Fans & Ventilation
A Practical Guide

The practical reference book and guide to fans, ventilation and ancillary equipment with a comprehensive buyers' guide to worldwide manufacturers and suppliers

W T W (Bill) Cory

First published 2005

ISBN 0-080-44626-4

A CIP catalogue record for this book is available from the British Library

Published by Elsevier in association with Roles & Associates Ltd

ROLES &associates

ELSEVIER

Amsterdam Boston Heidelberg London New York Oxford Paris
San Diego San Francisco Singapore Sydney Tokyo

Foreword

The word "fan" covers a wide variety of machines, from small table fans recognised by everybody to huge industrial fans consuming hundreds or thousands of kilowatts. Fans are very important to many industries since, for almost all human activities, there is a need to move or replace air.

The most obvious and well-known use of fans is in ventilation for comfort, which also includes air conditioning. However this is only a small part of fan applications. A list of such applications is extensive covering for example: mining, nuclear facilities, wood and paper production, textiles, computer rooms etc. For each there is a need to consider various aspects such as: correct design for specific requirements, best possible energy efficiency of the whole system, environmental influences (noise and vibration), personnel safety and global life-cycle costs.

A practical reference book about fans and ventilation is a welcome aid to all users who want to know practical information about fan design, selection and application and how these factors affect performance. The fact that *Fans & Ventilation* is written by Bill Cory ensures it is of high quality, and contains a substantial amount of practical and up to date information in this fast moving field of technology.

Bill Cory is currently Chairman of the Eurovent Working Group 1 "Fans" for many years. He was also President of AMCA from 2002 to 2003 and the most active member of ISO Technical Committee 117 "Industrial Fans". The list of the documents and Standards he has prepared, or participated in the preparation of, is impressive.

We have no hesitation in recommending *Fans & Ventilation*.

Sule Becirspahic

Director of Operations
Eurovent/Cecomaf

Dedication

This book is dedicated to the memory of my wife

Eleanor Margaret Cory, née McHale

She was born on 23 January 1933, we married on 26 July 1958 and she died on 8 November 2004.

Eleanor, not by any means a Dumbo (she passed her School Certificate when this meant something), sacrificed her career for mine. She gave me two lovely daughters and was a constant source of encouragement, advice and support.

To use modern parlance – I loved her to bits! Perhaps I should have told her this more often.

About the author

W T W (Bill) Cory, DEng, MSc, CEng, FIMechE, MCIBSE, MIAgrE FRSH, MIIAV

W T W (to his enemies!) or "Bill" (to his friends!) Cory first brought a light to his mother's eye on 4 October 1934. A bouncing 9lb. 5oz., he has been a heavyweight from that time on! The product of a boat builder's son and a farmer's daughter, he is unsure if it is salt or soil that he has had in his mouth ever since. He hopes it is one of the two!

Bill's career spans more than 50 years in the ventilation and fan manufacturing industries. He started his working life with Sturtevant Engineering Company Ltd and then continued with several companies, assuming positions of increasing responsibility. He joined Keith Blackman Ltd in 1976, becoming Technical Director in 1979. In 1984, when Woods of Colchester Ltd absorbed Keith Blackman Ltd, he was appointed Technical Director of the combined company and was responsible for the whole engineering staff. He retired from the Board of Woods in 1999 at the age of 65, but was retained by the company as a consultant. Members of staff say that they now see a lot more of him than previously! In 2001 Woods became a part of the Fläkt Woods Group.

Bill received his early technical education at Manchester College of Science and Technology and Northampton College of Advanced Technology, and the National College of Heating, Ventilation, Refrigeration and Fan Engineering. He gained a Master of Science degree in acoustics by distance learning from Heriot-Watt University in 1990 and in 1992 was admitted by London South Bank University, as its first Doctor of Engineering.

Bill Cory still serves on various AMCA and BSI committees dealing with ventilation and fans. He also leads the UK delegation to the corresponding ISO and CEN committees. He is a past member of the Council of the Institution of Mechanical Engineers, and past chairman of its Eastern Region as well as a past chairman of its Fluid Machinery Committee.

Bill is chairman of a number of technical committees and serves on the boards of various colleges and is a past president of Colchester Engineering Society. He has long been active in AMCA, HEVAC and FMA affairs and is a past chairman of FETA's Technical Management Committee. He was a director of AMCA from 1996 - 2004 and its President in 2002 - 2003 — the first non-North American to be so recognised. Recently he has become chairman of Eurovent Technical Committee WG1-Fans.

Bill Cory has presented over 50 papers to various technical institutions including the Institution of Mechanical Engineers, Chartered Institution of Building Services Engineers, Institution of Agricultural Engineers, Institution of Acoustics etc. He has given lectures to universities in Cagliari, Cairo, Helsinki, Sheffield, South Bank and Southampton. The subjects covered include fan performance measurement, fan acoustics, tunnel ventilation, condition monitoring, crop drying, natural ventilation etc.

Personal acknowledgements

This book has been based on a career of 50 years in the air moving industry during which I have benefited from the many friendships I have made.

Firstly I remember George Henry Gill of The Sturtevant Engineering Company who fired my enthusiasm for fans and Joseph Dunning, its Works Manager, who made sure I applied myself to becoming an engineer. I remember also William Osborne of the then National College of HVR & Fan Engineering who started me on a belated academic career. I learnt much from him which is incorporated in this book.

Of more recent years I have gained much from discussions with Prof Dr-Ing Hans Witt on explosion proof fans. I am also very grateful to Prof Richard Matthews of London South Bank University with whom I have collaborated on the design of mixed flow fans and tunnel ventilation.

Dr Ron Mulholland, Chief Engineer of Howden group Technology is a dab hand with the production of computer-generated illustrations which he has translated from my "back of a fag packet", dodgy sketches!

I wish also to say a special thank you to Mr Paul Wenden, Product Marketing Director, Fläkt Woods Ltd for providing many of the illustrations and who has also permitted me to use much material given in my papers to learned societies, and which were subsequently published by my then employer Woods of Colchester (now Fläkt Woods Ltd).

I thank Mr Steve Barker who produced many of the drawings for Chapters 1, 9 and 11, and a special thank you to Mrs Pauline Warner, my excellent secretary for many years, who produced the manuscript for some of the early chapters.

Finally, I would like to thank Ketty and Richard Tomes of Roles & Associates Ltd for their magnificent work in transforming many of my awful hand-drawn illustrations and editing much of my badly written manuscript and notes; creating, in my view, a work of art!

How to get the best from your fans

Using this book

Written specifically for fan users, *Fans & Ventilation* is intended to provide practical information about the outline design selection and installation of fans and how these affect performance. Fans & Ventilation is not intended to be a textbook on ventilation and air conditioning; rather it seeks to address the problems that exist at the interface between fan manufacturers and users. It is aimed at everyone who has technical problems as well as these wanting to know who supplies what, and from where.

Fans & Ventilation can be used in a variety of ways depending on the information required. For specific problems it is probably best used as a reference book. The detailed contents Section at the front of the book combined with the Reference index, Chapter 25, at the end, will simplify finding the appropriate topic. The introduction to the start of each Chapter will also provide valuable guidance. The bibliography Section at the end of many Chapters also provides useful references and suggestions for further reading.

As a textbook though, *Fans & Ventilation* may be read from cover to cover to obtain a comprehensive understanding of the subject. Of course, individual Chapters may be studied separately.

Chapter 1 covers the history of fans and details the various generic fan types. The properties of gases and gas flow are then discussed in the other early Chapters. The book then follows a logical pattern with Chapters 4 to 10 covering topics such as: performance standards, ducting systems, and flow regulation, constructional features, fan arrangements and bearings. Chapter 7 also provides useful information on fan materials and the stresses induced in the various parts of a fan. These stresses can be subject to mathematical analysis and an introduction is given to the methods used.

Chapters 11 to 13 are devoted to drives, couplings and prime movers. Noise and vibration are considered extensively as well as quality assurance, installation, fan economics and finally fan selection considerations, in Chapter 20, which are all clearly aimed at the user

Chapter 21 provides some fan applications illustrating the diversity of fan design and uses, showing there are many uses for fans outside the traditional areas. It also endeavours to demonstrate some of the sizing rules and features which should be included.

The Classification guide to manufacturers and suppliers, Chapter 24, is an invaluable and important part of the book. It summarises the various fan types, covering their differing styles, sizes and basic principles of operation. All definitions are in accordance with ISO 13349:1999 (BS 848-8:1999).

The guide has been categorised in a particular way to impose strict boundary limits on fan types and the operating conditions available, with the specific aim of simplifying the choice of supplier from the users' point of view.

The Classification guide includes most fan types, followed by ancillary products and services. Trade names are comprehensively listed too. It is preceded by the names and addresses and contact details of all companies appearing in the classification guide. These are listed alphabetically by country.

It is however strongly recommended that direct contact with the relevant companies is made to ensure that their details are clarified wherever necessary.

Heat
Cool
Vent

Contents

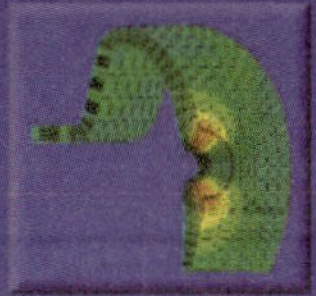

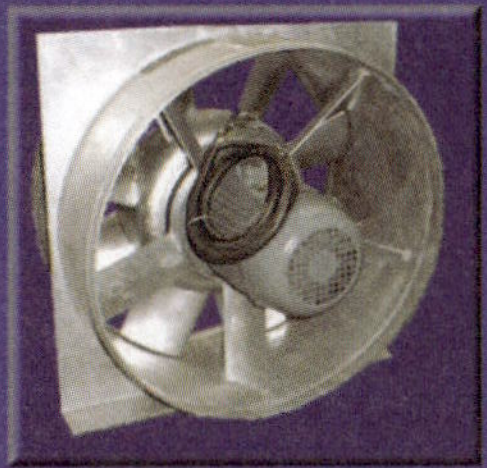

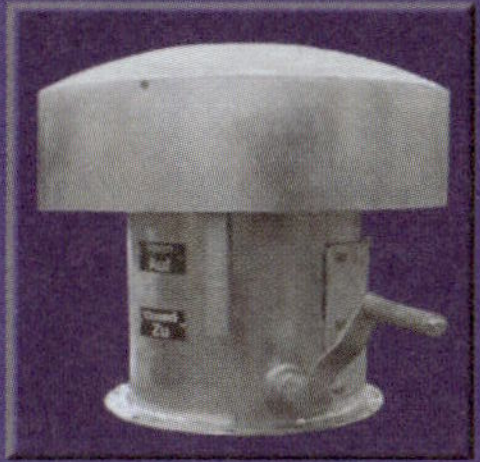

THE NEW POWERGRIP® GT3 BELT
Expands the limits of your ideas
Contemporary machine designs require advanced power transmission solutions. With the next generation synchronous rubber belt PowerGrip® GT3, Gates is one step ahead, providing drive designs not yet imagined. This technical tour de force transmits up to 30% more power than previous generation belts. PowerGrip® GT3 is available in 2, 3, 5, 8 and 14 MGT pitches and runs on existing drives, requiring no adaptation of the system. When you think the impossible, think Gates, the perpetual technology leader.
THE DRIVING FORCE IN POWER TRANSMISSION
Gates Power Transmission Ltd - Tinwald Downs Road - Heathhall - Dumfries DG1 1TS - Tel : 01387 / 24 20 21 - Fax : 01387 / 24 20 10
E-mail : ptindustrial@gates.com - Web Site : www.gates.com/europe/pti.
Gates®
A Tomkins Company

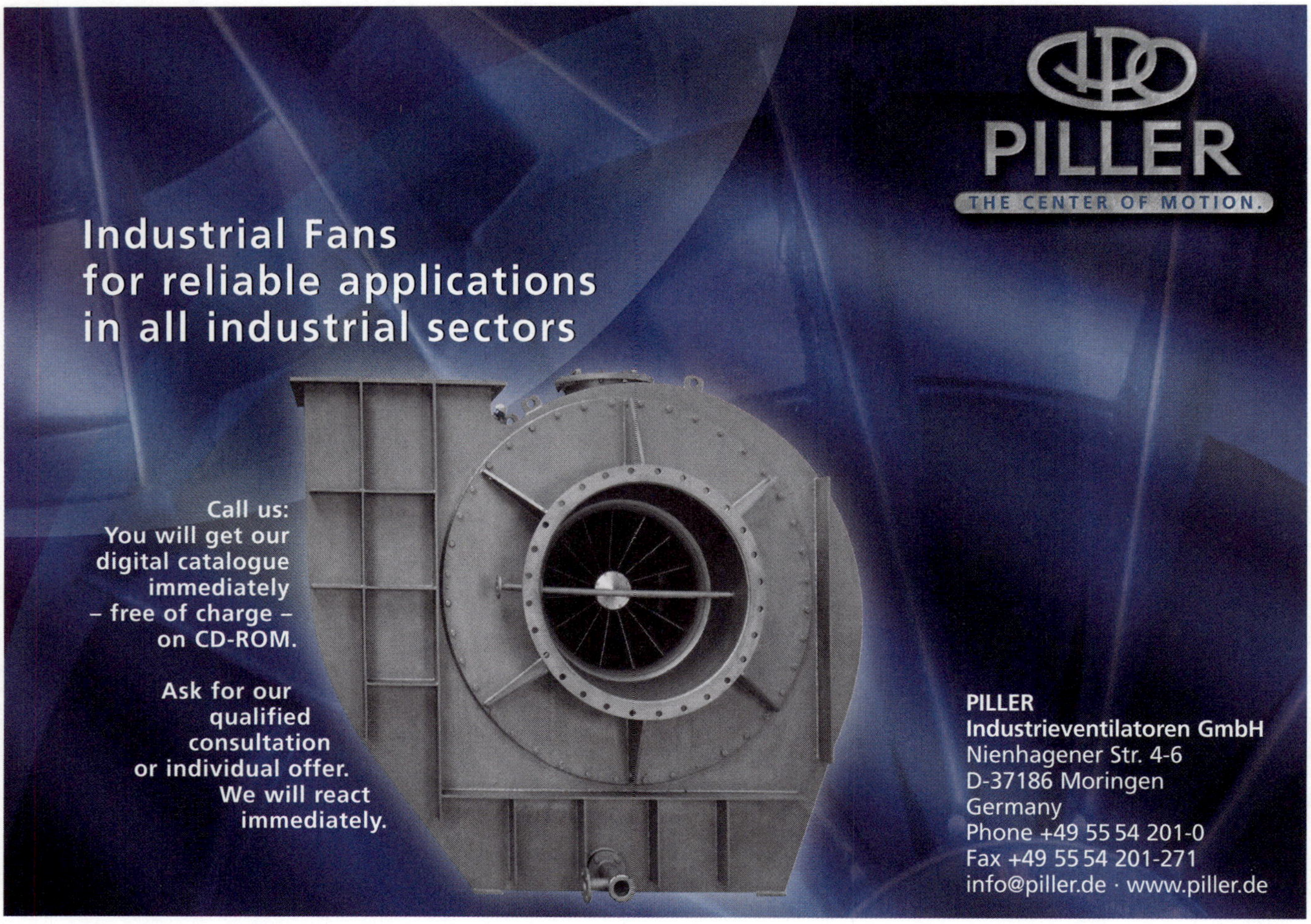
PILLER
THE CENTER OF MOTION.
Industrial Fans
for reliable applications
in all industrial sectors
Call us:
You will get our
digital catalogue
immediately
– free of charge –
on CD-ROM.
Ask for our
qualified
consultation
or individual offer.
We will react
immediately.
PILLER
Industrieventilatoren GmbH
Nienhagener Str. 4-6
D-37186 Moringen
Germany
Phone +49 55 54 201-0
Fax +49 55 54 201-271
info@piller.de · www.piller.de

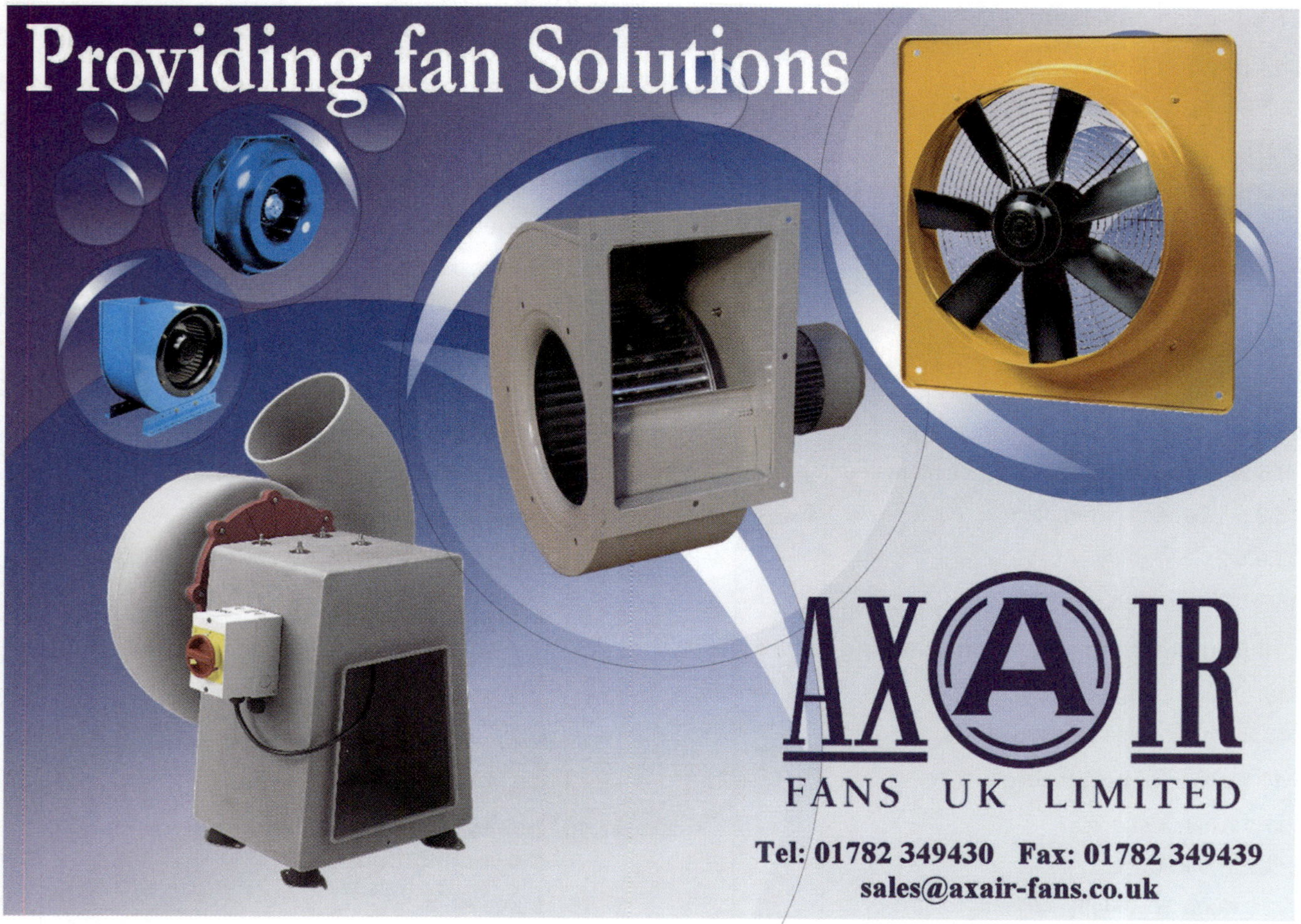
Providing fan Solutions
AXAIR
FANS UK LIMITED
Tel: 01782 349430 Fax: 01782 349439
sales@axair-fans.co.uk

LEADERFAN
A Trusted Name in
Air Moving Products for
over 50 years
www.leaderfan.com
Leader Fan Industries Ltd.
Tel. 416.675.4700 • Fax. 416.675.4707
Toronto, Ontario, Canada

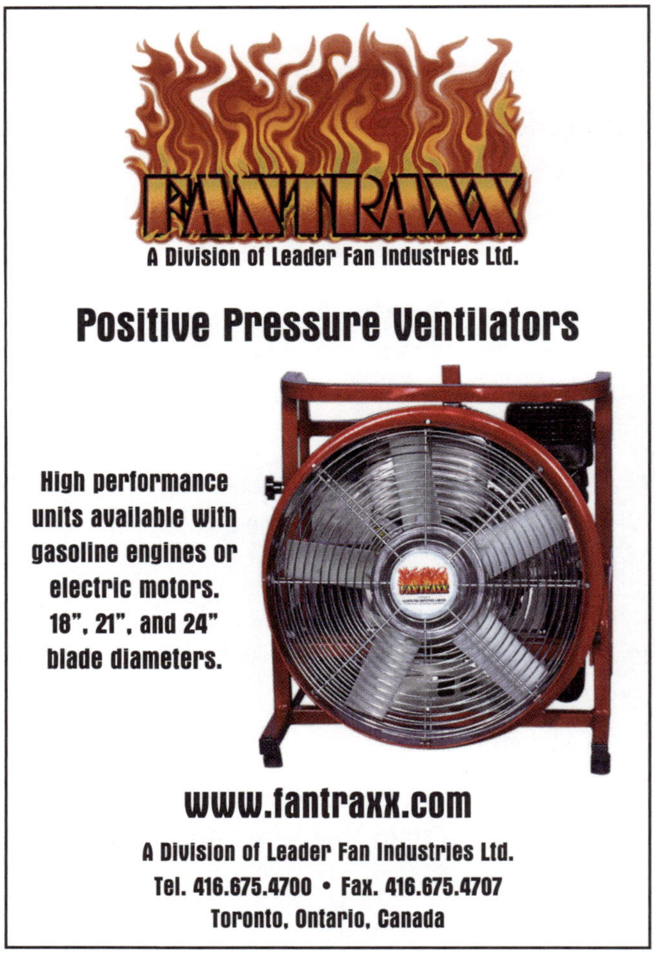
FANTRAXX
A Division of Leader Fan Industries Ltd.
Positive Pressure Ventilators
High performance units available with gasoline engines or electric motors. 18", 21", and 24" blade diameters.
www.fantraxx.com
A Division of Leader Fan Industries Ltd.
Tel. 416.675.4700 • Fax. 416.675.4707
Toronto, Ontario, Canada

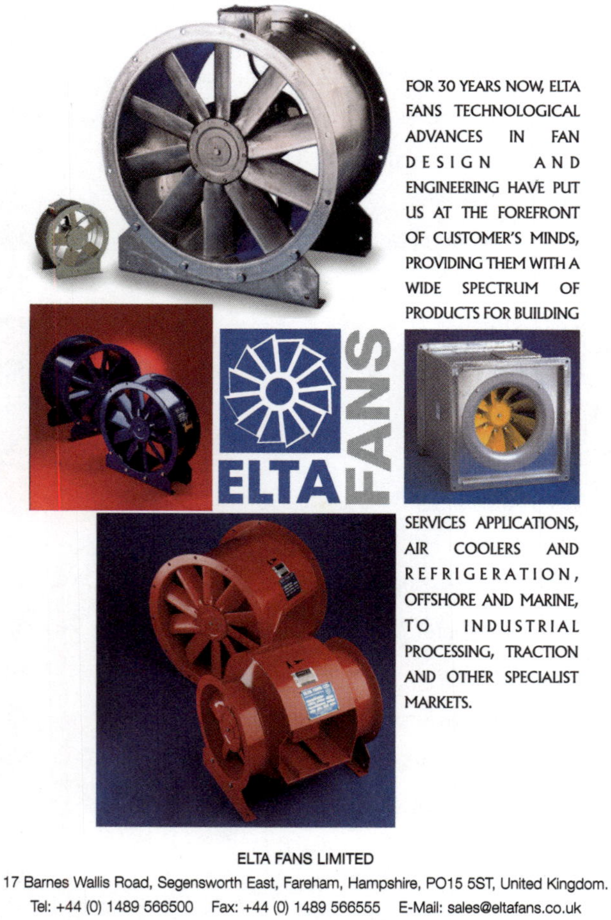
FOR 30 YEARS NOW, ELTA FANS TECHNOLOGICAL ADVANCES IN FAN DESIGN AND ENGINEERING HAVE PUT US AT THE FOREFRONT OF CUSTOMER'S MINDS, PROVIDING THEM WITH A WIDE SPECTRUM OF PRODUCTS FOR BUILDING
ELTA FANS
SERVICES APPLICATIONS, AIR COOLERS AND REFRIGERATION, OFFSHORE AND MARINE, TO INDUSTRIAL PROCESSING, TRACTION AND OTHER SPECIALIST MARKETS.
ELTA FANS LIMITED
17 Barnes Wallis Road, Segensworth East, Fareham, Hampshire, PO15 5ST, United Kingdom.
Tel: +44 (0) 1489 566500 Fax: +44 (0) 1489 566555 E-Mail: sales@eltafans.co.uk

COOPER GIVE FANS MORE UPTIME AND LESS DOWNTIME
Cooper Benefits
Reduce Down Time,(Planned or Unplanned).
Increase Maintenance Productivity.
Keeps Contaminants Out.
Can Increase Application Life.
Promotes Workplace Safety.
Cooper Features
Split to the Shaft Bearing. Superior Sealing.
Delivering Results in...
Trapped Positions. Hostile Environments.
Cooper will save your business money and improve bottom line finances. Our split roller bearings are proven in fan applications throughout the world across a range of heavy industries.
COOPER®
Cooper Customer Service Centres
UK, Africa, Asia, Australasia and the Middle East: Cooper Roller Bearings Co Ltd, Wisbech Road, King's Lynn, Norfolk PE30 5JX, England. Tel: +44 (0) 1553 763 447. Fax: +44 (0) 1553 761 113
Europe and South America: Cooper Geteilte Rollenlager GmbH, , Oberbenrader Str 407, 47704 Krefeld, Germany. Tel: + (49) 2151 713 016. Fax +(49) 2151 713 010
Locate Your Nearest Distributor by Visiting CooperBearings.com

1 Fan history, types and characteristics

In an age when political correctness has become the state religion, it is perhaps courting disaster to tell a joke about our fellow human beings. That it might be interpreted as racist by the professional do-gooders is doubly worrying. However, as a man of English-Scottish ancestry and with Welsh-Irish wife I feel impervious to such slings and arrows.

"Excuse me, my good man", said an Englishman lost in the wilds of Ireland. "Can you tell me the way to Ballykelly?" "If I were you, sir, I wouldn't start from here."

A perfectly correct and helpful answer. It's just the same with the fan world. We shouldn't have started when and where we did. But the die was already cast and a line from there to the present day shows us the path we trod. There were numerous setbacks and diversions, but an extension of that line, shows us the direction to the future. If we have studied that history, we may even avoid making the same mistakes twice, and will not have to suffer the old "Codger" in the corner saying "We tried that in 1961 and it didn't work".

To maintain the interest of those who like to classify and define, the Chapter continues with a description of the various fan types in what is hopefully a logical progression. It describes the shape of the characteristic curves, but the reader's patience will be rewarded in the Chapters that follow.

Contents:

1.1 Introduction

It is inevitable that the content of this chapter will reflect the personal experiences, and indeed preferences, of the author. Apologies are, therefore, proffered in advance to those companies whose products are conspicuous by their absence. The privilege of all historians is to be able to "slant" the investigations to suit their own individual prejudices - and I am no exception.

Mechanical fans are a particularly mature product - they have been around, and running most of the time, since at least the sixteenth century. Engineers will be the first to acknowledge that nothing is new, and most of the major design principles had been established by the early twentieth century. We, who have followed, have merely improved, tinkered with, or fitted theories to that which our fathers invented. We are but pygmies, standing on the shoulders of giants.

To appreciate the present and future developments, it is essential to know something of the past. Where we have come from gives us a direction as to where we might go in the future. It may also help to explain why there are so many different types of fan. The reasons for their existence are invariably that they met a customer need. Whilst managing directors may complain that they have half a million models in their manufacturing range, the chief engineer may reflect that if he or she were to meet all the requirements of flowrate, pressure and efficiency in the presence of hot, erosive and/or corrosive gases then an even larger range might be desirable.

1.2 Ancient history — "Not our sort of fan"

Few people ever pause to think that fan making is one of the oldest crafts in the world and that it dates back to the earliest times of which we have any clear record. The use of fans was already well established in the earliest Egyptian civilizations. This is made clear by the ancient bas reliefs in the British Museum, which depict women carrying feather fans. There is further evidence of the fact in the Cairo Museum, where there still exists the remains of a fan found in the tomb of Amenhotep, who died as far back as 1700 BC.

The royalty and notabilities of the ancient dynasties undoubtedly regarded fans as being one of their necessary accessories and throughout the centuries fans have continued to be quite important requisites in civilized life. The early fans, of course, were mainly carried in the hand by women and used for giving motion to the air for cooling the face. Originally they were all of the fixed type, made of feathers or of cloth or paper stretched on a framework of bamboo. Folding fans originated in Japan and were exported from there to China.

With the spread of civilization westwards, fans gradually became an accepted feature of social life in Europe. In the days of the Roman Empire they were a recognised item in bridal outfits. From Rome, fans spread to other countries, and by the 14th century they were generally in use in the European courts. By this time, however, a change had taken place in the purpose for which fans were used. They were no longer carried solely for the original purpose of fanning the face. They had become aids to feminine deportment. They were fashion accessories, used to accentuate feminine grace and aids to feminine wiles. Women used them to convey messages to their admirers by means of a conventional code of signals.

From then onwards, fans continued to be essential items in feminine equipment on all formal occasions. The centre of manufacture in the 17th century was Paris. But fans were also being made, to a considerable extent, in England. The revocation of the Edict of Nantes drove the French fan makers to this country, and by the middle of the 17th century, fan making was a well established trade. In fact, the fan makers sent a petition to Charles II protesting against the imports of fans from India.

The manufacture of ladies' fans reached its height in the 18th century. The craft had then become definitely an art. Being essentially feminine, fans lent themselves to extremely artistic treatment. They were made from ostrich feathers, fine parchment, taffeta, silk or fine lace mounted on ivory as well as on cane, and embellished with mother-of-pearl and precious metals. In the Victoria and Albert Museum and the South Kensington Museum, in London, there are large numbers of French, English, German, Italian and Spanish fans. See Figure 1.1.

Figure 1.1 A beautiful example of an 18th century fan

In more recent years, ostrich feather fans have been used not merely as a feminine accessory but as the sole covering of fan dancers. Fans of the feminine type had become so firmly established in the 17th and 18th centuries as necessary requisites for women, that The Worshipful Company of Fan Makers was concerned solely with the artistic side of fan making.

1.2.1 The advent of mechanical air movement using "air pumps" and fires

It has to be recognised that it is pure chance for the same word to be used for the contrivance behind which an oriental lady hides her face and the present day rotary machine for delivering a current of air. Only the Anglo-Saxon creates such confusion.

In Finland another form of confusion is found by the use of the word "puhallin" (a wind instrument) which covers both a trombone and a propeller fan. Of course, no such difficulty exists when using the French or German languages as "ventilateur" or "Ventilator" are more precise in their meaning and are unambiguous. All that is necessary is to define whether they are "powered" or "natural". The ladies with their "éventail" or "Fächer" are unlikely to be misunderstood.

The need for having some mechanical means of moving air for industrial and cooling purposes had been realized for many centuries. Punkahs were used in India hundreds of years ago. In its earliest form the punkah consisted of a large swinging flap covered with wet straw.

The first means of providing a forced draught of air was the bellows. It is believed that bellows of a primitive type were used in Egypt for assisting the combustion of fires as far back as 400 BC. In India a simple form of bellows made from goat skins was used for iron smelting in the very early ages.

The origin of the word bellows was blast-baelig - a blow bag. In the 11th century the first part of the name was dropped and in the 16th century the word baelig had become first belly, then bellies, and finally bellows. Bellows were almost the only means of blowing air until the 17th and 18th centuries, when blowing machines were developed. These consisted of a piston, cylinder and valve for moving air. In 1851, a double-acting blowing engine of tremendous size was used in Dowlan's Iron Works. This had a cylinder of 3.7 m diameter, the piston stroke was

A—DRUM. B—BOX-SHAPED CASING. C—BLOW-HOLE. D—SECOND HOLE. E—CONDUIT. F—AXLE. G—LEVER OF AXLE. H—RODS.

Figure 1.2 Georgius Agricola's reference to bellows and crude fans

3.7 m, the machine moved 21 cubic metres of air per second, and developed a pressure of 30 kPa.

Perhaps the earliest reference to mechanical ventilation was by Georgius Agricola in his book *De Re Metallica*, first published in 1556. He described the use of bellows and crude fans (Figure 1.2) in German underground metal mines in a manner which makes one assume that they were then well established. These early fans were, of course, made of wood with radial paddle vanes fitted to a spindle which rotated in a casing. Thus they were the first centrifugal fans and were rotated by animals, men or water mills.

It is interesting to observe that Agricola's book was translated from the Latin in 1912, by Herbert Clark Hoover, President of the United States of America. These days Presidents and less than humble engineers have more than enough trouble with English, let alone a foreign and dead language!

Much of the early history of fans is inextricably linked with that of mines, but up to about 1860, their ascendancy over other solutions was not, by any means, certain. John Smeaton (1724-1792) used reciprocating pumps for exhausting the foul air from coal mines in Northern England. In 1813 John Buddle (1773-1843) wrote to the Sunderland Society describing the methods which he had used in the collieries of North East England for generating the necessary air currents and thus the prevention of accidents from "firedamp". His exhausting piston pump had been installed in Hebburn Colliery in 1807.

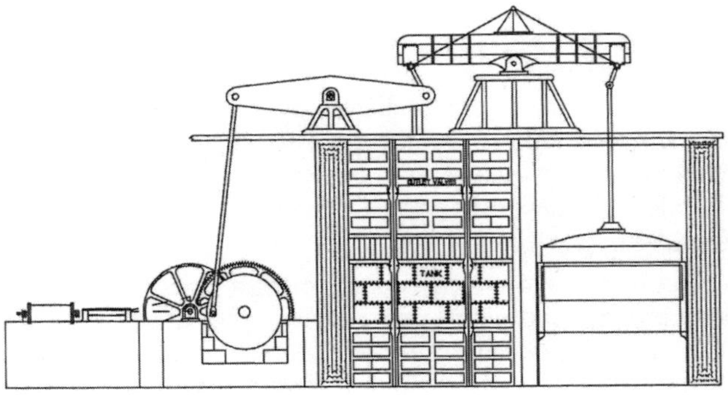

Figure 1.3 The Struve ventilator

Buddle also stated that "the standard air-course, or current of air, which I employ in the ventilation of collieries under my care, abounding in inflammable gas, equals from 5400 to 7200 cubic feet per minute". Allowing for the factor of exaggeration always present in any engineer's claims, we may note that 2.55 to 3.4 m^3/s (for those too-long metricated) is an exceedingly small amount and that nowadays flows 100 times as great would be considered necessary in such mines.

In addition to Smeaton's and Buddle's air pumps, other large machines working on the same principle were developed and one of the most successful of these was the Struve ventilator. William Price Struve of Swansea developed an air pump which employed circular air pistons shaped like bells or gas holders (Figure 1.3).

Generally each machine employed two of these which were moved up and down by means of a steam engine, the lower edge of this bell dipping into a circular water trough. This arrangement prevented leakage past the pistons. Each piston works as a double-acting pump. The air from the mine entered the space above and below the piston by means of a multitude of inlet valves and opens discharge valves, through which the exhaust air enters the atmosphere. These ventilators worked as exhausters and were connected to the top of the upcast shaft. In some cases ventilating pressures of 1.25kPa were produced.

The first Struve ventilator was installed at Eaglebush Colliery, South Wales and began to work in February 1849. The upcast shaft was 55 metres deep and the quantity of air circulated was 26.5 m^3/s at an average pressure of 0.9kPa. About a dozen of these machines are said to have been installed, the largest of which was erected by the Rhuabon Company in North Wales, the pistons of which were 7.6m diameter. The quantity of air produced by this machine was up to 28.3 m^3/s. All these machines suffered from slow piston speeds. Upkeep to retain their efficiency proved to be rather excessive, the valves requiring much maintenance with consequent stoppage of the machine. The useful effect reported for some of these machines was in the region of 50%.

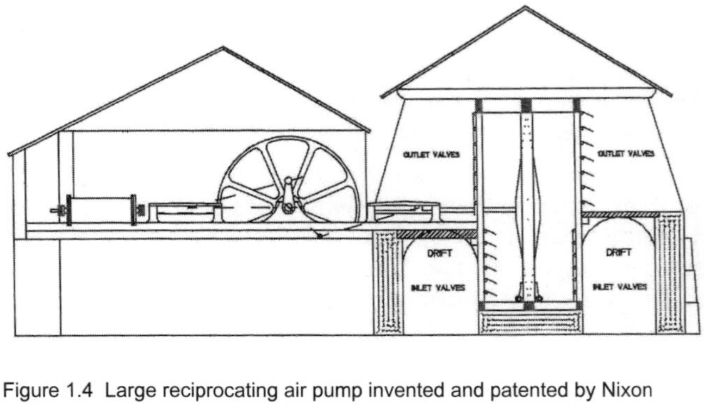

Figure 1.4 Large reciprocating air pump invented and patented by Nixon

Another type of large reciprocating pump was invented and patented by Nixon in 1861 (Figure 1.4). The first of these was installed at Navigation Collieries, Mountain Ash, South Wales; this was a horizontal machine having two rectangular shaped wooden pistons, each 9.1m long by 6.7m high, which ran on small wheels along rails in the wooden cylinders. The stroke of the pistons was 1.83m and when the machine ran at 6½ strokes per minute, it delivered air at the rate of 44 m^3/s.

The air enters the machine through flap valves and leaves through discharge valves. In Nixon's machine it was not possible to have water seals on the piston and leakage past the piston was a difficulty. The movement of the pistons was actuated by a steam engine. Two of these machines were installed in South Wales. Nixon's ventilator was said to have a useful effect of about 46% when in good condition. Having a multitude of small valves, it required careful maintenance if leakage was to be kept at a minimum.

To overcome the objections of the reciprocating air pumps of slow piston speed and much valve maintenance, rotary air pumps were invented and constructed. They consisted of vertical drums revolving eccentrically within a cylindrical chamber. By the revolution of the drum in a cylinder housing, spaces of varying capacity were formed causing the air to enter from the upcast shaft and by further movement of the drum, the return air was discharged into the atmosphere.

The Lemielle ventilator, which was extensively used in the ventilation of Belgian collieries, from about the middle of the 19th century, was one of the most successful of these rotary machines. Several were exported to England, starting the ventilation export trade. An example was that installed at Page Bank Colliery in about the year 1860. The drum was 4.6m diameter and 9.8m high and worked in a casing 6.9m in diameter. The useful effect reported by the North of England Institute Committee on Mechanical Ventilators for this machine was 23.4%. A further type of rotary air pump was that invented by Cooke, but very few of this type of ventilator were installed, and little is known of their design.

Perhaps the most alarming method of mine ventilation was to place a furnace at the bottom of the upcast shaft. By burning coal (what else?) a current of air to support the combustion was induced through the mine (Figures 1.5 and 1.6). The "stack-effect" of a deep mine meant that the pressure developed was then greater, and the method could not be used in shallow mines. Even so, a furnace was only capable of developing about 750 Pa and Buddle had to use "split ventilation" - dividing the workings into a number of parallel circuits to reduce the system resistance.

Many collieries favoured furnace ventilation around the mid 19th century as both air pumps and fans were considered to be

Figure 1.5 Early example of furnace at surface for ventilation of a mine

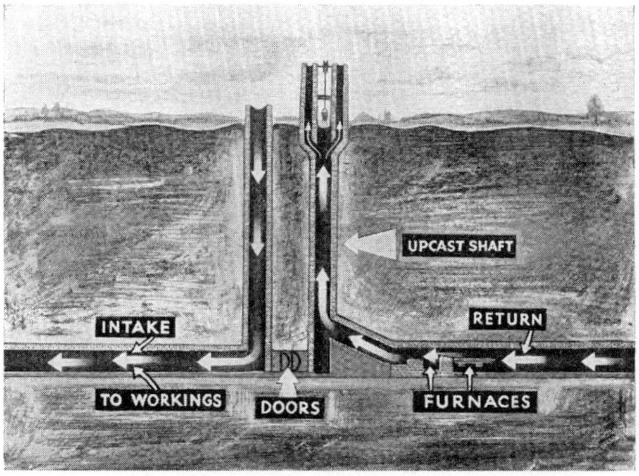

Figure 1.6 Early example of furnace underground for ventilation of a mine

unreliable. Just as mechanical ventilation was improving, a UK government select committee (1852), with that lateness of report and lack of accuracy that has always characterized politicians, stated that "any system of ventilation depending on complicated machinery is inadvisable, since under any disarrangement or fracture of its parts the ventilation is stopped, or becomes less efficient". It took a further 60 years before the UK Coal Mines Act of 1911 recognised that this problem could be easily overcome by having a running and standby fan. The committee also stated "that the two systems which alone can be considered as rival powers are the furnace and the steam jet".

Experiments soon proved that steam jets were extremely inefficient and were incapable of producing the larger flowrates of air required due to increasing colliery outputs, and the larger amounts of firedamp (methane) therefore being emitted. Furnaces could, however, cope and Nicholas Wood, (the backer of, and collaborator, with George Stephenson in the early development of railways) showed in tests at Hetton Colliery on 13th November 1852, that three furnaces at the bottom of the upcast shaft circulated 106 m^3/s with an underground ventilating depression of 486 Pa.

Even as late as 1946, Copy Pit and Clifton Colliery near Burnley had underground ventilating furnaces with chimneys belching out smoke for no apparent reason. Nobody would have suspected that these chimneys were in fact about 275 m high. The outlets were known locally as cupolas and can only have survived for so long as the mines were non-gassy.

1.2.2 Early mine ventilation fans

After the fans employed in German metal mines, described by Agricola, their use went into decline for almost 250 years. It was not until 1827 that a mine ventilating fan was re-introduced to a colliery near Paisley, Scotland. This had a number of inclined blades fixed to a vertical shaft rotating within a circular casing. The fan was fitted over the top of the upcast shaft and air was drawn through it and discharged to atmosphere. It could be argued that this was the first axial flow fan.

At the same time many mines in France and Germany experimented with fans working on the Archimedean screw principle, but these failed, not only from a lack of knowledge of the aerodynamic theory, but also because the metallurgy of the time did not permit them to run at the speeds necessary for an acceptable flowrate and pressure.

Attention therefore turned again to the centrifugal fan. The impeller of this was inherently stronger whilst the pressure developed was augmented by the centrifugal force applied to the air, in addition to the blade action. Lower rotational speeds, within the capacity of a typical steam engine, enabled useful duties to be performed.

In 1849 an open running 6 m diameter radial-bladed centrifugal fan with vertical shaft was installed at Gelly Gaer Colliery in South Wales. The engineer responsible for its design was William Brunton (1777-1851) who had been trained under Boulton and James Watt at the Soho Foundry, Birmingham. Not unnaturally the fan was directly driven through a crank from a steam engine. A model was shown at the Great Exhibition of 1851, held in Hyde Park, London.

In 1851, James Nasmyth (1808-1890), the inventor of the steam hammer, read a paper to the British Association at its meeting in Ipswich. He described a double inlet radial-bladed centrifugal fan again directly driven by a steam engine. His theory was put into practice in 1854 at Abercarn Colliery, South Wales. This fan had an impeller diameter of 4.12m and ran at 60 rev/min for a duty of 21.25 m^3/s against 125 Pa. Subsequently a larger fan of 4.57m diameter running at 80 rev/min was installed at Skiar Spring Colliery, Elsecar, Yorkshire, UK.

One of the most successful centrifugal fans of the mid 19th century was that designed by Theophile Guibal (1814-1888), (Figure 1.7). The fan, installed at the Jean Bart Colliery, was first described in *L'histoire generale des Techniques aux P.U.F.*, in 1859. Guibal was born in Toulouse and educated in Paris. At the time of his invention he was Professor of the Exploitation of Mines at the University of Mons, Belgium.

Many of the early fan designers had believed that an extract fan did not require a casing, but that the air should have a free and unrestricted access to the atmosphere. Guibal was the first to show that a casing was desirable and to develop the expanding evasée to slow down the air before discharge. By 1870 nearly 150 of these fans had been installed in Belgium, France and the United Kingdom with diameters varying from 4.8m to 15.5m and flowrates from 14 m^3/s to 100 m^3/s at depressions of 125 Pa to 1500 Pa.

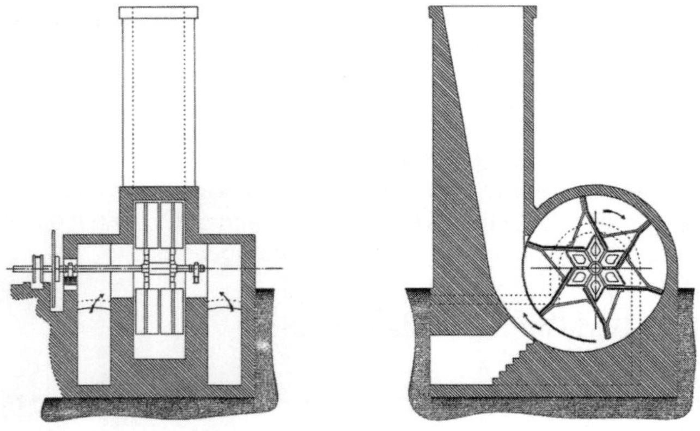

Figure 1.7 Guibal's successful centrifugal fan

In 1863 Christian Schiele of Manchester, England, patented an improved fan, which was developed in small sizes for blowing cupolas and in larger sizes for the ventilation of mines. His fan had a strongly built iron impeller which could rotate at much higher speeds. The blades were backward inclined and discharged into a gradually increasing volute. The consequences of these improvements were a much reduced size and capital cost for a given duty, which made it popular with the accountants, if no-one else (Figure 1.8).

J. R. Waddle of Llanelli, South Wales, introduced his first fan in 1864 at Bonville's Court Colliery. It replaced a furnace at the mine which had burnt 10 tonnes of coal per week to produce a flowrate of 4.72 m^3/s against 48.5 Pa. The fan was 4.88m diameter and circulated 14.16 m^3/s against 436 Pa. To some extent

Figure 1.8 Schiele's improved centrifugal fan

this fan was old fashioned when introduced, as it was open running, (Figure 1.9).

The impeller, however, had backward curved blades (Figure 1.10) and a tapered shroud so that it was extremely strong and had a non-overloading power characteristic. Fans of this type

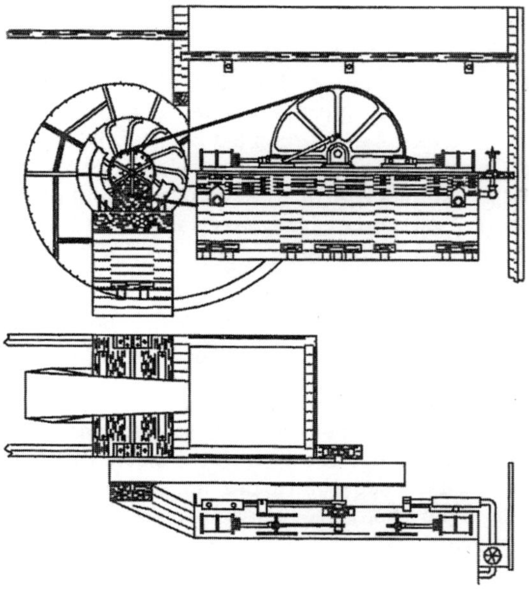

Figure 1.9 Waddle's open running fan

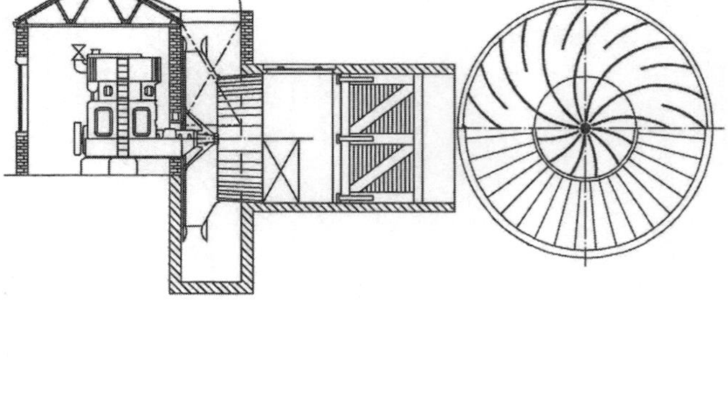

Figure 1.10 Cross-sections of Waddle's fan — with backward curved impeller blades

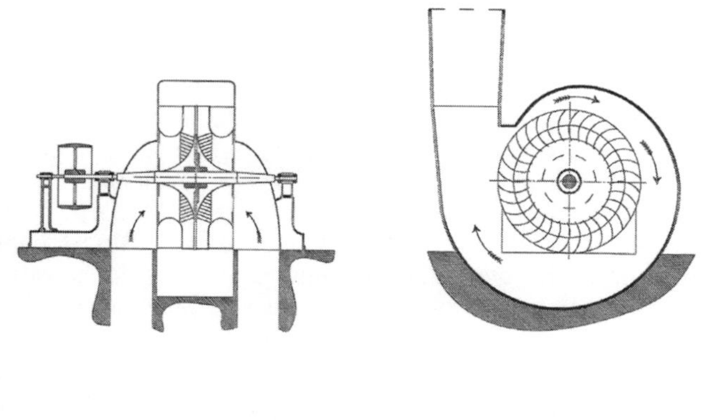

Figure 1.11 Cross-section through Professor Ser's fan

were built in diameters from 3.0 m to 15.5 m. Later examples from about 1890 were designed for higher peripheral speeds e.g. 5.5 m diameter at 300 rev/min), permitting a significant reduction in size for a given duty. They were widely used throughout South Wales and the rest of the United Kingdom, including the mines of Cory's Navigation Collieries, the reason for mentioning them here!

Professor Ser of the Ecôle Centrale de Paris designed his first fan in 1878, the theory being published in the *Memoires de la Société des Ingenieurs Civils*. Usually constructed in double inlet form it had 32 forward curved blades either side of the centreplate. These were of constant width but axially inclined (Figure 1.11).

The Capell fan was designed around 1883 by the Rev George Marie Capell, a graduate of Oxford University, and an Anglican priest. He said "it is now getting known that the life of a small fan, fast running, if the fan be properly constructed and balanced, is longer than that of the ponderous constructions of past times". In this he was putting into words what was being practised in France and Germany. His fan (Figure 1.12) was unique in the design of the impeller which essentially consisted of two concentric parts each having six backward curved blades either side of the centreplate. The inner and outer parts were separated by a drum having six port holes designed to have a total area equivalent to that of the impeller eyes. The inner part was unshrouded. As a peak efficiency of 70% was achieved, it may be deduced that the power of prayer exceeds that of the Mechanics of Fluids!

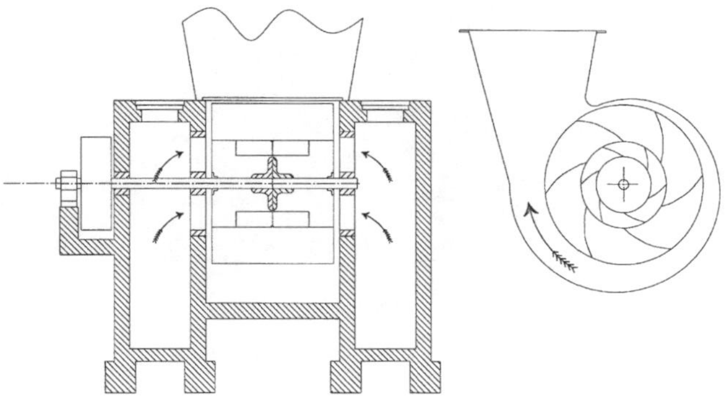

Figure 1.12 Cross-section through the Capell fan

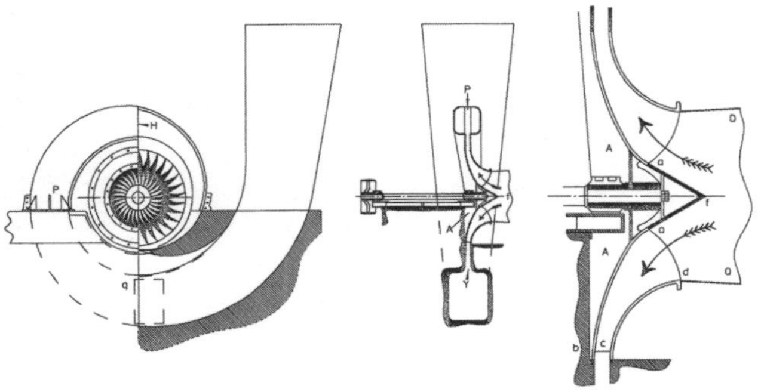

Figure 1.13 Cross-sections through Rateau's fan

Rateau's fan (Figures 1.13 and 1.14) of the late 1880s has sometimes been called the first mixed flow unit. In reality, however, it is perhaps best described as having compound blading with a truly axial inlet and centrifugal outlet, working in a complex volute having a gradually increasing cross-section. The blading was carefully designed for minimum shock losses and an efficiency of 80% was claimed.

The Guibal, Ser, Capell and Rateau fans were all subject to exhaustive practical tests. A detailed report by the Belgian Commission entitled *Les Ventilateurs des Mines* was published in the *Revue Universelle des Mines*, Vol.20, (1892), thus starting us along that perilous path of standardized methods of test, certification of performance and contract qualification.

The Mortier diametral Fan (Figure 1.15) was perhaps the first tangential or cross-flow fan. It was manufactured by Louis Galland at Chalon-sur-Saone, France. Efficiencies in excess of 70% were indicated by Charles Innes in his book *The Fan* (1916), perhaps suggesting that all is not progress. Later ver-

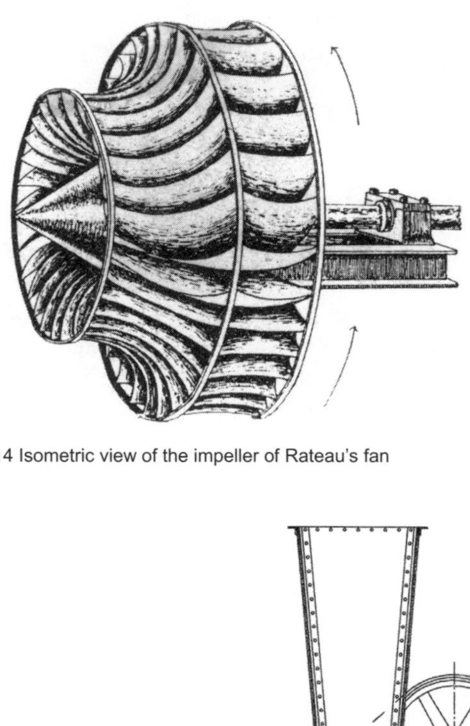

Figure 1.14 Isometric view of the impeller of Rateau's fan

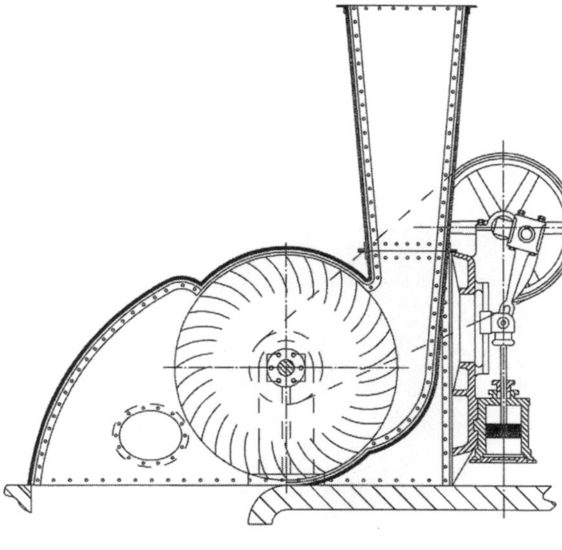

Figure 1.15 The Mortier diametral fan - perhaps the first tangential or cross-flow fan?

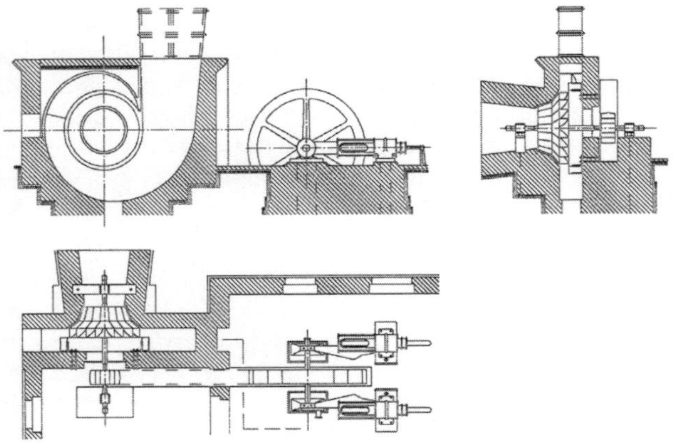

Figure 1.16 Pelzer Dortmund fan cross-sections

Figure 1.17 Impeller of Pelzer Dortmund fan

sions incorporated a movable section of scroll for flowrate control.

The Pelzer Dortmund fan (Figures 1.16 and 1.17) had twelve curved vanes designed for shock-free entry and with a radial discharge. It was the first to be manufactured in varying widths according to the fan flowrate and pressure development required.

Sam Davidson, who had left the shores of his native Ulster for the Assam tea plantations in 1864, was perhaps the next notable name in the fan industry. Dissatisfied with the crude and slow methods of withering and drying the tea leaf over open charcoal fires, he developed a cylindrical drying machine. In the development of his dryer, one feature was noted as a stumbling block to further progress. It relied on the natural draught induced by the furnace chimney.

Figure 1.18 Impeller of Davidson's multivane fan

Positive pressure from a fan was seen as the means of improving the drying rate. By a process of trial and error, and with an absence of any scientific instrumentation, he developed the forward curved bladed multivane impeller (Figure 1.18) patented in 1898. Witnessing the test of a tea drying machine fitted with one of these fans, a planter friend remarked "Why it's just like the Sirocco wind that blows off the desert". Sir Samuel Davidson, as he later became, immediately adopted the word as his trademark, and the fan was used widely for mine ventilation.

In all fans of the multivane type, in which the blades are axially long compared with their radial depth, there is a tendency for the air to "fill" the blade towards the backplate and for the side closest to the shroud to actually draw in air in a recirculatory mode. This was noted by Davidson, during his experiments and many of his early units were provided with an intermediate shroud to counterbalance the effect. BF Sturtevant, in his ordnance fan, provided the blades with cup-shaped indentations (Figure 1.19). These sought to prevent the air slipping to the back of the impeller. Perhaps more importantly, the blades were stiffer and could run at peripheral speeds approaching 503 m/s.

James Keith (1800-1843) started a fine engineering dynasty. His son George (1822-1912) was Provost, or Mayor, of his home town, the Royal Burgh of Arbroath, Scotland from 1889-1895. His grandson, also James, was renowned for the introduction to his workforce, and the world, of the eight hour working day. The resultant book, *A New Chaper in the History of Labour* was a best seller in 1893. To engineers, however, his important introduction was the Keith fan impeller of 1908 where

Figure 1.19 Impeller of B F Sturtevant's ordnance fan

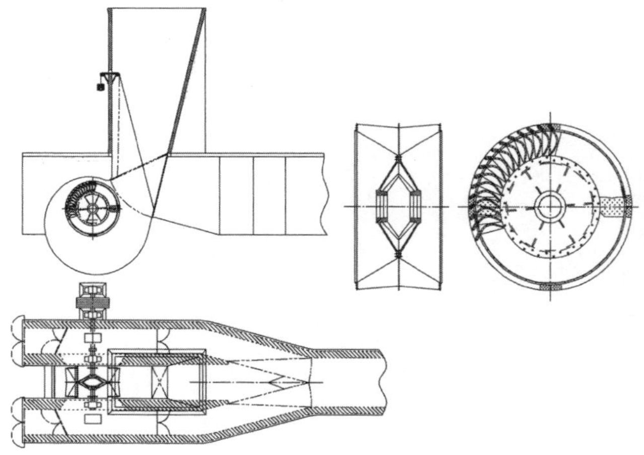

Figure 1.20 Cross-section through a Keith mine fan together with impeller detail

Figure 1.21 Keith mine fan during installation

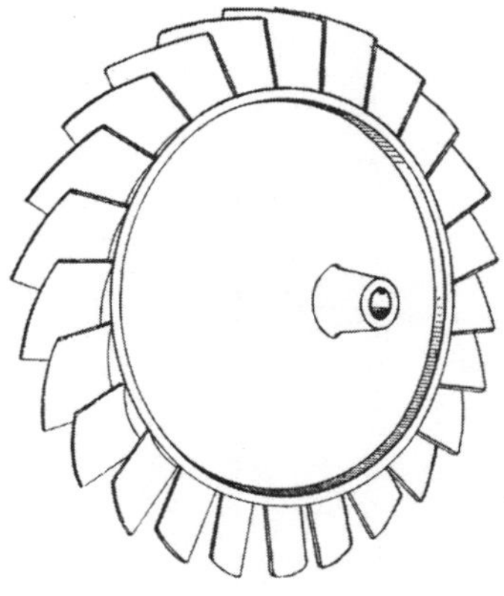

Figure 1.23 Rateau's axial impeller design

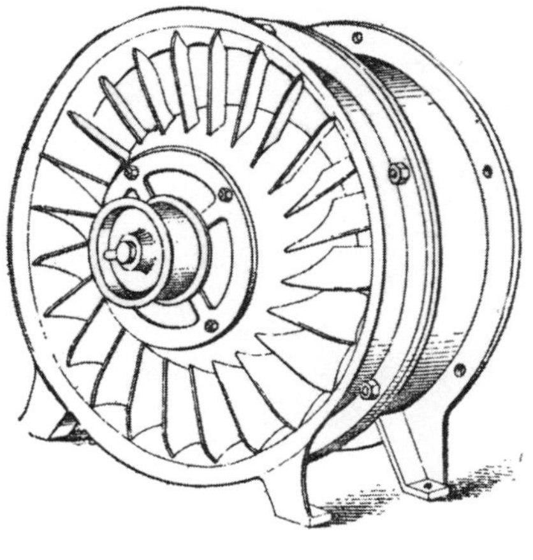

Figure 1.24 Rateau's horizontal cased axial flow fan

the external diameter was larger at the inlet or shroud side (Figures 1.20 and 1.21).

The peripheral speed was, therefore, higher here and in consequence the inductive effect was greater. A more even discharge of air across the blades was claimed whilst the nearly triangular shape gave great strength to resist centrifugal stresses and obviated the need for supplementary internal stays.

Another approach to the problem was in Waddle's Turbon fan (Figure 1.22). As with his backward-bladed fan, he adopted a novel, if not idiosyncratic approach. Instead of the impeller being built up from a large number of shallow blades of considerable axial length, rings were pressed by dies and made to interlock with each other. The corrugated rings were secured between the backplate and holding rings by means of stay bolts. The manufacturers claimed great torsional strength, the possibility of reverse running and that the cellular construction of the air passages resulted in the air being taken hold of more effectively. As an afterthought they also claimed that it was "silent running", which must have puzzled those still clinging to the belief that if it didn't make a noise, it wasn't doing much.

Turbon fans were made in sizes up to 2.54 m diameter which at 300 rev/min produced 1500 Pa fan static pressure and volume flowrates up to 280 m^3/s in double inlet form. The width was varied to suit the flowrate required and peak efficiencies of 75% were claimed.

Rateau applied his mind to the design of an axial flow fan. To achieve the high pressures required he developed a high hub to tip ratio unit (Figure 1.23) with steel vanes fixed to the rim of a slightly conical hub manufactured from cast iron. Upstream guide vanes were employed in the horizontal cased version (Figure 1.24) whilst the vertical version had a spiral admission chamber giving a contra-rotating entry (Figure 1.25). After the air left the impeller it was wholly axial and its velocity was decreased in a diffuser section.

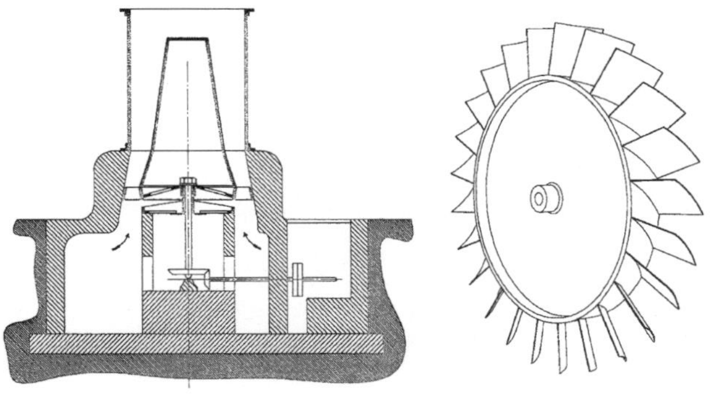

Figure 1.25 Vertical version of Rateau's axial flow fan

Figure 1.22 Waddle's Turbon fan

Whilst all this feverish activity for improving the fan was taking place, some clung to the methods of the past. Walker Brothers of Wigan, near Manchester, in the UK, sought to meet the wishes of the conservative engineers by producing the "Indestructible" fan (Figure 1.26). A good name can sell the most out-of-date product especially when the advertisers extolled the virtues of its strong construction. Aerodynamically however, all was not well, as it came complete with an "anti-vibration shutter", the blades discharging into a V-shaped aperture in the damper.

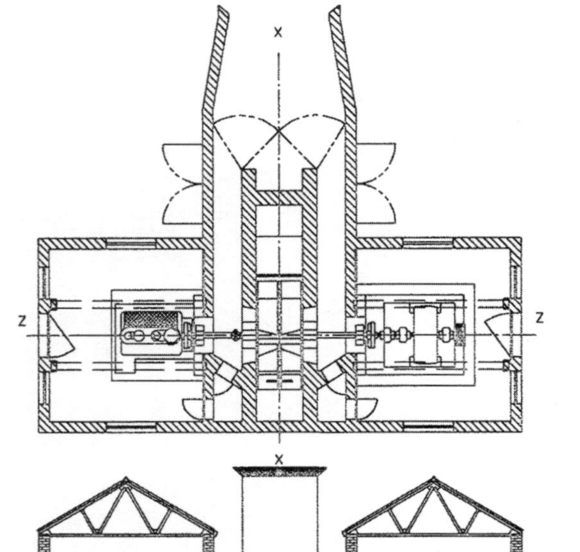

Figure 1.26 Cross-sections through Walker's so-called "Indestructible" fan

1.2.3 The dawn of tunnel ventilation

It was a natural progression from mines to tunnels. Many of the early tunnels were beset with ventilation problems during their construction. Those experienced by Marc and Isambard Brunel during their work on the first Thames tunnel are known from our school history lessons. The need for permanent ventilation did not become apparent until the 1870s and the use of the already established manufacturers of mine fans was an obvious solution.

One of the Great Western Railway of England's pioneering achievements in the field of civil engineering was the building of the 4½ mile long tunnel beneath the River Severn estuary. At the time of its construction it was the world's longest underwater and the first to connect two countries - England and Wales.

Work commenced in 1873 and the inaugural goods train ran through on the 9th January 1886, carrying South Wales coal bound for the metropolis. Passenger traffic did not commence until the December, awaiting the construction of some connecting lines, thus proving that the Channel Tunnel is unique in nothing.

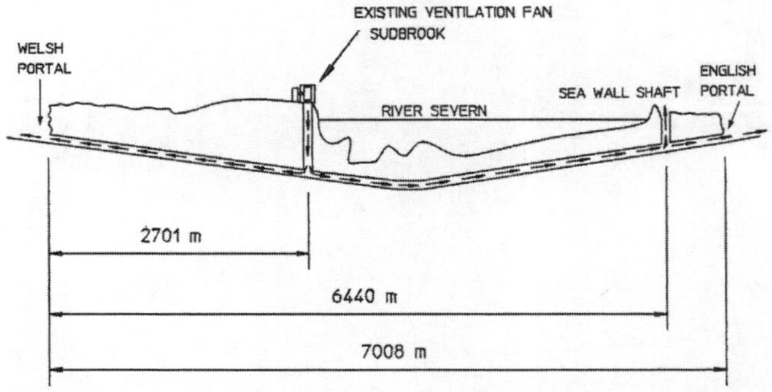

Figure 1.27 River Severn Estuary tunnel

During construction, following the death from inflammation of the lungs of two men who had been working in one of the headings, a Guibal fan having an impeller diameter of 5.5 m and a width of 2.1 m was installed. This was fitted to the top of the new pit shaft at Sudbrook (Figure 1.27). When the tunnel was completed a larger Guibal fan having an impeller diameter of 12.2m and a width of 3.7 m was installed for permanent ventilation. This was steam engine driven, the supply being from three Lancashire boilers each 2.1 m diameter by 7.9 m long. The maximum rotational speed of the fan was 60 rev/min, but less than half this was stated to be sufficient for normal operation.

Whilst the contractor, Thomas A. Walker, claimed that the appliances for ventilating the tunnel had proved to be thoroughly efficient, the inspecting officer, Colonel F. H. Rich noted that "the means of ventilation are ample, but did not act well when I made my inspection". Whatever the rights or wrongs, the Guibal fan did not last and was subsequently replaced by a Walker Indestructible fan with a capacity of 27.3 m³/s against 210 Pa fan static pressure. The characteristic curve (Figure 1.28) shows that this was not well-matched to the system and an operating efficiency of less than 40% was achieved. Nevertheless, apart from conversion of the original steam engine drive to electric motor, the unit continued to operate in its original form until very recently. Perhaps the name was well earned after all.

With the steam locomotive as the only proven and practical form of motive power, the idea of a long sub-aqueous railway tunnel raised acute problems of ventilation. Hence the first Mersey rail proposal envisaged pneumatic propulsion, a single carriage, fitting the bore like a piston, being alternatively sucked and blown through the tunnel between terminal air-locks.

This Mersey Pneumatic Railway was authorized by an Act of Parliament June 1866, but it failed to win support so, a more or-

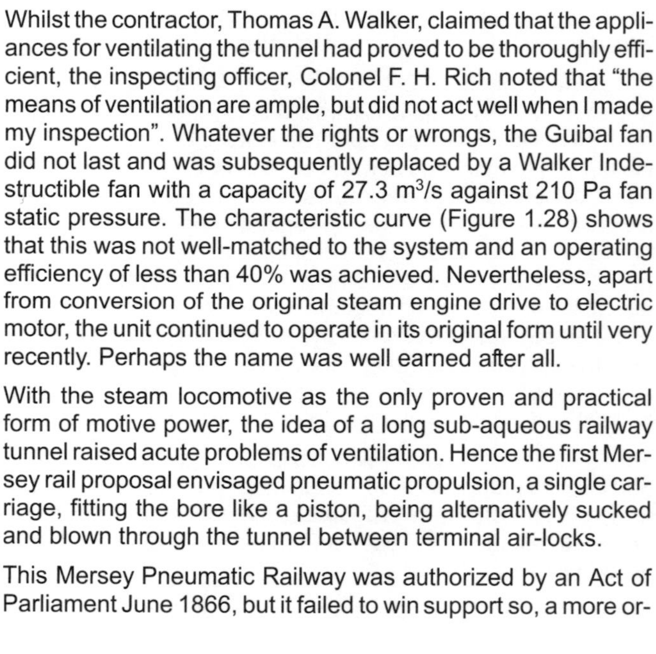

Figure 1.28 Fan characteristic curve for the Guibal fan

Figure 1.29 A Liverpool fan building

thodox scheme was substituted using condensing locomotives. The name was changed to the Mersey Railway Company and in 1871 it was authorized to make connections with main line railways on both banks and formally opened on the 20th January 1886 by the Prince of Wales.

Despite the use of giant steam-driven ventilating fans of Guibal design, but manufactured by Black Hawthorn, the tunnel had the dubious distinction of possessing the foulest atmosphere of any underground railway. There were two fans 12.2 m diameter x 3.7 m wide and two fans 9.1 m diameter x 3 m wide. It was claimed that the total extract was 274 m^3/s. It is interesting to speculate however, that as the fans were effectively in parallel, unless the smaller fans were operating at 33% greater speed, there could well have been a mismatch in pressure characteristics. The tunnel had a ruling gradient of 1 in 27, leading to the locomotives having to work very hard. It is scarcely surprising that as early as 1903 the line was electrified and steam locomotives banished from the tunnel forever.

1.2.4 The first Mersey road tunnel

Ventilation of road tunnels became of importance with the development of the internal combustion engine and the consequent carbon monoxide pollution. The Mersey road tunnel was conceived in the 1920s as an infrastructure improvement which, in a time of high unemployment, would give work to many. It was designed with a state-of-the-art ventilation system to reduce the carbon monoxide concentration and to maintain visibility. The fan stations still dominate the Liverpool skyline, along with the Liver building, and the Anglican and Catholic cathedrals. Many claim that the fan buildings, are, however, of the greatest architectural merit (Figures 1.29 and 1.30).

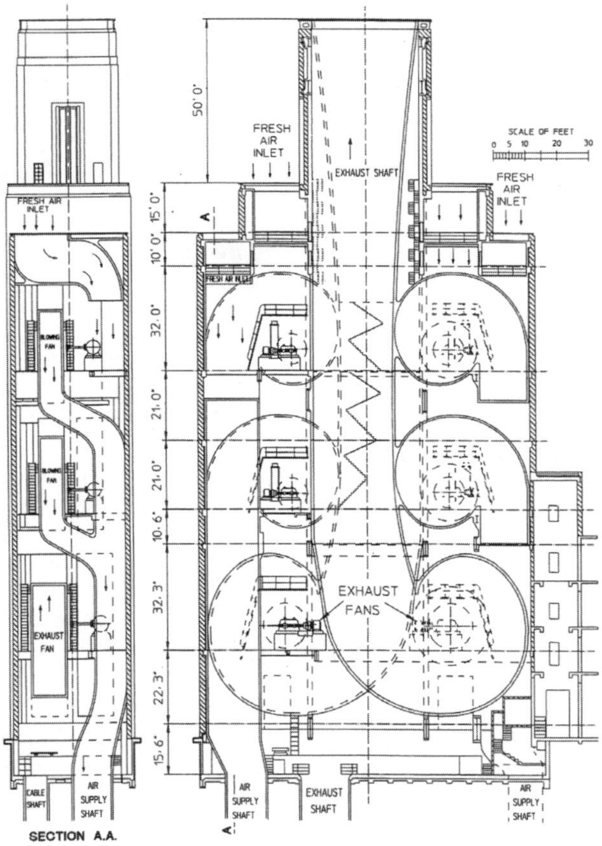

Figure 1.30 Section through a Liverpool fan station

Figure 1.31 Walker's "Indestructible" impeller

Figure 1.32 Walker's "Indestructible" fan

The nearest fan manufacturers to the tunnel, capable of constructing units of an appropriate size were Walker of Wigan and Sturtevant with a head office in London, but, importantly, a main works at Denton near Manchester. Each made bids and were so unlike each other as to cause the tunnel authorities much anguish. Walker offered its Indestructible design (Figures 1.31 and 1.32) - what else?

Sturtevant at that time had a French Chief Engineer named Lebrasseur. He designed a new backward curved bladed centrifugal fan which by appearance was the progenitor of today's modern fans and which for performance was far in advance of those currently available (Figures 1.33 and 1.34). The design, known in Sturtevant parlance as the GV/M was in reality the Grande Vitesse-Mersey thus showing an early French predilection for the use of these words. Unable to make up their minds, the authorities split the contract between the two companies, but not before the GV/M had proved its efficiency of greater than 80% on a test tunnel 46 metres long and with a cross-section 3.7 m x 3.7 m. The blowing fan tested had a capacity of 82 m^3/s.

Thirty fans in total were installed, duplicated to give running and standby capacity. The total operating supply flowrate was about 1917 m^3/s and that for extract 1211 m^3/s. It is of interest to note that the Walker Indestructible fans had impellers about twice the diameter of the Sturtevant GV/M type, but operated at a maximum speed of only 62 rev/min. All these fans have been operating almost continuously since 1934 and in 1994 celebrated their 60th anniversary.

Figure 1.33 The Sturtevant GV/M backward curved bladed centrifugal fan with temporary steel casing for test purposes

Figure 1.34 The Sturtevant GV/M backward curved bladed centrifugal fan with final concrete casing on site

1.2.5 Mechanical draught

It had been known for centuries that the output of a blacksmith's forge could be increased by the use of a bellows. Later small centrifugal fans were substituted as a labour saving device. As pressures were relatively high for the flowrate, narrow designs were developed incorporating cast iron casings. That produced by Beck and Henkel of Cassel, Germany is shown in Figure 1.35 and is an early example of a unit used not only for forge blowing but also cupolas producing cast iron. The complexity of the design must be admired as a high example of the iron founder's art, and creates a sense of envy for what we cannot do today - the cost would be enormous.

Another German fan of considerable interest is the Geneste-Herscher design (Figure 1.36) which gained first prize at the Paris Exhibition of 1900. We can see that, although of the forward curved bladed centrifugal type, considerable attention was paid to the form of the inlets whilst the volute had a rectangular cross section uniformly increasing to the outlet.

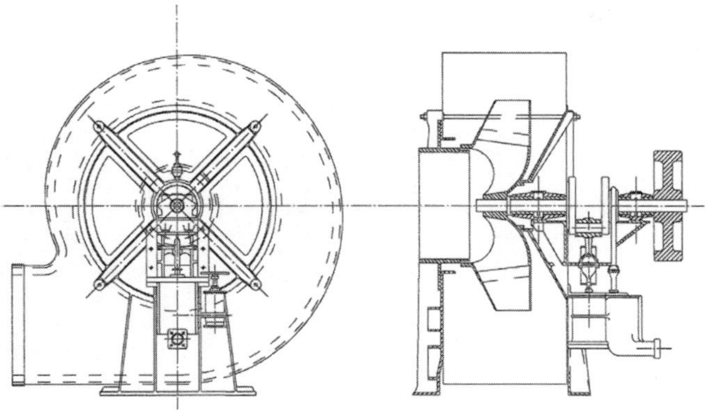

Figure 1.35 The Beck and Henckel centrifugal fan

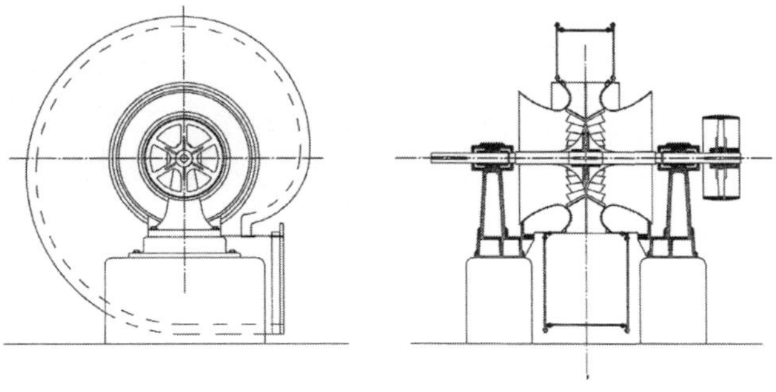

Figure 1.36 The Geneste-Herscher centrifugal fan design

We now come to another giant of the fan world, James Howden. Starting in 1854 as a consultant to the flourishing shipbuilding and engineering industry around Glasgow, he soon appreciated the need for improvements to engines and boilers. By 1881 he had developed and sold a range whose efficiency and output exceeded anything available.

During this period the concept of supplying air to a boiler under pressure from a fan so that it could also pass through a pre-heat section, to extract heat from the flue gases, emerged. Trials were carried out in the winter of 1882/3 and by the following year had been demonstrated on a refitted ship. Until then, trans-Atlantic steamers had to augment their driving force with sails, as with Brunel's *Great Eastern*, or had to proceed via Iceland and/or Newfoundland for refuelling. Now the resultant improvement in fuel efficiency and power enabled them to reach New York non-stop.

From this stage his company developed and the forced draught business increased to such an extent that it dominated all its activities. Boiler making eventually ceased in favour of fans for their marine forced draught system. By 1926 the system had been developed to the extent that land-based water tube boilers incorporating air pre-heaters, with forced and induced draught fans were operating with complete success.

James Keith's company had also manufactured boilers and naturally followed Howden's example in marine usage. A modified design of the patented impeller was applied to the forced ventilation of engine rooms. That for the *Lusitania* was manufactured in 1912 and is shown in Figure 1.37. Is it still at the bottom of the sea?

1.2.6 Air conditioning, heating and ventilation

There is considerable evidence that prehistoric man used fire to produce heat for his comfort. Native Americans also used open

Figure 1.37 James Keith's patented impeller

fires within their wigwams and allowed the products of combustion to escape through the hole at the apex, at the same time inducing fresh air.

Medieval Europeans developed fireplaces so that the smoke could be guided up a chimney, resulting in a stack effect which improved combustion and provided room ventilation. Of course, the Romans had done much better 1500 years before by constructing flues within the walls of their buildings to give the first central heating.

Public buildings were some of the first to employ a mechanical system of ventilation. Perhaps as a consequence of the large amounts of hot air produced, the Houses of Parliament in London were provided with a supply and extract system of ventilation as early as 1836. In the unlikely event of heating being necessary, air was drawn through steam coils adjacent to the fan. The air was also washed with water sprays and cooling could be achieved by the use of ice.

The more humble beginnings of building ventilation, however, started with the propeller fan which is believed to have originated in the United States. Perhaps times were hard, or the English considered gullible, for Lucius Fisher, Walter Burnham and James Morgan Blackman, all of Illinois, moved to the United Kingdom and formed the Blackman Air Propeller Ventilating Co. Ltd on 10th September 1883.

A number of propeller fan designs were produced in those early years, each having completely different blades, apparently conceived on the basis of "try anything once". All were designed for belt drive, usually from overhead line shafting. By 1891, however, an electrical direct drive version was available. A prototype produced at the time when the Tottenham factory closed is shown in Figure 1.38.

Perhaps even more interesting was the patented version produced by Blackman's engineer Mr Water, which had no separate motor either coupled to the spindle or belted to a pulley. The fan was its own motor, its periphery being the armature, its frame the field magnets and the commutator occupying the place of the pulley. Was this the first inside-out motor driven fan, albeit in a DC form?

This fan was stated to work at "a moderate speed consistent with sound and economical practice ... and all noise and risk of vibration is reduced to a minimum". By 1896 *Electrical Review*

Figure 1.38 The Blackman propeller fan prototype

was waxing lyrical in its description. The long extract that follows, is interesting for its language, if nothing else:

> ***Fresh air by electricity***
>
> *Of the many beneficent purposes to which electricity is applied, none can be more conducive to the comfort and health of the community than its use for driving ventilating fans; and it is with pleasure that we observe the rapidly increasing number of electrically driven fans that are being installed for the removal of all kinds of disagreeable fumes, such as the appetising(?) odours that arise from the kitchen, and the unhealthy products of gas burners* [incandescent and otherwise]. *Enquiries made at the London office of the Blackman Ventilating Company,* (the name had soon been shortened) *and an inspection of some of the installations of their well-known fans, has convinced us that a wide field is being opened up, and one that will form a valuable addition to the central load.*
>
> *Not only in the larger public buildings such as the Houses of Parliament, the Stock Exchange, Hotels Cecil, Metronome, Holborn Restaurant, etc. are electric Blackmans* (note the use of a name – just like Hoover) *freely used for ventilating the dining and smoking rooms, kitchens, and billiard rooms, but many leading club-houses, hotels and private residences are thus fitted.*
>
> *The wood cut* (Figure 1.39) *shows the electric Blackman with peripheral motor, as fixed to the upper part of a window. A considerable number of these latter are at work, some of them on windows of the most highly finished rooms in London, and the effect is in every way satisfactory.*
>
> *The stuffiness which was once a characteristic of the apartments on board ship is in many cases a thing of the past - electric fans are fixed in the dining saloons, drawing fresh air through them and forcing it away when practicable through the cooks' galleys, thus preventing the odours of cooking from penetrating various parts of the vessel, and preventing many an attack of* mal de mer, *the sleeping apartments are also ventilated.*
>
> *It is interesting to note that Messrs Siemens Brothers have six Blackman fans, direct coupled to Siemens motors, on board their cable ship* The Faraday, *and on its late trip up the Amazon, although the voyage was a most trying one, yet not a single case of yellow fever occurred, and the crew were able to take their meals in the dining saloons, and sleep in their berths, while on previous similar occasions they were driven to eat and sleep on deck.*
>
> *Speaking of ship ventilation reminds us that the Czar of Russia has followed the example of Her Majesty the Queen by having his magnificent yacht ventilated in this way.*

Early development of heating, ventilation and air conditioning was held back by the lack of authentic design data. Not only

Figure 1.39 The electric Blackman with peripheral motor

was it impossible to calculate the heating or cooling load, but little was known of equipment capacity, so that they could not be matched.

To proceed beyond the empirical methods, closely guarded by the few companies in the trade, it was necessary to develop the scientific principles involved. Thus was born the American Society of Heating and Ventilating Engineers which had its first annual meeting in 1895. It was followed by the Institution of Heating and Ventilating Engineers (UK) in 1897. In 1904 the American Society of Refrigerating Engineers was founded whilst the Swedish Heating, Ventilating and Sanitary Engineers Association commenced operations in 1909. All these organizations were active from the start in producing performance standards and in publishing records of research and applicational experience.

The expression "air conditioning" is believed to have been first used by S. W. Cramer who presented a paper on humidity control of textile mills to the National Cotton Manufacturers Association (USA) in 1907. The measurement and control of the moisture content of textiles was known as "conditioning" in the trade, so that the means of circulating humid air to achieve the desired textile moisture content was a natural extension.

Air conditioning was recognised as a branch of engineering in 1911 when Dr Willis H. Carrier presented his two papers *Rational Psychometric Formulae* and *Air Conditioning Apparatus* to the American Society of Mechanical Engineers. From thereon the use of fans for the air conditioning and ventilation of buildings was rapid. Until that time very large buildings had to have a "light well" at their centre so that not only could all rooms have access to natural light, but they could also be ventilated by opening the windows.

Now architects were released from this consideration. It is tempting to think that skyscrapers could not have reached their present size without fans. By the mid 1920s there were many centrifugal fan manufacturers producing standardized ranges of forward and backward curved types. Selection by multi-rating tables was common but it was H F Hagen of the B F Sturtevant Co. of Massachusetts who was the first to devise an ingenious graphical method under US Patent No. 1358107.

Figure 1.40 The Stork aerofoil backward bladed centrifugal impeller

1.2.7 Developments from the 1930s to the 1960s

In the late 1930s, Stork Brothers of Hengelo, in the Netherlands, introduced its aerofoil backward bladed centrifugal fan (Figure 1.40) which enabled efficiencies in the high 80s% to be achieved over a considerable portion of the characteristic. It coincidentally produced a reduction in noise levels.

In 1955 tests by Professor Sorensen had shown that the Schicht fan (Figure 1.41), produced by KKK of Frankenthal-Pfalz, Germany, could produce efficiencies in excess of 80%. Static pressure through the impeller remained constant and was only increased by retardation in the diffuser section. Due to the accelerated flow velocity, shaped blades were unnecessary, and the fan capacity was, therefore, unchanged by deposits, rust or erosion. In consequence the fan has been widely used for induced draught applications, control being by means of a radial vane inlet damper.

Aerex Ltd evolved a series of axial flow fans for mine ventilation using an impeller having patented blades of fabricated stainless steel, hollow formed to true aerofoil section. The blades could have their pitch angle changed at the periphery without entering the hub. Both up and downstream guide vanes were used. Fans were often arranged for horizontal drive through vee-belts from a side mounted motor and an integral outlet bend/diffuser was fitted. Many such fans were supplied to South Africa for use in gold and coal mines. A typical example for Wankie Colliery is illustrated in Figure 1.42.

The Axcent mixed flow fan was originally patented by Keith Blackman Ltd in 1958 (Figure 1.43) and was claimed to combine the advantages of both the axial and centrifugal types. With its steep pressure/flowrate characteristic and non-overloading power curve, its performance was more akin to a two stage axial fan. Subsequently improved versions have been produced with fan static efficiencies in excess of 70% and noise

Figure 1.41 The Schicht fan

Figure 1.42 An Aerex axial flow fan

Figure 1.43 An Axcent mixed flow fan

levels comparable with centrifugal fans. Such fans are widely used offshore for the ventilation of oil rig platforms in the North Sea. Their ability to maintain almost constant airflow under strong contrary winds has been as much valued as their low mass and compact dimensions.

1.2.8 More recent tunnel ventilation fans

Perhaps the notable feature of more recent tunnels has been the almost universal use of axial flow fans. The development of high duty aluminium alloys for the aircraft industry has meant that the tip speeds necessary for reasonable pressure development make the axial fan highly competitive. Flexible in design and much more compact, it can be installed horizontally, vertically or at any angle such that duct runs can be considerably simplified.

One of the early users of appreciable numbers of these fans was London Transport which has over 325 kilometres of tube railway. Generation of heat arises naturally from the continuous input of energy from train operation. There is a steady rise in the temperature of the air over a number of years due to the heat build up in the clay surrounding the tunnels which has to be corrected by ventilation.

Many fans for this usage were vertically mounted (Figure 1.44) and driven from a vertical motor through Vee-belts. The fans had to operate against widely fluctuating system pressures due to the piston effects of approaching or receding trains. They were designed with relatively low pitch angled blades to give a rising pressure characteristic back to zero flow, guarding against flow reversal. A number of manufacturers supplied these in sizes around 2.5 m diameter.

Probably the most recent usage of centrifugal fans for tunnel ventilation in Europe was in the late 60s by the Greater London Council. Both the Hyde Park Corner and Strand underpasses

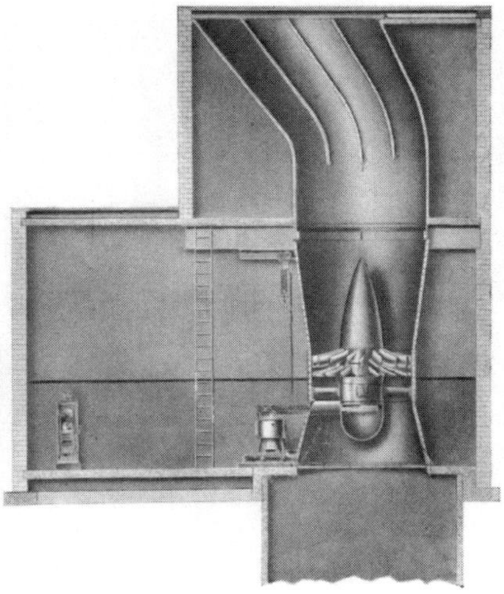

Figure 1.44 Vertically mounted tunnel ventilation fan

used backward bladed aerofoil fans. The former incorporated 8 Carter Howden 2.4 m diameter units (Figure 1.45). The Strand underpass, which was a conversion of the old Kingsway tram tunnel, used two 1.8 m double inlet double width fans, although these have subsequently been replaced by axial flow fans.

The first Mersey (Queensway) tunnel had been engineered on the grand scale and in 1925 no-one would have believed that it would ever reach vehicular saturation point. During its first year it handled over 3 million vehicles and by 1959, this had risen to 11 million vehicles. The original ventilation system could no longer cope and in 1964 additional axial flow fans were installed. Traffic continued to rise and in 1968 no less than 17 million vehicles were handled with 60,000 in one day.

Planning for a second (Kingsway) tunnel began in 1958 and this was opened in 1971. Ventilation was by the same upward semi-transverse system as used in the first tunnel with supply at the rate of about 0.3m^3/s per metre run. The blowing shafts were offset from the line of the tunnel whilst the adjacent exhaust shafts were positioned directly above (Figure 1.45). Ventilating stations were over their respective shafts, behind the promenade at Seacombe, and on the inland side of Dock Road. Both stations were surmounted by evasees which whilst not so meritorious as the ventilating stations of the first tunnel, nevertheless are noteworthy landmarks. They could even be said to have a 70s-style "pipe of peace" affinity with the Roman Catholic Cathedral, scurrilously known as "Paddy's Wigwam".

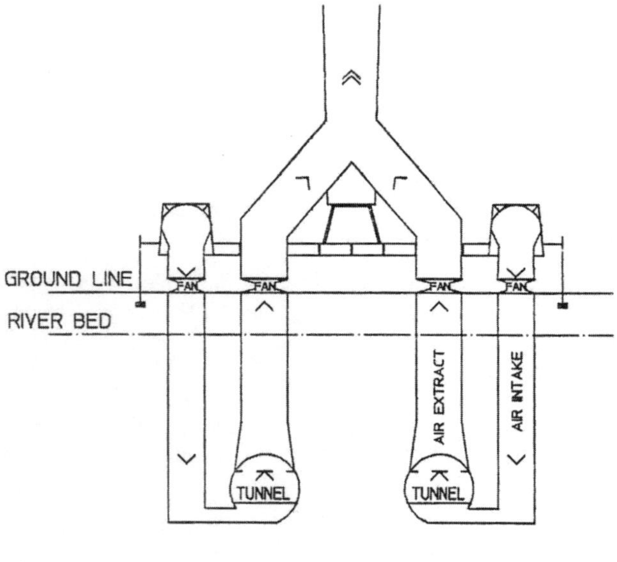

Figure 1.45 Mersey (Kingsway) tunnel ventilation system

Each tunnel tube is ventilated by two supply and two extract fans, one of each on either side of the river. An additional complete standby fan is linked together with the operating fan on bogies having traversing drives and carried on rails (Figures 1.46 and 1.47).

In operation, one fan is held in the surface position in line with the ventilation shaft, whilst its partner rests over a maintenance pit. In the event of failure, the fans automatically traverse to bring the standby into operation. Perhaps in imitation of the original tunnel, the order was split between Aerex and Davidson.

Each fan is driven through a 90° reduction gearbox coupled to a low speed induction motor. Fan speed is controlled by carbon monoxide monitors in the tunnels. Supply fans are 5.2 m diameter and have a duty of about 350 m^3/s against 750 Pa at 129

Figure 1.46 Mersey (Kingsway) tunnel Aerex fan

Figure 1.47 Mersey (Kingsway) tunnel Davidson fan

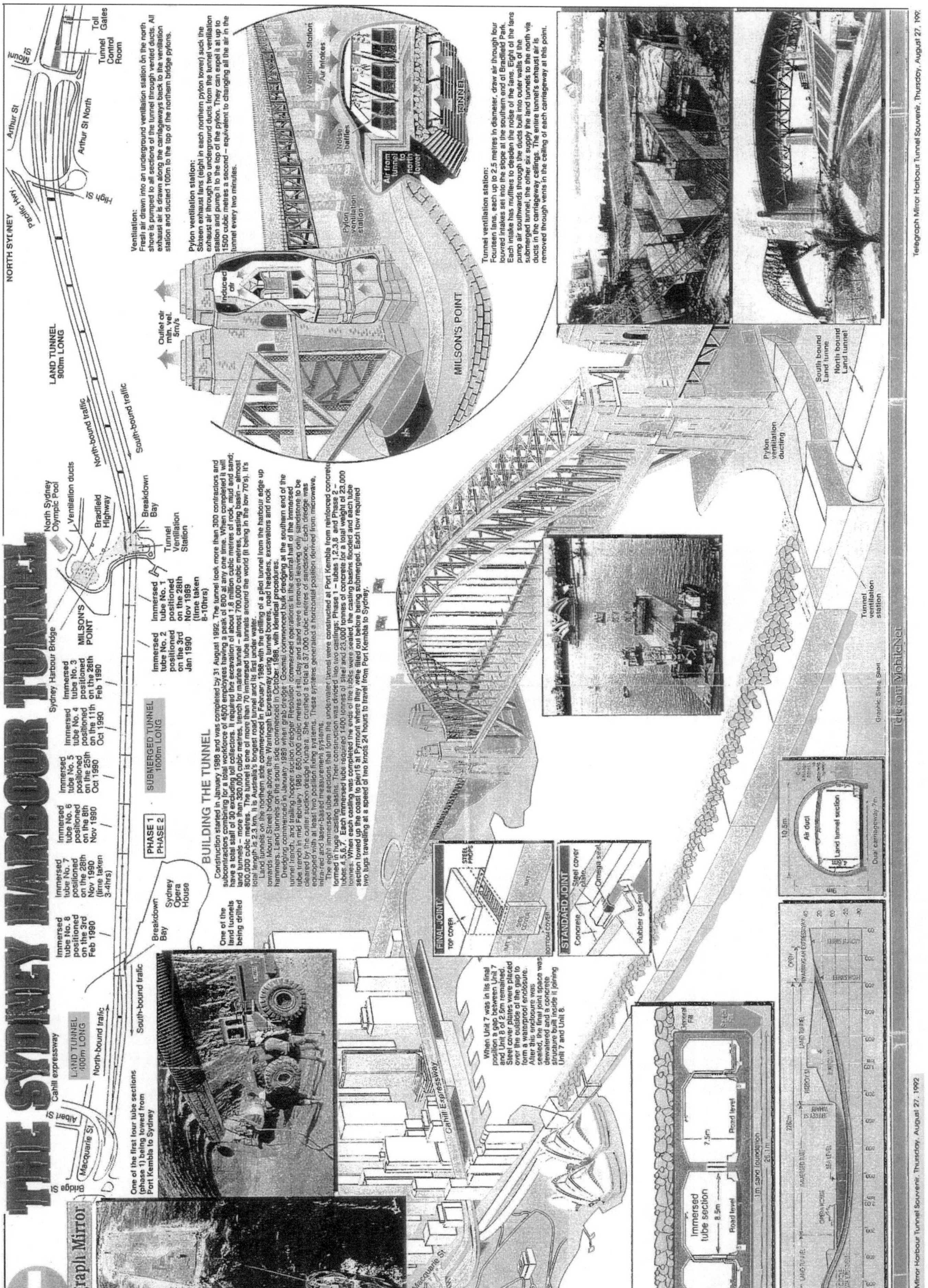

Figure 1.48 Article from the Australian *Telegraph Mirror Harbour Tunnel Souvenir,* illustrating the novel approach taken
Courtesy of The News Ltd, Sydney, Australia

rev/min absorbing 306 kW. The extract fans are 6.1m diameter and have a duty of about 387 m^3/s against 200 Pa at 245 rev/min for a power of 107 kW.

The Ahmed Hamdi tunnel is a 1640 metres long, two lane, two way road tunnel beneath the Suez Canal at El Shallufa, approximately 10 miles north of Suez, in Egypt. The ventilation is a fully transverse system supplying air through ducts under the road and extracting through the false ceiling which forms the extract duct. A total of 16 two stage fans, 1.9 m diameter, were installed in two extract and two supply fan chambers. The system was designed to reduce the carbon monoxide level to 250 ppm maximum and the diesel smoke level to 20% Westinghouse maximum.

Equipment had to withstand sand and dust storms and an ambient temperature of 45°C. It was also necessary for equipment to withstand a temperature of 250°C for one hour before breakdown. In the event of a fire, supply fans would be reversed and all 16 two stage fans would be extracting smoke from the tunnel.

To cater for the enormous increase in cross harbour traffic over the famous Sydney Harbour Bridge, Australia, and to relieve the subsequent heavy congestion on the bridge approach roads, it was decided that a tunnel should be constructed underneath the natural harbour. A newspaper article from the Australian *Telegraph Mirror* of 27th August 1992, illustrates the novel approach taken, see Figure 1.48.

The tunnel is 2.3 km in length. Two of the main requirements were that the supply fans had to be capable of running in reverse in an emergency and all fans be rated for smoke extract. Each of the fans has a duty of 53 to 103 m^3/s. (Figure 1.49). The testing programme was one of the most comprehensive ever, covering flowrate and pressure, power measurements, sound levels, bearing vibration, X-raying of all impeller components, high temperature tests at 200 °C for 2 hours, impeller strain gauged for centrifugal and fluctuating stress, and 24 hour run tests with reversals.

In Hong Kong, a number of tunnels (Eastern Harbour, Junk Bay, Lion Rock, Tates Cairn, MTR Island Line, etc) have been built to link the island to the mainland for both road and rail traffic. Some of these have been characterized by increasing fan capacity as traffic density has increased.

Figure 1.49 Supply fan for the Sydney Harbour tunnel

The Eastern Harbour crossing is but one of many and is a combined road (2.1 km) and rail (6 km) tunnel in one immersed tube which links Cha Kwo Ling near Kwun Tong on the Kowloon Peninsula with Quarry Bay on Hong Kong Island. The equipment was designed to cover normal tunnel ventilation, dilution and extraction of smoke and gases in a road tunnel through both overhead and low level side ducts. Emphasis was placed on the suitability of fans and associated acoustic treatment material being capable of working in high temperatures and in a hazardous environment. Fresh air is supplied from ventilation buildings located at each end of the tunnel, using 20 2.5 m diameter axial type fans. During emergency conditions 10, 2.8 m diameter exhaust fans operate to extract smoke. The environment is maintained by an intelligent computer control system. A total of some 180 fans in varying sizes are used.

Piston effects from moving trains in the Channel Tunnel cause the fans to operate over an extensive range of the fan characteristic. This calls for aerodynamic stability from windmilling to flow reversal, with a continuously rising and power limited fan characteristic. These criteria apply for both forward and reverse modes. The axial fans selected for both normal (NVS) and supplementary (SVS) ventilation (Figure 1.50) are hydraulically actuated with controllable blade pitch in motion. There are four 2 m diameter NVS axial fans having a capacity of 89 m^3/s and four 4 m SVS axial fans (Figure 1.51) each with a capacity of 300 m^3/s.

All these fans are aerodynamically stabilized by means of the Axico anti-stall ring which introduces two chambers, one on either side of the impeller, providing stable flow conditions and continuously rising fan characteristics in both flow directions. When in the stall region, the separated and highly turbulent flow is removed from the main flow annulus and entered into the stabilizing peripheral ring-shaped duct just upstream of the impeller blades.

1.2.9 Longitudinal tunnel ventilation by jet fans

This system of ventilation was first tried in Italy about 40 years ago. Ventilation cost is greatly influenced by the section length between access points at which fresh air may be supplied and polluted air exhausted. Longitudinal ventilation systems without ducts, in which the whole of the required airflow moves through the tunnel at constant velocity have become increasingly popular. To provide a positive longitudinal pressure difference, jet fans (Figure 1.52) are suspended from the tunnel roof and blow in the same direction as the traffic (normally one way) though they are often capable of reversal according to traffic density or for emergency smoke ventilation.

The lower fan efficiency can often be more than offset by the reduction in the pressure required due to the absence of a ducting system. Tunnels with lengths exceeding 1 km in length become increasingly difficult to ventilate by this method, as the tunnel air velocity becomes excessive. Hybrid systems of longitudinal and extract ventilation have, therefore, been developed.

Many hundreds of kilometres of road tunnel in Italy have been ventilated by the longitudinal induction method, including the Naples Tangenziale, the Lecco-Colico Super Strada around Lake Como, and the Frejus IV tunnel. The method has also been used in the UK for tunnels on the M25 London Orbital Motorway, the A55 North Wales Expressway and the A20, A27 and A38 trunk roads.

Barcelona, the principal commercial city in Spain, staged the 1992 Olympic Games. In order to relieve the current and anticipated congestion, the government built a new 12 km expressway, almost 3 km of which is underground in cut-and-cover tunnels.

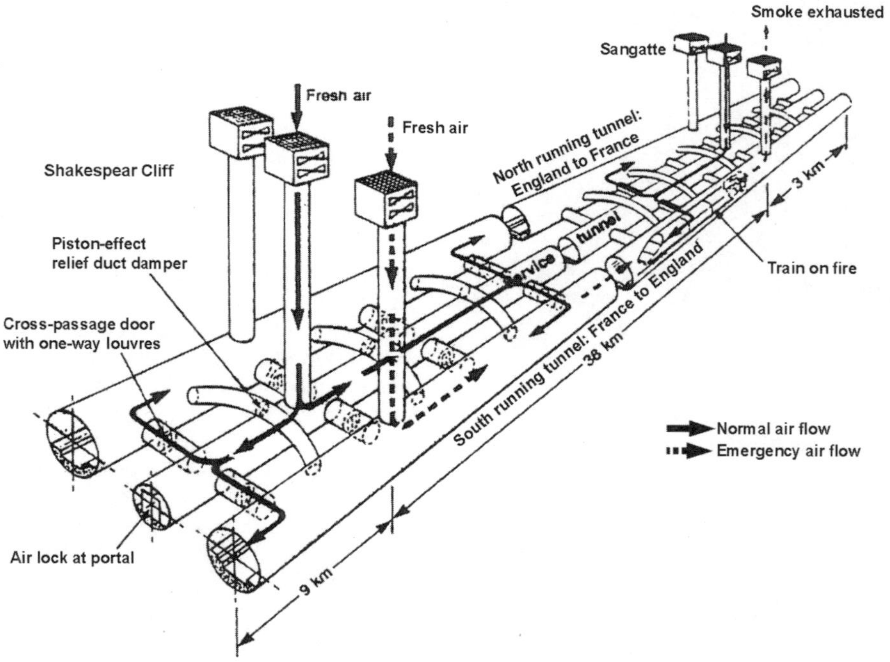

Figure 1.50 The Channel Tunnel ventilation system

Figure 1.51 4.5 m SVS axial flow fans for the Channel Tunnel

Figure 1.53 Purpose-designed jet fans

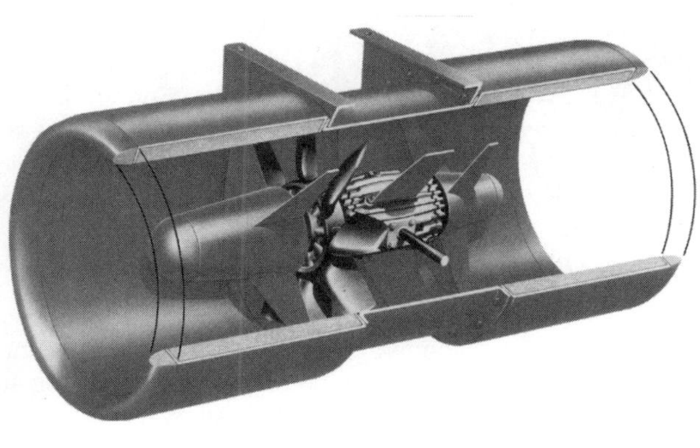

Figure 1.52 Typical jet fan

Comprising five tunnels, four single way and one for two way traffic, ventilation in these tunnels was designed on the longitudinal system, using main and jet fans. The longest tunnel, Vallvidrera, at 2.5 km includes three shafts, each having a 2.8 m aerofoil axial flow fan for smoke venting only. A "Galeria" provides a means of escape and 10 fans of 610 mm diameter, 2 speed, maintain pressure across each door to prevent smoke passing through. 30 purpose-designed jet fans of 1.6 m diameter and truly reversible (Figure 1.53) are grouped in 5 rows of 3, 15 each at either end of the tunnel. In the 4 remaining shorter tunnels a total of 710 mm uni-directional jet fans are used.

One of the strategic plans for the regeneration of London's old docklands area, made redundant by the sea container revolution, was the provision of the 1.6 km Limehouse Link road. This is believed to be the most expensive ever constructed on a per length basis.

A major challenge was to design a ventilation system which could deal with a disaster such as a 50 MW fire as well as the pollution caused by very heavy traffic flows. Other factors included the effect of noise on nearby residents and traffic control in the tunnel. The road was designed for a maximum of 1800 vehicles per hour per lane for free flowing traffic and ventilation is achieved by a system of 128, 710 mm jet fans mounted at intervals across the tunnel roof in groups of four (Figure 1.54).

Air is propelled in the direction of the traffic flow and then exhausted at the portals through grilles in the roof of the tunnel. From there it is ejected through exhaust chimneys by 8 2.8 m diameter and 4 x 1.5 m diameter axial fans (Figure 3.55) mounted on the roofs of the service buildings. The ventilation system was complex because of the road junctions.

Extensive computer modelling studies were carried out in order to analyse fire and smoke control in the case of fire or accident. For this exercise, the tunnel ventilation system is divided into

Figure 1.54 Jet fans used in the Limehouse Link road

Figure 1.55 Axial exhaust fans for the Limehouse Link road

six areas and the size of blaze anticipated is equivalent to a medium-size petrol tanker catching light. The level of ventilation has to be balanced between allowing people to move with safety and the need to blow the smoke away.

1.2.10 The rise of the axial flow fan

When reading the previous sections of this Chapter, it will have been noted that the years since World War II have been characterized by the rapid development of axial flow fans. This has been due in no small part to the efforts of Woods of Colchester Ltd – now part of the global Fläkt Woods Group. No other industrialised country manufactures such a high proportion of axial flow fans (well over 50% of the total). A brief history of this company therefore seems appropriate.

When the late Mr Maurice Woods came to Colchester in 1909, he had previous experience of operating an electrical generating station in Hampstead, London. His main interest was in the design and development of electrical machines and so he set up his company with premises at the Hythe and a total workforce of 6 people.

At that time electrical voltages and frequencies throughout the United Kingdom were far from standardized and there was considerable scope for small manufacturers to provide the many special machines required. Although the majority of motors were wound for DC supplies, Mr Woods built up his business and reputation by competently producing AC single phase machines for 100v 100Hz, 300v 400Hz and even 105v 77Hz AC. It was not long before the motors were being applied to ceiling and propeller fans so that by the 1930s, electrically-driven fans were the sole product (Figure 1.56).

Figure 1.56 Early production of electrically driven Woods fans

In 1947 the first standardized range of axial flow fans were introduced, these having sand cast constant chord, constant pitch aluminium impellers. It is believed that this range was the first to be manufactured on a batch production basis.(See Figure 1.57.)

Figure 1.57 First batch production of axial flow fans

By 1958, contra-rotating two stage axial flow fans were introduced, using many of the same components but with the second stage having opposite handed blades). By this means, the rotational energy of the air from the first stage was recovered. Instead of twice the pressure being developed, this was increased to three times. Many applications previously furnished with centrifugal fans could now be provided with these units which were more compact, cheaper and had a reduced starting load on the supply.

The performance of an aeroplane propeller can be changed by rotating the blades, such that their pitch angle is altered. In 1963 this technology was adapted by Woods, in its first range of Variable Pitch in Motion fans.

1.3 Definitions and classification

1.3.1 Introduction

In the early years of fans the design and manufacturing engineers were too busy making the things work to worry overmuch about definitions and classification. Once they had become established, however, these topics proved irresistible to academics and administrators. They have occupied their minds ever since.

It was not until 1972 that Eurovent produced its document 1/1 which gave agreed terms and definitions for fans and their components. This document was subsequently adopted by ISO and became ISO 13348. The content of this is described in more detail in this Chapter and in Chapters 9 and 11.

It should be apparent that classification can sometimes prove restrictive. Again the analogy with automobiles will indicate likely difficulties – estate cars had to become "people carriers" MPVs and stationwagons to sufficiently describe what was available.

Even the definition of what exactly is a fan has proved difficult for the industry to accept. The differences between it and a compressor are still the subject of much argument.

1.3.2 What is a fan?

We have seen from Section 1.2 that fans are built in all shapes and sizes. They run from the very lowest to high speeds. Their performances are just as different. Whilst it may be obvious, let us therefore have a general definition, on which hopefully we can all agree, of what we are talking about. That enshrined in Eurovent 1/1 and ISO 13348 is as follows:

> "A fan is a rotary-bladed machine which receives mechanical energy and utilizes it by means of one or more impellers fitted with blades to maintain a continuous flow of air or other gas passing through it and whose work per unit mass does not normally exceed 25 kJ/kg."

All very interesting, you may declare. But what exactly does it mean and why the need for an upper limit to the work per unit mass? The definition which follows is coloured, of course, by the texts which the author has read, and by his experiences over the years:

> "A fan is a rotary-bladed machine which delivers a continuous flow of air or gas at some pressure, without materially changing its density".

The words have been carefully chosen. Our sort of fan is not something for old-fashioned ladies to hide behind — thus the requirement for rotary motion. The flow is continuous into, through and out of the unit. Thus we can distinguish a fan from a positive displacement machine with pistons, vanes or lobes where the flow pulsates. A maximum pressure rise or density change has to be included to differentiate between fans and compressors. ASME, in its performance test Code PTC11 says that the boundary is "rather vague".

AMCA/ASHRAE in Standard 210/51 state that "the scope has been broadened by eliminating the upper limit of compression ratio". Nevertheless, a boundary exists somewhere.

ISO/TC117 has proposed that a maximum absolute pressure rise of 30% should be adopted. This equates to 30 kPa when handling standard air. For any others not yet fully metricated, this is about 120 ins water gauge. However, there are machines which we would recognise as fans developing pressures up to 240 ins water gauge or 60 kPa. Equally there are machines recognizable as compressors developing less than 6 kPa.

The prime function of a fan is, therefore, to move relatively large volumes of air at pressures sufficient to overcome the resistance of the systems to which they are attached. A fan's aerodynamic performance in terms of the pressure it generates as a function of flowrate, and how efficiently this is done, is what differentiates one fan type from another.

For any specific duty of flowrate and pressure rise, an infinite number of fans of varying types could be offered. Figure 1.58 shows an end elevation of their impellers. Apart from these variations in impeller design, the units could be of small diameter running at high rotational speed or conversely larger fans at low speeds. The selection of an appropriate fan will be influenced by space availability, driving method, noise limitations, aerodynamic and mechanical efficiency, mechanical strength and even, alas, capital cost and lead time. The manufacturer invited to tender may not have the optimum design within his manufacturing programme and this will lead to less than ideal solutions.

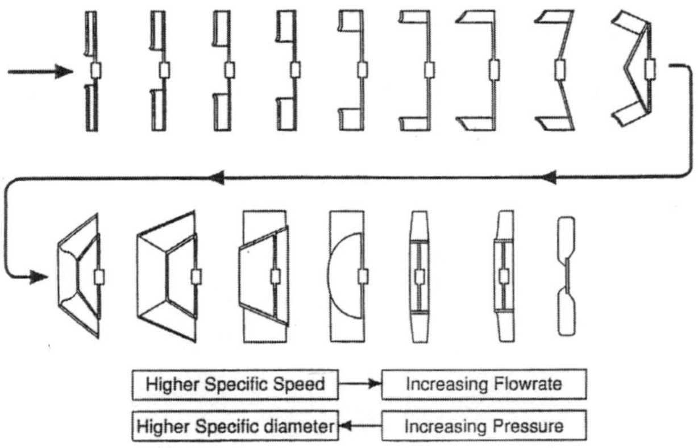

Figure 1.58 End elevation of impellers showing variation with flowrate and pressure

It will be noted that Figure 1.58 essentially indicates a continuous range of aerodynamic designs from low flowrate/high pressure through to high flowrate/low pressure. There is a continuing increase of inlet area available to the air from the narrow centrifugal fans through to the propeller fans where the total swept area is open to the flow. Whilst the main generic types may be identified as shown in Figure 1.59, there are in fact no definite boundaries between the types and there are many intermediate types which have been designed or are possible.

There is, as has been previously stated, a variety of fan designs, but practically and for the sake of *Fans & Ventilation*, we may identify five generically different types (Figure 1.59) characterised by their impellers and the flow through them:

a) **Propeller or axial flow** where the effective movement of the air is straight through the impeller at a constant distance from its axis. The major component of blade force on the air is directed axially from the inlet to outlet side, the resultant pressure rise being due to this blade action. There is also, of course, a tangential component which is a reaction to the driving torque and the air, therefore, also spins around the impeller axis. Suitable for high flowrate to pressure ratios.

b) **Centrifugal or radial flow** where the air enters the impeller axially and, turning a right angle, progresses radially outward through the blades. As the blade force is tangential, the air tends to spin with these blades. The centrifugal force resulting from the spin is thus in line with the radial flow of the air, and this is the main cause of the rise in pressure. According to the blade inclination or curvature, there may also be an incremental pressure rise due to the blade action. Suitable for a low flowrate to pressure ratio.

c) **Mixed or compound flow** where the air enters axially but is discharged at an angle between say 30° and 80°. The impeller blading extends over the curved part of the flow

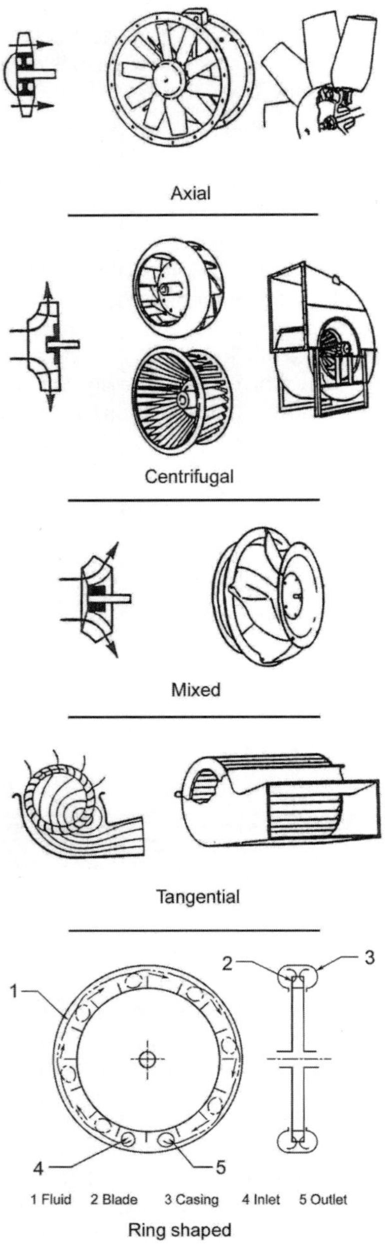

Figure 1.59 The five main generic fan types

path, the blade force having a component in the discharge direction as well as the tangential component. The pressure rise is thus due to both blade and centrifugal action. Intermediate in flowrate and pressure rise between the centrifugal and axial.

d) **Tangential or cross flow** in which a vortex is formed and maintained by the blade forces and has its axis parallel to the shaft, near to a point on the impeller circumference. The outer part of this vortex air is "peeled" off and discharged through an outlet diffuser. Whilst similar in appearance to a centrifugal impeller, the action is completely different, an equal volume of air joining the inward flowing side of the vortex. Thus air has to traverse the blade passages twice. Suitable for very high flowrates against minimal resistance.

e) **Ring-shaped in which the circulation of air or gas in a toric casing is helicoidal**. The rotation of the impeller, which contains a number of blades, crates a helicoidal trajectory which is intercepted by one or more blade, depending on the flowrate. The impeller transfers energy to the air or gas and is usually used for very low flowrates.

1.4 Fan characteristics

A fan's performance cannot easily be described by a single figure. Thus it differs from a motor car, which for many years was specified by its horsepower, under a known set of conditions e.g., RAC or DIN etc.

There are two quantities which are of interest to the user – the volumetric flowrate and the pressure rise. Both quantities vary over a wide range, but they do have a fixed relationship with each other. The best way of defining this relationship is to plot a characteristic curve on graph paper. Ideally it will be plotted at a fixed rotational speed, although for some direct driven fans an "inherent-speed" curve may be desirable.

Almost invariably the volumetric flowrate is plotted along the baseline (the x axis) whilst the fan pressure is plotted as the ordinate or y axis. This is the minimum amount of information which would be given. Other performance characteristics such as absorbed power, efficiency and noise level can also be added as further ordinates. Examples of these are shown against specific blade forms in Section 1.6 and onwards.

The peak efficiency of the fan can always be found at a specific point or duty on the curve. Where efficiencies are also added as curve information, this is easily identified as the "best efficiency point" (b.e.p.). As operation here gives the lowest power consumption of a particular design, it is desirable from an energy efficiency viewpoint. It usually achieves the added benefit of the lowest possible noise level for that particular design.

Fans can however be operated at other points on their characteristic curves, where, for example a smaller fan at higher speed can be selected, albeit at lower efficiency and higher noise level. These duties will be to the right of the b.e.p. In like manner a fan, which is oversized, will to the left of b.e.p. when the fan could be "stalled" with increased noise and vibration and unsteady flow. In the case of axial fan sit could even result in inadequate cooling of the electric motor and/or motor overloading.

1.5 Centrifugal fans

1.5.1 Introduction

Apart from the effects of varying blade widths and inlet areas, other differences in fan characteristics are attributable to differences in blade shape. In the Sections which follow, diagrams are included to show the impeller configuration and typical characteristic curves are also included.

1.5.2 Forward curved blades

These impellers first became popular at the end of the 19th Century and almost superseded all other types. A diagrammatic representation of the impeller is shown in Figure 1.60. They are considerably smaller for a given duty than all other designs.

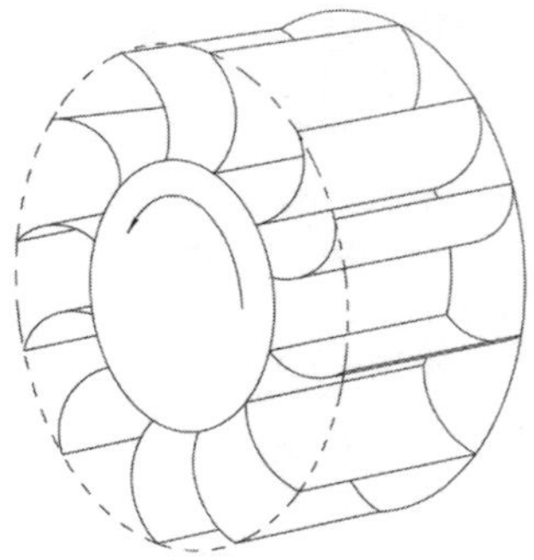

Figure 1.60 Forward curved impeller

Flowrate can be as high as 2.5 times that of the same size of backward-bladed fan. This is now seen to be not necessarily an advantage since casing losses, which are a function of velocity, will therefore be about six times a great. Thus even with an impeller total efficiency approaching the theoretical optimum of about 92%, the overall fan total efficiency would still be down to about 75%.

Such fans are now only used where space is at a premium, as they will be the most compact. Due to their smaller size they are usually cheaper, although the differences are much reduced with the greater possibility for automated manufacture of backward bladed fans. Nevertheless the scope for improvement has been appreciated and current designs achieve static efficiencies of 63% and total efficiencies of 71% at even lower speeds.

It will be noted that the performance curve has discontinuities due to stall and/or recirculation (see Figure 1.61). A large margin over the absorbed power is necessary where the system resistance cannot be accurately determined, or where it is subject to variation, to take account of the rising power characteristic.

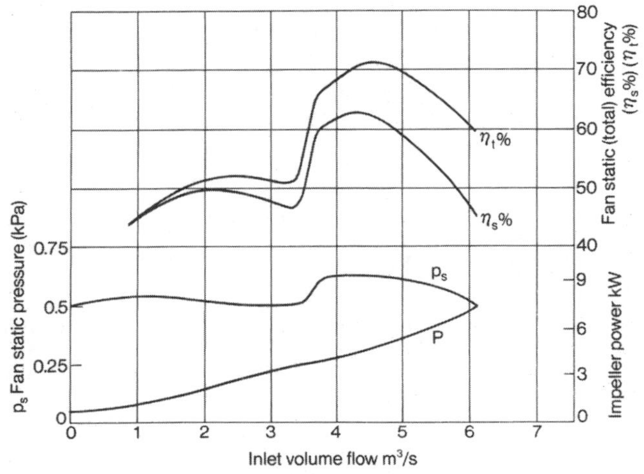

Figure 1.61 Forward curved fan — typical characteristic curves

The impeller has a large number of shallow blades in widths from 0.25 to 0.5D and runs at lower tip speed for the duty. Structural considerations have in the past limited the pressure development to about 1 kPa, but the narrower widths are now suitable for pressures up to 14 kPa.

Apart from low-pressure ventilation requirements, these fans are widely used for mechanical draught on shell-type boilers, oil burners, furnace recirculation etc.

1.5.3 Deep vane forward curved blades

These blades are considerably stronger than the conventional forward curved, being triangulated. They can thus run at higher speeds developing high pressure. A more detailed impeller drawing is shown in Figure 1.62, which perhaps explains why there is some reduction in flowrate. Nevertheless a more stable pressure/flowrate curve is produced (Figure 1.63) albeit with a moderate peak efficiency.

1.5.4 Shrouded radial blades

This useful design is represented diagrammatically in Figure 1.64 and can handle free flowing dust-laden air or gas. The impellers have the ability to deal with higher burdens than the backward inclined type. They are somewhat more efficient (up to 65% static) than the open paddle and also able to run at higher rotational speeds and thus develop higher pressures. The blades are inherently strong, as centrifugal forces have no bending effect. They are also simple and in sizes up to 900 mm can be easily flanged for rivetting and spot welding.

Blades are largely self-cleaning and are easily cleaned. Such fans are suitable for moderate free-flowing granular dust burdens.

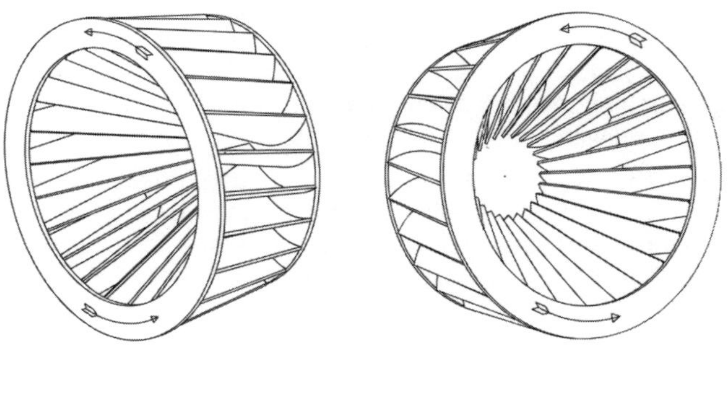

Figure 1.62 Deep vane forward curved impellers

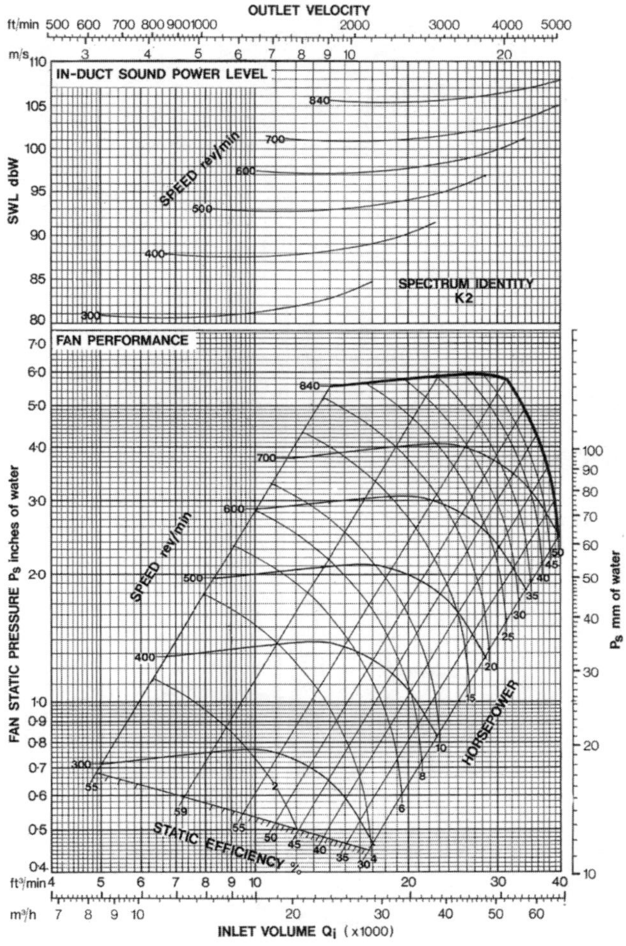

Figure 1.63 Deep vane forward curved fan — typical characteristic curves

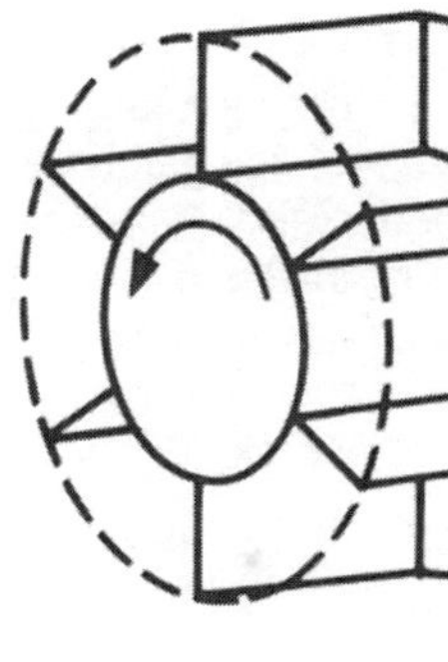

Figure 1.64 Shrouded radial impellers

It should be noted that the power rises continually towards free air (zero pressure) and a reasonable margin is necessary over the absorbed power, unless the system pressure can be accurately assessed. As the impeller has a backplate, wear is concentrated on this, but casing wear is correspondingly reduced compared with the open paddle.

Because of its characteristics, the shrouded radial impeller is widely used on gas streams having a significant dust burden, for example induced draught on rotary driers for the quarry and roadstone industries. A typical characteristic curve is shown in Figure 1.65.

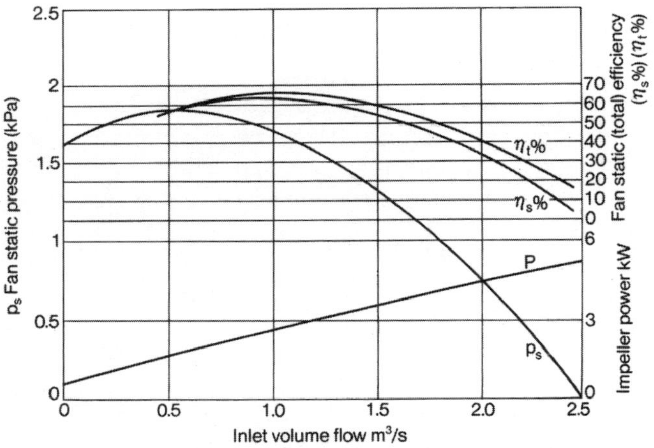

Figure 1.65 Shrouded radial fan — typical characteristic curves

1.5.5 Open paddle blades

This open paddle blade design is represented diagrammatically in Figure 1.66.

Figure 1.66 Open paddle impellers

This is the impeller for heavy dust burdens in excess of those possible with the shrouded radial. Its efficiency is only moderate (up to 60% static) but it is suitable for high temperatures. As there are no shrouds or backplates, the blades are free to expand. Standard units may therefore be used with gases up to 350°C, but special alloy wheels can be designed for the very highest temperatures.

It will be seen (Figure 1.67) that the pressure characteristic is stable over the whole range of flows but that the power rises continuously with flow. Open paddle fans are manufactured in various widths, where casing inlet and outlet areas are virtually equal. The narrower units are also suitable for high pressure applications such as direct injection pneumatic conveying.

1.5.6 Backplated paddle impellers

These are shown diagrammatically in Figure 1.68.

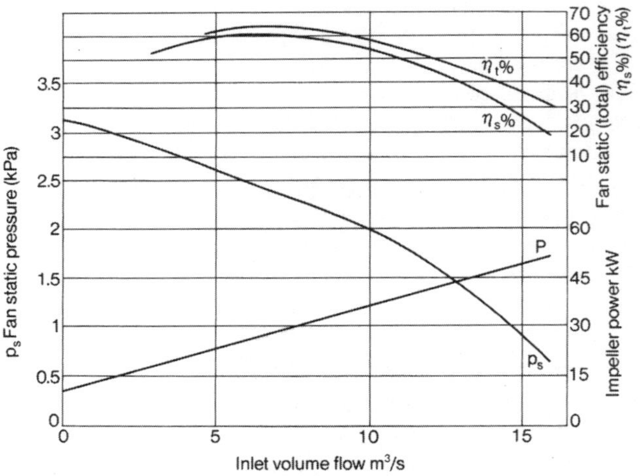

Figure 1.67 Open paddle fan — typical characteristic curves

Where the solids are fibrous in character, e.g. wool, paper, or wood shavings, there is tendency for them to wrap round the shaft of an open paddle and clog the unit. The backplate obviates this possibility. All characteristics are generally as the open paddle, except that the backplate paddle need to run about 3% faster taking approximately 6% more power for duties in its optimum range.

1.5.7 Radial tipped blades

The radial tipped blade design is represented diagrammatically in Figure 1.69.

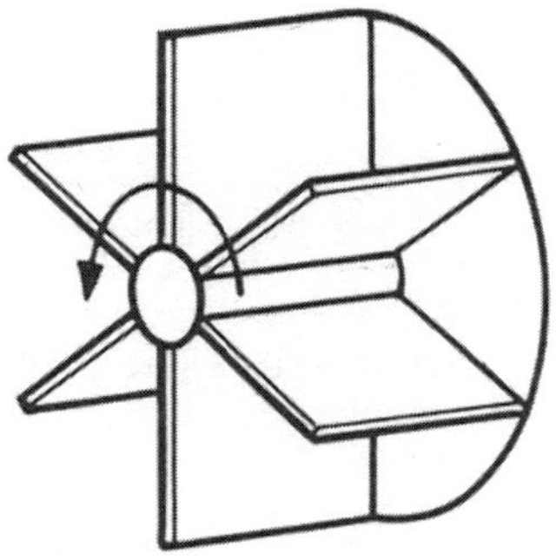

Figure 1.68 Backplated paddle impellers

This blade form is used as an alternative to the shrouded radial. Generally there is an increased number of blades and the heel of these is forward curved to reduce shock losses. The efficiency and flowrate are therefore improved for a given size, but the characteristics are otherwise similar. Fan static efficiencies up to 73% are possible.

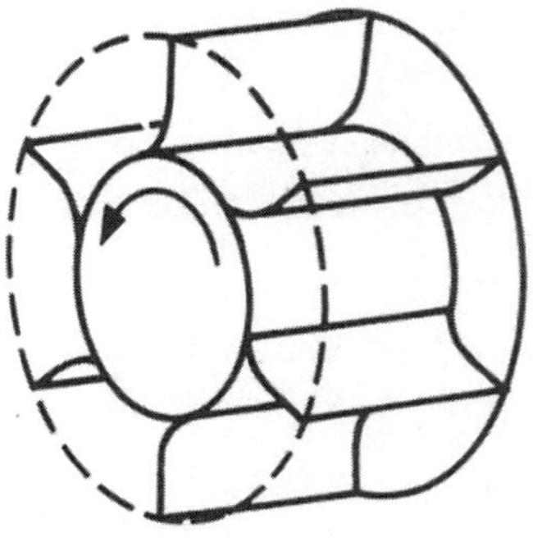

Figure 1.69 Radial tipped impellers

The units are widely used for induced draught on water tube boilers where low efficiency dust collectors are incorporated. Dust burdens similar to those of the shrouded radial, in Section 1.5.4 are acceptable.

1.5.8 Backward inclined blades

The impeller of these is represented in Figure 1.70.

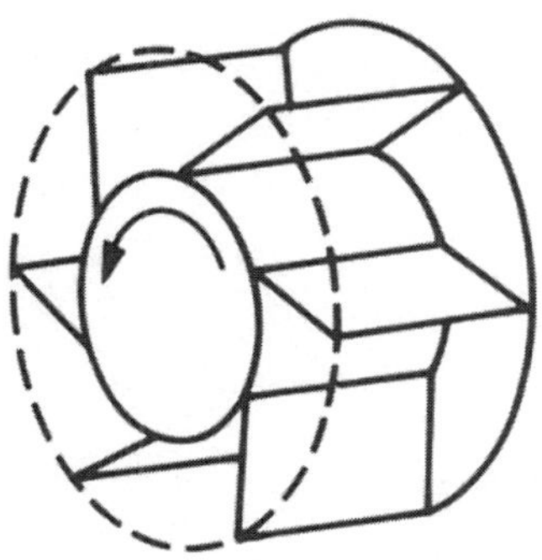

Figure 1.70 Backward inclined bladed impeller

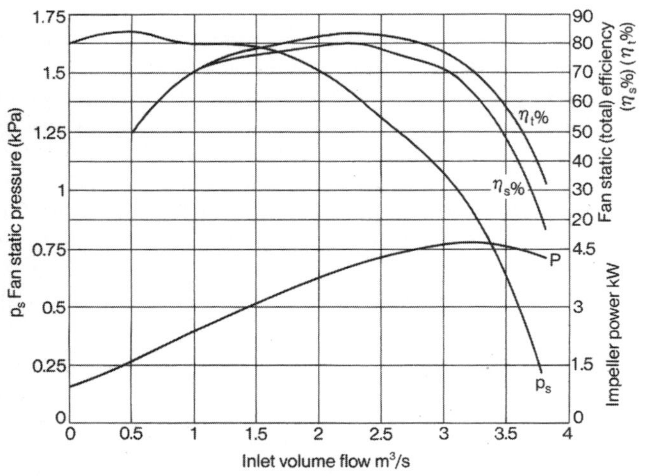

Figure 1.71 Backward inclined fan — typical characteristic curves

These may be considered at the "maids of all work". Due to their simplicity the blades lend themselves to simple methods of construction, at a moderate price, and they can easily be flanged for rivetting and spot welding up to size 900 mm. The design is of the high-speed type making them suitable for direct connection (Arrangement 4 and 8 for many duties).

Fan static efficiencies up to 80% peak have been achieved with the medium widths using the very latest aerodynamic knowledge. The wider fans have the additional advantage of a non-overloading power characteristic so that, with correct motor selection, the fan may operate over its complete constant speed pressure-flow curve. In its working range, the curve is also comparatively steep, so that large variations or errors in system pressure will have a smaller effect on flow rate. (See Figure 1.71).

The blades are self-cleaning to a certain degree and are in any case easy to clean because of their single plate flat form. They are therefore suitable for free-flowing granular dust burdens or moisture-laden air. In the absence of special factors, this impeller is the recommended form for all applications including commercial and industrial ventilation systems, low and high velocity air conditioning, the clean side of collectors in dust extract systems, fume extraction, etc.

Standard fans are available for operation at gas temperatures up to 350°C and special units employing high temperature alloys can be custom-manufactured for gases up to 500°C. In general terms, the narrower the impeller, the fewer the number of blades and the greater the blade outlet angle. Both these factors are conducive to the acceptance of higher dust burdens but counter-balanced to a certain extent by boundary layer effects and higher abrasive velocities.

1.5.9 Backward curved blades

These impellers are shown in Figure 1.72 and are preferred for certain applications where there may be disadvantages in the use of the backward inclined type. Due to the curvature, the blade angle at inlet can be made steeper for a given outlet angle. This generally enables shock losses to be kept low, whilst the curvature itself develops a certain degree of lift. It is therefore possible to arrange such fans with a pressure curve continually rising to zero flow.

They can be extremely stable, with none of the "bumps" in their curves found with other types, and most suitable for operation

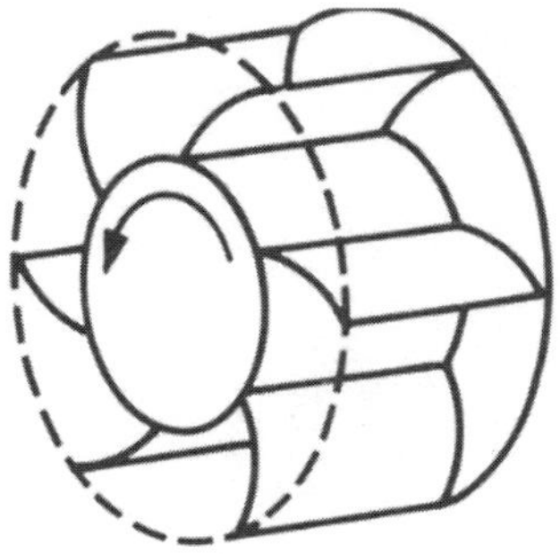

Figure 1.72 Backward curved bladed impeller

in parallel on multi-fan plants. With the special blade curvatures now used, efficiencies exceed 82% static, approaching those attained by aerofoil bladed fans.

The steeper inlet angle also results in a stronger blade, which can rotate at higher speeds. This is offset to a large extent, however, by the need to run at higher speeds for a given duty as compared with the backward inclined type. They are also more expensive as, unless complex press tools are used to "stretch" the metal, the blades cannot be flanged for rivetting or spot welding and have to be arc welded in position.

The curvature of backward curved blades (concave on the underside of the blades) is inclined to encourage the build-up of dust. As the impeller in its rotation tends to develop a positive pressure on the working convex face of the blade and negative effect on the underside, dust can lodge within the camber. This becomes more pronounced on the narrowest fans where the

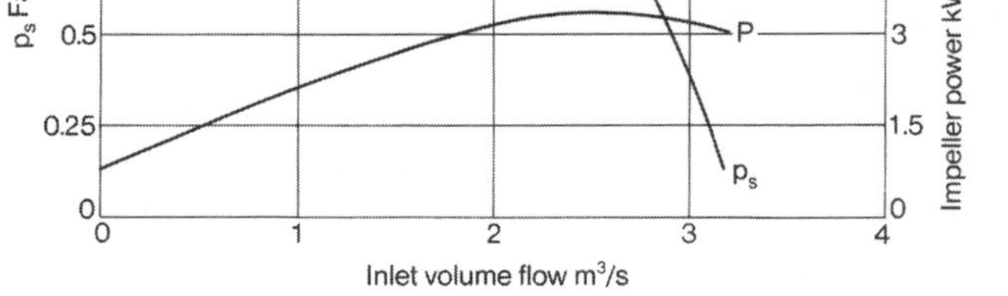

Figure 1.73 Backward curved fan — typical characteristic curves

camber is substantial and the chord is very much shorter than the developed blade length. The wider units have less curvature, although the effects are offset by the shallow outlet angles.

Generally backward curved impellers are not so suitable for high temperature operation, as differential expansion between blades and shrouds can be severe inducing additional stresses. Gas temperatures should therefore be limited to 350°C. Other advantages are the same as those of the backward include type, including a relatively steep pressure characteristic and non-overloading power curve. (See Figure 1.73).

1.5.10 Reverse curve blades

These blades are backward curved at their tips but forward curved at the heel (see Figure 1.74). Characteristics are generally similar to the backward curved type with the same limitations to their use. Shock losses at entry to the blade passages is reduced however and a slightly higher efficiency maintained outside the range of the b.e.p.

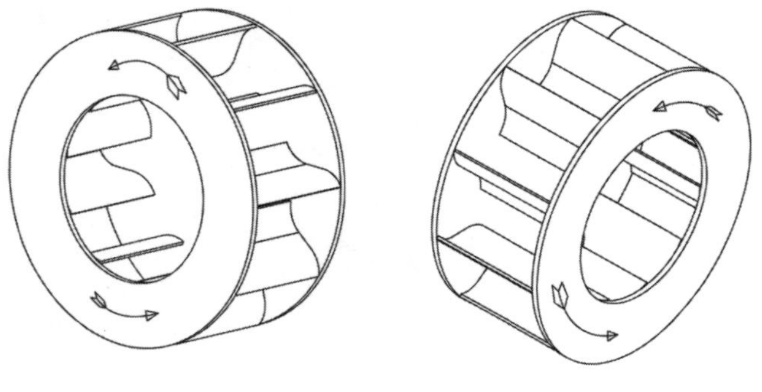

Figure 1.74 Reverse curve bladed impeller

1.5.11 Backward aerofoil blades

The impeller is shown in Figure 1.75. The blades produce lift forces, which counteract inter-blade circulation without requiring precise angles. Thus smooth flow conditions are maintained over a considerable portion of the characteristic.

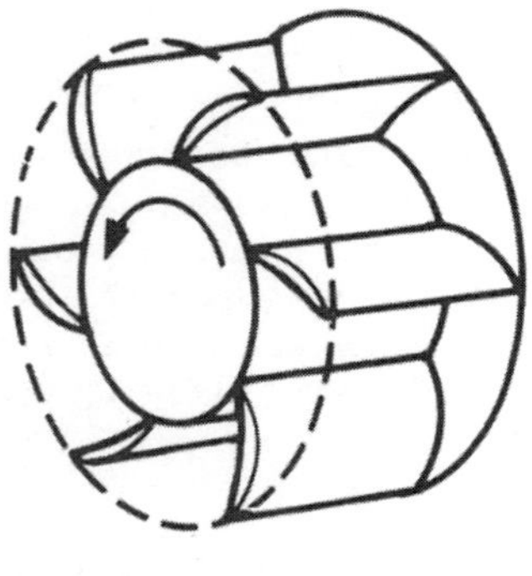

Figure 1.75 Backward aerofoil bladed impeller

Pressure losses in the impeller are thus reduced, as are those in the casing volute. Fan static efficiencies up to 88% have been achieved and total efficiencies of 91% are possible. An efficiency of at least 80% can be achieved over 40% of the volume flowrate at a given speed. It will be appreciated that at low flows the blades are stalled, resulting in a discontinuity in the pressure curve, which is not always acknowledged. (Figure 1.76).

Aerofoil should be used on low dust burdens, since particles penetrating the hollow welded blades can produce imbalance. Similar problems can arise with free moisture. Although precautions can be taken, such as solid nosing bars for dust or foam filling for moisture, the backward inclined is preferred for

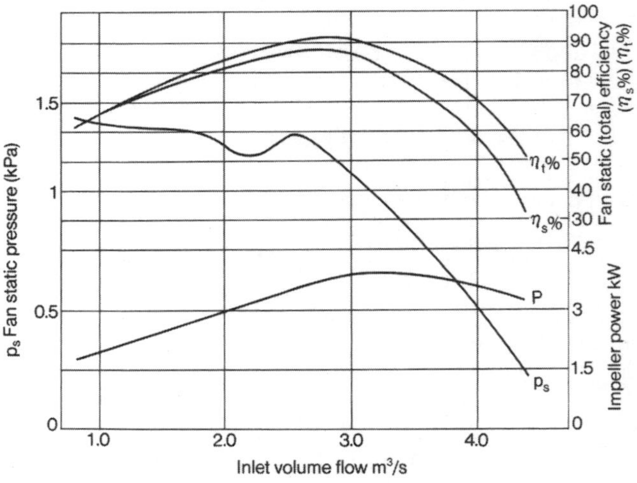

Figure 1.76 Backward aerofoil fan — typical characteristic curves

these applications, (see Section 1.5.8). Erosion of the blade noses will in any case reduce the efficiency. High temperatures may require "pressure relief" for the air trapped within the blades.

Whenever operating costs are of paramount importance, as when large powers are involved and where there is continuous operation at high load factor, the aerofoil is to be preferred. In general the advantages are not significant for fans below size 1000 mm. Aerofoils may also be necessary when increased duty is required from existing power lines: in many cases the power saved may allow a smaller motor to be installed so that the overall cost is the same. In other cases the additional fan price may be recovered in energy cost differences long before expiry of the period allowed for amortizing plant costs.

1.5.12 General comment

For all duties, the higher initial cost of backward bladed fans can usually be recouped many times over during the life of the unit, as the energy consumption will often be reduced by 25% compared with forward curved fans. Driving motors will also be smaller, and as the fans have a non-overloading power characteristic only a small margin is necessary over the absorbed power.

1.6 Axial flow fans

1.6.1 Introduction

Axial flow fans have developed rapidly since the Second World War due to the creation of a range of high strength aluminium alloys. These permit running at the rotational speeds necessary to produce worthwhile pressure. Axial fans adhere closely to classical theory and require less "know-how" than centrifugal fans. They may be placed in three general classifications according to how the flow is constrained:

- **Ducted fan** where the air has to flow through a duct thus encouraging it to enter and leave the impeller in an almost axial direction.
- **Diaphragm or ring mounted fan** where the air is transferred from one relatively large air space to another.
- **Circulator fan** where the impeller rotates freely in an unrestricted space. Examples are pedestal or ceiling fans.

1.6.2 Ducted axial flow fans

The various components possible in a ducted axial flow fan are shown in Figure 1.77. Not all the elements are present in a particular fan and the terminology for the various types is as follows:

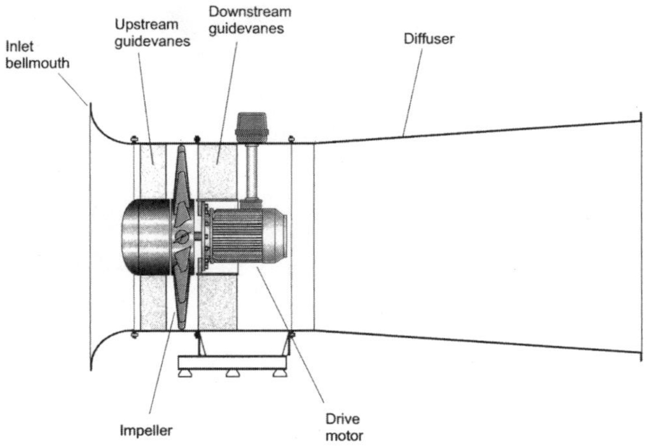

Figure 1.77 Components of a ducted axial flow fan

1.6.2.1 Tube axial fan

The tube axial fan is a fan without guide vanes and comprising only the impeller and casing. Fairings up and downstream of the impeller may be fitted. Such fans are usually selected for pressures up to about 750 Pa. (See Figures 1.78 and 1.79). Blades may have adjustable pitch at rest to cater for varying flowrates.

Figure 1.78 Examples of tube axial fans

1.6.2.2 Vane axial fan (downstream guide vanes – DSGV)

This is an axial fan with guide vanes downstream of the impeller to recapture the rotational energy and thus give a high pressure development and a higher efficiency. (See Figures 1.80 and 1.81).

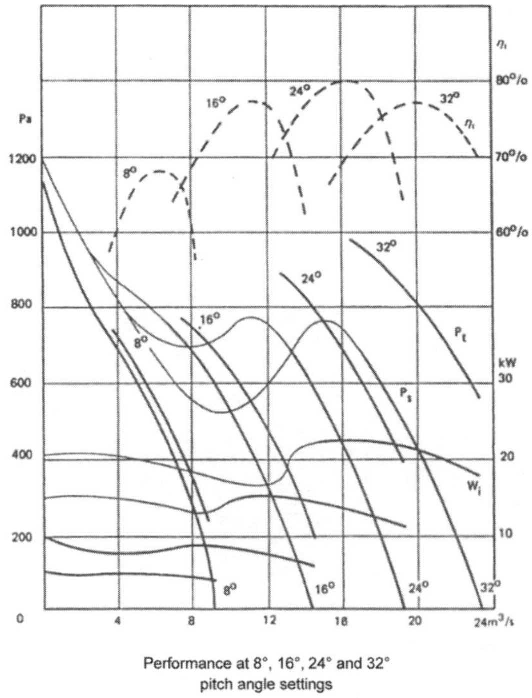

Figure 1.79 Tube axial fans — typical characteristic curves

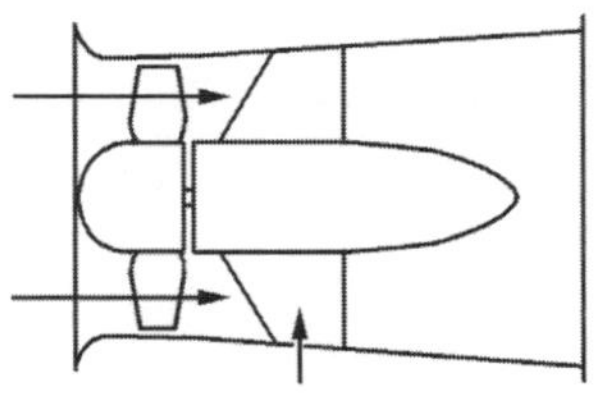

Figure 1.80 Vane axial fan (DSGV - downstream guide vanes)

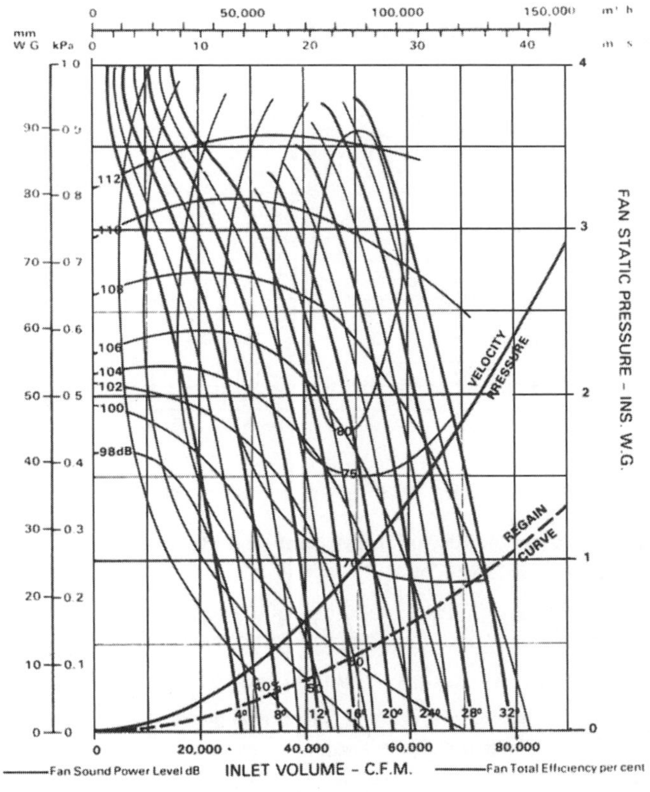

Pitch Angle	4	6	8	10	12	14	16	18	20	22	24	26	28	30	32
Peak B.H.P.	10	11.5	13	15.5	17.5	20	22.5	25.5	28	31.5	34	37.5	42	45	50

Air Density 0.075 lbs/ft³ Max Free Discharge 78,000 CFM Max Duty 64,000 CFM @ 2.5 ins SWG

Figure 1.81 Vane axial fan (DSGV) — typical characteristic curves

1.6.2.3 Vane axial fan (upstream guide vanes – USGV)

This is an axial fan with guide vanes upstream of the impeller. Pre-rotation of the air in the opposite direction to the impeller rotation means that lift forces, and hence the fan pressure are increased. The impeller removing the swirl pressure development can be higher than the corresponding DSGV fan albeit with a narrowing of the flow range. (See Figures 1.82 and 1.83).

Figure 1.82 Vane axial fan – (USGV upstream guide vanes)

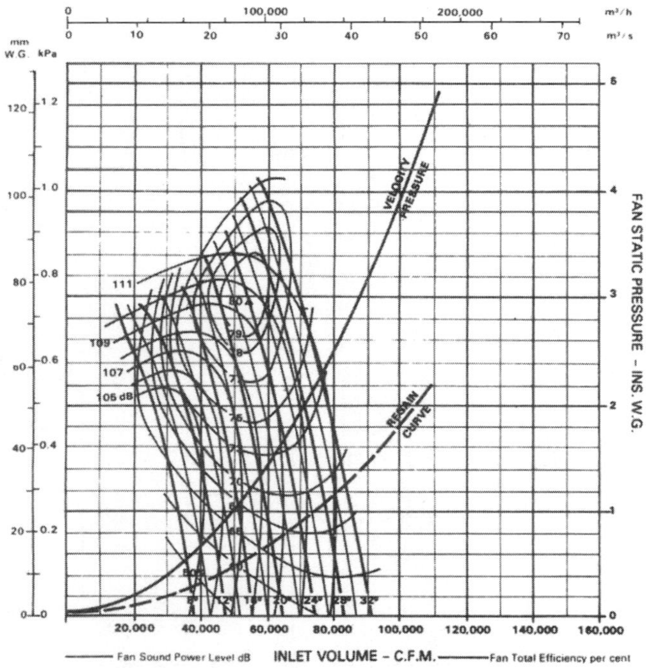

Pitch Angle	8	10	12	14	16	18	20	22	24	26	28	30	32
Peak B.H.P.	13.9	17.1	20	24	28	32	37	41	46	53	60	67	75

Air Density 0.075 lbs/ft³ Max Free Discharge 90,000 CFM Max Duty 70,000 CFM 2.8 ins SWG

Figure 1.83 Vane axial fan – (USGV) — typical characteristic curves

1.6.2.4 Vane axial fan (upstream and downstream guide vanes - U/DSGV)

By careful design the advantages of the two previous designs can be optimised to give the highest possible efficiency.

1.6.2.5 Contra-rotating axial flow fan

The contra-rotating type which has two separate impellers of opposite hand arranged in series, invariably with separate motors rotating in opposite directions. By this means, swirl from the first impeller is removed by the second impeller. The rotational energy is recovered and converted into useful static pressure. Thus instead of twice the single stage fan pressure being developed, this approaches three times that of a single stage tube axial fan (See Figures 1.84 and 1.85) Pitch angles are generally

Figure 1.84 Contra-rotating axial flow fan

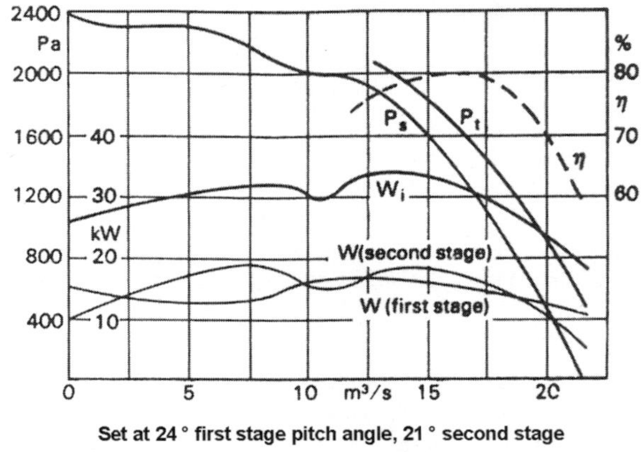

Figure 1.85 Contra-rotating axial flow fan— typical characteristic curves

adjusted so that each impeller takes equal power around the best efficiency point. This automatically secures an output flow free from swirl.

1.6.3 Blade forms

Whilst the variety of blade forms available for centrifugal fans is considerable, not nearly the same range is available in axial flow fans (Figures 1.86).

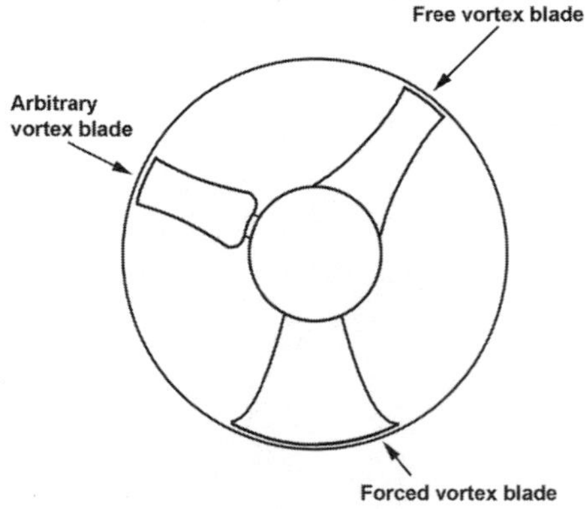

Figure 1.86 Axial flow impellers — variety of blade forms

The blades may be designed to three principles:

1.6.3.1 Free vortex

Each element of the blade performs equal work. A condition of radial equilibrium exists and the axial velocities over the blades are virtually constant. The blade chord at the tip is usually reduced whilst the twist near the hub can be substantial.

1.6.3.2 Forced vortex

The work performed by the blades is maximized at their tips leading to large tip chords when compared with the roots of the blades.

1.6.3.3 Arbitrary vortex

Intermediate between the two above.

Most axial fans are of an arbitrary vortex design to a greater or lesser extent. Blades have to be cut away near to their roots so that they do not interfere with each other. A truly forced vortex design would require minimum tip gaps between blades and the casing. Weight would also increase towards the periphery leading to greater centrifugal stresses.

1.6.4 Other types of axial flow fan

1.6.4.1 Truly reversible flow

Reversal of the direction of rotation of an axial fan reverses the direction in which the air flows. The performance of guide vane fans in reverse is extremely poor, but non-guide vane and contra-rotating fans will deliver 60% to 70% of the forward volume flow when reversed on a given system. The reduction is due to the fact that the aerofoil sections are operating tail-first and have their camber (curvature) in the wrong direction.

A truly reversible impeller can be built from standard parts by rotating every other blade through 180°. Half will then be running nose-first and half tail-first, the volume flow being about 85% of normal in each direction. A more recent innovation has been to design blades with two top surfaces (Figure 1.87) when the performance can be over 92% of normal in each direction.

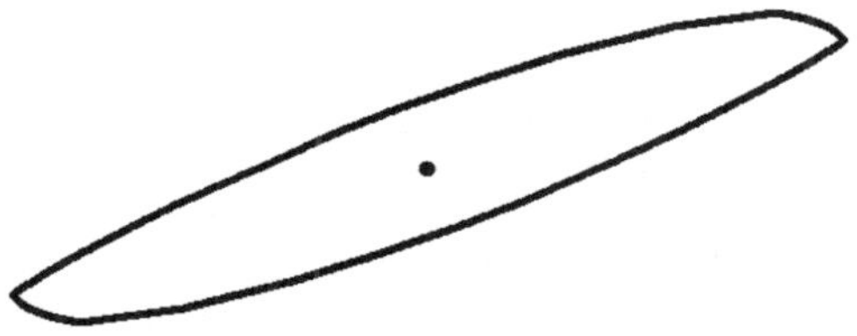

Figure 1.87 Truly reversible flow blade section

1.6.4.2 Fractional solidity

Impellers can be assembled on a standard hub by omitting some of the blades. Mechanical balance must, of course, be preserved, but there is no need for the blades to be evenly spaced. Peak pressure is reduced and the best efficiency point (b.e.p.) moves to a lower pressure and volume so that the speed must be increased for a given duty.

This can be an advantage when the impellers are directly driven by electric induction motors. Such motors have better efficiency and lower cost at higher speeds - a point which can be particularly significant with large low speed fans. Figure 1.88 shows the performance range of fans with 12 left-or-right-handed adjustable pitch blades, which could be assembled with 10, 9, 8, 6, 4, 3 or 2 blades, and multi-staged.

1.6.4.3 High pressure axial fans

These are designed with hub diameters between 50% and 70% of the impeller diameter, compared with 30% to 40% for a general-purpose range of competitive cost. Aerodynamically this reduces the pressure limitation set by the slow-moving roots of the blades. Mechanically the short blades can be made far stiffer so that the impeller can be run at higher tip speeds without danger of flutter. The ratio of the annular flow area to the total blade area decreases, making guide vanes or contra-rotation essential to recover the increased swirl energy. A typical fan is shown in Figure 1.89. Its performance is shown in Figure 1.90.

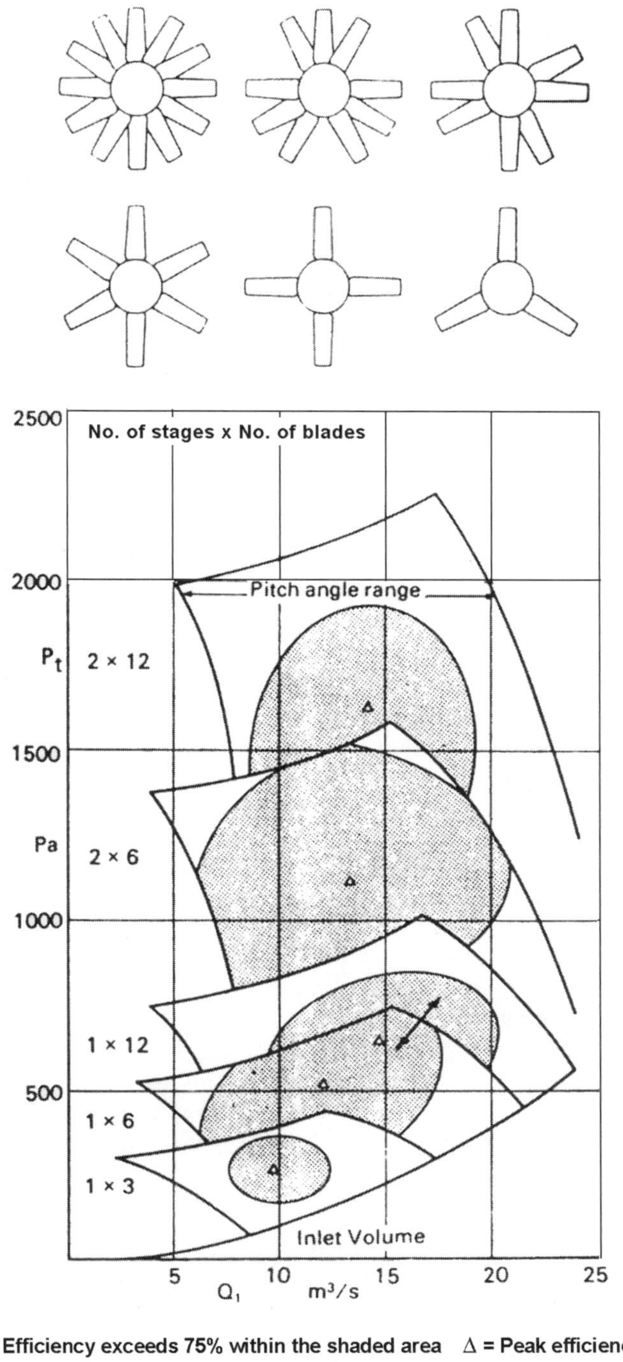

Figure 1.88 Performance range of fans available

Figure 1.89 Typical high pressure axial fan

1.6.4.4 High efficiency fans

When the power absorbed is measured in hundreds of kilowatts, every effort is made to achieve high efficiency. Among

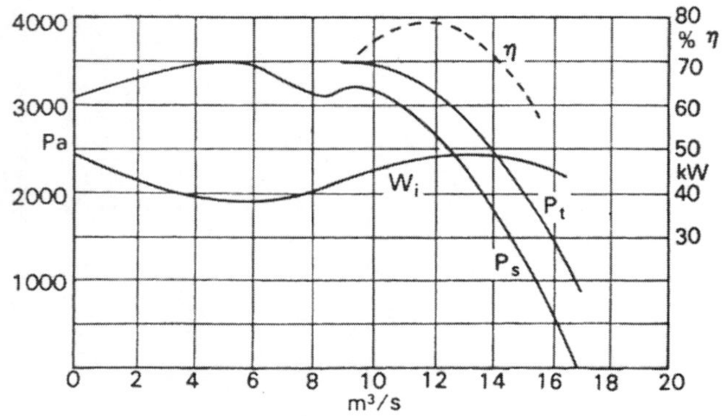

Figure 1.90 Typical high pressure axial fan performance curves

features distinguishing such designs from the general-purpose types are:

a) Hub diameters of 50% or more to improve the aerodynamic balance of the design from blade root to tip.

b) Blade form designed specifically for the required duty. When die forming is not justified, this entails increased labour to provide a good surface finish.

c) Aerofoil-section guide vanes, again designed specifically for the required duty.

d) Careful streamlining of the annulus passage, and fairing of bearing supports or other obstructions. This may entail moving the driving motor right out of the casing, introducing the necessary transmission elements to the impeller.

e) Space for a long tail fairing following the impeller hub and guide vanes to maximize fan total pressure by conversion of annulus velocity pressure.

f) Space for a long gradually expanding diffuser to minimize outlet velocity pressure, and maximize fan static pressure.

These measures may raise the peak fan total efficiency to 90%, compared with 80% for a good general-purpose model at optimum duty. Figure 1.91 show the constructional arrangement and Figure 1.92 shows typical performance curves.

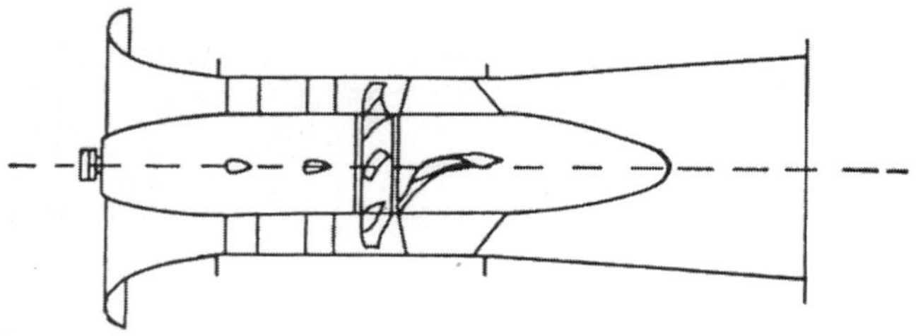

Figure 1.91 High efficiency axial fan — construction

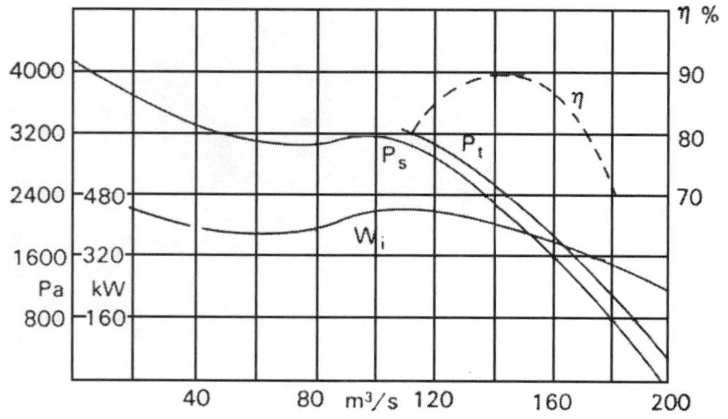

Figure 1.92 High efficiency fan — typical performance curves

1.6.4.5 Low-pressure axial fans

These are available in very large sizes for volumetric flowrates from 50 m³/s upwards at fan static pressure from 100 to 200 Pa. As an example they may be applied singly, discharging from the top of evaporative cooling towers, or in multiple, circulating air across extensive banks of heat exchange tubes.

Hubs are small and the blades long and few in number - three, four or six. Blades were at one time made of timber, but are now of hollow glass-reinforced polyester or similar. Mouldings or hollow aerofoil sections from steel or aluminium sheet are more usual. Guide vanes are unnecessary. (See Figure 1.93.)

Figure 1.93 Low-pressure axial fans

1.7 Propeller fans

1.7.1 Impeller construction

These may be regarded as a special type of axial fan designed to operate without a casing, the impeller being situated in a hole in a wall or partition. The fans are simple low cost units with broad bladed impellers usually formed from sheet metal. The blades are shaped to operate with an orifice flow pattern, deflecting the air with the minimum flow separation or vortex formation. Design techniques make use of flow visualization with stroboscopically viewed smoke trails.

1.7.2 Impeller positioning

The blade form is usually optimised for pressure differences across the partition from zero to about 100 Pa. Above the designed pressure the flow pattern changes drastically. The outlet jet assumes an expanding conical form with reverse circulation at its core as sketched. Towards zero volume flow, discharge is radially outwards, and the centrifugal mechanism is now responsible for pressure development.

Propeller fans are quiet and effective for ventilation purposes, both supply and exhaust. They are also used for unit heaters and similar applications where some resistance is encountered. For these an experimental matching of the fan and the unit is important since the pressure development and the flow pattern over the heat exchanger are very dependent on the blade and orifice plate positions.

1.7.3 Diaphragm, ring or bell mounting

As more of the impeller projects on the outlet side of the orifice, the free flow volume falls, because the inlet orifice flow no longer covers all the blade. At the same time the pressure at low flow rises because more blade is exposed on the outlet side for centrifugal action. The free flow can be substantially increased by rounding the orifice edge or fitting a rounded inlet ring. (See Figure 1.94 for the variants).

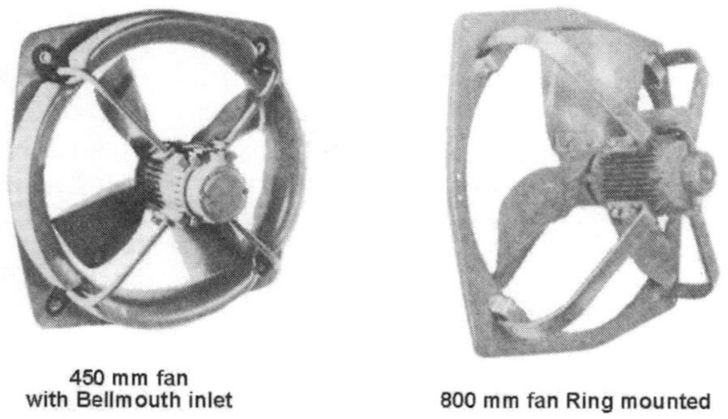

Figure 1.94 Examples of bellmouth or ring mounting

This is because the vena contracta is expanded and less velocity pressure is required for a given volume flow. Moderate pressure performance is also helped, but high pressure development is impaired. If the rounding is enlarged into a true bellmouth and a short tunnel formed around the impeller, the fan becomes in effect an axial fan, and is better served by an aerofoil section impeller.

1.7.4 Performance characteristics

The impellers of propeller fans are almost invariably mounted on the shaft of the driving motor. The air flow cools the motor, which can be totally enclosed to keep out dust. The impeller power rises rather sharply if the volume flow is drastically restricted, and the motor could be over-heated, particularly if on the downstream side, where centrifugal flow starves it of cooling air. However, propeller fans are not often used in systems where such excessive resistance could arise. Typical performance curves are shown in Figures 1.95 and 1.96.

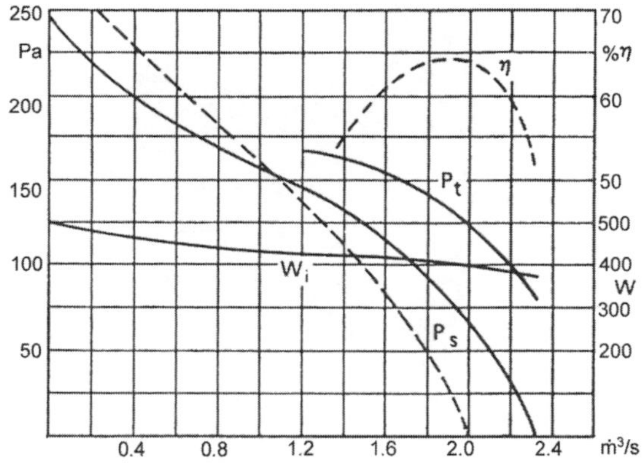

Effect on Ps of omitting inlet ring shown thus: – – –

Figure 1.95 Typical performance curves of ring mounted propeller fan

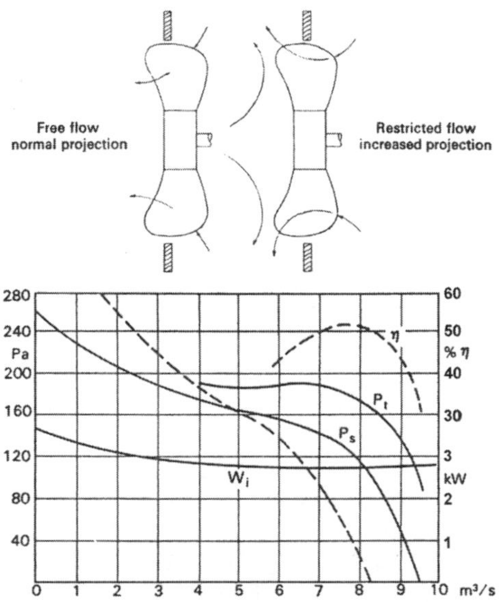

Effects on Ps of increasing downstream projection of impeller shown thus: – –

Figure 1.96 Typical performance curves of plate mounted propeller fan

1.8 Mixed flow fans

1.8.1 Why the need - comparison of characteristics

The suitability of a particular type of fan for a duty depends more on the relationship between the performance parameters than on their absolute values. This is especially true where there are limits to the size of the unit, and/or where the maximum speed is specified. In Section 1.3.2 the concepts of specific speed and diameter are discussed, and it is noted that there is an area for mixed flow fans between the two traditional types. This type has not been commercially available to any extent until recently.

For HVAC applications, there is a region for which neither centrifugal nor axial fan is ideal but for which a mixed flow fan can be designed. For the centrifugal fan to be of an acceptable size it has to be selected at efficiencies away from its peak; the axial fan has to have a high hub to tip ratio and/or has to be multi-staged to achieve the pressure.

Mixed flow fans should not be confused with in-line radials. Their casing diameter is generally smaller and they run at a spee d intermediate between axials and centrifugals.

1.8.2 General construction

The main elements of a true mixed flow fan are seen by reference to Figures 1.97 and 1.98, similar to a vee belt driven vane axial or in-line radial. The major difference is in the impeller, which is generally of fabricated construction. Both the front shroud and backplate are at an angle, so that the air follows a

Figure 1.97 Typical belt driven mixed flow fan

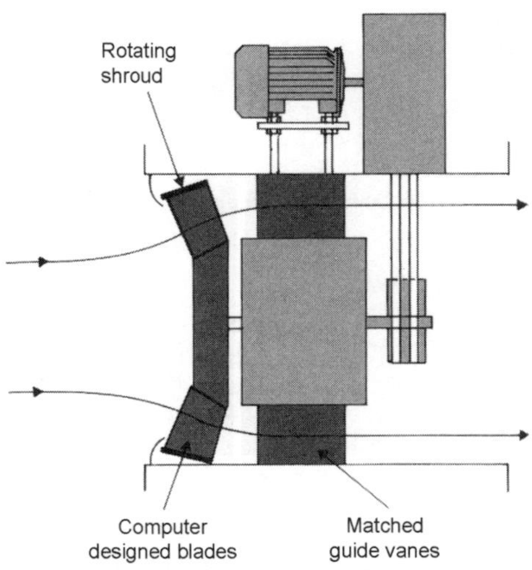

Figure 1.98 Cross-section through belt driven mixed flow fan

path somewhere between axial and centrifugal flow. It will be noted that the casing is just slightly larger than the impeller outside diameter.

1.8.3 Performance characteristics

Performance is intermediate between an axial and centrifugal of the same impeller diameter. A non-stalling characteristic is achieved and the power/flowrate curve is non-overloading. Pressures up to 2 kPa are possible with standard construction. A typical performance curve is shown in Figure 1.99.

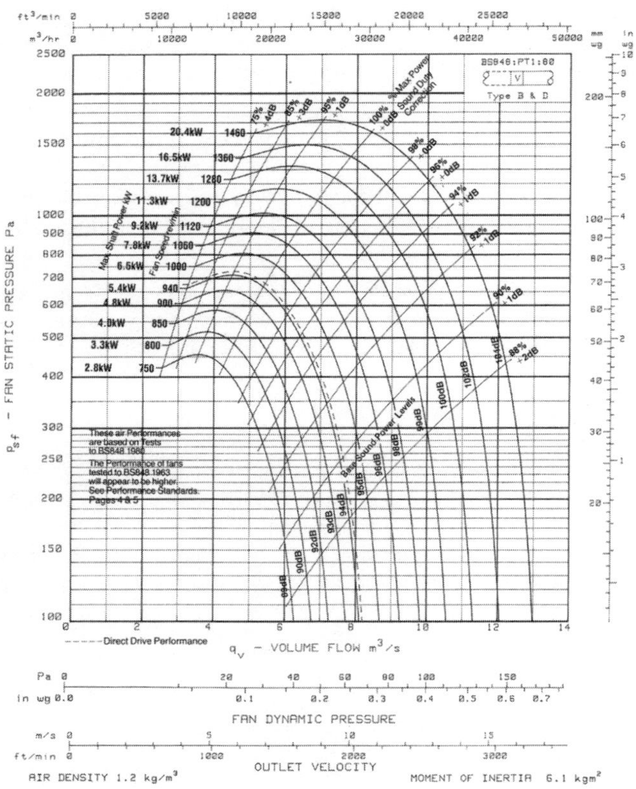

Figure 1.99 Typical performance curves — Mixed flow fans

1.8.4 Noise characteristics

The linear sound power level for this fan is intermediate between an axial and centrifugal if all are selected for best efficiency and the same duty. The centrifugal generally has a falling noise spectrum with frequency, whilst the axial peaks in the octave band containing the blade passing frequency. A mixed flow fan has a noise spectra somewhat similar to the centrifugal except that there is a marked reduction in the 63 and 125 Hz bands making silencing more easy.

1.9 Miscellaneous fans

1.9.1 Cross flow fans

The fluid path through the impeller is in a direction essentially at right angles to its axis both entering and leaving the impeller at its periphery. The impeller otherwise resembles a multivane forward curved centrifugal but has no side entries. Flow is induced by a vortex formed within the impeller.

Apart from structural considerations, there are no limitations on width so that it may be used to give a wide stream from a small diameter e.g., as in a unit heater. See Figures 1.100 and 1.101.

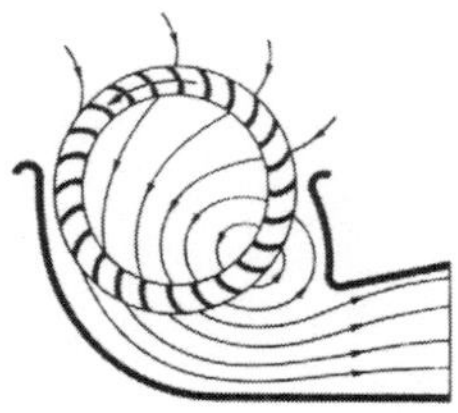

Figure 1.100 Cross-flow fan

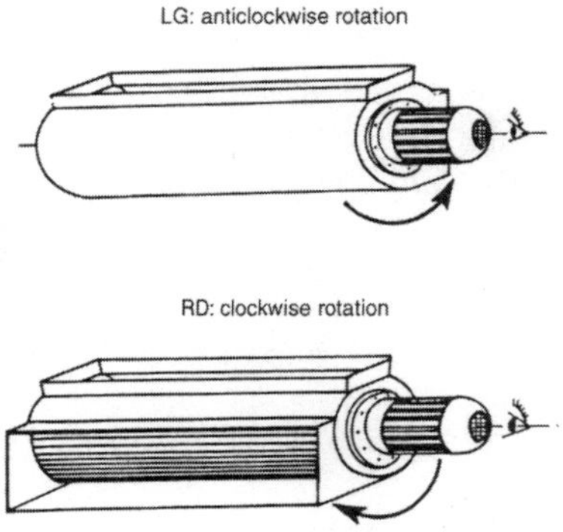

Figure 1.101 Rotation of cross flow fans

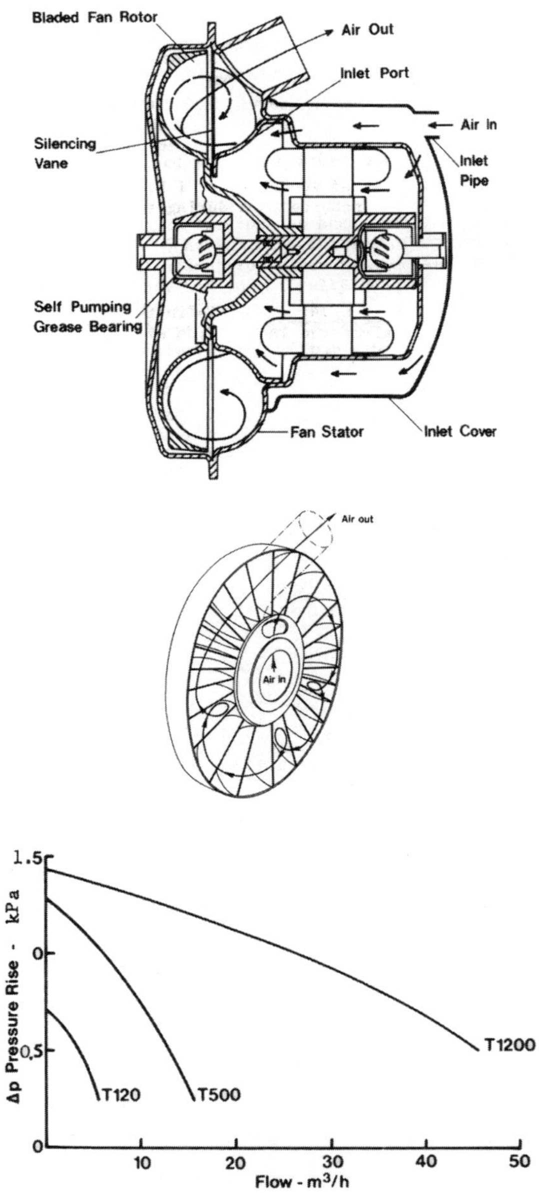

Figure 1.102 Toroidal fan, airflow pattern and performance curve

1.9.2 Ring shaped fans

The circulation of air in the toric casing is helicoidal. Rotation of the impeller, which has a number of blades, creates this helicoidal trajectory, which is intercepted by one or more blades depending on the flowrate.

The impeller transfers its energy to the air or gas in the manner shown in Figure 1.102 and is best used for very low flows at high pressure albeit at only a moderate efficiency.

1.10 Bibliography

The Fan Museum, 12 Crooms Hill, Greenwich, London SE10 8ER, UK, Tel: 020 8305 1441

De re Metallica, Georgius Agricola, Courier Dover Publications, Paperback version,1912, ISBN 0486600068

The Fan: Including the Theory and Practice of Centrifugal and Axial Fans, Charles H. Innes, Manchester Technical Publishing Co., 1904. vi, 252, [4]pp.

Rational Psychrometric Formulae, Paper by Dr Willis Carrier, (ASME, 1911).

Apparatus For Treating Air, U.S. Patent No. 808897 issued January 1906.

ISO/DIS 13348 *Industrial fans — Tolerances, methods of conversion and technical data presentation.*

Eurovent 1/1 – 1984 *Fan Terminology.*

AMCA Standard 210/ASHRAE Standard 51, *Laboratory Methods of Testing Fans for Rating.*

ISO/TC 117 *Industrial fans: Standardization in the field of fans used for industrial purposes including the ventilation of buildings and mines.*

2 The properties of gases

For all those whose knowledge of physics is sketchy, and who wondered if the gas laws would ever be useful in their lives – here is the answer. Boyle, Charles and Dalton have centre stage. Without these you can never completely understand the underlying rules of the fan engineer.

Essential information about those properties of air and other gases which must be known for fan selection are given in this Chapter. In the case of hazards, guidance is detailed regarding legislation and safety standards.

Contents:

2.1 Explanation of terms

2.1.1 Introduction

Gases, together with liquids and solids, are our names for the various forms in which substances naturally occur. Thus we speak of the gaseous state, the liquid state and the solid state. Sometimes we call these the three phases of a substance. Gases and liquids are often grouped together as fluids. Fluids differ from solids in that they readily take up the shape of the container in which they are placed.

A solid body subjected to a small shear force undergoes a small elastic deformation and returns to its original shape when the force is removed. When subjected to larger shear force the shape may be permanently changed due to plastic deformation.

A fluid, when subjected to an arbitrarily small shear force undergoes a continuous deformation. This happens regardless of the inertia of the fluid. For a fluid the magnitude of the shear force and the speed of deformation are directly related. In a solid body it is the deformation itself, which is related to the shear force.

A fluid may be either a liquid or a gas. A gas differs from a liquid in that it will expand to completely fill the container. A gas at conditions very close to boiling point or in contact with the liquid state is usually called a vapour. Fluids are compressible; gases being much more compressible than liquids.

A substance can exist in all three states. A typical example of this is ice, water and steam. When ice is heated at constant pressure, the ice converts to water at the melting point and to steam at the boiling point. If the steam pressure is increased at constant temperature, the steam converts to water at the saturation (vapour) pressure.

Solid particles can be suspended in a gas. Such a combination, gas plus particles, is very common in dust control, pneumatic conveying etc. When the particles distribute themselves evenly through the gas, we speak of a homogeneous mixture. When concentration gradients occur, we speak of a heterogeneous mixture.

Gases display greatly varying properties. For the purposes of fans and fan systems, the following characteristics of gases should generally be known:

- density
- relative humidity and liquid content
- viscosity
- compressibility
- temperature and changes of state
- chemical composition and solid content

2.1.2 Changes of state

2.1.2.1 Boiling point

The boiling point is the temperature at which a liquid converts to vapour or gas at a particular local pressure. The boiling point is usually stated at the standardised atmospheric pressure, 101.325 kPa. The boiling point of water at this pressure is 100° C. The boiling point of all liquids is heavily dependent upon pressure.

2.1.2.2 Melting point

The melting point is that temperature at which a substance changes from the solid to the liquid state and also solidifies from a liquid to a solid. The melting point in most substances is pressure dependent only to a very limited degree and this is especially true of substances, which are normally gases at ambient conditions e.g., air, nitrogen, oxygen etc.

2.1.3 Ideal gases

A gas consists of a large number of molecules, each of which has a random motion. These molecules are very small and very close together with the scale being such that for all practical purposes a gas can be considered continuous and uniform. The behaviour of a gas is a function of the average distance between the molecules, compared to the size of molecule. If the molecule can be considered small compared to the average distance between molecules, then the potential energy arising from the mutual attraction of the molecules may be ignored and the gas can be considered an ideal or perfect gas. The important properties of an ideal gas at rest are density and pressure.

2.1.4 Density

The density of a gas is defined as the total mass of the molecules in a unit volume.

Thus in SI units density is specified in kg/m^3.

2.1.5 Pressure

Since the molecules are in continuous motion, they are always colliding with other molecules or the solid surfaces of their container. In a perfect gas, all these collisions are taken to be perfectly elastic i.e. when a molecule strikes a solid surface, the surface experiences a force equal and opposite to the time rate of change of momentum of the rebounding molecule. This force causes the gas to exert an overall pressure on the container or other immersed body.

This force per unit area is defined as the pressure, the units in the SI system being Pa (Pascals) 1 Pa = 1 N/m^2. In a fluid at rest the pressure acts normally to the solid surface.

2.2 The gas laws

2.2.1 Boyle's law and Charles' law

The kinetic energy of the molecules increases with increasing temperature. The important effects of this fact are given in Boyle's law and Charles' law, which state that the volume of a perfect gas varies inversely with absolute pressure and directly with absolute temperature, respectively.

The total effect is more properly stated by the equation of state:

$$P = \rho R T \qquad \text{Equ 2.1}$$

where:

P	=	pressure
ρ	=	density
R	=	gas constant
T	=	absolute temperature

In the design of the majority of fan systems, the gas may be considered as incompressible without introducing significant error. The normal boundary, between the assumption that the gas is incompressible or that it is compressible, as accepted in ISO 5801 is for pressures up to 2 kPa. In many calculations, therefore, the air density may be considered constant and the absolute pressure is directly proportional to the absolute temperature.

Since an ideal gas is assumed to be composed of molecules, which are very small perfect spheres, and the collisions of these

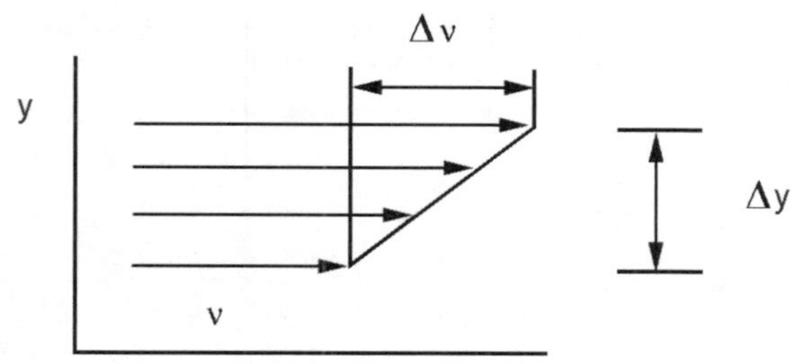

Figure 2.1 Definition of viscosity

molecules with one another and solid boundaries are assumed to be elastic, an ideal gas can only exert pressure normal to a surface. Thus, no frictional force exists in any ideal gas, even if strong velocity gradients exist. All gases, however, consist of molecules, which do not behave as elastic spheres, and thus no gas is truly ideal. Real gases are capable of exerting pressure parallel to the surface of a body, which is moving with respect to the gas. The magnitude of the force parallel to the surface is used to define an important property of real gases – viscosity. The effects of viscosity on the behaviour of real gases causes resistance to flow; the resistance is proportional to the velocity gradients, which exist in the gas.

2.2.2 Viscosity

The absolute viscosity (m) is defined as the shearing stress for a unit rate of change of velocity. It has the units of Newton-sec per metre squared in the SI system. The shearing stresses are proportional to the ratio of absolute viscosity to density, called kinematic viscosity.

Viscosity (the ability to flow) is a property of fluids (both liquids and gases) treated under the heading of rheology. The work rheology derives from the Greek "rheos" meaning flow.

Between two layers of fluid flowing at different speeds, a tangential resistance, a shear stress, is developed because of molecular effects. We say that the shear stress is caused by the internal friction of the fluid or conversely that the fluid transmits shear forces by reason of its internal friction.

A liquid in motion is continuously deformed by the effects of these shear forces. The magnitude of the stress depends on the rate of shear deformation and the sluggishness of the liquid, i.e. the viscosity.

Viscosity is defined for flow in layers, laminar flow, by Newton's law of viscosity and is illustrated diagrammatically in Figure 2.1.

$$\tau = \mu \frac{\Delta v}{\Delta y} \quad \text{Equ 2.2}$$

where:

τ = shear stress (N/m^2)

μ = dynamic viscosity (kg/ms)

Δv = change in viscosity (m/s)

Δy = distance between layers (m)

In viscous flow equations the dynamic viscosity divided by the density of the liquid is given the symbol v. This parameter is called kinematic viscosity.

$$v = \frac{\mu}{\rho} \quad \text{Equ 2.3}$$

where:

v = kinematic viscosity (m^2/s)

μ = dynamic viscosity (kg/ms)

ρ = density (kg/m^3)

The SI unit for kinematic viscosity is 1 m^2/s.

2.2.3 Atmospheric air

Atmospheric air is a mixture of gases, water vapour and impurities (both solid and gaseous). The proportions of the important constituents for dry air at sea level are given in Table 2.1. This table may be considered representative of air at all the altitudes usually experienced in fan engineering.

Constituent	Chemical symbol	% by Volume	% by Weight
Nitrogen	N_2	78.09	75.52
Oxygen	O_2	20.95	23.15
Argon	Ar	0.93	1.28
Carbon dioxide	CO_2	0.03	0.04
Also traces of helium, hydrogen, krypton, neon, ozone etc.			

Table 2.1 Constituents of atmospheric air

Table 2.1 shows that air is primarily a mixture of nitrogen and oxygen,(both of which are diatomic gases) with molecular weight calculated from the average constituents.

For purposes of uniformity, standard air has been defined as air with a density of 1.2 kg/m^3 and an absolute viscosity of 18.19 x 10^{-6} Pa.s. This is substantially equivalent to air at a temperature of 20 °C, 50% relative humidity and a barometric pressure of 101.325 kPa. The ratio of specific heats, (γ), is taken to be 1.4, which is the expected value for a perfect diatomic gas.

The temperature and barometric pressure of atmospheric air vary widely with weather conditions and geographical location, most noticeably altitude. In order to simplify design, a standard atmosphere has been defined. This gives the atmospheric pressure, temperature and therefore, density with altitude. (See Table 2.2.)

Altitude m	Temperature °C	Atmospheric pressure kPa	Gas density kg/m^3
0	15.00	101.32	1.230
100	14.35	100.13	1.215
200	13.70	98.94	1.201
300	13.05	97.77	1.189
400	12.40	96.61	1.177
500	11.76	95.46	1.166
600	11.11	94.32	1.155
700	10.46	93.20	1.145
800	9.81	92.08	1.134
900	9.16	90.98	1.123
1000	8.51	89.88	1.112
1100	7.86	88.80	1.102
1200	7.21	87.72	1.091
1300	6.56	86.66	1.080
1400	5.90	85.61	1.069
1500	5.25	84.56	1.058
1600	4.60	85.53	1.047
1700	3.95	82.50	1.037
1800	3.30	81.49	1.026
1900	2.65	80.49	1.016
2000	2.00	79.49	1.006
2100	1.35	78.51	0.996
2200	0.70	77.54	0.986
2300	0.53	76.57	0.976

Altitude m	Temperature °C	Atmospheric pressure kPa	Gas density kg/m³
2400	-0.60	75.62	0.967
2500	-1.25	74.68	0.957
2600	-1.90	73.74	0.948
2700	-2.55	72.82	0.938
2800	-3.20	71.91	0.929
2900	-3.85	71.00	0.919
3000	-4.50	70.11	0.909
3100	-5.15	69.23	0.900
3200	-5.80	68.35	0.890
3300	-6.46	67.48	0.880
3400	-7.11	66.62	0.871
3500	-7.76	65.77	0.862

Table 2.2 Standard atmospheric data versus altitude

The density of atmospheric air is also a function of the humidity. Although the change in density with humidity is not large, it is often significant and air system designers should be cognizant of these changes. Remember that increasing humidity lowers the density since water vapour is lighter than dry air. The density of saturated air for various barometric and hygrometric conditions is shown in Table 2.3.

Dry-bulb temp °C	Density of saturated air for various barometric pressures and dry bulb temperatures — kg/m³					
	Barometric pressure kPa					
	97	98.5	100	101.5	103	104.5
-2.0	1.244981	1.263273	1.282390	1.302927	1.324194	1.340401
-1.0	1.239396	1.258667	1.277753	1.297353	1.319731	1.335505
0.0	1.234260	1.254012	1.272975	1.292141	1.315018	1.330532
1.0	1.229423	1.249325	1.268119	1.287163	1.310140	1.325506
2.0	1.224768	1.244618	1.263236	1.282324	1.305166	1.320447
3.0	1.220207	1.239902	1.258360	1.277553	1.300147	1.315376
4.0	1.215680	1.235188	1.253510	1.272800	1.295123	1.310307
5.0	1.211147	1.230483	1.248697	1.268037	1.290121	1.305254
6.0	1.206587	1.225792	1.243921	1.263247	1.285157	1.300224
7.0	1.201994	1.221119	1.239179	1.258431	1.280239	1.295225
8.0	1.197375	1.216468	1.234459	1.253595	1.275367	1.290260
9.0	1.192743	1.211838	1.229752	1.248752	1.270533	1.285328
10.0	1.188116	1.207227	1.225045	1.243920	1.265728	1.280428
11.0	1.183512	1.202631	1.220330	1.239113	1.260938	1.275553
12.0	1.178948	1.198047	1.215603	1.234343	1.256148	1.270693
13.0	1.174432	1.193466	1.210866	1.229616	1.251342	1.265837
14.0	1.169963	1.188879	1.206131	1.224925	1.246506	1.260970
15.0	1.165527	1.184277	1.201420	1.220251	1.241632	1.256073
16.0	1.161092	1.179644	1.196770	1.215560	1.236712	1.251125
17.0	1.156606	1.174968	1.192231	1.210795	1.231747	1.246101
18.0	1.151991	1.170232	1.187875	1.205877	1.226746	1.240975
19.0	1.146325	1.164887	1.182780	1.200987	1.222584	1.237641
20.0	1.141813	1.160033	1.78197	1.196304	1.217665	1.232675
21.0	1.137279	1.155335	1.173591	1.191607	1.212804	1.227740
22.0	1.132735	1.150742	1.168962	1.186898	1.207980	1.222830
23.0	1.128188	1.146207	1.164311	1.182174	1.203177	1.217939
24.0	1.123646	1.141691	1.159639	1.177435	1.198380	1.213061
25.0	1.119111	1.137164	1.154946	1.172681	1.193576	1.208190
26.0	1.114582	1.132601	1.150234	1.167912	1.188756	1.203320
27.0	1.110055	1.127983	1.145503	1.163126	1.183912	1.198445
28.0	1.105523	1.123300	1.140754	1.158323	1.179039	1.193559

Dry-bulb temp °C	Density of saturated air for various barometric pressures and dry bulb temperatures — kg/m³					
	Barometric pressure kPa					
	97	98.5	100	101.5	103	104.5
29.0	1.100978	1.118548	1.135988	1.153503	1.174134	1.188656
30.0	1.096404	1.113730	1.131206	1.148664	1.169195	1.183730
31.0	1.091787	1.108856	1.126408	1.143808	1.164226	1.178775
32.0	1.087106	1.103942	1.121596	1.138932	1.159230	1.173786
33.0	1.082339	1.099014	1.116769	1.134037	1.154213	1.168756
34.0	1.077460	1.094100	1.111930	1.129122	1.149185	1.163679
35.0	1.072440	1.089240	1.107079	1.124186	1.144155	1.158549
36.0	1.067247	1.084478	1.102216	1.119229	1.139139	1.153361
37.0	1.061846	1.079865	1.097342	1.114250	1.134151	1.148108
38.0	1.056198	1.075460	1.092459	1.109249	1.129210	1.142784

Table 2.3 Density of saturated air at various temperatures and barometric pressures

2.2.4 Water vapour

Whilst the gaseous constituents of air may be considered to be essentially constant, the amount of water vapour contained within the air can vary enormously. The properties of moist air are dependent on the relative amounts of water vapour and dry air. The state of an air-water vapour mixture is completely defined by specifying its pressure, temperature and humidity.

2.2.5 Dalton's law of partial pressure

Dalton's law states that each component of a gas mixture exerts a pressure that is determined by the volume and temperature of the mixture regardless of the other constituents involved. The pressure of each of the components is called its partial pressure.

2.3 Humidity

2.3.1 Introduction

With no water vapour present, the partial pressure of the air must equal the barometric pressure. When water vapour is added it exerts a certain pressure regardless of whether or not the air is present. The saturated condition exists when the actual vapour pressure is equal to the vapour pressure of the pure liquid at the same temperature.

Partially saturated air contains vapour that is superheated, that is the temperature of the mixture and therefore that of the vapour is higher than the saturation temperature for the existing vapour pressure.

2.3.2 Relative humidity

The relative humidity (rh) of an air-water vapour mixture is defined as the ratio of the vapour pressure existing compared to the vapour pressure at saturation for the same dry-bulb temperature. This is also equal to the ratio of the mole fractions under the same conditions.

rh is usually express as a percentage but occasionally as decimal (less than unity).

2.3.3 Absolute humidity

Absolute humidity (ah) is the actual weight of water vapour existing per unit weight of dry air or gas. It is usually expressed in kg water vapour per kg of dry air.

The humidity of an air-water vapour mixture is frequently expressed as either % relative humidity or by giving the wet-bulb depression.

2.3.4 Dry bulb, wet bulb and dew point temperature

Unless otherwise specified, the temperature of an air-water vapour mixture is that temperature which is indicated by an ordinary or dry-bulb thermometer. This dry-bulb temperature is the temperature of both the air and the water vapour in the mixture. The wet-bulb temperature may be determined by submerging a water-covered bulb in the air-water vapour mixture until equilibrium is obtained.

The wet-bulb temperature will be lower than the dry-bulb temperature as long as evaporation continues. If no evaporation is possible, the mixture is saturated and the wet and dry-bulb temperatures for this condition will be identical. the dew point temperature of an air-water vapour mixture is the saturation temperature corresponding to the absolute humidity of the mixture.

The dew point temperature may also be considered as that temperature at which condensation begins when the mixture is gradually cooled.

2.3.5 Psychrometric charts

Thermodynamic properties of dry and moist air have been tabulated by a number of authorities including CIBSE (The Chartered Institute of Building Services Engineers) in the United Kingdom. However, a chart presentation of the data is preferable, especially where this can encompass all the values of humidity from completely dry air, through fractional humidities, to completely saturated air.

Such charts can be drawn for a number of temperature ranges but that for normal atmospheric air is shown in Figure 2.2. To quote CIBSE Guide C:

The chart has been constructed using two fundamental properties specific enthalpy and moisture content as basic, linear co-ordinates. Other physical properties are not then shown as linear scales. The 30 °C dry bulb line has been constructed at right angles to lines of constant moisture content and the scale of specific enthalpy inclined obliquely to the vertical scale of moisture content. In this way lines of constant dry bulb temperature are approximately vertical, diverging slightly each side of 30 °C and the traditional appearance of the chart preserved.

The wet bulb values plotted are those read from a sling, or ventilated, psychrometer but lines of percentage saturation are plotted instead of relative humidity. Within the comfort zone there is little practical difference between percentage saturation and relative humidity and, of course, the difference diminishes as the saturated or dry states are approached.

2.4 Compressibility

2.4.1 Introduction

All gases are compressible but this can generally be neglected for fan systems where the pressure above atmospheric is less than 2.5 kPa.

It may be noted for example that water is about 100 times more elastic than steel and about 0.012 times as elastic as air. Compressibility is very temperature dependant and slightly pressure dependant. Any values used must related to the operating conditions. Classically, compressibility is expressed in terms of the bulk modulus defined by the relationship:

$$\text{Compressibility} = \frac{1}{k} \qquad \text{Equ 2.4}$$

$$k = \rho\frac{\Delta p}{\Delta \rho} = \frac{\Delta r}{\Delta \nu} \qquad \text{Equ 2.5}$$

where:

k	=	bulk modulus of the gas (N/m^2)
p	=	pressure (Pa)
ρ	=	density of the gas (kg/m^3)
ν	=	volume of gas (m^3)
Δ	=	change of magnitude

The change in volume due to a change in pressure can be calculated directly from the definition:

$$\Delta\nu = -\frac{\nu\Delta\rho}{k} \qquad \text{Equ 2.6}$$

where the minus sign indicates that the volume decreases with increasing pressure.

2.4.2 Gas data

Densities and specific volumes of air and many other common gases are readily available for a wide range of pressures and temperatures. However compressibility data for gases other than dry air may be difficult to obtain. Nevertheless the data for air may usually be accepted without serious error.

2.4.3 Acoustic problems

Compressibility can also be of importance in acoustic problems.

The acoustic velocity, or wave speed is directly related to the bulk modulus and compressibility. If acoustic resonance occurs in the ductwork the acoustic velocity must be known to effect a successful cure. Acoustic resonance can be a very serious problem creating destructive ducting vibrations and large pressure pulsations. The acoustic velocity calculated from the bulk modulus applies to pure clean gas. If the gas has solid particles, the acoustic velocity will be greatly reduced from the theoretical value. Testing may be the only approach to find the true value.

2.5 Hazards

2.5.1 Introduction

The work "hazard" is in common use in the English language and it must be defined here to show the context in which it is used hereafter.

Hazard:

A physical situation with a potential for human injury, damage to property, damage to the environment or some combination of these.

It can be seen from the definition that three distinct types of effect are considered but in some cases one hazard may lead to others. Fire, for example, can be a serious health hazard.

A hazardous substance:

A substance, which, by virtue of its chemical properties, constitutes a hazard.

2.5.2 Health hazards

The fan user must consider the effects of the gas and its solid content, on the health of the operators and employees. Most countries have legislation limiting the exposure of employees to substances judged to be hazardous. If the gas to be handled is

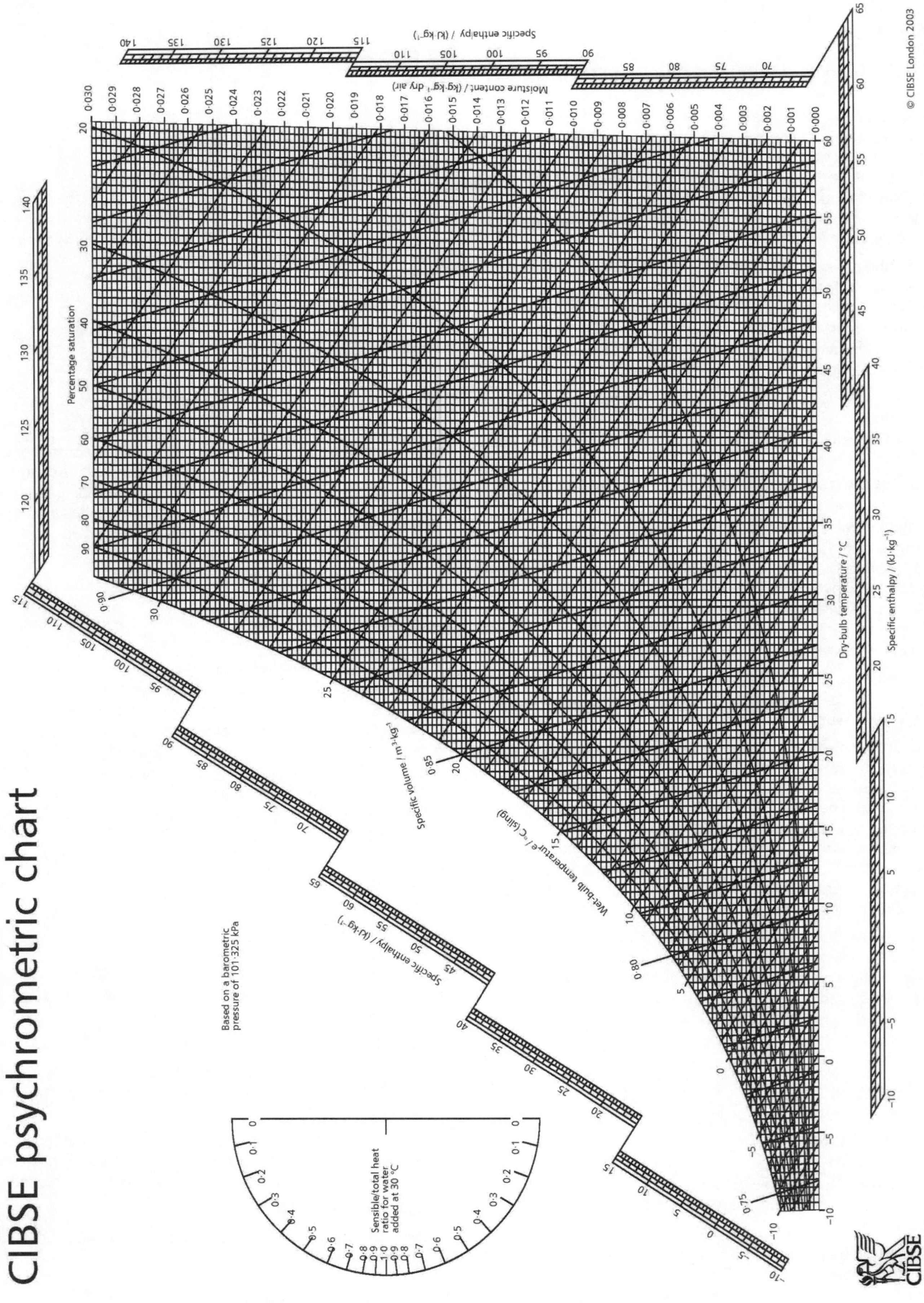

Figure 2.2 Psychrometric chart

Courtesy of CIBSE (Chartered Institution of Building Services Engineers) — Reproduced from CIBSE Guide C: Reference data

listed in local regulations the fan manufacturer must be informed. The type of health hazard must be specified.

Another health hazard, sometimes not recognised as such, is noise. Some countries have regulations stipulating the acceptable noise levels and exposure times. It must be remembered that the fan duty will largely determine the fan noise. High pressure fans will be noisier than low pressure fans. System resistance should therefore be kept as low as possible.

Some fan types are inherently noisy. Large equipment, in general, is noisy. Noise levels can be attenuated by fitting acoustic enclosures. However, these tend to drastically diminish the maintainability of the equipment by hindering access. In some instances, costly acoustic enclosures have been removed at site and scrapped in order to achieve acceptable access. One easy solution to this hazard is to declare certain areas "Ear Protection Zones".

2.5.3 Physical hazards

Physical hazards include fire and explosions as well as corrosion and temperature. The degree of risk attached to the hazard is dependant upon the properties of the gas and the solid content.

The fan equipment itself may pose a physical hazard. Within the European Union, the Machinery Directive, 89/392/EEC, and the amending directive 91/368/EEC, which came into force on 1st January 1993, place the responsibility for safety on the machine designer. The machine must be designed to be safe in all aspects:

- installation
- commissioning
- operation
- maintenance

If the designer is unable to devise a completely safe machine the areas of concern must be documented and recommended precautions communicated to the user. Because this is a legal requirement in all EU countries the machine designer may not be relieved of the obligations by a third party. A supporting European 'C' type standard is EN 14461.

2.5.4 Environmental hazards

We are becoming more aware of the limitations of our environment. The Earth's resources and waste-disposal capabilities are finite. Stricter limitations will be imposed gradually on the amount of pollutant which can be released, while the list of pollutants will become longer. The fan user must be aware of the full consequences of leakage of gas from the fan and installation.

The environment can be considered in two separate identities:

- local
- global

If the site is surrounded or close to a town what risk is likely to the population, structures or habitat in the event of a failure? In the global sense, what are the likely cumulative effects of product leakage?

2.5.5 Installation hazard assessment

The user and system designer are in full possession of all the relevant available facts regarding the gas and the installation. Any assumptions made should be passed to the fan manufacturer and identified as such. The user must assess the risks attached to all the possible hazards and decide what, if any, leakage is acceptable. Gas properties reviewed during the assessment should include:

Auto ignition point	The temperature about which a substance will start to burn without an ignition source being necessary.
Flash point	The lowest temperature at which a gas will burn if an ignition source is present.
Atmospheric boiling point	The temperature at which any liquid will boil at atmospheric pressure. 101.325 kPa.
Vapour specific gravity	Specific gravity is the ratio of a vapour's density to air at standard conditions, atmospheric pressure. 101.325 kPa.

The nature of the hazards will also dictate the type of duct connections to be used. Spigot, flat-face, flanged, raised-face flanged, ring-type joints. Process upset conditions must be considered as part of the assessment. Upset conditions which last for more than one or two hours may have a significant impact on pump and ancillary equipment selection.

The physical location of the fan, indoor or outdoor, will decide the behaviour of the leakage once outside the fan. Will any vapour cloud quickly disperse on a breeze which always blows over the un-manned site or will a manned enclosed fan house gradually build up a dangerous concentration of gas? Only the user can assess these questions and specify the necessary precautions.

It is the responsibility of the user to define exactly what the fan is intended to do. It is the responsibility of the fan manufacturer to supply equipment to meet the required performance.

2.6 Bibliography

CIBSE (The Chartered Institution of Building Services Engineers), 222 Balham High Road, Balham, London, SW12 9BS UK. Tel: (+44) 020 8675 5211, Fax (+44) 020 8675 5449.

CIBSE Guide C: 2001, ISBN 0750653604.

ISO 5801:1997, *Industrial fans — Performance testing using standardized airways.*

The Machinery Directive 89/392/EEC, as amended by Directives 91/368/EEC, 93/44/EEC and 93/68/EEC. Implemented in the UK by the Supply of Machinery (Safety) Regulations 1992 and the Supply of Machinery (Safety) (Amendment) Regulations 1994.

3 Air and gas flow

It is an unfortunate fact that the relationship between academics and engineers in the ventilation industry is less than perfect. The former produce theories from their research which rarely get transferred to industry. In like manner, the latter may install plant for which the working data is never relayed back to academia. Worse, balancing subcontractors endeavour to put right the mistakes made in design. Even worse, system pressures are "guessed" (with a large safety margin). Too many fan enquiries specify 500 Pa for exhaust systems and 1500 Pa for the supply air handling unit.

We have not completely finished with basic theory and it is necessary to introduce the work of three further scientific giants — Newton, Euler and Bernoulli. An appreciation of their work is essential as they give the foundations for fan engineering.

This Chapter on air and gas flow, therefore attempts to bring the science and engineering together for the mutual benefit of the two sides. It emphasises, it is hoped, the need for more auditing of actual plant. How does actual performance compare with design intentions? Certainly there is a need to feed back the actual site data to the universities and a company's data bank.

Contents:

3.1 Basic equations

3.1.1 Introduction

Fan engineering has, over the years, developed a certain mystique in the development of its "Laws" and basic equations. It should however be recognised that, as with other specialities, Newton's Laws of Motion are followed and the subject, in reality, is merely a branch of Applied Mechanics. Delving into the subject a little more deeply, we may deduce that the great majority of design work and of the operation of fans is encompassed by the Mechanics of Fluids. It is therefore imperative that we understand some of the basic concepts of air and gas flow and their applications as outlined in the following Sections.

3.1.2 Conservation of matter

Conservation of matter or the continuity equation is merely a mathematical statement that, during a flow process, matter is neither created nor destroyed.

Thus the mass flow in a fluid element (assuming no leakage to outside) remains constant i.e.,

$$\rho_1 A_1 v_1 = \rho_2 A_2 v_2 \qquad \text{Equ 3.1}$$

where:

ρ_1	=	air or gas density at position 1 (kg/m^3)
ρ_2	=	air or gas density at position 2 (kg/m^3)
A_1	=	cross-sectional area at position 1 (m^2)
A_2	=	cross-sectional area at position 2 (m^2)
v_1	=	air or gas velocity at position 1 (m/s)
v_2	=	air or gas velocity at position 2 (m/s)

In the particular case of flows where the pressures are less than about 2.0 kPa, air and many other gases may be treated as if they were incompressible. Thus $\rho_1 = \rho_2$ i.e., the density of the air/gas remains constant and

$$A_1 v_1 = A_2 v_2 \qquad \text{Equ 3.2}$$

3.1.3 Conservation of energy

The principle of the conservation of energy is encapsulated within the First Law of Thermodynamics, which states that, in a non-nuclear process, energy cannot be created or destroyed.

We may also say that when a system undergoes a thermodynamic process, the net heat supplied is equal to the net work done. This law is based on the work of Joule, who found by experiment a "mechanical equivalent of heat".

3.1.4 Real thermodynamic systems

In a real system there are inevitably losses such that the conversion process is less than 100% efficient.

The Second Law of Thermodynamics therefore states that:

> It is impossible for a system to produce net work in a thermodynamic cycle if it only exchanges heat with sources /sinks at a single fixed temperature.

This Law is based on a principle proposed by Clausius. He stated that heat flows unaided from hot to cold but cannot flow, unassisted, from cold to hot. Lord Kelvin used the proposal to show that work may be completely transformed into heat. However, only a proportion of heat could be transformed into work. If a gas is heated at constant volume there will be no work done but the energy level of the gas will be increased thus:

$$Q = mc_v(T_2 - T_1) \qquad \text{Equ 3.2}$$

$$= m(U_2 - U_1)$$

where

Q	=	heat transferred (kJ)
m	=	mass of gas (kJ)
c_v	=	specific heat capacity at constant volume (kJ/kg.k)
T_2	=	final absolute temperature (k)
T_1	=	initial absolute temperature (k)
U_2	=	final specific internal energy (kj/kg)
U_1	=	initial specific internal energy (kJ/kg)

Note: There is no degree symbol associated with the absolute temperature. Absolute temperatures in Kelvin can be converted to degrees Celsius by subtracting 273.15.

Specific heat capacity is normally abbreviated to specific heat. It is easy to see that specific internal energy, U_1 is equal to the product c_v and the absolute temperature, internal energy is an intrinsic property of a gas and is dependent upon the temperature and pressure. In this case it would have been possible to use degrees Celsius to obtain the same result.

However it is worthwhile working in absolute temperatures consistently to avoid problems with rations. If a gas is restrained and applied at constant pressure there will be work done, thus:

$$Q = mc_v(T_2 - T_1) + W \qquad \text{Equ 3.3}$$

$$= mcp(T_2 - T_1)$$

$$= m(h_2 - h_1)$$

so that:

$$W = m\left[(h_2 - U_2) - (h_1 - U_1)\right]$$

also

$$W = p(V_2 - V_1)$$

and:

$$h = U + pv$$

where:

W	=	work done (kJ)
c_p	=	specific heat capacity at constant pressure (kJ/(kg.K))
h_2	=	specific enthalpy (kJ/kg)
h_1	=	specific enthalpy (kJ/kg)
p	=	absolute gas pressure (kPa)
V_2	=	final gas volume (m^3)
V_1	=	initial gas volume (m^3)
v	=	gas specific volume (m^3/kg)

Absolute pressures are gauge pressures plus 101.325 kPa. The International Standard Atmosphere, at sea level, is 101.325 kPa. The actual local sea level atmospheric pressure is not constant and will vary with the weather by +/- 4%. some locations which experience severe weather conditions may experience larger variations. The atmospheric pressure will reduce at altitudes above sea level.

Enthalpy is an intrinsic property of a gas and is dependent upon the temperature, pressure and volume. The total enthalpy in a system, H, is the product of gas mass, m, and the specific enthalpy, h. Equation 3.3 can be rewritten as shown in equation 3.4 when it is known as the Non-flow energy equation. U is the product of m and u.

Note: The specific heat capacities, c_v and c_{p1} are variables not constants. The values for dry air, not real air, at atmospheric pressure and 275 K are 0.7167 and 1.0028; at 1000 K the values increase to 0.854 and 1.411.

$$Q = (U_2 - U_1) + W \qquad \text{Equ 3.4}$$

For heat to be transferred into or out of a system a temperature differential must exist. The general equation for heat transfer by conduction is thus:

$$Q = \frac{ka(T_h - T_c)}{L} \qquad \text{Equ 3.5}$$

where:

q	=	energy transfer (kW)
k	=	thermal conductivity (kWm/(m²K))
a	=	area (m²)
T_h	=	hot absolute temperature (K)
T_c	=	cold absolute temperature (K)
L	=	length of conductive path (m)

The thermal conductivity, k, will not be a simple value based on the boundary material. The conductivity value used must take account of the inside and outside boundary layer films and, if necessary, an allowance made for the reduction in conductivity due to surfaces being coated with deposits or modified by corrosion.

It will be appreciated that the rate of heat transfer due to conduction is proportional to the temperature differential. If the heat source cools as transfer proceeds it will take an infinite length of time to transfer all the heat available providing there are no losses. Energy losses usually occur via convection and radiation and by heating the system as well as the gas. Perfect systems are massless; only the mass of the working fluid is considered.

Entropy is another intrinsic property of gases. Entropy is very unusual when compared to other gas properties; entropy only changes when heat transfer occurs. Entropy is not dependent upon temperature, pressure or volume. A change in entropy is defined as:

$$ds = \frac{dQ}{T} \qquad \text{Equ 3.6}$$

where

ds	=	change in entropy (kJ)
dQ	=	heat transfer (kJ)
T	=	absolute temperature (K)

The units for specific entropy, s, are kJ/(kg.K). Values of intrinsic properties: u_1 h_1 s; are quoted in gas tables and appear on the axes of gas charts. It is very important to verify the base temperature of printed data before starting calculations. Some gases use 0 °C and some, like refrigerants, use – 40°C.

3.1.5 Bernoulli's equation

Consider an elemental tube in which flow is entirely parallel to the boundaries. For simplicity assume it to have constant cross-section area of δa (although it can be shown it is not essential to do so).

The forces on the element may be equated to the rate of change of momentum. In the direction of flow, the forces are: due to change in pressure:

$$p\delta A - (p + \delta p)\delta A = -\delta p \delta A$$

due to change in height above some datum:

$$-\rho g\ \delta s \sin 0\ \delta A = -w\delta H\delta A$$

Rate of change of momentum in direction of flow

$$= \rho\delta A v(v + \delta v) - p\delta A v^2 = p\delta A v\delta v$$

thus

$$-\delta p\delta A = p\delta H\delta A = p\delta A v\delta v$$

and rearranging

$$v\delta v + \frac{\delta p}{\rho} + g\delta H = 0$$

which in the limit becomes

$$vdv + \frac{dp}{\rho} + gdH = 0$$

On integration, this gives

$$\frac{v^2}{2} + \int \frac{dp}{\rho} + gH = \text{constant} \qquad \text{Equ 3.7}$$

H is measured from any arbitrary datum, and any change of datum results in a change in H and an equal change in the constant of integration. If the air is considered as incompressible, which is acceptable for fan pressure below about 2.0 kPa, then equation 3.7 reduces to

$$\frac{v^2}{2g} + \frac{p}{\rho g} + H = \text{constant, known as Total Head} \qquad \text{Equ 3.8}$$

Although strictly only applicable to flow along a stream tube of an ideal frictionless fluid, equation 3.8 is often used to relate conditions between two sections in a practical system of flow through a duct. If the mean total head is measured at the two sections, it will be found that the value at the downstream section is less than that at the upstream section. This is due to resistance to flow between the sections and the difference in head is known as loss of total head. When making measurement however, it is customary to use gauge pressure, i.e. pressures greater or less than atmospheric pressure.

Considering two sections, subscript 1 referring to the upstream section and subscript 2 referring to the downstream section, then

$$\frac{v_1^2}{2g} + \frac{p_{at1} + p_1}{\rho g} + H_1 = \frac{v_2^2}{2g} + \frac{p_{at2} + p_2}{\rho g} + H_2 + \Delta H \qquad \text{Equ 3.9}$$

where ΔH is the loss of total head between the two sections. This may be rewritten

$$\frac{v_1^2}{2g} + \frac{p_1}{\rho g} = \frac{v_2^2}{2g} + \frac{p_2}{\rho g} + \Delta H + \left(H_2 - H_1 - \frac{p_{at1} - p_{at2}}{\rho g}\right) \qquad \text{Equ 3.10}$$

Now, if p_{at} represents the atmospheric pressure at a height H above some datum, and $p_{at} + \delta p_{at}$ at a height H + δH above the same datum, and a column of air of cross-section A is considered,

$$p_{st}A - (p_{at} + \delta p_{at})A = \rho g A(H + \delta H) - \rho g A H$$

from which

$$-\delta p_{at} = \rho g \delta H \qquad \text{Equ 3.11}$$

If ρg remains constant, then equation 3.8 may be rewritten

$$\frac{p_{at1} + p_2}{\rho g} + H_2 + H_1$$

and inserting this in equation 3.10 gives

$$\frac{v_1^2}{2g} + \frac{p_1}{\rho g} = \frac{v_2^2}{2g} + \frac{p_2}{\rho g} + \Delta H \qquad \text{Equ 3.12}$$

Multiplying throughout by ρg gives the equation in terms of pressure:

$$\tfrac{1}{2}\rho v_1^2 + p_1 = \tfrac{1}{2}\rho v_2^2 + p_2 + \Delta p \qquad \text{Equ 3.13}$$

or

$$p_{t1} = p_{t2} + \Delta p$$

In equation 3.13, p_1 and p_2 are known as the static pressures at the two sections and may be positive or negative according to whether the absolute pressure is greater or less than the ambient atmospheric pressure which, as stated above, is the arbitrary datum or zero to which static pressure is generally referred.

The sum of static pressure and velocity pressure $\left(p + \tfrac{1}{2}\rho v^2\right)$ is known as the total pressure p_T. Although in many cases the air density remains substantially constant, this may not be so where the height between two parts of a system is considerable, or if there is a temperature gradient.

Equation 3.13 shows that the resistance of a system of ducting expressed as a pressure loss for a particular flow rate, is equal to the difference between the total pressures at the two ends of the system. In practice the use of this equation to calculate the resistance of a system is complicated by the fact that the velocity nearly always varies considerably between the centre and the duct walls, although the static pressure, except near bends, is often sensibly constant across a section.

In determining the pressure loss it is not correct to calculate the velocity pressure component of the total pressure from the expression:

$$\tfrac{1}{2}\rho v_m^2$$

where:

v_m	=	the mean velocity and is equal to Q/A
Q	=	the volume flow
A	=	cross-sectional area of the airway

Strictly speaking, and neglecting any variations in the static pressure p across the section, the mean velocity pressure must be calculated from the kinetic energy per unit time divided by the volume flow per unit time, that is, in a circular duct:

$$p_{v(mean)} = \int_O^R \tfrac{1}{2}\rho v \times v^2 \times 2\pi r \, dr \div \int_O^R v \times 2\pi r \, dr$$

or

$$p_{v(mean)} = \tfrac{1}{2}\rho \int_O^R v^3 r \, dr \div \int_O^R vr \, dr \qquad \text{Equ 3.14}$$

In most cases where equation 3.13 is used, the error due to the incorrect method of calculating $p_{v\,(mean)}$ is allowed for by an experimentally determined loss factor or coefficient for the form of velocity distribution it is hoped will be encountered. It will be assumed here that $p_{v(mean)}$ is based on the simple calculation in conjunction with this factor.

3.2 Fan aerodynamics

3.2.1 Introduction

It is not the intention of this book to give detailed data for the aerodynamic design of fans. As has been said elsewhere, it rather seeks to inform both manufacturers and users of the information necessary at their common interface, so that correct choices are made to their mutual advantage.

Nevertheless, it is of value to cover the basics of the theory, to show what is and is not possible, It will also show the background to Chapter 1 and explain how those characteristic curves match with the fundamental fluid mechanics. A detailed design guide could be written and it would certainly require a similar number of pages to this volume, to do the subject justice.

3.2.2 Elementary centrifugal fan theory

To fully understand therefore, Sections 1.5 and 1.6 in Chapter 1, dealing with fan characteristics, Chapter 5, Section 5.6 on system effect factors and Chapter 6 on flow regulation, some knowledge of the elementary theory is essential.

For the sake of simplicity the analysis which follows is not mathematically exact and further assumes that the air or gas is incompressible.

A centrifugal fan receives air or gas at the impeller eye and delivers it to the casing volute at high velocity by imparting rotational energy. The kinetic energy produced by the impeller is converted into pressure energy within the volute. Fan efficiency therefore depends on how much kinetic energy is produced, how low the impeller losses can be kept, and how well this kinetic energy is converted into potential energy (or static pressure) within the casing.

Considering the velocity triangles in Figure 3.1, the work done on the gas by the impeller will be the energy difference between exit and entry in the direction of rotation.

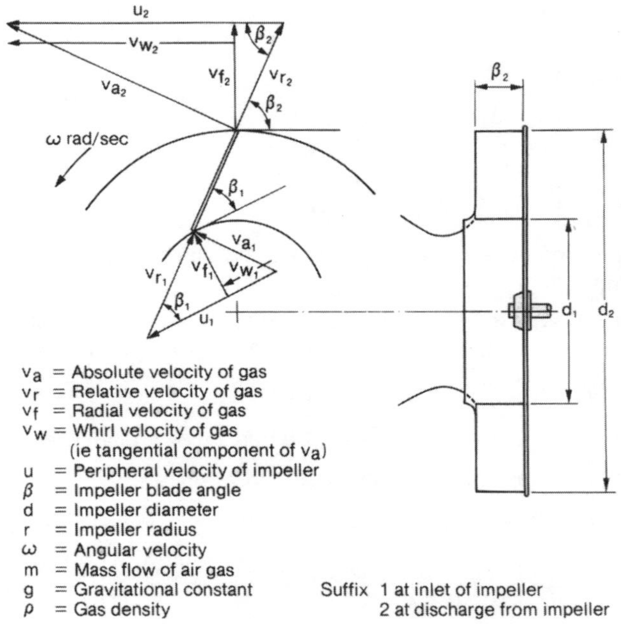

Figure 3.1 Theoretical flow pattern in a centrifugal fan impeller with backward inclined bladed impeller

Energy in air at impeller exit

= torque x angular displacement

= rate of change of (tangential momentum x radius x angular displacement)

= tangential momentum x radius x angular displacement

= $m \, v_{w2} \, r_2 \, \omega$

in like manner the energy in air at impeller inlet

= $m \, v_{w1} \, r_1 \, \omega$

Now $r_1\omega = u_1$ and $r_2 \, \omega = u_2$

Energy given to the air by the impeller

= $m\left(v_{w2} \, u_2 - v_{w1} u_1\right)$

The theoretical or Euler head H developed by the impeller is defined as the height to which the same weight of gas could be raised by an equal amount of work.

Thus:

$$mgH = m\left(v_{w2}\, u_2 - v_{w1}\, u_1\right)$$

or

$$H = \frac{1}{g}\left(v_{w2}\, u_2 - v_{w1}\, u_1\right)$$

In fan work it is usual to know the pressure developed $(p = \rho g H)$ and therefore $p = \rho\left(v_{w2}\, u_2 - v_{w1}\, u_1\right)$.

Under normal circumstances at the design duty, the air will enter the easiest way, i.e. radially and then $v_{w1} = 0$.

Thus:

$$p = \rho\, v_{w2}\, u_2 \qquad \text{Equ 3.15}$$

Considering the impeller in cross-section with a width at its tip of b_2, it may be said that the volume of air or gas delivered per unit time $Q = \pi\, d_2\, b_2\, v_{f2}$.

Now the impeller blades at the outlet may be either:

a) Backward inclined (straight, curved, or aerofoil) as in Figure 3. 1.

when

$$u_2 = v_{w2} + v_{f2} \cot \beta_2$$

or

$$v_{w2} = u_2 - v_{f2} \cot \beta_2$$

Now, as:

$$p = \rho\, v_{w2}\, u_2 \text{ and } Q = \pi d_2\, b_2\, v_{f2}$$

$$p = \rho\, v_{w2}\left(u_2 - \frac{Q}{\pi d_2 b_2} \cot \beta_2\right) \qquad \text{Equ 3.16}$$

This theoretical characteristic is a straight line with a downward slope.

b) Radial (straight shrouded, open or backplate paddle, or radial tipped) as in Figure 3.2 when $v_{w2} = u_2$ and $v_{f2} = v_2$

$$p = \rho u_2^{\,2} \qquad \text{Equ 3.17}$$

This theoretical characteristic is a horizontal straight line.

c) Forward curved as in Figure 3.3 when

$$v_{w2} - u_2 = v_{f2} \cot\left(180° - \beta_2\right)$$

$$p = \rho u_2\left[u_2 + \frac{Q}{\pi d_2 b_2} \cot\left(180° - \beta_2\right)\right] \qquad \text{Equ 3.18}$$

This theoretical characteristic is a straight line with an upward slope.

It will be seen that for a given speed of rotation and a given pressure, the volume flow rate is dependent on the width of the impeller and the blade angle. Reputable centrifugal fan manufacturers will have many different width ranges with varying blade numbers and outlet blade angles to meet all duties economically. All these theoretical characteristics are shown in Figure 3.4.

The theoretical pressure will be reduced by the following factors, the aim of the fan engineer being to keep them to a minimum:

Relative rotation losses

In addition to the normal flow of fluid within the impeller, the inertia effect of the fluid causes a rotation of the fluid relative to the impeller. Also, when the impeller is mounted between bearings due to the effect of the rotating shaft, the fluid will have a definite

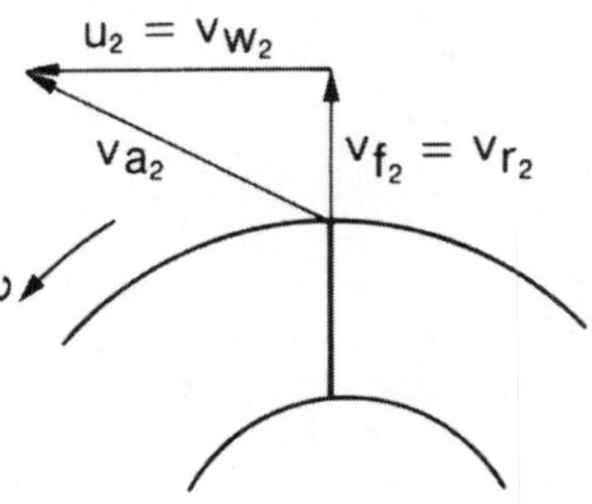

Figure 3.2 Theoretical flow pattern at impeller outlet for radial blades

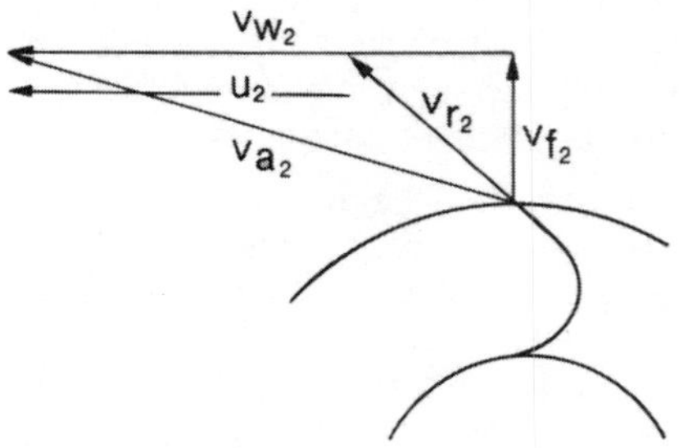

Figure 3.3 Theoretical flow pattern at impeller outlet for forward curved blades

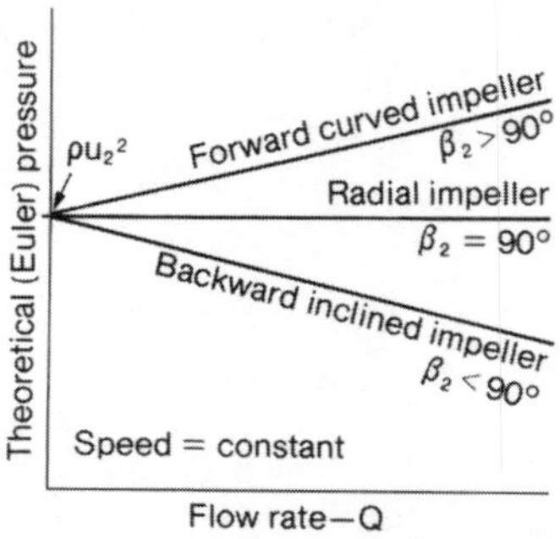

Figure 3.4 Theoretical p-Q characteristics for different values of impeller discharge angle

tangential whirl velocity at entry to the impeller blade. Both of these factors reduce the pressure that the fan is capable of producing, but they do not affect the efficiency.

Friction losses

These are caused by gas friction and also include volute losses. (The volute is that part of the fan which converts velocity energy into pressure energy. This is normally achieved by arranging the discharge channel so that the cross-sectional area gradually increases, thus reducing the flow velocity.)

Shock losses

Losses arise at entry to, and exit from, the impeller blade because the blade angles are only correct for the design duty. On both sides of this shockless flow condition losses will occur.

Other losses

- **Leakage**: occurs from discharge to suction and through the shaft entry hole.
- **Disc friction**: due to the rotation of the impeller shroud and backplate within the gas.
- **Mechanical losses**: caused by the bearing friction and friction at any shaft seal. These losses differ from those of the previous three groups in that whilst they affect the overall efficiency they do not alter the basic fan characteristic.

The actual characteristic, with its losses are shown for a backward inclined impeller in Figure 3.5. Actual against theoretical

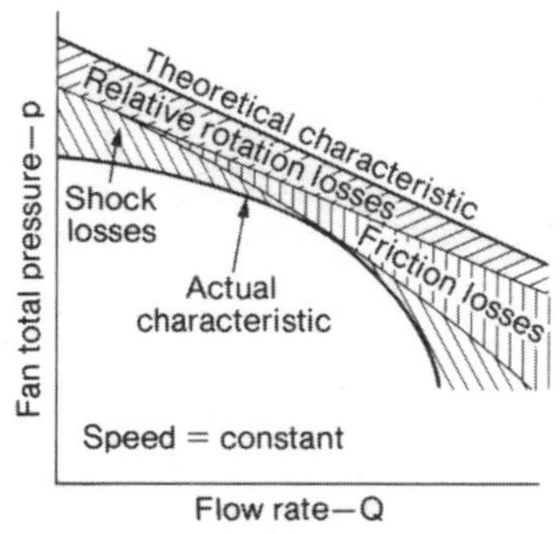

Figure 3.5 Deviation of actual fan characteristics for impeller having backward inclined vanes

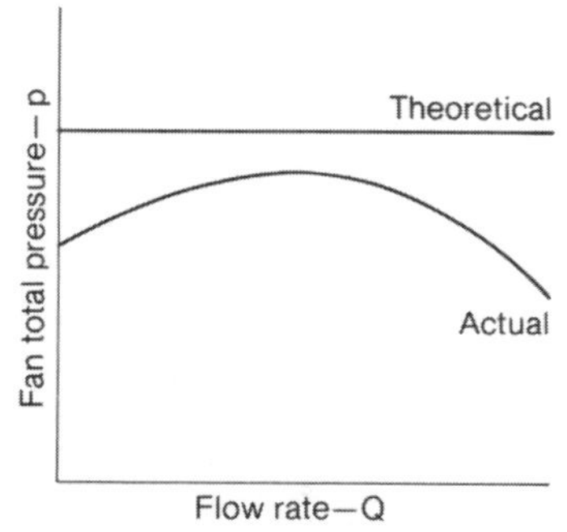

Figure 3.6 Characteristics for radial blade fan

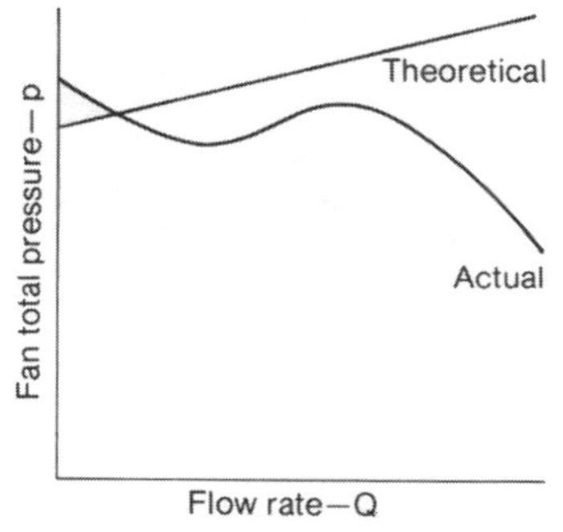

Figure 3.7 Characteristics for forward curved fan

characteristics for radial and forward curved fans are shown in Figure 3.6 and Figure 3.7 respectively.

Important Note

It must be emphasised that all the above assumes straight flow into the impeller eye and consideration of the equations will show that if this is not the case then the pressure developed will be reduced.

Variable inlet vanes purposely use this fact to impart swirl in the direction of rotation. This can be progressively increased by closure of the vanes with a corresponding reduction in the pressure developed. There will of course be some additional friction losses. Further information is given in Chapter 6, Section 6.5.

More importantly, from the system designer's viewpoint, it will be seen that if straight flow into the fan inlet is not achieved due to poor inlet connections, then the fan will not develop its test pressure. Insufficient straight ducting on the fan inlet side, sagging flexible connections, absence of straighteners in bends, and too tight bends can all be responsible. Where fans are mounted in plenum chambers there must be a sufficient distance from the fan inlet(s) to the chamber walls for the same reason.

Often the system designer is himself short of space. He may then have to provide less than ideal connections. A section on system effect factors (Chapter 5, Section 5.4) has therefore been included and this will enable the designer to make such allowances as are necessary in specifying the fan duty so that the required flow may be achieved.

3.2.3 Elementary axial fan theory

Figure 3.8 shows an axial flow fan blade section at some particular radius, with its associated velocity triangles. The air enters the impeller axially with a velocity $v_1 = v_{m1}$, and leaves with velocity v_2.

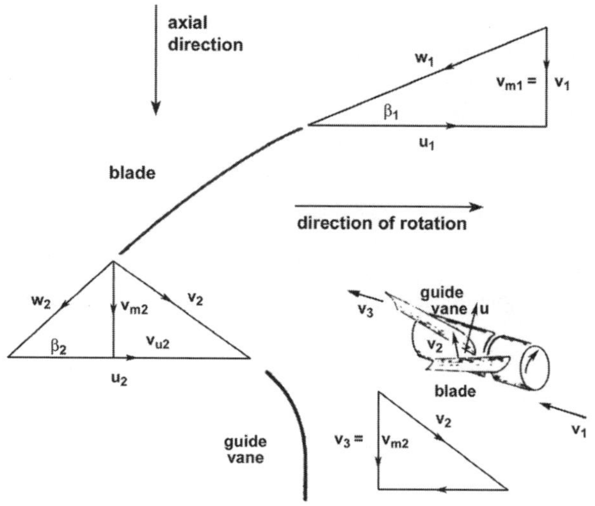

Figure 3.8 Axial flow blade velocity triangles

The shape of the triangles is almost identical with those of a backward bladed centrifugal fan, but it should be noted that $u_1 = u_2$, and $v_{m1} = v_{m2}$. The total pressure developed is given by the same equation as for a centrifugal fan, namely, $\rho u_2 v_{u2}$, v_{u2} being the rotational component of v_2. It should be noted that the expanded form of Euler's equation no longer includes a forced vortex component since $u_1 = u_2$ at each radius.

The theoretical characteristics may be derived since:

$$v_u = u - v_m \cot \beta_2$$

$$p = \rho u v_u = \rho u^2 - \rho u v_m \cot \beta_2$$

$$= \rho u^2 - \rho u \cdot \frac{4Q}{\pi d_2^2 \left(1 - \nu^2\right)} \cdot \cot \beta_2 \qquad \text{Equ 3.19}$$

Where ν equals hub to tip ratio D_1/D_2. The characteristics are shown in Figure 3.9, and are seen to be very similar to those for a backward bladed centrifugal fan, apart from the stall point.

It is usual to design a blade to give the same axial velocity and pressure development at each radius, in which case

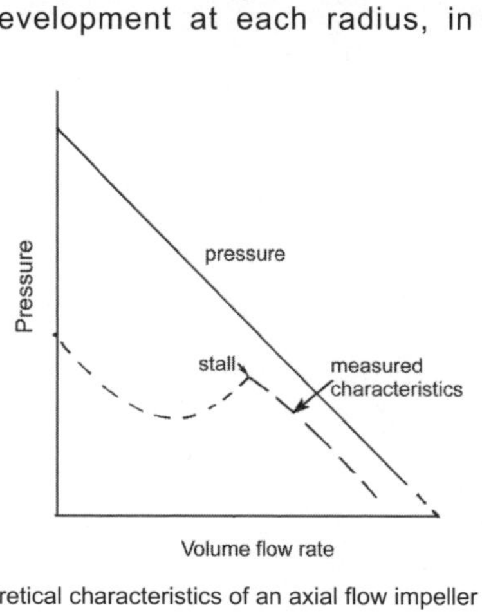

Figure 3.9 Theoretical characteristics of an axial flow impeller

$p = \rho\omega r v_u$ = constant, or $r v_u$ = constant. This will be seen to be the condition for a free vortex and permits radial equilibrium of forces on the fluid. It is necessary to have increased blade angles at the hub section to achieve the higher values of v_u at the smaller radius. Departures from free vortex designs have therefore been made, which limit the blade chord adjacent to the hub. These develop less pressure in this region and are known as arbitrary vortex designs.

Alternative forced vortex designs are also available, where maximum pressure development takes place at the tips of the blades. For good efficiency the tip gap needs to be kept to an absolute minimum.

Since the air leaving the impeller has a rotational component of velocity, v_u, there is a loss of total pressure of

$$\tfrac{1}{2}\rho\left(v_2^2 - v_m^2\right) = \tfrac{1}{2}\rho v_u^2 \qquad \text{Equ 3.20}$$

if the rotational energy is allowed to be dissipated along the duct system. Downstream guide vanes may be fitted to reduce the velocity to v_m and thereby regain static pressure equal to $\tfrac{1}{2}\rho v_u^2$.

Even so, many commercial designs are produced without guide vanes to reduce costs, these being known as Tube Axials. The resulting loss in efficiency is relatively unimportant at low fan power.

It is possible to avoid rotational energy loss by having a guide vane upstream of the impeller which pre-rotates the entering air in a direction opposite to that of the impeller rotation. The impeller is designed to do sufficient work on the air to remove this rotation (Figure 3.10).

Then,

$$p = \rho u_2 v_{u2} - \rho u_1 v_{u1} = 0 - \rho u_1(-v_{u1}) = \rho u v_{u1} \qquad \text{Equ 3.21}$$

and, at the design point,

$$v_{u1} = v_m \cot \beta_1 - u$$

$$\rho u v_{u1} = \rho u v_m \cot \beta_1 - \rho u^2 \qquad \text{Equ 3.22}$$

Another type, the contra-rotating fan, makes use of air leaving an impeller with rotation to enter a second impeller rotating in the opposite direction. This second impeller acts in a similar manner to that of an upstream guide vane fan, as can be seen from the velocity triangles, in Figure 3.11. There, the inlet and outlet velocity triangles for each impeller have been combined into a single diagram, made possible since v_m and u are the same in each case. Each impeller develops the same pressure if u and v_u for each are the same, and the air is discharged axially, that is:

$$p = 2\rho u v_u \qquad \text{Equ 3.23}$$

A similar arrangement, with both impellers running in the same direction, is possible by using guide vanes between the impellers. Whilst this obviates the need for opposite handed impeller, a large angular deflection of the air is necessary. Very careful design of these intermediate guide vanes is required to ensure that flow separation does not occur.

3.2.3.1 Use of aerofoil section blades

As with centrifugal fans, the air passing through an impeller constructed with sheet metal blades will not follow the blade profile very accurately unless the number of blades is infinite. Since aerofoil data is available, it is possible to predict the performance of an axial flow fan more accurately if blades of aerofoil profile are used. The velocity triangles for such blades are shown in Figure 3.12 and are seen to differ from those previously considered only by the addition of a mean relative velocity vector, $w_\infty = \tfrac{1}{2}(w_1 + w_2)$ to which the blade section is inclined at its angle of attack, α. The mean blade angle is β, with an effective blade angle (blade air angle) between vectors of w and u of $\beta - \alpha$.

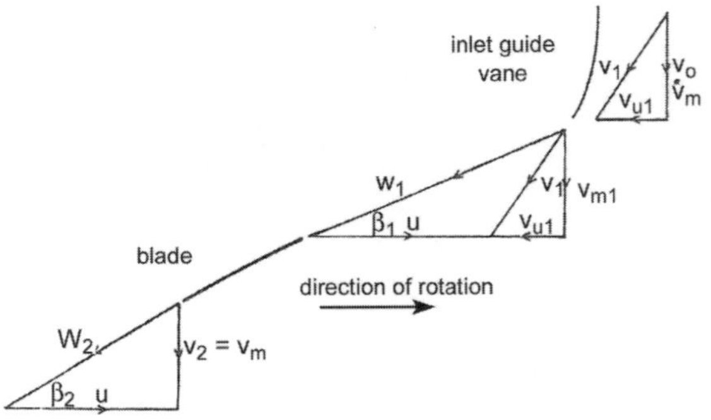

Figure 3.10 Axial flow blade with upstream guide vane

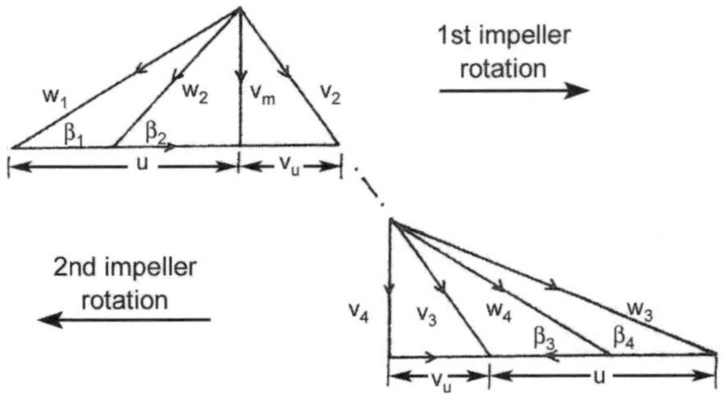

Figure 3.11 Contra-rotating fan velocity triangles

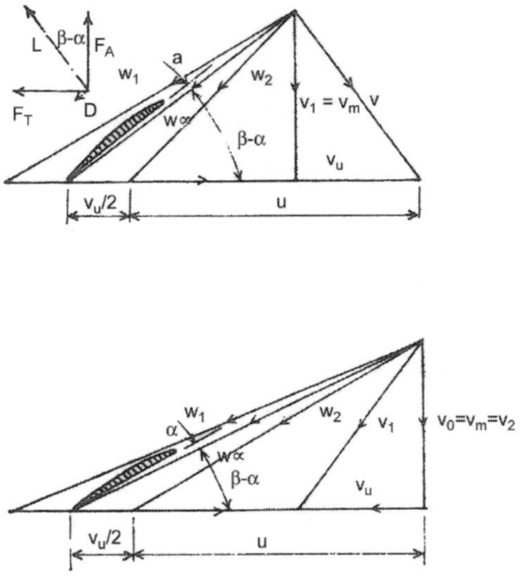

Figure 3.12 Use of aerofoil section axial flow blades

The static pressure difference across the impeller may be found, since

$$p = \rho u v_u = p_{t1} - p_{t2} = p_2 + \tfrac{1}{2}\rho v_2^2 - \left(p_1 + \tfrac{1}{2}\rho v_1^2\right)$$

$$= p_2 - p_1 + \tfrac{1}{2}\rho\left(v_2^2 - v_1^2\right)$$

Static pressure difference,

$$p_2 - p_1 = \rho u v_u - \tfrac{1}{2}\rho\left(v_2^2 - v_1^2\right)$$

$$= \rho u v_u \pm \tfrac{1}{2}\rho v_u^2$$

$$= \rho v_u\left(u \pm \tfrac{1}{2} v_u\right) \qquad \text{Equ 3.24}$$

where the negative sign refers to the downstream guide vane impeller, and the positive sign to the upstream guide vane impeller.

This pressure difference over the impeller swept area may be equated to the axial thrust due to the aerodynamic lift forces L on the blades

$$F_A = L\cos(\beta - \alpha) = (p_2 - p_1)\cdot 2\pi r dr$$

for an element of blade.

If there are z blades, each of chord c,

$$zc\cdot dr\cdot C_L\cdot \tfrac{1}{2} r w_\infty^2 \cos(\beta - \alpha) = \rho v_u\left(u \pm \tfrac{1}{2} v_u\right) 2\pi r\cdot dr$$

and writing blade spacing, $s = 2\pi r/z$

and substituting

$$u \pm \tfrac{1}{2} v_u = w_\infty \cos(\beta - \alpha)$$

$$\tfrac{1}{2}\frac{c}{s}\cdot C_L w_\infty = v_u$$

or

$$\frac{v_u}{w_\infty} = \tfrac{1}{2} C_L \frac{c}{s} \qquad \text{Equ 3.25}$$

The above simplified blade element theory, whilst adequate for exploratory design, ignores the effect of drag. To consider more fully the forces on the aerofoils it is necessary to equate the thrust force F_A, which is due to static pressure rise less any pressure loss, to the axial force due to the lift and drag.

3.2.4 Elementary mixed flow fan theory

The mathematics of mixed flow fans becomes even more complex than that given in Sections 3.2.2 and 3.2.3, as there are both axial and centrifugal components to the airflow. In general, however, it can be said that characteristics similar to the backward bladed centrifugal are achieved. The Euler theory still "reigns"!

3.3 Ductwork elements

3.3.1 Introduction

In the design of a ductwork system it is the practice to add the resistance of all the elements in the index leg together, to determine the total (or static) pressure loss. The fan must develop this pressure at the design flowrate. The system and fan will then be in harmony. (See Chapter 4.)

The resistance of duct fittings and straight ducting is invariably determined from the Guides produced by CIBSE or ASHRAE. Both bodies have a similar approach and treat the pressure losses as a function of the local velocity pressure. This function is usually regarded as a constant and thus the loss becomes:

$$p_L = k_F \times \frac{1}{2}\rho v^2 \qquad \text{Equ 3.26}$$

where:

p_{Lf} = pressure loss (Pa)

k_F = constant

ρ = local air density (kg/m^3)
(usually taken as standard 1.2)

v = local velocity (m/s)

Whilst this may be reasonably true in the normal working range, it is important to know that k_F has a Reynolds Number dependence and that at low Reynolds Numbers k_F can increase enormously, whilst in fully turbulent flow, if ever attained, the value could be less.

There are very few textbooks which even admit this variation. The only one of note is Idelchik's *Handbook of Hydraulic Resistance* which gives a very detailed exposition of the subject and is noteworthy for its comprehensiveness. Miller's *Internal Flow Systems* is also recommended.

It might be thought that the topic is somewhat esoteric, but it is suggested that with the increasing use of inverters and other variable flow devices, it is important to know that at high turndown ratios, the system resistance curve diverges ever more from the oft quoted $p_L \propto Q^2$. Thus power absorbed is not $\propto$ fan speed N^3, even if there were no bearing, transmission and control losses.

In like manner, the loss in straight ducting is usually quoted as

diameter d m	Average velocity v m/s	Reynolds No $Re = \frac{\rho v d}{\mu}$	Relative roughness $\frac{k}{d}$	Friction factor f	Flow quality
0.1	2.5	16492	0.0015	0.0076	Tr
	5	32985		0.0067	
	10	65970		0.0063	
	15	98955		0.0059	
	20	131940		0.0057	
0.25	2.5	41231	0.0006	0.006	Tr
	5	82463		0.0055	
	10	164926		0.005	
	15	247388		0.0048	
	20	329851		0.0047	
0.315	5	103903	0.00048	0.0051	Tr
	10	207806		0.0047	
	15	311710		0.0046	
	20	415613		0.0045	
	25	519516		0.0044	
0.63	5	207806	0.00024	0.0043	Tr
	10	415613		0.0042	
	15	623419		0.0039	
	20	831226		0.0038	
	25	1039032		0.0036	
1	5	329851	0.00015	0.0039	Tr
	10	659703		0.0037	
	15	989555		0.0036	
	20	1319406		0.0035	
	25	1649258		0.0034	
2	10	1319406	0.000075	0.0033	Tr
	15	1979109		0.0032	
	20	2638812		0.0031	
	25	3298516		0.003	
	30	3958218		0.00295	
2.5	15	2473887	0.00006	0.00295	Tr
	20	3298516		0.0029	
	25	4123144		0.00285	
	30	4947773		0.0028	
	40	6597031		0.0028	

Table 3.1 Friction factors versus duct size and velocity

Note 1: Values apply to standard air

Note 2: All values are in the transitional range

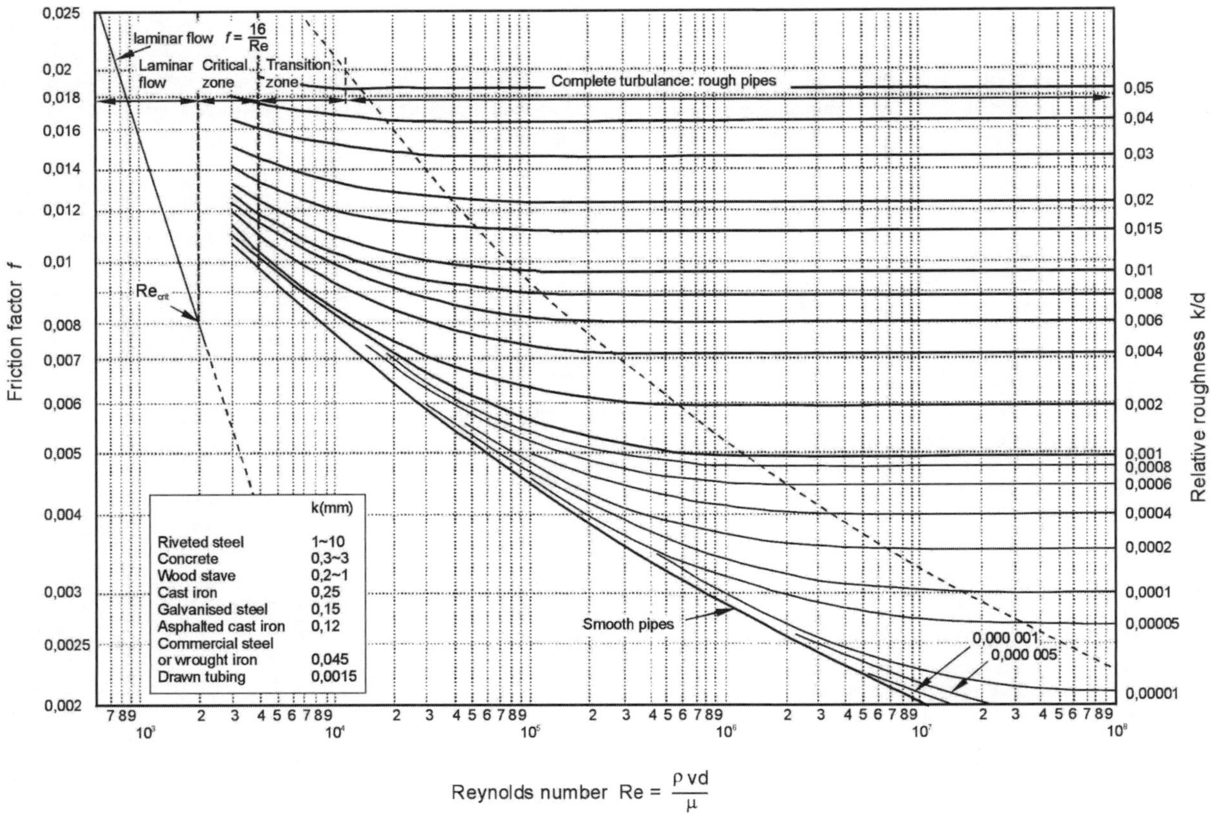

Figure 3.13 Friction factor versus Reynolds number - Moody chart

$$p_{Ls} = \frac{fL}{m} \cdot \frac{1}{2}\rho v^2 \qquad \text{Equ 3.27}$$

and $\frac{fL}{m}$ is taken to be a constant k_s

where:

L	=	length of straight duct (m)
m	=	mean hydraulic depth (m)
	=	$\frac{d}{4}$ for circular cross-sections
f	=	friction factor

Again, as L and m are constants and f is assumed to be constant, the loss is taken to be

$$p_{Ls} = k_s \cdot \frac{1}{2}\rho v^2 \qquad \text{Equ 3.28}$$

And thus another problem is created, for f is **not** a constant but rather a function of absolute roughness and Reynolds Number.

The Moody chart shown in Figure 3.13 shows that in the transitional and lower zones $f \neq$ constant, and that again, as flow enters the critical zone there are significant increases in f, then a sudden drop, before climbing again in the laminar zone.

Referring now to Table 3.1, this covers the range of sizes and velocities encountered in HVAC practice. Assuming an absolute roughness applicable to g.s.s. (galvanised sheet steel), it can be seen that in all these cases the flow is transitional. The relative roughness and friction factor therefore vary enormously as shown. Thus with decreasing flow, and therefore velocity, the reducing velocity pressure is partially offset by the increase in f.

A system resistance curve is likely to be of the form shown in Figure 3.14 although for most HVAC systems the flow at which instability occurs is very close to zero flow. For mine ventilation, where the size of roadways can be considerable and the

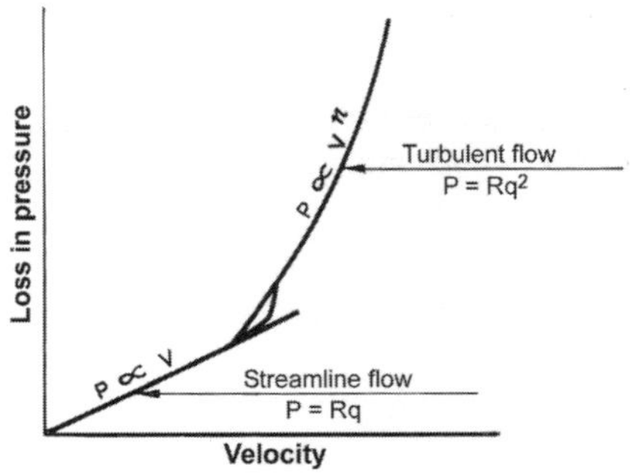

Figure 3.14 A system resistance curve

Reynolds Number is higher, this shape of system resistance curve has been recognised for at least 50 years. Somewhat later in Section 3.4.1 it will be shown how the formula has been tailored to fit the facts by reducing the index of v velocity from 2 down to 1.9 or even less.

Vigilant readers of this text will have detected that the author is somewhat cynical and he would suggest that it hardly seems worth the struggle to reach the truth, if there is any! Better to go back to basics. In this computer age, it should be possible to develop a programme to give the correct f for the velocity, diameter and roughness. Whether the effort is applauded, however, may still be debatable.

Norman Bolton at NEL, East Kilbride, was responsible for a programme of work which measured the resistance of supposedly identical ducts and fitting from three different manufacturers. The variation in pressure loss p_L was enormous, thus proving that quality is everything. It also suggests that so-called balancing of systems is not enough and that, to use "management speak", a full system audit should be carried out.

The results should be fed back into the company design database. Some aspects of ductwork design are rarely mentioned in textbooks. The Sections which follow are a mixture of basic fluid dynamics and practical "nous".

3.3.2 Diffusers

These are attached to the discharge or fan outlet and are used to improve the fan static pressure of medium to high pressure fans. They may also be used in a system at the end of the outlet duct to atmosphere. A change from high to low velocity is accompanied by a conversion from velocity pressure to static pressure. There is always a loss in this conversion such that the total pressure is never the same before and after the diffuser. Efficiency of conversion E is never 100%.

E = percentage converted of difference in initial and final velocity pressures.

True efficiency of conversion in the diffuser itself depends almost entirely on the angle of taper. If however, the diffusing taper is followed by a length of straight ducting 4 to 6 diameters long, then there is some additional conversion after the taper. In such cases, the overall efficiency, as determined by test, is related to the angle of the taper and the area ratio.

The included angle of a jet of air which is confined by the walls of a duct is about 7°. If the taper is more than this then the flow leaves the walls and dead areas result.

An unconfined jet of air in free space has an included angle of about 3°, but the jet is spread by the induction of secondary air so that the actual included angle increases to about 16°. This can be seen from smoke photographs of air jets, the results being summarised in Figures 3.15 to 3.17.

The static regain, or increase in static pressure in the larger duct

$$p_{sr} = \cdot E\left(p_{v1} - p_{v2}\right) \qquad \text{Equ 3.29}$$

where:

$\cdot E$ = efficiency of conversion expressed as a decimal

It is generally more convenient to calculate the regain from the initial velocity pressure, and to make allowance for the difference by an area term i.e.:

$$p_{sr} = \cdot E\left[1 - \left(\frac{A_1}{A_2}\right)^2\right] p_{v1} \qquad \text{Equ 3.30}$$

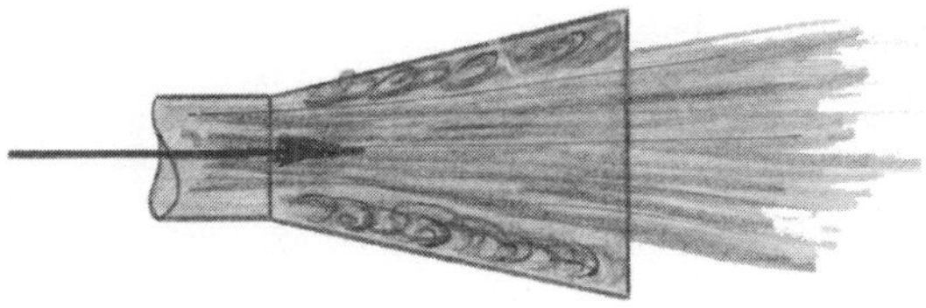

Figure 3.15 Short diffuser at large angle

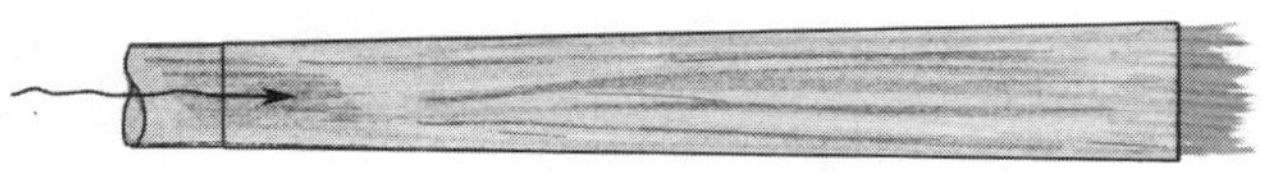

Figure 3.16 Very long diffuser at small angle

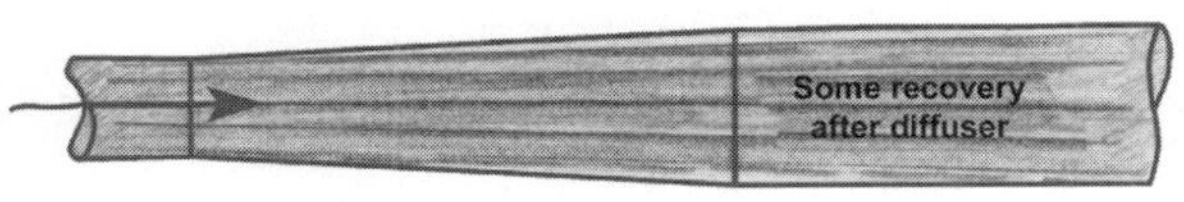

Figure 3.17 Normal diffuser followed by duct

where:

p_{v1} = initial velocity pressure (Pa)

p_{v2} = final velocity pressure (Pa)

A_1 = initial cross-sectional area of diffuser (m^2)

A_2 = final or outlet cross-sectional area of diffuser (m^2)

$\cdot E$ = efficiency of conversion expressed as a decimal

At 100% theoretical efficiency of conversion:

$$p_{s1} + p_{v1} = p_{s2} + p_{v2}$$

or

$$p_{s2} = p_{s1} + \left(p_{v1} - p_{v2}\right)$$

so

$$p_{s1} + p_{v1} = p_{s1} + \left(p_{v1} - p_{v2}\right) + p_{v2}$$

but

$$p_{v2} = p_{v1} \times \left(\frac{A_1}{A_2}\right)^2$$

because velocity pressure is inversely proportional to area^2.

Then

$$p_{s1} + p_{v1} = p_{s1} + \left[p_{v1} - p_{v1}\left(\frac{A_1}{A_2}\right)^2\right] + p_{v2}$$

$$= p_{s1} + p_{v1}\left[1 - \left(\frac{A_1}{A_2}\right)^2\right] + p_{v2}$$

So, the static regain p_{sr} or addition to the initial fan static pressure p_{s1} is the term $p_{v1}\left[1 - \left(\frac{A_1}{A_2}\right)^2\right]$ which is exactly the same as $\left(p_{v1} - p_{v2}\right)$.

As the efficiency of conversion is never 100%, the actual regain will be:

$$\cdot E p_{s1}\left[1 - \left(\frac{A_1}{A_2}\right)^2\right] \qquad \text{Equ 3.31}$$

From this a combined factor F may be obtained from the value:

$$\cdot F = \cdot E\left[1 - \left(\frac{A_1}{A_2}\right)^2\right]$$

and the regain p_{sr} is then $\cdot F \times p_{s1}$.

Values of F from experiment are plotted with included angle of taper and various ratios of $\frac{A_1}{A_2}$. Those from the tests of Kratz and Fellows are the most reliable, see Figures 3.18 - 3.20.

It is impossible to include factors for every possible design of diffuser. Those given are for circular cross-section diffusers. If the cross-section is square or rectangular then the efficiency is somewhat less for a given included angle. It is suggested that an average reduction of 5% or .05 in $\cdot$ E is a suitable allowance.

Diffusers for steel plate industrial fan outlets often transform from rectangular or circular cross-section. Draw the view each way and estimate the mean included angle. Then use the values for a circular design. If the design is critical and has to be passed by a performance test, it is wise to be on the safe side with the factor.

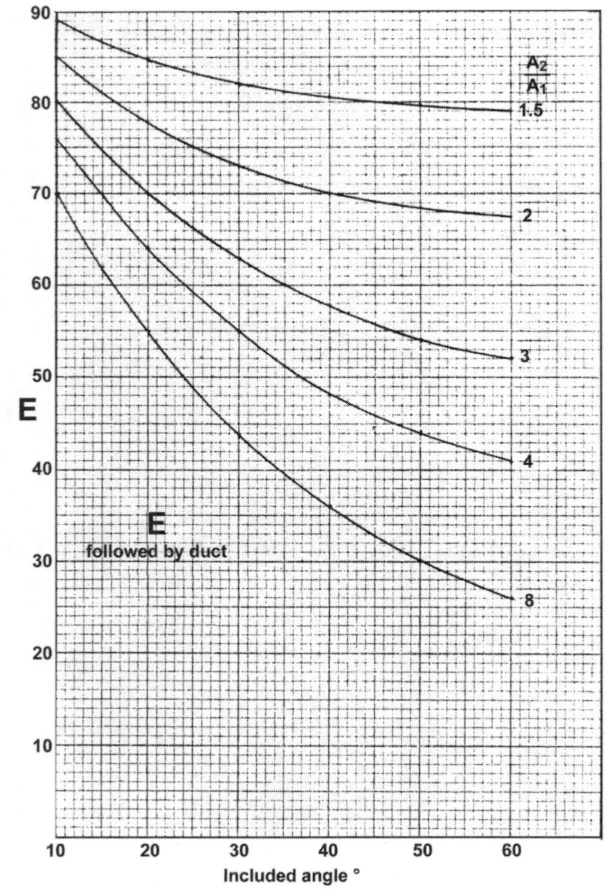

Figure 3.18 Diffuser efficiency versus included angle and area ratio

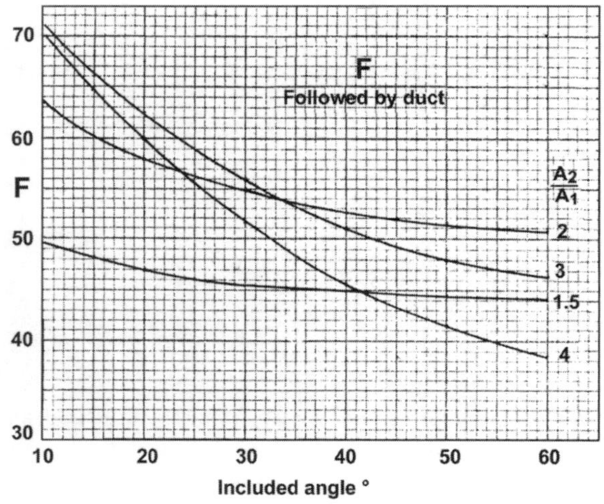

Figure 3.19 Regain in a diffuser followed by a duct ($p_{sr} = \cdot F \times p_{s1}$)

In an exhaust system which has a diffuser fitted on the fan discharge direct to atmosphere, any gain due to this must be subtracted from the calculated resistance of the system. A fan to deal with the required flowrate at this nett resistance is then selected.

If a diffuser follows immediately after a bend in the system, the full recovery will not be achieved. It is then prudent to use about 0.7 x ·F. With one diameter of straight duct between the bend and the diffuser use 0.8 x ·F. If 5 diameters of ducts are between the bend and the diffuser then the full values of ·F as shown on the curves may be used.

Air velocity has some effect on efficiency. When this initial velocity is very high e.g. above 37.5 m/s there is some loss in efficiency but no definite data is available.

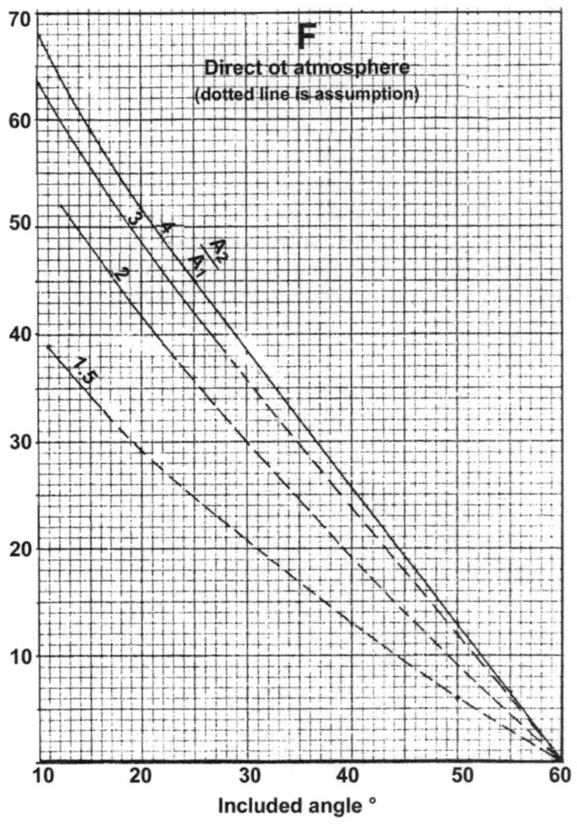

Figure 3.20 Regain in an open-ended diffuser ($p_{sr} = \cdot F \times p_{s1}$)

The loss of pressure in a diffuser due to imperfect conversion may be calculated by the same method using (1- ·E) for the loss factor.

Thus

$$p_{sl} = (1 - \cdot E)p_{s1} \times \left[1 - \left(\frac{A_1}{A_2}\right)^2\right] \qquad \text{Equ 3.32}$$

It is important to remember that different factors are required for a discharge direct to atmosphere compared to one with a following duct. One should also note that on forward curved bladed centrifugal fans, and indeed on many other modern designs using a shield or tongue piece in its outlet, there is already an allowance for some gain in the catalogue tables or characteristic curves. The listed performance is based upon some regain by expansion from the nett throat area to the area of duct equal to the full discharge connection size. (See Figure 3.21.)

The static pressure is based on readings taken at some distance from the fan outlet and includes this gain. Hence a diffuser cannot add much to the performance. If for reasons of duct design an expander is used, it is customary to ignore any possible gain.

There is a practical limit to the final diameter of a diffuser fitted to a fan outlet if followed by a ducting system. The larger its final diameter, the more expansive is the ducting which follows. It is all a question of economics and the life cycle i.e. initial cost versus running costs. The effect of the diffuser is to reduce the power absorbed by the fan. This saving must be considered in

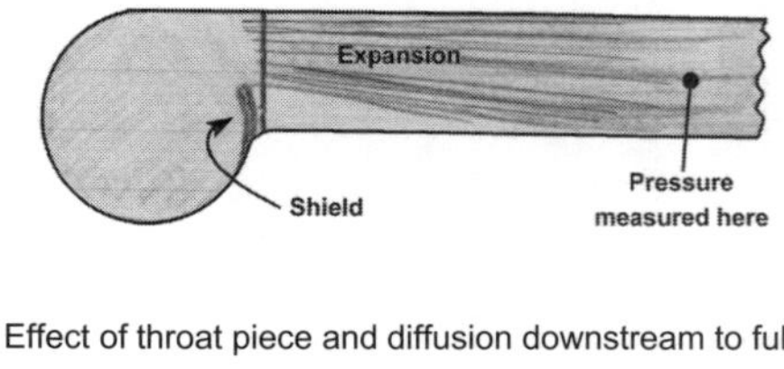

Figure 3.21 Effect of throat piece and diffusion downstream to full outlet duct area

relation to the cost of the ventilation system as a whole, including the fan, motor and ducting system.

On centrifugal fans for mine ventilation a diffuser is invariably fitted. It has a taper on one side only. The discharge is direct to atmosphere, at which point the static pressure above ambient is zero. Thus the static pressure at the fan outlet is negative and as much below atmospheric pressure as the velocity at that point converted into static pressure. A gain in pressure with the final at atmospheric or zero gauge must start from a negative. This negative pressure is transferred to the fan inlet and the fan is selected for the required flowrate at the calculated resistance of the system less the static regain.

3.3.3 Blowing outlets

When air is discharged from an outlet, the perimeter of the airstream is slowed down by contact with the surrounding air, which is induced onto the primary airstream as secondary air. The jet in consequence expands with distance from the outlet.

As stated previously, an unconfined jet of air in free space has an included angle of about 3°, but due to the induction of secondary air, its "spread" is increased to around 14 – 16°. American tests on air blasts from circular outlets are shown in Figure 3.22, the results being plotted for the percentage of the initial velocity on the centre line at distances measured in diameters.

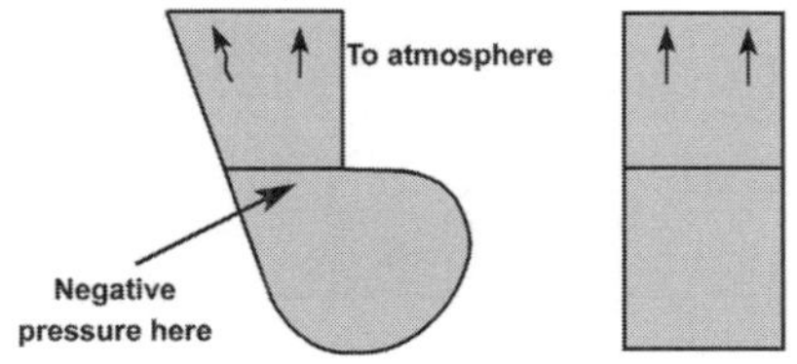

Figure 3.22 Diffuser fitted to a centrifugal mine fan

In the 1950s, Sturtevant Engineering Company made tests on three blast outlets of virtually the same cross-sectional area:

229 mm diameter

203 mm square

and 254 mm x 165 mm

When plotted on a basis of initial velocity on the centre line against $\frac{\text{distance m}}{\sqrt{\text{outlet area m}^2}}$ the results were virtually on the same

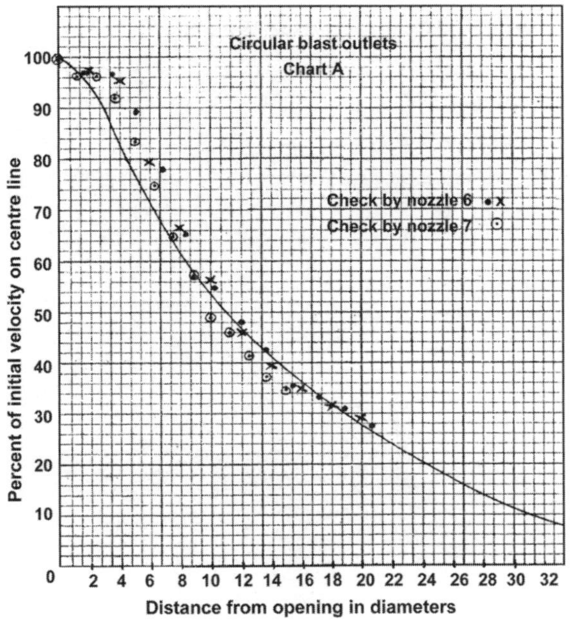

Figure 3.23 Circular blast outlets

curve and were in reasonable agreement with the American tests shown in Figure 3.23.

Hence, provided the long side of the rectangle is not more than 1.5 x the short side, the chart may be used by taking the cross-sectional area of the square or rectangular outlet and converting it into the corresponding equal area circular outlet.

Table 3.2 is a numerical equivalent of Figure 3.23.

Percentage of initial centre line velocity	$\frac{\text{Distance m}}{\text{Diameter m}}$	$\frac{\text{Distance m}}{\sqrt{\text{Area m}^2}}$
90	3.0	3.38
80	4.4	4.95
70	6.25	7.05
60	8.5	9.6
50	11.0	12.4
40	14.5	16.4
30	19.0	21.4
20	24.0	27.1
10	31.0	35.0
5	36.0	40.6

Table 3.2 Circular blast outlets

For example:

for 50% of the initial centreline velocity, the distance in metres from the outlet

= diameter of outlet m x 11

or

$$= \sqrt{\text{outlet area m}^2} \times 12.4$$

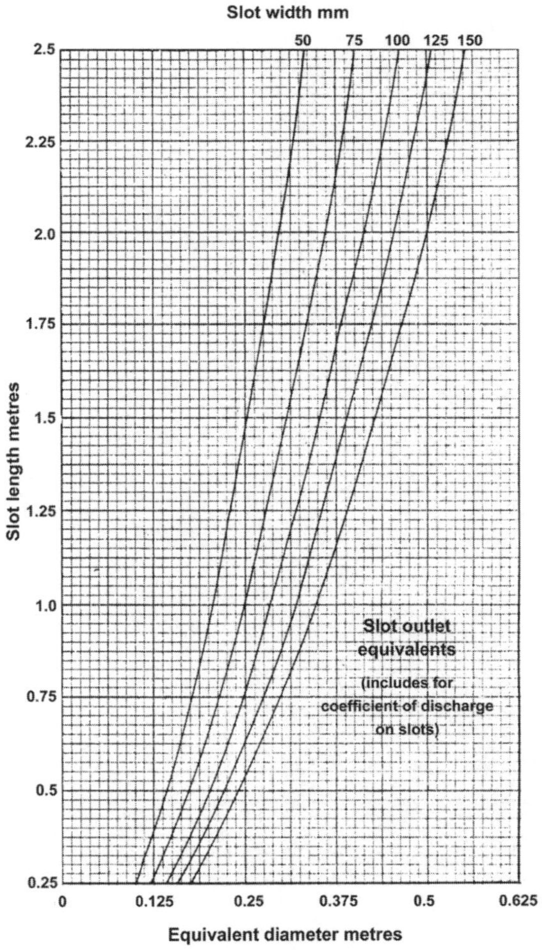

Figure 3.24 Slot outlet equivalents

Tests have also shown that the ratio of centreline velocity to average velocity is about 3.0 irrespective of outlet size, shape or initial velocity between 10 and 50 diameters from the outlet. In industrial ventilation, maximum velocity is usually the important factor from the viewpoint of draughts on persons. More recent data has suggested that the "blow" is to much greater distances, but practical experience suggests otherwise. It may be that this data is based on theory unsupported by actual site tests.

Narrow slot outlets require a different approach as the rate of fall in velocity with distance is greater. Figure 3.24 shows the equivalent diameter in metres against slot length for various slot widths, and is based on American data.

For example a slot on 762 mm x 76 mm is equivalent in performance to a 216 mm circular outlet. Its throw may then be determined from Figure 3.23 in the normal way. No practical confirmation has been made for all the combinations and it would be wise to restrict its use to slots of less than 76 mm long. Figures 3.25 and 3.26 show tests on a 914 mm x 38 mm slot, which suggest some caution.

Oscar Faber and John Kell used multiple nozzles to introduce ventilating air from high level in the original ventilation system at

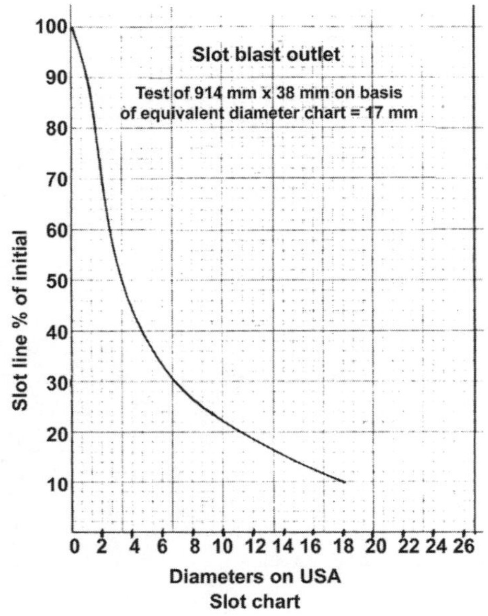

Figure 3.25 Slot blast outlet

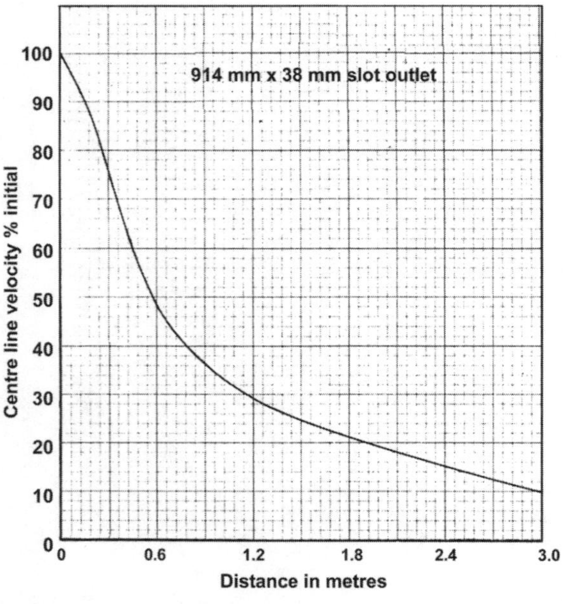

Figure 3.26 914 mm x 38 mm slot showing throw

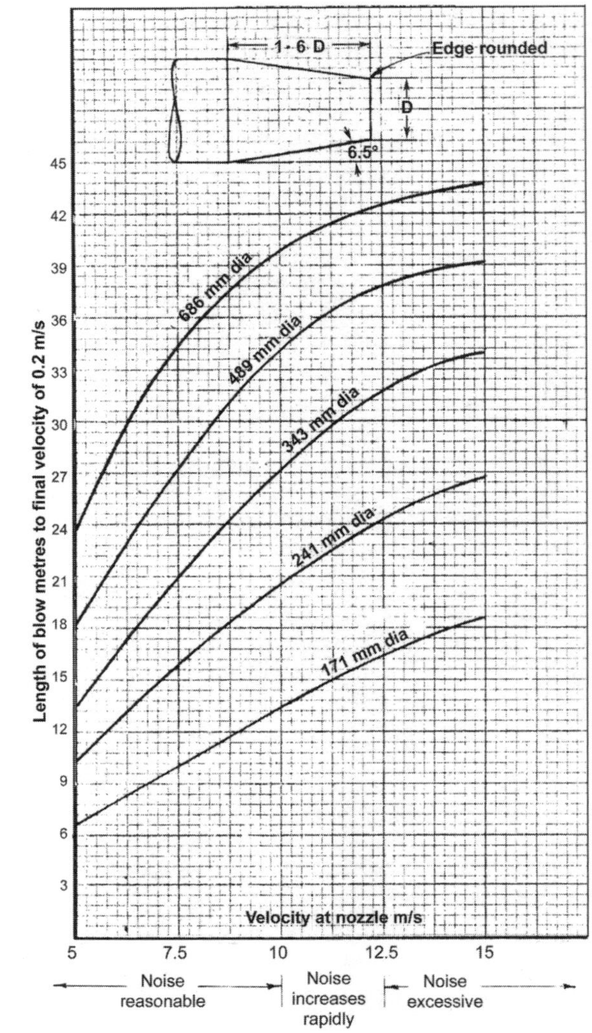

Figure 3.27 Faber's tests on round nozzles

the construction of the Earls Court exhibition main hall in London (311500 m^3 and 23000 people maximum).

The published tests results for which this nozzle scheme was designed are shown in Figure 3.27. These results are in general agreement with the practical experience of many engineers.

With Faber's design of nozzle fixed on the end of a short duct from a main duct, the static pressure required, as measured in the branch duct is given by:

$$p\,(\text{Pa}) = 0.535\,(\text{vel m/s at nozzle mouth})^2 \qquad \text{Equ 3.33}$$

At an area ratio of 0.535 the value of $K \simeq 1.06$. Readers may like to do the mathematical manipulation to justify the formula.

3.3.3.1 Punkah louvres

Another application of nozzles is in the cabin ventilation of ships where even today, Punkah louvres are used. These have the advantage that they can be swivelled to vary the direction of blow, to suit the particular preferences of the occupiers (see Figure 3.28).

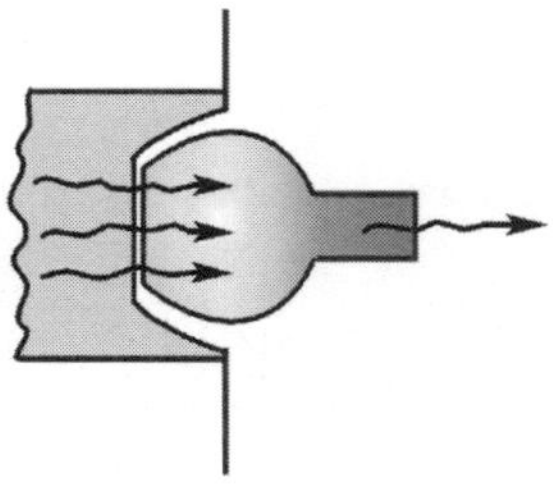

Figure 3.28 Punkah louvre

A standard range has been developed over the years in accordance with Table 3.3.

Dia of outlet mm	Dia of "ball" mm	m^3/s at 125 Pa	K
25	50	0.007	1600
37.5	75	0.016	700
50	100	0.028	400
62.5	125	0.049	228
73	150	0.064	175

Table 3.3 Performance details for Punkah louvre range

The makers claim that the pressure loss $Pa = \left(m^3/s \times k\right)^2$

3.3.2 Grilles

The length of "throw" from an air supply grille is important in design to avoid draughts. Throw is usually defined as the distance from the grille to where the air velocity has fallen to 0.25 m/s. This velocity should be achieved at not less than 2 to 2.14 m above the floor. Modern grilles are manufactured to a number of proprietary designs for which it is best to consult the manufacturers for recommendations as to the best type for a particular application.

One design has multi-deflecting vanes approximately 6 mm deep x 6 mm centres. These may be set to the required angle at the manufacturers or may be adjusted on site by bending the vanes with a special tool. Grilles with straight deflecting vanes generally produce the maximum throw for a given entering air velocity, but other types are available which produce a wide spread of the air with less risk of complaints from draughts.

With deflecting vanes, the air velocity is increased after leaving the grille. It is obvious from Figure 3.29 that width B is less than A, becoming more reduced as the angle of deflection is increased.

Figure 3.29 Reduction of width of issuing airstream with increased angle of deflection

3.3.2.1 Sizing of grilles on blowing systems

High level (above 3 m minimum height): The basis of selection is normally to obviate the noise caused by air impingement on the vanes. The maximum velocity on the entry side of the grille will depend on the application and the following maximum values are suggested:

- Board rooms, private offices etc 3.5 m/s
- General offices 5.0 m/s
- Industrial applications 7.5 m/s

Low level: The basis of selection is to achieve good comfort conditions for the occupants without noticeable draughts. If the entering air is at a temperature below that in the room i.e. a cooling application, then a minimum height of 2 m is suggested. The maximum velocity on the outlet side of the grille should be:

- Occupant very near the grille 0.5 m/s
- Generally with private offices 2.5 m/s

As stated, "angling" the vanes produces a greater outlet velocity than that normally on the inlet side. Multiplying the selected velocity by the appropriate factor in Table 3.4 will provide the entering air velocity.

Vane angle degrees	0	10	20	30	40	50
Factor	1	0.98	0.94	0.86	0.76	0.7

Table 3.4 Factor for the entering air velocity

At the design stage it is usual to assume a mean angle and factor of around 0.85.

The resistance may be determined as the leaving velocity pressure. Normally it is preferable not to spread the air vertically in industrial applications (and indeed in some offices with exposed steelwork) as there is a risk of hitting beams at ceiling height, or of blowing cooled air too rapidly down into the occupied zone.

Grilles fitted at the top of riser ducts in walls may have several horizontal deflectors behind the vanes. These may then be set to assist the air in turning.

3.3.2.2 General notes on blowing outlets

In factory heating with warm air on the overhead plenum system, the outlets are typically from 150 mm to 275 mm diameter at 3 m to 3.8 m above floor level with an average velocity of 5 m/s to 6 m/s (see Figure 3.30). The sophistication of outlet grilles is rarely merited other than for aesthetic reasons.

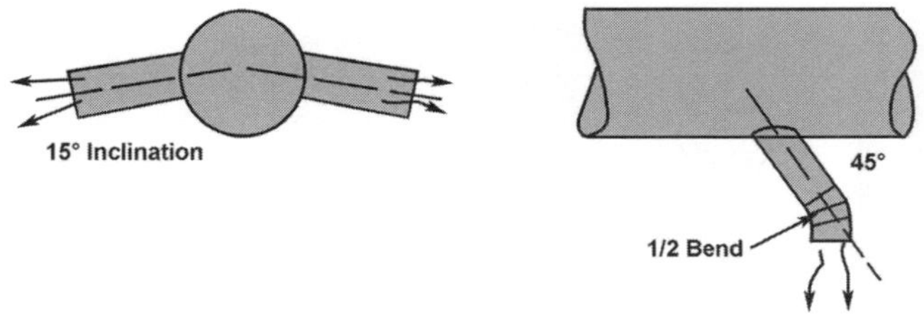

Figure 3.30 Typical outlets for factory plenum heating system

For general ventilation of factories, similar outlets may be used. Common examples are in the ventilation of very hot workrooms such as those for pressing and finishing of garments, laundry ironing rooms, etc. Tapered outlets have been used in some installations (see Figure 3.31). An outlet velocity of 3.75 m/s to 5 m/s has been found satisfactory.

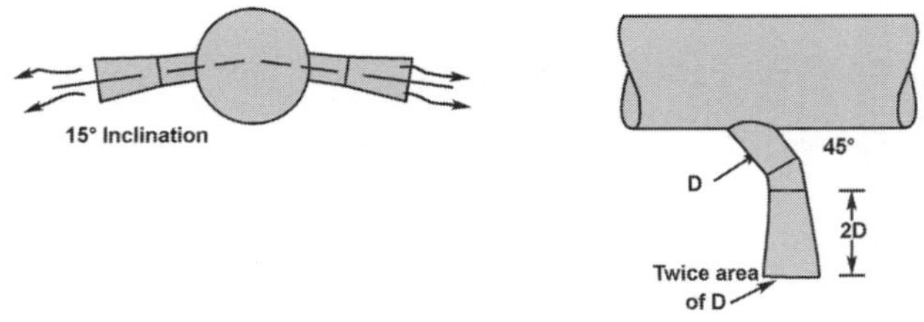

Figure 3.31 Tapered outlets for factory general ventilation

Cold air douche plants are often supplied for applications such as steel rolling mills, tin-plate rolling and glass furnaces. Here the operators are subjected to high radiant heat. As well as the copious quantities of beer which some plants allow, drop ducts from the main duct are positioned to blow cool air onto the workers! The air does not have to be cooled artificially, but is merely external atmospheric air. The outlet velocity is usually about 3.75 m/s (see Figure 3.32).

Textile conditioning plants have outlets on drop pipes from the main duct and are fitted with special diffusing outlets. These are spaced at intervals to cover the area of the room to be conditioned. Many different designs of outlet are available. Two of the

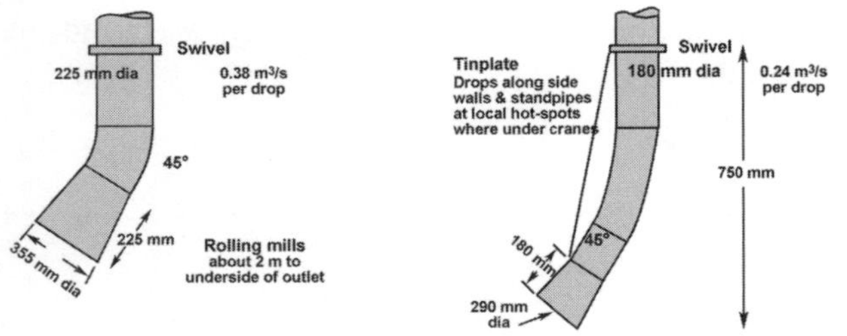

Figure 3.32 Outlets for cold air douche plants

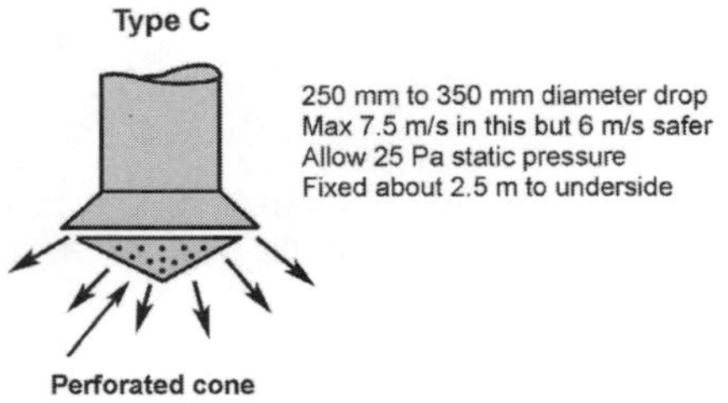

Figure 3.33 Type "C" outlet for textile air conditioning

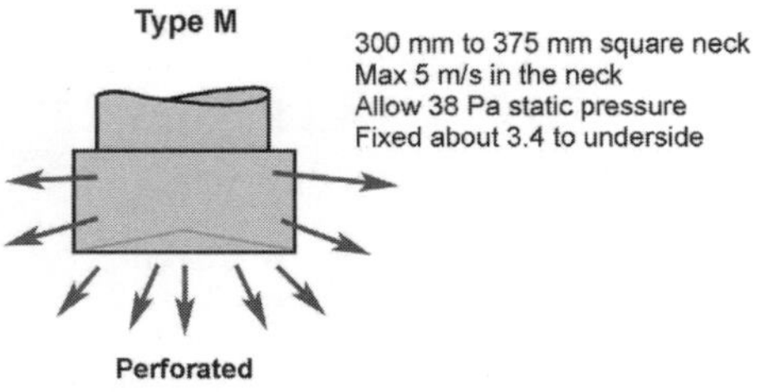

Figure 3.34 Type "M" outlet for textile air conditioning

simplest, which have proved satisfactory, are shown in Figure 3.33 and 3.34.

In factories with very high ceilings, the plenum warm air system is often fitted with drop pipes, or down corners, fixed adjacent to stanchions. These discharge the warm air nearer to floor level (see Figure 3.35). The drops may be from 200 mm to 280 mm diameter splitting into two outlets about 750 mm above the floor. Velocity should be 3.75 m/s to 5 m/s.

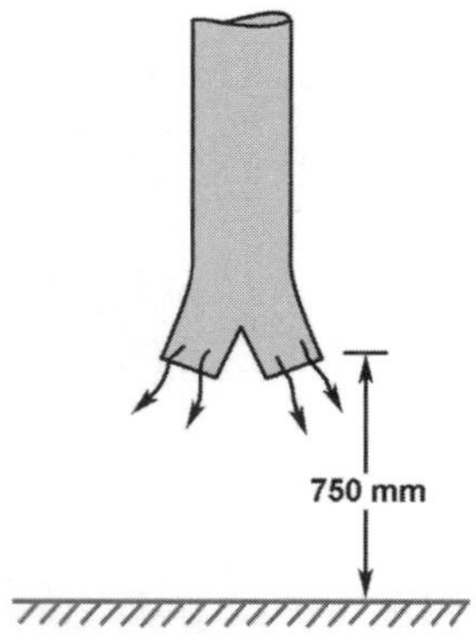

Figure 3.35 Drop pipes for warm air

3.3.4 Exhaust inlets

Consider the case of air exhausted by a very small point source (Figure 3.36). We can assume a sphere with a surface area of A m^2 at any radius r from the point of extract.

Let v = velocity m/s at radius r, assumed to be equal over the sphere.

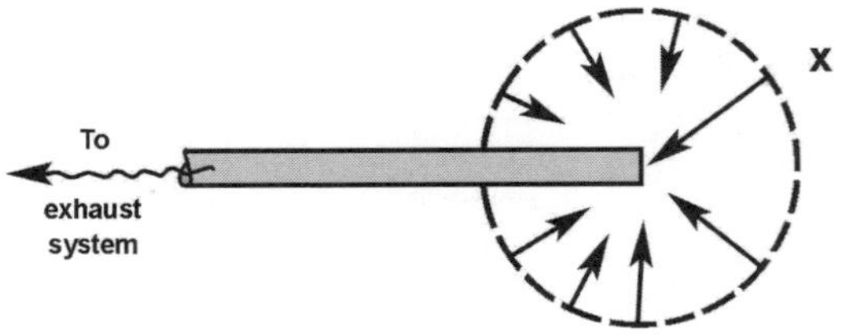

Figure 3.36 Extract from a point source

The extract volumetric flowrate $Q\ m^3/s = A \times v$

but $A = 4\pi r^2 = 12.57r^2$

So v m/s at any radius $r = \dfrac{Q}{12.57r^2}$

In actual practice the extract is not from a point source and the flow is not completely the same from all directions. In 1932 Dalla Valle investigated an open ended duct freely suspended in space, and found that the centre line air flow relationship was:

$$Q = v\left(10r^2 + A\right) \qquad \text{Equ 3.34}$$

where:

v	=	velocity measured on centre line (m/s)
r	=	distance from open end (m)
A	=	area of open ended duct (m^2)

Hence

$$v = \frac{Q}{10r^2 + A} \qquad \text{Equ 3.35}$$

The actual extract is shown in Figure 3.37.

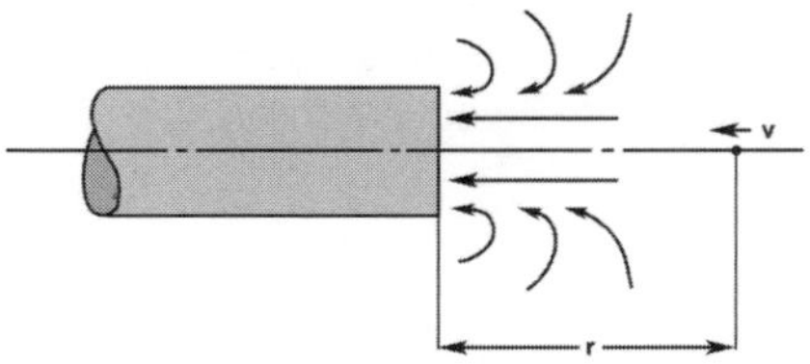

Figure 3.37 Actual extract from open ended duct

Laboratory and site tests have confirmed the general correctness of the equation.

To take the example of a circular exhaust opening having the following dimensions

Face area	=	0.093 m^2
Velocity	=	0.5 m/s
At distance	=	0.61 m

then $Q = 0.5\left(10 \times 0.61^2 + 0.093\right) = 1.907\ m^3/s$

If the same velocity was required at 1.22 m

then $Q = 0.5\left(10 \times 1.22^2 + 0.093\right) = 7.489$

It will be noted that Q is proportional to slightly less than the distance squared.

Note also that the velocity v varies directly as Q irrespective of the face velocity into the opening.

These points emphasise that when extracting dust, the hood must be as close as possible to the source of production and that to increase the velocity at a given distance must involve an increase in Q. The limitation on Q is of course due to economic factors. If velocity is insufficient to extract the dust effectively, it might be thought that a reduction in the size of the opening for a given volumetric flowrate would increase its "pulling power", but this is not so.

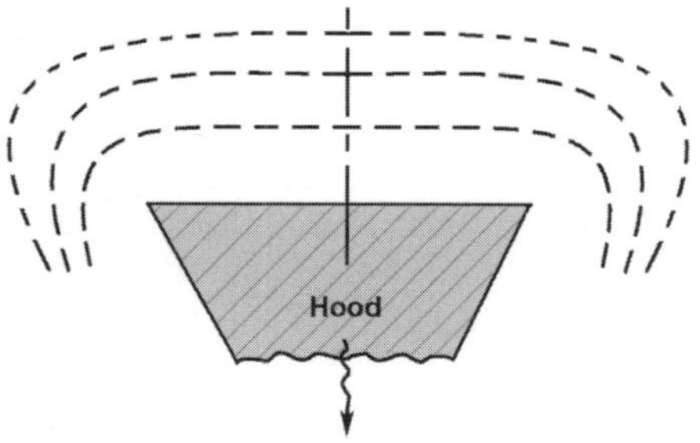

Figure 3.38 Flattening of velocity contours at hood face centre

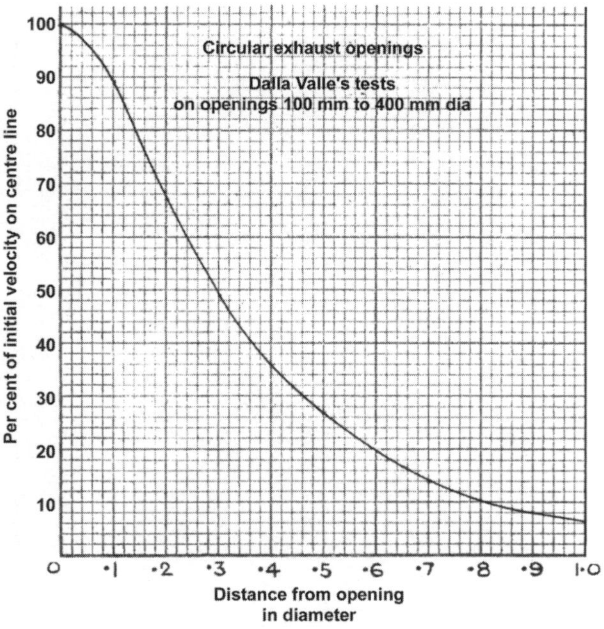

Figure 3.39 Circular exhaust openings (Dala Valle's tests on 100 mm to 400 diameter)

The centreline velocity is a useful guide in practice. In normally shaped hoods as used in dust collecting, the velocity contours are flat in the regions opposite the main portion of the hood (see Figure 3.38).

The graphical representation of Dalla Valle's tests is shown in Figure 3.39.

3.3.4.1 Comparison of blowing and exhausting

It is important to realise the great difference in effect of distance from the opening when comparing blowing and exhausting. At 31 diameters from the opening, 10% of the initial velocity is still maintained when blowing, but the distance for this 10% velocity contour is only 0.8 diameters when exhausting, see Figure 3.40.

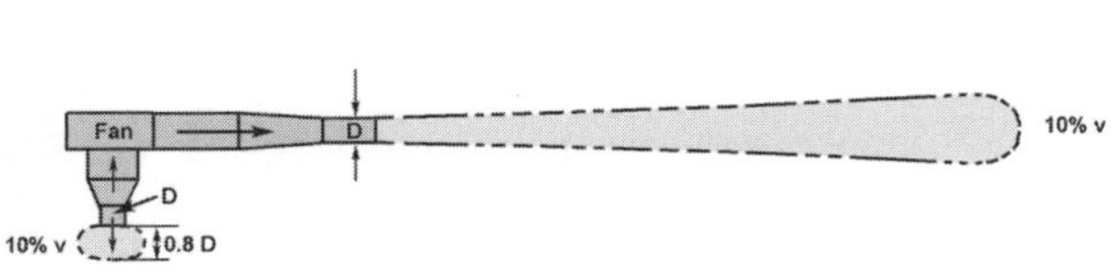

Figure 3.40 Comparison of velocity contour distance when blowing and exhausting

3.3.4.2 Airflow into exhaust opening for dust extract

When air enters a duct through a hood having any shape other than that of a perfect bellmouth, a vena contracta is formed (see Figure 3.41).

This is a point where unwanted air velocity is attained i.e. velocity above that needed in the duct to carry away the dust.

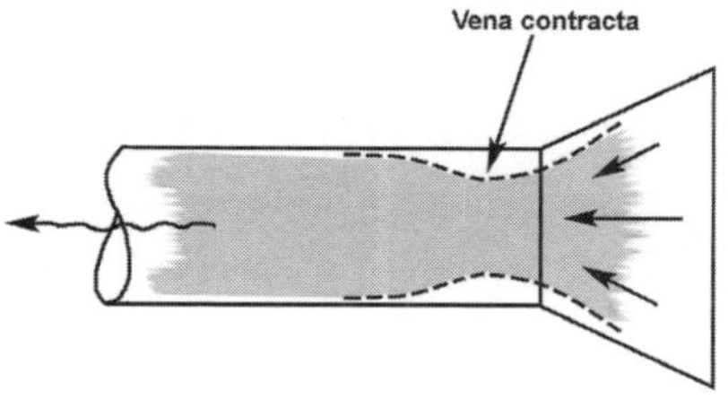

Figure 3.41 Formation of a vena contracta

The acceleration of the air to this excess velocity requires pressure and is shown as residual static pressure at the point where the airstream fills the duct at normal velocity. Normal duct velocity in average dust extract systems is from 16 to 23 m/s.

A common method of measurement in the USA is to drill one or more holes preferably as small as 1.6 mm diameter, free from burrs on the inside, at one duct diameter from the throat for all tapered entrances. For open ended or flanged ended openings, the hole is drilled at these duct diameters from the end. (See Figure 3.42.)

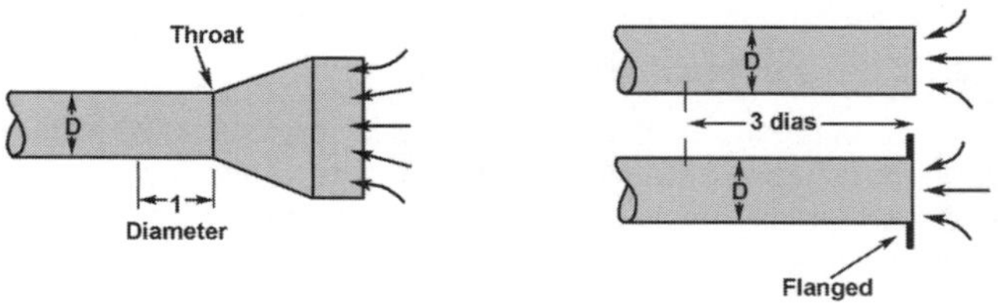

Figure 3.42 Positions of tappings for flow measurements

Press a rubber tube, connected to a pressure gauge, tightly against the hole and read the static depression.

Then

$$Q = 1.29\sqrt{p_s} \times A \times C_e$$

at normal temperatures and barometric pressures.

where:

Q	=	extract flowrate (m^3/s)
p_s	=	static depression (Pa)
A	=	cross-sectional area of duct at point of measurement (m^2)
C_e	=	coefficient of entry, which varies from 0.6 to 0.98 in commercial work

C_e also varies to some extent with velocity, see Figure 3.43

$$\therefore C_e = \frac{Q}{1.29\sqrt{h_s} \times A}$$

$$= \frac{1}{1.29\sqrt{h_s}} \times \frac{Q}{A}$$

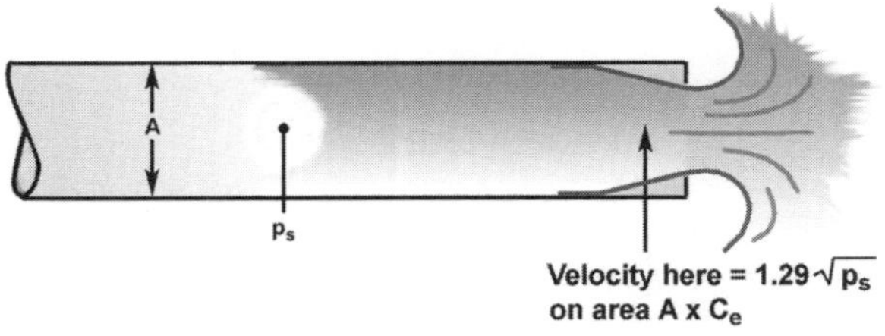

Figure 3.43 Measurement of extract flowrate

But $\frac{Q}{A}$ is the velocity in the duct after the vena contracta and equals $1.29\sqrt{p_v}$ at normal temperature and barometric pressure where p_v is the velocity pressure in the duct.

So,

$$C_e = \frac{1}{1.29\sqrt{p_s}} \times 1.29\sqrt{p_v} = \frac{1.29}{1.29} \cdot \frac{\sqrt{p_v}}{\sqrt{p_s}}$$

or

$$C_e = \sqrt{\frac{p_v}{p_s}} \qquad \text{Equ 3.36}$$

With

p_v = mean velocity pressure in duct after vena contracta (Pa)

p_s = static side hole depression taken in position specified to be clear of the vena contracta

Many tests have indicated that on a given extract opening, the value of C_e increases with velocity, indicating some Reynolds Number dependence.

3.3.4.3 Loss of pressure in hoods

The loss of pressure in an exhaust inlet is very much dependent upon its shape. It is mainly due to the contraction of the airstream which results in an increase in velocity at that point. In a bell mouthed entrance (Figure 3.44) there is virtually no contraction of the entering airstream. To create a flow of say 20 m/s at A or a velocity pressure of 250 Pa requires a static depression of 250 Pa in the duct.

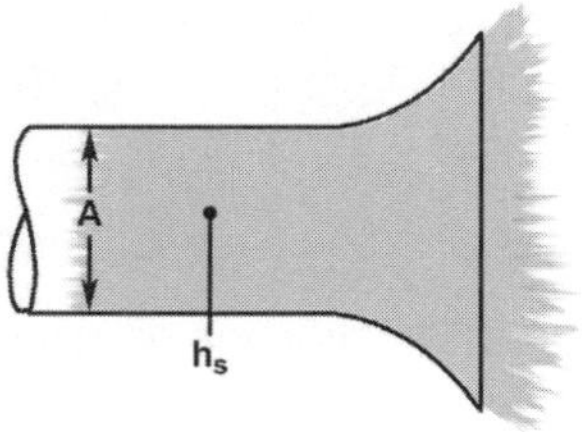

Figure 3.44 Bell mouthed inlet

Thus if there are no losses:

$p_s = p_v$

When there is a contraction of the entering airstream then:

$p_s = p_v + p_L$

where:

p_v = velocity pressure in the duct (Pa)

p_L = extra static depression for the increased velocity (Pa)

or

$p_L = p_s - p_v$

The value of p_L relative to the velocity pressure in the duct is:

$$\frac{p_s - p_v}{p_v}$$

But C_e as already shown = $\sqrt{\frac{p_v}{p_s}}$

or $\sqrt{p_v} = C_e\sqrt{p_s}$

or $p_v = C_e^2 p_s$

Substituting for p_v in the formula for relative p_L:

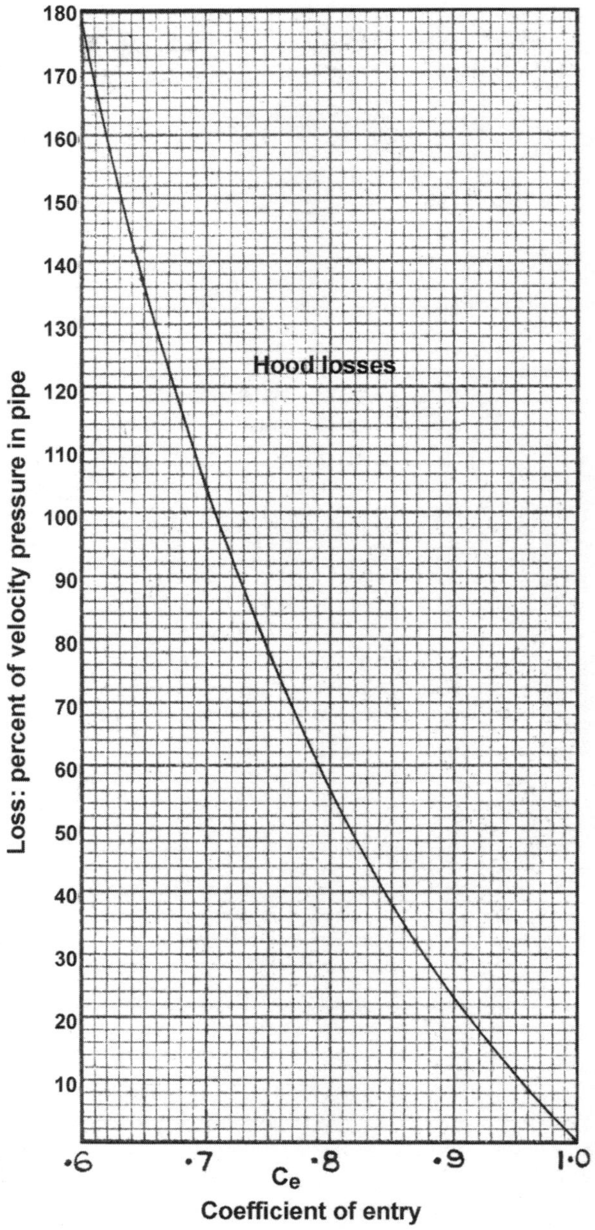

Figure 3.45 Hood losses

$$p_L = \frac{p_s - C_e^2 p_s}{C_e^2 p_s} = \frac{p_s(1 - C_e^2)}{C_e^2 p_s} = \frac{1 - C_e^2}{C_e^2}$$

Or as a percentage of the velocity pressure in the duct.

$$p_L = 100\left(\frac{1 - C_e^2}{C_e^2}\right) \qquad \text{Equ 3.37}$$

Figure 3.45 shows this in graphical form for values of C_e from 0.6 to 1.0.

In practice, the estimation of this loss is required in the design of dust extracting plant. It is generally possible to estimate the value of Ce from some similar known example. In especial cases a model may be made and checked by a laboratory test.

Typical values of Ce are given in the paragraphs which follow. It may be appreciated that absolute accuracy in the figure is not required and is in fact impossible to achieve at the estimation stage. Results of tests have been given to three decimal places, but a rounded approximate figure may be all that is necessary.

Note: p_L represents the mean facing tube reading as usually taken on the inlet side ducting of the fan. It is the equivalent of the resistance depression up to the point of measurement, but must be a mean over the area of flow.

3.3.4.4 Values of coefficient of entry C_e

Typical values of C_e are shown in Figures 3.46 to 3.52.

Diameter mm	C_e
100	·765
150	·745
200	·735
250	·725

At about 20 m/s in duct. C_e is less at lower velocity

Figure 3.46 C_e for plain open ended duct

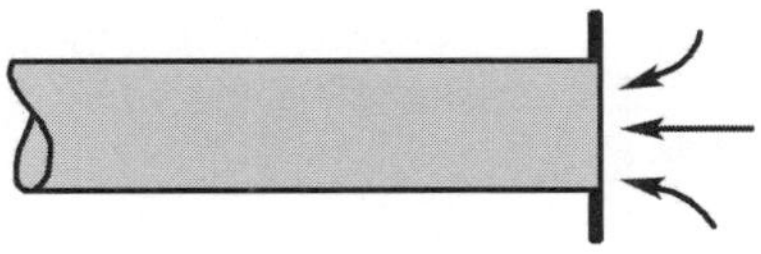

Diameter mm	C_e
100	·86
150	·855
200	·835
250	·81

At about 20 m/s in duct. C_e is less at lower velocity

Figure 3.47 C_e for flanged open ended duct

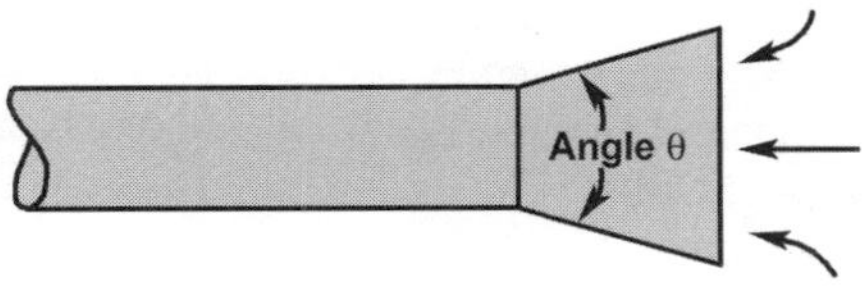

Included angle θ	Round		Square	
	Pipe 200 mm dia or larger	Pipe less than 200 mm dia	Pipe 200 mm square or larger	Pipe less than 200 mm square
20°	·96	·91	·94	·89
40°	·99	·93	·95	·90
60°	·98	·92	·94	·88
75°	·96	·91	·93	·87
90°	·95	·90	·91	·85
120°	·91	·86	·87	·82
150°	·86	·82	·83	·79

Figure 3.48 C_e for tapered hoods

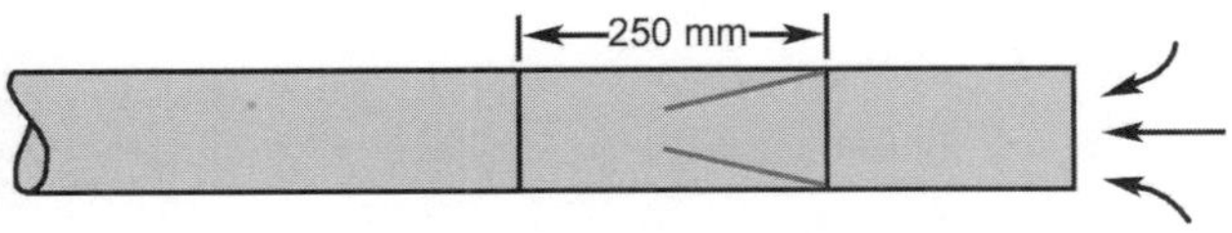

Pipe dia mm	Mouth square	C_e
100	100	.75
200	200	.748

Figure 3.49 C_e for square mouthed ducts

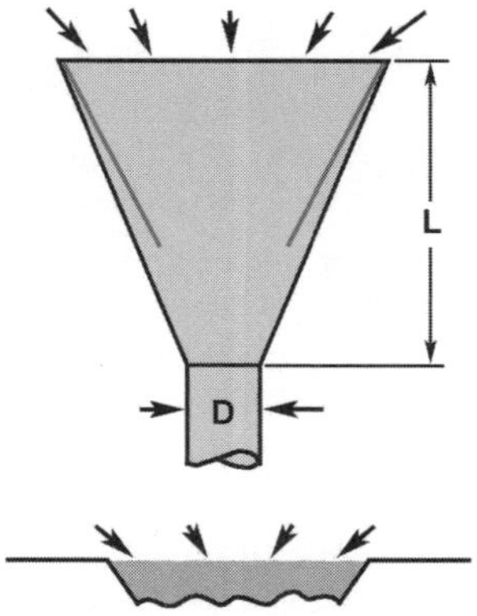

			Open	Flanged
D mm	Mouth mm	L mm	C_e	C_e
75	75 x 225	150	·809	·849
100	100 x 300	200	·854	·865
125	125 x 375	225	·814	·844
200	200 x 600	300	·785	·804

Figure 3.50 C_e for rectangular hoods ratio 1 : 3

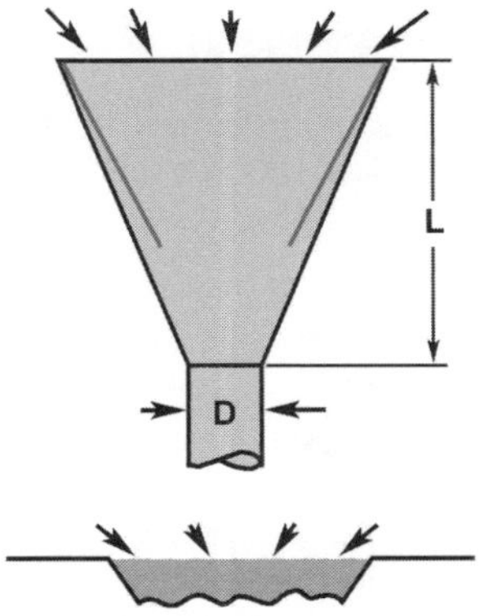

			Open	Flanged
D mm	Mouth mm	L mm	C_e	C_e
100	100 x 125	125	·849	·89
150	190 x 250	175	·838	·865

Figure 3.51 C_e for rectangular hoods ratio 3 : 4

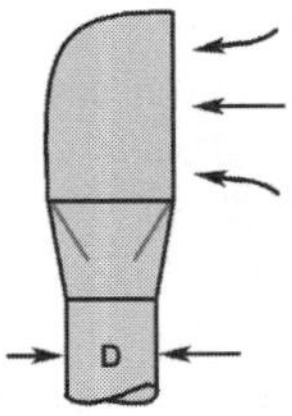

D mm	Hood mm	C_e
150	Plain with 165 x 50 wheel	·80 without wheel
		·725 with wheel

For average hoods including the obstruction of grinding or buffing wheel take C_e at ·71

Figure 3.52 C_e for grinding wheel hoods

3.3.4.5 General notes on exhausting

Detailed designs for hoods to suit most applications may be found in the standard design manuals produced by machinery manufacturers and also in *Industrial Ventilation* published by ACGIH®.

A point to note is that, in all dust extract work, the hood should be fitted to enclose the source of the dust as much as possible, whilst in fume extract the hood should be reasonably close to the area of evolution. These considerations should be clear from a basic study of air flow into exhaust openings.

In dust extract, the principle is to so design the hood that the particles are thrown from the point of generation directly into the throat of the hood. Grinding wheels, as an example, may be revolving with a peripheral velocity of 30 m/s.

For grinding wheels the throat velocity (see Figure 3.53) should be about 5 m/s to 5.5 m/s with a duct velocity from 17.5 m/s to 20 m/s. Normal duct sizes vary from 75 mm to 180 mm diameter and are generally standardised by manufacturers for their own types and size of wheels.

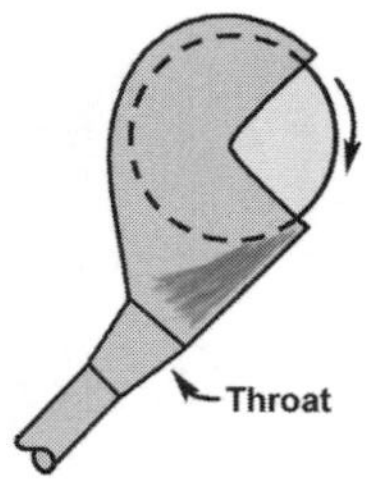

Figure 3.53 Grinding wheel hood showing throat and maximum enclosure

Overall system resistances for complete dust extract systems are typically in the range of 1000 Pa to 1500 Pa although extensive systems may reach higher values.

The cutters on wood working machinery, such as planers and moulders must be hooded so that the chips are thrown directly into the throat of the hood.

The air velocity into the opening of the hood around the cutters may be from 5.5 m/s to 8 m/s. The duct velocity is typically from 20 m/s to 22.5 m/s on chips and from 16 m/s to 17.5 m/s on sawdust. The duct connections to each hood range from about 75 mm to 180 mm diameter. System resistances are typically from 1250 Pa to 1500 Pa but for larger more extensive systems, could be higher.

For extract from spray booths the velocity into the open side of the enclosure may be from 0.5 m/s to 1 m/s with a general average of 0.75 m/s to 0.8 m/s.

In fume hoods over appliances the velocity into the actual opening may be as low as 0.25 m/s up to 1 m/s. The duct connection to the hood may have a velocity of 7.5 m/s to 10 m/s with the main ducting sized for 12.5 m/s to 15 m/s. Plant overall resistance with fume discharged direct to atmosphere can be as low as 250 Pa to 325 Pa. However, the addition of fume collection equipment to give a clean discharge to atmosphere can add considerably to this figure.

For further information on dust and fume hoods refer to Chapter 21, Section 21.7.

Sizing of extract grilles for HVAC plant

The sizing of extract grilles is very similar to the method described for those used for supply air. Again, due to the wide range available it is recommended that the manufacturer should be consulted. In general, noise becomes an important factor and the "throw" does not arise. The maximum velocity of the entering air is suggested to be as follows:

- Boardrooms and private offices 3.5 m/s
- General offices 5.0 m/s
- Industrial applications 7.5 m/s

The velocity of the air entering the grille is affected by the vane angle. For straight vane grilles, multiply the selected velocity by the factor in Table 3.5 to obtain the upstream velocity after the vanes.

Vane angle degrees	0	10	20	30	40	45
Factor	1	0.98	0.94	0.86	0.76	0.71

Table 3.5 Factors for straight vane grilles

For most straight vane grilles it is usual to allow a mean factor of 0.85. The resistance is assumed to be equal to the velocity pressure at the air entry. Sizing is based on the upstream velocity (i.e. immediately after the grille).

3.4 Friction charts

Charts have been published in various text books or the guides of the major institutions and societies which produced results without the need for tedious calculations. In former times they gave the frictional resistance in ins.w.g. per 100 ft of straight duct. Reading from volumetric flowrate in ft^3/min across horizontally to the duct diameter line in ins., a vertical line projected down to the bottom scale gave the friction. The velocity in ft/min could also be determined.

More recent versions have been converted to the SI units with flowrates in m^3/sec, duct diameters in mm or m, velocities in m/s and friction in Pa/m.

All these published charts look very similar, especially if the same units are used. Before accepting any particular version it is wise to check at small and large diameters to see what differences are present. Note especially that there will be areas of the chart which are close to the stated formula whilst at the extremities they are less accurate. Some charts show the preferred areas shaded. That shown as Figure 3.54 is as good as any.

The author's reaction is that, in an age of computers, it is just as easy to return to the classical formula, inserting the value of frictional coefficient appropriate to the relative roughness and Reynolds Number as obtained from the Moody chart in Figure 3.13. Table 3.1 has been compiled for a range of duct sizes and velocities. It will be noted that for all the velocities encountered in ventilation systems the flow quality is in the transitional zone where f is not a constant. The variation of f for both a constant duct size and a constant velocity is considerable.

3.4.1 Duct friction

The friction loss of straight ducting is not usually the most important element in determining the resistance of a ventilation system. Why then has so much effort been expended over the years in producing equations for its determination?

The classical equation is:

$$p_{Ls} = \frac{fL}{m} \times \frac{1}{2}\rho v^2 \qquad \text{Equ 3.38}$$

where:

p_{Ls} = pressure loss in a straight duct (Pa)

f = dimensionless friction factor

L = length of straight duct (m)

m = hydraulic mean "depth" (m)

ρ = air density (kg/m^3)

v = mean air velocity (m/s)

The hydraulic mean depth is defined as:

$$m = \frac{A}{P}$$

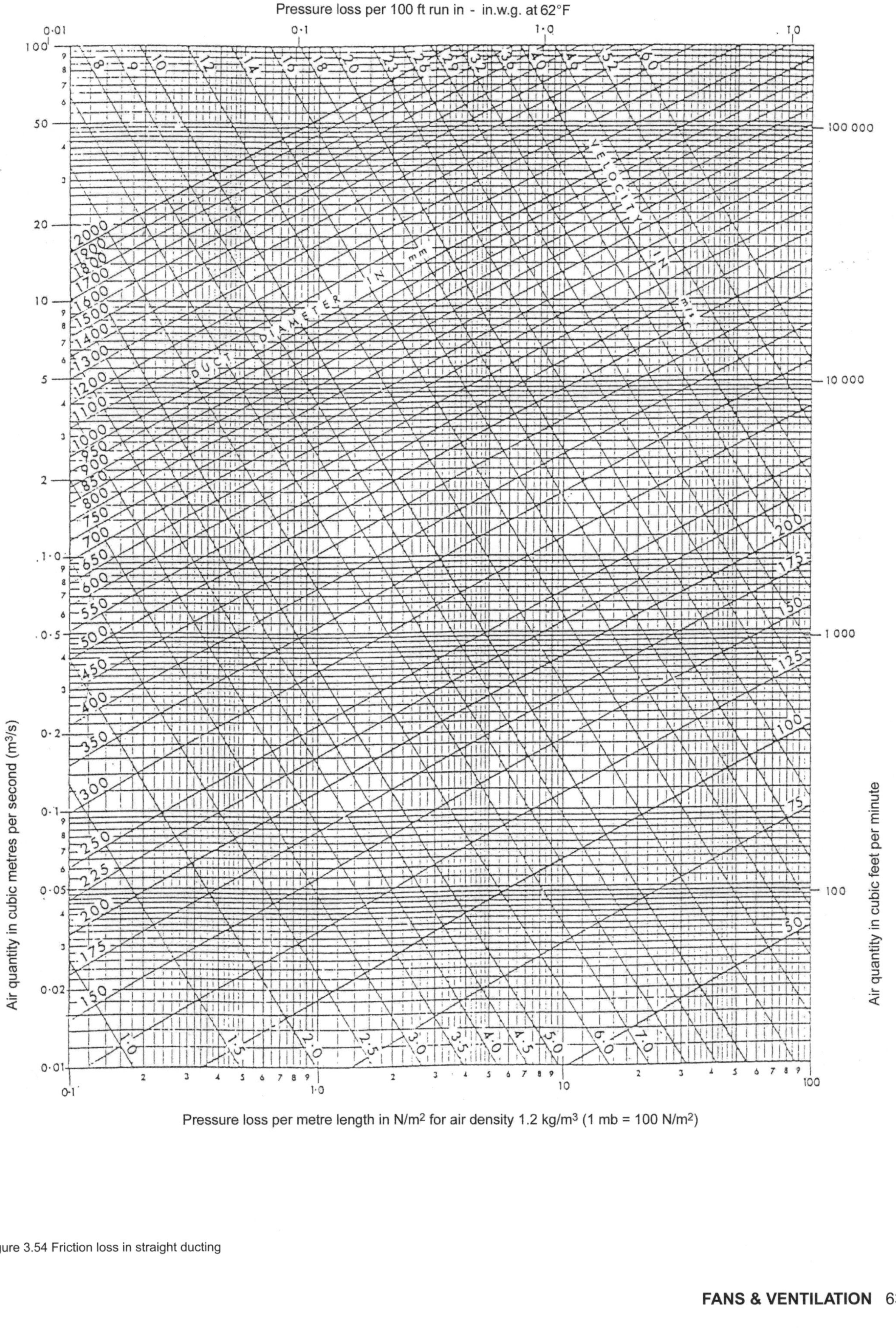

Figure 3.54 Friction loss in straight ducting

where:

A = cross-sectional area of duct (m^2)

P = perimeter of duct (m)

For a circular cross-section $A = \frac{\pi d^2}{4}$ and $P = \pi d$

where:

d = diameter of duct

Thus:

$$m = \frac{A}{P} = \frac{\pi d^2}{4} \div \pi d = \frac{d}{4}$$

and

$$p_{Ls} = \frac{4fL}{d} \times \frac{1}{2}\rho v^2 \qquad \text{Equ 3.39}$$

Here it should be noted that in American and some German texts, the pressure loss is defined for a circular duct and their formula becomes

$$p_{Ls} = \frac{fL}{d} \times \frac{1}{2}\rho v^2$$

No difficulty should be encountered provided one realises that their values of f, the friction factor, are four times the value, to compensate.

It has frequently been assumed that f is a constant and this leads to the conclusion that:

$$p_L \propto Q^2$$

This is very far from the truth, especially at low velocities. In fact $f \propto fn$. Re and the relative roughness of the duct. The relationship is best shown on the Moody chart in Figure 3.13. Numerous formulae have been produced to make the necessary corrections to the classical equation, these usually resulting in an index to v of less than 2 and an index to d of more than one.

As stated earlier, due to the numerous formulae having been produced, the author will content himself with examples from the pre-SI units era.

In the 1930s the then ASHVE (American Society of Heating and Ventilating Engineers, now ASHRAE), put forward the following empirical formula for the American market:

$$p_{Ls} = \frac{0.75fL}{d^{1.31}} \times \left(\frac{v}{4005}\right)^{1.84}$$

At about the same time the then IHVE (Institution of Heating and Ventilating Engineers, now CIBSE), was giving its formula for the British user as:

$$p_{Ls} = \frac{0.0001577\,L}{d^{1.269}} \times v^{1.852}$$

ASHVE gave suitable charts for the coefficient of friction, whilst this was included within the IHVE equation. In both of these formulae:

p_{Ls} = frictional resistance (ins.w.g.)

L = length of straight duct (ft)

d = diameter (ins)

v = mean air velocity (ft/min)

There were some differences in the air density assumed, the American data being for dry air at 70°F and 29.92 ins Hg barometric pressure whilst the British values were based on the then standard air at 60 °F, 29.53 ins. Hg and 60% relative humidity. For "average" sheet metal construction IHVE specified an addition of 20%.

For any other air density, the pressure loss due to friction at the same air velocity was obtained by multiplying the "standard" density value by:

$$\left(\frac{\rho}{0.0751}\right)^{0.852}$$

where:

ρ = air density at stated conditions lb/ft^3

It will be noted that in both American and British formulae, the friction was shown to vary as 1.84 to 1.852 the power of the velocity. Most practical engineers, however, continued to calculate friction losses as varying as the square of velocity. Provided the changes in velocity on a given system were relatively small (say less than 10%), the error was negligible and likely to be less important than variations due to manufacturing tolerances. Also, the friction loss was taken as directly proportional to air density, again without serious error.

The fact that the Fan Laws defined similar variations in fan performance was an added advantage. Indeed such assumptions were in order, because the calculated values can never be more than estimates, due to the inexact knowledge of constructional roughness, covered, as already noted by a 20% addition. Normal roughness does not necessarily mean bad workmanship, but essential constructional features such as circumferential joints which at that time were as many as 40 per 100 ft. Nevertheless, the variations in calculated resistance from the ASVE 1930s data to the most recent formulae, of more than 30% can never be justified. It has not, however, deterred the researchers, and Loeffler's formula of the 1980s, whilst showing similarities with the historical formula has increased the velocity index to about 1.9.

The formulae for galvanised steel ducts with an absolute roughness of:

$$\varepsilon = 0.0001524 \text{ m } (0.0005 \text{ ft})$$

$$p_L = a\frac{L\,Q^{1.921}}{D^{5.066}}$$

where:

$a = 1.717$ E-02 (for SI units)

or

$a = 3.534$ E-09 (for Imperial units)

where:

p_L = total pressure loss (Pa or in. wg)

Q = flow rate (m^3/s or cfm)

D = duct diameter (m or ft)
(or equivalent diameter of rectangular ducts)

L = duct length (m or ft)

To repeat, duct friction is usually a very small item in the overall resistance of a typical ventilation plant. In a dust extract or wood refuse collection plant, the frictional resistance is usually much higher as the air velocity in such systems is also higher.

3.5 Losses in fittings

We have seen that, over a limited working range, the pressure losses in both straight ducting and fittings are a function of the velocity pressure. It is therefore possible to equate the two and to state the loss at fittings in equivalent diameters of straight duct.

3.5.1 Bends

In the case of bends it is important to note that much American data is categorised on the basis that the radius of a bend is to its centreline. British practice is usually to give the inside radius. When looking at data, make sure you are comparing like with like.

The loss of pressure in a bend following by further straight ducting is less than if it discharges to atmosphere. In the former case there is some recovery in the expansion of the airflow to the full duct diameter. (See Figure 3.55.)

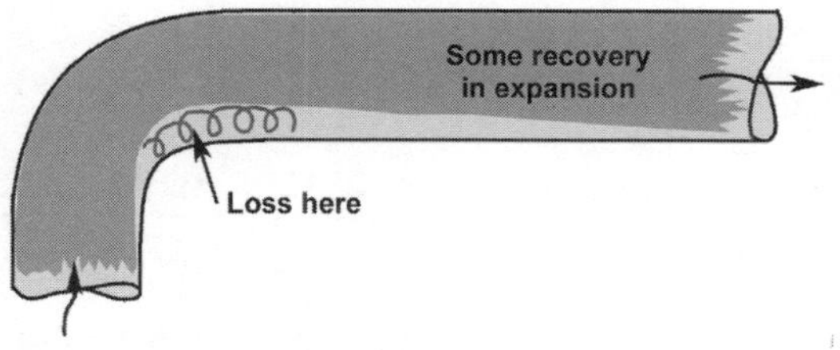

Figure 3.55 Recovery in duct after a bend

It should be noted that the factors are in diameters. For example a 355 mm diameter single radius 90° bend is equivalent in resistance to 9 diameters of straight duct. Its equivalent length in metres is then $\frac{355}{1000} \times 9 = 3.2$ metres. When dealing with rectangular bends, the equivalent is taken on the "way" of the bend i.e. on dimension W (see Figure 3.56). Two 90° bends of exactly the same cross-section will have different pressure losses according to the "way". One is an easy bend and the other a hard bend.

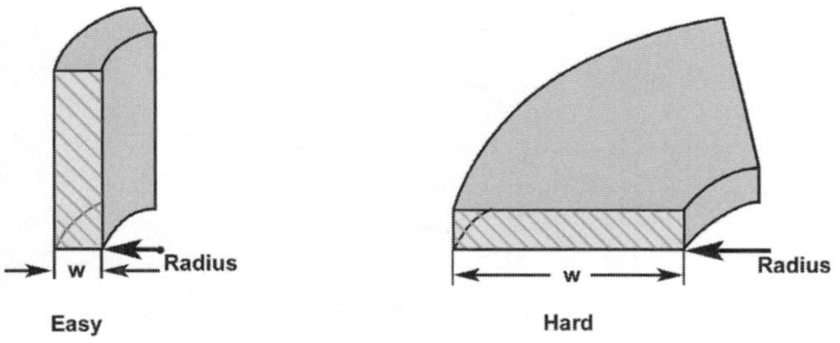

Figure 3.56 Easy and hard bends

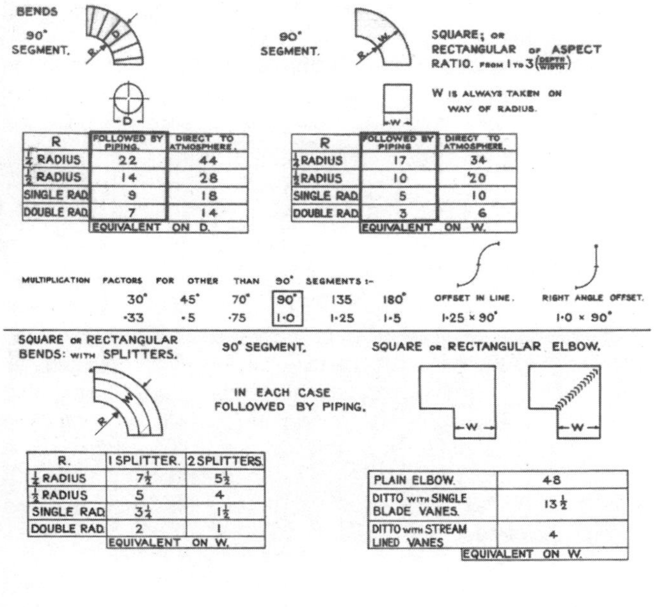

R	FOLLOWED BY PIPING.	DIRECT TO ATMOSPHERE.
¼ RADIUS	22	44
½ RADIUS	14	28
SINGLE RAD	9	18
DOUBLE RAD	7	14
	EQUIVALENT ON D.	

R	FOLLOWED BY PIPING	DIRECT TO ATMOSPHERE.
¼ RADIUS	17	34
½ RADIUS	10	20
SINGLE RAD	5	10
DOUBLE RAD	3	6
	EQUIVALENT ON W.	

R.	1 SPLITTER.	2 SPLITTERS.
¼ RADIUS	7½	5½
½ RADIUS	5	4
SINGLE RAD.	3¼	1½
DOUBLE RAD.	2	1
	EQUIVALENT ON W.	

PLAIN ELBOW.	48
DITTO WITH SINGLE BLADE VANES.	13½
DITTO WITH STREAM LINED VANES	4
	EQUIVALENT ON W.

Figure 3.57 Duct resistance equivalent lengths for bends

The hard bend throws the air to one side as it turns the corner and so causes higher resistance. This loss can be reduced by the inclusion of splitters. Figure 3.57 gives equivalent lengths in diameters for a number of different bends, including those with splitters.

The equivalent length in diameters is based upon the assumption that one velocity pressure is lost in 55 diameters of ducting. Or to be precise the equivalent of one velocity pressure is lost in frictional resistance. Extensive tests have been made on bends of various designs and their losses measured.

These were then converted into fractions of velocity pressure. This factor is then independent of velocity over a limited working range. For example a bend with a resistance of 50 Pa at 10 m/s (velocity pressure 60 Pa) therefore has a loss factor of 0.83. As resistance may be taken as the square of velocity over this limited range, at 20 m/s the loss would be 200 Pa and the velocity pressure would be 240 Pa and the loss factor would still be $\frac{200}{240}$ i.e. 0.83.

To repeat, it is convenient in estimating the resistance, or pressure loss of a ducting system to calculate assuming that bends are equivalent to so many metres of straight ducting.

3.5.1.1 Reducing the resistance of awkward bends

When ducting is to be arranged in large buildings it is often impossible to find the space to incorporate bends of a reasonable radius. It is then possible to insert vanes or splitters to reduce the pressure loss. See Figures 3.58 to 3.61.

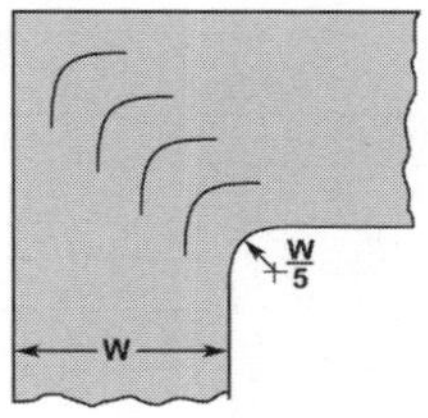

Figure 3.58 Bend with splitters

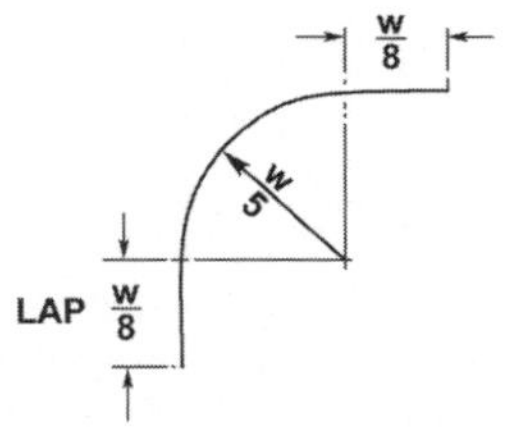

Figure 3.59 Detail of splitter

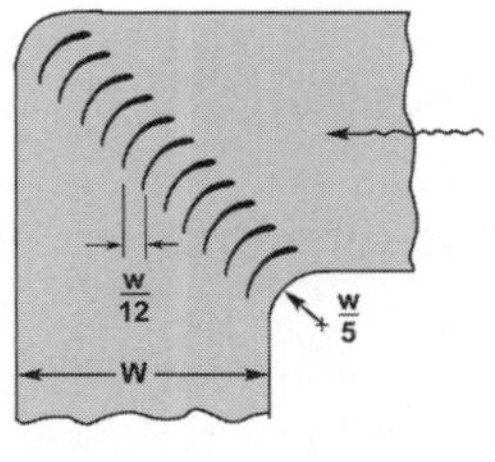

Figure 3.60 Bend with aerofoil section vanes

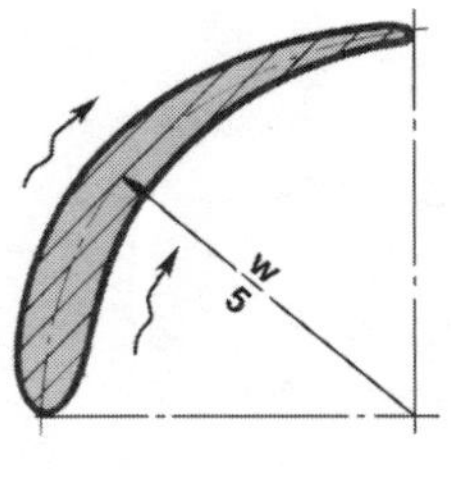

Figure 3.61 Detail of aerofoil section vane

The aerofoil section vanes are cast aluminium and are less liable to be noisy that sheet metal splitters. They also result in a lower pressure loss i.e.

sheet metal splitters p_{Lb} = 0.24 x velocity pressure

aerofoil section vanes p_{Lb} = 0.11 x velocity pressure

An alternative design of splitter which encompasses the complete bend may also be used. This effectively divides the bend into a number of parallel sections for which the dimensions are known. The loss for these may then be calculated and the highest value used. See Figures 3.62 and 3.63.

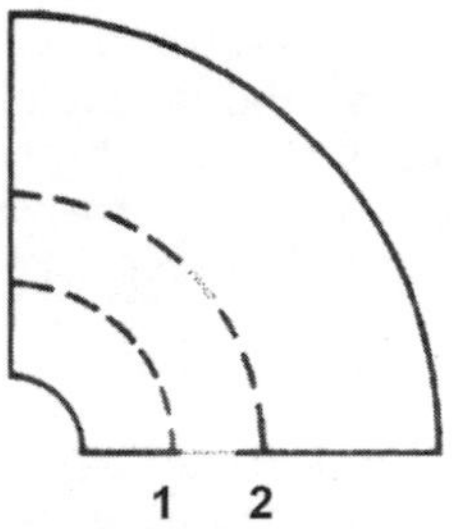

Figure 3.62 Splitter radius in radiused rectangular bends

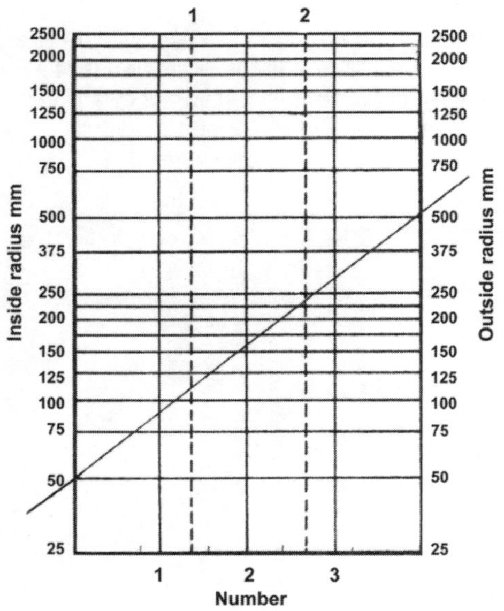

Figure 3.63 Chart for determining position of splitters

For example, as shown by the line drawn across the chart in Figure 3.63, a bend has an inner radius of 50 mm and an outer radius of 500 mm. If there were 2 splitters, these would be positioned at radii of 112 mm and 230 mm.

If there were 3 splitters, these would be positioned at radii of 90 mm, 160 mm and 260 mm.

3.5.2 Branches and junctions

These may be treated in the same way as bends and equivalent lengths calculated or measured from tests. Care must be taken to ensure that the "way" of the junction is recognised and also to note the direction of airflow, (i.e. whether blowing or exhausting) see Figures 3.64 and 3.65.

3.5.3 Louvres and grilles

These are best treated as the loss being a function of velocity pressure. Typical figures are shown in Figure 3.66. The manufacturers will however, have figures obtained from tests and should be consulted when chosen.

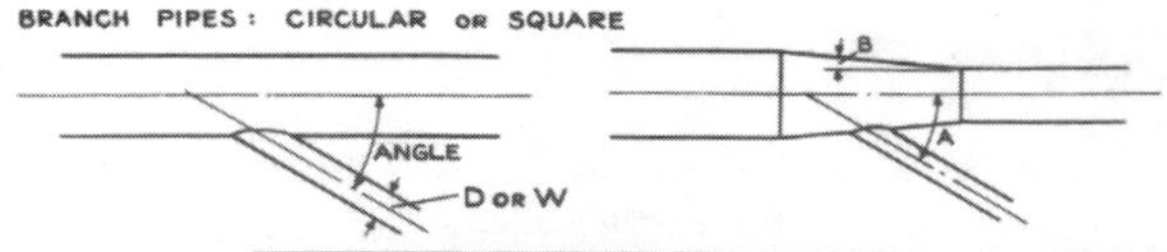

ANGLE OF BRANCH	EQUIVALENT ON D OR W
15°	5
30°	9
45°	12
60°	21

IF TAKEN OFF TAPER THE ANGLE = A−B

LOSS DUE TO TURNING THE AIR IN BLOWING PLANTS & IMPACT OF BRANCH AIR ON MAIN STREAM IN EXHAUST PLANTS

Figure 3.64 Duct resistance equivalent lengths for branches and junctions

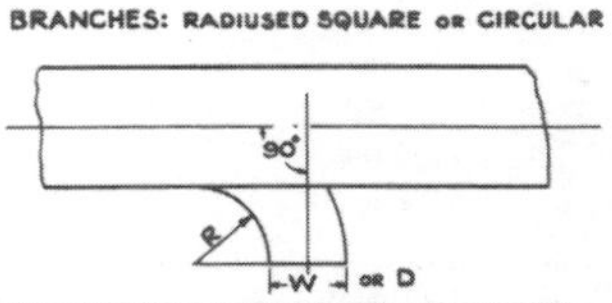

RADIUS	EQUIVALENT ON W OR D
¼ RADIUS.	23
½ RADIUS.	15
SINGLE RAD.	12
DOUBLE RAD.	9

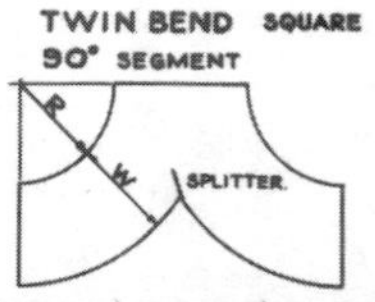

RADIUS	EQUIVALENT ON W
¼ RADIUS.	20
½ RADIUS.	13
SINGLE RAD.	8
DOUBLE RAD.	5

Figure 3.65 Duct resistance equivalent lengths for branches and junctions

Air inlet louvres with wire mesh screen:

Loss: Pa

Face velocity m/s	90% free area	70% free area
4	5	15
4.5	7.5	20
5	10	25
5.5	12	30
6	15	35

Grilles: Latttice type

Loss: Pa

	Face velocity m/s			
Free area	2	3	4	5
50	15	33	55	90
60	10	23	40	60
70	8	18	30	45
80	5	13	23	35

Ornamental

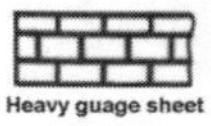

Heavy guage sheet metal

	Face velocity m/s			
65% F A	2	3	4	5
Grille only	2	5	10	18
With vanes	8	20	33	53

Perforated plate

	Face velocity m/s			
F A	2	3	4	5
3 mm 33.75%	8	18	32	50
10 mm 56.5%	5	10	18	28

Figure 3.66 Pressure losses in louvres and grilles

3.5.4 Expansions and contractions

These are best treated by k factors as listed according to the particular type (see Section 3.3.2 on diffusers). Note that contractions normally have a very low total pressure loss provided the included angle is less than 45°.

$$p_{LEC} = k \times \text{velocity pressure}$$

3.5.5 Square or rectangular ducting

Until recently special tables or charts were not available for the resistance of square or rectangular ducting. Even now, the few charts which are do not show all the combinations of width and depth desirable. Accordingly, an equivalent table may be used (Table 3.6). This table shows the size of round ducting which is equivalent in frictional resistance to a square, or any rectangular duct, when passing the same volumetric flowrate of air. The air velocity, of course, is not the same.

For example, from Table 3.6 a 535 mm x 280 mm rectangular duct is equivalent to a 405 mm diameter duct. A 405 mm diameter duct is equivalent to a 370 mm square duct.

Round duct mm dia	Square duct mm x mm	Rectangular duct - depth mm d						
		230	305	380	455	535	610	685
		Width mm w						
155	140	90	75	-	-	-	-	-
180	160	115	100	75	-	-	-	-
205	185	155	125	100	90	-	-	-
230	210	205	155	125	115	100	90	-
255	230	230	180	155	125	115	100	90
305	280	370	265	215	190	165	140	125
355	325	495	355	280	240	215	190	165
405	370	660	470	370	315	280	255	215
455	420	865	620	470	395	345	305	265
510	465	1030	760	585	485	420	355	330
560	510	1245	940	710	585	510	430	380
610	560	1450	1120	840	685	595	520	455
660	605	-	1270	1015	815	710	610	535
710	650	-	1485	1180	965	815	710	610
760	700	-	-	1370	1120	940	785	710
815	745	-	-	1525				815
865	795	-	-	-				915
915	840	-	-	-	-			1015
965	885	-	-	-	-	1575	1295	1170
1015	935	-	-	-	-	-	1475	1295

Table 3.6 Equivalent dimensions of round, square and rectangular ducts for equal friction and flowrate

Note: Sheet metal duct design is always an approximation. In the smaller sizes the dimensions have therefore been rounded to the nearest 5 mm. In the larger sizes some rounding has also been made.

The basis of Table 3.6 is:

$$\text{Round duct diameter} = 1.265 \times \sqrt[5]{\frac{(dw)^3}{d+w}} \qquad \text{Equ 3.40}$$

It is usual to design the system in the first instance on the basis of circular ducting and then to convert it into the equivalent rectangular cross-section. In many cases the depth may be kept constant for constructional reasons e.g. where the ducting is in a void above a false ceiling.

3.5.6 Non g.s.s. (galvanised steel sheet) ducting

The friction loss in ducting manufactured in other materials is best obtained from the absolute roughness and relating it to its size to calculate the relative roughness and hence the friction factor. Aluminium and PVC ducts will then be seen to have lower friction. Spiral wound ducting may have a higher friction depending on the smoothness of the internal surface.

In the past, large ducts in public buildings were often built into the masonry fabric and finished with glazed tiles. This was when the pressure loss due to friction was as good as that for g.s.s. Those days are unlikely to return, but underground air ducts are still used in applications such as grain drying. In these cases the approximate "correction factors" given in Table 3.7 may be used.

Surface	Average correction factor to g.s.s value
Smooth cement	1.2
Rough concrete	1.4
Good brickwork	1.5
Acoustic lining	1.5

Table 3.7 Correction factors for other materials

3.5.7 Inlet boxes

The author, during his perhaps too long a career, has come across many instances where the ductwork manufacturer has provided an inlet box to the fan, to give side entry. Reference to Chapter 5, Section 5.6 shows that this leads to a "system effect" such that the fan no longer gives its rated catalogue performance. The fan requires a fully developed symmetrical air velocity profile free from swirl at its inlet.

It is the system designer's responsibility to provide this. Where an inlet box entry cannot be avoided, it should preferably be or-

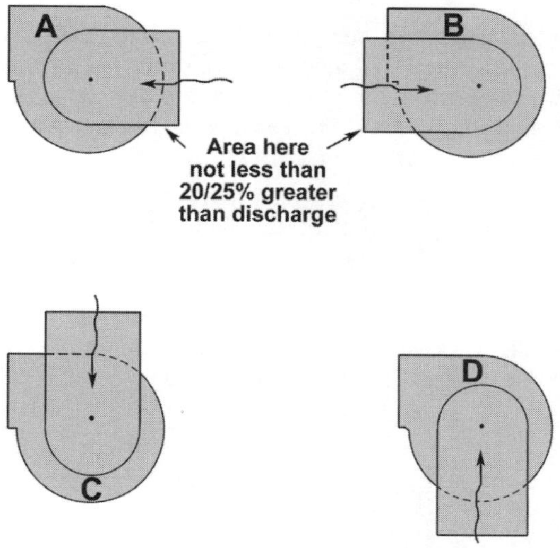

A Minimum loss: $0.5 \times p_v$ in inlet diameter

B Maximum loss: $1.0 \times p_v$ in inlet diameter
Avoid this as it may cause slow surge

C & D $0.75 \times p_v$ in inlet diameter

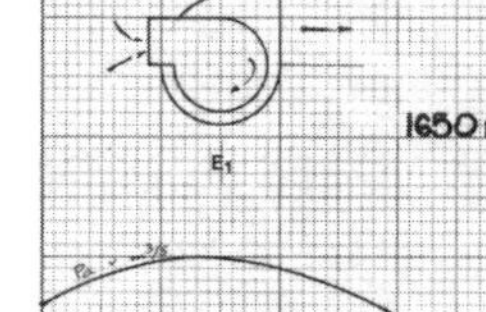

Spiral inlet box:

Volume is seriously reduced

Avoid this type

Figure 3.67 Fan inlet box losses

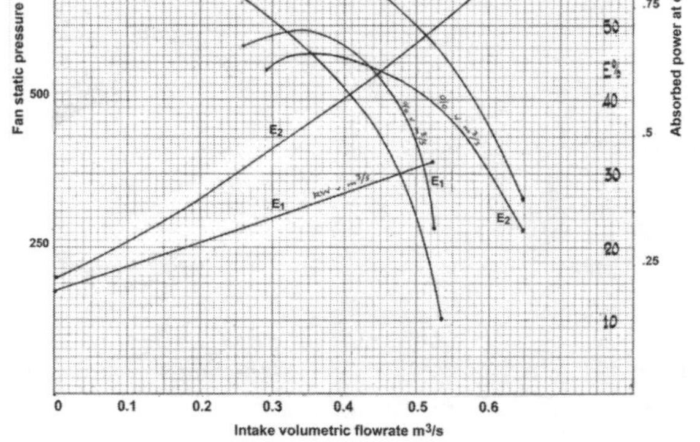

Figure 3.68 Spiral inlet box losses

dered from the fan manufacturer. The manufacturer will usually be able to supply a box which is tapered to suit and has an internal swirl baffle. If having said all this, the ductwork designer still wishes to be responsible for their supply he should be aware that simple box pressure losses can be very dependent on their orientation. Figure 3.67 gives some information of a very approximate nature — the loss is also very dependent on the fan design. The spiral design type E should be avoided at all costs — the volumetric flow is seriously reduced. See Figure 3.68 which is a typical example.

3.5.8 Discharge bends

At the fan discharge, due to centrifugal forces, the air velocity at the outer extremity of the casing (furthest from centreline) is higher than that at the other end of the discharge (nearest to centreline). This is even more so when the fan casing is fitted with a shield or tongue piece.

Bends fitted directly to the fan outlet flange therefore receive a distorted air velocity profile. It is always good practice to have at least 4.5 equivalent diameters of straight duct on the fan outlet to allow for good diffusion. Where this cannot be accommodated, the bend loss will be greater than normal. The approximate effect is given in Figure 3.69. More comprehensive information is given in Chapter 5, Section 5.6 and AMCA Publication 201.

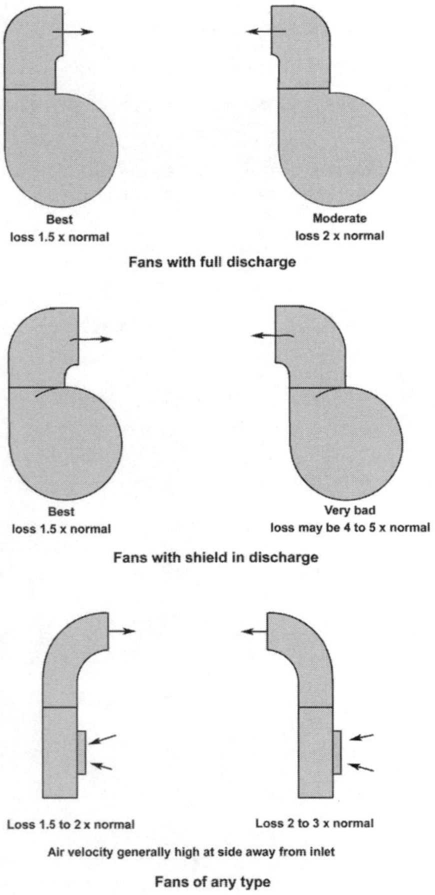

Figure 3.69 Discharge bend losses

3.5.9 Weather caps

These are not nearly so common nowadays. They should however be fitted at the final discharge to atmosphere where this is vertically up. The ducting must be protected from the ingress of rain. In former times, they were known as "Chinamen's hats" – a descriptive term with no racial connotations!

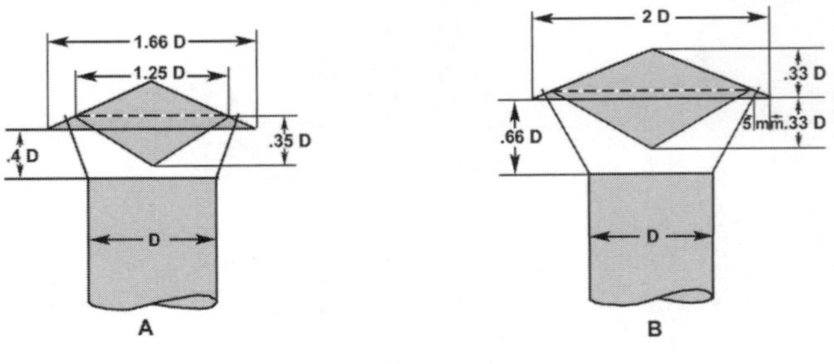

Figure 3.70 Proportions of weather caps

The smaller the diameter of this cap, the lower it must be fitted to the duct end to prevent rain ingress. But the lower it is fixed, the greater its resistance. The resistance is also affected by the design of the inverted cone.

Two accepted designs have been tested as shown in Figure 3.70. Design B is American and rather high in the gap. For British weather conditions it should probably be fixed slightly lower, when the pressure loss should not exceed 0.25 x p_v in the duct.

The static pressure loss for Design A is 1.0 x p_v

The static pressure loss for Design B is 0.2 x p_v

If the velocity is high in the discharge duct to atmosphere, as in dust collecting systems, a tapered diffuser should be fitted before the weather cap.

As an example:

Consider a straight duct discharging at 20 m/s i.e. p_v = 240 Pa. The loss in a cap to Design A would then be 1.0 x p_v = 240 Pa. Now assume a tapered duct is fitted with an included angle of say 7° and an area ratio of 1.75 to 1, which is a reasonable design.

From Figure 3.20 (in Section 3.3.2 on diffusers) with discharge direct to atmosphere and interpolating to 7°, the static regain is about 0.5 x p_v = 120 Pa. Resistance of the weather cap is reduced due to the lower velocity, which at 1.75 x area of duct would be about 11.4 m/s with a velocity pressure of 78 Pa, which would probably offset the frictional resistance of the length of tapered duct.

In general it may be taken that by the use of a taper with a larger weather cap, the discharge resistance in such cases can be eliminated with consequent saving in absorbed power.

3.6 Air duct design

There are two essentially different principles used in the design of air ducts.

- Graduated velocity with duct friction per metre maintained constant
- Velocity maintained approximately constant

The graduated velocity method is used for ventilating plants and as duct sizes are reduced in mains and branches, the velocity is also reduced, maintaining friction approximately constant per metre. This results in economy of power consumption of the fan.

In industrial schemes the initial velocity at fan discharge may be relatively high, but in public building schemes the duct velocity is limited by noise, which is an initial factor. Not only must air noise in ducts be eliminated, but the design of all sections of the plant must be for low resistance in order that a slow speed, quiet-running fan can be installed.

The velocity maintained approximately constant method is used for pneumatic collecting plants as the requirement is to provide the velocity to keep the particles in suspension throughout the system. Too high a velocity means excessive resistance with consequent high power consumption. Too low a velocity means a risk of choking at bends etc, with consequent complaints. Velocity may be varied slightly in different branches according to the ideas of the designer, but in principle the basis is constant velocity.

3.6.1 Blowing systems for H & V

3.6.1.1 Design schemes

Round piping

1. Make a line diagram, or isometric of proposed run of ducts with all branches and outlets shown. On this diagram mark the volumes of air to be delivered by each outlet and the totals to all branches from main duct.
2. For general industrial schemes the piping is sized on the basis of 1.6 Pa/m. In very extensive layouts i.e. with distribution ducts up to 120 m – 150 m long, it may be increased to around 3.3 Pa/m.

Initial velocity in the duct system will vary from 10 to 11.5 m/s in relatively small layouts, up to 18.5 to 21.5 m/s in extensive industrial systems.

It is important to note that when the system is for a public building, such velocities cannot be used because of air noise in ducts and in the noise generated by the fan. When quietness is essential the maximum air speed in ducts should be kept between 6 to 8.5 m/s, and for less important cases it may be 8.5 to 11.5 m/s.

General: Subdivide the main duct and branches by tapering down after air outlets with reasonable compromise. Included angle of tapers between 2.5° and 10°.

Too many tapers should be avoided, and with small "pops" (say 150 mm dia.) 3 may be taken on each section without reduction. With larger "pops" either a single outlet, or a pair, is usual practice.

Note that the sizing of the ducts on the basis of construction friction per metre does not in itself ensure the flow of the calculated volumes in the various branches. Balancing of resistance is necessary as described later. Size the duct to the nearest 3 mm in smaller sizes to nearest 6 mm in larger sizes.

Rectangular piping

1. Make a line diagram of system with volumes indicated exactly as in scheme A.
2. Assess the sizes of ducts as round piping
3. Convert these round duct diameters into equivalent rectangular by the Equivalent chart, in Table 3.6, which shows sizes for equal friction at equal volume. One side of rectangular duct, such as the depth, is kept constant in many cases, or at least so far as is reasonable.

General: The size of the fan, and hence its discharge dimensions, are not known at this stage. The initial area of main duct is not necessarily equal to the fan discharge, but of course should never be less in area.

If a fan supplies a main duct which immediately branches into two directions, it is usual to come from the discharge in a rectangular duct of same area. Then divide into two with each area proportional to the respective air volumes. Finally, transform from these on each side to the area decided in the duct layout assessment.

An adjustable splitter damper is desirable at the junction as flow from the fan discharge is generally uneven.

3.6.1.2 Duct resistance calculation

The design basis of friction per metre will be known at this stage. Prepare a scale layout diagram.

1. Examine this scale diagram and decide which is the longest run from fan discharge to remote air outlet. The equivalent length of this longest run is the actual length in metres, as measured from the diagram, plus the equivalent length in metres for each bend in this run, plus the equivalent length of any junction.

Values for bends, junctions etc. are given in this Chapter and also in CIBSE and ASHRAE guides.

As mentioned in Section 3.5.1, if the piping is rectangular it is important to note the "way" of each bend and to use correct dimension to work out the equivalent length. Ignore any resistance set up by duct tapers.

2. If the total equivalent length of the longest run calculated in metres is L, then duct frictional resistance is

   $L \times \text{friction Pa/m} = \text{Pa}$

3. If ducts are not in galvanised sheet steel, use the correction factor as given in Table 3.7.
4. Add an extra 25% of duct resistance only as a margin for balancing. Do not include resistance of heaters, washers, coolers, filters etc. in this addition as these should be known more accurately.

3.6.1.3 General notes

It will be appreciated that when a duct is sized on equal friction per metre, the velocity is gradually reduced from the fan to the remote end of the system. Hence it might be expected that there would be a gain in static pressure due to this reduction.

It is normal to neglect any such gain, and this was advised by ASHVE. Some engineers allow a regain of half the difference in initial and final velocity pressure in the longest run of duct. This is deducted from the calculated frictional resistance.

Actually, as will be shown, the pressure changes in a duct system are extremely complicated, and cannot be assessed with accuracy in commercial work. Experience over years has shown that the simple method as given will provide a reasonable approximation to the actual working resistance when installed.

Before the design is finally approved it is necessary to check the overall resistance of the plant. This includes duct resistance (with margin), addition for any special type of air outlet or grille, fresh air inlet louvres, filters, heaters, etc.

If the calculated overall resistance is found to be excessive for the particular type of system, it would then involve too high a fan speed. Noise in operation must be considered, and also the power absorbed by the fan, both of which are related to overall resistance.

If resistance is too high, then either redesign the ductwork for lower velocity or increase the area of filters, heaters, etc. to reduce their resistance.

Overall resistance values depend upon local conditions and experience is necessary to judge. Table 3.8 may be used as a guide noting that these values may currently be viewed as low. However, in an energy conscious world we should be endeavouring to reduce system resistance.

Type of system	System resistance
Factories:	
Heating only	200 Pa to 300 Pa
H & V with washer	300 Pa to 750 Pa
Public buildings:	
Ventilation only	100 Pa to 250 Pa
H & V	150 Pa to 300 Pa
H & V with washer	200 Pa to 350 Pa
HVACR with noise control	1000 Pa to 1500 Pa

Table 3.8 Typical static pressure loss in various systems

3.6.2 Exhaust ventilation systems for H & V

Velocity increases towards fan.

3.6.2.1 Industrial schemes

Design ducts on equal friction per metre, allowing 12.5 to 15 m/s in duct at fan inlet. Calculate resistance as described and add 25% margin.

Public buildings

Quietness is the important factor. With ducts connected to fan inlet box (i.e. without other plant items) design for 6 to 7.5 m/s in main.

If quietness is vital, keep down to 7.5 m/s in main at fan inlet, and 2.5 to 4 m/s in branches.

Even at low velocity internal acoustic treatment of ducts may be necessary.

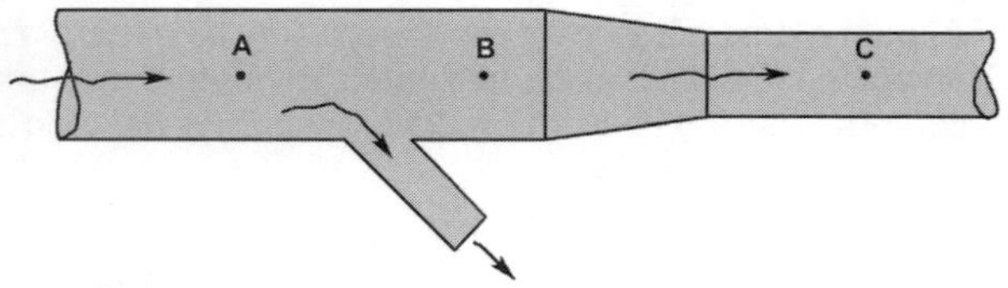

Figure 3.71 Effects of tapers and outlet pops

In Figure 3.71 the flow of air at B is less than at A by the amount passed through the outlet pop. Hence the velocity at B is less than at A and so a static pressure regain results. In passing from B to C there is a fall in static pressure as flow is restricted by the taper.

Many of these take-off outlets and tapers occur in a duct system and, as already shown, the effects are neglected in the generally accepted method of calculating duct resistance. Hence it is obvious that these gains and losses must cancel approximately because long experience has shown tat the accepted method is satisfactory.

Duct tapers

Tests show that at normal velocities, the static loss in a taper is relatively small but the aggregate of many taper in along main can be a considerable item in the resistance.

For example, in a long duct on an installation at a textile mill it was estimated that the tapers represented about 700 Pa. The IHVE guide at that time gave the loss in a taper as 0.2 x velocity pressure in the small end, and on this basis the main at the mill calculated at 840 Pa. In contrast, American sources gave the loss as 0.04 to 0.05 of the velocity pressure in the small end, and this caused confusion. In fact, this was the loss in total pressure.

In a duct taper with included angle up to 10°, the conversion of velocity is complete and the loss occurs in the duct immediately after the taper due to slight turbulence at the walls in regaining the full flow area, see Figure 3.72. Friction loss is negligible.

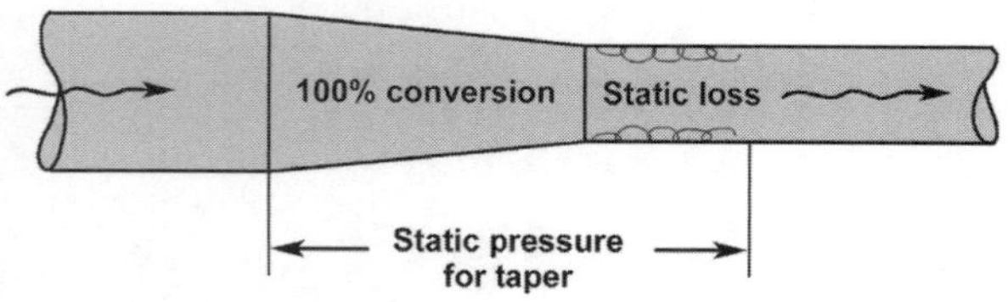

Figure 3.72 Loss in contractions

Analysis of tests shows that it is not satisfactory to calculate the static pressure loss from a single factor multiplied by the velocity pressure in the small end. In some cases this gives results which are less than the difference in velocity pressure at the entrance and exit.

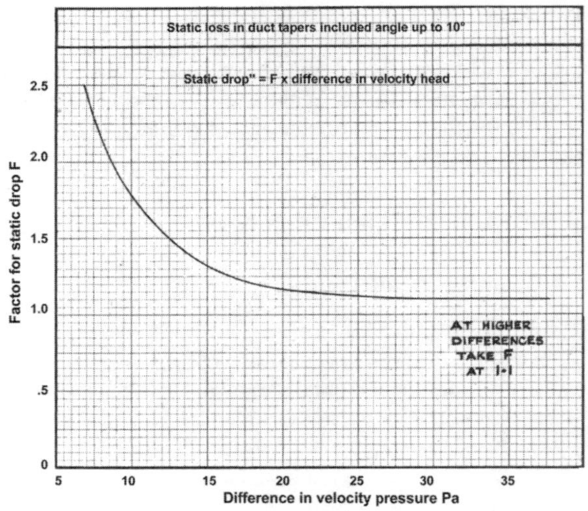

Figure 3.73 Static loss in duct tapers

Tests have shown that the best approximation for practical work is given by a variable factor multiplied by the difference in velocity head in the taper. This loss of static pressure becomes greater in proportion when the differential is very small, as shown in Figure 3.73.

3.6.2.2 Take-off regain

In the normal design of ducting, the pop or take-off is not fitted on the taper, but is at the end of the duct before taper. Experiment has shown that the regain of static pressure is very much higher than would be expected. It varies with velocity to some extent and is greater when velocity is very low in the duct. The regain is estimated as a percentage of the difference in velocity head before and after the take-off.

For practical purposes it is suggested that average values may be taken at:

90% when ducts are designed at 0.82 Pa/m

82.5% when ducts are designed at 1.63 Pa/m

75% when ducts are designed at 2.45 Pa/m

It is of course risky to overestimate this regain in a commercial calculation.

3.6.2.3 Effect of change in volume

Whilst small variations in air volume passed through a duct system may be calculated as the square, it is not advisable to try this when the change is very considerable. The change in regain and taper losses from a very underloaded condition to an overloaded condition might upset the accuracy of this estimation. There will also be a change in Reynolds Number.

Large changes in flow on a given system are unusual, but cases are known where ducts have been installed to deal with some future condition in a factory. One must also remember that some systems with a variable fans speed can cope with a 10 : 1 reduction.

3.7 Balancing

Owing to the fall in pressure in the length of the main duct, the air outlets at the initial end will deliver more air, and those at the extreme end less air, than the mean if the system is not balanced.

3.7.1 Unbalanced system example

In a blowing system with round piping designed on the percentage system there are 2 similar lines of ducts each with 14 outlets. No balancing adjustments were provided, and tests were made for the air volume flowrate discharged. The results are given in Table 3.9.

Outlet pop No.	Line 1 m^3/s	Line 2 cfm
1	0.107 (+ 12% on Mean)	0.093 (+ 19% on Mean)
2	0.106	0.091
3	0.104	0.089
4	0.103	0.088
5	0.101	0.085
6	0.099	0.083
7	0.098	0.080
8	0.094	0.078
9	0.093	0.077
10	0.091	0.073
11	0.088	0.070
12	0.086	0.067
13	0.083	0.065
14	(74% of No 1) 0.079 (-17% on Mean)	(66% of No 1) 0.061 (-22% on Mean)
	1.332	1.100

Table 3.9 Variation in flow on a typical unbalanced system

Methods of balancing

In blowing systems the connection of an air outlet pop to the main is made at 45°. The development of the hole in the main to attach the pop is of a peculiar shape. Its total length is 1.41 x diameter of pop and is shown in Figure 3.74.

Design balancing is based upon alteration of the length of this hole, producing restriction as required and gradually adding resistance to the outlets from those at the extreme end of the main to those near the fan. See Figure 3.75.

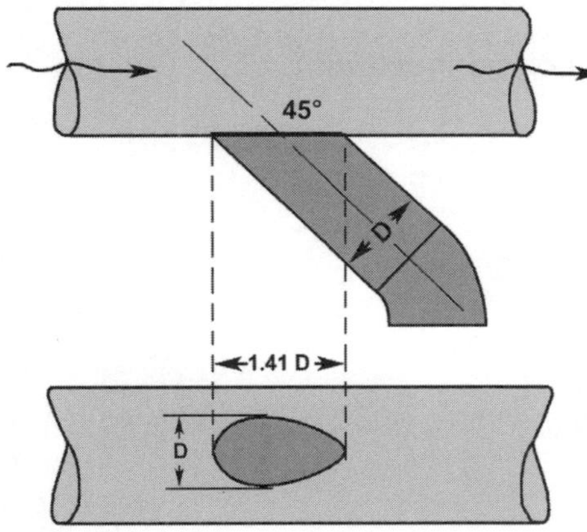

Figure 3.74 Hole in main duct for branch

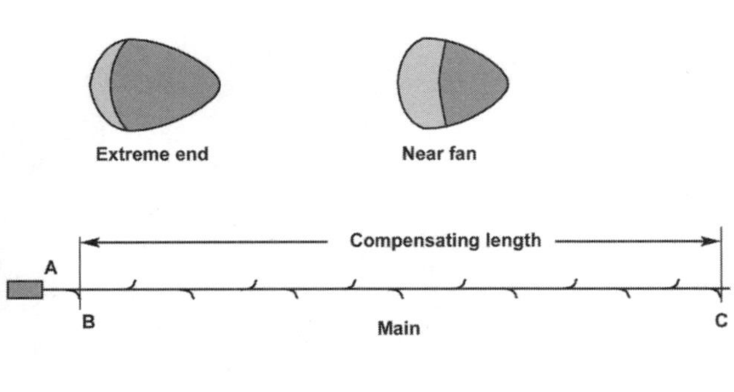

Figure 3.75 Restriction to balance resistances

To balance the system, the resistance from A to B must be equal to that from A to C. the difference is the friction of the main between B and C and this must be compensated by adding resistance at B. This has been checked by experiments as will be seen later.

3.7.2 Balancing scheme

1. On a line diagram of the duct system with its outlets, mark the length in metres from the extreme outlet to each of those before it, in the direction towards the fan, i.e. the extreme outlet is O, and first outlet is the length of main in metres between it and the extreme outlet.
2. Convert these lengths in metres into diameters of pop. If the pop is 150 mm diameter or 0.15 m, and length is 21.3 metres for example, this is 142 diameters.
3. From the balancing chart in Figure 3.76, read against each value in diameters the opening length in percentage of pop diameter. For example, if 140 diameters, the resistance equivalent is 76.5% of pop diameter.

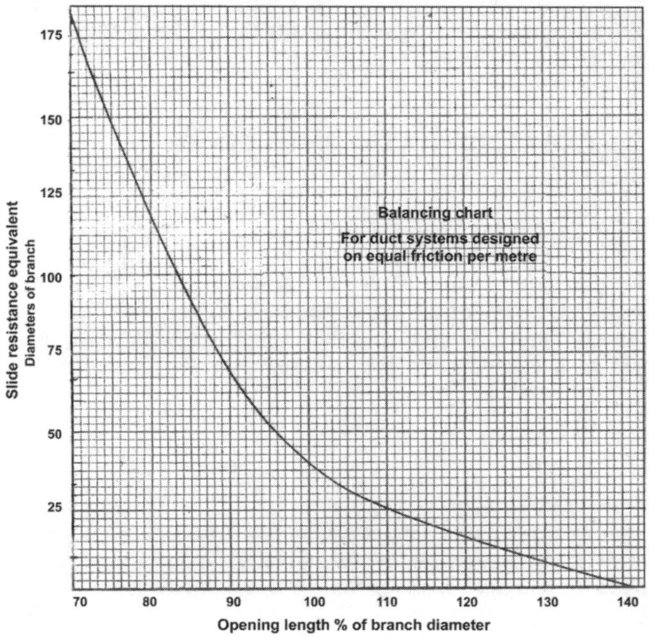

Figure 3.76 Balancing chart

The Table in Figure 3.77 shows three examples worked for a length of ducting with pops of 150 mm, 250 mm and also of graduated diameter from 215 mm to 250 mm diameter.

4. These lengths of opening are then specified on the working drawing for ducting as millimetres, and are usually shown alongside for ducting as millimetres, and are usually shown alongside the pop in a circle thus: (225)

   Work to nearest 5 mm.

3.7.3 Balancing tests

Experiments made some time ago by Sturtevant Engineering Company Ltd showed that as the main duct static resistance is increased, equivalent to various lengths of main duct, an exactly similar addition of static resistance had to be inserted into the branch to maintain flow constraint.

They first set up conditions to represent a branch at the extreme end of a main. The air velocity was 7.5 m/s in both the main and the branch. There was no control resistance in the branch.

Conditions were then created to represent the first branch in a system with the main velocity 20 m/s and the branch velocity 7.5 m/s.

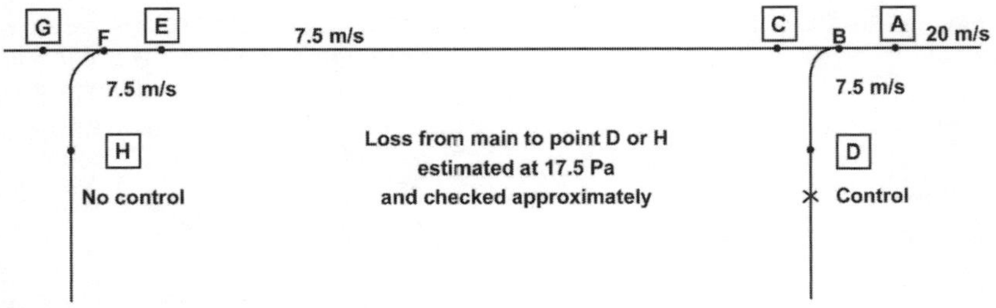

Test:	A	B	C	D	E	F	G	H	Resistance of main B - F	Control resistance added D - H	Difference in pressure velocity A - C	Regain A - C	% Regain
1	50	117.5	145	100					50	50	106	95	90
2	150	222.5	250	205	70	67.5	90	50	155	155	106	100	94
3	275	345.5	375	325					275	275	106	100	94

Figure 3.77 Results of balancing tests

The results can be seen in Figure 3.77

As the air velocity attained the branch value at the entrance to the branch, this regain must be passed into the main air stream and is returned in the regain from A to C. The volume flow in the branch was measured by a Venturi.

3.8 Notes on duct construction

3.8.1 Dirt

Provide cleaning doors (slides in smaller ducts) for all supply systems, even after a filter. Dirt deposits 30 mm thick have been found in ducts. Even after good filtration beware of blowing air directly onto a wall or ceiling, as dirty marks will appear in time.

3.8.2 Damp

If underground ducts are proposed, make enquiries as to the nature of the ground. Ordinary concrete is not waterproof and is porous. In heating plants with underground ducts there has been trouble with attainment of temperature due to evaporation of moisture in ducts.

Waterproof cement rendering will obviate trouble against normal drainage:

1 part cement, 3 parts washed sand; mix with soapy water (50 g soft soap per litre). The free lime in the concrete combines with the alkali in the soap forming a calcium compound which fills the pores in the concrete.

3.8.3 Noise

Beware of drumming of rectangular ducts, particularly the top surface. Round ducts are free from this trouble.

If rectangular ducts are used they must be very amply stiffened or "cross-folded".

In public buildings the ducts were formerly made in builders' works as they were less likely to cause trouble. Square corners were unavoidable to get into the space provided, but the addition of vanes reduced the pressure loss.

No sharp edges of any form should be left on which air is blown. No splitters of light gauge which might vibrate. Edges turned over to air flow.

Beware of noise from room to room with short connections on a main duct. See Figure 3.78.

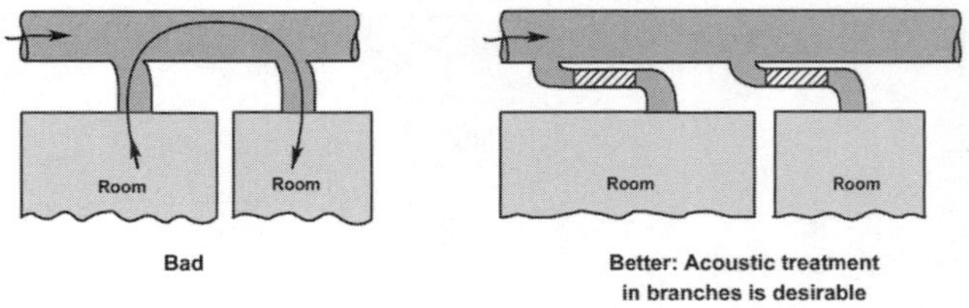

Figure 3.78 Noise from room to room

3.8.4 Inlet and discharge of fans

Transmission of noise to ducts is obviated by rubberised canvas connections, 150 mm to 225 mm clear space. Treatment with shellac after fixing is sometimes advocated.

3.8.5 Temperature control

When temperature control is required it will be necessary to insulate builders' work ducts internally to reduce the lag.

3.8.6 Branch connections

Examples of these are shown in Figure 3.79. ASHRAE advocate method A rather than B.

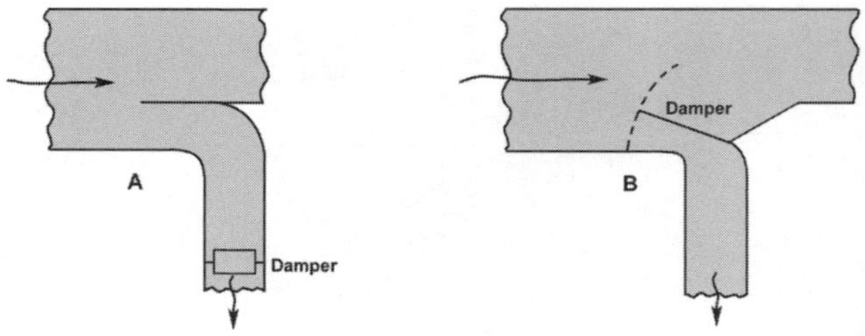

Figure 3.79 Branch connections

3.8.7 Fire damper

When a duct must pass through a fireproof wall, a special damper has to be fitted

6 mm thick for small ducts,

9 mm thick for large ducts

The fitting of a fire damper is illustrated in Figure 3.80.

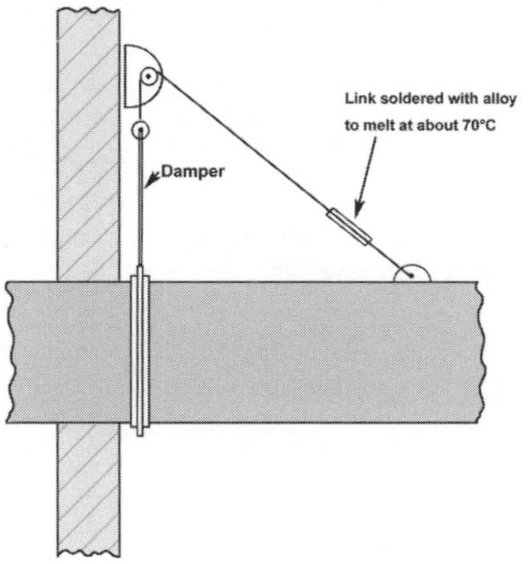

Figure 3.80 Fitting of fire damper

3.8.8 Adjustment of damper at outlets

These may be fitted as slot and slide, or hit and miss slides adjusted by poking through the grille. Examples are shown in Figures 3.81 and 3.82.

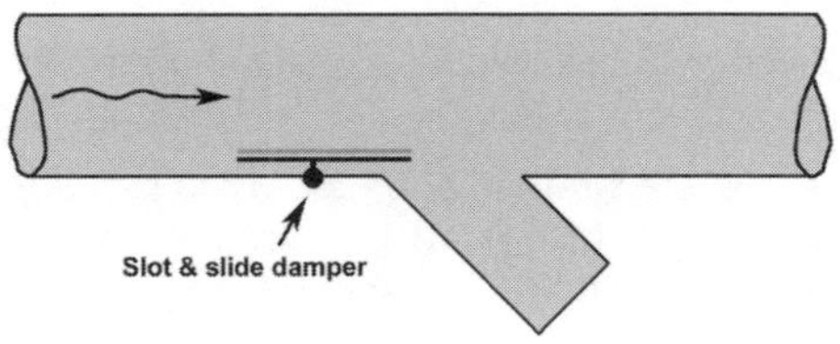

Figure 3.81 Duct outlet slide

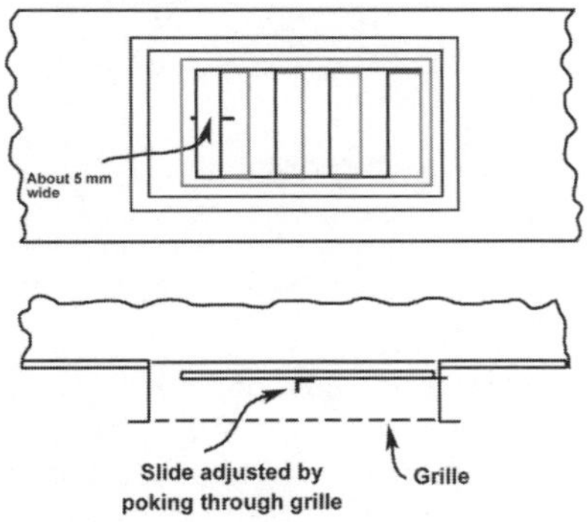

Figure 3.82 Hit and miss slide

3.9 Duct design for dust or refuse exhaust

Long experience has decided the most suitable diameters of the connections to exhaust hoods for all the usual machines to which dust or refuse collection is applied. These standards are available from machine manufacturers or system designers.

The velocity necessary to provide adequate margin for the suspension of the particles in the airstream is also known for most types of dust or refuse. Table 3.9 shows some examples.

Machine	Duct velocity m/s
Grinding wheel dust	23
Buffing wheel dust	20
Sawdust, dry	18
Wood chips, normal machines	20
Wood chips, high speed machines	23
Wood sand papering machines	12/13

Table 3.9 Duct velocities for types of dust or refuse

The range of air velocity used by engineers is from about 12 to 25 m/sec, but 18 to 23 m/s covers the usual requirements. For unit collectors or individuals grinding or buffing machines, lower velocities are common in the short connecting pipes e.g. 18.5 m/s for grinders and 17 m/s for buffing machines.

Many plants are at work successfully which were designed for constant air velocity in all mains and branches. Some designers vary the velocity in a system in different branches according to the types of machines connected. For example, in a wood refuse plant the branches to sawdust-producing machines may be designed for 18 m/s; with those to chip-producing machines at 20 to 23 m/s, and with all mains at a nominal 20 m/s. This may vary slightly in mains due to approximations for duct diameters to the nearest 5 mm.

3.9.1 General notes

In an extensive woodworking plant, a separate system may be installed to deal with the saws, as sawdust can be sold. Another separate system deals with planers and moulders etc., the chips collected being discharged to a boiler or a refuse destructor.

Wood sandpapering machines should be handled by a separate plant, or as individual units, as this dust is extremely fine and it requires a textile filter to collect.

Grinding machines and buffing machines should no be connected to the same exhaust plant. Sparks from grinding might ignite lint from the buffs with risk of fire.

When a woodworking machine has multiple connections, e.g. a four-cutter or six-cutter moulder, it is important to keep in mind the effect of it being out of service with blast-gales (dampers) on connections closed. This might result in too low a velocity in the main to carry the refuse from other machines still in service on this section. Actually, when the material is in the main, the minimum carrying velocity is considerably less than those mentioned, say 75% of normal, and this allows some latitude. Experience is the only guide in difficult cases.

3.9.2 Design scheme

On an outline plan of the factory, mark the positions of machines with their exhaust points and sizes according to the schedule. Lay out a suitable run for ducting, noting that branches in an exhaust system enter the main at 30°, or through patent junctions with almost parallel entry.

From the diameter of connection and selected velocity calculate the flow or obtain this from a manufacturer's data. The diameter of the main is then calculated in its graduated sizes as branches enter, from selected velocity and total flow at any given point. Work to the nearest 10 or 5 mm in main sizes. This alters the selected velocity slightly and the final figure is used for friction calculation.

3.9.3 Calculation of resistance

1. Estimate entry loss at the hood most remote from the fan.
2. Calculate the approximate equivalent length in metres of this most remote branch from hood to main. That is, the length of straight piping plus equivalent length in metres for bends.
3. Branch loss at entry to main from B to A for exhaust systems is less than in blowing. (Figure 3.83.)

   Now total up the equivalent length of branch, estimate its friction loss in mm w.g. Add entry loss from item 1.

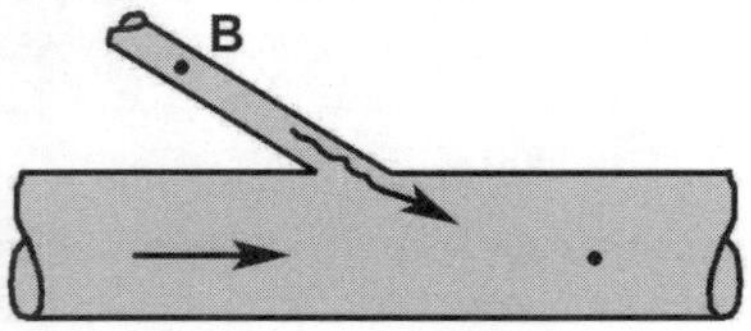

45°	30°	15°	Patent junction
9	4	2	1

Figure 3.83 Loss in equivalent diameters of B

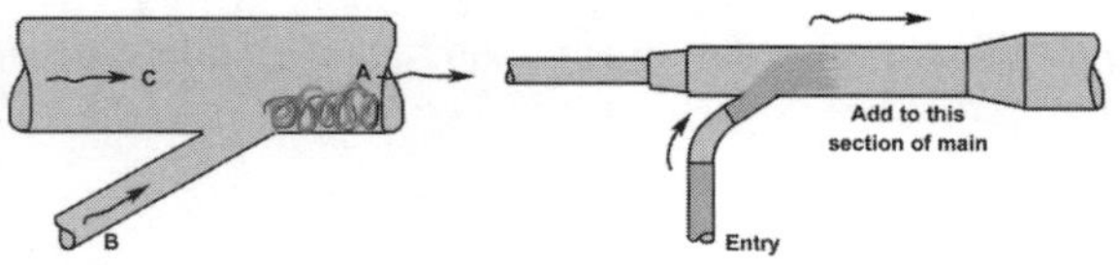

Figure 3.84 Entry of air from a branch

4. Measure each length of main between the entry of branches and allow for any addition from item 5. Neglect tapers and include as straight duct.

5. The entry of air from the branch, if at an appreciable angle to main, causes a loss in the main from C to A due to turbulence, and is shown in Figure 3.84. A summary is given in Table 3.10. (Note this is in diameters of A and **not** B.) Estimate this loss at each branch of entry and add to the friction of the section of main following any given point of entry.

Diam A / Diam B	45°	30°	15°	0° Parallel junction
	Loss in diameters of A			
1	7	4	1	Neglect
1½	6	3	1	Neglect
2	5	2½	¾	Neglect
2½	4	2	½	Neglect
3	3	1½	½	Neglect

Table 3.10 Loss in branch in diameters of A (from step 5.)

6. Add values of steps 1, 2, 3, 4 and 5 and mark the total on the diagram at each point of entry. It may be conveniently shown in a square thus: [50]

7. The complete frictional and turbulent resistance of the suction main is entered at the fan inlet as suction side resistance depression. If velocity pressure is added, it is then static suction, but most performance tables for fans are based upon fan static pressure and so this is the figure required when dealing with the fan speed etc.

Note: The resistance depression to be set up by the fan must include the separating apparatus. In wood refuse systems a cyclone separator is used and is always on the discharge side of the fan. Hence, to the resistance depression on the suction side from step 7, must be added to the frictional resistance of the discharge duct with its bends, and the resistance of the cyclone separator. The latter will normally have a resistance of 35 to 50 mm.

In dust systems either a cyclone or a textile bag filter may be used as decided by experience of the particular application. These may be installed on either suction or discharge side of the fan. If on the suction side, the resistance depression must be added to step 7, plus the resistance of the discharge duct on the fan with its weather cap.

If on the discharge side, then the resistance of the piping, together with that of the cyclone or bag filter added to step 7, will represent the fan static pressure.

3.9.4 Balancing of dust extract systems

Balancing of the system is the adjustment of resistance so that in the example in Figure 3.85 the resistance from remote hood at A to the fan inlet at B is approximately equal to the resistance of the branch near the fan from C to B. If not balanced, C would exhaust too much air and A too little, as compared with that to meet designers requirements.

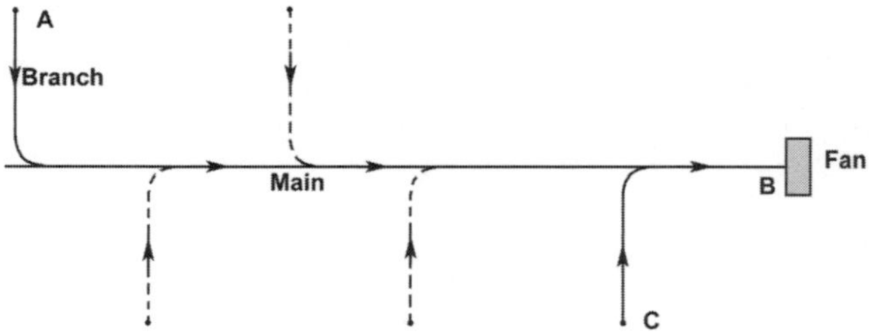

Figure 3.85 Example of dust extract system balancing

Any artificial resistance put into the circuit must be of such nature that dust, sawdust or woodchips cannot build up on it to cause a blockage. An orifice in a plate inserted between a pair of flanges in branch C could be used to impose artificial resistance for balancing, but it would probably build up and cause a choke.

Experience has shown that when the air is carrying material, the best restriction is in the form of a conical piece, see Figure 3.86, inserted into the end of a branch where it joins the main.

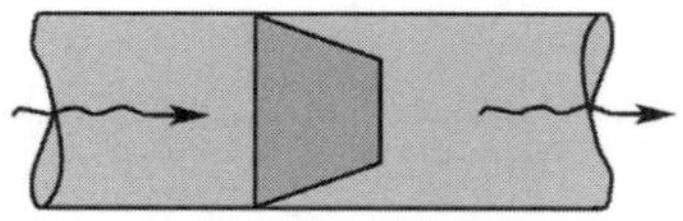

Figure 3.86 Internal conical piece for balancing

Material passes easily through this and the desired added resistance is attained by a suitable diameter of the small end of the cone.

The cone is inserted in the inlet of its patent junction with the main, and has an included angle of 30° to 40°. If a relatively small reduction is required, say 5 mm or less than branch diameter, then the end of the branch itself is closed to the required dimension and inserted into its junction with the main.

If the velocity were exactly equal throughout the entire system this balancing would involve only the question of so much added resistance. As mentioned, there may be some differences in velocity in branches and in the main, due to the ideas of the designer.

So balancing is worked on static suction depression and when these are equal in the branch and in the main at any given point of entry, the system is balanced. All branches are, of course, treated as required. See the formula illustrated in Figure 3.87.

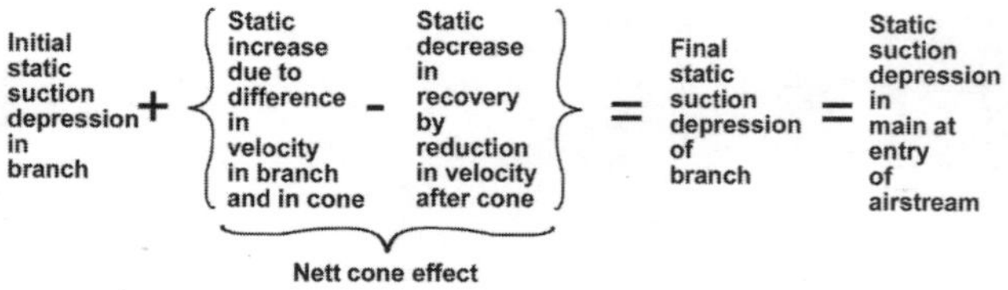

Figure 3.87 Effect of cones in branches

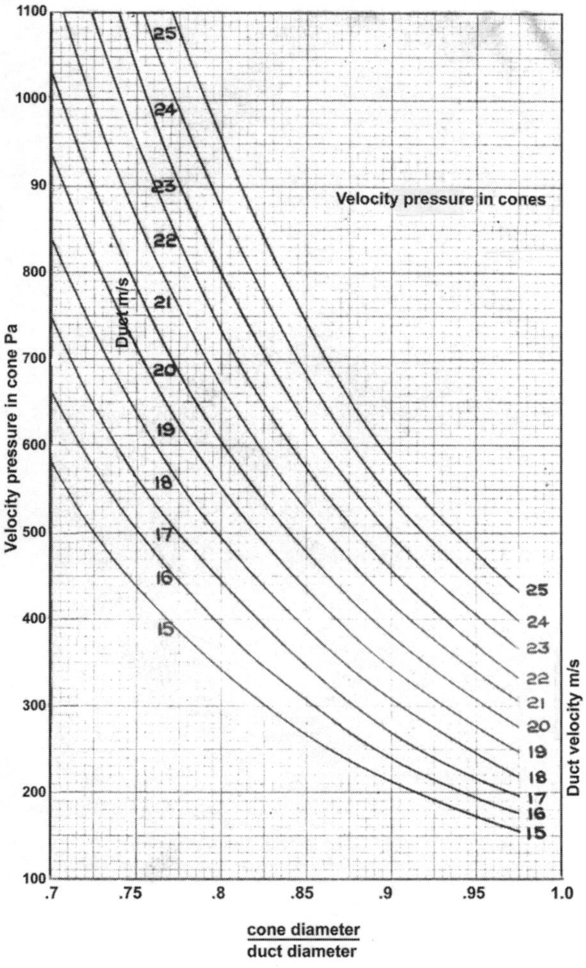

Figure 3.88 Velocity pressure in cones

If a cone is inserted in a long length of piping there is considerable recovery, as measured by tests. When inserted in the junction, the air leaving is in a turbulent state, and any recovery is balanced by a loss. Experience of results on the method of calculation described has indicated that any recovery may be neglected.

The cone inserted in a branch must have the same net effect as the difference in static depressions. As no recovery is assumed after the cone, this difference is equal to the increase in velocity pressure from the branch to the mouth of the cone. If the initial velocity pressure in the branch is known then the final velocity pressure at the mouth of the cone has to be v_{pi} + difference in depressions. The required diameter of the mouth of the cone to produce this velocity pressure is given in Figure 3.88.

From the required additional pressure, read across to branch velocity m/s and then down to value of $\frac{\text{cone diameter}}{\text{duct diameter}}$.

3.10 Bibliography

CIBSE (The Chartered Institution of Building Services Engineers), 222 Balham High Road, Balham, London, SW12 9BS, UK, Tel: (+44) 020 8675 5211, Fax: (+44) 020 8675 5449 Web: www.cibse.org.

ASHRAE (American Society of Heating, Refrigerating and Air-Conditioning Engineers Inc.), 1791 Tullie Circle, N.E., Atlanta, GA 30329, USA. Tel: (404) 636-8400, Fax: (404) 321-5478 Web: www.ashrae.org, Email:ashrae@ashrae.org.

Handbook of Hydraulic Resistance, I E Idelchik, Begell House Publishers Inc., 2001 ISBN 1567000746.

Internal Flow Systems (2nd completely revised edition) Edited by D S Miller, BHR Group Ltd, 1996 ISBN 0947711775.

NEL (National Engineering Laboratory), East Kilbride, Glasgow, G75 0QU, UK, Tel: 01355 220222 Fax: 01355 272999, Email: info@nel.uk, Web: www.nel.uk.

Heating and air conditioning of buildings, Oscar Faber and John Kell, *IHVE Journal*, March 1938, (Work on the construction of the 12 acre Earls Court Exhibition building in London, involved conducting full- scale tests on the special ventilating jet nozzles).

Heating and air conditioning of buildings 9th Edition, Faber and Kell, Edited by P L Martin, 20 December 2001, Butterworth-Heinemann Ltd, ISBN 075064642X.

Studies in the design of local exhaust hoods, Dalla Valle, J.M., & Hatch, T., ASME Transactions, Vol. 54, 1932.

Industrial Ventilation: A Manual of Recommended Practice, 24th Edition, ACGIH® (American Conference of Governmental Industrial Hygienists), 2001 ISBN: 1882417429.

Simplified Equations for HVAC Duct Friction Factors, J J Loeffler, ASHRAE Journal.

AMCA 200-95, *Air Systems.*

AMCA 201-02, *Fans and Systems.*

AMCA 203-90, *Field Performance Measurement of Fan Systems.*

Fan Application Guide 2nd Edition, FMA (HEVAC).

Fan and Ductwork Installation Guide 1st Edition, FMA (HEVAC).

4 Fan performance Standards

Until very recently there were more than 12 national Codes for fan testing, incorporating over 70 specific duct arrangements. However, three international Standards, ISO 5801, ISO 5802 and ISO 13347 for specifying the aerodynamic and noise performance of fans have received considerable attention. As they alone embody the latest agreements within ISO, their virtues have been extolled in many quarters. Nevertheless, misunderstandings as to their intent and accuracy are apparent.

This Chapter outlines the reasoning behind the various decisions made, how fan performance Standards may be compared and corrects current misunderstandings. ISO Standards are discussed and the differences with previous Standards explained. Shortcomings in the latter have been identified and are rectified.

Contents:

4.1 Introduction

Until the early 1920s, the methods for testing the aerodynamic performance of industrial fans were legion. It is no exaggeration to say that these were determined by the various manufacturers according to their own beliefs, prejudices or downright commercial considerations. At about that time, ASHVE (the American Society of Heating & Ventilating Engineers, a forerunner of ASHRAE – the American society of Heating, Refrigerating and Air-conditioning Engineers) in the USA and IHVE, (a forerunner of CIBSE – the Chartered Institution of Building Services Engineers) in the United Kingdom both set up fan standardisation committees, which produced recommendations for the conduct of such tests and the calculation methods to be used. Subsequently, these recommendations were incorporated into the appropriate national Standards.

The situation worsened however, as other organisations believed that they had to issue documents if they were not to be left out of the "race". It seemed that we had simply exchanged one set of problems for another, as ever more organisations felt impelled and qualified to issue their own versions of a fan Standard. Not only did ASME issue its own Standard in the USA but the FMA (Fan Manufacturers Association) in the UK, recognising the deficiencies of the then British Standard, also issued its own code in 1952. Many other National Standards bodies had by then joined the game so that by the 1960s proliferation had made the matter worse than ever.

Into the chaotic situation which existed, ISO stepped with great confidence. It set up Technical Committee TC117 in 1963 to discuss the formulation of an International Standard which could be agreed to be all the major industrial fan manufacturing nations. It started off with considerable optimism and after various excursions along the way (see Section 4.1.3) eventually settled into a dull routine where each nation sought to protect its own Code at the expense of all the others. Eventually, it dawned that compromise was essential if work was to be completed this side of the grave!

You may well ask "Why all the fuss?" Does it come as a surprise to know that not all those national Codes were of the same technical merit, and serious discrepancies could result? A few years ago the company for whom the author was then working, carried out a series of tests on one particular fan to various standards. The supposed differences in performance (see Figures 4.2 and 4.3) were alarming. In fact, of course, nothing should have changed. If efficiencies had been plotted, then, with an unchanged fan absorbed power, these should have been proportional to the fan pressure. This latter is very much a convention (see Section 4.1.3). It should be noted that one of the fans was a tube axial type with appreciable outlet swirl. How this swirl energy is treated can have an appreciable effect on the results. The diffusion at the fan outlet can also be important and how much velocity pressure is converted into useful static pressure may be dependent on the length of such ducting.

This is a world which endeavours to preach the value of free trade. Increasingly, it has had to accept the fact of globalisation. As a contribution to harmony between nations, it is essential that valid comparisons can be made between different companies (and nation's) products. Only if they are all tested to the same standard test code is this possible.

The fan engineer works under the disadvantage of handling a fluid, which cannot be seen or directly weighed. If necessary a pump flow could be determined by catching the water in a bucket. The engineer does not have the possibility of determining airflows in that way. Furthermore, in the "real" world, air travels in three dimensions and is turbulent. If one is making measurements under actual installation conditions, it is therefore desirable to take a great many measurements of velocity and direction.

This is the thinking behind ANSI/ ASME PTC 11-1984, which is a Code, developed in the USA, for determining performance under operating conditions of large fan units such as those required for mechanical draught in central power stations.

It normally requires the use of a calibrated 5 hole pitôt tube combined with a temperature sensor, as shown in Figure 4.1. A traverse is taken directly on the fan discharge and the many measurements of pressure (total and static), direction (pitch and yaw) and temperature (wet and dry-bulb) are then integrated to obtain the total flow and pressure. This normally requires the aid of a computer to reduce the otherwise tedious hand calculations.

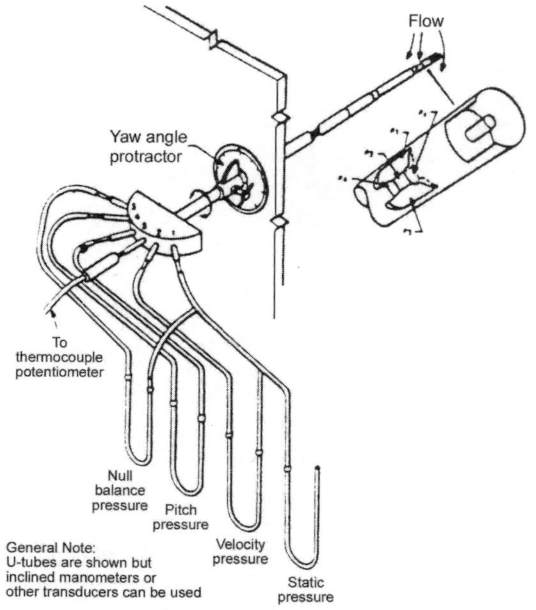

Figure 4.1 View of 5 hole Pitot tube

In Australia, the Standard AS 2936-1987 adopted a similar philosophy, but permits the adoption of a simplified 3 or 2 hole yaw meter. These methods have no devices for straightening the swirling airflow, but determine the velocity in the actual direction of flow. Vectoring is then applied to obtain the mean axial flow velocity and hence the volumetric flowrate.

It should be noted that these methods do not necessarily give an accurate result for the fan static pressure. Due to diffusion at the fan outlet, there will be an exchange of kinetic and static energy such that the maximum pressure may be developed at least 3 duct diameters from the fan outlet.

Whilst design programmes exist which can closely predict the performance of a fan, it is nevertheless essential to conduct tests to confirm them. Even the most advanced design techniques such as CFD (Computational Fluid Dynamics) require the input of empirically determined correction factors.

Fan tests will be conducted for one or more of the following reasons:

i) Tests carried out during the development of a product range to confirm the design programme

ii) Tests carried out to provide selection data for a catalogue (either paper copy or electronic)

iii) Acceptance tests at the manufacturers' works to confirm that a unit meets the customer's specification

iv) Acceptance tests on site to confirm that a unit meets the customer's specification and/or to confirm that the system resistance is correct or needs modification.

Laboratory tests are essential if the full characteristics of the fan are to be determined from zero flow (shut-off) to full flow (free delivery). Field tests are invariably limited to a particular duty point unless artificial resistance can be inserted into the circuit.

Until very recently there were more than 12 national Codes for fan testing, incorporating over 70 specific duct arrangements. However, three international Standards, ISO 5801, ISO 5802 and ISO 13347 for specifying the aerodynamic and noise performance of fans have received considerable attention. As they alone embody the latest agreements within ISO, their virtues have been extolled in many quarters. Nevertheless, misunderstandings as to their intent and accuracy are apparent.

This Chapter outlines the reasoning behind the various decisions made, how performance to other Standards may be compared and corrects current misunderstandings. These ISO Standards are discussed and the differences with previous Standards explained. Shortcomings in the latter have been identified and are rectified.

The aim is to collect, steady and generally organize the flow in a suitable test airway, and this is achieved in the various laboratory test methods. The major national Codes for fan performance permitted a fan to be tested in a number of ways. It has been calculated that there were over 70 distinct test methods in use. Many of the methods incorporated in ISO 5801 were taken from the American, British and French Standards. Not all of these are of the same technical merit, and it will come as no surprise that some discrepancies can still result. In a world committed to free trade as its contribution to harmony between nations, this is a little strange. Some may think that the differences in measured fan performance are not serious. This is not the case. It is a cause for joy that ISO 5801 is currently under review resulting, hopefully, in these differences being minimized and in a reduction of its present 232 pages.

4.1.1 Fan performance

The performance of a fan is affected by the connections made to its inlet and outlet. Ducting, where fitted, not only has a pressure loss, but can act as an impedance, modifying the flow into or out of the fan casing. In extreme cases it can prevent the development of a full velocity profile. Ideally the flow velocity vectors should be symmetrical and axially aligned (free from yaw) and without swirl or spin (pre or contra) if the fan is to develop its design duty.

4.1.2 The outlet duct

In many former test Codes, the outlet duct was simulated in either of two ways:

i) Using a parallel duct, usually of a similar area to the fan outlet, for those fan types where the outlet flow permits acceptable measurements of flow and pressure on the outlet side, e.g. centrifugal fans.

ii) For other fan types - and, in particular, axial flow fans without guide vanes - a short length of parallel ducting of the same size and shape as the fan outlet. This means that all measurements of flow and pressure were made on the inlet side.

No doubt the majority of tests carried out in accordance with. these Codes yielded comparable results, but discrepancies could arise (see Figures 4.2 and 4.3).

In each of the figures the same fan was tested to various Standards. For the tube axial fan i.e., without discharge guide vanes, not only is the peak pressure different according to the Code used, but there were considerable differences in the measured Fan Static Pressure over the working range of flowrates. For the centrifugal fan, the conditions at the fan outlet are critical, especially where a tongue piece is fitted and, as with a backward-bladed fan, the impeller is towards the back of the fan casing. The "increase" in performance and efficiency by adding a straight duct of the same cross-section as the outlet and only

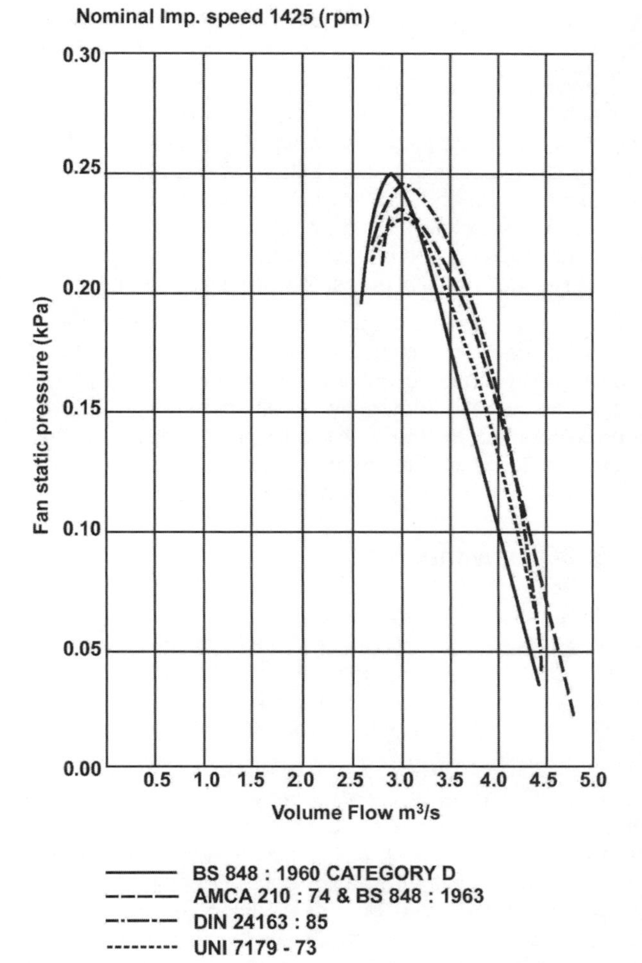

Figure 4.2 Performance of 610 mm tube axial fan to different national Codes

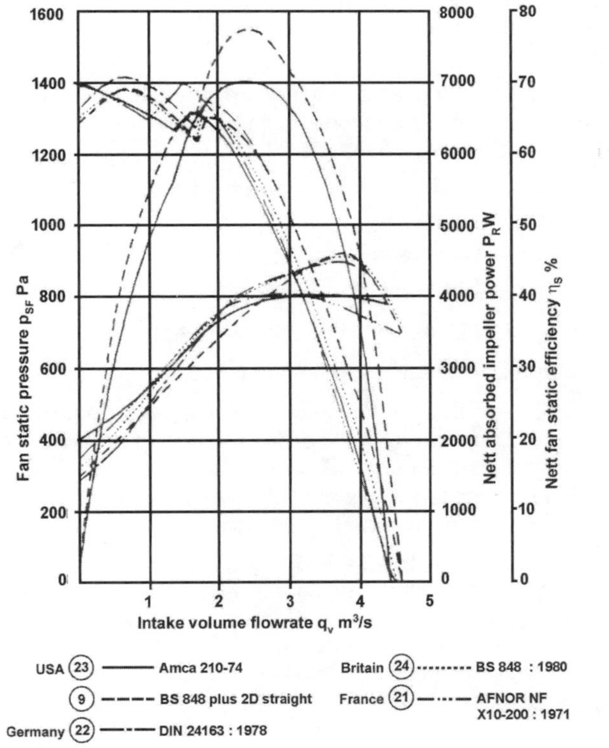

Figure 4.3 Performance of 630 mm backward inclined centrifugal fan to different national Codes

two equivalent diameters long before the circular outlet duct, will be noted.

The conventions for velocity pressure also differed. In Scandinavia and the Low Countries some companies used to present their data with a discharge loss, which was assessed for the different outlet duct configurations. This loss was referred back to the velocity in the annulus between the outside and hub/stator diameters. Essentially the efficiency calculated was an impeller /stator efficiency and did not encompass the overall fan. In such cases the Fan Total Pressure would be apparently higher.

The discharge loss was calculated for a constant diffuser or cone efficiency and was the same no matter what the impeller pitch angle. It comprised a conventional duct velocity pressure and an impact loss. Apparently, the effects of residual swirl had been discounted. Where such swirl was present, the discharge losses would be greater than calculated, and the available pressure from the complete fan would be reduced.

4.1.3 ISO conventions

The International Standards Organization Committee TC117 was charged over thirty years ago with the production of a mutually acceptable performance test Code. Many of the Committee arguments were fierce and some members adopted entrenched positions, which they were reluctant to abandon.

With the approach of the true European Common Market on 1st January 1993, a new sense of urgency developed, for it was intended that any resulting ISO Standard would be adopted as a CEN Standard by the so-called accelerated PQ procedure. Great Britain tried to anticipate the outcome of the deliberations by revising BS 848:Part 1 in a new edition published in 1980. There were some scares along the way, for at one stage the French and Belgian delegations were proposing that fan performance be reported as mass flowrate kg/s against specific energy J/kg! Of course, Great Britain was not completely correct and changes were taking place even at a final meeting in Florence on 5th May 1993. Nevertheless, agreement was reached and ISO 5801 was finally published in 1997. It has subsequently been adopted in its entirety by Britain, France and Italy with dual numbering e.g., in the UK it is also BS848 Part 1 : 1997.

A number of the concepts included are new to those not familiar with BS 848:1980. Attention should be drawn to the following, which are of major importance:

i) It recognizes that a fan will perform differently according to how it is installed.

   Type A with free inlet and outlet

   Type B with free inlet and ducted outlet

   Type C with ducted inlet and free outlet

   Type D with ducted inlet and outlet

It will be seen that the two alternative connections previously mentioned have been combined to give the four possible installation categories (Figure 4.4). In installations of type A, a partition in which the fan is mounted may support a pressure difference between the inlet and outlet sides.

ii) It allows considerable flexibility in the methods of measuring flowrate. Where these are based on the devices and coefficients described in ISO 5167 for orifice plates, nozzles and venturi tubes they will have equal validity. In effect all the devices given in BS 848:Part I: 1980 have been included with the addition of other test assemblies such as the French "caisson-reduit" used with an inlet or outlet orifice and the American (AMCA) multi-venturi chambers.

The coefficients of discharge for the British conical inlets have again changed. The 1963 version of BS 848 gave values up to 0.975 at high Reynolds numbers. In the 1980 edition this was

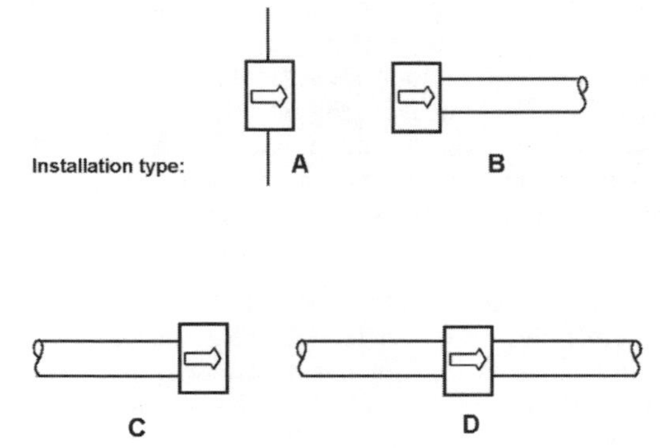

Figure 4.4 Fan installation categories

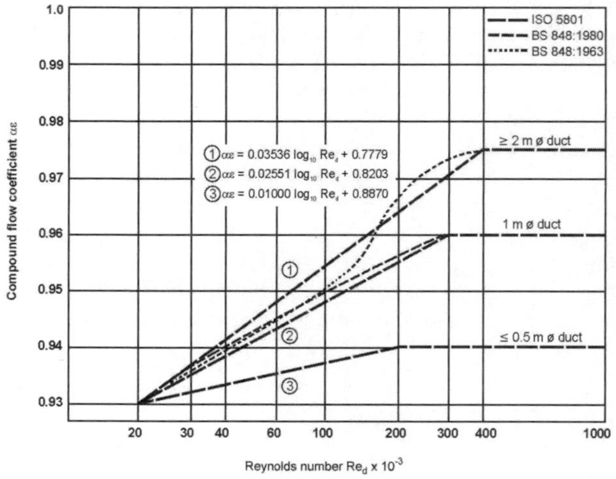

Figure 4.5 Compound flow coefficient of conical inlets

reduced to 0.96. A review of all the data collected by NEL and others suggested that the value should be diameter related and that boundary layer effects are present (Figure 4.5). Calibration of the inlet was always allowed and will continue. For those who practised this, no changes were therefore necessary.

However it is noteworthy that a number of companies have not changed the data in their catalogues for many years. With the various changes in Standards, in some cases dating back to 1952, these cannot be correct.

Pitôt-static tube traverses are permitted without calibration although these are now restricted to cylindrical ducts to minimize the uncertainty. The four major types - NPL modified ellipsoidal, CETIAT, AMCA and AVA may all be used at the low angles of pitch and yaw without correction.

iii) Fan pressure is defined as the difference in stagnation pressures at fan inlet and outlet, see equation 4.1. Below about 2.0 kPa this is virtually the same as the previously defined Fan Total Pressure. For the ventilation and air conditioning industry, therefore, no problems arise, although it will be noted that there will be less emphasis on fan static pressure.

$$p_F = p_{sg2} - p_{sg1}$$

$$p_{sgx} = p_x \left[1 + \frac{\kappa - 1}{2} M_{ax2} \right]^{\frac{\kappa}{\kappa - 1}} \qquad \text{Equ 4.1}$$

iv) It introduces the concept of "common parts" of the ducting adjacent to the fan inlet and/or outlet sufficient to ensure an accurate and consistent determination of fan pressure no matter what method of flow measurement or control is used. The dimensions of these parts have been specified such that the duct area must be closely matched to the fan

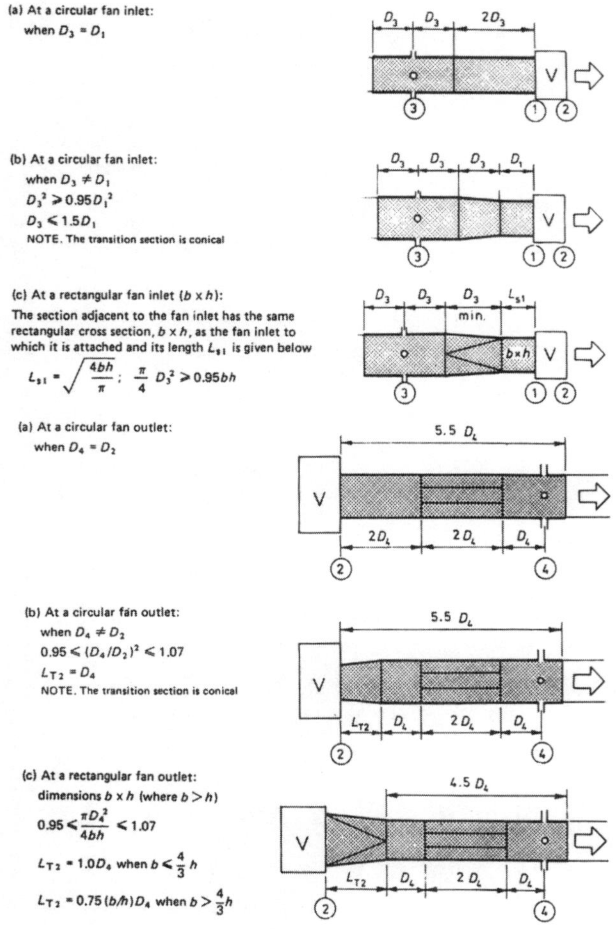

Figure 4.6 Common parts for ducting on fan inlet and outlet

inlet/outlet area as relevant, whilst their length is generally longer than those previously used. (Figure 4.6).

v) It specifies the use of a "conditioner" on the outlet of installation type B or D fans. This is designed to dissipate any swirl energy, which is not normally available for overcoming the system resistance.

vi) It defines the inlet and outlet areas of the fan as the gross areas inside the casing at the appropriate plane.

vii) Site testing is considered of sufficient importance to be transferred to a separate document (ISO 5802). The traversing techniques for a variety of duct cross-sections are detailed.

ISO considered it illogical and unacceptable for different fan types of the same installation category to need different test methods because of the differing outlet flow. Thus the necessity to devise an outlet simulation, which had the combined requirements of conditioning the flow to permit worthwhile measurements without severely hampering the fan by excessive pressure losses. These losses were likely to be an important part of the fan pressure determination and would be calculated on the basis of straight, fully developed flow.

Then arose another requirement for the common part — to match its actual increase in pressure loss in the presence of non-uniform and swirling flow to that corresponding to a long, straight uniform duct. This was considered a fair requirement, which would neither unduly penalize nor benefit a fan with such an outlet flow. Unfortunately politics intervened and some exceptions to this desirable situation continue to be permitted.

Compared with the use of the simulation, or common part, it can be stated that, in the presence of non-uniform and swirling flow from the fan outlet:

a) a short length of duct benefits the fan

b) a multi-cell straightener, as used in outlet side testing in other national Standards, tends to penalize the fan unduly.

4.1.4 Common parts of ducting

Satisfactory measurements of pressure cannot be taken immediately adjacent to the fan inlet or outlet and it is necessary to establish test stations some distance away, where the flow can be normalized.

The quantity measured at these stations is the static pressure, to which is added some conventional velocity pressure to obtain the effective total pressure. Oversized ducts can enhance fan performance whilst insufficient length can also result in inaccurate measurements of fan pressures. The common parts include a duct on the outlet side of the fan, having a length of five equivalent diameters to the pressure measuring point and incorporating a standardised flow straightener. Without such parts, different values of pressure can result according to the character of the airflow at the fan outlet. The velocity distribution at this point often contains considerable swirl. Even when free from swirl it is far from uniform. This results in an excess of kinetic energy or velocity pressure over the conventional allowance of $\frac{1}{2}\rho v^2$ caused by the proportionality of kinetic energy to the local value of rv^3 (mass flow x velocity pressure) so that the excess where v is high exceeds the deficit where v is low.

Now the non-uniformity of the axial velocity components diminishes as the flow proceeds down the duct and the excess energy reaches a minimum of a few percent of $\frac{1}{2}\rho v^2$ within a length equal to two or three duct diameters, but full uniformity is not reached until about 4.5 diameters, (Figure 4.7). Part of the original excess is lost, but part is converted into additional static pressure, the conventional velocity pressure remaining constant. This addition is available for overcoming external resistance, and in order to credit it to the fan, as it should be for type B and type D installations, it has been determined that the test station for outlet side pressure measurement should be more than five duct diameters from the outlet (Figure 4.8).

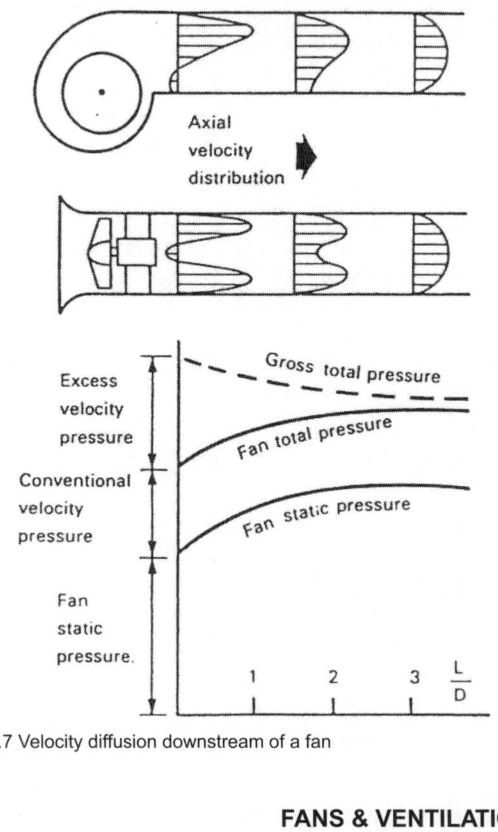

Figure 4.7 Velocity diffusion downstream of a fan

Country of Origin	Test Coded	Date	Ducted Outlet Simulation	Straightener	Figure No.	Comments
United Kingdom	BS 848	1980 1963	ISO common parts Duct	Etoile (8 radial vanes)	12-15 16	Equates with ISO 5801 within limited of uncertainty Fan will "benefit" compared with ISO 5801 if appreciable swirl is present.
Untied States	AMCA 210	1985	2D or 3D if test on inlet. Duct + straightener if tested on outlet	Multi-cell	17	Fan will benefit if inlet test method chosen. May be penalized if outlet method chosen – especially if velocity profile is poor and swirl is present.
France	AFNOR NFX10200	1971 1986	Straightener + diverging duct Outlet common part including straightener + diverging duct	Croisillon (vanes) Etoile (18 vanes)	18 19	Pressure may be overstated due to reduced number of straightener vanes and also because pressure is not measured at fan outlet area. Provided pressure is measured in common part, will equate with ISO 5801 within limits of uncertainty.
Germany	DIN 24163	1985	Duct	-	20	Fan benefits when there is swirl. Regretfully ISO recommendations have not been incorporated despite its recent date.
Italy	UNI 7179-73P	1973	Duct + straightener	Multi-cell	21	Outlet tests may be optimistic, due to increased duct size allowed where pressure is measured. This is partially offset by increased resistance of straightener

Table 4.1 A comparison of national Standards

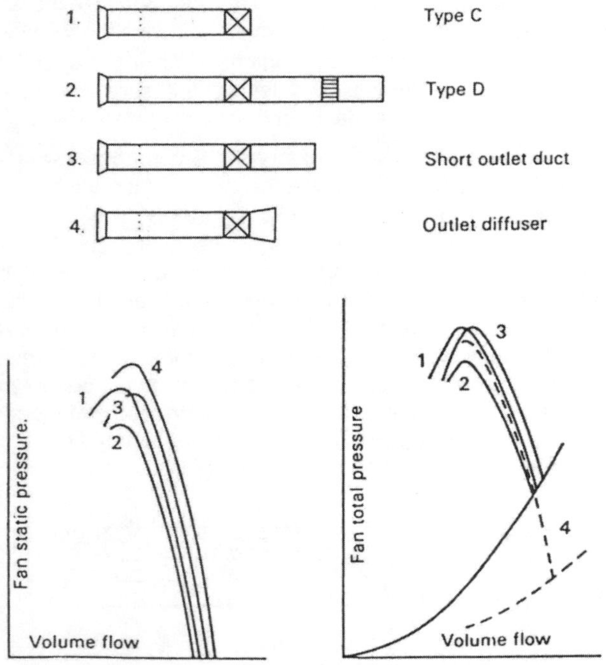

Figure 4.8 Fan characteristic with outlet swirl

A transition section may be used to accommodate a difference of area and/or shape but to minimize the effects of any change in aerodynamic impedance, it is specified that the duct area shall be within the limits of 5% less and 7% more than the fan discharge area. The dimensions of the transition are also specified to give a small valley angle.

4.1.5 National Standard comparisons

Figures 4.9 and 4.10 show the requirements in BS 848:1980 and ISO 5801 for the outlet duct simulation. Bearing in mind the difficulties concerning fans with non-uniform and swirling flow at the outlet, the effect of using various national Codes for testing such fans for installation categories B and D compared with ISO 5801 is shown in Table 4.1.

Common parts on the fan inlet are shorter and the pressure measurement station need be only three equivalent duct diameters from the fan inlet. This reflects the more regularized conditions, which apply on this side. For the same reasons, in an accelerating flow, a greater deviation in the upper limit of duct diameter is permitted. The lower limit is set at 5% less area of

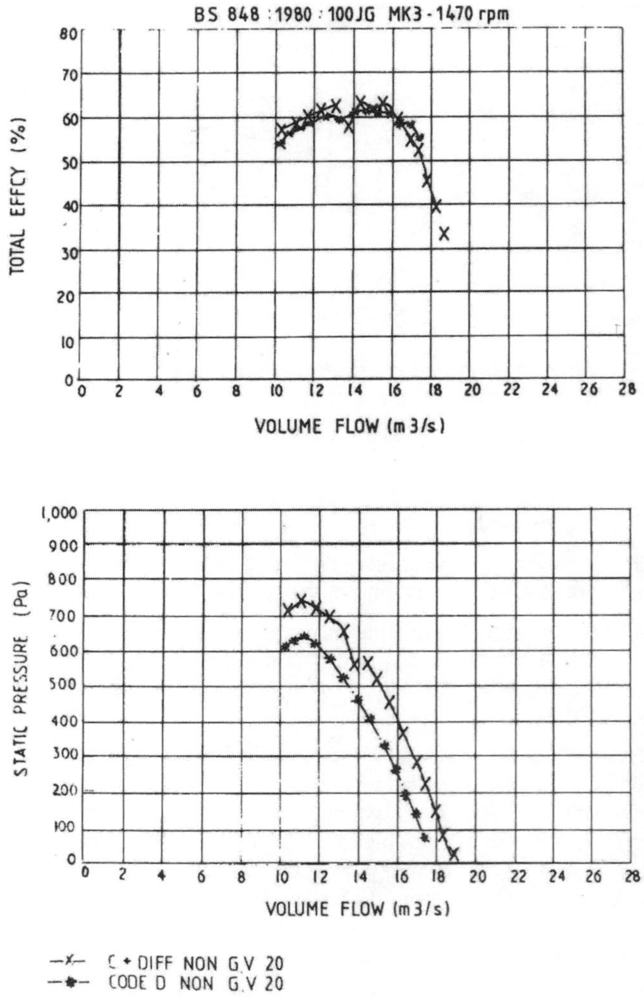

Figure 4.9 Effect of outlet connections on low pitch angle performance

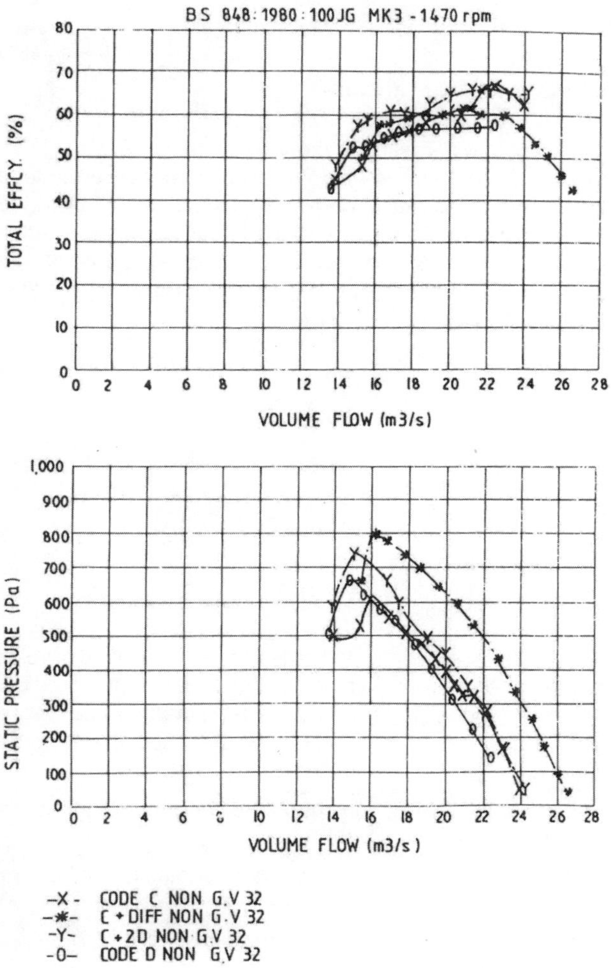

Figure 4.10 Effect of outlet connections on high pitch angle performance

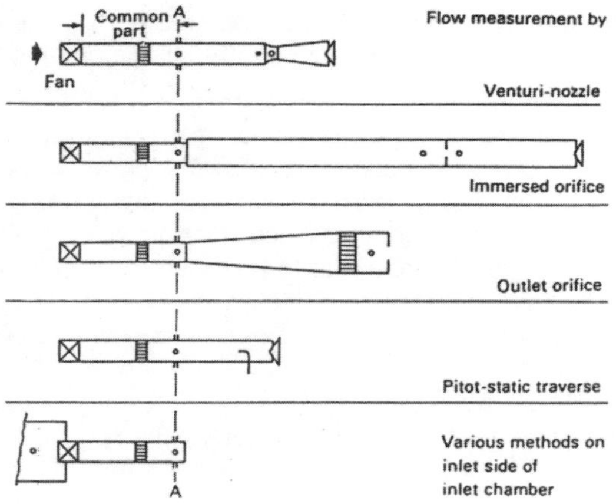

Figure 4.11 Principle of the common parts applied to type B test airways

duct to fan inlet. Again the transition angles are specified to minimize the effects of flow separation. The principle of the common parts applied to type B test airways is shown in Figure 4.11.

4.1.6 Flow conditioners

The swirl energy at the fan outlet is only recovered in a straight uniform duct if more than about 100 diameters long. In the presence of swirl, simple measurements of effective pressure or volume flow are impossible, and it must, therefore, be removed when tests are to be taken in a duct on the outlet side of the fan, to give information on performance. An effective flow straightener or conditioner will do this. If it removed just the swirl energy and no more, the minimum energy convention would be satisfied. However, the energy actually removed is very dependent on the combination of swirl pattern and straightener. Again, the need for an agreed standard outlet duct will be appreciated.

In practice, a fan with a lot of outlet swirl ought not to be selected for use with a long straight outlet side duct, because the friction loss in the latter will be substantially increased. Guide vanes should be fitted which will remove and recover (instead of removing and destroying) the swirl energy. The flow straightener will then just ensure that test conditions are satisfactory in the downstream duct: the relatively small outlet swirl components from centrifugal, guide-vane axial or contra-rotating fans will be removed without measurable disturbance to the performance.

The actual design of straighteners to be used in the standardised test ducts is therefore of great importance. It is appropriate to review the two types which were considered, and which are also used in ISO 7194.

a) The AMCA straightener is used only to prevent the growth of swirl in a normally axial flow, and does not improve asymmetric velocity distributions. It consists of a nest of equal cells of square cross-section and has a very low-pressure loss. Typical use is either side of an auxiliary booster fan where this is necessary to overcome the resistance of the airway when a complete fan characteristic is required. It is especially preferred adjacent to a flow-measuring device. This type of straightener is illustrated in Figure 4.12

b) The Etoile straightener is again designed to eliminate swirl and is of little use in the equalization of asymmetric velocity distributions. The eight radial vanes should be of sufficient thickness to provide adequate strength but should not exceed 0.007 D_4 for pressure loss considerations. This straightener has a similar pressure drop to the AMCA straightener, i.e., approximately 0.25 times the approach velocity pressure, but is also easier to manufacture. More importantly, it allows the static pressure to equalize radially as the air flows through it. This is not the case with the AMCA straightener, which can produce variations in the

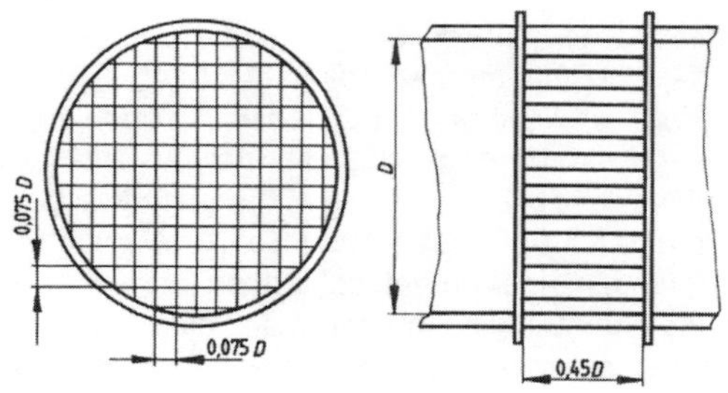

Figure 4.12 AMCA multi-cell straightener

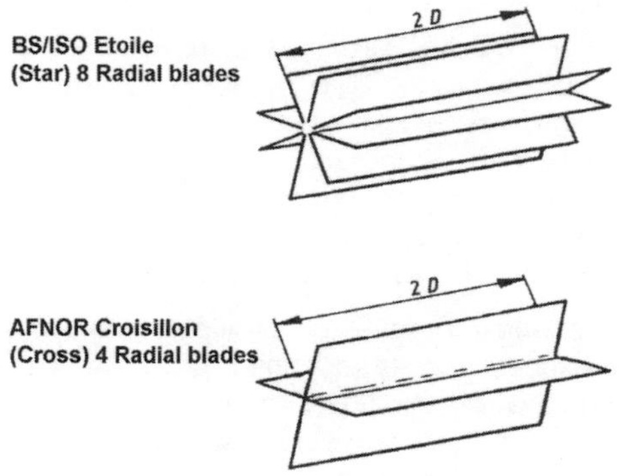

Figure 4.13 Etoile and Croisillon straighteners

static pressure across the duct downstream. The Etoile straightener is therefore preferred in the common duct on the fan outlet and is shown in Figure 4.13.

It should be noted that well designed centrifugal fans or axial and mixed flow fans with efficient outlet guide vanes will not be penalized at design duty by the incorporation of flow conditioners in the proposed test ducting. However, an axial flow fan without outlet guide vanes will be penalized by the 1980 Standard up to as much as 13 points on peak efficiency and over 20% on pressure. Centrifugal fans with poor outlet velocity profiles may also suffer. When operating away from the best efficiency point i.e. "off-design", residual swirl may be present in all types of axially ducted fans, such that the straightener will reduce the pressure developed.

4.2 Laboratory Standards

Various other Standards bodies throughout the world have also published fan test Codes. These are not necessarily of the same technical merit. In a global economy, where fans of various nations and manufacturers compete regularly, this can present problems where comparisons have to be made. It is almost impossible to make such comparisons where technical catalogues present data from tests to different national Codes.

In 1997 the first international Standard was published – ISO 5801 – and it is strongly recommended that this should be used in all competitive situations, both in customers, specifications and for acceptance tests. Even this is not enough. Many alternative methods are detailed in the Standard, which may give slightly different results. Preferably the same method should also be used for an appropriate installation category (See Section 4.4)

4.3 Determining the performance of fans in-situ

4.3.1 Introduction

The need to revise existing national methods of measuring the aerodynamic performance of fans under site conditions has been felt for some time. Hence, early in the life of ISO Technical Committee TC117, work commenced on a "stand-alone" document. Again the time for preparation has been extremely long, but compromises have been reached which enabled Standard ISO 5802 to be published. This is largely an amalgam of the French AFNOR X10-201 for siting of the velocity-area measuring points and BS 848:Part 1:1980 Section 3 relating to pressure, calculation, instrument calibration and uncertainties. Thus all the commonly encountered airway cross-sections are addressed together with relevant velocity-area methods.

4.3.2 Performance ratings

Catalogue rating tables and performance curves are produced from tests carried out according to the procedures specified for standardised airway conditions. In actual systems, however, it is rare for fans to be installed exactly reproducing those specified in the laboratory Standard. It will be remembered that ISO 5801 specifies "common parts" both upstream and downstream of the fan. These ensure a fully developed, swirl free and symmetrical velocity profile presented to the inlet.

The fan is enabled to develop its full potential and also to recover the excess velocity (dynamic) pressure at the fan discharge and convert it into useful static (potential) pressure. At the same time any useless residual swirl is removed. For these reasons, it is likely that the site performance will be degraded when compared with a laboratory test in a standardised airway. The magnitude of the difference may be considered as an indication of the quality of the system design.

4.3.3 Measuring stations

A major problem of testing in the field is the difficulty of finding suitable locations for making accurate measurements of flowrate and pressure. Wherever possible, the system designer should consider the provision of a suitable measurement station before manufacture. If this is not possible then temporary or permanent alterations to the ducting may be necessary to improve the accuracy of the test.

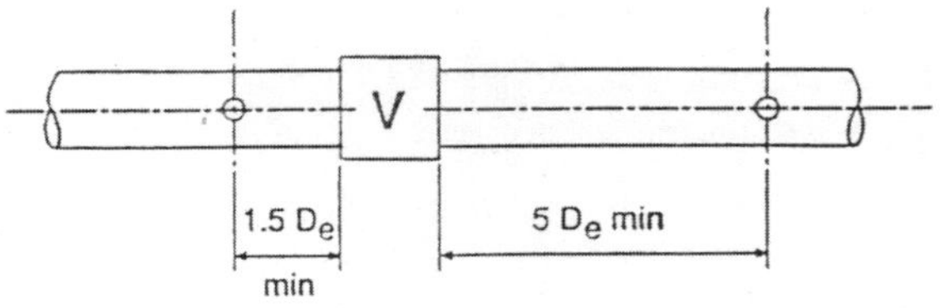

Figure 4.14 Location of pressure measurement planes for site testing

Most field tests will need to be carried out by some kind of velocity-area method using either pitôt-static tubes or anemometers. A traverse plane suitable for the measurements necessary to determine flowrate (Figure 4.14), would have the following attributes:

a) the velocity distribution should be uniform throughout the traverse plane
b) the flow streams should be at right angles to the traverse plane
c) the cross-sectional shape of the airway in which the traverse plane is located should be regular
d) the cross-sectional shape and area of the airway should be constant for some distance both upstream and downstream of the traverse plane
e) the traverse plane should be located to minimize the effects of leaks between the traverse plane and the fan.

A location at least five equivalent diameters downstream of the fan in a long straight uniform cross-section duct would provide ideal conditions for a pitôt traverse assuming a vane axial or centrifugal unit. For a tube axial a location upstream would be preferable to obviate the errors resulting from swirl. In all cases where the traverse plane has to be close to the fan, an upstream location is preferred. This will give a more acceptable velocity profile from symmetry, fullness and swirl-free points of view. It will also minimize the effects of leakage. In some installations with parallel flow paths it may be necessary to use more than one traverse plane and add their results.

4.3.4 Flowrate measurements

The Standard includes recommendations for the number and distribution of measurement points in the traverse plane when a velocity-area method is used. For circular ducts the measuring points are spread over a minimum of three diameters with at least three points per radius. The positioning may be to either log-Tchebycheff or log-linear rules (Figure 4.15).

Similar information is given for annular, rectangular (Figure 4.16) and other common regular shapes. Rules are also included for duct cross-sections, which do not correspond closely to any of the standard shapes.

Since the flow at a traverse plane is never absolutely steady, the velocity pressure measurements indicated by a pitôt-tube/manometer combination will fluctuate. Each measurement will, therefore, need to be averaged on a time-weighted

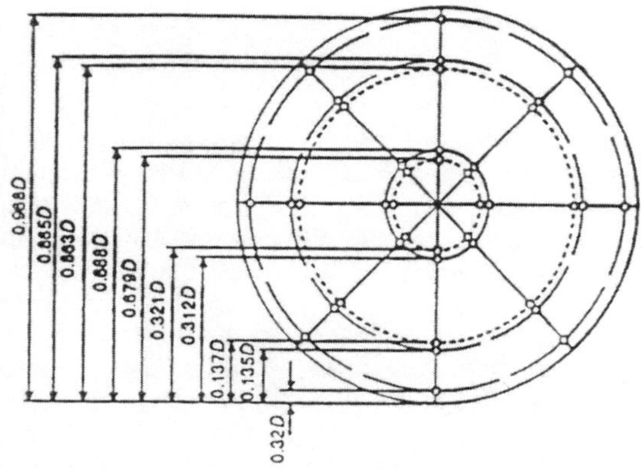

Figure 4.15 Siting of measuring points in a circular section with four diameters and three measuring points per radius

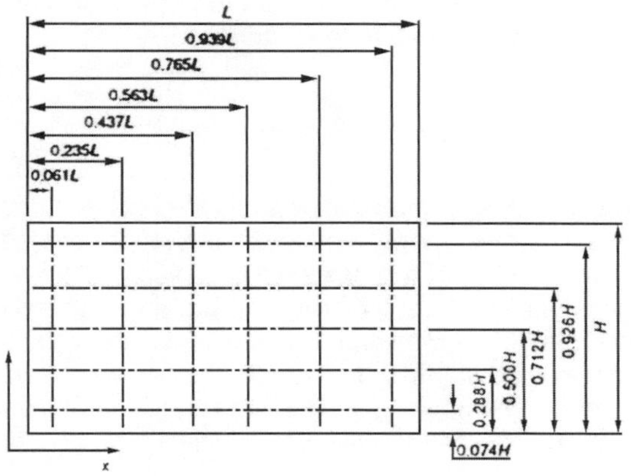

Figure 4.16 Rectangular section with six cross-lines and five measuring points per cross-line

basis. The four designs of pitôt-tube permitted in ISO 5801 are all considered primary instruments and may be used without calibration provided they are in good condition. They do not all have the same insensitivity to pitch and yaw. ISO 3966 and ISO 7194 indicate likely errors for each type under non-normal flow. The modified ellipsoidal head of the NPL design is preferred as it is the least sensitive to misalignment.

4.3.5 Pressure measurements

Care must be taken to ensure that static pressure measurements on both the inlet and outlet of the fan are taken relative to atmospheric pressure or to that existing within a common test enclosure. Under reasonably uniform flow, free from swirl and separation, four interconnected wall tappings may be used (Figures 4.17 and 4.18). As with ISO 5801, Fan Pressure is defined as the difference in stagnation pressures at fan outlet and inlet. At pressures less than about 2.0 kPa, this is virtually the same as the previously defined Fan Total Pressure.

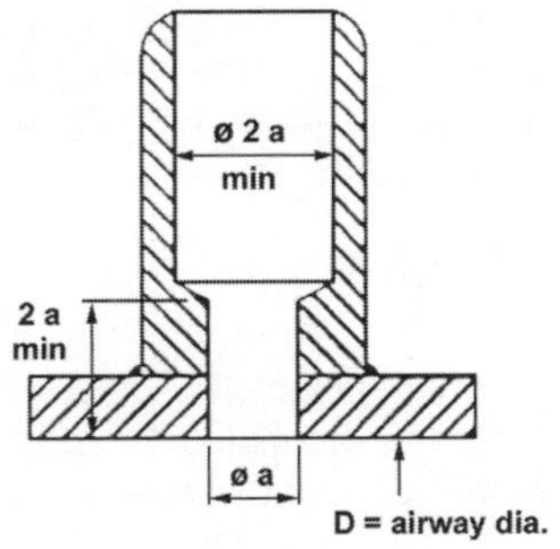

Figure 4.17 Construction of wall pressure tappings Note: a to be not less than 1.5 mm nor greater than 10 mm and not greater than 0.1 D

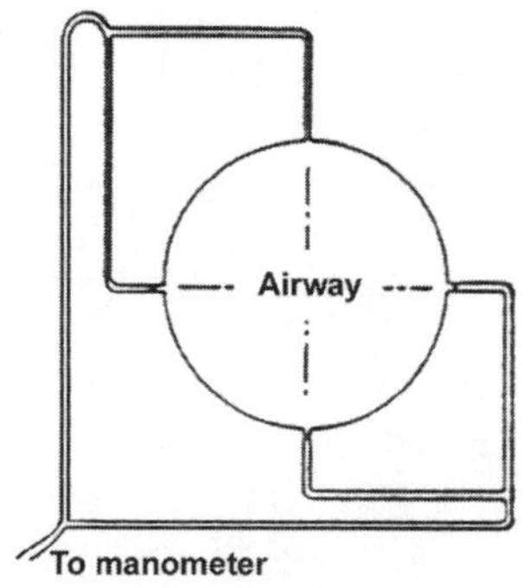

Figure 4.18 Tapping connections to obtain to obtain average static pressure in circular airway (Shown interconnected to single manometer)

4.3.6 Power measurements

The drive shaft power may be determined either directly through a torque meter or deduced from the electrical power input to the motor terminals and using the summation of losses method.

4.4 Installation category

The differences in fan performance according to installation category are as much a function of the fan type and design, as

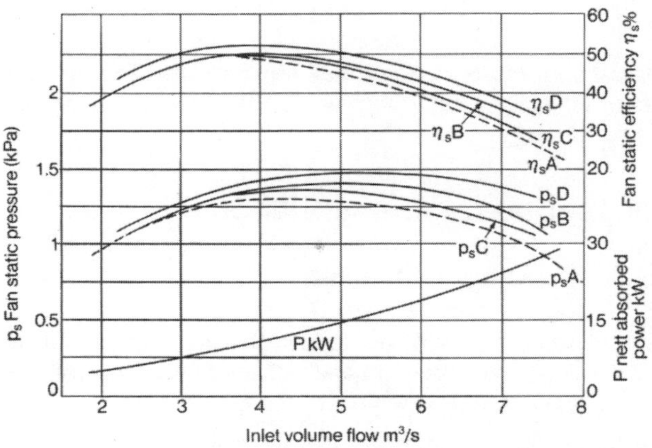

Figure 4.19 Typical performance curves for a forward curved centrifugal fan to different installation categories

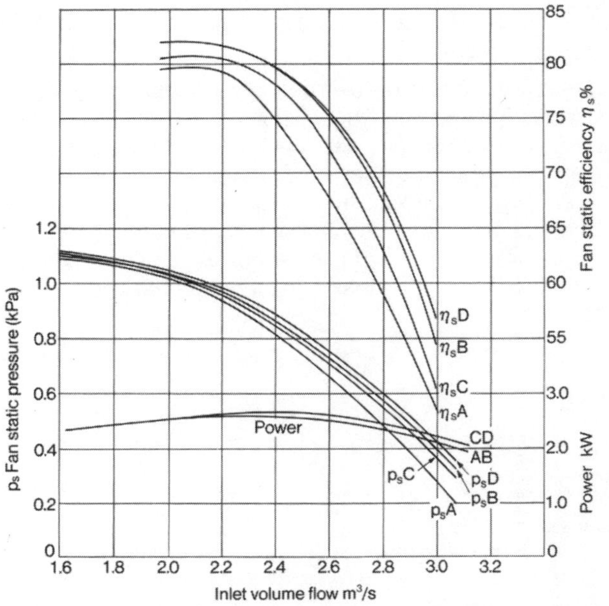

Figure 4.20 Typical performance curves for a backward inclined centrifugal fan to different installation categories

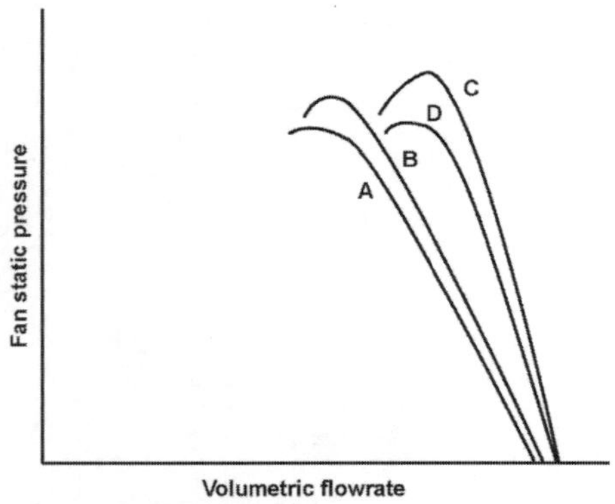

Figure 4.21 Typical performance curves for a tube axial fan to different installation categories

of the position of duty point on the particular characteristic curve. In practice the type B and type D characteristics for most fan types will be nearly the same at the best efficiency point, provided the fan is supplied, in its free inlet form, with a properly shaped entry cone or bellmouth. With the same proviso, type A performance will coincide with type C.

The essential difference remaining is that between free outlet and ducted outlet performance, which is significant for fans of all kinds though it diminishes as the ratio of fan velocity pressure to fan total pressure falls. It will also be affected by tongue pieces in a centrifugal fan outlet. In the latter, a length of ducting is desirable to enable some recovery of dynamic pressure to useful static pressure to be achieved from the distorted velocity profile.

Typical performance curves for a forward curved centrifugal, a backward inclined centrifugal and a tube axial fan are shown in Figures 4.19, 4.20 and 4.21.

4.5 Testing recommendations

4.5.1 Laboratory test stands

Tests for rating should be carried out on a duct system, with flow and pressure measurement and with instrumentation all meeting the requirements of ISO 5801.

It is not proposed to detail all the alternative set-ups, as there are a considerable number of these. The Standard totals 232 pages and has given the author many happy (?) hours of reading. Suffice it to say that the requirements are detailed and must be followed closely. However a typical duct arrangement is shown in Figure 4.22.

If a fan is provided with its own bearings it should be tested after a sufficiently extended "run-in" period. The inlet and outlet should be away from all walls. Free space should be sufficient to permit air to enter or leave the fan without setting up an unmeasurable resistance. The laboratory should be of sufficient volume to ensure that it is free from any air currents that could affect the performance. If it is necessary to discharge the air into another room, then make-up air will be needed.

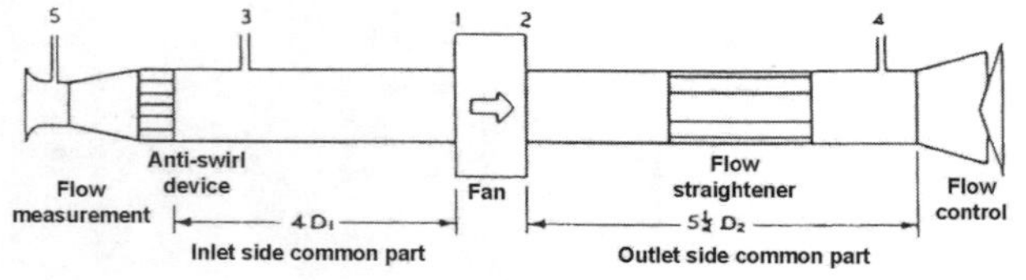

Figure 4.22 Typical example of a standardised test airway

4.5.2 Field tests

The use of standardised laboratory test stands in the field is usually impossible. Long lengths of straight ducting to "calm" the flow are rarely feasible whilst permanently installed flow measuring devices such as orifice plates, venturis etc., will have too high a pressure loss. All these lead to higher than necessary absorbed power.

Whenever the real installation differs from the idealized (and recommended) laboratory arrangement there will be a loss of fan performance due to the effects of swirl and/or distorted undeveloped velocity profiles. This is especially true where there are duct bends directly on the fan inlet and/or outlets. It is recommended to read AMCA 201 or *The Fan and Ductwork Installation Guide*, published by FMA (Fan Manufacturers Association). Both of these give information on how to calculate the magnitude of likely performance reduction.

4.5.3 Measuring flowrate

Fan flowrate can be expressed as either the volumetric flowrate in m^3/s or the mass flowrate in kg/s. If a laboratory test is to comply with ISO 5801 it is essential that readings are taken at the prescribed measuring planes and are downstream of any flow straightening device and at a sufficient distance to ensure flow calming. Many flow measuring devices are permissible within the Code e.g. orifice plates, inlet cones, venturi meters, multi-nozzles etc. All are valid provided the correct coefficients of discharge are used.

Pitot static tube traverses are permitted, but these are perhaps more dependent on operator skill. They are however often the only method possible on site. All types of pitôt head are permitted, but the writer would recommend the NPL modified ellipsoidal type, which is less susceptible to pitch or yaw errors.

4.5.4 Measuring fan pressure

Fan pressure is defined as the stagnation pressure at outlet minus the stagnation pressure at inlet. Up to about 2.0 kPa this is virtually the same as Fan Total Pressure. Care should be taken to ensure that the appropriate value is specified i.e. "total" or "static". This may depend on the data used for calculating the system pressure and therefore whether "velocity" pressure is included.

4.5.5 Measuring air density

Fan performance is a function of the air (or gas) density handled by the fan. It is therefore necessary to take such measurements of wet and dry bulb temperature, barometric pressure and even perhaps chemical composition so that the density may be calculated. It should be noted that standard air density is assumed to be 1.2 kg/m^3. This equates to dry air at 20°C and 101.325 kPa or to moist air at 16°C and 100 kPa and 50% RH, but these properties are not part of the definition.

4.5.6 Measuring fan speed

Rotational speed can be measured by various types of tachometer. A good accuracy is essential as fan performance is very sensitive to even small variations in speed. The fan laws (see Section 4.6) show that flowrate varies directly as the speed, pressure as the square of the speed and absorbed power as the cube of the speed.

4.5.7 Measuring absorbed power

Various prime movers can be used to drive a fan, but more than 99% are electric motor driven. To obtain good figures for absorbed power, it is necessary to at least use a calibrated motor where input volts and amperes can determine the output power. The so-called two-wattmeter method may also be used. For the highest accuracy, however, it is essential to use a dynamometer or torque meter.

4.5.8 Calibration and uncertainties

Instruments used for a fan test should be calibrated frequently and this calibration should be traceable back to National / International Standards. There will be uncertainties associated with any calibration correction and the measured quantities may have a random error, which may be superimposed on a systematic error. If measurements are repeated over a sufficient period of time then it may be possible to obtain the magnitude of the systematic error.

4.5.9 Test results

The results of a fan test should be expressed in terms of volumetric flowrate against fan pressure at a constant rotational speed. Fan absorbed power and fan efficiency may also be given. Inlet air or gas density is also essential.

A fan characteristic curve may be plotted for either the duty range or the full curve from SND (fully closed) to FIO (fully open).

4.6 Fan Laws

4.6.1 Introduction

It may seem like heresy to many fan engineers to question the validity of the so-called "Fan Laws". They are in fact approximations albeit, in many well defined situations, very close approximations. As they are so widely used without query or comment, it seems appropriate to look at their derivation.

When considering the performance of a series of fans, it is apparent that they can be made in a geometrically similar range of sizes and that they can be run at an infinite number of rotational speeds. They can also handle gases or air having varying physical properties — temperature, humidity, density, viscosity, and specific heats. For the manufacturer to test under all these varying conditions would be impossible and it is therefore desirable to be able to predict the performance of one fan in a series from tests made on another, perhaps with a variation also in speed and gas conditions.

4.6.2 The concept of fan similarity

To develop the Fan Laws requires that we appreciate the concept of similarity and recognize its limitations.

In geometry, we are aware that similar triangles have equal angles and the lengths of sides are in proportion. From this we are able to develop three complementary types of similarity:

- **Geometric similarity** in which two units have length dimensions in a constant ratio throughout and equivalent angles are equal.
- **Kinematic similarity** in which the dimension of time is added to length and all peripheral flow velocities at any point within a machine are in a constant ratio to the velocities at corresponding points of the similar unit.
- **Dynamic similarity** in which acceleration is introduced and the forces at corresponding points in the two machines also bear a constant relationship.

Whilst it might be thought that geometric similarity would be easy to achieve, it should be remembered that if strict adherence is necessary then this would require that metal thicknesses would have to be proportional, along with clearances, weld dimensions, fasteners etc. The exigencies of manufacturing methods and the commercial availability of the required elements dictate that this cannot be the case.

Surface roughness would also need to be proportional with size. Sheet metal roughness is almost constant over a range of thicknesses whilst welding protuberances etc., may well be a function of operator skill and quality control. Shaft diameters and the scantlings of impellers and other items are determined by the mechanical loads imposed such as centrifugal stresses, critical speeds, and fatigue stresses. This may result in the dimensions of such rotating parts diverging from those calculated by strict geometrical similarity.

Fortunately the effect of these differences is usually small and can be ignored in all but the most extreme cases. The relative

Critical dimensions	%
Impeller	
Blade tip diameter	± 0.25
Blade heel diameter	± 0.25
Blade chord & width	± 0.2
Blade profile (deviation from template)	± 0.2
Rim inlet diameter - formed Rim inlet diameter - machined	± 1.0 ± 1.0
Rim inlet curvature (deviation from template)	± 1.0
Peripheral run-out	± 1.0
Inlet	
Throat curvature (deviation from template)	± 1.0
Inlet/impeller rim clearance when running*	± 20.0
Inlet/Impeller setting when running*	± 10.0
Housing, inlet box(es), and all accessories	± 0.4
* Expressed as a percentage of actual clearances	

Table 4.2 Permissible divergences from strict geometrical similarity for a centrifugal fan

Critical dimensions	Pitch design	
	%	%
Impeller	**Fixed**	**Variable**
Blade tip diameter	± 0.25	+ 0.125 - 0.25
Hub diameter	± 0.375	± 0.125
Blade chord length	± 0.1	± 0.1
Blade profile	± 0.1	± 0.1
Blade angle of twist	+ 2.0 °	± 1.5 °
Blade angular setting	± 0.1 °	± 0.5 °
Blade tip clearance when running*	± 20.0	± 20.0
Casing		
Impeller casing	± 0.2	± 0.2
Inlet box, inlet bell and discharge casing	± 0.4	± 0.4
Angular setting guide vanes	± 2.0 °	± 2.0 °
Axial setting of guide vanes	± 0.2	± 0.2
Accessories	± 0.4	± 0.4
* Expressed as a percentage of actual clearances		

Table 4.3 Permissible divergences from strict geometrical similarity for an axial fan

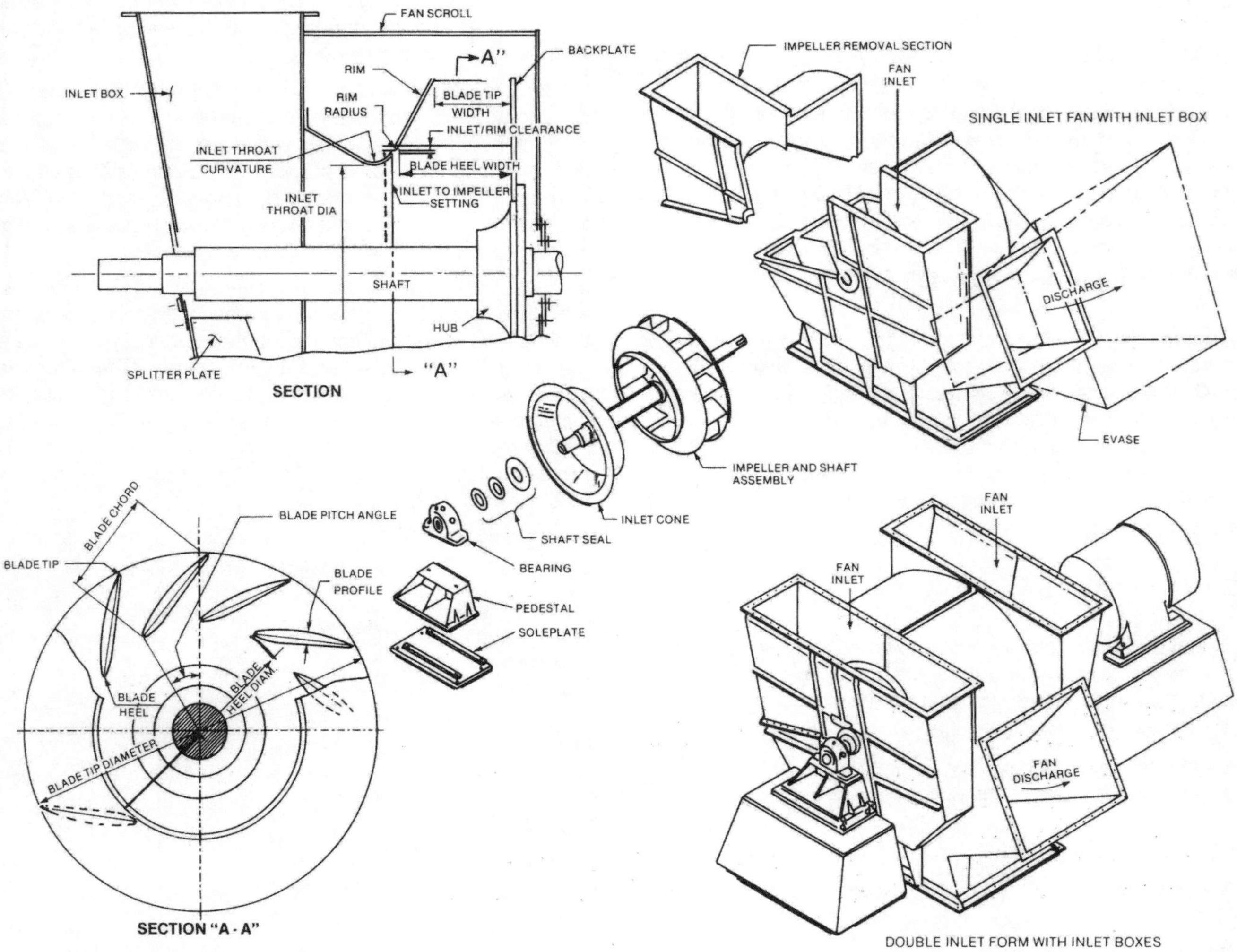

Figure 4.23 Terminology and critical dimensions for centrifugal fans

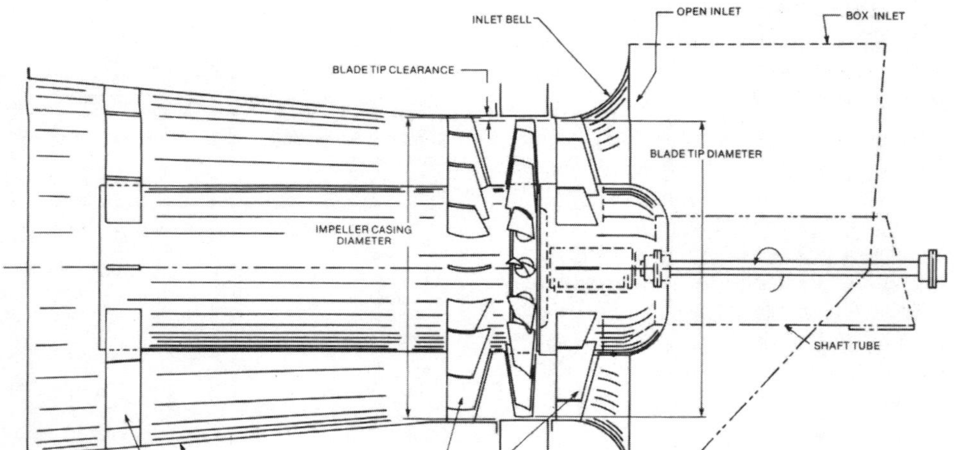

Figure 4.24 Terminology and critical dimensions for axial fans

clearances between different parts of the fan can also vary but these may be of great importance and should be eliminated by both careful design and by quality control at the manufacturing stage.

Figures 4.23 and 4.24 give the terminology and show those dimensions which are critical. These, together with Tables 4.2 and 4.3 have been abstracted from AMCA 802. They give recommendations for maximum divergences of these critical dimensions from strict geometrical similarity without invalidating the "Fan Laws" used in performance prediction, within the stated uncertainties of the method.

One of the requirements of dynamic similarity is that Reynolds numbers be equal at all corresponding points in the two fans - model and predicted. Differing cross-sectional areas within the impeller blade passages and into and out of the casing, dictate that Reynolds number vary considerably. It is, therefore, both customary and convenient to refer to a single arbitrary figure based on the impeller tip diameter D and the peripheral velocity at this point πND together with the air or gas properties at the fan inlet - mass density ρ and viscosity μ.

Thus fan Reynolds number

$$Re_F = \frac{\rho}{\mu}\pi ND^2$$

Changes in Re_F can be the result of varying N or D or both. By altering only N, any size effects that might accompany a change of D can be eliminated. Tests by Phelan suggest that there is a threshold limit for Re_F for each and every fan design below which increasing deviations from the fan aerodynamic laws occur.

The approximate threshold limits for various designs are given in Table 4.4. It will be noted that the lowest limiting value is for the paddle fan where, due to its simple design, flow is highly turbulent throughout the flow passages. More sophisticated designs have higher threshold values indicating that flow is in the transitional region, until speeds are reached at which most of the passages are hydraulically rough. Shock losses follow the Fan Laws and are independent of Reynolds number but are less with the increasingly efficient designs.

Type of Fan	Impeller design	Re_F Threshold Fan Reynolds number
Centrifugal	Radial	0.4×10^6
	Forward curved	0.8×10^6
	Backward inclined	1.0×10^6
	Backward curved	1.5×10^6
	Backward aerofoil	2.0×10^6
Mixed flow	Compound curvature	2.0×10^6
	Meridional acceleration	2.5×10^6
Axial	High hub/tip ratio	2.5×10^6
	Low hub/tip ratio	3.0×10^6

Table 4.4 Approximate threshold fan Reynolds numbers for different types of fan

For dynamic similarity Mach numbers in the test and predicted fan must be the same, which is unlikely unless they develop the same pressure. When operating at high pressures, above say 2.0 kPa, the air or gas may no longer be considered incompressible and a compressibility coefficient has to be introduced into the simplified form of the Fan Laws. This coefficient is a function of the polytropic exponent n and the absolute pressures at fan inlet and outlet.

The assumption of a polytropic process between the fan connections as defined by total pressures is in itself only an approximation of what actually happens inside the fan. It is, however, adequate for predictive purposes.

To simplify any analysis, it is again convenient to specify a single fan Mach number based on the peripheral velocity of the impeller blade tips when compared with the speed of sound C as defined by the air or gas density at the fan inlet. Thus:

$$Ma_F = \frac{\pi ND}{C} = \frac{\pi ND}{\sqrt{\gamma Rt}}$$

where

C	=	speed of sound (m/s)
R	=	gas constant (287 J/kg.°K)
t	=	absolute gas temperature (°K)

From compressibility effects, variations in Ma_F produce no deviation from the simple fan laws unless they approach a value of around 0.3.

This value may appear lower than anticipated, but it should be recognised may well indicate a local value within the blade passages approaching 1.0. Critical conditions can then develop resulting in a "choking" effect where there is a limitation on the flowrate. It is not usually a problem unless the blade passage is highly obstructed. Figure 4.25, also abstracted from AMCA 802 gives allowable variations in Ma_F.

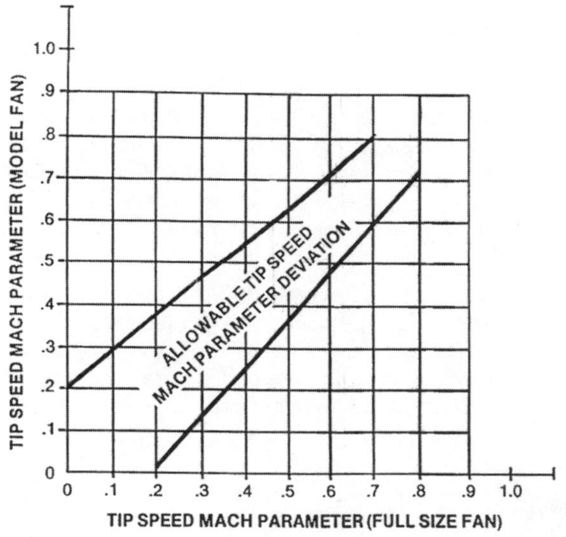

Figure 4.25 Allowable variations in fan Mach numbers

4.6.3 Dimensional analysis

The capacity of a fan "Q" is dependent on:

Capacity	Q (m^3/s)
Fan size	D (m)
Fan speed	N (rev/s)
Gas density	ρ (kg/m^3)
Gas viscosity	μ (Pa.s)

Thus:

$$Q \propto fn\,(D, N, \rho, \mu)$$

or

$$Q \propto D^a N^b \rho^c \mu^d$$

If we assign to each of the physical properties detailed above the fundamental units of mass M, length L and time T we then have:

$$L^3T^1 \propto fn\left(L, T^1, ML^{-3}ML^{-1}T^1\right)$$

or

$L^3 T^{-1} \propto L^a T^b M^c L^{-3c} M^d L^{-d} T^{-d}$

Equating indices we have:

for	M : 0	= c + d	or c = -d
for	L : 3	= a-3c-d	or a = d + 3c + 3
or	a	= d-3d + 3	or a = 3-2d
	T : -1	= -b-d	or b = 1-d

Thus:

$$Q \propto D^{3-2d} N^{1-d} \rho^{-d} \mu^{d}$$

or:

$$Q \propto ND^3 \left(\rho \frac{ND^2}{\mu}\right)^{-d}$$

The formula can be altered to $Q \propto ND^3 \left(\pi\rho \frac{ND^2}{\mu}\right)$ without affecting its validity as π is a constant, and if we note that πND = fan tip speed u then it will be seen that the term in brackets has the form $\rho \frac{uD}{\mu}$ i.e. some sort of Reynolds number.

This is a dimensionless quantity. For reasonable variations in this fan Reynolds number, its effects will be small. ISO 5801 requires that the test condition is within the range 0.7 to 1.4 times the fan Reynolds number for the specified duty.

Provided that these limits are met then:

$$Q \propto ND^3 \qquad \text{Equ 4.1}$$

It is anticipated that this "Law" would be accurate to at least the catalogue tolerances of ISO13348. In general if the test fan Reynolds number is lower than the specified fan Reynolds number, then the law will be pessimistic, whilst if the test number is higher than the duty number the results of the calculation will be optimistic.

At very "high" numbers (test and duty) i.e. above the so-called threshold number for a particular design (see Table 4.4), the effects may be ignored but the dangers of predicting the performance of a small and /or high-speed fan are apparent. These effects have been noted as being especially serious with high efficiency fans, e.g. aerofoil bladed centrifugals.

In like manner we can calculate the fan pressure (static or total).

The pressure of a fan p is dependent on the same quantities and thus :

$$p \propto fn\,(D, N, \rho, \mu)$$

or

$$p \propto D^a N^b \rho^c \mu^d$$

Pressure has the dimensions of force (mass x acceleration) per unit area and using dimensional analysis we have:

$$ML^{-1} T^{-2} \propto fn\,(L, T^{-1}, ML^{-3}, ML^{-1} T^{-1})$$

or

$$ML^{-1} T^{-2} \propto L^a T^{-b} M^c L^{-3c} M^d L^{-d} T^{-d}$$

Equating indices we have:

M : 1 = c + d   or c = 1-d

L : -1 = a-3c-d   or a = 3c + d-1

or

a = 3-3d + d-1   or a = 2-2d

T : -2 = -b-d   or b = 2-d

Thus:

$$p \propto D^{2-2d} N^{2-d} \rho^{1-d} \mu^{d} \qquad \text{Equ 4.2}$$

or:

$$p \propto N^2D^2 \rho \left(\rho \frac{ND^2}{\mu}\right)^{-d}$$

or:

$$p \propto N^2D^2 \rho \left(\pi\rho \frac{ND^2}{\mu}\right)^{-d}$$

Again the function in brackets is in the form of the fan Reynolds number and with the same provisos we may say that:

$$p \propto \rho N^2 D^2 \qquad \text{Equ 4.3}$$

The fan power absorbed W is proportional to Q x p and therefore:

$$P \propto ND^3 \times \rho N^2 D^2$$

or

$$P \propto \rho N^3 D^5 \qquad \text{Equ 4.4}$$

Note: Capital P is for power whilst small p is for pressure.

It must be emphasised that these simplified laws apply to a specific duty point of Q, p and P. As $P \propto Q \times p$, the efficiency of the unit will remain unchanged. When the fan is applied to a system we cannot change the speed N without altering *all* the quantities.

Just as fans have laws, which govern their behaviour, so have systems. The usual fan system consists of a number of fittings such as bends, grilles, transformation pieces, junctions, etc. Between these will be lengths of straight pipe or ducting.

The pressure loss in fittings, assuming a constant friction loss factor κ:

$\propto$ velocity pressure

$\propto v^2$

$\propto Q^2$ as $v = \frac{Q}{\text{cross-sectional area}}$

In like manner the pressure loss in straight ducting

$$\propto \frac{fLv^2}{m}$$

where:

f	=	friction factor
L	=	length of duct
v	=	air/gas velocity
m	=	mean hydraulic depth
		$\left(\frac{\text{cross-section area}}{\text{perimeter}} \text{ of duct}\right)$

Unfortunately the friction factor is **never** a constant over the complete fan characteristic. For many ventilation systems we are in the transitional zone between laminar and fully turbulent flow. The index for v may be nearer 1.8 even at the design flow rate. It will fall to 1.0 at zero flow. However, this would upset all those people who for years have been declaring that, on a given system, as $Q \propto v$, we may say that the loss in straight ducting and fittings is also $\propto Q^2$. Thus overall $p \propto Q^2$ and a system line may be plotted on the fan characteristic accordingly, see Figure 4.26. This is only strictly correct for flows varying by about 20% from design (see Chapter 5 and 6).

A change in fan speed alters the point of operation from A to B i.e. along the system curve. This is because, as previously shown in the Fan Laws, for a given fan and system $Q \propto N$, $p \propto N^2$ and therefore $p \propto Q^2$ for the fan as well, but only if f remains constant, or nearly so. It should be repeated that this system

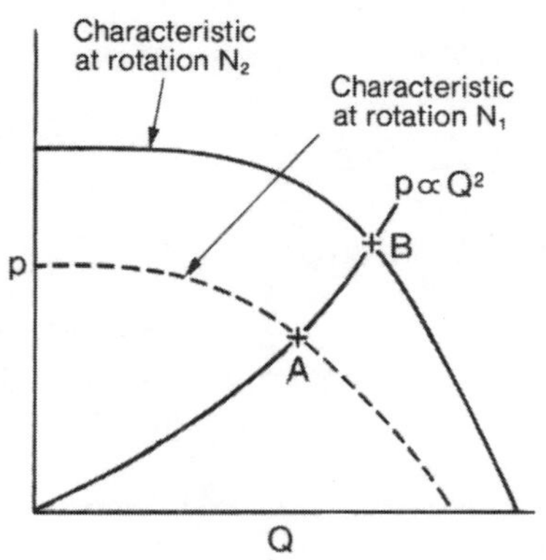

Figure 4.26 Fan and system characteristics

law is only valid for speed changes of about 20%. Over this value the divergence in the value of f becomes too great.

Thus if a fan is applied to a system and its speed is changed from N_1 to N_2.

$Q \propto N$ i.e. $$Q_2 = Q_1 \times \frac{N_2}{N_1} \quad \text{Equ 4.5}$$

$p \propto N^2$ $$p_2 = p_1 \times \left\{\frac{N_2}{N_1}\right\}^2 \quad \text{Equ 4.6}$$

$P \propto N^3$ $$P_2 = P_1 \times \left\{\frac{N_2}{N_1}\right\}^3 \quad \text{Equ 4.7}$$

An increase of 10% in fan rotational speed will therefore increase volume flow Q by 10%, pressure developed p by 21% but power absorbed P by 33%, assuming air/gas density is unchanged. Unless large motor margins over the absorbed power are available, therefore, the possibilities of increasing flow by speed increase are usually limited.

At the same speed and gas density, a fan of a different size, but geometrically similar, will have a performance as given below:

$Q \propto D^3$ i.e. $$Q_2 = Q_1 \times \left(\frac{D_2}{D_1}\right)^3 \quad \text{Equ 4.8}$$

$p \propto D^2$ i.e. $$p_2 = p_1 \times \left\{\frac{D_2}{D_1}\right\}^2 \quad \text{Equ 4.9}$$

$P \propto D^5$ i.e. $$P_2 = P_1 \times \left\{\frac{D_2}{D_1}\right\}^5 \quad \text{Equ 4.10}$$

In a range of fans to ISO 13351, where the size ratio averages 1.12, the approximate increase per size will therefore be 40% on capacity, 25% on pressure, and 76% on power.

At the same tip speed and gas density, N_1, D_2 will equal N_2D_2

now $$Q_2 = Q_1 \times \frac{N_2}{N_1} \times \left(\frac{D_2}{D_1}\right)^3$$

but then $$\frac{N_2}{N_1} = \left(\frac{D_1}{D_2}\right)$$

$\therefore$ $$Q_2 = Q_1 \times \left\{\frac{D_2}{D_1}\right\}^2 \quad \text{Equ 4.11}$$

also $$p_2 = p_1 \times \left\{\frac{N_2}{N_1}\right\}^2 \times \left(\frac{D_2}{D_1}\right)^2 \quad \text{Equ 4.12}$$

$\therefore$ $$p_2 = p_1$$

and $$P_2 = P_1 \times \left\{\frac{N_2}{N_1}\right\}^3 \times \left\{\frac{D_2}{D_1}\right\}^5 \quad \text{Equ 4.13}$$

$$P_2 = P_1 \times \left\{\frac{D_2}{D_1}\right\}^2$$

Thus in a series of fans sized to ISO 13351 (a Renard R20 series) at constant tip speed and gas density, the approximate increase per size will be 25% on both capacity and power for the same pressure. The speed will be reduced by 11 %.

In the above analysis, we have assumed that:

- The air is incompressible - a reasonably accurate assumption at fan pressures up to about 2.0 kPa - and that air / gas velocity triangles at inlet and outlet retain similarity after a speed change. As an alternative the change in k_p from test conditions to specified duty should not exceed ± 0.001.
- Velocities are substantially below the speed of sound and there are no Mach number effects
- $\frac{\text{fan tip speed}}{\text{velocity of sound}} < 0.25$, say (see Figure 4.21)
- Changes of Reynolds number are maintained within the limits shown.
- Relative roughness of fan parts remain unchanged with variation in size.

If all these effects were included in our dimensional analysis additional variables would be introduced and the mathematics complicated accordingly. The overall fan laws would then become:

$$Q \propto ND^3 \; (Re_F)^a (Ma_F)^b k_p^{\,c} \Delta^d \quad \text{Equ 4.14}$$

$$p \propto N^2D^2 \; (Re_F)^e (Ma_F)^F k_p^{\,g} \Delta^h \quad \text{Equ 4.15}$$

$$P \propto N^3D^5 \; (Re_F)^j (Ma_F)^k k_p^{\,1} \Delta^m \quad \text{Equ 4.16}$$

where:

Re_F fan Reynolds number = $\frac{\pi \rho N D^2}{\mu}$

Ma_F fan Mach number = $\frac{\pi N D}{\sqrt{\gamma R t}}$

$= \frac{\text{fan tip speed}}{\text{velocity of sound}}$

compressibility coefficient ≃

$$\frac{2 + 2z(r-1)}{2 + (z+1)(r-1)}$$

where:

$z = \frac{(\gamma - 1)P}{\gamma Qp}$

r = absolute pressure ratio across fan

γ = ratio of specific heats (1.4 for air)

R = gas constant (287 J/Kg. °K)

t = absolute gas temperature (°K)

Δ = relative roughness

$= \frac{\text{absolute roughness of component}}{\text{impeller diameter}}$

The calculation of r is dependent on whether the fan is ducted on the inlet and/or outlet.

The velocity of sound in air at sea level and 20°C (293°K) = 344 m/s.

Care must be taken to use N in rev/s in the calculation of fan Reynolds and Mach numbers.

Relative roughness should not normally be of interest except when predicting the performance of a very small fan from tests on a larger unit, or where impeller scantlings are varied substantially.

Further information on the above is given in a number of advanced textbooks, e.g *Cranfield Series on Turbomachinery*. It is important to note however that the exponents a, b, c, etc are peculiar to a given design of fan and probably a given duty point. Work is being carried out in many research establishments to establish them. Usually they only need to be known when it is important to achieve the duty within very close tolerances i.e. within 2%.

Approximate Reynolds numbers and absolute roughness effects are typically combined in manufacturers data. Those for a medium pressure backward inclined centrifugal fan are shown in Figure 4.27. The effect of fan Reynolds numbers on the peak static efficiency is shown in Figures 4.28.

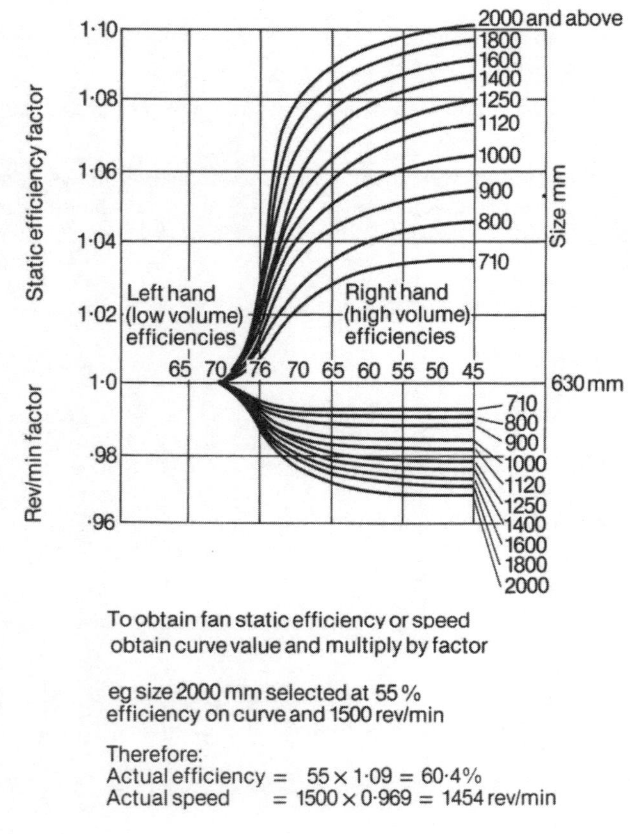

Figure 4.27 Effects on medium width centrifugal fan with backward inclined impeller

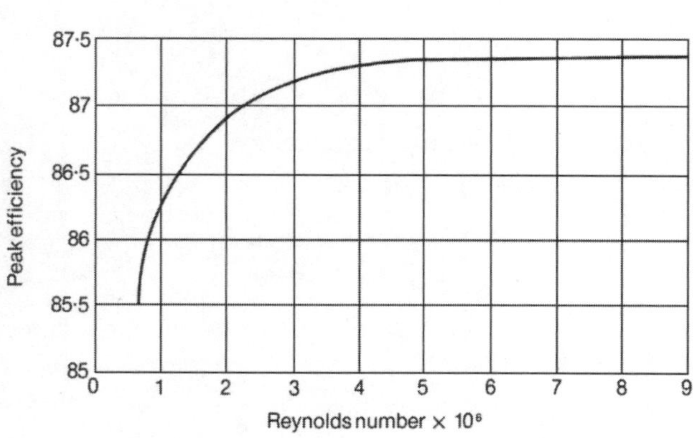

Figure 4.28 Reynolds number effects on the peak static efficiency of aerofoil bladed fans

4.7 Specific values

4.7.1 Specific speed

The specific speed of a fan at a given duty is the speed at which a geometrically similar or homologous fan would have to run to give unit flowrate and unit pressure at the same point of rating (assumed same efficiency) when handling air or gas of unit density. Thus by manipulating the fan laws

$$N_s = \frac{NQ^{0.5}\,\rho^{0.75}}{p^{0.75}\,k_p^{\;0.25}} \qquad \text{Equ 4.17}$$

If SI units were used then N_s (and N) should be in rev/s.

4.7.2 Specific diameter

Specific diameter D_s is the impeller diameter of the geometrically similar or homologous fan for which the specific speed has been calculated.

$$D_s = \frac{D_p^{\;0.25}}{\rho^{0.25}\,Q^{0.5}\,k_p^{\;0.25}} \qquad \text{Equ 4.18}$$

4.7.3 Composite charts

Reference to Figure 4.29 show that it is possible to plot all of a manufacturer's product range on a single chart. Specific diameter and efficiency have been plotted against specific speed. It will be seen that the specific speed at maximum efficiency is a unique value for a particular design.

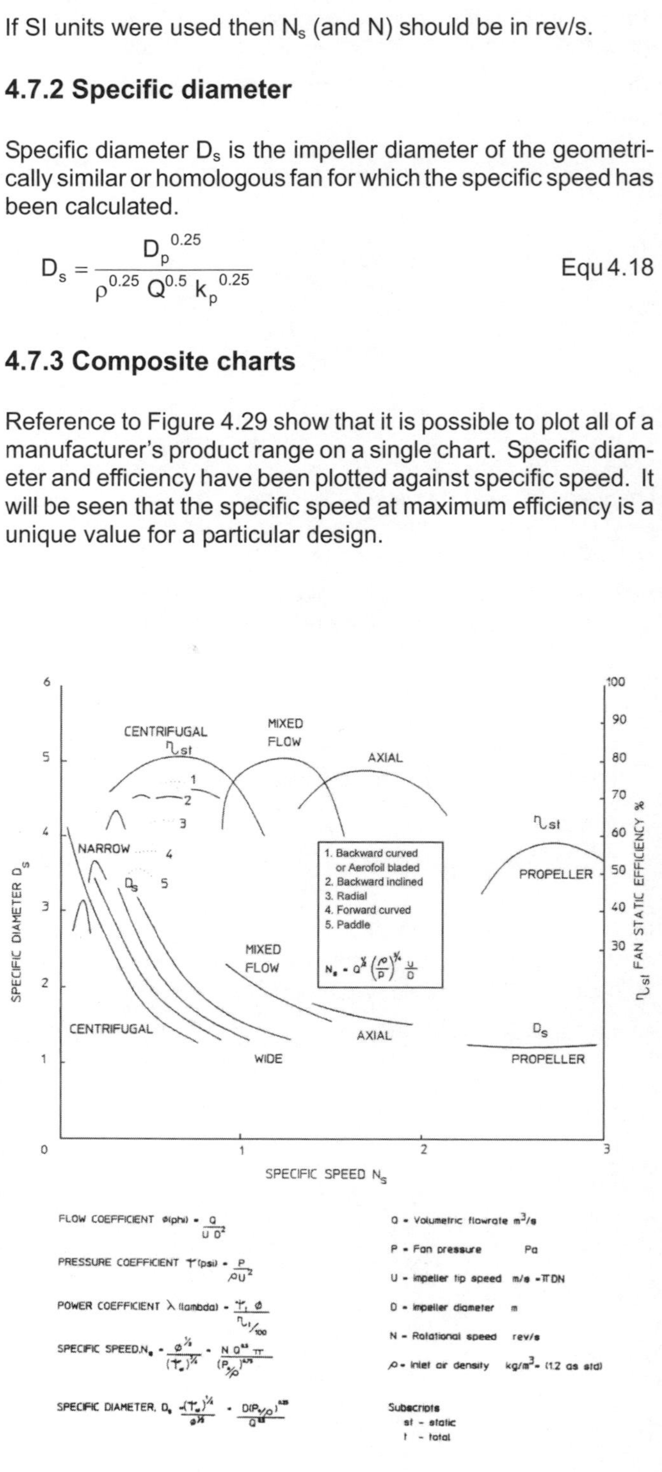

Figure 4.29 Specific diameter and efficiency against specific speed for a range of fans

Use of such charts is useful in both the selection and design of fans. The manufacturer can identify gaps in his range if adequate coverage of all duties is to be achieved.

4.8 Bibliography

ANSI/ASME PTC 11-1984, *Fans: Performance Test Codes.*

AS 2936-1987, *Industrial fans - Determination of performance characteristics (known as the SAA Fan Test Code)* superseded by: AS ISO 5801-2004 : *Industrial fans - Performance testing using standardized airways* identical to ISO 5801:1997.

ISO 13347-1:2004, *Industrial fans — Determination of fan sound power levels under standardized laboratory conditions — Part 1: General overview.*

BS 848-1:1997, *Fans for general purposes. Performance testing using standardized airways.*

DIN 24163-3, *Fans; performance testing of small fans using standardized test airways.*

ISO 7194:1983, *Measurement of fluid flow in closed conduits — Velocity-area methods of flow measurement in swirling or asymmetric flow conditions in circular ducts by means of current-meters or Pitot static tubes.*

ISO 3966:1977, *Measurement of fluid flow in closed conduits — Velocity area method using Pitot static tubes.*

The Measurement of Airflow, E. Ower and R.C. Pankhurst, Pergamon Press, Oxford 1977.

Pressure-probe methods for determining wind speed and flow direction, D.W. Bryer and R.C. Pankhurst, NPL (National Physical Laboratory).

AMCA 01, *Fans & Systems.*

The Fan and Ductwork Installation Guide, UK Fan Manufacturers Association, (HEVAC).

AMCA 203, *Field Performance Measurement of Fan Systems.*

Axial Flow Fans and Compressors: Aerodynamic Design and Performance (Cranfield Series on Turbomachinery Technology), A.B. McKenzie, Ashgate Publishing Ltd.

ISO 5801:1997, *Industrial Fans — Performance testing using standardised airways.*

ISO 5802:2001, Industrial Fans — Performance testing in-situ.

A study of the influence of Reynolds Number on the performance of centrifugal fans, J.J. Phelan, S.H. Russell and W.C. Zeluff, ASME Paper No. 78-WA/PTC-1, 1978.

BS7405 :1991, *Selection and application of flowmeters for the measurement of fluid flow in closed conduits.*

AMCA Publication 802, *Industrial process/power generation fans — Establishing performance using laboratory models.*

5 Fans and ducting systems

A theme of this book has been that the fan and its system interact. Performance is not solely the responsibility of the fan manufacturer or the system designer. Each has his own tasks in achieving that harmony, when the two are in balance.

Fans and their ducting systems have to be in balance i.e. the system resistance (or back pressure of a system) and the fan pressure are equal. This normally only occurs at one volumetric flowrate if the fan characteristic has a negative slope and the system characteristic is rising.

A system will have a number of components each of which will have a pressure loss which is a function of the velocity of air or other gas which is flowing through it.

It is essential to realise that the capacity of a fan is not fixed, but is determined to a great extent by the system which is attached. Hence this concept is continually repeated in many of the chapters.

This Chapter looks at the problems in more detail and perhaps emphasises the need for continual dialogue between fan and system engineers. Buying fans through a purchasing department committed to spending the fewest bucks is fraught with danger.

But ductwork designers appear to know little of system effect factors – an aim of this Chapter is therefore to rectify that deficiency. Hopefully, it will lead to the reader looking for the other references given.

Contents:

5.1 Introduction

Just as fans have laws which govern their behaviour, so too have their systems. Fan systems can be an assembly of ducts, filters, coolers, heaters, dampers, louvres, terminal devices, screens etc. Alternatively, it might be a boiler, economiser, pre-heater chimney stack and associated flues. Yet again, it could be a dryer, heater and ducting or a dust collector, hoods and ducting. The variety of systems is virtually endless, but some of the more popular are described in more detail in Chapter 21.

Most systems draw air, or some other gas such as flue gas, from one space and discharge it into another. The means of producing this air movement in a controlled fashion is by the use of a fan with its prime mover.

5.2 Air system components

A typical air system will contain one or more of the following components:

- System inlet
- Distribution system
- Fan and prime mover
- Control apparatus
- Conditioning apparatus
- System outlet

These are shown in Figure 5.1 taken from AMCA 200-95.

5.2.1 System inlet

An air system will usually include a device such as a louvre, filter, mesh screen or guard, grille etc., where the air or gas enters the system. These elements are necessary for personnel safety as well as to preclude the entry of rain, dust and other unwanted materials which we do not wish to collect.

Some of these items may be an architectural feature such that their appearance may be of more importance than their functional efficiency as they may be visible from the exterior of a building.

5.2.2 Distribution system

This will be made up of the straight ducting, bends, junctions, diffusers and reducers. It will be purpose-designed to convey the air or other gas from the system inlet(s) to the system outlet(s). In certain cases, enclosed spaces in the structure such as plenum chambers or other enclosures above ceilings may be used to confine the flow. Holes in walls may also direct the air.

5.2.3 Fan and prime mover

A fan is necessary to produce a pressure difference between the inlet and outlet of the system such that the required flow of air or gas is passed. The fan must be correctly designed and selected to produce the requisite flowrate against the specified pressure differential for satisfactory system operation.

Different fan designs produce different flowrates against different system pressures. The absorbed power will be a function of these two properties and the fan efficiency. Their variation with time may also affect prime mover selection. For consideration of the factors involved see Chapter 1, which not only gives typical characteristic curves but also the history of how these differences arose.

5.2.4 Control apparatus

In most air systems it is desirable to regulate or control the flowrate according to some external requirement. This might be the variation in ambient conditions through the year, the reduction of a boiler output, the change of drying capacity according to stock moisture content etc, etc.

Control and regulation of the flowrate through the system is usually in response to some monitoring signal such as air velocity, pressure, temperature or humidity. It may also be desirable to regulate the flowrate in the individual branches of the ducting according to whether they are in use or not. Examples of this would be the individual rooms of a hotel air conditioning system, the extract points of a wood refuse extract system or the outlet connections of a multi-boiler induced draught plant, etc.

Control devices such as dampers function by increasing or decreasing their pressure loss and thus reducing or increasing the flowrate. Variable inlet vanes act on the air or gas entering the fan to give a controlled amount of pre-swirl. This reduces the amount of work carried out and thus the pressure developed by the fan.

In recent years the control of the fan, by varying the rotational speed of the prime mover, has become ever more popular especially with the introduction of inverters with induction motors. Chapter 6 gives a résumé of the methods used including other types of variable geometry designs of fan.

5.2.5 Conditioning apparatus

Most ventilation systems are designed to take the air or other gas from the inlet and change its condition before discharging it at the outlet. These changes could be:

- Altering its temperature by passing through a heater or cooler

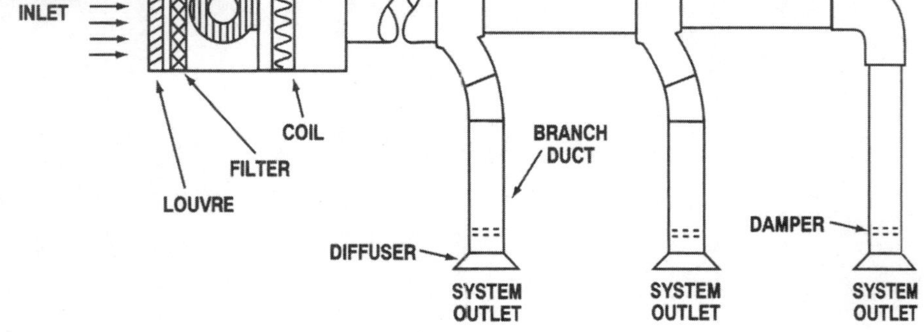

Figure 5.1 Typical fan system

- Altering its humidity by passing through a dryer or washer
- Altering its solids content by passing through a filter or dust collector

Many conditioning devices require an outside energy source such as hot water, or electrical resistance for a heater, or chilled water for a cooling coil. Other apparatus such as filters or cyclonic dust collectors are passive and have no external energy connection. All such apparatus however has a pressure loss, increasing the fan pressure requirement and therefore having an important effect on the fan selection and the absorbed power.

5.2.6 System outlet

A ventilation system usually terminates with a special component at the end of each of the outlets. This component may be a simple wire mesh screen, a ceiling diffuser or a special grille. In many cases these may incorporate control devices such as dampers and/or mixing boxes. In air conditioning, the distribution requiring careful outlet positioning and diffusers to achieve the desired air motion and temperature conditions.

5.3 System curves

Just as fans have characteristic curves, so also do systems.

It has been shown that fan performance cannot be adequately described by single values of flowrate and pressure. Both quantities are variable, but have a fixed relationship with each other.

This relationship, demonstrated in Chapter 1, is best described graphically in the form of a fan characteristic. Volumetric flowrate is normally plotted along the base with the fan pressure, absorbed power and efficiency as ordinates. Such characteristic curves are specific to:

- a given fan design and size (usually based on impeller diameter)
- impeller rotational speed
- air/gas conditions (temperature, barometric pressure, humidity, chemical composition and, therefore, gas density)

Chapter 2 showed how to calculate the system pressure caused by the resistance of a system to the required volumetric flowrate. The resistance can also be plotted along the base with the system pressure as ordinate. For a specific system the pressure for a number of points may be calculated and these points would be joined be a curve — the system characteristic. Again, it is specific to the air/gas conditions. In general, the more air required to be circulated, the more pressure required. As noted in Section 5.2, a typical system will comprise a number of components connected by a ducting system comprising straight ducting, bends, junctions, etc.

The head loss in metres of fluid flowing in straight ducting:

$$h_L = \frac{fL}{m} \times \frac{v^2}{2g} \qquad \text{Equ 5.1}$$

where:

f	=	friction factor
L	=	length of duct (m)
v	=	air/gas velocity (m/s)
m	=	mean hydraulic depth

$$= \frac{\text{cross-sectional area}}{\text{perimeter}} = \frac{m^2}{m} = m$$

For a circular cross-section duct:

$$m = \frac{\pi d^2}{4} \div \pi d = \frac{d}{4}$$

Head loss may be converted to pressure loss for:

$$h_L = \frac{p_L}{w} = \frac{p_L}{\rho g}$$

or

$$p_L = h_L \rho g$$

or

$$p_L = \frac{fL}{m} \times \frac{1}{2}\rho v^2 \qquad \text{Equ 5.2}$$

Note: In some literature, mostly of German or American origin, p_L is defined in terms of circular cross-section ducting, i.e.

$$p_L = \frac{fL}{d} \times \frac{1}{2}\rho v^2 \qquad \text{Equ 5.3}$$

As $m = \frac{d}{4}$, the value of f has to be 4 times larger in this literature, for in the UK

$$p_L = \frac{4fL}{d} \times \frac{1}{2}\rho v^2 \qquad \text{Equ 5.4}$$

If we define $v = \frac{Q}{A}$, and if we assume that the flow is fully turbulent, then we may also assume that f is a constant, then

$$p_L \propto Q^2$$

In like manner, the pressure loss in fittings

$$= k \times \frac{1}{2}\rho v^2$$

Again if we assume fully turbulent flow, k may be taken as a constant and

$$p_L \propto \frac{1}{2}\rho v^2$$

$$\propto v^2$$

$$\propto Q^2$$

Thus overall $p_L \propto Q^2$ and the system line may be plotted accordingly.

If we draw both fan characteristic and system characteristic to the same scales of flowrate and pressure, they may be plotted on the same grid.

The intersection of the two curves will be the point of fan operation on that particular system, again assuming the same gas conditions for each (see Figure 5.2).

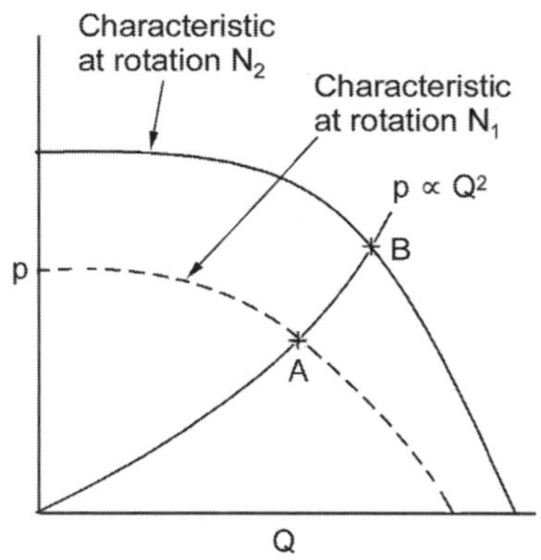

Figure 5.2 Elements in a typical air system

Note that:

Q	=	flowrate through duct of fitting (m^3/s)
w	=	weight of gas per unit volume (kg m/s^2)
ρ	=	density of air or gas (k/m^3)
A	=	cross-sectional area of duct (m^2)
N	=	fan rotational speed (rev/s or rev/min)
W	=	absorbed fan power (W or kW)

A change in fan speed alters the point of operation from A to B ie along the system curve. This is because, as shown in the "Fan Laws", (Chapter 4), for a given fan and system:

$Q \propto N$

$p \propto N^2$

and $\therefore p \propto Q^2$ for the fan as well.

Thus if a fan is applied to a system and its speed is changed from N_1 to N_2 then:

$Q \propto N$

$$\text{ie } Q_2 = Q_1 \times \frac{N_2}{N_1} \qquad \text{Equ 5.5}$$

$p \propto N^2$

$$\text{ie } p_2 = p_1 \left(\frac{N_2}{N_1}\right)^2 \qquad \text{Equ 5.6}$$

$W \propto N^3$

$$\text{ie } W_2 = W_1 \times \left(\frac{N_2}{N_1}\right)^3 \qquad \text{Equ 5.7}$$

An increase of 10% in fan rotational speed will therefore increase volumetric flowrate Q by 10%, pressure developed by the fan and the system pressure by 21%, but power absorbed W by 33%, assuming air/gas density is unchanged and that the friction factor for straight ducting and fittings remains virtually constant.

Unless large motor margins over the absorbed power are available, therefore, the possibility of increasing flowrate by a speed increase are usually limited unless substantial over-design is incorporated. Speed increase also leads to increased stresses within the fan impeller (and other parts) also $\propto N^2$.

Most importantly, it has been assumed that the friction factor f is also constant. Whilst this is almost true for small changes in duct velocity, it is not true for large changes.

Reference to the Moody chart in Chapter 3, Figure 3.13, shows that this is not the case in the laminar and transitional zones. Only in the fully turbulent zone is it remotely close to the truth. In general f increases in all systems from design flow down to near zero flow where, by definition, the flow is laminar.

Thus p_L is **not** $\propto Q^2$ over a wide range of flows and thus:

$$Q_2 \neq Q_1 \times \frac{N_2}{N_1}$$

$$p_2 \neq p_1 \times \left(\frac{N_2}{N_1}\right)^2$$

$$W_2 \neq W_1 \times \left(\frac{N_2}{N_1}\right)^3$$

for a fan and system.

The fan "law" still applies to the fan alone at a near constant fan efficiency. It does not however apply to the attached system, over a range of volumetric flowrates greater than say 10%. Where the fan speed is reduced over a turndown ratio of say 10:1 (e.g. with inverter control), the expected power savings $\propto N^3$ will **not** be achieved as claimed in many catalogues.

Table 3.1 in Chapter 3, shows the Reynolds numbers for a range of duct sizes and air/gas velocities. The corresponding friction factor for straight smooth ducting is shown as taken from the Moody chart, (Chapter 3, Figure 3.13), for typical galvanized sheet steel ducts. f is far from constant and is in fact a function of Reynolds Number and relative roughness.

It is a similar situation for duct fittings. Whilst the pressure loss through these is normally assumed to be

$$p_L = k \times \frac{1}{2}\rho v^2$$

where k is a constant, it is known that k in fact varies with the duct Reynolds number.

The supporting experimental evidence for this statement is sparse, although the work of Idelchik and Miller, is perhaps the most valuable. Turbulence in a right angled circular bend leads to dead areas as shown in Figure 5.3, with a resultant value for k typically as detailed in the graph in Figure 5.4.

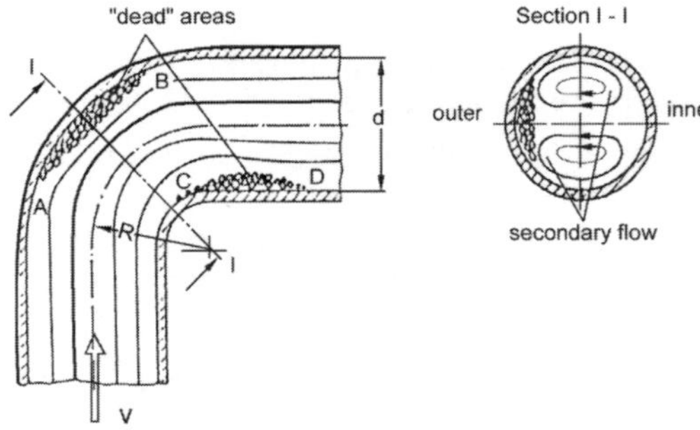

Figure 5.3 Cross-section through a right angled circular section bend showing "dead" areas

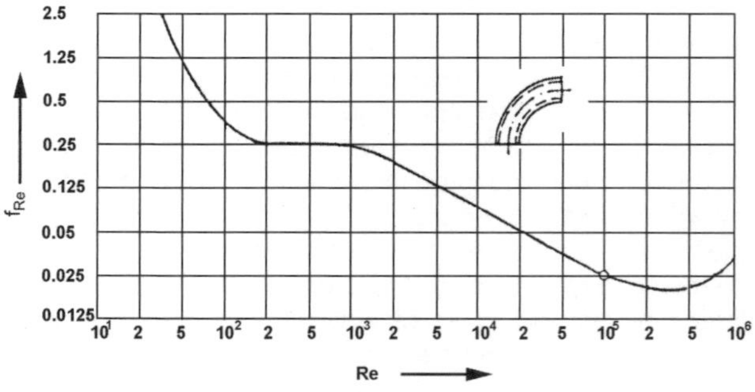

Figure 5.4 Values of k against Reynolds number

It will therefore be appreciated that for a typical system $p \propto Q^n$ where $n < 2$. Typically it will be between 1.7 and 1.9. For systems incorporating absolute filters and little else, $n \rightarrow 1$. For the flow through granular beds such as grain, n will lie between 1.25 and 1.4 according to its variety and moisture content.

There will be very few systems where the flow is fully turbulent and consequently $f \neq$ a constant.

There will always be a flowrate where there is a change from transitional to laminar. At this point it is likely that the system pressure will increase. In all systems the velocity index will change from around 1.8 down to 1.0 with decreasing flow. A real system pressure curve is likely to be as shown in Figure 5.5.

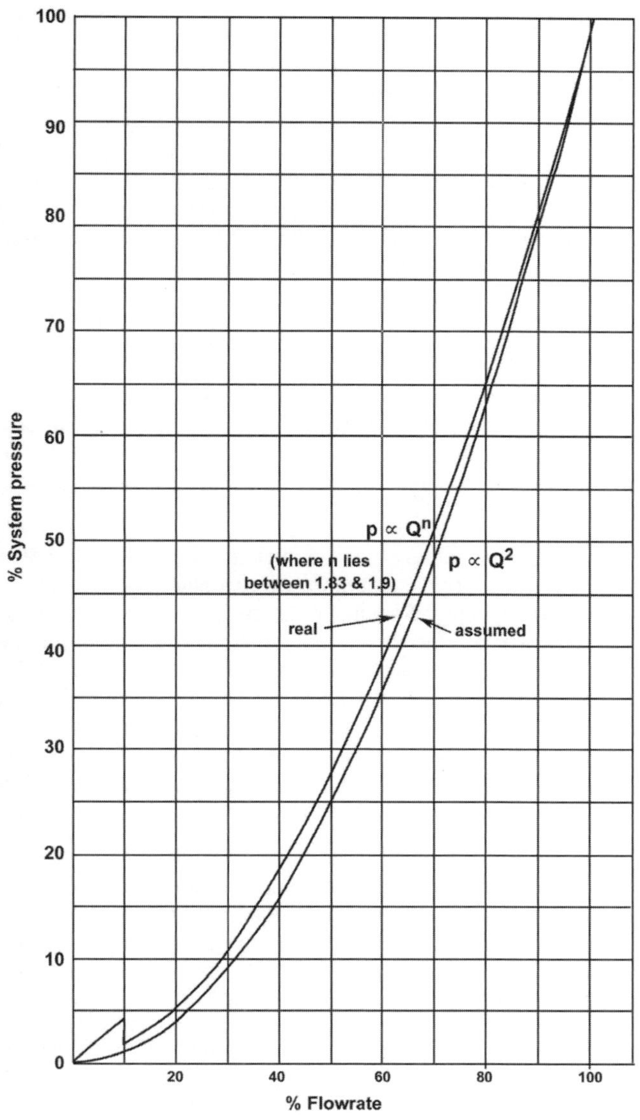

Figure 5.5 Real system pressure curve

The transition point will vary from one system to another according to the amount of laminar flow present due to low velocities at filters etc. Only pneumatic conveying plant, dust exhaust and high velocity air conditioning are likely to have flows which are fully turbulent. These effects should be recognized especially when speed control is included. To repeat, fan efficiency **will** change and power absorbed will **not** vary as N^3. Power savings are therefore likely to be somewhat less than claimed e.g. between N^2 and $N^{2.5}$. At very high turn down ratios, the savings will be even less.

It will be noted that the index for Q is continually varying and is not a fixed value. For small plants, the index appears to tend to smaller values – certainly below the 1.9 or thereabouts quoted by Loeffler et al.

It will however be concluded that a square law relationship assumed in applying tolerances to performance data as called for in AMCA 211 and ISO 13348 (catalogue fans) is perfectly valid for small variations of 3% or even 5% of flowrate.

The curve assumes standard air, and if there is a variation in temperature and /or barometric pressure along the duct run then the curve becomes even more complex to calculate. Such cases are not unknown. Again, it should be emphasised that much lower indices are to be expected in grain drying, fuel beds, etc.

5.4 Multiple fans

5.4.1 Fans in a series

As an approximation it may be said that when fans are connected together in series then, at any give volumetric flowrate, each fan adds its corresponding fan total pressure to the combined output with its corresponding power. In actual practice there is a slight loss in pressure in the connections between the stages.

In more exact work it should be noted that the total pressure of the combination is equal to the sum of the fan total pressures of the individual units minus the losses in the interconnecting duct. Thus the fan static pressure of the combination is equal to the total pressure of the first stage plus the static pressure of the second stage there being only one velocity pressure lost at the final outlet. With high pressures compression becomes important. The second stage will receive its air at a density increased by the pressure of the first. Due to this increased density its pressure development will be correspondingly greater, together with its absorbed power.

For normal commercial requirements, series operation is in use mainly for air supply to furnaces, which require a relatively high pressure at a small air flow. Two stages meet most needs, but a larger number of stages may be used for applications such as industrial vacuum cleaning, pneumatic conveying etc.

A test on a Sturtevant 2 stage STI type fan is shown in Figure 5.6 and the results are show in Table 5.1.

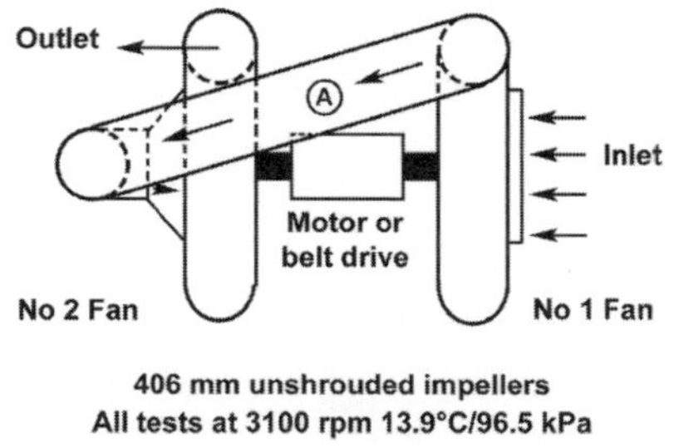

Figure 5.6 Example of test on Sturtevant 2 stage STI type fan

Item	Fan static pressure at discharge Pa	Volumetric flowrate m^3/s	Absorbed power Nett kW	Fan static pressure at "A" Pa
No 1 fan alone	3275	0	0.276	-
	3139	0.024	0.350	-
	2665	0.092	0.667	-
	1183	0.211	1.133	-
No 2 fan alone	3338	0	0.350	-
	3176	0.024	0.388	-
	2740	0.093	0.735	-
	1203	0.213	1.156	-
No 2 fan with inlet bend	3301	0	0.283	-
	3089	0.024	0.291	-
	2354	0.086	0.623	-
	872	0.182	0.940	-
Pair of fans as sketched	6676	0	0.723	3276
	6153	0.033	0.902	3064
	4359	0.118	1.670	2018
	1318	0.224	2.267	461

Table 5.1 Results of test on 2 stage fan

5.4.2 Fans in parallel

For a given system total pressure the volume delivered by the combination is the sum of the individual units at the same fan static pressure. This is only strictly true where the two fans are connected to a chamber.

If the fans blow directly into a common duct then neglecting losses, the volume delivered by the combination for a given total pressure is the sum of the volumes delivered by the individual fans at the same fan total pressure.

Multivane forward curved bladed fans are not usually suitable for parallel operation due to the shape of the fan curves. The stall of low volumetric flowrates means that there may be as many as three flowrates, where the fan pressure is the same.

Because of the pronounced peak in the pressure/volume curve, where there is any possibility of large and rapid fluctuation in system resistance, a forward curved fan selected at any pressure Q above the dotted line (see Figure 5.7) can be unstable. If, for any reason, the flow drops the point of operation can move from something normally around B to C where the fan head is slightly less. The change in volume may have been small and the system back pressure will have stayed almost unaltered. Thus the system pressure will be in excess of the fan pressure causing the flow to decrease rapidly back to A. Since the back pressure is still above the shut-off pressure a reversal of flow can occur.

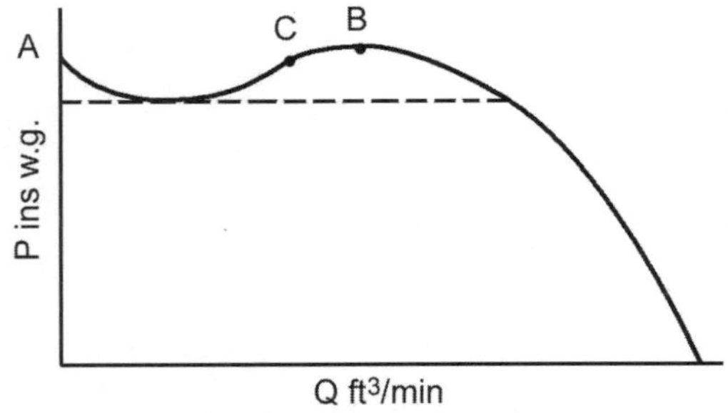

Figure 5.7 Characteristic of forward curved fan showing instability

The system is then at a standstill and the system pressure (which we assume is $\propto Q^2$) now drops below the shut-off pressure. Volume flow increases and the operating point moves up the curve past the equilibrium point. It then comes back and may tend to overshoot, thus repeating the cycle.

Such behaviour is accentuated at higher pressures, on long duct runs or when the fan discharges into a chamber of large dimensions. The instability is often not found during normal fan performance tests as these conditions do not then exist.

It will be seen that the practice of selecting over-large fans for a system to reduce the outlet velocity can be extremely dangerous. It may even lead to operating points to the left of the peak pressure B which should be avoided under all circumstances.

It is usually necessary to operate identical fans together to ensure that each does an equal share of the work.

5.5 Fan installation mistakes

There are two possible mistakes when fan impellers are installed on site:

- Incorrect rotation — due to the motor wired for running in the wrong direction
- Wrong hand — applying to impellers with blades of either forward or backward type. This may be due to transposed impellers in a pair of handed fans or to insertion the wrong way round of a double inlet impeller or to a wrong handed impeller sent in error.

5.5.1 Incorrect rotation

This is common particularly for fans with the impeller mounted directly on the motor shaft extension. In this arrangement, with ducts fitted on inlet and discharge of fan, it is not easy to see any rotating part. Observation has to be made on the shaft as seen down the gap between the motor and the fan. This mistake can arise when the erector leaves the job before it is wired. Many people think that if a fan runs in the wrong direction it will "blow from where it should suck", which is of course not true.

It is important to note that in some installations the reduced flow due to incorrect rotation is not obvious to the customer. Hence if the job is wrong and not checked he may not complain but in time will be dissatisfied with the work. Examples from experience will illustrate this.

In a sawdust collecting plant a backplated paddle fan handled 1.65 m^3/s with incorrect rotation and actually worked in a poor manner. When corrected the flowrate was 2.41 m^3/s. Other sawdust collecting plants have given similar results. A paddle bladed centrifugal fan was installed for handling exhaust from paint spraying booths with a textile bag filter on the discharge. It was put into operation, with another similar plant, with incorrect rotation. They worked this way for some time until a visit was made and the fault noted. The volumetric flowrate was 2.029 m^3/s as compared with 3.303 m^3/s when corrected, see Figure 5.8.

The only means of checking by the customer was the feel of the air entering the booths. It was designed for a face velocity of 0.825 m/s but in the wrong fan rotation was about 0.5 m/s. As 0.5 m/s is common for cut-price work, it is easy to see that a customer might never complain, although not satisfied.

Narrow cast iron centrifugal fans are liable to this mistake. A 225 mm fan on a small job handled 0.035 to 0.038 m^3/s in the wrong rotation and 0.069 m^3/s when corrected.

A cast iron fan with forward curved bladed impeller handled 81% of specified flow with power about the same either way.

One case is known of a cast iron fan which had been running in the wrong direction for seven years before it was noticed!

On forward curved multivane fans the wrong rotation is obvious as the flow is so much reduced and cannot fail to be noticed. The same applies to wide backward bladed fans, (see Figure 5.9).

Very narrow backward inclined bladed fans installed for blowing might not be noticed. In Figure 5.10, a 760 mm diameter type 30/25 fan which was designed of duty on 0.66 m^3/s (140 cfm) against 7.47 kPa (30 ins. swg) handled about 0.52 m^3/s (1100 cfm) at virtually the same power consumption. This is based on the system resistance following a square law relationship $p \propto Q^2$. The customer is interested in the flowrate handled and not in the pressure set up, this flow being judged by very rough observation in many cases.

With wide backward bladed fans a wrong handed impeller, with rotation correct, cannot fail to be noticed owing to the effect on power. For example, a double inlet backward curved bladed fan had its impeller inserted the wrong way by the erector. When the customer started up after the erector had left, he reported 5 times the normal power with the starter impossible to keep in. It will be seen that the effect on flow of the wrong hand is very slight, but the power characteristic is altered completely, because it has become, in effect, a forward curved impeller. See Figure 5.11.

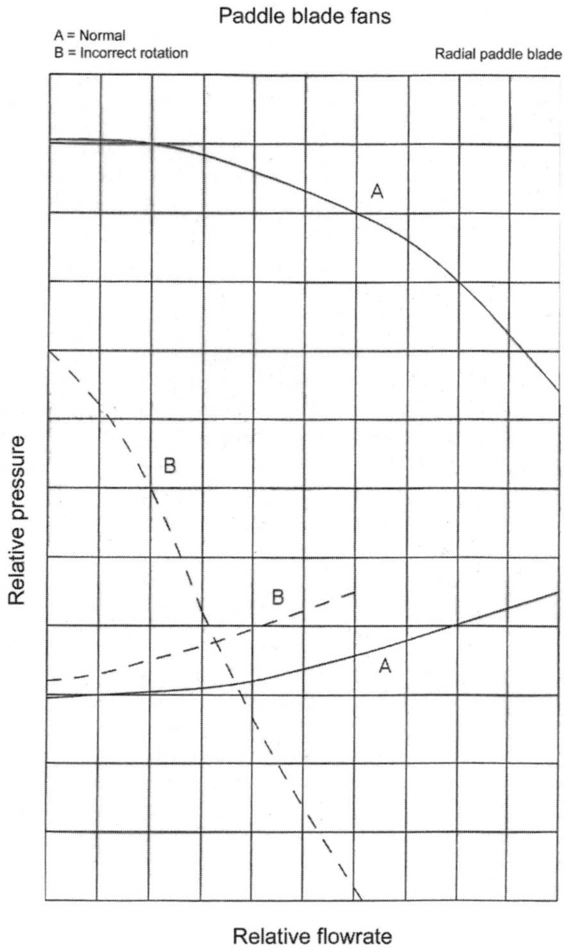

Figure 5.8 Paddle bladed fan with correct and incorrect rotation

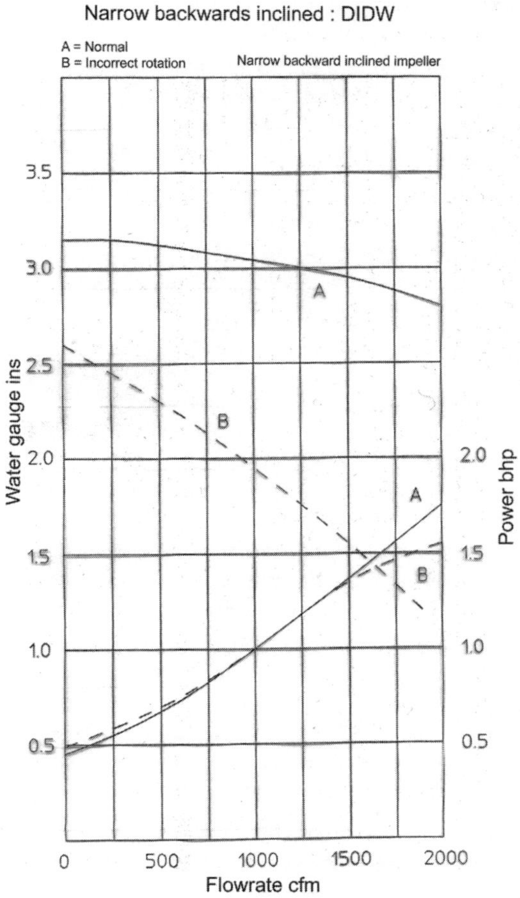

Figure 5.10 Narrow backward bladed blowing fan with correct/incorrect rotation

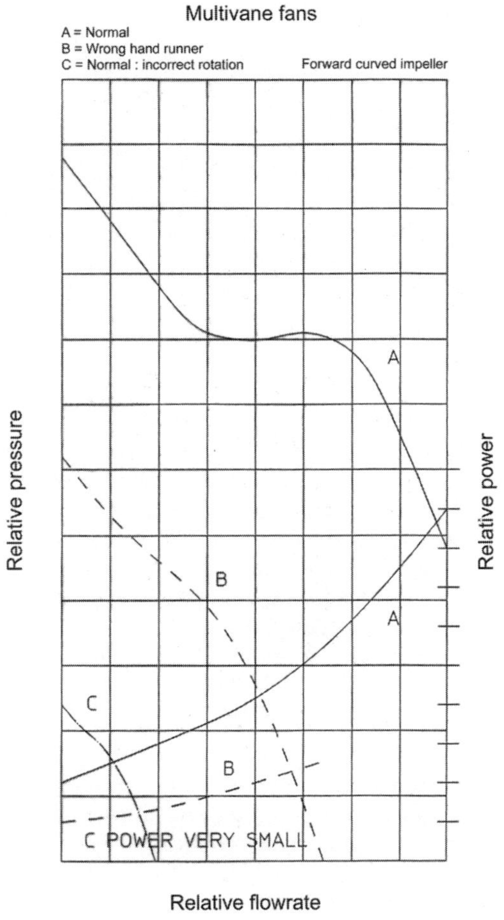

Figure 5.9 Forward curved multivane fans

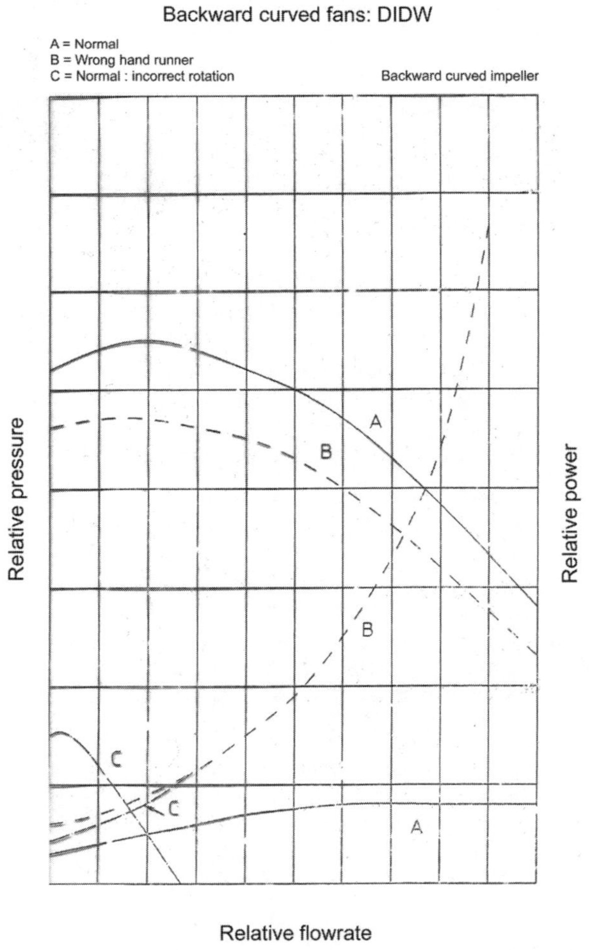

Figure 5.11 DIDW backward curved fans with installation errors

5.5.2 Wrong handed impellers

Paddle bladed fans can normally be left out of this consideration as if put in the wrong way it means that the spider is in front of the blades instead of behind. This will reduce flow to some extent but not seriously.

With forward curved bladed fans a wrong handed impeller with the rotation correct should not fail to be noticed by its results. It might just pass, however, as flowrate in average cases could be down to around 63%, with less power absorbed.

Note: Fans of the backplated paddle type for wood refuse collection usually have greater clearance at the throat of the casing, and in the wrong rotation will handle relatively more air than normal paddle bladed fans. This is confirmed by experience.

5.6 System effect factors

It has been known for may years that the ducting adjacent to a fan can have a considerable effect on the air flowrate. This applies to both the fan and ductwork itself.

Reference to Chapter 3, shows that a fan will only achieve its optimum performance when the flow at the inlet is fully developed with a symmetrical air velocity profile. It must also be free from swirl. On the fan discharge a similar situation is present. There is a need for the asymmetric profile at the discharge to diffuse efficiently and again reach a fully developed state.

In the case of fans with an inline casing, e.g. axial and mixed flow fans, there is also the possibility of residual swirl, especially if operating away from the design i.e. best efficiency point. In the case of tube axial fans, the problem can be especially severe with swirl existing up to almost 100 diameters of ducting. The only solution is to incorporate a flow straightener, which destroys the swirl, or guide vanes which can recover the swirl energy.

The system designer should therefore remember that a good arrangement of the ductwork is one that provides the above conditions at the inlet and outlet of the fan. It is his responsibility to make sure that they exist.

Ductwork engineers have been heard suggesting that due allowance should be made for less than perfect connections in fan catalogues. But how bad should they be? The reduction in flowrate for some particularly notorious examples has reached more than 60%. The first attempt in the UK at providing advice was given in the Fan Manufacturers' Association *Fan Application Guide* of 1975. It has subsequently been translated into French, German and Italian by Eurovent. This however, was purely subjective – what was good, bad or indifferent.

In the USA, AMCA published the first edition of Publication 201. This attempted to give a number of ductwork examples and quantified the effect as an additional immeasurable pressure loss. It was based on some experimental evidence back up be experience. This basis is not strictly correct as it assumes that the "loss" is proportional to the velocity pressure squared. Whilst reasonably acceptable in the working range of a fan, it is less accurate close to the shut-off (static non delivery) or at the other end of the fan characteristic (free inlet and outlet).

In January 1988 the UK Department of Trade and Industry approved a grant covering 40% of the cost of a project to establish by experimental measurement at NEL (National Engineering Laboratory), the effect of commonly used, fan connected ductwork fittings on fan aerodynamic performance. These would be installed in conjunction with a number of different fan types. The results were subsequently published in abbreviated form by the FMA in 1993 as its *Fan and Ductwork Installation Guide*.

The ductwork designer is strongly recommended to obtain these publications. They deserve the widest possible readership.

Hopefully there would not then be so many bad examples to amuse the cognoscenti.

For the benefit of those anxious to know more immediately, the following paragraphs are appended. These are based on AMCA 201 which is much easier to use in practice.

5.6.1 Inlet connections

Swirl and non-uniform flow can be corrected by straightening or guide vanes. Restricted fan inlets located too close to walls or obstructions, or restrictions caused by fans inside a cabinet, will decrease the usable performance of a fan. The clearance effect is considered a component part of the entire system and the pressure losses through the cabinet must be considered a system effect when determining system characteristics.

Installation type D fans (the Series 28 standard) have been tested with an inlet cone and parallel connection to simulate the effect of a duct. Figure 5.12 shows the variations in inlet flow which will occur. A ducted inlet condition is as (i), the unducted condition as (iv), and the effect of a bell mouth inlet as (vi). Flow into a sharp edged duct as shown in (iii) or into an inlet without a smooth entry as shown in (iv) is similar to flow through a sharp edged orifice in that a vena contracta is formed. The reduction in flow area caused by the vena contracta and the following rapid expansion causes a loss which should be considered a system effect.

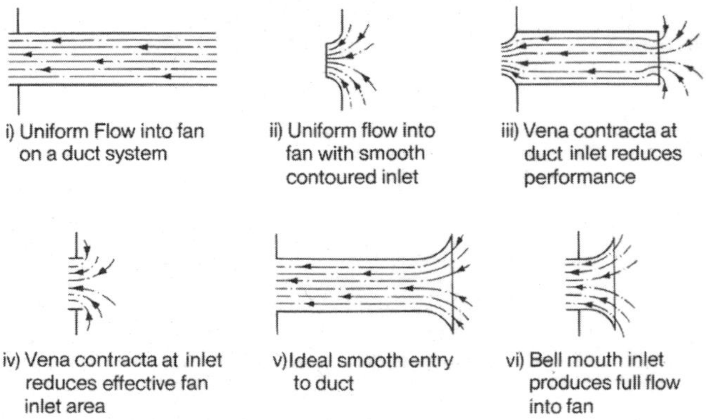

Figure 5.12 Typical inlet connections for centrifugal fans

Wherever possible fans with open inlet-installation types A or B should be fitted with bell mouths as (vi) which will enable the performance as installation types C or D to be realised.

If it is not practical to include such a smooth entry, a converging taper will substantially diminish the loss of energy and even a simple flat flange on the end of a duct will reduce the loss to about one half of the loss through an unflanged entry. The slope of transition elements should be limited to an included angle of 30° when converging or 15° when diverging. Where there is additionally a transformation from rectangular to circular; this angle should be referred to the valley.

5.6.1.1 Non-uniform flow

Non-uniform flow into the inlet is the most common cause of deficient fan performance. An elbow or a 90° duct turn located at the fan inlet will not allow the air to enter uniformly and will result in turbulent and uneven flow distribution at the fan impeller. Air has weight and a moving air stream has momentum and the air stream therefore resists a change in direction within an elbow as illustrated.

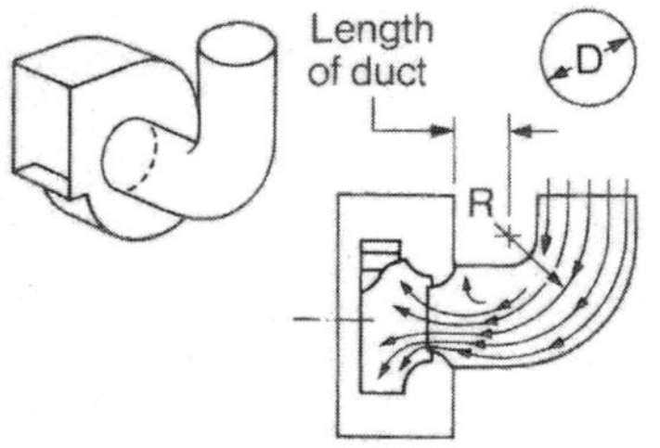

R/D	No duct	2D duct	5D duct
0.75	1.5	0.8	0.4
1.0	1.2	0.7	0.3
2.0	1.0	0.6	0.2
3.0	0.7	0.4	0.2

Figure 5.13 Systems effects expressed as velocity pressures. Non-uniform flow into a fan from a 90° round section elbow, no turning vanes

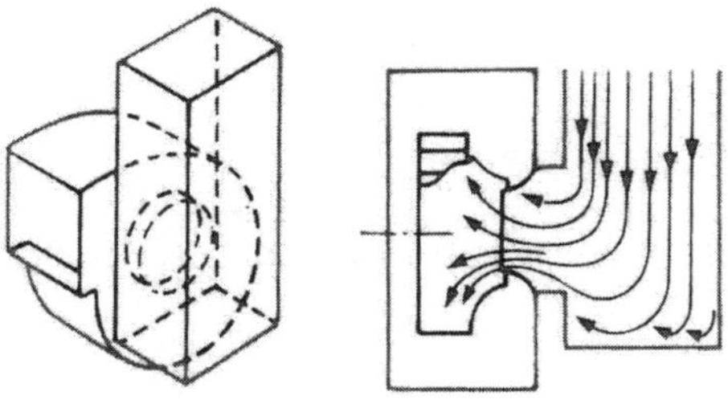

Figure 5.14 System effects expressed as velocity pressures. Non-uniform flow into a fan from a rectangular inlet duct

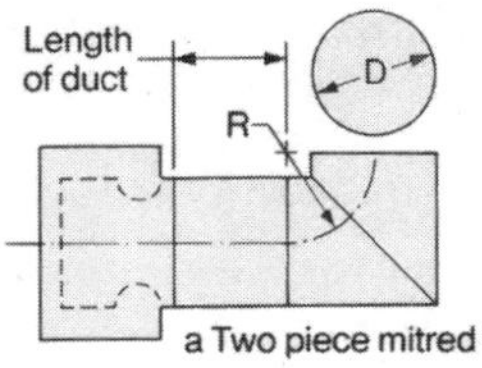

R/D	No duct	2D duct	5D duct
—	3.0	3.0	1.0

a Two piece mitred 90° round section elbows—no vanes

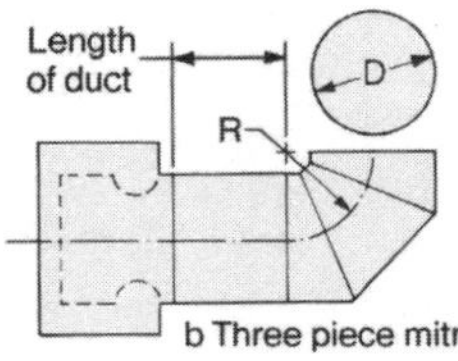

R/D	No duct	2D duct	5D duct
0.5	2.5	1.5	0.8
0.75	1.5	1.0	0.5
1.0	1.2	0.7	0.3
2.0	1.0	0.5	0.3
3.0	0.8	0.5	0.3

b Three piece mitred 90° round section elbow—no vanes

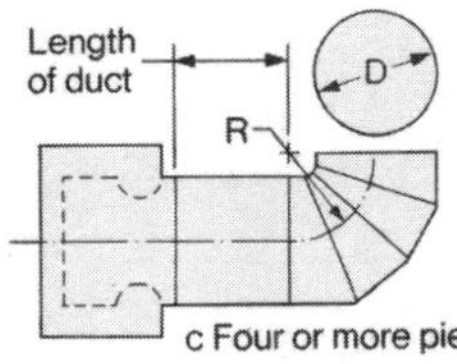

R/D	No duct	2D duct	5D duct
0.5	1.8	1.0	0.5
0.75	1.5	0.8	0.4
1.0	1.3	0.7	0.3
2.0	1.0	0.5	0.3
3.0	0.7	0.4	0.2

c Four or more piece mitred round section elbow—no vanes

$$D = \frac{2H}{\sqrt{\pi}}$$

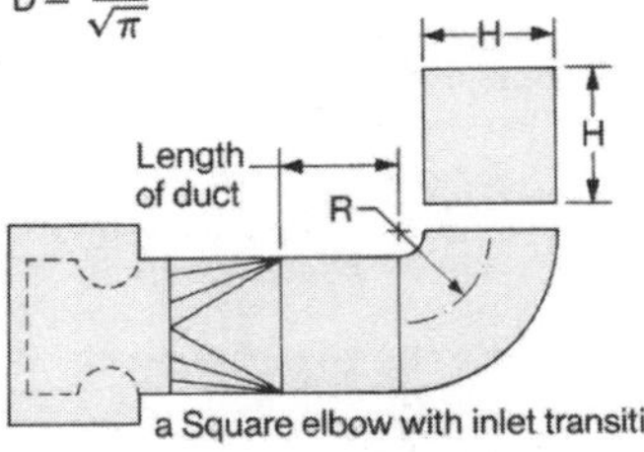

R/H	No duct	2D duct	5D duct
0.5	2.5	1.6	0.8
0.75	2.0	1.5	0.7
1.0	1.5	0.7	0.3
2.0	0.8	0.5	0.2

a Square elbow with inlet transition—no turning vanes

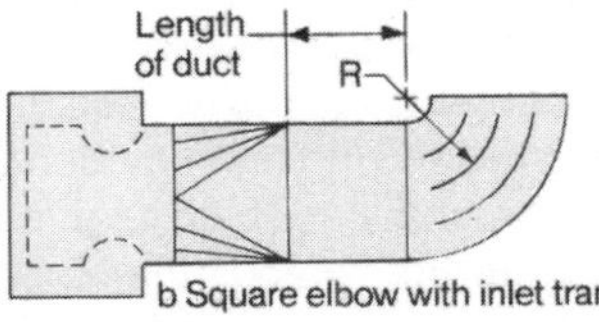

R/H	No duct	2D duct	5D duct
0.5	0.8	0.5	0.3
1.0	0.5	0.3	0.2
2.0	0.3	0.3	0.1

b Square elbow with inlet transition—three long turning vanes

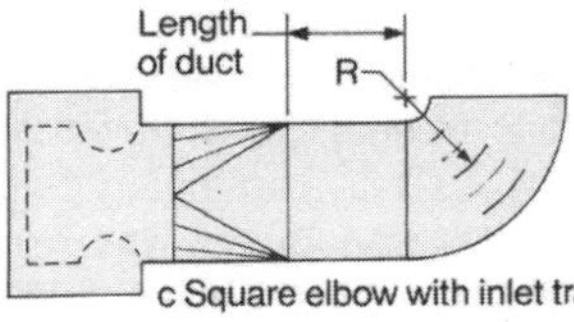

R/H	No duct	2D duct	5D duct
0.5	0.8	0.5	0.3
1.0	0.5	0.3	0.2
2.0	0.3	0.3	0.1

c Square elbow with inlet transition—short turning vanes

Figure 5.15 System effects of ducts of given radius /diameter ratios expressed as velocity pressures

The systems effects for elbows of given radius diameter ratios are given in Figures 5.13 to 5.15. These losses only apply when the connection is adjacent to the fan inlet and are additional to the normal loss. In Figure 5.14 the reduction in capacity and pressure for this type of inlet condition are difficult to tabulate. The many differences in width and depth of duct influence the reduction in performance to varying degrees. Such inlets should therefore be avoided. Capacity losses of 45 % have been observed. Existing installations can be improved with guide vanes or the conversion to square or mitred elbows with guide vanes. In Figure 5.15 the inside area of the square duct (H x H) is equal to the inside area circumscribed by the fan inlet spigot. The maximum included angle of any converging element of the transition should be 30°, and for a diverging element 15 °.

Note that when turning vanes are used and there is a reasonable length of duct between the fan inlet and elbow, the effect on fan performance is low. If the straight exceeds 6 diameters, the effect is negligible. Wherever a right angle on the fan inlet is necessary, it may be preferable to use our own design inlet boxes which incorporate anti-swirl baffles and for which the performance is known.

5.6.1.2 Inlet swirl

Another cause of reduced performance is an inlet duct which produces a vortex in the air stream entering a fan inlet. An example of this condition is shown in Figure 5.16.

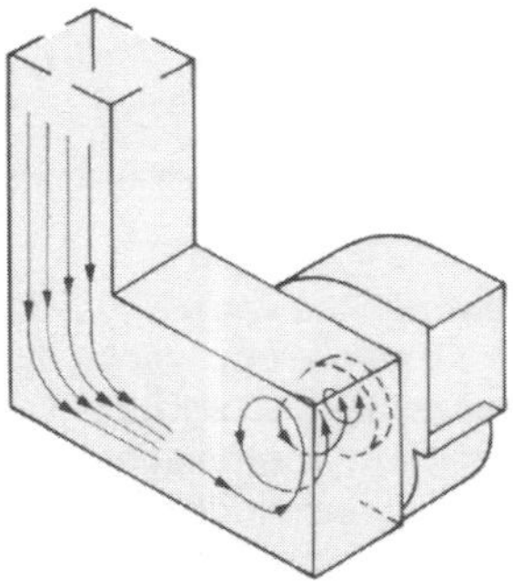

Figure 5.16 Loss of performance due to inlet swirl

The ideal inlet duct is one which allows the air to enter axially and uniformly without swirl in either direction. Swirl in the same direction as the impeller rotation reduces the pressure-volume curve by an amount dependent upon the intensity of the vortex. The effect is similar to the change in the pressure-volume curve achieved by inlet vanes installed in a fan inlet which induce a controlled swirl and so vary the volume flow. Contra-swirl at the inlet will result in a slight increase in the pressure volume curve but the horsepower will increase substantially.

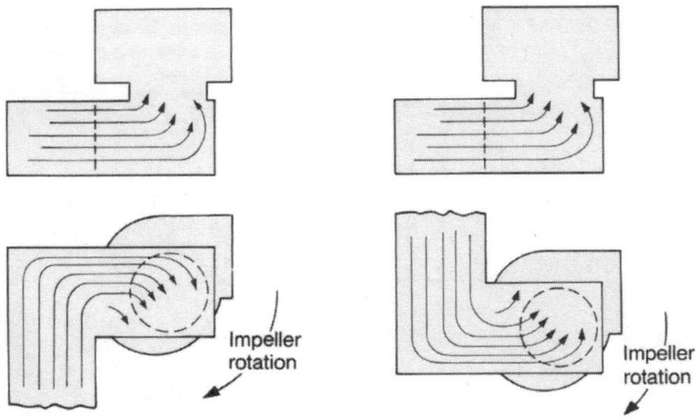

Figure 5.17 Examples of duct arrangements which cause inlet swirl

Inlet swirl may arise from a variety of conditions and the cause is not always obvious. Some common duct connections which cause inlet swirl are illustrated in Figure 5.17, but since the variations are many, no factors are given.

Wherever possible these duct connections should be avoided, but if not, inlet conditions can usually be improved by the use of turning vanes and splitters.

5.6.1.3 Inlet turning vanes

Where space limitations prevent the use of optimum fan inlet connections, more uniform flow can be achieved by the use of turning vanes in the inlet elbow. Many types are available from a single curved sheet metal vane to multi-bladed aerofoils. (See Figure 5.18.)

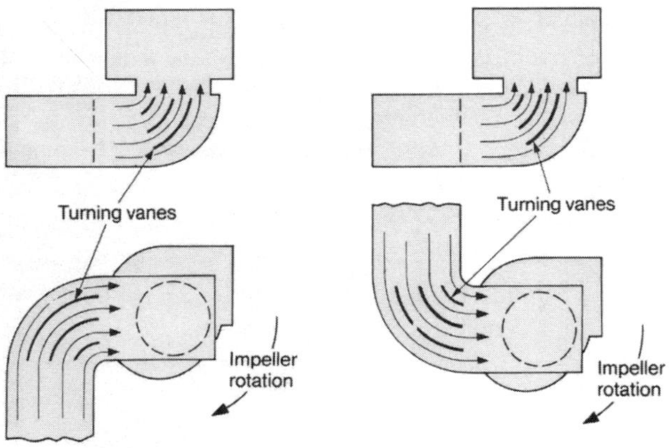

Figure 5.18 Pre-swirl (left) and contra-swirl (right) corrected by use of turning vanes

The pressure loss through the vanes must be added to the system pressure losses. These are published by the manufacturer, but the catalogued pressure loss will be based upon uniform air flow at entry. If the air flow approaching the elbow is non-uniform because of a disturbance further up the system, the pressure loss will be higher than published and the effectiveness of the vanes will be reduced.

5.6.1.4 Straighteners

Air flow straighteners (egg crates) are often used to eliminate or reduce swirl in a duct. An example of an egg crate straightener is shown in Figure 5.19.

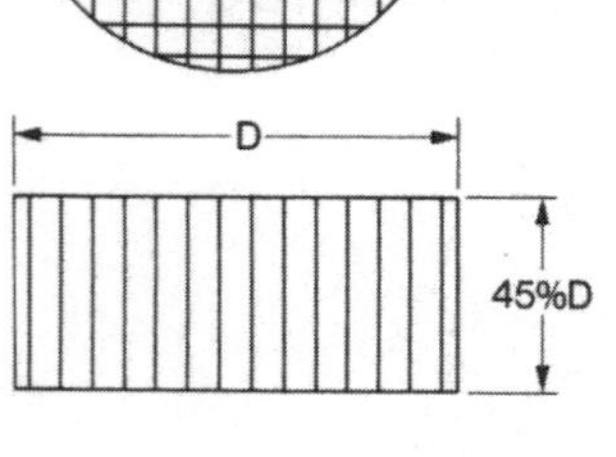

Figure 5.19 Example of egg crate air flow straightener

5.6.1.5 Enclosures (plenum and cabinet effects)

Fans within air handling units, plenums, or next to walls should be located so that air flows unobstructed into the inlets. Performance is reduced if the distance between the fan inlet and the enclosure is too restrictive. It is usual to allow one-half of the inlet diameter between enclosure wall and the fan inlet.

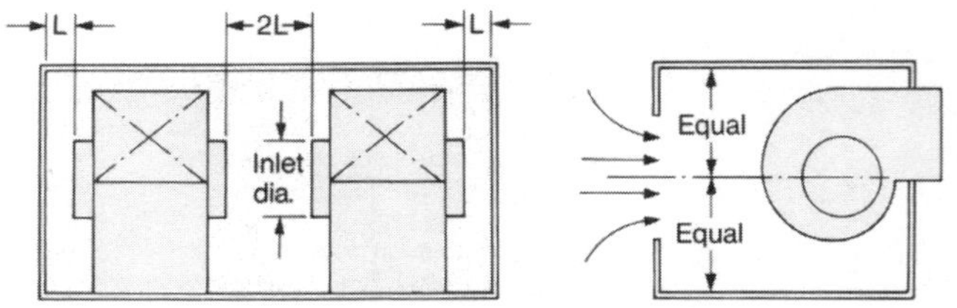

L Distance inlet to wall	System effect additional fraction of velocity pressure at inlet
0.75 x dia. of Inlet	0.2
0.5 x dia. of Inlet	0.4
0.4 x dia. of Inlet	0.5
0.3 x dia. of Inlet	0.8
0.2 x dia. of Inlet	1.2

Figure 5.20 System effects of fans located in common enclosures

Multiple DIDW fans within a common enclosure should be at least one impeller diameter apart for optimum performance. Figure 5.20 shows fans located in an enclosure and lists the system effects as additional immeasurable velocity pressure.

The way the air stream enters an enclosure relative to the fan also affects performance. Plenum or enclosure inlets of walls which are not symmetrical to the fan inlets will cause uneven flow and swirl. This must be avoided to achieve maximum performance but if not possible, inlet conditions can usually be improved with a splitter sheet to break up the swirl as illustrated in Figure 5.21.

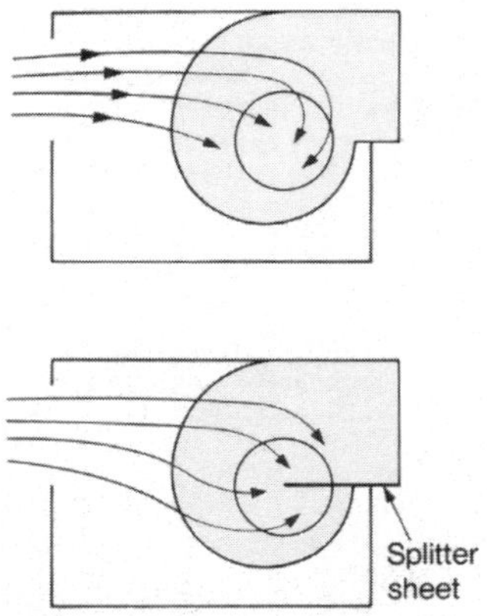

Figure 5.21 Use of splitter sheet to break up swirl. Above, enclosure inlet not symmetrical with fan inlet: preswirl induced. Below, flow condition improved with a splitter sheet: substantial improvement would be gained by repositioning inlet symmetrically

5.6.1.6 Obstructed inlets

A reduction in fan performance can be expected when an obstruction to air flow is located in the plane of the fan inlet. Structural members, columns, butterfly valves, blast gates, and pipes are examples of more common inlet obstructions. Some accessories such as fan bearings, bearing pedestals, inlet vanes, inlet dampers, drive guards, and motors may also cause obstruction. The effects of fan bearings as in Arrangements 3 and 6 are given in Figure 5.22. For these and other examples refer to the manufacturer as they are not part of AMCA 201.

Inlet obstructions such as bearings and their supports reduce the performance of a fan. The loss takes the form of reduction of volume and pressure, the power usually remaining constant. On single inlet fans Arrangement 3 and DIDW fans Arrangement 6, bearings are mounted near the inlet venturi(s). The free passage of air into the inlet(s) is thus affected. Wherever possible Arrangement 1 fans should therefore be selected.

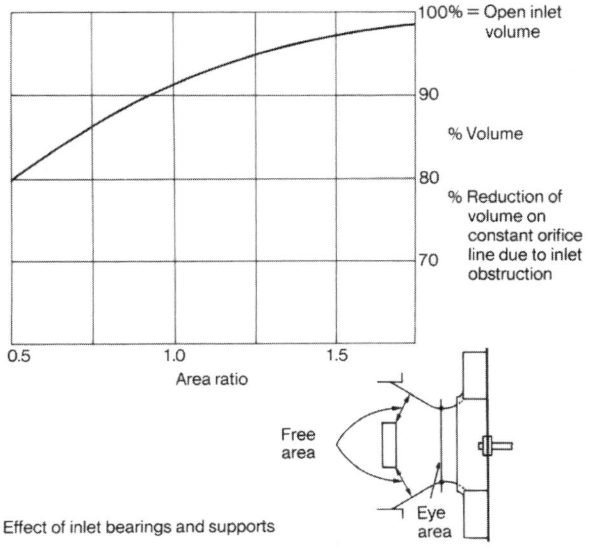

Figure 5.22 Loss of performance caused by obstruction by inlet bearings and supports

A measure of this loss is given in Figure 5.22, the degree of obstruction being assessed from the ratio

$$\frac{\text{Minimum free area at plane of bearings}}{\text{Free area at plane of impeller eye}}$$

where the free area is taken to mean the minimum area through which the air has to pass between the bearing and the wall of the venturi. The effect on performance is given as a reduction in volume below that which would be attained by the equivalent open inlet Arrangement 1 or 4 fan having no bearing obstruction, then taken as a percentage reduction down a constant orifice line.

Figure 5.23 gives the compensation necessary in the fan selection process to attain the required performance when using the normal open inlet curves. This adjustment can be either by:

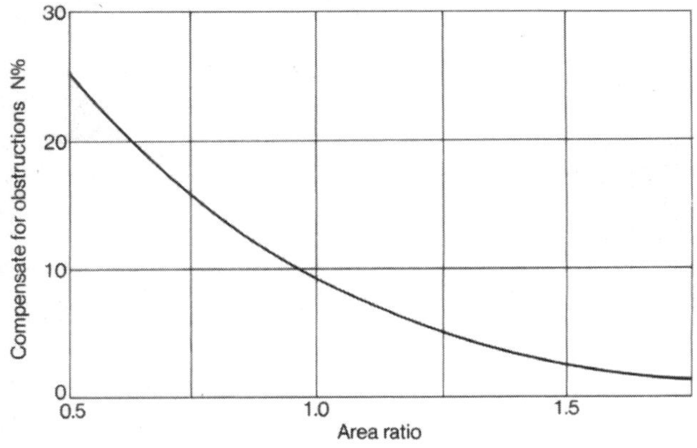

Figure 5.23 Compensation in fan selection required, using open inlet curves

- Increasing the running speed by N% after the fan has been selected

- Increasing the volume by N% and the pressure as the volume squared before the fan is selected.

The power taken by the fan with inlet bearings will be approximately the same as a fan with open inlet, at the same speed. It will thus be necessary to increase the power for a given duty by N^3 % (see Figure 5.23).

5.6.1.7 Drive guards obstructing the inlet

Arrangement 6 fans may require a belt drive guard in the fan inlet. Depending on design, the guard may be located at the plane of the inlet, or it may be "stepped out". Depending on the location of the guard, and on the inlet velocity, the fan performance may be significantly affected by this obstruction.

It is desirable that a drive guard in this position has as much opening as possible to allow maximum flow to the fan inlet. However, the guard design must comply with applicable Health & Safety Act requirements.

System effect factors for drive guards situated at the inlet of a fan may be approximated as 0.4 x inlet velocity pressure where 5 % of the fan inlet area is obstructed increasing to 2.0 x inlet velocity pressure where it is 50%.

5.6.2 Outlet connections

The velocity profile at the outlet of a fan is not uniform, but is shown in Figure 5.24. The section of straight ducting on the fan outlet should control the diffusion of the velocity profile, making this more uniform before discharging into a plenum chamber or to the atmosphere.

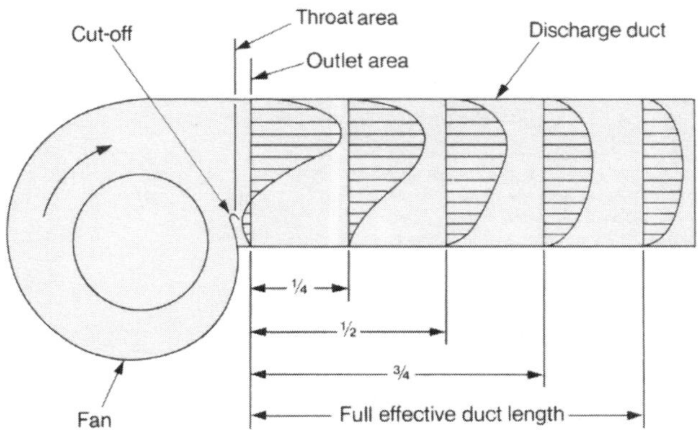

Figure 5.24 Velocity profile at fan outlet (see also Figure 5.25)

Alternatively, where there is a ducting system on the fan outlet, the straight ducting is necessary to minimise the effects of bends, etc.

The full effective duct length is dependent on duct velocity and may be obtained from Figure 5.25.

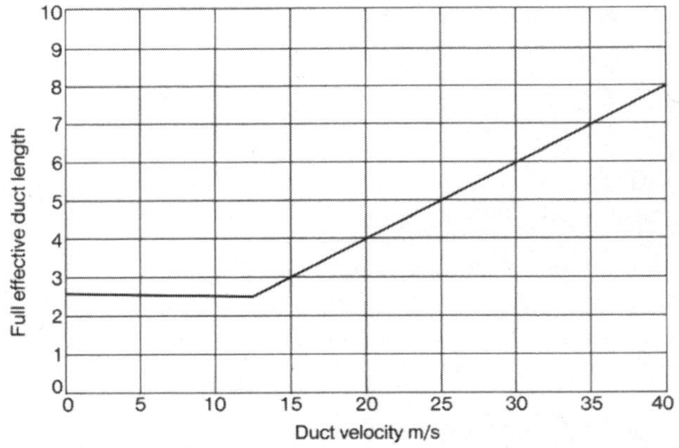

Figure 5.25 Full effective duct length expressed in equivalent duct diameters

If the duct is rectangular with side dimensions a and b, the equivalent duct diameter equals $\sqrt{\frac{4ab}{\pi}}$.

The effect of outlet bends depends on their orientation relative to the fan and also on the ratio of throat area to outlet area is

Throat area / Outlet area	Outlet elbow position	No outlet duct	1/8 effective duct	1/4 effective duct	1/2 effective duct	Full effective duct
0.4	1	3.0	2.5	2.0	0.8	No system effect
	2	5.0	4.0	2.5	1.2	
	3	6.0	5.0	3.0	1.5	
	4	6.0	5.0	3.0	1.5	
0.5	1	2.0	1.5	1.2	0.5	No system effect
	2	3.0	2.2	1.7	0.8	
	3	4.0	3.0	2.2	1.0	
	4	4.0	3.0	2.2	1.0	
0.63	1	1.5	1.5	1.0	0.3	No system effect
	2	2.0	1.5	1.2	0.5	
	3	3.0	2.2	1.7	0.8	
	4	2.5	2.0	1.5	0.7	
0.67	1	0.7	0.5	0.3	0.2	No system effect
	2	1.0	0.8	0.5	0.3	
	3	1.5	1.2	0.8	0.3	
	4	1.2	1.0	0.7	0.3	
0.8	1	0.8	0.7	0.4	0.2	No system effect
	2	1.2	1.0	0.7	0.3	
	3	1.5	1.5	1.0	0.3	
	4	1.5	1.2	0.8	0.3	
0.88 - 0.89	1	0.7	0.5	0.3	0.2	No system effect
	2	1.0	0.8	0.5	0.3	
	3	1.2	1.0	0.7	0.3	
	4	1.0	0.8	0.5	0.3	
1.0	1	1.0	0.8	0.5	0.3	No system effect
	2	0.7	0.5	0.4	0.2	
	3	1.0	0.8	0.5	0.3	
	4	1.0	0.8	0.5	0.3	

Table 5.2 System effect factors for outlet elbows for SISW fans

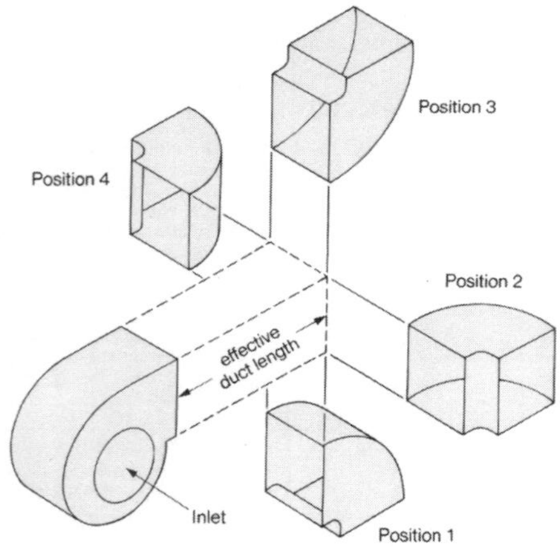

Figure 5.26 Outlet duct elbows

shown in Figure 5.26 and Table 5.2 gives the system effect factors for SISW fans. (For DIDW fans use the appropriate multiplier from the following: Elbow Position No 2 x 1.25, Elbow Position No 4 x 0.85, Elbow Positions No 1 & No 3 x 1.00.)

The use of an opposed blade damper is recommended when volume control is required at the fan outlet and there are other system components, such as coils or branch takeoffs downstream of the fan. When the fan discharges into a large plenum or to free space a parallel blade damper may be satisfactory.

For a centrifugal fan, best air performance will be achieved by installing the damper with its blades perpendicular to the fan shaft; however, other considerations may require installation of the damper with its blades parallel to the fan shaft. Published

throat area / outlet area	SP multiplier
0.4	7.5
0.5	4.8
0.63	3.3
0.67	2.4
0.8	1.9
0.88	1.5
0.89	1.5
1.0	1.2

Table 5.3 Pressure loss multipliers for volume control dampers

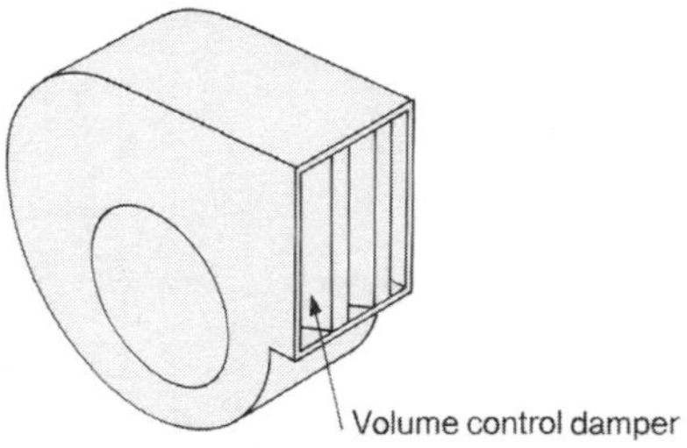

Figure 5.27 Volume control damper installed at fan outlet

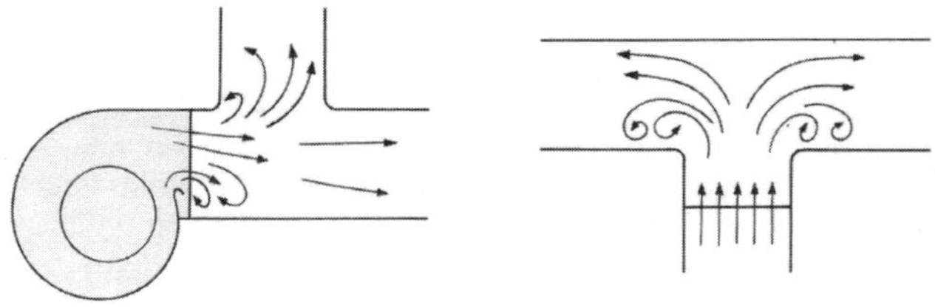

Figure 5.28 Branches located too close to fan

pressure losses for control dampers are based upon uniform approach velocity profiles.

When a damper is installed close to the outlet of a fan the approach velocity profile is non-uniform and much higher pressure losses through the damper can result, see Figure 5.27. The multipliers in Table 5.3 should be applied to the damper manufacturer's catalogued pressure loss when the damper is installed at the outlet of a centrifugal fan.

Where branches are fitted on the fan outlet, a section of straight is especially important, see Figure 5.28. Split or duct branches should not be located close to the fan discharge. A straight section of duct will allow for air diffusion.

5.7 Bibliography

AMCA Publication 200-95, *Air Systems*

Handbook of Hydraulic Resistance, I E Idelchik, Begell House Publishers Inc., 2001 ISBN 1567000746.

Internal Flow Systems (2nd completely revised edition) Edited by D S Miller, BHR Group Ltd, 1996 ISBN 0947711775.

Simplified Equations for HVAC Duct Friction Factors, J J Loeffler, ASHRAE Journal, January 1980.

AMCA 211-05, *Certified Ratings Programme - Product Rating Manual for Fan Air Performance.*

ISO/DIS 13348, *Industrial fans - Tolerances, methods of conversion and technical data presentation.*

Fan Application Guide, 2nd edition, FMA (HEVAC).

Fan and Ductwork Installation Guide 1st edition, FMA (HEVAC).

AMCA 201-02, *Fans and Systems.*

6 Flow regulation

This Chapter reviews a number of the factors affecting the efficient utilisation of energy in fans, their systems, their prime movers and especially their flowrate controls.

It is useful therefore to re-examine the fundamentals and it is hoped that the resulting conclusions may be of value to system designers, users and energy managers.

No one method of flow regulation is applicable to all applications. How the system resistance varies with flow, and whether there is a fixed element, very much determines the choice.

It is important to emphasise that no one method is applicable to all systems. Whilst speed control of induction motors by inverters is currently the most popular, there are situations where, because of a fixed element in the system resistance, other methods are more appropriate. This Chapter gives the necessary information.

Contents:

6.1 Introduction

Energy costs rose considerably during the 1970s following a succession of crises affecting the Middle East oil-producing nations. Despite a temporary respite in the 1980s following a rapid increase in North Sea oil production, and the discovery of other sources, this escalation continued in the 1990s. In the 21st century there is also now a "green" issue to be faced in the realization that continued burning fuels is leading to ever increasing levels of CO_2 in the upper atmosphere. Global warming is now largely accepted as a possible threat to mankind.

For all these, and many other reasons, the spotlight of efficiency has been directed to the reduction in energy consumption of all types of machinery, but none more so in fluid or turbo-machinery such as fans, pumps and compressors. Such concerns need not — indeed should not — be solely altruistic. The savings in running costs can usually justify a small increase in first cost, even for the humble fan.

If "carrot" is not enough, however, we have in some areas to contend with a little "stick". Recent changes to the UK's building regulations, for example, encourage the installer to design air conditioning or mechanical ventilation systems to meet defined energy targets. We even have to contend with a new found enthusiasm for "natural" ventilation. Preference will in any case be given to plants which incorporate efficient means of flowrate control such that supply and demand can be more closely matched all times.

This Chapter reviews a number of the factors affecting the efficient utilisation of energy in fans, their systems, their prime movers and especially their flowrate controls.

All or some of the following strategies should be considered.

a) Ensure that the plant is only in use when required.

b) Use some form of capacity control to match the flow to requirements.

c) Prime movers to be of high efficiency and matched to demand.

d) Keep plant and motors in largest possible units, consistent with a) and b) above.

e) Reduce system resistance to a minimum.

f) Make no unnecessary energy conversions.

None of these strategies should give rise to any surprise amongst practising fan engineers. We are, however, in an advertising age where the advantages of high efficiency motors and of inverter controls have been trumpeted to the disadvantage of the others. It is useful therefore to re-examine the fundamentals and it is hoped the resulting conclusions may be of value to system designers, users and energy managers.

6.2 The need for flowrate control

Every fan is selected and installed for a given flowrate and system pressure, but there will be many occasions when the demand will not be at this design maximum. Boiler induced draught units will have to cope with gas flows varying according to the amount of fuel being burned and therefore the boiler output. A fan on a grain drying installation will have to blow through more crop as the harvest progresses. On a ventilation plant there may be differences between winter and summer duties whilst on VAV (variable air volume) systems the fan capacity and system requirements must be continuously balanced.

For these and many other examples, the fan manufacturer needs to provide or advise on capacity control systems. Before considering specific cases, it is necessary to determine how the system demand may vary as on this will depend not only the best method of control to use, but also which type of fan is most suitable. To operate the fan at a higher rate than necessary is wasteful in energy. Whilst increasing the initial cost, fan control systems will usually more than pay for themselves over the life of the fan.

The manner in which fan demand may vary can be categorised categorized as follows although combinations of these are also possible.

6.2.1 Constant orifice systems

In these the plant remains unchanged, but the air flowrate through it may need to vary. When there are fixed elements such as straight ducting, bends, takeoffs etc., and the flow is fully turbulent, then we may apply normal systems resistance "laws".

Thus system pressure $p_s \propto$ (flowrate Q)^2. If the capacity control is to maintain its efficiency constant then as fan power

$$P \propto Q \times p_s$$

$$P \propto Q \times Q^2$$

or $P \propto Q^3$ Equ 6.1

As fan capacity $Q \propto N$ it will be seen that speed variation is the optimum solution provided that power source efficiency can also remain constant over the range required. With AC electric motors, good efficiencies are maintained down to about 50% power (i.e. 80% fan flowrate).

It should again be noted (see Chapter 5) that whilst a system may be fully turbulent for the design flowrate and just below this figure, this will **not** be the case for high turn down ratios. Inevitably, flow will become laminar as zero is approached. Then

$$P \propto Q \times p$$

$$P \propto Q \times Q^n$$

or $P \propto Q^{n+1}$ Equ 6.2

where n varies continuously from just less than 2 at the design flow down to 1 at zero flow.

Fan speed and efficiency will also vary.

6.2.2 Parallel path systems

Here the airflow may vary, but pressure required remains virtually constant. Examples which come to mind are mechanical draught systems where one fan may cater for more than one boiler. As boilers are shut off or started up according to demand, so the gas flowrate will vary. Provided the common ductwork is short, however, the pressure drop through each boiler and therefore through the system will remain unchanged. If the capacity control is to maintain a constant efficiency then as

power $P \propto$ flowrate Q x fan pressure p_s

i.e. $P \propto Q$ Equ 6.3

Similar situations can arise in central extract systems where dampers in parallel ducting legs may be shut according to whether a machine is or is not operating and therefore emitting dust or fumes. With an on-floor grain drying plant, the floor area to be ventilated will increase as the harvest progresses but at constant grain depth and drying rate the pressure demand would remain unchanged.

6.2.3 Series path systems

The airflow needs to remain fairly constant, but pressure required will vary. For a fan ventilating a tunnel during construction, the air requirements at the working face will remain constant, depending only on the number of men working and air required to cool or supply the machinery. The length of ducting taken to a fresh airsource will, however, increase as the work progresses. If the fan control is to maintain a constant efficiency then as

power $P \propto$ flowrate Q x fan pressure p_s

i.e. $P \propto p_s$

Similar situations can arise in drying plants with bottom ventilated bins where pressure will increase with the depth of bed.

6.2.4 Variable air volume (VAV) systems

In a VAV system, as applied to the air conditioning of a building environment, the airflow rate to each separate room or occupied space is varied both individually and continuously. Thus the instantaneous cooling demands of a room may be satisfied. Such a system is shown in Figure 6.1 and consists of a central unit (1), ducting (2), flow variators (3) and supply air terminals (4). Each flow variator is controlled by a room thermostat (5) and demands a constant pressure in the ducting. This is maintained by the pressure transducer (6) which controls the fan flowrate by altering fan speed, inlet guide vane angle, disc throttle position, impeller pitch angle or such other method of flow variation as installed.

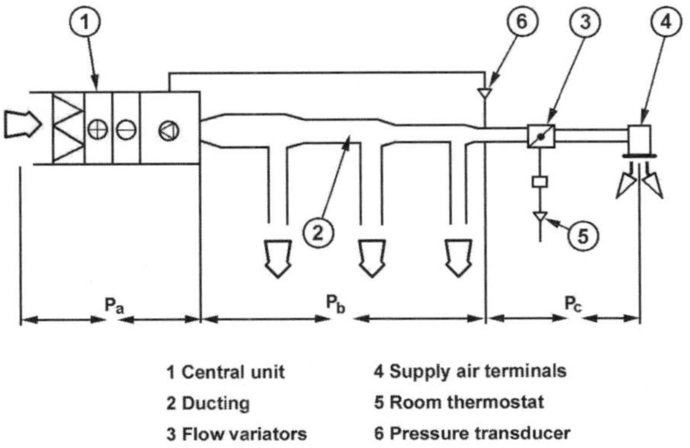

Figure 6.1 Variable air volume (VAV) system

The system pressure required may be divided into three main parts:

p_a: Pressure loss in the air handling unit, which varies generally as something less than the square of the fan air flow (any filters p_f may be $\propto Q$) $p_a \propto Q^2$

p_b: Frictional pressure loss in the ducts, which varies as something less than the square of the air flow. $p_b \propto Q^2$

p_c: Constant pressure loss across the flow variator. This can amount to between 10% and 50% of the total pressure loss in the system. $p_c = c$

Reference to Figure 6.2 shows that the resulting system curve of "orifice" is far from the usual square law relationship where $p_s \propto Q^2$. When assessing the suitability of the fan we must, therefore, consider that the resultant

$$p_s = (p_a + p_b + p_c) \propto Q^2 + c \qquad \text{Equ 6.4}$$

Even this is not the complete truth. For the reasons given in Chapter 5 and Section 6.2.1

$$p_s = (p_a + p_b + p_c) \propto Q^{n+1} + c \qquad \text{Equ 6.5}$$

It must be emphasised that no type of fan flowrate control is applicable to all installations. The type selected will depend on the

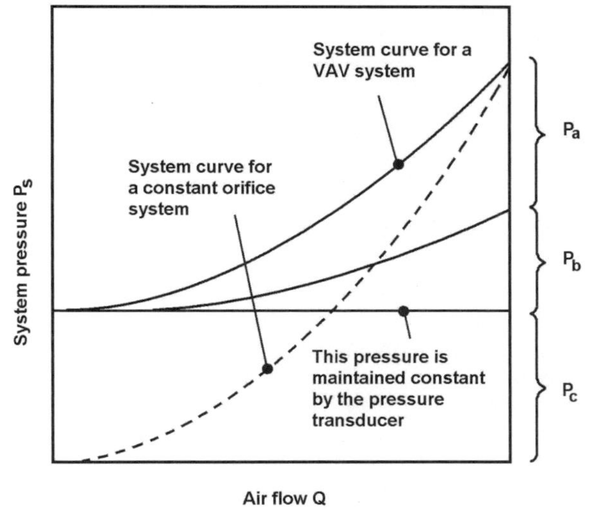

Figure 6.2 System pressure in a VAV system

turndown ratio required, how the system resistance varies and the presence of contaminants or high temperatures. Where the system has high values of fixed resistance elements, variable speed solutions will not operate to best advantage. With reduction in fan speed, the fan may develop insufficient pressure to satisfy system requirements. Some of the features and advantages/disadvantages of the various designs are detailed in the following Sections.

6.3 Damper control

The reduced efficiency accepted, dampers offer a low first cost method of controlling flowrate. They are easily adjusted and additional space is often minimal as they are inserted in the existing duct layout. They are manufactured in all types of material according to the gas constituents and temperature. They can be positioned either in the inlet or outlet duct, this being determined by fan type and characteristics.

Since a damper operates by adding resistance to the system or by "destroying" fan pressure, its only effect upon fan power is to move the operating point nearer to the closed condition. With the wider backward bladed fans, this may have little or no effect on power absorbed as the power characteristic is virtually constant (non-overloading) over the working range. With rising pressure a characteristic of closed conditions it also means that the amount of pressure to be dissipated across the damper is ever increasing. The overall efficiency can then be very low.

For the narrower backward bladed fans and for other blade designs where the power absorbed reduces significantly at lower flowrates, an outlet damper is a reasonably economical control situation. With wide forward curved bladed, or multivane fans where the pressure characteristic is flat or even reducing to zero flow, the amounts of pressure to be dissipated across the damper are reduced and the fan/damper combination is reasonably efficient. It can, therefore, be recommended where system resistance and power absorbed are sufficiently low to justify the use of the multivane.

6.3.1 Parallel blade dampers

The free area through the damper is not substantially reduced until the blades have been turned through a considerable angle. The quadrant arm, therefore, has to move through a large arc for a small reduction in fan capacity. This means that such a damper may best be installed on systems requiring flows between 70% and 100% of full capacity. The greater the number of blades, the more movement is necessary for a given flow reduction. Its sensitivity enables predetermined lever settings to give good repeatability of flowrate.

It is more readily manufactured for rectangular ducts and thus is mainly used on the outlet of centrifugal fans. It may, however, be used on fans fitted with a boxed inlet when a degree of pre-swirl (and power saving) is achieved. (See Section 6.5.3.)

6.3.2 Opposed blade dampers

These act in the same way as parallel blade dampers, but alternate blades are made to turn in the opposite direction. The free area through the damper reduces more proportionally with the blade angle. Flowrate reduction is thus almost directly proportional to the angular movement of the damper control arm. Again, this type is normally restricted to the rectangular outlets of centrifugal fans, as the complexity involved in the sealing and leakage of circular units makes this variant too expensive.

The dampers are also used when it is necessary to maintain an even distribution of air immediately downstream of the damper, due to the proximity of branch take-offs etc.

6.3.3 Single blade swivel dampers

These are a very simple form of control similar in operation to the parallel bladed type. They can be easily manufactured in circular or rectangular cross-section and thus may be easily positioned on the inlet or outlet of both centrifugal and axial flow fans. Although less movement of the damper arm is needed, sensitivity is also reduced and their use should be restricted to systems requiring flow regulation between 50% and 100% of full flowrate.

It should also be noted that at low flow rates considerable distortion to the velocity profile can result. Under these circumstances their use adjacent to the inlet of both axial and centrifugal fans may be detrimental.

6.3.4 Guillotine dampers

These consist of a single plate which can move from one side of the duct to "cut-off" the airflow. Most widely used for fan isolation, they should only be used for flow regulation after careful consideration. The velocity profile will be considerably distorted to one side and damage to the fan can eventually result. Where a tongue piece is positioned in the outlet of a centrifugal fan, or where the impeller is asymmetrically placed, special care must be taken or there may be a zero effect.

Radial inlet vanes are considered later, for whilst they also act as a damper, their major intention is to promote pre-swirl into the fan inlet. Thus, they materially affect the fan geometry and an additional power saving results.

A comparison of the flowrate control for various types of damper is shown in Figure 6.3. It must be emphasized that this is approximate only. In fact, it is specific to a particular fan and ducting system. The general trends however may be taken as indicative.

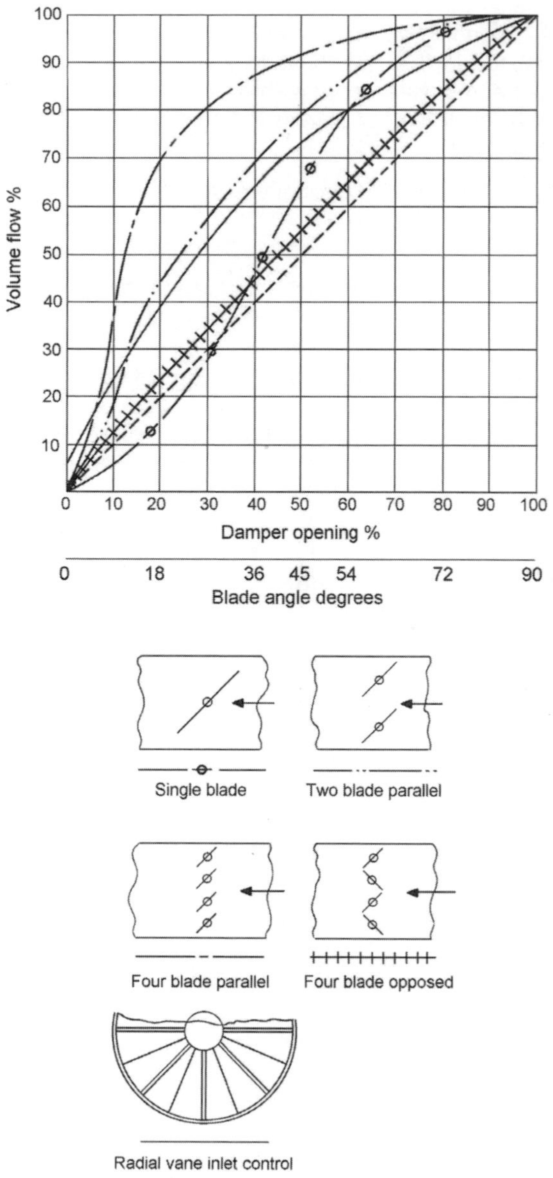

Figure 6.3 Approximate effect of damper blade opening on flowrate (constant system)

6.4 Variable speed control

Where high efficiency fans, such as centrifugal units fitted with backward inclined, backward curved or aerofoil impellers or premium efficiency downstream guidevane axial flow fans are installed on constant orifice or series path systems, then reduction in flowrate by varying the speed is preferred. In this way full advantage can be taken of the fan characteristic without sacrificing the inherent low energy demand. It should be noted that speed variation is not usually suitable for parallel path systems due to the reduction in pressure developed ($P_s \propto Q^2 \propto N^2$) with decreasing flow.

Whether speed variation can be used on VAV systems will depend on the fixed element of system resistance due to the flow variator. Where this is 10% of the total fan pressure at maximum duty it is acceptable, but at 50% the variation in flowrate possible will probably be unacceptable.

Suitable prime movers for variable speed include:

- AC electric induction motors with inverter drive
- Slip ring and commutator type AC electric motors
- DC electric motors
- Variable vee belt drives with AC electric motors
- Steam turbine and reciprocating motors

Multi-speed dual wound or pole changing electric motors can be used when the operating requirements are clearly defined. For example there may only be specific winter and summer, or continuous and overload, duties to be met. In conjunction with damper control, a wider duty variation is possible, and this combination is often a very simple solution to the problem.

Where continuously variable control down to about 50% design flowrate is required, the economy achieved by slip couplings of the eddy current, scoop control fluid, or powder type may be indicated. There is the additional advantage of improved starting by gradually "letting in" the fan inertia. In all such cases close consultation between system designer, fan manufacturer, and coupling manufacturer, is necessary to achieve the best results in energy saving.

A steam turbine drive with a gearbox to give optimum matching of fan and turbine speeds is usually the most efficient. It is only considered for industrial applications, however, where a suitable steam supply is available.

Table 6.1 below shows typical overall drive efficiencies for a 15 kW input at ¼, ½, ¾ and full speed.

Prime mover and control	Range	Usage	Typical overall efficiency			
			¼ N	½ N	¾ N	N
DC motor with AC input through thyristor control	Speed and torque adjustable over range	Speed reduction required at full torque	70	86	88	89
AC motor with variable vee rope drive	Speed adjustable down to ⅙ N. Torque increases with decreasing speed	Infrequent requirements for speed reduction	70	80	83	85
AC motor with inverter voltage and frequency control in rotor circuit	Speed adjustable down to 1/12 N.	Torque reduces substantially with speed	50	60	77	85
AC slip-ring motor with resistance control in rotor circuit	Full speed control only possible down to ½ N.	Minor speed adjustment or easy starting as control losses substantial	22	45	67	89
AC motor with slip coupling	Adjustable over whole range with electronic equipment and tachometer generator	Limited speed reduction and easy starting	20	41	62	82
Steam turbine and gearbox with variable supply	Adjustable over whole range but requires suitable steam supply	Good speed reduction but requires gearbox to match optimum turbine speeds	70	86	88	90

Table 6.1 Typical overall drive efficiencies

It is not always realised that centrifugal impellers of backward bladed design, whilst shown on performance data as having a smooth continuous characteristic of pressure against flow, often have a small order discontinuity close to their peak pressure point. This discontinuity usually increases with impeller width and is the result of a rotating stall "cell" between adjacent blades. Manufacturers try to obtain the maximum airflow from a given casing size by incorporating wide impellers. This results in the performance being obtained with the smallest space envelope.

For a given resultant pressure rise there is a relationship between the blade inlet and outlet radii. The inlet cone throat diameter is dictated by the blade inlet diameter. Thus there is an optimum width of impeller for the correct inlet throat area/impeller inlet blade area ratio. An increase in this value will result in the impeller being susceptible to inlet disturbance and the resultant discharge airflow may contain disturbing pulsations. These can be difficult to deal with, and the downstream ducting may become "live" to low frequency vibrations.

The area of instability is shown in Figure 6.4 which also indicates a typical VAV system curve. If speed control is used as a means of modulation then entry into this unstable area is inevi-

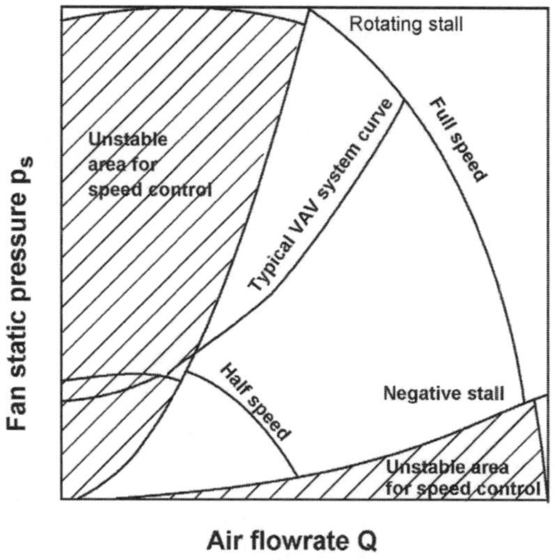

Figure 6.4 Instability with speed control of wide backward bladed centrifugal fan

table. This is often overlooked with the availability of low cost (but lower efficiency) prime movers.

6.5 Variable geometry fans

The possibilities for varying the fan geometry are limitless. Many exotic methods have been tried on both centrifugal and axial flow fans. In all systems, the intention is to vary the inlet/outlet velocity flow triangles. At inlet, pre- or contra swirl of varying amounts may be induced by the use of variable angle radial vane inlet controls. They have been used extensively for over forty years with backward inclined or aerofoil bladed centrifugal fans where they have proved particularly successful, and also with axial flow fans where the simplicity of a non-rotating control has been desired. The operating range at high efficiency with axials is, however, somewhat narrow. Mixed flow fans are becoming more popular and again this method of control is widely used.

For axial flow fans, the alteration of impeller blade pitch angle at rest has been available for many years but over the last two decades the means of varying the pitch angle in motion has extended from the high technology mine ventilation and mechanical draught installations into the more humble HVAC plant. This is now seen to be an extremely efficient and versatile form of control, rivalling the inverter drive on constant orifice systems. It also gives useful power savings on variable flow/constant pressure and constant flow/variable pressure systems.

Other less popular methods of centrifugal fan control have consisted of variable angle impeller tips and a rotating plate attached to the impeller backplate which can vary its axial position and, therefore, the impeller blade width. A cylindrical drum moving axially over the impeller periphery to achieve the same result has been more extensively used in North America.

Some of the most popular types are now described in a little more detail.

6.5.1 Radial vane inlet control (RVIC)

The full pressure development of a fan is achieved only when the air enters the impeller eye axially and without swirl. If the air or gas entering the fan is already spinning in the direction of impeller rotation, the fan will develop less pressure.

Both flowrate and power absorbed will thus be reduced. It is the purpose of this control to induce pre-rotation. In effect, it alters the design pressure/flow characteristic whilst largely maintaining the fan's efficiency. Thus the power consumption can be considerably reduced with lowering fan capacity.

Radial vanes are most effective with backward bladed high flowrate fans where the pressure curve rises considerably above the duty condition, the power is non-overloading, and the impeller inlet velocity vectors are of such a magnitude that they can be materially affected. With other types, especially the forward curved, power savings are not nearly so great and often only marginal.

The relationship between control arm movement and flowrate reduction is intermediate between the two previous types. Such dampers should never be used on direct pneumatic conveying or high dust burden extract systems as they require many parts within the air/gas stream subject to erosion and/or corrosion.

A typical performance characteristic for a backward aerofoil centrifugal fan is shown in Figure 6.5. Superimposed are the effects of the various types of system and thus the energy savings achieved. It should again be noted that a typical VAV system will have a system characteristic intermediate between the parallel path and constant orifice systems. When the fixed element of system resistance is a large proportion of the total, then the power savings will approach those for parallel paths, whilst if it is small, then the power saving will be similar to that for a constant orifice system.

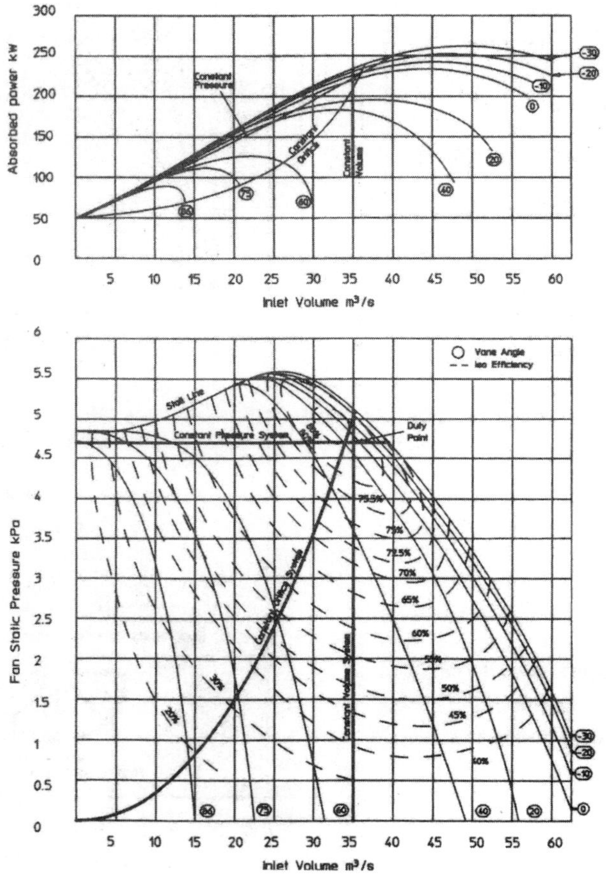

Figure 6.5 Typical performance characteristics of aerofoil bladed centrifugal fan fitted with a RVIC

The mechanical design of the vanes and particularly the mechanism can cause problems because of the need for continuous maintenance and greasing. This is due to the high friction and corresponding high operating torque required for the operation of the actuating mechanism.

This mechanism usually comprises an external ring and a number of actuating levers, one lever for each vane (see Figure 6.6). The vanes are supported by a larger hollow collar at the centre to allow the fan shaft to pass through. The ring is nor-

Figure 6.6 RVIC with external operating gear

mally supported by a number of rollers. The actuating levers are connected to the ring via double links to overcome the great differences between the paths of the levers and the external ring.

It is such mechanical problems which have resulted in doubt as to their reliable operation for VAV systems, especially as the fans are often of double inlet design necessitating cross linkage between the two assemblies.

As with speed variation, when considering the use of RVICs as a means of control, then the resultant area of instability may lead to problems. In "wider" impeller designs this area can be large, see Figure 6.7. This has lead some to claim that one should not consider their use if a flowrate of less than 50% of design is required. Below this ratio simple damper control would have to take over, with its resultant inefficiency.

However, by correct impeller/RVIC design selection, the area of instability can be very small with modulation over the entire VAV system curve totally stable. Normally a turndown to 20% can be achieved with a single speed drive motor and this is generally

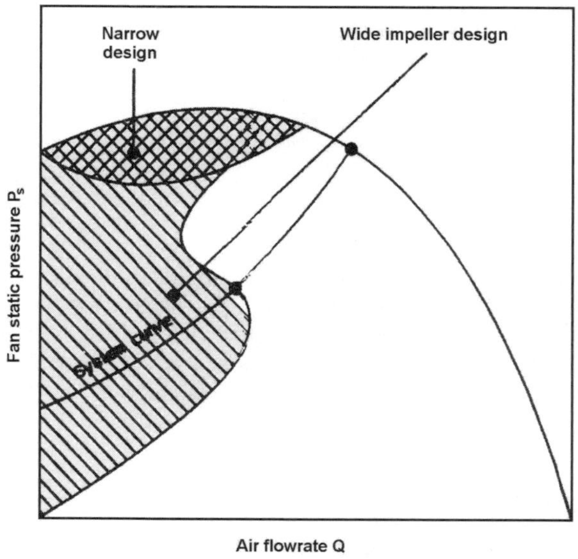

Figure 6.7 Instability with radial vane inlet control of backward bladed centrifugal fans

sufficient for VAV system use. By using a two speed fan, operation down to 10% of design is feasible.

It should be noted that due to the relatively large clearances necessary at the centre support, zero flow is impossible and even with complete closure there will be a leakage of up to 8%.

As well as inducing pre-swirl, the RVIC imposes an additional and increasing resistance as the vanes approach full closure. This is the explanation for the corresponding reduction in efficiency, as this loss of energy is then attributed to the fan/RVIC combination. RVICs are very expensive and the price for two fitted to a double inlet fan can even exceed the price of the bare fan itself.

Controls incorporating an internal mechanism can be less expensive (Figure 6.8) but are usually limited to clean dry air applications.

Figure 6.8 RVIC with internal operating mechanism

6.5.2 Semi-circular inlet regulator

First introduced by Davidson & Co of Belfast, this is a very much simplified device for imparting swirl to the air entering the inlet of a centrifugal fan. It consists of a split circular plate in which the top and bottom halves swing in opposite directions (Figure 6.9) and thereby induce the required circular motion to the incoming gas stream. Extremely simple in concept and therefore cheaper to produce, it is only slightly less efficient than the RVIC.

Figure 6.9 Davidson semi-circular inlet regulator

6.5.3 Differential side flow inlet control

Where a centrifugal fan has to be fitted with an inlet box for side air entry, the possibility for incorporating a simplified method of flowrate control is apparent. If the box is fitted with a set of parallel bladed dampers then these can impart pre-swirl (Figures 6.10 and 6.11). Thus a power saving almost as good as a RVIC can be achieved, (Figure 6.12).

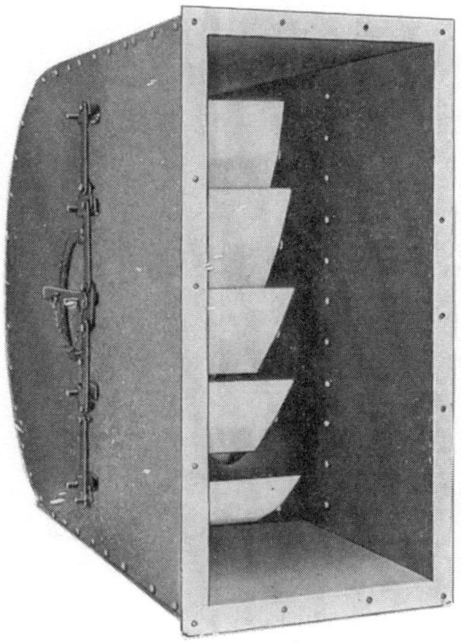

Figure 6.10 Inlet box incorporating side flow control

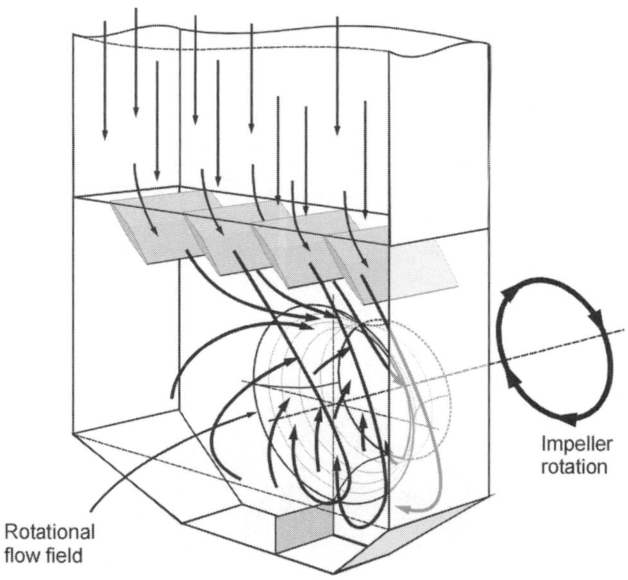

Figure 6.11 Flow path of air with differential side flow inlet control

6.5.4 Disc throttle

The unit comprises a profiled circular plate supported co-axially within a centrifugal impeller. It is described in UK Patent 2,119,440B. It is necessary for the inner edges of the blades to be parallel to the impeller axis so that a close clearance can be maintained with the periphery of this disc throughout its movement. The plate is carried by an axially extending shaft which projects outwards through the inlet venturi and is moved axially by means of an actuator of any convenient kind. The actuator is

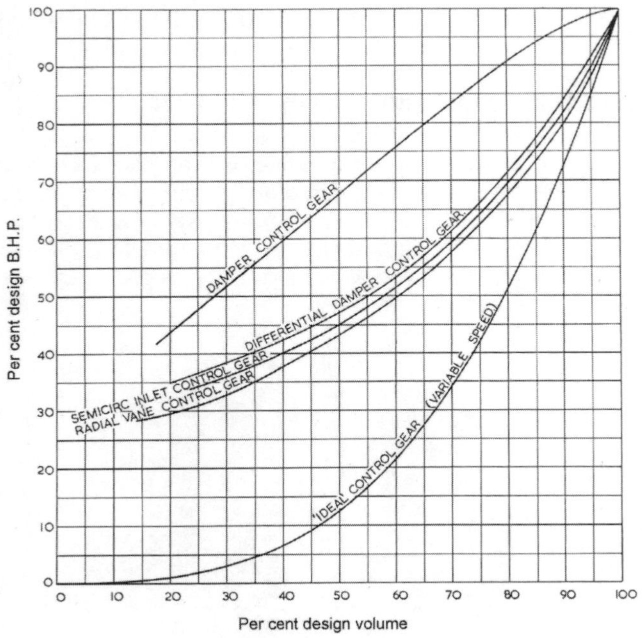

Figure 6.12 Power absorbed by various types of fan control

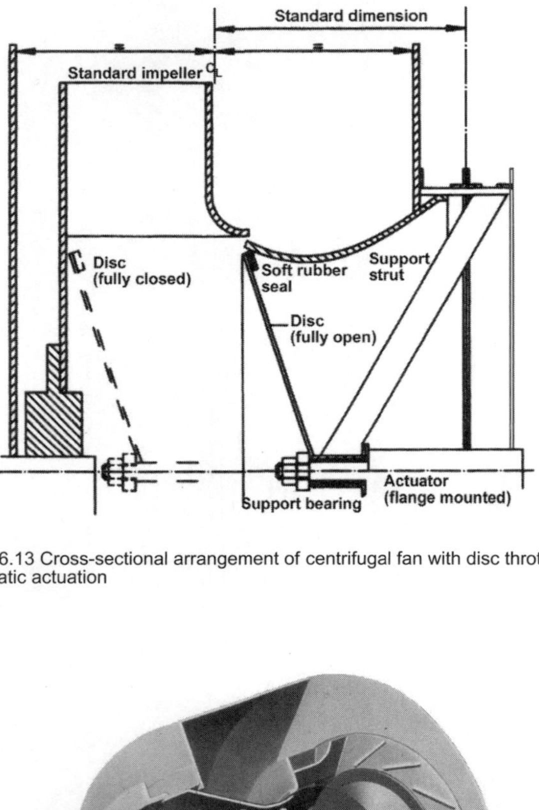

Figure 6.13 Cross-sectional arrangement of centrifugal fan with disc throttle for pneumatic actuation

supported from the fan casing by suitable brackets or rods and where the travel is particularly long, an additional sliding bearing may be incorporated to support the shaft. A cross-section of the arrangement is shown in Figure 6.13 and the general layout is shown in Figure 6.14.

Movement of the rod alters the position of the disc axially with respect to the impeller's blades and this effectively controls the flowrate by varying the active width of the blades. The disc does not rotate and it will be seen that there are, therefore, a minimum of moving parts. This produces an inexpensive device, and a high efficiency is maintained for a considerable turndown. A soft rubber ring can be attached to the outer edge of the disc so that when the damper is withdrawn up to the venturi, the inlet flowrate is almost zero.

Conversely, with the plate close to the impeller backplate, the flow is at a maximum and almost the same as that for a fan without a disc throttle.

This, therefore, permits the control to be used with very wide impellers to achieve the maximum flowrate from a given space envelope, without the risk of entering the stall range. Its simplicity and effectiveness has been optimised with the development of a special range of impellers having dimensions calculated to make the best possible use of the disc throttle.

The control offers a substantial energy reduction compared with conventional dampers. There is also an additional power saving compared to radial vane inlet controls. With the damper plate acting on width, operation is unaffected by blade shape and these may, therefore, take many of the forms commonly used in centrifugal fans, such as backward inclined, backward curved, aerofoil, shrouded radial and radial tipped.

Flowrate control is substantially linear over a wide range. Even forward curved bladed fans may be fitted when an additional power saving over normal dampers is made, albeit small, in contradistinction to the radial vane inlet control.

Again, with narrower width high pressure impellers, the power savings become less but the other advantages outlined remain. The disc throttle is a competitive solution to many centrifugal fan flowrate control problems.

As the effective width of the impeller is narrowed, there is still a small stall point at each setting until at about 1/3 effective width this can no longer be detected. The unstable area for disc throttle is therefore very unlike the RVIC (see Figure 6.7) and is shown in Figure 6.15.

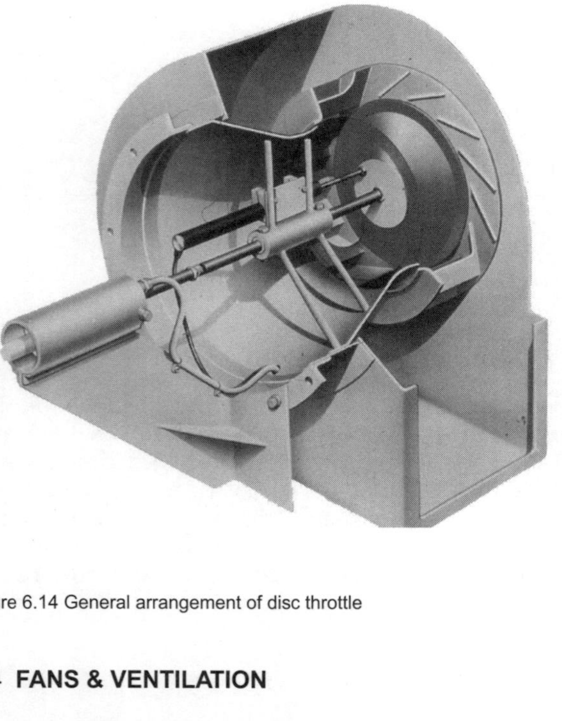

Figure 6.14 General arrangement of disc throttle

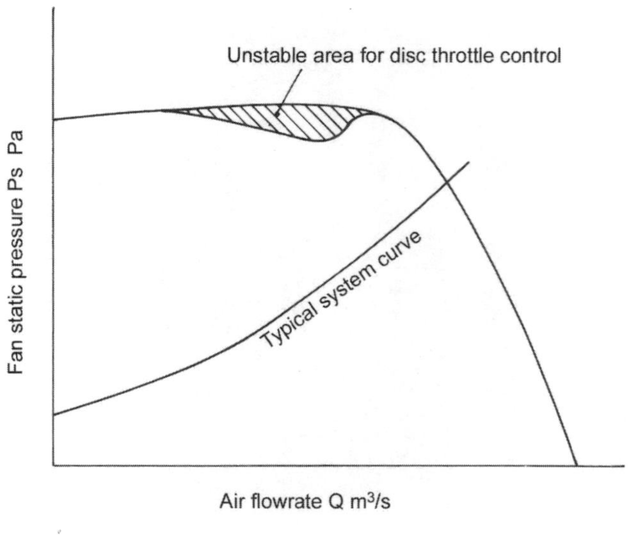

Figure 6.15 Instability with disc throttle of wide centrifugal fans

6.5.5 Variable pitch-in-motion (VPIM) axial flow fans

One of the most important parameters in the design of any turbo machine is the angle which the outer edges of the blades make with the tangent of the peripheral motion. As this angle is increased, so the volume flowrate will also increase, and this applies to axial, mixed flow or centrifugal fans. At the same time the pressure, which is a function of the swirl, remains substantially constant.

It will, therefore, be seen that if the pitch of the blades of an axial flow fan could be altered in motion, then an effective method of volume control would be available. The technology to do this already existed with the aircraft propeller, albeit where the number of duty hours was considerably less than the humble ventilating fan. Nevertheless, over the last few years, the systems necessary have been simplified to enable a sufficiently reliable fan to become available for normal HVAC applications.

As previously stated, only variable pitch axial fans can adequately meet the needs of constant orifice, constant flowrate or constant pressure systems. The energy savings made have been amongst the highest achieved, and a reasonably good efficiency is maintained over a turndown of 4:1. The aerodynamic performance of such fans is, of course, similar to normal adjustable pitch-at-rest axials and a typical characteristic is shown in Figure 6.16. The centrifugal force on an individual fan blade can be considerable and is a function of the blade weight and its rotational speed. For a typical application, this force can be as much as 600 times the dead weight. These forces are usually resisted by anti-friction bearings of the ball or roller type. Such bearings have a lower capacity under the virtually static conditions prevailing, and in the early days failure was not uncommon. With increasing experience, however, the problems have been overcome.

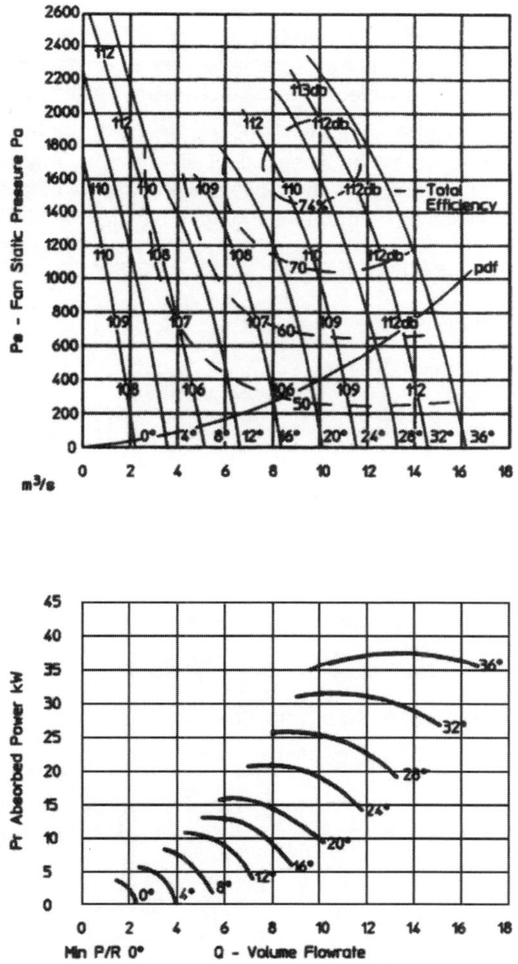

Figure 6.16 Characteristic curves for 710 mm diameter VPIM axial flow fan at 2950 rev/min and handling standard air

Levers at the base of each blade convert the equal pitch angle adjustment into axial movement of a sliding member within the impeller hub. This may be controlled in a number of ways:

a) By movement of pneumatic bellows against a spring as shown in Figure 6.17. The bellows are expanded by compressed air through a rotary air seal onto a shaft extension.

b) By an actuator (either pneumatic or electric) giving axial movement through levers to the stationary race of a ball thrust bearing, the revolving race being coupled to the sliding actuator within the hub.

An alternative pneumatic arrangement is shown in Figures 6.18 and 6.19, with an overall fan assembly shown in Figure 6.20.

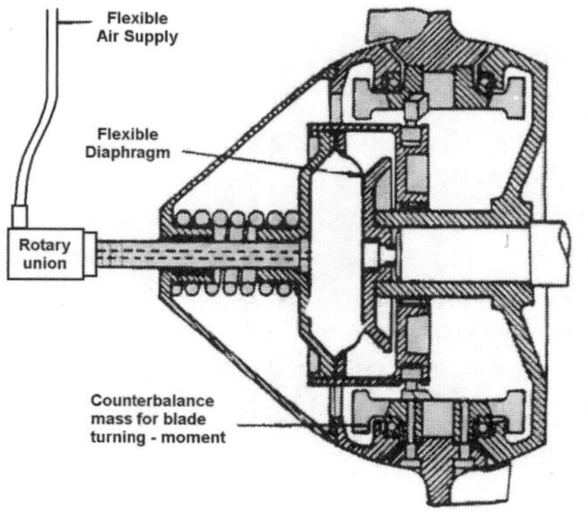

Figure 6.17 Cross-sectional arrangement of hub mechanism for VPIM axial flow fan (compressed air operation)

Figure 6.18 Sideways view of alternative form of pneumatically operated VPIM axial flow fan

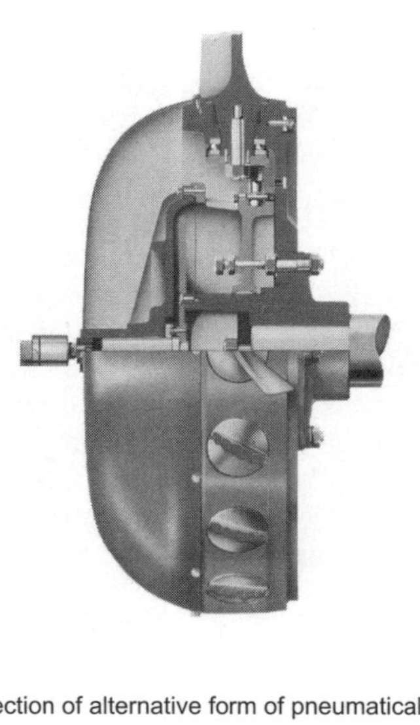

Figure 6.19 Cross-section of alternative form of pneumatically operated VPIM axial flow fan

Figure 6.20 General arrangement of VPIM axial flow fan

In all cases, when the fan is running, a force must be applied to each blade to maintain the required pitch angle or it would rotate to a position near zero pitch angle where the centrifugal forces on it were in balance. Weights are sometimes attached to the blade root, at right angles to the blade pitch, to produce a counterbalancing moment and thus reduce the actuating force necessary. In the event of compressed air supply failure, the flowrate will, of course, revert to minimum unless some alternative is available.

6.6 Conclusions

The advantages of maintaining a good fan efficiency across the range of operating points are clear - low running costs which can lead to the additional capital cost being recouped in a very short period of time - often less than two years. A high efficiency impeller may not necessarily be more expensive as, with a reduction in internal losses, the fan may even be reduced in size for a specific duty.

In an age of aggressive marketing, care must be taken to read beyond the advertising "blurb". No form of flowrate control is applicable to all types of system and the user must distinguish between the different types of system. Speed control by the use of inverters with induction motors is not a universal panacea.

Graphs of the type shown in Figures 6.12 and 6.21 are common, but attention is again drawn to some of the assumptions made and to the fact that they are only applicable to fully turbulent constant orifice systems, where $P \propto Q^3 \propto N^3$. It must be appreciated that they are approximate and that they refer to specific items of equipment. The full cubic power saving is never achieved in practice. The general conclusions are, however, valid.

In the analysis, the backward bladed fan has an assumed static efficiency of 80%, whilst for the forward curved and variable pitch axial, this is 60% both at the design flowrate. The differences would be smaller if both axials and centrifugals were selected on a total pressure basis as recommended in the fan test standards ISO 5801/2. Special attention is drawn to the use of wide backward bladed centrifugal fans with 2 speed (dual wound 4/6 pole shown) motors and disc throttle dampers. This is a relatively cheap installation rivalling more sophisticated methods in its control efficiency.

DC motors with thyristor control surpass all others, but AC motors with inverter drives are almost as efficient and much more reliable. Both enable high efficiency centrifugal fans to match the power savings of variable pitch axial flow fans.

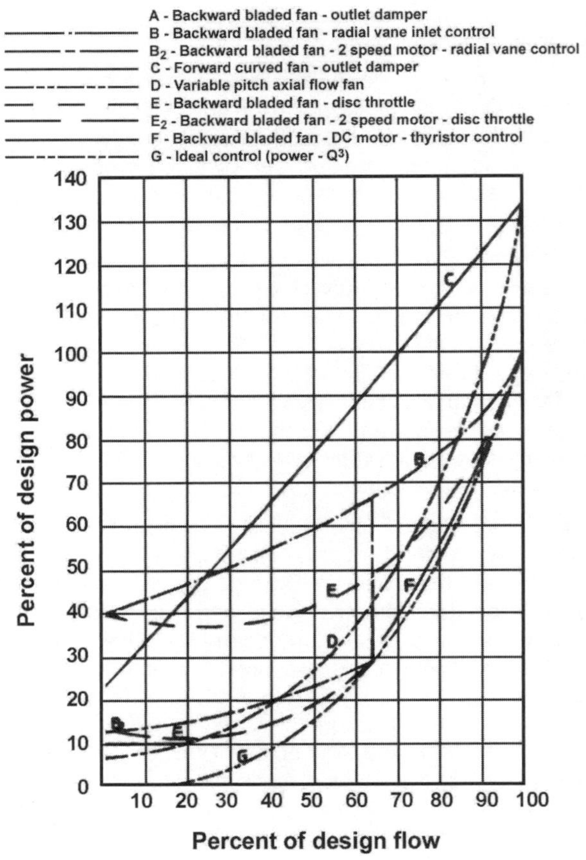

Figure 6.21 Power savings for damper and speed control

Speed control, whilst the preferred method for constant orifice or fixed systems, and also usable in many constant flow systems, is not applicable to constant pressure systems.

You would expect a fan manufacturer to say it, but more care should be devoted to selection of appropriate equipment. Where comparisons are to be made on the basis of absorbed power, certification schemes such as those provided by AMCA and Eurovent become necessary. Performance data needs to be independently validated.

Remember that:

$$P(\text{Power input})\ \text{kW} = \frac{Q \times p_f}{\eta_f \times \eta_m \times \eta_t \times \eta_c}$$

where:

Q	=	flowrate (m^3/s)
p_f	=	fan (total) pressure (kPa)
η_f	=	fan (total) efficiency (decimal)
η_m	=	motor efficiency (decimal)
η_t	=	transmission efficiency (decimal)
η_c	=	control efficiency (decimal)
P	=	input power (kW)

The need to avoid unnecessary energy conversions is obvious, and direct drive fans should be considered wherever possible.

ETSU, BRESCU and their more recent successors, and others can take justifiable pride in the manner in which they brought to public attention, the reduction in running costs by changing from normal to high efficiency motors, when a saving of perhaps 5% can be made. How much greater would be the savings if the many fans with impeller efficiencies of 50 to 60%

were changed for units having efficiencies of greater than 75%, and if appropriate fan regulators were fitted which were matched to their systems.

There is, of course, one foolproof method of saving power. Don't leave a fan idling! Switch it off when it is not doing any useful work.

A particular example of this technique may be found in some bulk storage grain drying plants. Here the fan is controlled by a hygrostat and can only be run when the ambient air has a moisture content below the equilibrium moisture content of the grain, thus permitting some useful drying to take place without the need for auxiliary heat.

6.7 Bibliography

Centrifugal fans, UK Patent 2,119,440B, 1983-11-16, W T W Cory, Patent granted 1985.

ETSU, (Energy Technology Support Unit), set up by the UK government in 1974. Superceded by Future Energy Solutions (Part of AEA Technology), PO Box 222, Didcot, OX11 0WZ, UK, Tel: 0870 1906374, Fax: 0870 1906318.

BRESCU, Building Research Energy Conservation Support Unit. Replaced by BRESEC (British Research Establishment Sustainable Energy Centre) in the UK; Tel: 0870 1207799, e-mail brescuenq@bre.co.uk, www.bre.co.uk.

7 Materials and stresses

Whilst the fan industry has been characterised throughout this book as "mature", there has nevertheless been a revolution over the last few years in its use and selection of materials. The axial flow fan owes its increasing popularity to the availability of lighter materials which have reduced the centrifugal stresses to acceptable levels. Invented at the beginning of the nineteenth century, it did not prove a manufacturing success until after the 2nd World War. The aircraft industry had developed the aluminium alloys which were just what the fan industry wanted! This has been followed by the increasing use of engineering plastics.

For centrifugal fans pre-galvanised sheet has become an accepted norm for light duty fan casings, often of lock-formed construction. Aluminium alloys and even plastics have been introduced for impellers. At the other end of the duty scale, nickel and titanium alloys have extended the peripheral speeds and hence pressures that fans are able to achieve.

This Chapter does not seek to be a comprehensive textbook on materials. Rather it seeks to point those interested to the right sources of information. The stresses induced in the various parts of a fan can be subject to mathematical analysis and an introduction is given to the methods used. With the advent of specialised computer programmes, however, some readers may be tempted to think that a knowledge of first principles is unnecessary. It is hoped that these paragraphs will disabuse them of such thoughts!

Contents:

7.1 Introduction

The modern fan consists of many parts which may be made from a number of materials. The choice of these will be determined by their cost, ease of manufacture and mechanical attributes. Increasingly, also, appearance may have some effect – especially where the fan is in the public eye.

Whilst the rotating parts of all fans will be subject to centrifugal forces, the resultant stresses may determine the thickness or scantlings of their components.

At the present time 3 material groups are in the ascendant:

- Sheet steels and cast irons
- Sheet and cast aluminium alloys
- Engineering plastics and composites

For the sake of analysis, however, we may make a more coarse definition of metals or non-metals and these are discussed in Section 7.3.

7.2 Material failure

Whilst engineers may argue over the way that materials fail, it has to be recognised that there is no universally accepted definition of the manner in which this occurs. Figure 7.1 shows the generally accepted points on the journey to failure.

The initial stage is usually a straight line relationship where stress is proportional to the extension. The graph then curves slightly to the yield point, following which there is irreversible plastic flow. The ultimate tensile strength is the maximum point at which the crack initiates. There is then a propagation stage where a crack develops until finally the material breaks.

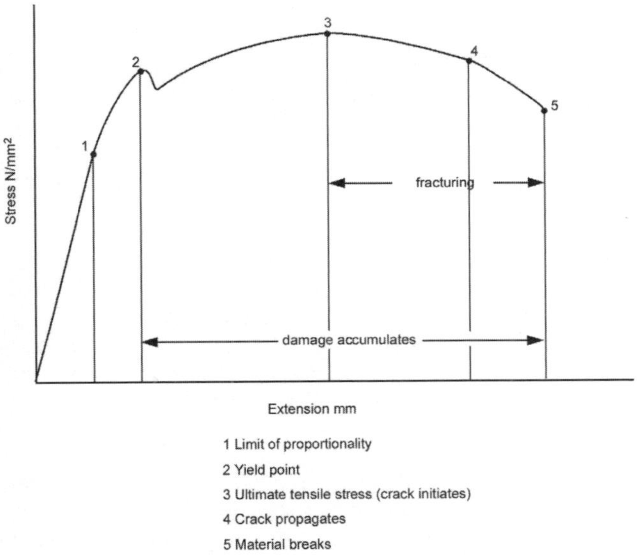

Figure 7.1 Typical phases of failure of a metal

7.3 Typical metals

7.3.1 Metal structure

All metal are recognised as having a crystalline structure. The crystals are geometrically regular in shape. The molecules are attracted to each other by "binding forces" which are non-directional and encourage these molecules to take up a regular shape. Whilst all solids have some tendency to become crystalline, metals are likely to form the most regular and packed arrangement.

Where impurities are present, the crystals like to form around them. The metallurgist tries to improve the strength of the material, by controlling the order of the metal crystals and introducing other elements necessary to improve some particular property desired for the alloy.

7.3.2 Carbon steels

Small percentages of carbon are introduced into steel to improve its strength. At the same time this may reduce its ductibility and weldability. Approximate physical properties are as shown in Figure 7.2.

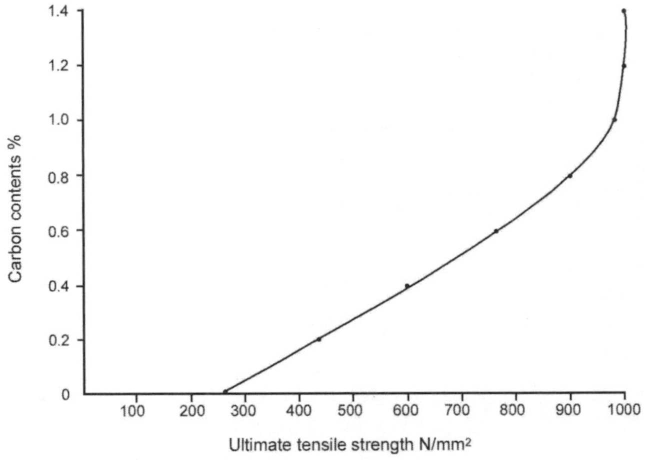

Figure 7.2 Typical strength of steel with varying carbon contents

Typical properties of such steels are shown in Table 7.1

Type	Low carbon steel	Structural steel	Steel casings	Machined part steel
% Carbon	0.1	0.2	0.3	0.4
% Manganese	0.35	1.4	-	0.75
Yield stress N/mm²	220	350	270	480
Ultimate tensile stress N/mm²	320	515	490	680

Table 7.1 Carbon content versus strength of steels

7.3.3 Low-alloy and alloy steels

Low-alloy steels have small amounts of chromium, magnesium, molybdenum and nickel to increase certain physical properties. Alloy steels have an even larger percentage of these elements, together with silicon, vanadium and others to give increased strength and hardness.

7.3.4 Cast irons

These are iron and carbon alloys which have somewhat more than 2% carbon. They may be subdivided into grey and white varieties.

7.3.4.1 Grey cast iron

These types have a grey appearance with a structure of ferrite, pearlite and graphite. The latter exists as either flakes or spheres. Nodular or spheroidal graphite cast iron is obtained by adding magnesium which helps the graphite to form spheres. This material is widely used for the hubs of centrifugal fan impellers.

7.3.4.2 White cast iron

This material is hard and brittle due to its structure of cementite and pearlite. It is difficult to machine and is therefore used for wear resisting components. In the past it has been used for cast scroll segments of mill exhausters.

7.3.4.3 Malleable cast iron

These are forms of cast iron which are heat treated to improve their ductility whilst retaining their high tensile strength. Three types are usually recognised:

Whiteheart — which is heated with an iron compound to give a ferrite outer skin and a ferrite/pearlite core

Blackheart — which is soaked at high temperature to break down the cementite and then slowly cooled to produce ferrite and graphite.

Pearlite — very much the same as Blackheart, but cooled faster to produce a higher strength

7.3.5 Stainless steels

This term describes a group of steel alloys containing over 11% chromium. There are four main categories, which in turn may be subdivided into many different proprietorial and generic grades.

Austenitic — which contain 17 to 25% chromium combined with 8 to 20% nickel and/or magnesium and other trace alloying elements. They are easily weldable due to the low carbon content and in their raw state are non-magnetic. Magnetism can however, be induced by heavy working. Good strength is combined with high corrosion resistance.

Ferritic — again have a high chromium content greater than 17% together with medium carbon content and small quantities of molybdenum and silicon. Good corrosion resistance rather than high strength and generally non-hardenable. Magnetic.

Martensitic — have a high carbon content up to 2% and a low chromium content generally around 112%. Difficult to weld. Magnetic.

Duplex — grades contain both austenitic and ferritic phases. High tensile strength at normal temperatures is combined with good corrosion resistance due to the addition of trace elements. Weldable, but becomes brittle above 300 °C.

7.3.6 Non-ferrous metal and alloys

This term is used for all those metals or alloys which do not contain iron as the base element. Apart from copper they are rarely used in a pure form and hence the term alloy is often more appropriate. Some typical properties of these alloys are given in Table 7.2.

Main constituent	Ultimate tensile strength N/mm^2	Typical alloys
Aluminium	100 to 500	duralumin, silumin
Copper	200 to 1100	Brasses, cupronickels, aluminium & tin bronzes, gunmetal
Magnesium	150 to 340	
Nickel	400 to 1200	Monel®, Inconel®, Hastelloy®, Nimonic®
Titanium	400 to 1500	TiCu, TiAl, TiSn
Zinc	260 to 360	A, B, ZA12

Table 7.2 Properties of non-ferrous alloys

7.3.6.1 Aluminium alloys

These are widely used in the fan industry where lightness combined with strength is desired. Whilst pure aluminium is relatively weak, the addition of small quantities of other elements can increase its strength and hardness enormously. Mechanical properties can also be improved by work hardening.

There are now a very large number of aluminium alloy grades available in both casting grades and sheet form.

Axial flow fan blades and hubs are frequently cast in grades such as LM6 and LM31. Readers are referred to relevant British, European and International standards for further information.

Centrifugal fans can have impellers and casings fabricated from relevant sheet grades, many of which are weldable. Again reference to standards is recommended.

The use of silumin, a grade containing about 12% silicon has especial properties for fans in explosive atmospheres. When subject to a grinding action, the material tends to fracture, before frictional deformation and heat can result.

7.3.6.2 Copper alloys

Whilst copper in its pure form may be used for electrical components, its alloys are of particular interest to the fan engineer. Thus brasses may be used as anti-spark features at the boundaries between close running, stationary and rotating parts (see Chapter 8). In this case admiralty brass, which has a small lead content, is particularly good. It has been widely used in fans for coal mines and offshore oil rigs. Some authorities, however, restrict the use of alloys containing lead and its acceptance should be verified.

The fans used for the ventilation of oil tanker holds have to be of intrinsically non-sparking design. In such cases the complete impeller may be made of aluminium bronze together with potential rubbing parts.

7.3.6.3 Magnesium alloys

Not used to any extent in the fan industry, due to their flammability. There may however be a use for them in certain special applications.

7.3.6.4 Nickel alloys

Nickel is commonly alloyed with copper, chromium and iron to produce a range of materials with high temperature and corrosion resistance. The Nimonics® and Hastelloy® have been extensively used for high temperature fans (in excess of 500 °C) whilst Monel® has been used for fan shafts, due to its ability to withstand shock loads (when dampers have to close in micro-seconds or large “lumps” pass through the fan).

7.3.6.5 Titanium alloys

Titanium may be alloyed with many other elements to produce a range of materials which are extremely light, strong and resistant to many corrosive gases and vapours. In consequence they may be used to produce a lightweight impeller which can rotate at high speed to produce high pressures. Anything is possible, so long as you can afford it!

7.3.6.6 Zinc alloys

Particularly useful for the production of small die cast parts, due to the ease of casting. Provided that stresses and shock loads are not high, then a zinc alloy may be acceptable.

7.4 Engineering plastics

7.4.1 Introduction

The use of plastics in the manufacture of fans has increased tremendously over the last two decades, especially in small units of all types. There has also been an increase in their use for the blades of large axial fans up to the very largest sizes.

The plastics used may be divided into three generic types:

- Thermoplastics
- Thermosets
- Composites

As their names imply, thermoplastic polymers can be re-softened by heating, in contra-distinction to thermosets where they cannot.

Many practical applications of plastics in the fan industry need to use composite grades to meet the necessary strength and durability requirements.

7.4.2 Thermoplastics

These are probably the most widely used group of plastic materials and include the following:

- ABS (acrylonitrile butadiene styrene)
- PVC (poly vinyl chloride)
- Polyethylene
- Polyamides (nylons)
- Polypropylene
- PTFE (polytetrafluoroethylene)

7.4.3 Thermosets

Whilst perhaps not used so widely in their solid form, they are nevertheless recognised as important for surface coatings and finishes. Examples of thermosets are:

- Alkyds
- Epoxies
- Polyesters
- Silicones

7.4.4 Composites

These are expected to be the group with the most exciting future. Not only has glass fibre been used as a strengthening agent, but there is now the possibility of using carbon fibres with even greater strength properties. Grades currently popular are:

- GRP (glass reinforced plastic)
- SMC (sheet moulding compound)

But there will be many more to come in the future.

7.4.5 Mechanical properties of plastics

These vary enormously, not only according to type, but also from one manufacturer to another. It is best to check with the suppliers of the appropriate grades and ascertain from them how their figures were obtained and also what supporting test work they can instance. Table 7.3 is therefore given as typical only.

One unfortunate property of plastics from a fan manufacturers' point of view is that even at temperatures just above ambient they are affected by "creep". Thus they are subject to extension (time dependent strain) under the most moderate stresses. It is therefore important to design for a known working life.

Plastic type	Ultimate tensile strength N/mm^2	Modulus of elasticity kN/mm^2
Epoxies	80	8
GRP	<180	<20
Nylon	60	2
Polyethylene	20	0.6
PTFE	14	0.3
PVC	50	3.5

Table 7.3 Typical mechanical properties of plastics

7.5 Surface finishes

Surface finish is an important aspect of fan appearance at the present time. Often fans are contained in plant rooms that are visible to the public.

Surface finish is also important in maintaining the underlying materials in good condition. There are numerous ways of protecting the surfaces of a machine.

Important to the successful completion of most surface finish systems is adequate preparation of the base material to ensure adhesion for an appropriate coating thickness.

The Swedish Standard SS 055900 has received wide acceptance with its Sa grades. These are given in Table 7.4. It is stated that this Standard is equivalent to ISO 8501-1.

Designation	Preparation
Sa 1	Light blast cleaning removing the worst millscale and rust
Sa 2	Blast cleaning to remove the majority of millscale and rust
Sa 2½	Thorough blast cleaning with some remaining surface staining
Sa 3	Blast cleaning to pure metal with no remaining surface staining

Table 7.4 Surface preparation grades

7.6 Surface protection

7.6.1 Introduction

To give the basic materials of a fan protection against temperature, corrosion and erosion or to improve its appearance, it is important to provide a good surface finish. The possibilities are endless, but may be considered under five basic headings:

- painting
- galvanising
- plating
- lining
- coating

These will now be discussed in a little more detail, although it is important to emphasise the necessity of discussion with a reputable supplier or specialist sub-contractor.

It is unfortunate that everyone seems to believe that he has a God-given right to specify his own unique finish. Thus the fan manufacturer may be burdened with non-standard paint systems or even unusual colours. The consequent increased workload in just substituting one paint for another has to be imagined — change of brushes or applicators, cleaning of pipelines etc., etc. Wherever possible, users are recommended to study Eurovent document 1/9 on the surface treatment of fans.

7.6.2 Painting

The number of paints in existence, and the methods by which they are applied, must total many thousands. Correct surface preparation, choice of paint system and careful application must all be right to give satisfactory protection and good appearance. Paints may be categorised into the following types:

- primers e.g. zinc phosphate or zinc chromate
- air drying e.g. alkyd resins, chlorinated rubbers or esters
- two pack e.g. epoxy or polyurethanes

Some of the types detailed above may be restricted in their use by local or national ordinances, especially where they are likely to end up being poured into the drains. There is a trend towards water-based paints, as apposed to oil or lead bases, for such reasons.

There are a number of national and international Standards which are relevant including BS 381, BS 5493 and BS 7079.

7.6.3 Galvanising

This is the term used for the coating of iron and steel components with zinc. It is probably a more robust surface than paint in protecting the underlying metal from corrosion. The initial bright finish (often enhanced by the inclusion of a small amount of aluminium in the molten zinc tank), however, rapidly "dulls" in service.

The coating is usually defined by its weight per unit area in accordance with the grades specified in BS 729. See also ISO 1459 and subsequent revisions. Weights can vary from around 300 to 800 g/m^2, the heavier coatings being applicable to thicker materials, or where the ambient atmosphere is aggressive e.g. an oil refinery close to the sea.

7.6.4 Plating

Perhaps the most commonly recognised plating is that using chrome. Not only can it give an excellent surface and appearance, but it also gives a measure of protection against many adverse environments. Many plating systems are quite complex and have a layer of copper beneath the chrome.

Nickel can also be used for electroplating and the relevant standards for both materials are BS1224, ISO 1456 and ISO 1458.

7.6.5 Lining

Perhaps more popular in the past than nowadays, is the lining of industrial fans with thick rubber to all surfaces in contact with the gas being handled. The lining is applied in sheets up to about 6 mm thick to the casing of either cast iron or sheet steel. Impellers also may be lined. These are usually of the open paddle bladed type, although it is possible with shrouded types provided sufficient clearance is maintained at the interface between the shroud lip and the fan inlet cone.

There are two main types of rubber used:

- Natural rubbers for ambient temperatures where the air/gas is oil-free
- Synthetic rubbers such as nitryl, butyl or neoprene for gas temperatures up to 120 °C and/or when fumes are present.

Both natural and synthetic rubbers are available as hard or soft grades. The hardness scales used are the Shore scale or the IRHD (International Rubber Hardness Degrees).

Hard rubber or ebonite is 60-80 Shore D scale or 80-100 IRHD. Soft rubber (India rubber if natural) is 40-80 Shore A scale or 40-80 IRHD. For further information see ISO 7619.

The design of rubber-lined components is especially important to ensure that there is adhesion (see BS 6374). The procedure requires that:

- metal surfaces are shot blasted to Sa 2½
- adhesive is applied to all the surfaces to be lined
- rubber sheets are manually laid with overlapping joints
- rubber is vulcanised by heating to 120 °C using steam or hot water.

Testing of the lined components is essential to ensure their integrity and the following are commonly specified:

- Spark testing at 20kV to guarantee continuity
- Rap testing with a special hammer to check the adhesion between rubber and metal
- Hardness testing using a hand-held gauge to measure this hardness and to ensure that the vulcanising process has been completed.

Other materials which have been used for thick lining of fans include many other polymers and organic materials. PVC and other plastics have also been employed.

7.6.6 Coating

The term coating is used to apply to thin coating perhaps of only 150 μm thick. Typically these are of glass-like appearance and are baked on. To ensure continuity they require all sharp edges to be rounded and welds ground smooth over and above the requirements of Sa 3.

7.7 Stressing of centrifugal impeller

7.7.1 Introduction

When designing a centrifugal impeller it is important to be able to calculate the stresses induced, to ensure the selection of the correct materials. Such impellers may be considered to comprise four elements:

- shroud
- backplate
- blades
- hub

On the above the shrouded and backplate may be considered as discs. Centrifugal forces act on these discs, as well as the blades and hub. The loads imposed by the air/gas on the impeller are invariably small when compared with those due to rotation. The latter, of course, become especially important in high pressure fans when peripheral speeds are high.

Any element of the disc will have three stresses acting on it, namely, radial, tangential, and axial. The latter is quite small and is neglected. Fundamental equations to determine the radial stress R and the tangential stress T produced in the disc were derived by Dr. A. Stodola in his book on steam and gas turbines.

These equations are based upon the following assumptions:

a) The disc is symmetrical about a plane perpendicular to the axis of rotation.

b) The disc thickness varies only slightly, so the slope of the radial stresses toward the plane of symmetry is negligible.

c) The stresses are uniformly distributed over the cross section.

In applying the basic equations, it is necessary to express the shape of the profile by some mathematical equation or have the profile closely approximate it. For very special applications, a single equation may be used; e.g., the De Laval constant strength disc. However, for general work the disc is usually divided into a number of sections having some particular shape such as conical rings, constant thickness rings, hyperbolas, etc., and then the stresses in these sections are found.

The method using parallel sided, constant thickness "flat" shrouds or backplates can give especially accurate results and is described below. It is perhaps one of the important reasons for using flat shrouds, as well as making blade shapes simpler. However, because of its simplicity and adaptability to any disc shape or load condition it has been widely used for all types of impeller.

7.7.2 Sum and difference curves

The method uses the sum S and the difference D of the tangential T and radial R stresses, as applied to parallel-sided discs, i.e.:

$$S = T + R$$

$$D = T - R$$

For the special case of a constant thickness disk, Stodola's basic equations reduce to

$$S = (1+\nu)\frac{\rho}{2}\left[K_1 - u^2\right] \qquad \text{Equ 7.1}$$

$$D = (1-\nu)\frac{\rho}{4}\left[u^2 + \frac{K_2}{u^2}\right] \qquad \text{Equ 7.2}$$

where:

$K_1 = \dfrac{4b_1E}{(1-\nu^2)}$

and

$K_2 = \dfrac{8\omega^2 b_2 E}{(1-\nu^2)}$

ν = Poisson's ratio, or the ratio of the strian perpendicular to a force to the strain in the direction of the force (0.3 for steel)

ρ = density of the material (kg/m^3)

u = tangential velocity (m/s)

ω = angular velocity (rad/s)

E = modulus of elasticity (N/m^2)

b_1 & b_2 = constants depending upon the stress conditions at the bore and rim

For a disc rotating at a given speed, the only variables for any given radius are K_1 and K_2. Hence, arbitrary values of K_1 and K_2 may be assumed, and the values of S and D may be plotted against the tangential velocity u. In this way the chart shown in Figure 7.3 is obtained.

By means of the chart, the tangential and radial stresses at any radius in a parallel-sided disc can be found. As K_1 and K_2 are constants, any pair of curves which will satisfy the given stress conditions at the bore and rim will also give the values of S and D at points between. The correct pair is chosen by trial and error.

It should be noted that there is a degree of approximation in these curves which were originally calculated for tangential velocities in ft/s and stresses in lbf/in^2. They have been converted to SI units without altering the original shapes – hence the unusual scales.

To illustrate, assume a parallel-sided disc rotating, at 5000 r.p.m. has inside and outside diameters of 140 mm and 565 mm. There is no external load at either bore or rim, i.e., the radial stress is zero at these two radii. The corresponding peripheral velocities are 36 and 146 m/s respectively, and the S and D curves should intersect on both these lines.

By trial it may be seen that the only pair of curves which do this on Figure 7.3 intersect at approximate stresses 38 N/mm^2 at the bore and at 143 N/mm^2 at the rim. The values of K_1 and K_2 used in plotting these two curves were the correct ones for this particular case. The values of the radial and tangential stress at any point along the disc can then be found.

$$R = \frac{S-D}{2}$$

$$T = \frac{S+D}{2}$$

7.7.3 Discs of any profile

The sum and difference curves may be used for an impeller of any profile by approximating its shape with a number of constant thickness sections. These imaginary parallel sided sections will have different widths. In the transition from one section to the next it is assumed that the radial stress varies inversely with the thickness and the change in tangential stress equals the change in the radial stress times the Poisson's ratio for the material.

7.7.4 Effect of the blades

The impeller blades, because of the centrifugal force acting upon them, increase the stresses induced in the shroud and the backplate but since these stresses are not continuous they do not contribute to their strength. The additional stress due to this dead load may be cared for by the following procedure through the use of the sum and difference curves.

a) The vanes are divided into a number of imaginary lengths, generally extending between the points of transition of the imaginary parallel-sided rings making up the impeller.

b) The centrifugal force of each length is found from:

$$F = \frac{Wu^2}{r} \qquad \text{Equ 7.3}$$

where:

W = mass of the length (kg)

u = peripheral velocity of the approximate centre of gravity of the length (m/s)

r = radius of the approximate centre of gravity of the length (m)

F = centrifugal force (N)

c) The additional radial stress R′ due to this load may be considered to act at the outer side of the inner ring of the step. It equals the total force for all the vanes, zF, divided by the circumferential area of the outer side of the inner ring, i.e.,

$$R' = \frac{zF}{\pi t'd} \qquad \text{Equ 7.4}$$

d) After the change in radial stress ΔR at the step is found, the additional external radial stress R′ is subtracted from it before the change in the sum and difference curves is found.

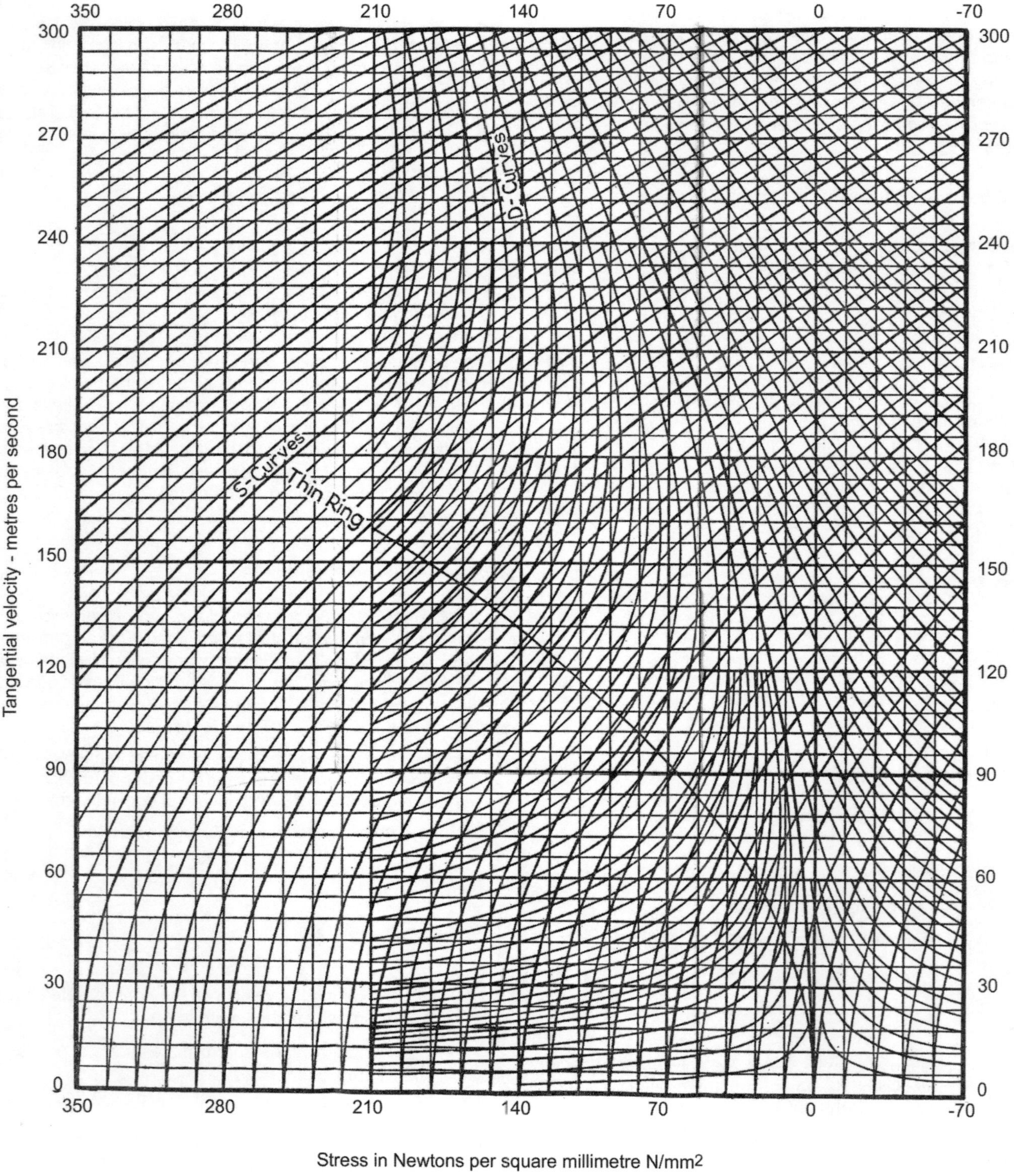

Figure 7.3 Sum and difference curves for stresses in discs

e) The rest of the procedure is the same as that outlined in the previous Section.

If the impeller has a shroud and backplate, it may be assumed that each carries an equal share of the dead load. For wide impellers, this may be nearer to allocating $\frac{2}{3}$ to the backplate and $\frac{1}{3}$ to the shroud.

7.7.5 Speed limitations

Rearrangement of equations 7.3 and 7.4 shows that the maximum hoop stress in the shroud or backplate f_h is:

$$f_h \propto N^2\left(ad_2{}^2 + bd_1{}^2\right) \qquad \text{Equ 7.5}$$

where:

d_2	=	outside diameter of shroud or backplate
d_1	=	inside diameter of shroud or backplate
N	=	rotational speed
a & b	=	constants for a particular design

It will be seen that the smaller d_1, the lower this stress. Thus from a strength point of view, with lower flowrates and higher fan pressures, the inlet diameter to the impeller shroud should be reduced.

We should also realise that from an aerodynamic viewpoint, an oversize impeller inlet may lead to a rapid change from low velocity to high velocity in the blade passages with consequent losses. Thus the narrowing of the width of standard fans by just changing the blade and casing width is to be avoided wherever possible.

7.7.6 Impellers not made of steel

The sum and difference curves plotted as Figure 7.3 are for steel. For any other material, a new chart could be plotted, but it would be quite laborious.

An inspection of the equations shows that the only factors involving the material are its density and Poisson's ratio ν since the chart is plotted with assumed values of K_1 and K_2. Approximate values of these properties, as taken from handbooks for common impeller materials are given in Table 7.5.

Material	Density ρ kg/m^3	Poisson's ratio ν
Steel	7833	0.30
Brass	8719	0.33
Aluminium	2768	0.33
Cast iron	7086	0.27
Bronze	8525	0.35

Table 7.5 Typical densities and Poisson's ratios for common metals

A value of Poisson's ratio of 0.30 may be used for all these materials without a great error. If this is done the values of S and D or stress will be directly proportional to the material densities. i.e. the stress scale is compressed or extended in that ratio.

Thus, an impeller of any common material may be calculated as if it were made of steel, but the resulting radial and tangential stresses must be reduced in the ratio of ρ/7833 where ρ is the specific weight of the impeller material.

It will be noted that, whilst aluminium alloys are very much lighter than steel, their yield stress may not reduce to the same extent. Thus it is possible to design impellers manufactured from a suitable aluminium alloy, which can rotate faster and generate greater fan pressures than the equivalent manufactured in steel.

7.7.7 Stresses in the fan blades

The fan blades may be considered as uniformly distributed loaded beams with rigid supports (encastré ends) at the backplate and shroud. They are subject to a maximum bending moment of $\frac{WL}{12}$ where W is the total distributed load on the blade, which comprises the centrifugal force and the pressure difference across the blade. The centrifugal force is by far the greater and the forces due to the pressure difference may be ignored.

Considering an element of blade width 6, thickness t and length dl as shown in Figure 7.4 the force normal to the element dF′ will be:

$$dF' = dF\ \cos\beta = b\,t\,dl\,\rho_m\omega^2 + \cos\beta \qquad \text{Equ 7.6}$$

where:

ρ_m = density of blade material (kg/m^3)

To achieve a consistent result in SI units, b, t and r will all need to be measured in metres, with $\omega = \frac{2\pi N}{60}$ rad/sec and N in rev/min.

The maximum bending moment M:

$$M = \frac{dF'b}{12}$$

$$= \frac{b^2t}{12}\rho_m\omega^2\ r\ dl\ \cos\beta$$

The section modulus Z:

$$Z = \frac{tdl}{12} \div \frac{t}{2}$$

$$= \frac{t^2dl}{6}$$

Thus the maximum bending stress =

$$= \frac{b^2\rho_m\omega^2\cos\beta}{2t}\ \text{N/m}^2 \qquad \text{Equ 7.7}$$

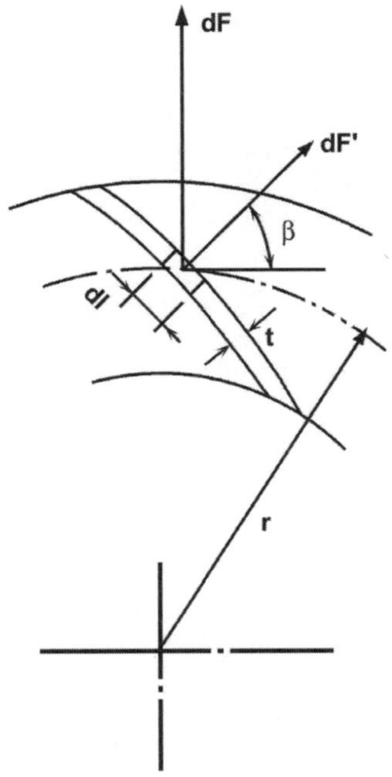

Figure 7.4 Stresses in an element of a rotating centrifugal fan blade

If the blades are welded to the shroud and backplate, the stress in the weld will be:

$f = 2m\,\omega^2\, r\, Ll \cos 45°$ for a double fillet weld

where:

- l = width of the weld (m)
- m = mass of the blade (kg)

The strength of a weld is taken to be that at the weld "throat" i.e. l cos 45 ° x weld length L.

For a riveted impeller, we are interested in the shear stress in the rivets which will be:

$$f_{sr} = \frac{m\,w^2\,r}{Za}$$

where:

- m = blade mass (kg)
- Z = number of rivets
- a = cross-sectional area of a rivet (m^2)

7.7.8 Finite element analysis (FEA)

All that has been said so far assumes an impeller with a relatively flat shroud, a constant thickness backplate and simple blades. Where these do not exist and there is an appreciable slope to the shroud, complex blade forms and backplates stiffened with cones, the calculations become too complex. In any case the structure is statically indeterminate.

With the advent of PCs and their ability to handle Finite Element Analysis programmes, however, the problem has largely "gone away". It is now possible for junior engineers to obtain accurate results of stress without really understanding what is happening.

Back in the 1970s there were valid concerns with the quality of these programmes. Now, with what seems like limitless computer power, the FEA has been linked to 2D and 3D CAD packages. Automatic mesh generators have been developed which take a CAD defined volume and fill it with tetrahedral elements, thus dividing the impeller into a number of very small elements as in Figure 7.5.

But beware - all problems have essentially been reduced to that of a cantilever beam – loads applied at one end and constraints at the other. Invariably the constraint has been modelled as a fully encastré support – something that is impossible to achieve in practice.

Note: There are however many good FEA programmes, which can provide balanced loading and minimal constraint. Make sure yours is one of them!

7.8 Stressing of axial impellers

7.8.1 Introduction

Axial flow fan impellers will also be subject to centrifugal forces and thus the various elements will be "stressed". As in most cases the blades are "cantilevered" and only supported at the end adjacent to the hub, fluctuating stresses are more important. These are due to aerodynamic forces and vary according to the duty position on the fan characteristic. Fatigue is therefore the important criteria in determining the life to failure.

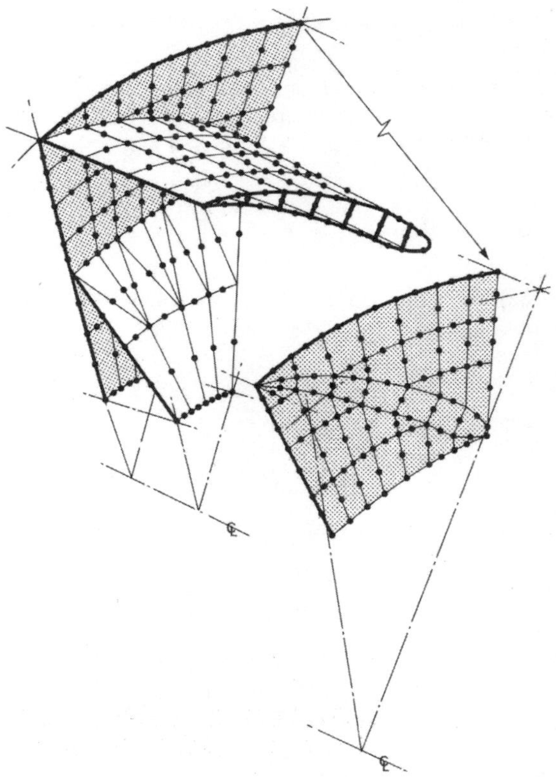

Figure 7.5 Detailed finite element mesh for a backward aerofoil impeller

7.8.2 Centrifugal loading effects

True aerofoil blades vary in section along their length. It is preferable for the centroids of each section to lie on a radial line, when the stress at the blade root will be:

$$\frac{\rho_m \omega^2}{A_o}\int_{r_1}^{r_2} A(r)\cdot r\,dr \qquad \text{Equ 7.8}$$

where:

- A_o = cross-sectional area of blade root (m^2)
- A_r = cross-section area of any element (m^2) at radius r (often function of r)

The static pressure difference across the blade swept area and the torque combine to give a bending moment on each blade. These should be resolved along the blade to give a bending moment at the blade root normal to a neutral axis for which the section modulus is least. The section modulus may be found by drawing an enlarged aerofoil section, dividing it into a number of strips. A summation of these will give:

$$I = \sum dA \times y^2$$

Beyond this it is difficult to particularise as each design will be unique. General equations as for centrifugal fans are not usually possible.

7.8.3 Fluctuating forces

Apart from out-of-balance, the only readily perceived cause of a fluctuating force has been due to aerodynamic effects and these are magnified at unstable parts of the fan characteristic curve.

In the design of any axial flow impeller, it is therefore necessary to ascertain the magnitude of not only the centrifugal stresses that are imposed, but also the fluctuating stresses. The ratio of these will lead to a determination of the operational life. During the last fifty years, vast strides have been made in the advance of metallurgy, particularly as it relates to the use of non-ferrous alloys. Many of these were developed for the aircraft industry and have a considerable increase in tensile strength, but most importantly, a greater resistance to fatigue.

The use of such new alloys, however, often presents problems in the methods required in the foundry, heat treatment, forge or machine shops. If the full advantages are to be obtained, it is essential that the design engineer is aware of the characteristics of the material being used and how they will be down-rated according to the manufacturing processes involved.

For complete success a three-stage design and testing programme is preferable with appropriate iterations as necessary between each of these stages:

- Finite Element Analysis
- photo-elastic coating tests
- strain gauging

7.8.3.1 Finite Element Analysis

As with centrifugal impellers, it is not proposed to give a detailed description of the methods used for axial machines. Suffice it to say that such programmes are readily added to CAD systems and are now considered essential if we are to be aware of the highest stress points in a blade or hub, examples of which are shown in Figures 7.6 and 7.7.

Figure 7.7 shows the stress resulting only from the centrifugal loading and on this must be superimposed the fluctuating stress caused by aerodynamic and other effects. At the present time these are not easily susceptible to mathematical evaluation and it is best to deduce them experimentally. Nevertheless, a fatigue crack will start initially at a point of high stress concentration such as a keyway, toolmark, oil hole, start fillet, inclusion, change of section or any other "stress raiser". The FEA and CAD programmes assist in the identification of such problem areas and lead to modifications which will improve the design.

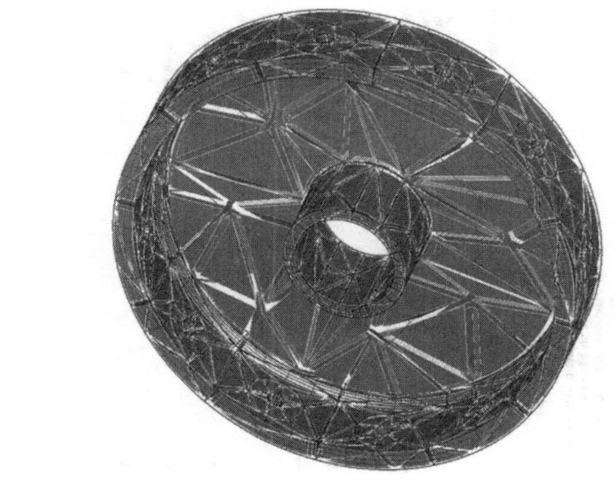

Figure 7.6 FEA mesh of a hub

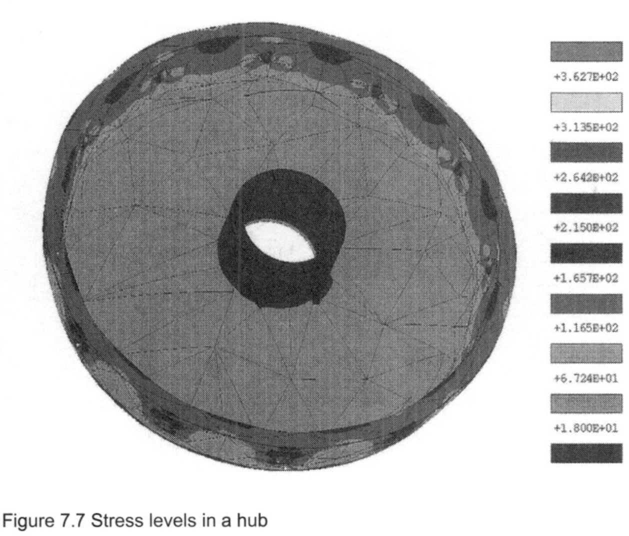

Figure 7.7 Stress levels in a hub

7.8.3.2 Photoelastic coating tests

In any FEA programme assumptions have been made and, for complete confidence, these should be validated (see Section 7.7.8). Photoelasticity is therefore used to both confirm the overall stress distribution and to enable the high stress points to be immediately identified.

When a photoelastic material is subjected to a load and then viewed with polarised light, coloured patterns are seen which are directly related to the stresses in the material. The colour sequence observed starts at black, (zero) and continues through yellow, red, blue-green, yellow, red, green, yellow, red green with increasing stress and repeating. The transition between the red and green colours is known as a "fringe". The number of fringes increases in proportion to the increase in stress and is illustrated in Figure 7.8.

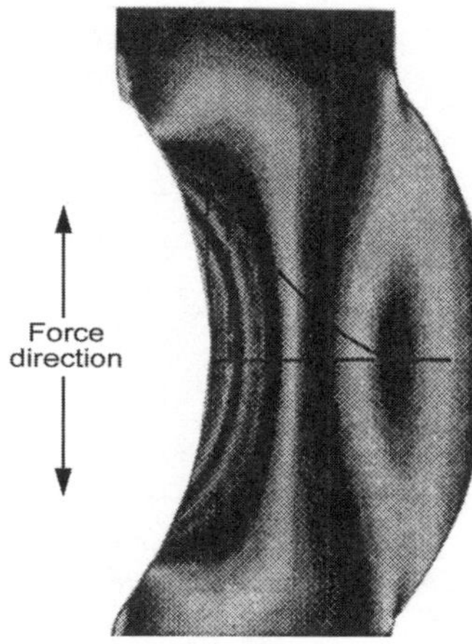

Figure 7.8 Photoelastic stress patterns

7.8.3.3 Strain gauge techniques

Whilst photoelastic methods can give quantitative results, strain gauge techniques are preferred, as these also permit the measurement of the fluctuating stresses, so important in the assessment of the fatigue life of the component. High stress points in an impeller blade or hub, as identified in the Finite Element Analysis and confirmed by the photoelastic tests, should then be fitted with strain gauges.

Stress in a material cannot, of course, be measured directly and must be computed from other measurable parameters. We, therefore, use measured strains in conjunction with other properties of the material to calculate the stress for a given loading. Bonded resistance strain gauges are normally used (Figure 7.9) these being cemented to the blade, hub or other part as re-

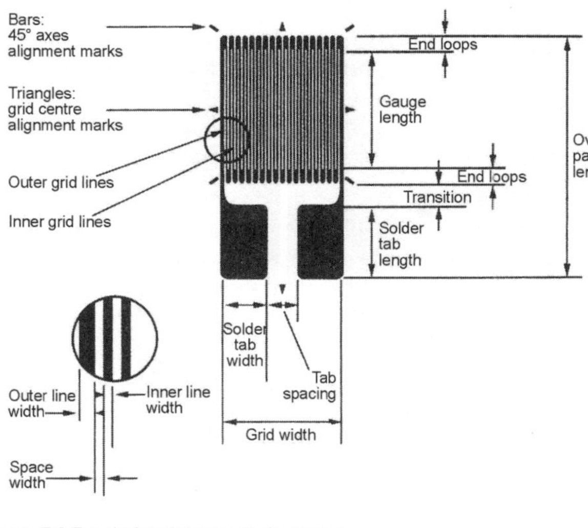

Figure 7.9 Bonded resistance strain gauge

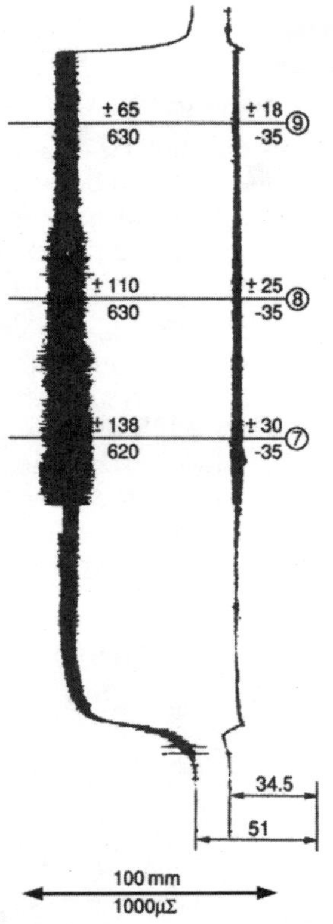

Figure 7.10 Strain gauge trace

quired. An initial unstrained gauge resistance is used as a reference measurement. When the fan is run, a change in resistance will occur which can be equated to the strain. The variation in the strain, due to fluctuating forces, can be seen on the trace produced. It is necessary to assess this value as it is far from constant (Figure 7.10).

7.8.3.4 Fatigue

Failure under low cycle fatigue is rapid. It is easily recognised and is usually due to the rotational frequency coinciding with the natural frequency of the component. With a blade, it is common to "tap" it with a hammer and measure the acoustic emission and analyse its frequency. It is a simple matter to rectify by local stiffening. Such failures are especially rapid in the "stall" region.

There will however be many other resonances over the whole frequency spectrum which can be captured by the acoustic emission. These resonances become ever closer at increasing frequencies and lead to high cycle fatigue.

The term fatigue is used to describe the failure of a material under a repeatedly applied stress. The stress required to cause failure, if it is applied many times, is, of course, much less than that necessary to break the material in a single "pull".

As previously stated, fatigue causes many of the failures of axial impeller rotating parts and it is, therefore, necessary to design against this eventuality. To repeat, in an impeller there will be a mean stress, due to centrifugal loading, and a fluctuating stress imposed on this, due to aerodynamic effects.

Experience has shown that for satisfactory correlation with actual behaviour in service, full size blades and hubs should be tested in conditions as close as possible to those encountered during service. Some basic information can however be obtained from simple laboratory tests.

A Roell-Amsler vibraphore resonant frequency machine can be used establish the fatigue strength of the aluminium alloys used. Test samples are cast as shown in Figure 7.11 and these are then subject to high cycle fatigue at various mean stress levels an at variously defined numbers of stress reversals (cycles). A tensile test is also carried out on one of the run-out fatigue specimens in order to give a tensile strength value and thus permit all the data to be plotted on a Goodman diagram (Figure 7.12). Figures 7.13 and 7.14 give typical impeller and impeller hub stresses versus LM25-TF fatigue data. LM25-TF is a heat treated aluminium alloy frequently used for hubs and clamp-plates. It is interesting (and very informative) to compare the as cast data with that published for smooth specimens.

Examination of the fracture surfaces of the failed specimens has shown that in the majority of cases, failure initiates from defects, however minute, in the aluminium casting. It has also been noted that the larger defects correspond to the lower fatigue lives.

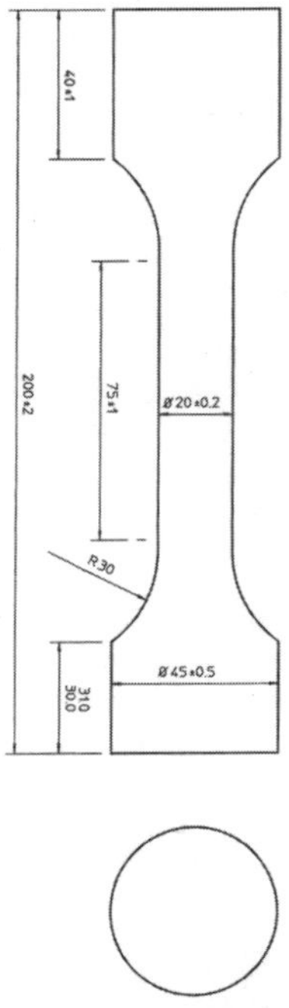

Figure 7.11 As cast specimen for fatigue testing

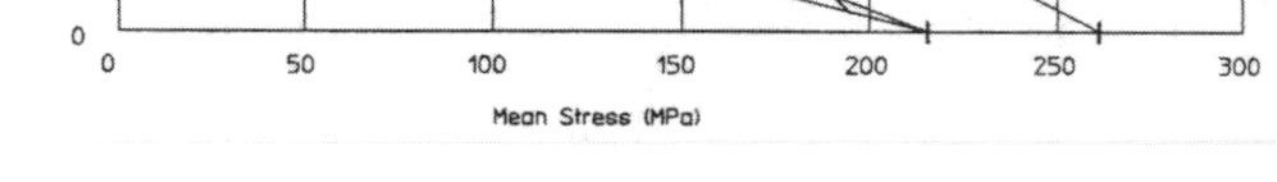

Figure 7.12 Goodman diagram for LM25-TF cast aluminium

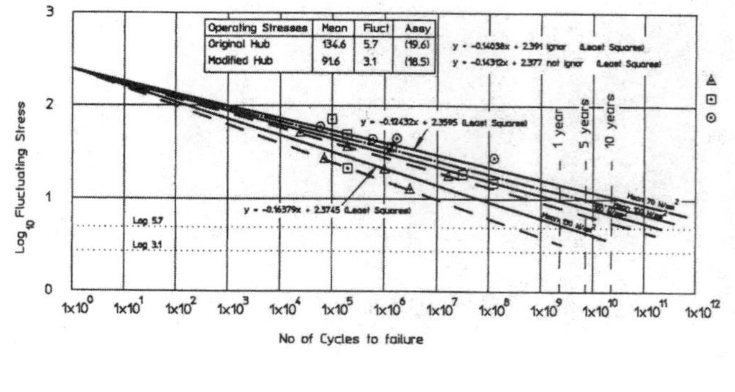

Figure 7.13 Typical impeller hub stresses versus LM25-TF fatigue data

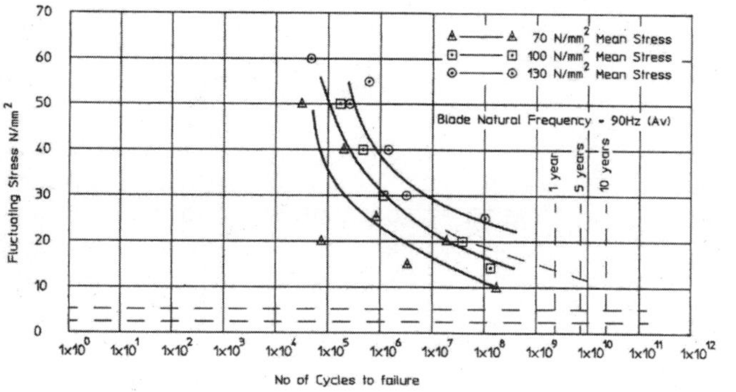

Figure 7.14 Typical impeller stresses versus LM25-TF fatigue data

7.8.3.5 Fracture mechanics

This is a relatively new subject which looks at the fracture toughness of cast materials and their rates of fatigue crack growth. This type of research has enabled fan manufacturers to determine design rules which specify acceptable defect sizes under combinations of steady and fluctuating stress. The tests are carried out in accordance with BS 6835:1988 and ASTM E647.

Figure 7.15 is an example of the results obtained from LM25-TF.

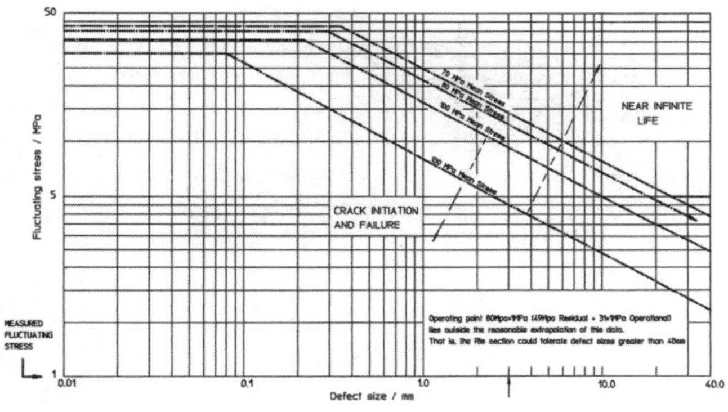

Figure 7.15 Defect size and stress in rim of LM25-TF hub

7.8.3.6 Performance and fluctuating stress curves

It is convenient during the performance (rating) tests of a fan to also measure the fluctuating stress at various flow rates. From these tests, some interesting conclusions have been deduced.

Whilst the fluctuating stress generally increases towards the stall point at that particular impeller blade pitch angle, the maximum is not necessarily coincident with the stall (Figure 7.16). Furthermore, whilst different aerofoil shapes may give similar aerodynamic results, this does not apply to the fluctuating stress values. For new ranges of metric axial flow fans and also for large special purpose tunnel ventilation units, manufacturers have developed improved sections (Figure 7.17) which have reduced fluctuating stress values away from the stall point.

Note especially, that in reverse rotation high maxima can occur on the negative slope of the characteristic - what would otherwise be assumed to be an acceptable operating point for this condition. Note also that maximum fluctuating stresses generally increase with increasing pitch angles (Figure 7.18). Truly reversible sections have also been developed which not only give virtually the same airflow in each direction (tube axial), but also have extremely low values of fluctuating stress across the whole performance characteristic (Figure 7.19).

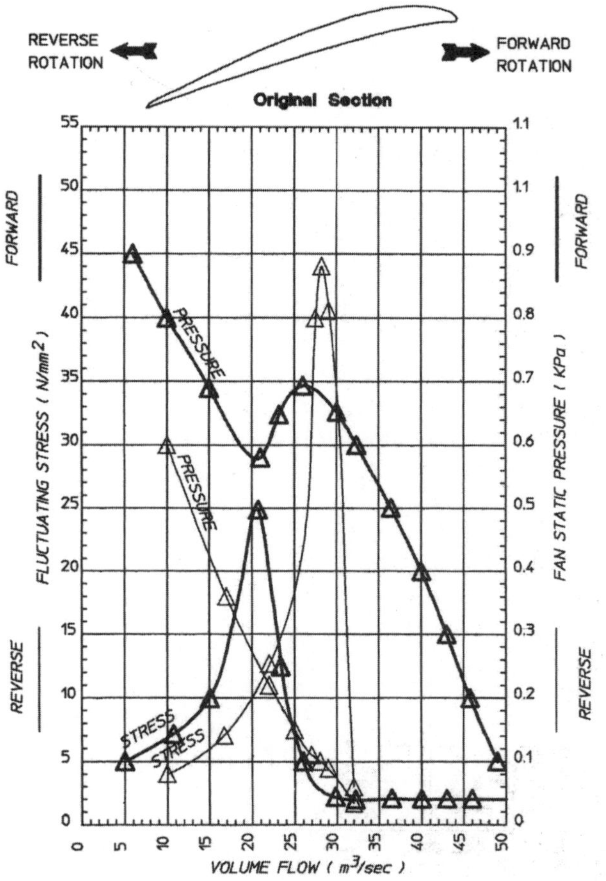

Figure 7.16 Göttingen design blades – pressure and fluctuating stress against flowrate

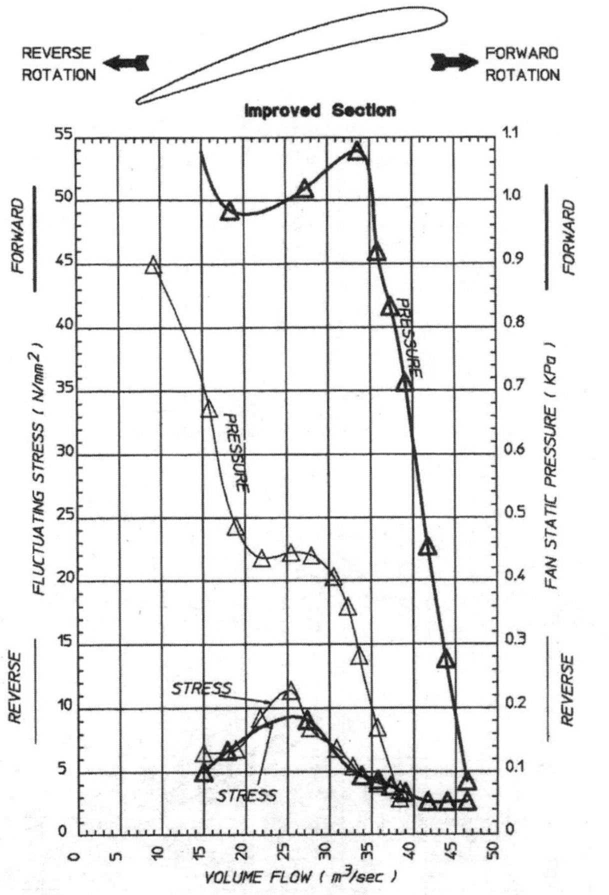

Figure 7.17 NARAD design blades - pressure and fluctuating stress against flowrate

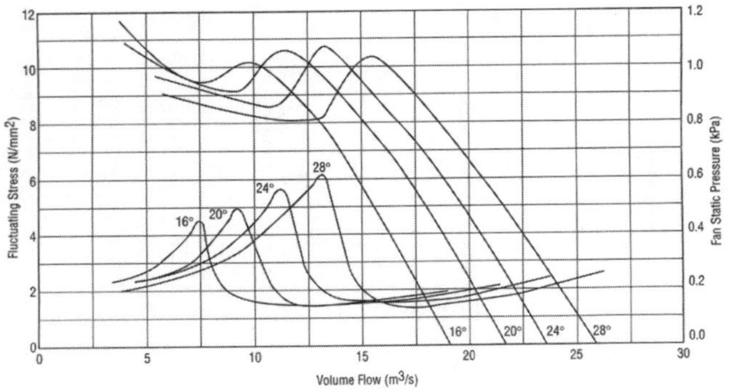

Figure 7.18 NARAD design blades – pressure and fluctuating stress against flowrate with varying pitch angles

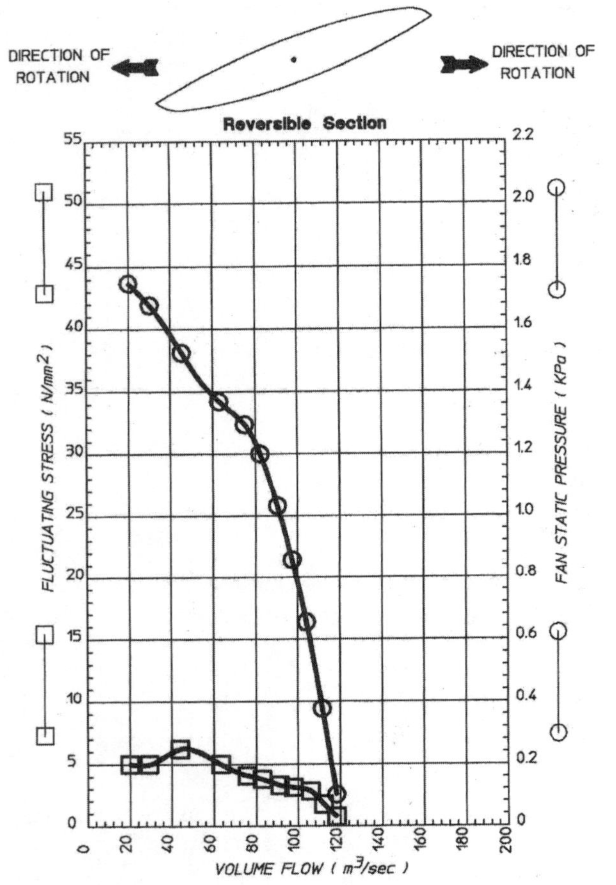

Figure 7.19 Reversible design blade - pressure and fluctuating stress against flowrate

7.8.3.7 Conclusions

The techniques described in this Section can act as a powerful tool for obtaining the same integrity with axial flow fans as has been achieved over many years with centrifugal fans.

It is essential that a design and testing procedure is adopted which recognises that a major cause of failure in axial impellers is due to insufficient knowledge of the fatigue criteria and how they are affected by casting quality. Close co-operation between design and production departments is necessary to ensure that the stated operating life is achieved. Constant vigilance is, nevertheless, indicated with continual research to improve knowledge. Reference to Chapter 17, Section 17.6 may be useful for practical solutions and advice.

7.9 Shaft design

7.9.1 Introduction

The shaft of all types of fan may be treated as a beam carrying the impellers as point loads if the shaft is long, or as a thickening of the shaft if it is short. The bearings, especially if self-aligning, are treated as simple supports. Only in the old-fashioned sleeve bearings, where the journal might be 3 diameters long, was it possible to consider them as approaching rigid encastré supports.

The shaft must be considered for three different criteria and that giving the largest diameter must be taken as the basis of the design:

- Maximum sheer stress
- Maximum direct stress
- Critical speed

In order to carry out these calculations, it will be necessary to fix the type, size and position of the bearings (see Chapter 10). Where the fan is driven through vee belts (see Chapter 11) the belt tension will give an additional load which is used for calculating stresses. It should not however be used for critical speed determination as, unlike out-of-balance, it is unidirectional.

7.9.2 Stresses due to bending and torsion

Bending stresses result from the overhang effects of the impeller and from the moment produced by the belt pull in indirect drive units. Torsion results from the work done by the fan in rotating at the speed necessary to achieve the duty. If the system resistance is lower or higher than that specified, this will affect the power absorbed and thus the torque required. It may also affect the belt pull in indirect drive units and thus the bending stress.

Max direct stress f is:

$$f = \frac{16}{\pi d_s^{3}}\left(M + \sqrt{M^2 + T^2}\right) \qquad \text{Equ 7.9}$$

Max shear stress q is:

$$q = \frac{16}{\pi d_s^{3}}\sqrt{M^2 + T^2} \qquad \text{Equ 7.10}$$

where:

M = maximum bending moment

T = maximum torque

D_s = shaft diameter

All in consistent SI units.

The acceptable stresses will be determined by the shaft material, whilst the maximum bending moment and torque are determined by the arrangement of impeller, bearing centres and belt pull, etc.

It is essential to allow reasonable factors of safety on the maximum stresses attained to cater for the effects of unbalance, additional accelerating torque at start-up, fatigue, over tightened vee belts etc.

7.9.3 Lateral critical speeds

As the rotational speed of a fan is increased, it will be seen that at certain speeds the shaft may vibrate quite violently whereas at speeds above and below these it will run relatively quietly. The speeds at which these severe vibrations occur are known as the critical speeds of the rotating assembly.

If a unit operates at or near a critical speed, large amplitudes of vibration can be built up. Such a condition results in dangerously high stresses, possible rubbing of the impeller eye on the inlet cone, and large cyclical forces transmitted to the foundations. It is therefore important that there is a margin between the running and critical speed.

Many textbooks suggest that this margin should be a minimum of 20%. The author suggests however that for all non-symmetrical arrangements, i.e. all single inlet fans, the ratio of critical speed should be at least 1.5. This ratio is a measure of the shaft stiffness and determines the dynamic effect of unbalance. For a given system it can be shown that the eccentricity of the centre of gravity of the impeller is increased by 80% for a ratio of 1.5 but only 20% when the ratio is 2.5. The disturbing forces, which have to be resisted by the bearings, bearing supports and ultimately the foundations, increase in proportion to the eccentricities.

Where fans are handling large quantities of foreign matter and are thus subject to build-up, erosion, corrosion or temperature distortion, a minimum ratio of $\frac{Nc}{N}$ of 1.8 is recommended.

For double inlet fans, due to the symmetricity, the ratio for clean air fans may be reduced to 1.3.

Ratios close to 2 should however be avoided as they may coincide with the second harmonic of critical speed.

It can be shown that all critical speeds are:

$$N_c \propto \frac{d_s^{\,2}}{m^{0.5}} \times l^{1.5} \qquad \text{Equ 7.11}$$

where:

d_s	=	shaft diameter (m)
m	=	impeller mass (kg)
l	=	distance from impeller c.g. to supporting bearing (m)

The actual values will depend on the fan arrangement, bearing centres, overhang of impeller etc. This formula is therefore a simplification but does show which factors are of importance.

It should be noted that perfect balance of an impeller and shaft is impossible. There is always a residual unbalance however small. Rotation produces a centrifugal force of the mass centre which is balanced by the springing action of the shaft.

Below the first critical speed, the centre of gravity (c.g.)of the impeller and shaft assembly rotates in a circle about the geometrical centre, whereas above the first critical speed the shaft rotates about the c.g. This leads to extremely smooth running and is the "norm" for turbo-generators. There are now engineers advocating its use for large fans especially where the impeller is between bearings and the blockage effects of the shaft are severe. It does of course require that the fan rapidly accelerates through the critical speed.

The axis of rotation changes at the critical speed from the geometric centre to the centre of gravity. When the shaft rotates at critical speed the restoring force of the shaft is neutralised and the action is dynamically unstable, hence large amplitudes of vibration may occur.

7.9.4 Torsional critical speed

In addition to the lateral critical speeds described in Section 7.9.3 there are torsional critical speeds where two or more rotating masses are connected by a shaft. These must be avoided for trouble-free running.

As a fan impeller rotates, small torque impulses may develop and be transmitted to the shaft. They may be caused by slight misalignments, the passing of the impeller blades by the casing cut-off or tongue piece, or by rapid fluctuations in system resistance. If the frequency of these impulses coincides with, or is a multiple of, the torsional critical speed, then large amplitude oscillations may build up and a possible shear fatigue failure occur.

Most fan installations will have only two masses the fan impeller and the motor rotor for which the frequency F:

$$F = \frac{1}{2\pi}\sqrt{\frac{I_p E_s (J_1 + J_2)}{J_1 J_2 L}} \qquad \text{Equ 7.12}$$

where:

F	=	natural frequency (Hz)
I_p	=	polar moment of inertia of shaft (m^4)
		$= \frac{\pi d_s^{\,4}}{32}$
d_s	=	shaft diameter (m)
E_s	=	shear modulus of elasticity (Pa)
L	=	shaft length between masses (m)
J	=	mass moment of inertia = mr^2
m	=	mass of impeller or rotor (kg)
r	=	radius of gyration (m)

The formula becomes very much more complex for a stepped shaft.

7.10 Fan casings

It is the usual practice to strengthen with angle iron or flat bars, the large areas of metal forming the sides of centrifugal fan casings. This prevents the "drumming" of the relatively thin sheet. The areas of sheet metal or plate so formed may be treated as rectangles of sides a and b subjected to a uniform pressure and supported around its perimeter. Then the maximum bending stress f will be:

$$f = \frac{pa^2b^2}{2t^2\left(a^2 + b^2\right)} \qquad \text{Equ 7.13}$$

where:

t	=	thickness (m)
p	=	pressure (Pa)
a & b	=	dimensions (m)
f	=	stress (Pa)

The circumferential surfaces and also the casings of axial flow and other in-line fans may be considered as thin cylinders.

The direct stress will be due to the force $p \cdot 2 \cdot r \cdot l$ across the resisting section of area $2 \cdot t \cdot l$. l being the length of the casing. Thus the direct stress will be:

$$f = \frac{p \cdot 2 \cdot r \cdot l}{2 \cdot t \cdot l} = \frac{pr}{t}$$

Where an electric motor must be supported in an axial flow fan casing, this will often be suspended by tie rods or brackets, for which additional load the casing must be designed.

7.11 Mechanical fitness of a fan at high temperatures

The strength of metals and plastics varies according to their temperature. When handling air or gas at temperatures other

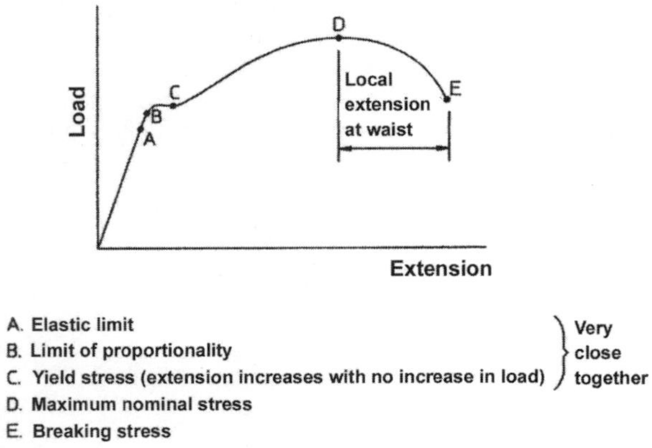

Figure 7.20 Stress/strain relationship for a typical steel

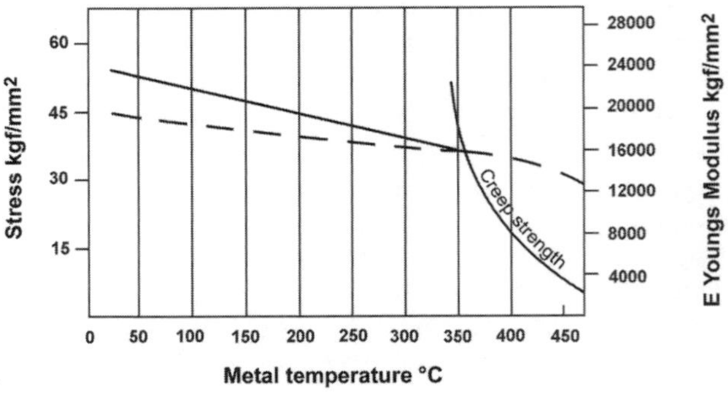

Figure 7.21 Reduction in fan running speed due to gas temperature

than ambient, the materials of construction may need to be de-rated from the values normally given in textbooks.

As noted in Chapter 8, Section 8.6.2, all elements of the fan must be satisfactory. Those within the gas stream are likely to take up the same temperature, but elements outside may take up a temperature somewhere between that of the gas stream and the ambient air around the fan.

It is important to note the stress/strain relationship for the typical steel used in the fan construction as shown firstly as Figure 7.1 and repeated as Figure 7.20 with more detail. This diagram is applicable to a given temperature. The general shape This diagram is applicable to a given temperature. The general shape of the relationship between load and extension however remains similar. At increased temperatures, the values of A, B, C, D and E all reduce together with the value of the extension to failure

$$\text{Stress} = \frac{\text{load}}{\text{cross-sectional area}}$$

$$\text{Strain} = \frac{\text{extension}}{\text{original length}}$$

In the past, factors of safety were applied to the ultimate stress (i.e. D) in determining the design stress. Nowadays, with the common use of Finite Element Analysis, it is frequently the case that a design stress within the elastic limit or yield is specified. Account must be taken of any shock loadings.

It should be noted that above 400°C creep stresses become important. At high temperatures under stress it is found that the ordinary condition of elasticity of metals changes to a state of viscous flow whereby continuous deformation or creep proceeds at slow rates. Above about 535°C **any** stress however small would cause continuous flow or creep in carbon steels. A molybdenum content is of value in reducing the rate of creep. **It is therefore necessary to decide a creep rate for reasonable impeller life**.

The choice of steel has to be carefully considered and must be related to the exact range of working temperatures. Stainless steel is not always the answer - some grades are **weaker** at high temperatures than carbon steels. The reduction in strength with temperatures of a typical carbon steel is shown in Figure 17.21, together with the variation in the modulus of elasticity.

Impeller — Forces acting on the impeller are centrifugal stresses (air forces generally negligible).

$$\text{Centrifugal force} \propto (\text{rev/min})^2$$

$$\text{Safe rev/min}_{\text{Temp}} =$$

$$= \text{Safe rev/min}_{20°C} \times \sqrt{\frac{\text{steel strength at temperature}}{\text{steel strength at } 20°\text{C}}}$$

e.g. at 315°C = 86% of rpm at 20°C

Shaft — Usually the most important factor affecting the shaft is its critical speed (i.e. whirling takes place).

$$\text{Critical speed } N_c = \frac{\text{constant}}{\sqrt{\text{deflection}}}$$

$$\text{deflection } \Delta = \frac{WL^3}{KEI}$$

All factors are constant except Young's Modulus E which falls with increasing temperature.

Therefore for the shaft:

$$\text{Safe rev/min}_{\text{temp}} =$$

$$= \text{Safe rev/min}_{20°C} \times \sqrt{\frac{\text{E at temperature}}{\text{E at } 20°\text{C}}}$$

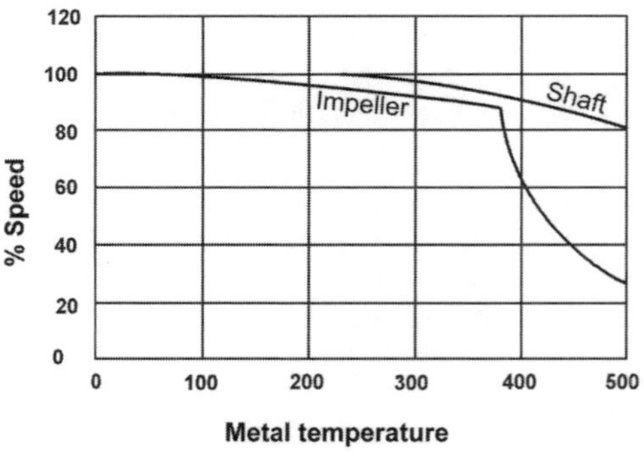

Figure 17.22 Reduction in fan speed due to metal temperature

Thus all factors may be combined on a single graph as shown in Figure 17.22. It will be seen that the impeller is usually the most important item. The drastic fall-off in safe operating speed for a carbon steel impeller above 400 °C will be noted.

7.12 Conclusions

The mechanical design of arduous duty fans can be extremely complex and is best left to the expert. Modern materials are not always fully documented and their limitations may be found only through (bitter) experience. Nevertheless, the application of principles from Strength of Materials and Theory of Machines can produce acceptable designs.

7.13 Bibliography

LM6 and LM31 — included in BS 1490:1988, *Specification for aluminium and aluminium alloy ingots and castings for general engineering purposes.* - Replaced by EN 1559-1:1997.

BS 1471:1972, *Specification for wrought aluminium and aluminium alloys for general engineering purposes - drawn tube.*

BS 1475:1972, *Specification for wrought aluminium and aluminium alloys for general engineering purposes – wire.*

BS 1490:1988, *Specification for aluminium and aluminium alloy ingots and castings for general engineering purposes.*

ISO/DIS 3522, *Aluminium and aluminium alloys — Castings — Chemical composition and mechanical properties.*

ISO 7722:1985, *Aluminium alloy castings produced by gravity, sand, or chill casting, or by related processes - General conditions for inspection and delivery.*

DIN 1725:1998, *Aluminium casting alloys.*

SS 055900 Edition: 3, *Preparation of steel substrates before application of paints and related products — Visual assessment of surface cleanliness — Part 1: Rust grades and preparation grades of uncoated steel substrates and of steel substrates after overall removal of previous coatings.* December 1988, SIS, Swedish Standards Institute, SE-118 80 Stockholm Sweden, Tel +46 8 555 520 10, Fax: +46 8 555 520 11, Email: sis.sales@sis.se, www.sis.se.

ISO 8501-1:1988, *Preparation of steel substrates before application of paints and related products — Visual assessment of surface cleanliness — Part 1: Rust grades and preparation grades of uncoated steel substrates and of steel substrates after overall removal of previous coatings.*

EUROVENT 1/9 - 2002, *Surface treatment for industrial fans.*

BS 381C:1996 *Specification for colours for identification, coding and special purposes.*

BS 5493:1977, *Code of practice for protective coating of iron and steel structures against corrosion.*

BS 7079-A3:2002, ISO 8501-3:2001, *Preparation of steel substrates before application of paints and related products. Visual assessment of surface cleanliness. Preparation grades of welds, cut edges and other areas with surface imperfections.*

BS 729:1971 *Specification for hot dip galvanized coatings on iron and steel articles.*

ISO 1459:1973 *Metallic coatings — Protection against corrosion by hot dip galvanizing — Guiding principles* Revised by: ISO 1461:1999 *Hot dip galvanized coatings on fabricated iron and steel articles — Specifications and test methods.*

BS 1224:1970 *Specification for electroplated coatings of nickel and chromium.*

ISO 1456:2003 *Metallic coatings — Electrodeposited coatings of nickel plus chromium and of copper plus nickel plus chromium.*

ISO 1458 : 2002 *Metallic coatings — Electrodeposited coatings of nickel.*

ISO 7619:1997, *Physical testing of rubber. Determination of indentation hardness by means of pocket hardness meters.*

BS 6374-1:1985 *Lining of equipment with polymeric materials for the process industries. Specification for lining with sheet thermoplastics.*

Steam and gas turbines, with a supplement on *The prospects of the thermal prime mover Vol 1,* Aurel Stodola, New York, P. Smith, 1945.

BS 6835-1:1998, *Method for the determination of the rate of fatigue crack growth in metallic materials. Fatigue crack growth rates of above 10^{-8} m per cycle.*

ASTM E647-00 *Standard Test Method for Measurement of Fatigue Crack Growth Rates.*

Centrifugal Pumps and Blowers, Austin H Church, Krieger Publishing Company, (June, 1972) ISBN 0882750089.

Fans: (In SI/Metric Units) William C. Osborne, Elsevier Science Ltd, 1977 ISBN 0080217265.

Centrifugal Fan Guide, W. T. W. Cory, Keith Blackman, 1980.

Axial Fan Impeller Integrity:Goodman Diagrams and Real-Time Radiography, W. T. W Cory, GEC REVIEW, Volume 9, No.3, 1994, page 154.

8 Constructional features

Fans have developed over a very long period of time and are therefore considered to be a "mature" product. As with automobiles, this means that there are remarkable similarities between the competing products of different manufacturers. Whilst the more cynical amongst us will put this down to blatant copying, it should also be recognised that once a buyer's specification is sufficiently detailed and has been established for a length of time, then the resulting solutions will also be remarkably similar.

Thus, just as all "super minis" in the car world look much the same, so it is with Category 1 fans. Just as all medium sized saloon cars exhibit considerable similarities, so do Category 2 fans. It is only with purpose-made fans to Category 3, that real differences become apparent. Of course, the mass-produced fan can be customised and various extras can be added — just as cars having alloy wheels, leather seats, air conditioning, satellite navigation, etc, etc.

This Chapter cannot describe all the options which are available. To repeat, the fan industry is a mature one. Often the options are the sole means of differentiation. Thus they proliferate ad nauseam. Those that are most popular (or appeal to the author) are described in the next few pages.

Contents:

8.9 Bibliography

8.1 Introduction

Centrifugal fans can be manufactured to various casing thicknesses and with various forms of construction according to usage. Thus at one extreme they can be handling clean air whilst at the other, air or gas handled can be at a temperature well above ambient and/or may contain substantial quantities of moisture and/or solids. It may also be at high pressure such that loadings on the fan casing and the associated ducting system are much higher than usually expected for a HVAC fan.

Connection to the ducting may be via flexible connections, or alternatively may be directly connected. In the latter case the fan has to withstand additional loads due to the dead weight of these connections. Where gases, or the surrounding ambient atmosphere, are at a high or low temperature, additional loading can result from the effects of expansion or contraction.

To ensure that the buyer can choose an appropriate form of construction, and to assist him in either specifying or recognising what he buys, ISO 13349, Section 5.3 gives a categorisation which is outlined in Table 8.1. This in no way indicates any form of grading but reflects current practice. Category 1, (Figure 8.1) is as valid for low pressure clean air applications as Category 3 is preferred for heavy industrial usage.

Category	1 (see Figure 8.1)	2 (see Figure 8.2)	3 (see Figure 8.3)
Usage	Light HVAC Clean air	Heavy HVAC Light industrial	Heavy industrial
Air/gas	Clean	Light dust or moisture	Dirty air/gas containing moisture and/or solids or high pressure or high power
Casing features (typical)	Lockformed, spot-welded or screwed construction Cradle or angle frame mounting	Lockformed, seam welded or fully welded construction. Semi-universal construction with bolted on sideplate	Fully welded fixed discharge
Casing thickness	<0.0025 D	> 0.0025 D	> 0.00333 D

Note: D is the impeller nominal diameter in millimetres

Table 8.1 Categorisation according to casing construction and thickness

This categorisation is particularly appropriate for centrifugal fans, as the great majority of axial flow fans are supplied for clean air, albeit some handle small amounts of entrained moisture. Nevertheless, there is no specific restriction to centrifugals. The special features detailed in the subsequent Sections may be limited to specific types of fan, which will be identified when appropriate. It is often difficult to differentiate between these special constructional features and the ancillaries described in Chapter 16.

Figure 8.1 Typical Category 1 fan

A distinction has been made that constructional features are part of the basic fan as manufactured, whilst ancillaries are bolt-on "goodies" which may or may not be supplied. Readers can enjoy themselves looking for the undoubted anomalies which arise!

8.1.1 Cradle mounted fans (centrifugal - Category 1)

These are very light duty fans for clean air applications. They are normally manufactured from pre-galvanized sheet steel and are either of lockformed or flanged and spot welded construction. The bearings are usually of the ball race type, grease packed for life. The casing volute is often supported in a cradle which can be bolted on to give different angles of discharge.

8.1.2 Semi-universal cased fans (centrifugal – Category 2)

This is best understood by reference to Figure 8.2. It will be noted that the casing "snail" consists of a scroll plate seam welded to the volute sides. Mild steel fabricated sideplates are bolted on at an outer pitch circle diameter such that they can be assembled to any of the standard angles of discharge, (see Chapter 9).

Figure 8.2 Typical Category 2 fan

Figure 8.3 Typical Category 3 fan

8.1.3 Fixed discharge cased fans (centrifugal – Category 3)

These fans are purpose made for a specific contract and have a fixed position for the casing outlet flange. They are usually of sheet steel welded construction and are most common for fans having impellers greater that 1000 mm diameter, (see Figure 8.3).

8.1.3.1 Horizontally split casings

Because of their size, fixed discharge fans may have to be split horizontally to facilitate transport and/or site assembly. The "split" comprises and angle flange terminating each half casing and these can then be bolted together (see Figure 8.4).

Figure 8.4 Typical large fan with casing split on horizontal centreline

8.1.3.2 Casings with a removable segment

Whilst a horizontally split casing facilitates transport and assembly, it may not be ideal for routine maintenance or for breakdowns. For vertically up (0°), top horizontal (90°) or any angular (45°, 315° etc.) discharges, it may require that the discharge ducting also be disassembled before the impeller/shaft assembly can be removed for maintenance. A removable segment (see Figure 8.5) overcomes this difficulty. The segment should be larger across its extremities than the impeller diameter.

Figure 8.5 Typical large fan casing with removable segment

8.2 Inlet boxes

Inlet boxes are provided to give air side entry to the fan inlet. This also permits the bearings to be mounted outside the airstream. The large flat faces of the box require adequate stiffening to prevent drumming. Pressure losses in boxed inlets can be substantial (see Chapter 3, Section 3.5.7) and for this reason are best supplied by the manufacturer as part of the fan. The proportions of the box and internal anti-swirl baffles are critical to performance and are very much dependent on the actual fan design.

Figure 8.6 DIDW fan with dual inlet boxes

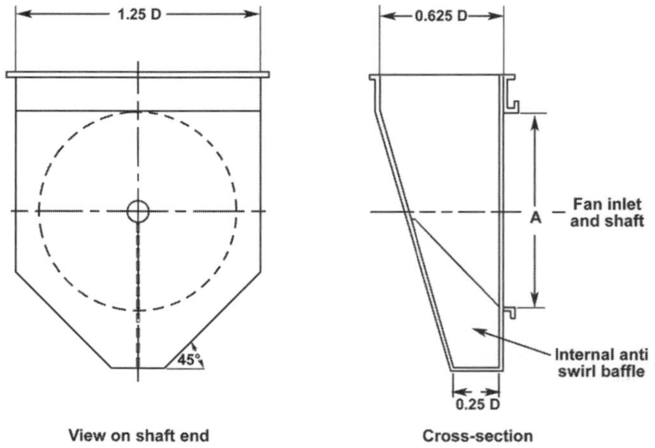

Figure 8.7 Proportions of an inlet box

They are designed to give minimum pressure loss in the working range and to ensure an absence of swirl at the impeller entry. A typical fan and inlet box is shown in Figure 8.6, whilst the proportions which have proved satisfactory for many fans are shown in Figure 8.7.

8.3 Other constructional features and ancillaries

For more detailed information refer to Chapter 16, and Figure 8.8 may be helpful.

8.3.1 Inspection doors

These permit examination of the fan impeller for material build-up or erosion. They are usually positioned on the scroll so that the impeller blades may be readily seen and cleaned. If positioned at a low level any dust may be easily removed.

Doors may occasionally, and additionally, be positioned on the volute sides to permit the shroud and/or backplate of the impeller also to be viewed and cleaned.

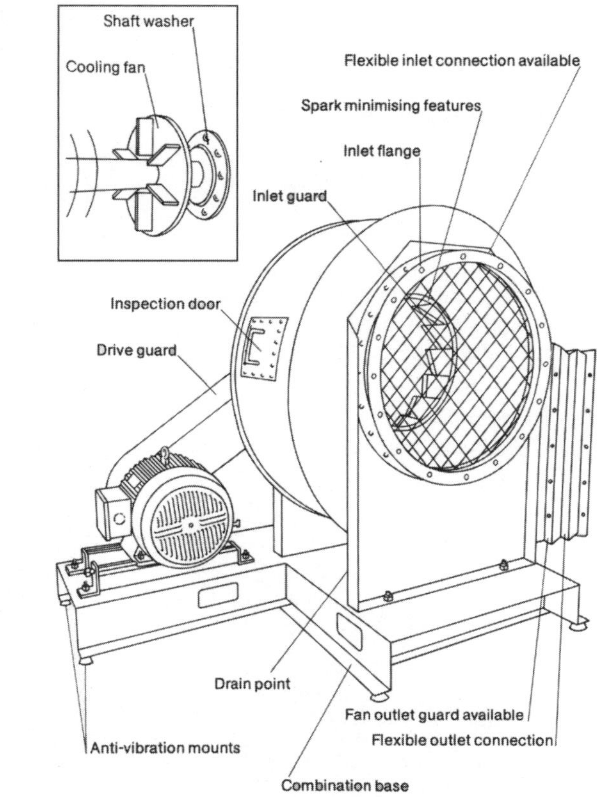

Figure 8.8 Constructional features and ancillaries for centrifugal fans

The inspection door usually consists of a steel plate positioned over a rectangular or circular hole in the casing. If positioned on the scroll, it must of course be rolled to match. Quick release fitting are **not** recommended – rather the door should be held by bolts and nuts, requiring a spanner to be used. Too easy a removal could be dangerous when the fan is running. The rotating impeller will be in close proximity and will be highly dangerous. It may even be advisable to have an electric interlock with the power supply, such that when the door is removed, the fan cannot run.

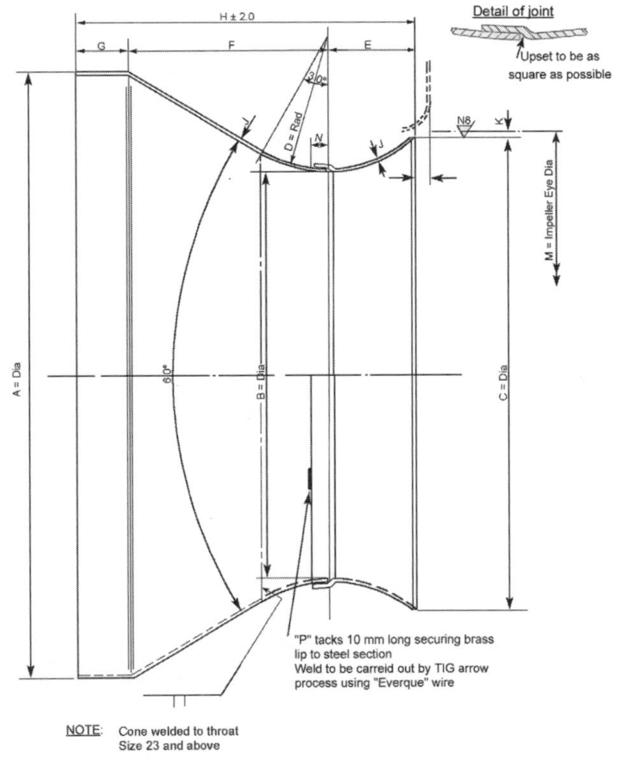

Figure 8.9 Inlet Venturi cone with anti spark features

8.3.2 Drain points

Where a fan is handling air contaminated with liquids or vapours, it is recommended that a drain point is positioned at the lowest point of the scroll. This may be screwed to accept piping or fitted with a closing plug.

8.3.3 Spark minimising features

A non-ferrous rubbing ring is attached to the inlet cone or Venturi, where the cone is adjacent to the eye of the impeller, and contact could take place, see Figure 8.9. A non-ferrous shaft washer is also necessary. These will minimise the possibility of incendiary sparks being produced. Such features are essential where explosive or inflammable gases or vapours are bing handled. The material pairings are especially important and are detailed in prEN14986.

8.3.4 Design of explosion proof fans

The ATEX Directive 94/9/EC of the European Union came into force at the end of June 2003. This placed obligations on both users and manufacturers of equipment, such as fans, which could be the cause of explosions. As a result CEN (Commitée Européen Normalisation) was mandated to produce prEN 14986. Not only does this give detailed recommendations on the spark minimising features, it also details other requirements concerning bearing selection, vee belt drives, clearances, material stresses, etc.

8.4 Gas-tight fans

There are three possible areas where leakage may take place:

- leakage of welds and seals in the casing
- leakage at static interfaces such as flanges and joints
- leakage at shaft seals (dynamic rotating interfaces).

8.4.1 Tightness of the casing volute

An almost absolute casing tightness can only be achieved between metallic materials when the components, such as the scroll and volute sides, are correctly and continuously welded together. This requires close inspection and quality control. It is normally carried out at the same time as the inspection of splitting flanges. The main areas of concern are the inspection door openings and any removable segments.

8.4.2 Static assemblies

This type of interface has to be capable of disassembly from time to time. The usual joint comprises plane surfaces. A very common method is to use an "O" ring of some elastic material between two flanges as shown in Figure 8.10. Blind holes are

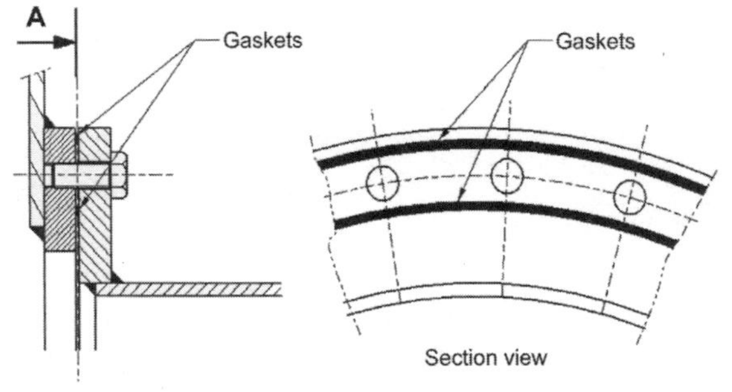

Figure 8.10 Common tightening methods for static assemblies

recommended and through holes with nuts and bolts should be avoided.

8.4.3 Absolute tightness

In practice absolute tightness can never be achieved, and there will always be some degree of leakage. However, something approaching zero leakage can be obtained through welding. The type of assembly shown in Figure 8.11 is difficult to disassemble and requires the welds at the periphery of the thin plates to be broken.

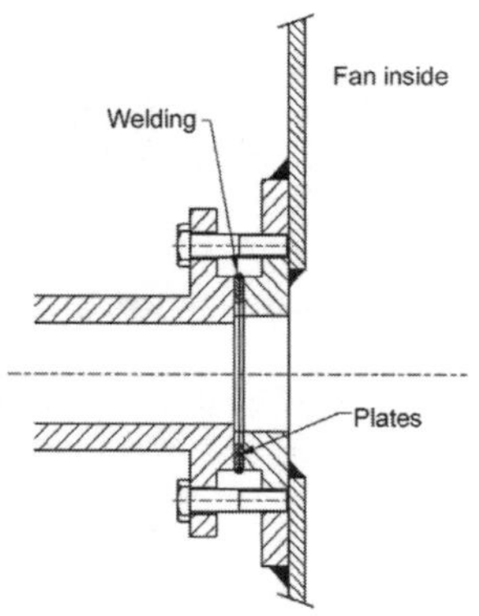

Figure 8.11 Welded flange with added plates

The bolting together of two surfaces such as flange faces, only provides a limited tightness even when the flanges have a high degree of surface finish and the bolts are "torqued-up" to a significant value.

8.4.4 Sealing without joints

In certain cases, it is possible to achieve a reasonable degree of gas tightness by using a knife edge plane contact as shown in Figure 8.12. This design requires that the geometry of the contact surfaces is very good and that the surface roughness is minimal.

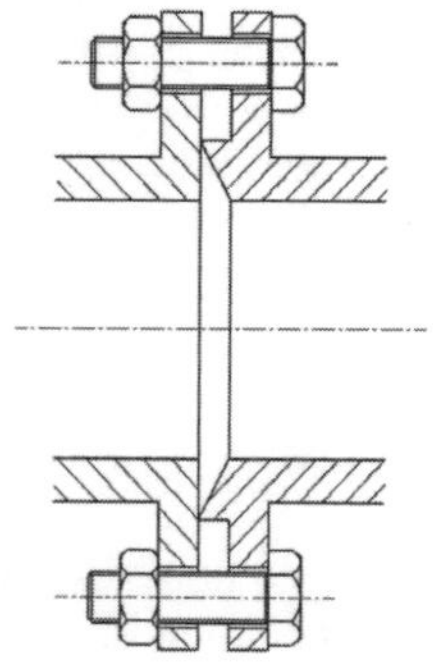

Figure 8.12 Knife edge plane contact

The example shown has a knife edge in contact with a plane surface. One of the two pieces should preferably be much more ductile than the other. This type of assembly should be restricted to parts less that about 100 mm for the maximum dimension.

8.4.5 Gaskets

With a sealing gasket, a high level of gas tightness can be achieved with less than perfect surface quality even on larger areas. The gasket material must have good elasticity, plasticity and low permeability. It must also have good resistance to the temperature, corrosiveness and erosiveness of the gas being handled.

8.5 Shaft seals

8.5.1 Near absolute tightness

It is possible to achieve a virtually leak proof fan by employing a direct driven fan having a flanged end shield motor. Even if gas escapes through the seal at the shaft extension, it is still contained within a totally enclosed motor housing. This should be naturally cooled and there is then no shaft seal at the non drive end.

Other methods may also be used for fans in the gas industry, see Figure 8.13, which shows a fan arranged with shaft seals and drive through a coupling.

Figure 8.13 Direct driven leak proof fan for the gas industry

8.5.2 Shaft closing washer

The shaft closing washer described in Section 8.3.3 as part of the spark minimising features may also be used as a simple seal. Provided it is made from a soft brass or similar, the hole can be of exactly the same diameter as the shaft. It will easily "run in" without causing any damage. Provided the ratio of critical speed to running speed is high, the shaft deflection is low and the balance grade better than G 6.3 (preferably G 2.5), elongation of the hole will be minimal.

8.5.3 Stuffing box

A box is filled with a soft packing, such as greased rope. This packing is compressed against the shaft by a gland. The gland is usually split as illustrated in Figure 8.14 and held in place by swivel bolts.

The gland tightness is critical – too tight and heat will be generated. There will also be a frictional power loss. If insufficiently

Figure 8.14 Components of split stuffing box and gland

tightened there will be considerable leakage. Maintenance is therefore greater than for other types.

8.5.4 Labyrinth seals

These are most commonly used and many variants exist. All however require a polished shaft, see Figure 8.15. The labyrinth ring is in two parts, typically stainless steel or PTFE.

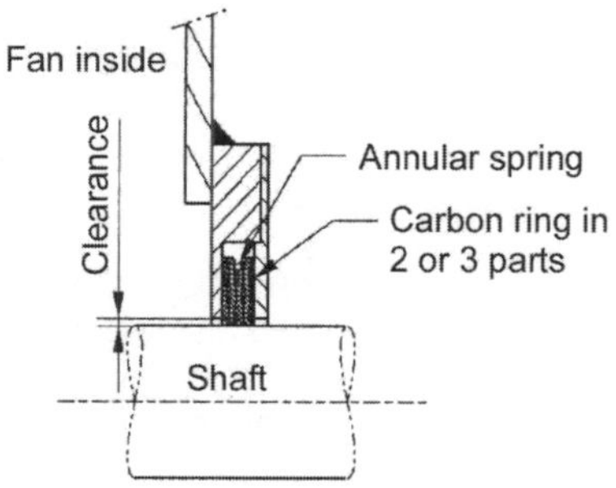

Figure 8.15 Labyrinth seal

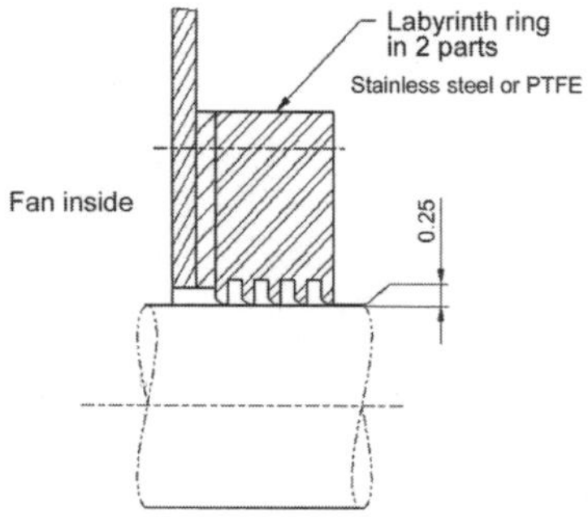

Figure 8.16 Labyrinth seal with annular springs

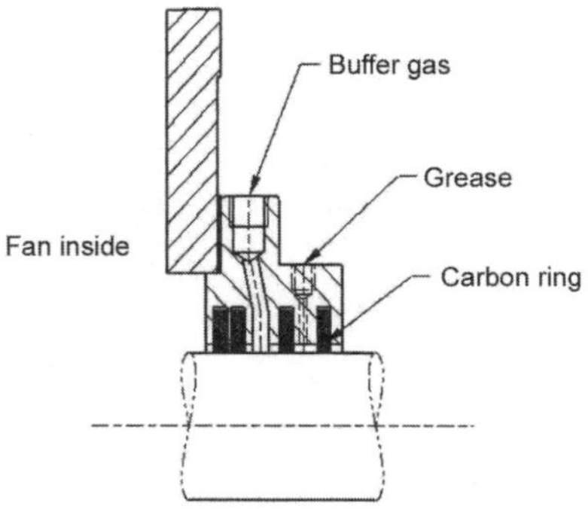

Figure 8.17 Labyrinth seal with floating bushing

Better tightness can be achieved with a floating bushing. The carbon rings are made in two or three parts which are kept closely to the shaft with annular springs (Figure 8.16). A floating bushing as shown in Figure 8.17 can also be used.

8.5.5 Mechanical seals

If the fan operates at a high pressure, ordinary packing may be unsatisfactory. Some form of mechanical seal must then be employed. A typical example is shown in Figure 8.18.

In this design a collar is attached to the shaft by a setscrew. The position of the collar causes the compression springs to exert a

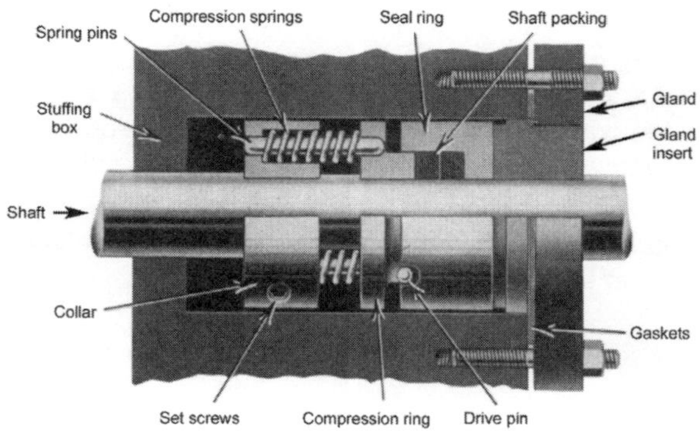

Figure 8.18 Section through a mechanical seal

force through the shaft packing to a seal ring. All the parts described above rotate with the shaft. The gland insert is fixed to the gland which is stationary, and hence rubbing takes place between this insert and the seal ring. By varying the number of gaskets between the gland and the box, the best setting for gas tightness and wear can be decided.

8.6 Fans operating at non-ambient temperatures

8.6.1 Calculation of the duty requirement

Whilst not exactly a special feature it is convenient at this point to say something about the calculation of the required fan performance.

When fans handle air or some other gas, which has a density differing from the standard 1.2 kg/m^3 then performance will vary in accordance with the Fan Laws (see Chapter 4). Thus at a constant volumetric flowrate, the pressure developed, the weight flowrate and the power absorbed will all vary directly with the density of the air or gas being handled. Fan efficiency remains unchanged.

A fan being essentially a "constant-volume" machine, it is necessary to know how the duty requirement has been calculated.

a) Fan flowrate must always be converted to the actual conditions at the fan inlet. Does the customer require the same volume or weight flow?

b) It is important to know under what conditions the fan pressure has been calculated. How will this vary with temperature?

c) Will the fan be required to start on cold air? Is there a need for dampers to assist?

d) Find out the maximum temperature reached during operation – there may be a heat build-up.

An understanding of these rules is important for correct fan selection, determining the correct operating speed where this is variable and also to determining the power consumption over the duty cycle.

8.6.2 Mechanical fitness at high temperature

The strength of metals and plastics varies according to their temperature. When handling air or gas at conditions other than ambient the materials of construction of the fan will therefore also vary from the values normally given in textbooks.

It is important to remember that all elements of the fan must be satisfactory:

a) Impeller

b) Shaft

c) Bearings

d) Casing

Elements within the air or gas stream are likely to take up the same temperature, but elements outside may take up a temperature somewhere between that of the gas stream and the ambient air around the fan.

For detailed methods of calculation to determine material suitability refer to Chapter 7.

8.6.3 Maintaining the effectiveness of the fan bearings

It is important that the "temper" of the balls or rollers is maintained. Normal greases are likely to break down at temperatures above about 90°C. For these two reasons it is essential to reduce the amount of heat which is transmitted from the gas stream, along the shaft to the first bearing. There are a number of ways in which this objective may be achieved.

a) The first and most important method is to add an auxiliary cooling disc to the shaft between the casing and inner bearing. With a simple aluminium bolt-on construction having six open radial blades this extends the operating gas temperature from 75 °C to a maximum of 350 °C as heat is dissipated from the shaft and the temperature at the bearing reduced to less than 90 °C, see Figure 8.19. A more sophisticated shrouded copper impeller has been used with d) below for gas temperatures up to 650 °C. This is just visible through the mesh in Figure 8.20.

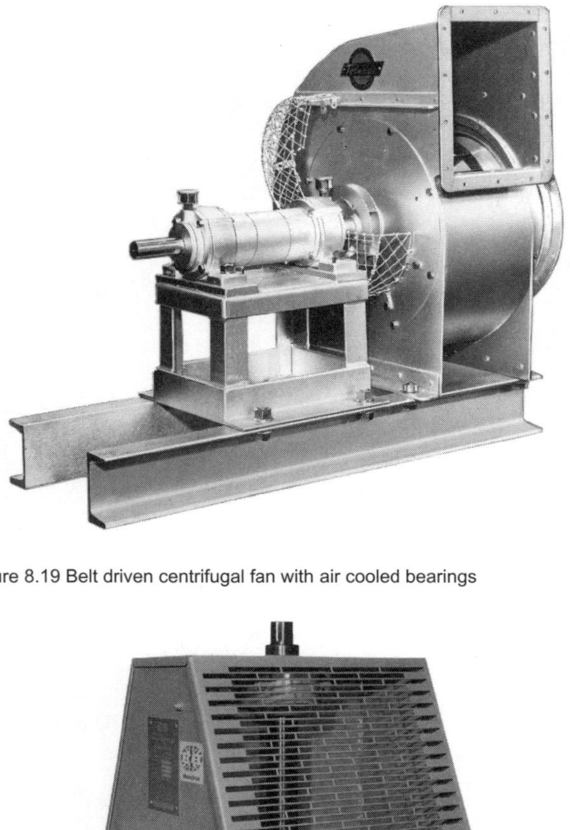

Figure 8.19 Belt driven centrifugal fan with air cooled bearings

Figure 8.20 Fabricated plug type fan with internal shrouded copper cooling impeller

Figure 8.21 Plug fan for the glass industry

b) At higher temperatures water-cooled sleeve bearings may be used. The water ensures that the oil lubricant does not become too thin and also that the white metal babbit does not melt. (See Chapter 10.)

c) Spacer couplings which make a heat "break" in the shaft may also be used above 400 °C. Shaft slots have also been used.

d) Insulated "plugs" on the drive side are typically used above 500 °C to minimise problems from radiated heat, (see Figure 8.21).

8.6.4 Increased bearing "fits"

Bearings are manufactured with various grades of clearance between the rotating elements and the raceways, the normal clearance being designated CN. Table 8.2 gives typical details of the grades available, it being noted that C3, C4 and C5 have clearances greater than normal. Whilst C3 bearings are commonly used where the product of bearing size in mm and rotational speed in rev/min exceeds 175 000 to dissipate frictional heat, C4 or C5 may be necessary with fans handling gases at up to 650 °C.

8.6.5 Casing features

These may require the ability to withstand loads externally applied at high temperatures due to the expansion of the customer's ducting. A preferable alternative is to provide high temperature flexible connections on the fan inlet and outlet and to ensure that clients separately support their ducting.

The casing itself will expand, growing up from its feet. As the pedestal will be cooler, this may destroy the clearances between inlet cone and impeller eye or shaft and shaft entry point. The growth is a function of temperature and size. Clearances of inlet cones and at shaft entry may then need to be increased above about 350°C. At temperatures above about 450°C it is common to support the fan casing near its centreline so that growth of all parts is radially outwards and clearances are not affected.

Where oxygen is present in the gases, "scaling" of a mild steel case will take place above 400°C at increasing rates to 500 °C where it becomes catastrophic. COR-TEN® steel and other

Bore diameter d		Radial internal clearance									
		C2		Normal		C3		C4		C5	
over	incl	min	max	min	max	min	max	min	max	min	max
mm		µm									
	6	0	7	2	13	8	23	-	-	-	-
6	10	0	7	2	13	8	23	14	29	20	37
10	18	0	9	3	18	11	25	18	33	25	45
18	24	0	10	5	20	13	28	20	36	28	48
24	30	1	11	5	20	13	28	23	41	30	53
30	40	1	11	6	20	15	33	28	46	40	64
40	50	1	11	6	23	18	36	30	51	45	73
50	65	1	15	8	28	23	43	38	61	55	90
65	80	1	15	10	30	25	51	46	71	65	105
80	100	1	18	12	36	30	58	53	84	75	120
100	120	2	20	15	41	36	66	61	97	90	140
120	140	2	23	18	48	41	81	71	114	105	160
140	160	2	23	18	53	46	91	81	130	120	180
160	180	2	25	20	61	53	102	91	147	135	200
180	200	2	30	25	71	63	117	107	163	150	230
200	225	4	32	28	82	73	132	120	187	175	255
225	250	4	36	31	92	87	152	140	217	205	290
250	280	4	39	36	97	97	162	152	237	255	320
280	315	8	45	42	110	110	180	175	260	260	360
315	355	8	50	50	120	120	200	200	290	290	405
355	400	8	60	60	140	140	230	230	330	330	460
400	450	10	70	70	160	160	260	260	370	370	520
450	500	10	80	80	180	180	290	290	410	410	570
500	560	20	90	90	200	200	320	320	460	460	630
560	630	20	100	100	220	220	350	350	510	510	700
630	710	30	120	120	250	250	390	390	560	560	780
710	800	30	130	130	280	280	440	440	620	620	860
800	900	30	150	150	310	310	490	490	690	690	960
900	1 000	40	160	160	340	340	540	540	760	760	1 040
1 000	1 120	40	170	170	370	370	590	590	840	840	1 120
1 120	1 250	40	180	180	400	400	640	640	910	910	1 220
1 250	1 400	60	210	210	440	440	700	700	1 000	1 000	1 340
1 400	1 600	60	230	230	480	480	770	770	1 100	1 100	1 470

Table 8.2 Typical radial internal clearance of deep groove ball bearings

proprietary grades, which have a copper content, scale at a slower rate. Information is available from the manufacturer on the rate for these and many other steels.

As an alternative, the casing may be "aluminised", which effectively eliminates the problem. Above about 570 °C stainless steel casings are usually necessary from scaling, strength and stability considerations.

It should be noted that scaling will not occur if the gases are inert e.g. nitrogen. Flue gases may be inert under conditions of perfect combustion, i.e. do not contain oxygen in its free form.

8.6.6 Lagging cleats

European legislation now covers the maximum safe temperature for surfaces which may come into contact with the hands or other parts of the human body. It may also be desirable for efficiency reasons to limit the amount of heat which may be dissipated from the casing. In these cases, lagging cleats should be added to assist in the anchoring of insulating materials.

8.6.7 Mechanical fitness at low temperature

There are no real problems with gas temperatures down to about -30 °C but allowance must be made for the power increase due to the higher air density. Below -40°C mild steel becomes increasingly brittle. It may be necessary to use an aluminium impeller or steel with high nickel content. Shafting should also be of nickel steel whilst bearing plummer blocks must be cast steel (not cast iron). Grease lubricants should be checked for suitability - they must not solidify or separate.

8.7 High pressure fans

Casings of high pressure fans need to be of sufficient thickness and strength to withstand the internal bursting pressure. This is normally calculated by determining the hoop stress in the scroll and the bending stress in the volute sides. Another consideration is the thrust load on the fan bearings. In a closed circuit fan, this can be considerable. There is also the attendant leakage at the shaft entry hole. The features detailed in Sections 8.6.1 and 8.6.2 reduce both the thrust and any outward leakage.

8.7.1 Scavenger blades

These are narrow (usually radial) blades attached to the rear of the impeller backplate and running in the space between the volute side and the impeller (see Figure 8.22). Air is induced at the shaft entry hole and an axial thrust developed in the opposite direction to that of the main impeller. The resultant axial load at the fan bearing can thereby be reduced to a very low figure albeit with an increase in absorbed power.

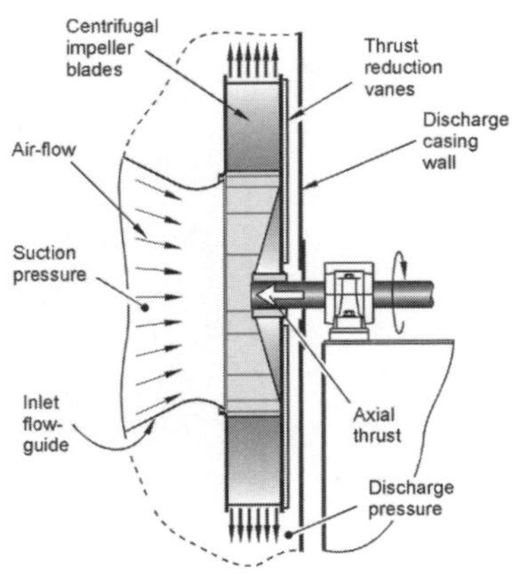

Figure 8.22 Cross-section of fan with scavenger blades

8.7.2 Pressure equalizing holes

These are small holes in the impeller backplate which allow a minimum quantity of air to pass through, thereby reducing the pressure difference between the space behind the impeller and the suction zone at its inlet, see Figure 8.23. Again the resultant axial load on the fan bearing is reduced, with a slight reduction in fan efficiency. Stresses in the fan backplate will increase and the holes may act as a stress raiser.

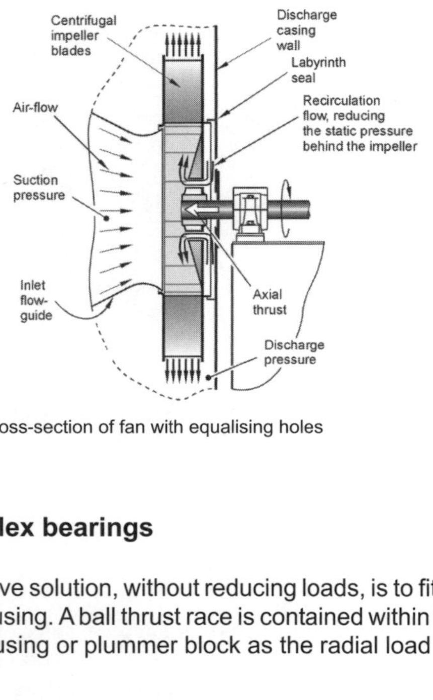

Figure 8.23 Cross-section of fan with equalising holes

8.7.3 Duplex bearings

An alternative solution, without reducing loads, is to fit a duplex bearing housing. A ball thrust race is contained within the same bearing housing or plummer block as the radial load bearing.

8.8 Construction features for axial and mixed flow fans

8.8.1 Features applicable

Many of the features described for centrifugal fans in Sections 8.1 to 8.6 inclusive, are also applicable to axial and mixed flow fans. Examples which readily come to mind are inspection doors (with the provisos detailed) and drain points. The latter may be used where the fan is at the lowest point of the system. Cooling discs may be used with bifurcated fans (see Section 8.9.4) where the air is above 75 °C and heat transmitted along the shaft could otherwise damage the motor.

Scavenger blades and pressure relief holes are not of course applicable but the reduced pressure development of these fans make them unnecessary.

8.8.2 Short and long casings

Tube axial fans may be provided with so-called "short" or "long" casings.

Short casings are normally used on fans at the entry (Installation Category B) or at the exit (Installation Category C) of the ducting system. They can also be used in non-ducted situations (Installation Category A). Access to the motor and impeller in all these cases is then easy. See Figure 8.24.

Figure 8.24 Short cased axial flow fan

The terminal box can be on the motor carcase in its normal position, noting however that there is some blockage to the airflow where this is along the motor body length. A terminal box on the motor endshield may be preferable for this reason.

Long casings are normally used on fans contained within a ducting system which has elements on both the fan inlet and outlet (Installation Category D). The fan casing will be sufficiently long to encompass the impeller and motor length, normally terminating in flanges, see Figure 8.25. An external terminal box is fitted, so that electrical wiring can be carried out without access to the motor, the fan manufacturer providing the wiring between this box and the motor terminals. This wiring is normally contained within rigid piping or a flexible conduit. Vane axial fans with downstream guide vanes and mixed flow fans are invariably provided with long casings.

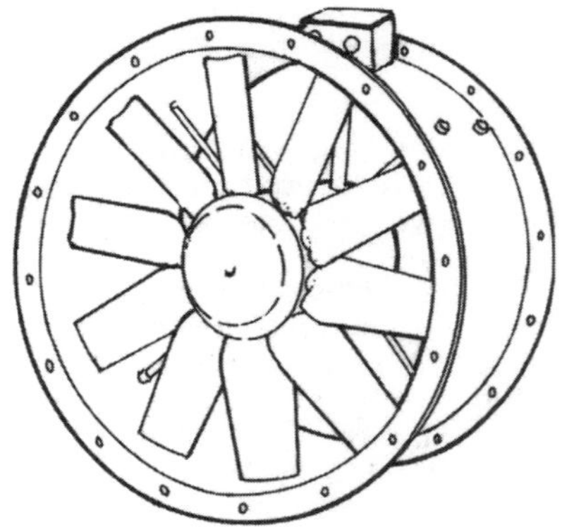

Figure 8.25 Long cased axial flow fan

8.8.3 Increased access casings for maintenance

There are a number of variants on this theme which are particularly popular for marine use and for kitchen extraction.

a) A short cased fan is manufactured with an external terminal box. The motor mounting arms are bolted on the inside of the fan casing, enabling the motor with impeller to be removed for overhaul while the casing remains in situ and any attached ducting does not need to be disturbed. An extension duct bolted to one of the fan flanges with a door

Figure 8.26 Marine fan with downstream duct section having large inspection doors

Figure 8.27 Marine fan with swing-out "Maxcess" casing

or doors gives an opening of 180 ° for this removal (see Figure 8.26).

b) Instead of the extension duct detailed above, a more simple "inspection" duct can be substituted. This is fitted with an access door of ample size for inspection, lubrication or cleaning. On larger sizes the door may be carried on hinges instead of being bolted on.

c) For the most arduous duties, the so-called "Maxcess" casing is preferred. Here the motor and impeller are mounted on a very large hinged door which can be swung out for access and maintenance, without disturbing any associated ducting. (See Figure 8.27.)

8.8.4 Bifurcated casings

Directly driven axial flow fans have their motors in the airstream, which can be both an advantage and disadvantage. Whilst the moving air cools the motor, if there is high temperature or corrosive elements present, then it is desirable for the motor to be outside. A bifurcated, or "split" casing is a solution. This is shown in Figure 8.28. The airstream is diverted either side of the motor compartment and then rejoins again downstream. Thus the motor is open to the cooler or cleaner ambient

Figure 8.28 True bifurcated axial flow fan

Figure 8.29 Bifurcated axial flow fan with one-sided motor compartment

atmosphere. True bifurcated fans can be installed vertically at high level in chimneys where the wind can blow through the motor compartment to give excellent cooling.

A variant on the true bifurcated fan is for the motor compartment to be only open to atmosphere on one side, see Figure 8.29. The blockage effect is less but requires a diversion plate to be fitted to encourage a cooling air path if a TEFV motor, as discussed in Chapter 13, is fitted.

8.9 Bibliography

ISO 13349:1999, BS 848-8:1999, *Fans for general purposes. Vocabulary and definition of categories.*

prEN 14986, *Design of fans working in potentially explosive atmospheres.*

ATEX DIRECTIVE 94/9/EC, *equipment and protective systems intended for use in potentially explosive atmospheres.*

9 Fan arrangements and designation of discharge position

The need for understanding between fan manufacturers and system designers is nowhere more apparent than in the nomenclature for describing the fan inlet and outlet orientation. The history of attempts at removing any possible misunderstanding is described with a few words, but the illustrations are of most importance. Someone once said that one good picture is worth a thousand words. For once the author was dumbstruck!

Contents:

9.1 Introduction

Over the years the need for understanding between manufacturers and their customers has determined that an agreed nomenclature for centrifugal fans and their components was absolutely essential. This applied to both positions of the outlet flange and the mechanical driving arrangements. Motor positions for indirect drives also had to be categorised. Whilst individual companies often had their own coding, this was not necessarily helpful in a competitive situation. Confusion could arise e.g., when one manufacturer's Arrangement 1 was designated Arrangement 3 by another.

9.2 Designation of centrifugal fans

9.2.1 Early USA Standards

Probably the first attempts at an industry wide standard were made by the US National Association of Fan Manufacturers in its Bulletin No 105 dating back to the 1930s. This bulletin covered the designation of the discharge of centrifugal fans, the position of inlet boxes, the arrangement of fan drives, and the standard designation of motor positions. The relevant diagrams for these designations are shown in Figures 9.1 to 9.4.

It is of interest to note that these standards have been used in the USA ever since, albeit with a few deletions and additions.

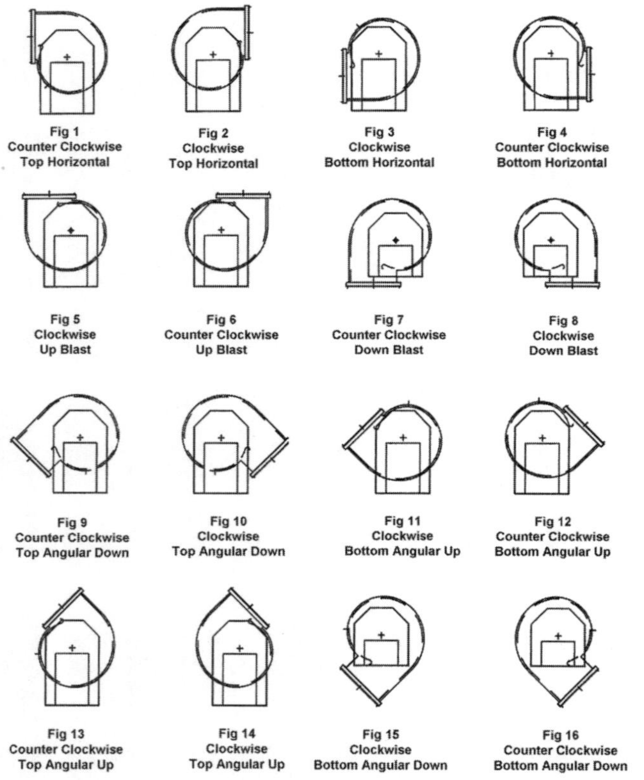

Figure 9.1 Standard designation of fan discharge

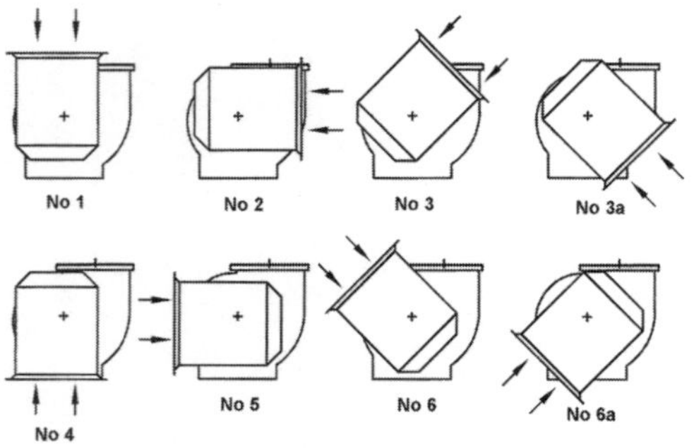

Figure 9.2 Designation of position of inlet boxes

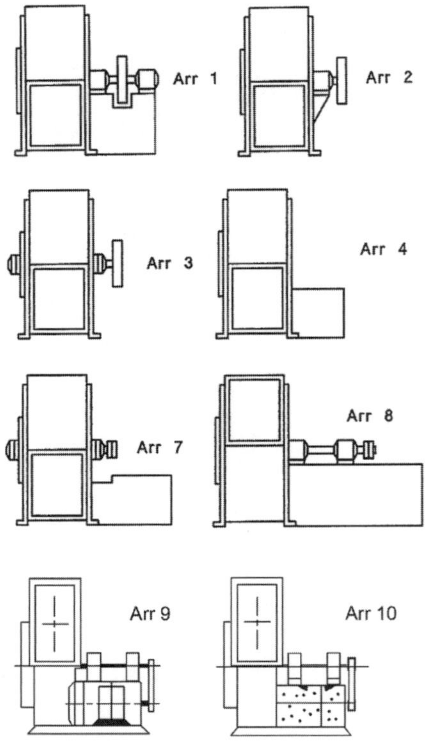

Figure 9.3 Standard arrangements of centrifugal fan drive (AMCA - USA)

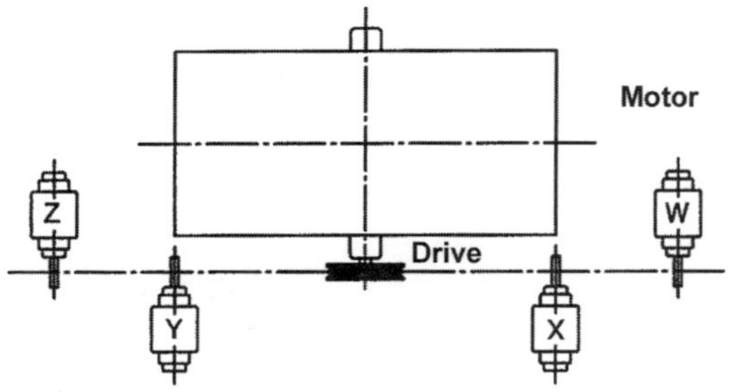

Figure 9.4 Standard designation of motor position

The NAFM has been succeeded by AMCA International, which has been influenced to some extent by the subsequent ISO standards.

9.2.2 Early British Standards

Early efforts at the standardisation of nomenclature for discharge position and arrangements of drive etc were largely based on these American standards, but with some significant improvements.

Instead of "clockwise" and "counter-clockwise" for rotation, "right-hand" and "left-hand" were the designations perhaps on the basis that a right-hand thread is screwed clockwise to tighten. The position of the outlet was given an angular designation starting at 0 for bottom horizontal and proceeding around the protractor i.e.

45	for bottom angular up
90	for vertical up
135	for top angular up
180	for top horizontal
225	for top angular down
270	for vertical down
315	for bottom angular down

Thus the designations become R0 or L0. R90 or L90 etc. These were standardised in both FMA 3:1952 and British Standard

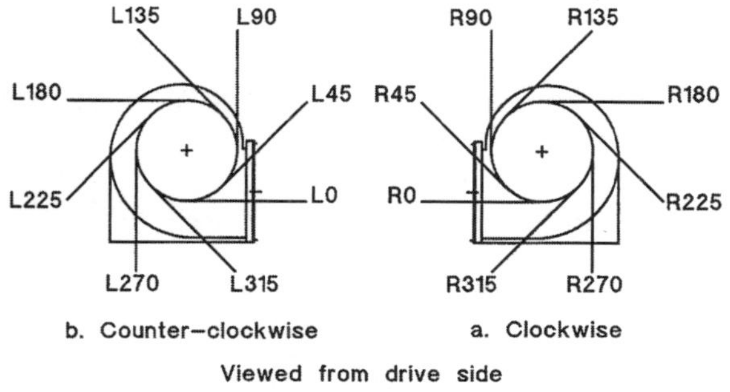

Figure 9.5 Standard designation of fan discharge (FMA and BSI - UK)

949:1939 and are best shown by reference to Figure 9.5. These designations were repeated in the 1963 and 1980 editions.

In like manner the designations for motor position were appended to FMA 3:1952 and BS 848:1963 and 1980. However, instead of the letters W, X, Y and Z, the letters B, C, D and A respectively were used, see Figure 9.6.

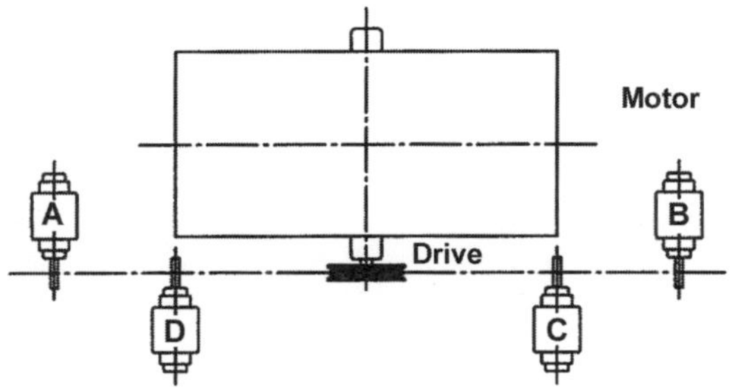

Figure 9.6 Standard designation of motor position (FMA and BS 848:1993)

9.2.3 European and International Standards

With the growing Europeanisation of the fan industry the 1980s witnessed a demand for a more widespread standard. Eurovent (The European Committee of Air Equipment Manufacturers) responded to this with document 1/1 of 1972. Whilst the British and American Standards were tabled as working documents, certain important changes were made in the interests of acceptability. These were:

- Rotation would be identified by the letters LG (signifying Left, Gauche or Links) and RD (signifying Right, Droite or Recht). Thus the 3 main European languages were all recognised.
- An angular position would be identified by a number showing the degrees, but starting at 0 for vertical up outlet instead of 0 for bottom horizontal.

As in all the preceding standards, these designations were to be taken when viewed along the axis of the fan on the driveside. It should here be noted that the driveside was identified as the side opposite the inlet for a single inlet fan, no matter what was the actual position of the drive. This was stipulated principally for those occasions where a single inlet fan had a direct drive motor fitted in the fan inlet. There are however other rare instances of indirect drive on the inlet side. For double inlet centrifugal fans the direction of rotation is determined when viewed from the driveside.

These outlet positions are shown in Figure 9.7 and having recently been accorded worldwide recognition in ISO 13349. It should be noted that intermediate positions may be identified by an appropriate figure for the angle of the outlet. For the user, it is necessary to discuss with the manufacturer exactly what is available, depending on the constructional methods. All angles from 180° to 225° may require special constructions at extra cost.

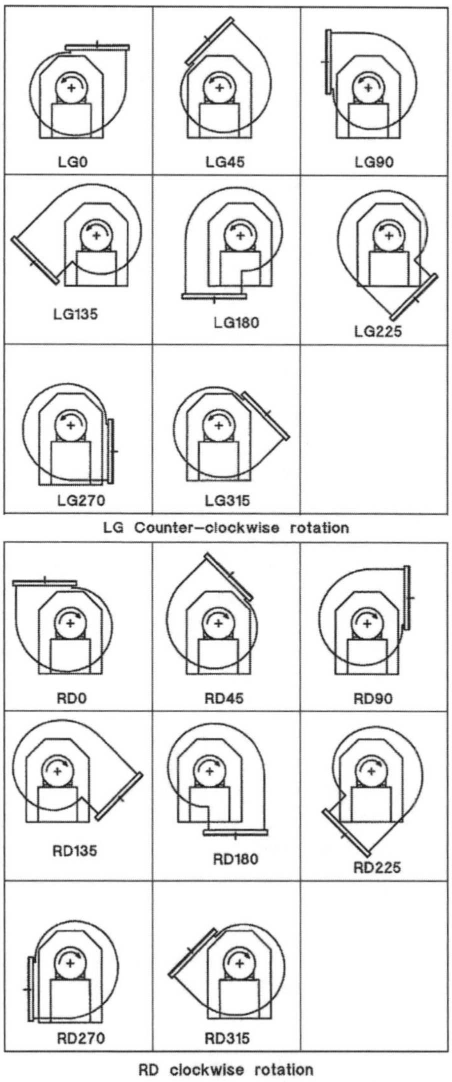

Figure 9.7 Standard designation of fan discharge (Eurovent and ISO)

The position of component parts of a centrifugal fan with volute casing are also standardised in Eurovent 1/1:1972 and ISO 13349 figure 20.

Whilst these diagrams indicate the angular position of a motor if mounted on the fan casing, they do not identify the alternative positions of a motor for an indirect drive (belt or chain) when at or near ground level. For these cases both Eurovent and ISO have adopted the American W, X, Y, and Z positions.

Fan specifiers are encouraged to specify ISO 13349 as this will obviate all possible ambiguities. However it has to be recognised that there are still some manufacturers using these earlier standards, albeit in diminishing numbers. For assistance in such cases, the following Table 9.1 of equivalents may be of help.

ISO 13349 Eurovent 1/1	BS 848 1939/63/80 FMA	AMCA Int. 99-2404	NAFM Bulletin 105 and early AMCA
LG or RD 0	L or R 90	CCW or CW 0	CCW or CW UB
LG or RD 45	L or R 135	CCW or CW 45	CCW or CW TAU
LG or RD 90	L or R 180	CCW or CW 90	CCW or CW TH
LG or RD 135	L or R 225	CCW or CW 135	CCW or CW TAD
LG or RD 180	L or R 270	CCW or CW 180	CCW or CW DB
LG or RD 225	L or R 315	CCW or CW 225	CCW or CW BAD
LG or RD 270	L or R 0	CCW or CW 270	CCW or CW BH
LG or RD 315	L or R 45	CCW or CW 315	CCW or CW BAU

Table 9.1 Equivalent fan discharge designations

Key:

CCW	=	Counter Clockwise
CW	=	Clockwise
UB	=	Up Blast
TAU	=	Top Angular Up
TH	=	Top Horizontal
TAD	=	Top Angular Down
DB	=	Down Blast
BAD	=	Bottom Angular Down
BH	=	Bottom Horizontal
BAU	=	Bottom Angular Up

9.2.4 European and International Standards for fan arrangements

Until the 1980s the standardisation of fan arrangements was largely non-existent. Each company continued to use its own designations. Regrettably a small number still do. At that time BSI launched work on BS 848 Part 8 and had reached the stage of a working draft. This included a section on fan arrangements and these largely followed North American standards as exampled in what had now become AMCA Standard 99-2404. Since the original NAFM Bulletin No. 105 however, Arrangements 5 & 6 , which required flanged (rigid) couplings had become obsolete and were no longer included. The BSI draft took advantage of this fact to use these two numbers for other purposes. Arrangement 5 was therefore proposed for direct drive without a motor supporting stool or pedestal, the motor being bolted to the fan casing by its flanged endshield. Arrangement 6 was utilised for the DIDW version of Arrangement 3 , which was restricted to SISW fans. There was certain logic in this – twice 3 equals 6!

Meanwhile UNI, the Italian standards organisation had also produced its standard UNI 7972 which had a very much more comprehensive range of fan arrangements, again using the American designations where possible.

At this point in time ISO determined that it would commence work on a "Vocabulary and definition of categories" which, as noted, was published as ISO 13349:1999, giving the drive arrangements for centrifugal fans. These are shown in Table 9.2.

9.3 Designations for axial and mixed flow fans

9.3.1 Direction of rotation

This is not normally of any great concern for the fan user except when obtaining spare parts. Sometimes, however, it may affect the magnitude of system effect factors. The manufacturer may need a code for determining the handing of impeller parts.

ISO 13349 specified that the rotation is determined from the side opposite the inlet, (see Figure 9.8).

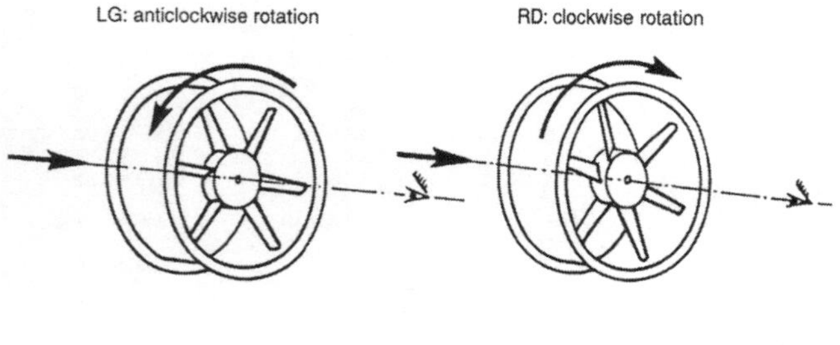

Figure 9.8 Direction of rotation of axial and mixed flow fans

9.3.2 Designation of motor position

These are best determined from Figure 9.9. The codes used in ISO 13349 are for horizontal and vertical axes.

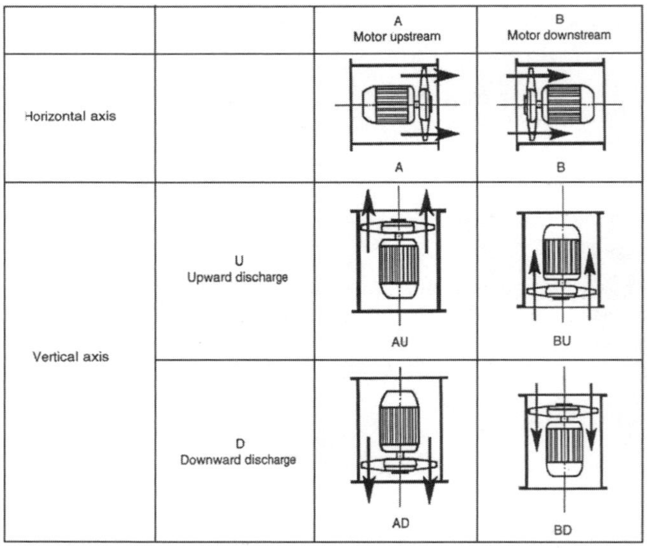

Figure 9.9 Designation of motor position for axial and mixed flow fans

9.3.3 Drive arrangements for axial and mixed flow fans

These also have been standardised in ISO 13349 and 99-2404 of 1998. The similarity with the corresponding centrifugal fans will be recognised.

A description of the driving arrangements is given in Table 9.3. It will be noted that not all arrangements available for centrifugal fans are applicable to axial and mixed flow fans.

9.4 Belt drives (for all types of fan)

Variously known as belt or rope drives, these are most commonly of vee section. For further information refer to Chapter 11.

Standard arrangements Nos. 1 2 3 6 9 10 11 12 13 14 18 and 19 are all applicable to belt drive of flat, classical or wedge form. It should be noted that the fan bearing nearest the pulleys and belts will be subject to a unidirectional radial load. This may limit the power which can be transmitted unless recourse is made to layshafts or pulleys between the bearings.

9.5 Direct drive (for all types of fan)

This description is limited to those designs where the fan impeller is directly mounted on the shaft extension of a suitable electric motor or other prime mover. The motor must be capable of supporting the weight of the fan impeller and also of resisting the end thrust produced by the pressure difference across the impeller. Standard Arrangement Nos. 4 5 15 and 16 are all applicable.

9.6 Coupling drive (for all types of fan)

This description is applicable to Arrangement Nos. 7 8 and 17 . A flexible coupling permitting limited misalignment is now normally used. The motor may be removed for maintenance purposes without disturbing the fan alignment. Arrangements 8 and 17 are particularly appropriate for large high powered fans and there are generally no limitations on the power to be transmitted.

Arrangement No.	Description	Motor position (see Figure 9.4)	Outline drawing
1	Single-inlet fan for belt drive. Impeller overhung on shaft running in 2 plummer block bearings supported by a pedestal.	—	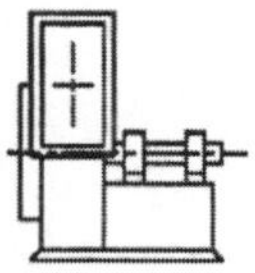
2	Single-inlet fan for belt drive. Impeller overhung on shaft running in bearings supported by a bracket attached to the fan casing.	—	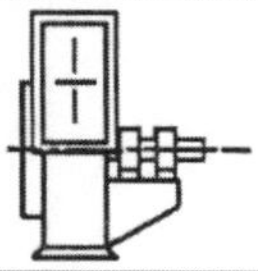
3	Single-inlet fan for belt drive. Impeller mounted on shaft running in bearings on each side of casing and supported by the fan casing.	—	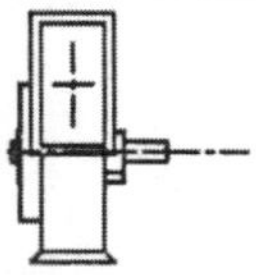
4	Single-inlet fan for direct drive. Impeller overhung on motor shaft. No bearings on fan. Motor supported by base.	—	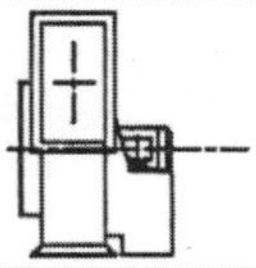
5	Single-inlet fan for direct drive. Impeller overhung on motor shaft. No bearings on fan. Motor attached to casing side by its flanged end-shield.	—	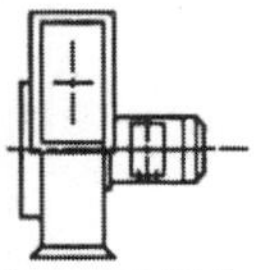
6	Double-inlet fan for belt drive. Impeller mounted on shaft running in bearings on each side of casing and supported by the fan casing.	—	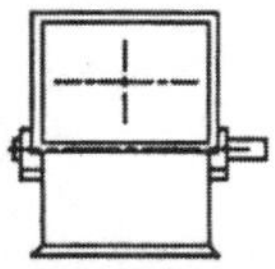
7	Single-inlet fan for coupling drive. Generally as arrangement 3 but with a base for the driving motor.	—	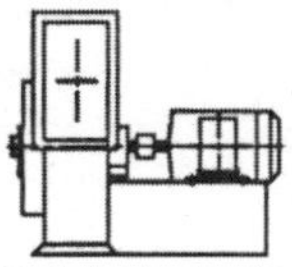
8	Single-inlet fan for coupling drive. Generally as arrangement 1 plus an extended base for the driving motor.	—	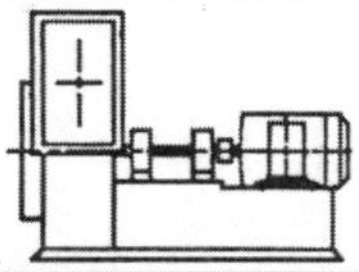
9	Single-inlet fan for coupling drive. Generally as arrangement 1 but with the motor mounted on the outside of the bearing pedestal.	W or Z	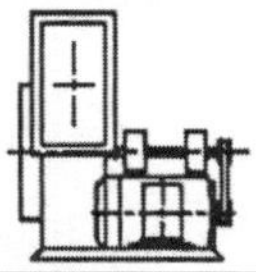
10	Single-inlet fan for belt drive. Generally as arrangement 1 but with the drive motor inside the bearing pedestal.	—	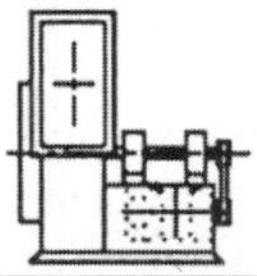

Table 9.2 Standard drive arrangements for centrifugal fans

Arrangement No.	Description	Motor position (see Figure 9.4)	Outline drawing
11	Single-inlet fan for belt drive. Generally as arrangement 3 but with the fan and motor supported by a common base frame.	W or Z (very rarely X or Y)	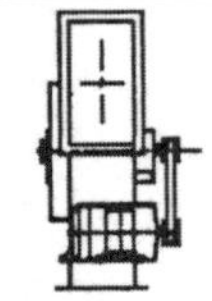
12	Single-inlet fan for belt drive. Generally as arrangement 1 but with the fan and motor supported by a common base frame.	W or Z (very rarely X or Y)	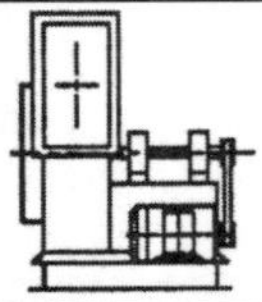
13	Single-inlet fan for belt drive. Generally as arrangement 1 but with the motor fixed underneath the bearing pedestal.	—	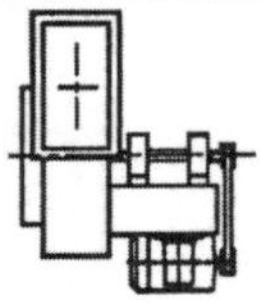
14	Single-inlet fan for belt drive. Generally as arrangement 3 but with the motor supported by the fan scroll.	—	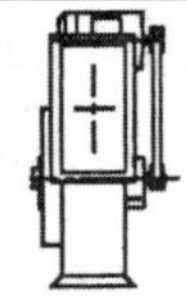
15	Single-inlet fan for direct drive. Driving motor in-set within impeller and fan casing.	—	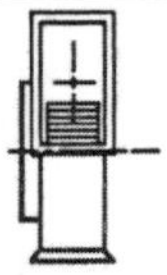
16	Double-inlet fan for direct drive. Driving motor in-set within impeller and fan casing.	—	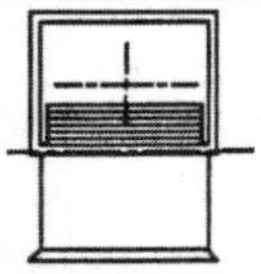
17	Double-inlet fan for coupling drive. Generally as arrangement 6 but with a base for the driving motor.	—	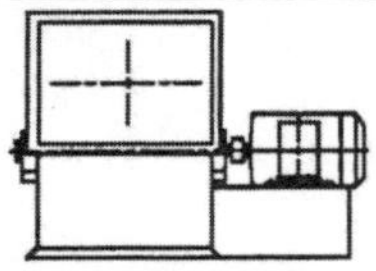
18	Double-inlet fan for belt drive. Generally as arrangement 6 but with a fan and motor supported by common base frame.	W or Z (very rarely X or Y)	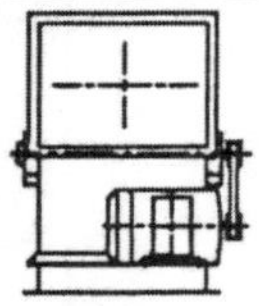
19	Double-inlet fan for belt drive. Generally as arrangement 6 but with the motor supported by the fan scroll.	—	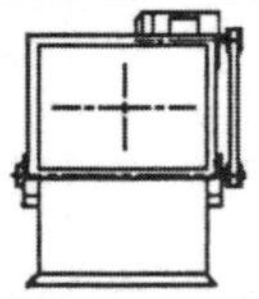
NOTE Arrangements 1, 3, 6, 7, 8 and 17 may also be provided with the bearings mounted on pedestals for base set independent of the fan housing.			

Table 9.2 Standard drive arrangements for centrifugal fans *(continued)*

Arrangement No.	Description	Motor position (see Figure 9.4)	Outline drawing
1	For belt drive. Impeller overhung on shaft running in 2 bearings, suitably supported.	—	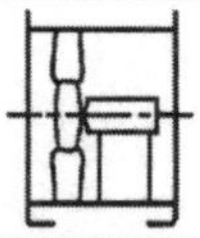
3	For belt drive. Impeller overhung on shaft running between bearings and supported by fan housing.	—	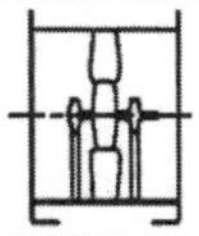
4	For direct drive. Impeller overhung on driving motor shaft. No bearings on fan. Driving motor base-mounted or integrally direct-connected.	—	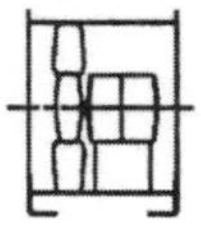
7	For coupling drive. Generally as arrangement 3 but with a base for the driving motor.	—	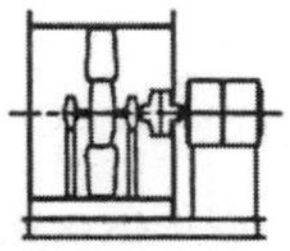
8	For coupling drive. Generally as arrangement 1 plus an extended base for the driving motor.	—	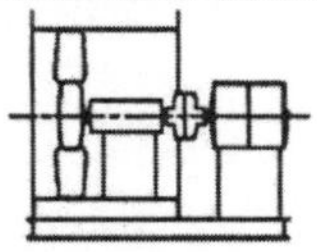
9	For belt drive. Generally as arrangement 1 but with a driving motor outside and supported by the fan casing.	—	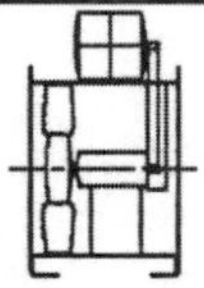
11	For belt drive. Generally as arrangement 3 but with fan and driving motor outside and supported by a common base frame	W or Z (very rarely X or Y)	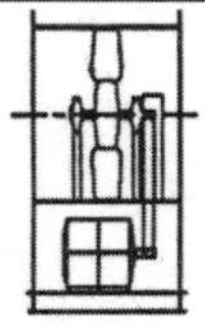
12	For belt drive. Generally as arrangement 1 plus an extended base for the driving motor.	W or Z (very rarely X or Y)	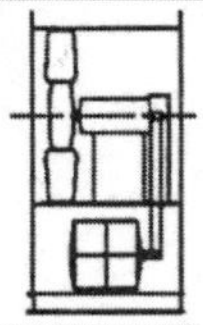

Table 9.3 Drive arrangements for axial and mixed flow fans

9.7 Single and double inlet centrifugal fans

Standard fans are usually manufactured as Single Inlet Single Width designated SISW or alternatively SWSI (especially in Northern America).

Where a large volumetric flowrate is required, a Double Inlet Double Width fan designated DIDW or alternatively DWDI may be used. At a given speed for a given diameter approximately twice the flow can be handled at the same pressure and efficiency.

For a given flowrate and pressure the DIDW fan will be approximately 70% of the size at the same efficiency. It will also run faster, permitting the selection of a cheaper motor.

9.8 Other drives

Around 99% of all fans incorporate electric driving motors. However petrol or diesel motors are used where electrical supplies are unavailable or perhaps where portability is desirable. In mechanical draught installations on steam boilers, the availability of steam has often encouraged the use of steam turbines. These, of course, are not limited to the set speeds of electric motors on AC supplies.

9.9 Bibliography

BS 848, ISO 13349, *Fans for general purposes. Vocabulary and definition of categories.*

UNI 7972:1980, *Ventilatori industriali. Classificazione e terminologia.*

AMCA 99-2404-03, *Drive arrangements for centrifugal fans.*

Eurovent 1/1, *Fan terminology.*

10 Fan bearings

Many types of bearings can be found on fans, of which rolling element and plain bearings are by far the most numerous and form the main part of this Chapter. More exotic bearings, for example air bearings and magnetic bearings, may be used for some very special applications and are briefly discussed.

Other factors which play an important part in the choice of bearings include thermal expansion and heat losses. Any fan when it operates will experience a temperature rise and this can give different amounts of expansion between the stator and rotor which in turn may impose additional forces on the bearings or a requirement to design the overall bearing system to compensate for such events. The load may in some cases contribute to the problem by its own shaft expansion. All bearings have some frictional losses which appear as heat and may require some bearing cooling. Lubrication plays an important part in maintaining bearing temperatures at an acceptable level and in some cases cooling of the lubricant may be essential.

Contents:

10.1 Introduction

Wherever there is rotating machinery there will be a need for bearings i.e. those components whereby forces are transmitted between solids which are moving relative to each other. It is at such interfaces that friction takes place, accounting in its turn for significant amounts of energy to be added to that required for the air power provided by a fan impeller.

It is also at these interfaces that wear occurs, with a consequential risk of malfunctioning and/or overcoming the effects of wear, not only on the impeller and stationary parts, but often more importantly on the fan bearings and shaft.

The change of lubrication from an empirical art to an exact science, now dignified with the title "Tribology" grew out of the studies of Beauchamp Tower. He reported to an Institution of Mechanical Engineers committee set up in 1879. Osborne Reynolds, that giant of Victorian engineers, analysed these results and in 1886 showed that in certain circumstances, the relative motion and convergent geometry could generate sufficient pressure to overcome the loads applied to a bearing and prevent the two surfaces from making physical contact.

10.1.1 General comments

There is a wide variety of bearing types used for fans of which plain and rolling element bearings are by far the most numerous and form the main part of this Chapter. More exotic bearings, for example air bearings and magnetic bearings, may be used for some very special applications and are briefly discussed.

Although the bearings essentially support and position the impeller, they may be called upon to withstand some of the other forces imposed by the driven load. The rotor weight will always act downwards whatever the motor attitude but the forces arising from the load, where applicable, may be in any direction and even vary according to the load conditions. The type of bearing selected will depend upon these conditions in addition to any limitations imposed by the environment. There is clearly a difference in the type of bearing used for impellers running horizontally or vertically. Except for some very small fans and fans intended to run with the shaft in any direction, particular attention may need to be paid to the choice of bearings.

Other factors which play an important part in the choice of bearings include thermal expansion and heat losses. Any fan, when it operates, will experience a temperature rise, or indeed may handle hot gases. This can give different amounts of expansion between the fan casing and bearing support structure, which in turn may impose additional forces on the bearings or a requirement to design the overall bearing system to compensate for such events. The fan may in some cases contribute to the problem by its own shaft expansion. All bearings have some frictional losses which appear as heat and may require some bearing cooling. Lubrication plays an important part in maintaining bearing temperatures at an acceptable level and in some cases cooling of the lubricant maybe essential.

The fan attitude, forces from the driven load, air or gas temperatures and site ambient conditions all affect the bearing reliability and life. In turn the maintenance requirements are determined by these factors and the type of bearing selected. Generally the manufacturer will fit bearings suitable for the specified requirements but customers may have a preference for a particular bearing type. For example, sometimes rolling element or plain bearings may be suitable and the customer has a preference based on his experiences.

This Chapter covers various aspects of bearing selection, bearing housings, operation, lubrication, life and maintenance. Monitoring bearing performance by means of auxiliary equipment to protect against failure is also discussed in Chapters 15 and 18.

10.1.2 Kinematic pairs

A machine has been defined as "an apparatus for applying mechanical power, consisting of a number of interrelated parts, each having a definite function". The parts in contact, and between which there is a relative motion, form a "kinematic" pair consisting of two solid bodies in contact. Lubrication is inevitably necessary for good operation. Often additional elements are included, for example, the balls or rollers and cage of a typical bearing race.

Kinematic pairs fall into two categories:

Lower, in which surfaces touch over a fairly large area whilst sliding, one relative to the other. These would include pistons, sleeve bearings and screws used for converting rotary to linear motion or vice versa.

Higher, in which there is only line or point contact between the surfaces and relative motion may be partly turning and sliding. Examples include wheels on rails, anti-friction (ball and roller) bearings, or gears and pinions.

The majority of modern fans are fitted with rolling element bearings. As design has become more advanced, parts have been expected to rotate at higher speeds leading to higher stress levels. It has become the norm to get "a quart out of a pint pot". In general this has favoured the increasing adoption of ball/roller, or anti-friction, bearings.

10.1.3 Condition monitoring

It is inevitable that in every decade there will be a theme to fascinate our political masters. Having survived the "white heat of the technological revolution" what now? Undoubtedly one of the contenders is our "business efficiency" and this is recognised as vital if we are to expand, or indeed survive, in an increasingly competitive world.

The use of CNC machinery for production; of computer systems in the design and accounts departments; and even of sophisticated marketing techniques in the sales office, all continue apace. Only recently has the efficient maintenance of machinery been recognised as a potential field for extra profit.

Condition monitoring techniques have frequently been introduced but have themselves been monitored for cost effectiveness. Companies have often wasted money on such systems but the losses have been ignored. Perhaps maintenance itself should be more closely investigated instead of being accepted as an inevitable overhead.

Mechanical methods of condition monitoring are of most interest where the fan has ball/roller bearings (higher pairs), although some can be of use in analysing the special problems of sleeve bearings. Chemical methods can be of value in all cases.

The cost of preventative maintenance programmes, involving periodic stopping, stripping down and re-starting of an installation, is becoming prohibitive. This is particularly so with capital intensive or even automatic plant. Various techniques have therefore been developed to determine the condition of fans whilst they are running, with the intention that only when there is an indication of impending damage or malfunctioning due to excessive wear, will they be stopped. These techniques may be conveniently grouped under two headings and some examples are given for each:

Mechanical

- Vibration analysis — For general monitoring of plant condition.

- Spike energy detection — Methods for early warning of bearing failure.
- Shock pulse measurements — Methods for early warning of bearing failure
- Kurtosis monitoring — Methods for early warning of bearing failure

Further information on these techniques as applied to fans is given in Chapter 15.

Chemical

- Spectrographic oil analysis programmes (SOAP)
- Heat detection and thermography
- Ferrography or particle analysis

Further information on these techniques as applied to fans is given in Chapter 18.

10.2 Theory

Once a fan designer has decided whether to use a lower (sleeve bearings) or a higher (anti-friction bearings) pair then the following results may be stated:

1) In a **lower pair**, the two surfaces conform to each other and contact will be dispersed over the whole of the nominal area of contact. However, practical surfaces are never completely smooth and true contact will be restricted to a limited number of peaks. A rough rule is that the true area of contact will be only about 0.1% of the nominal area, whilst the total area of the peaks in contact equals the total load on the surfaces divided by the "flow stress" of the material.
2) In a **higher pair**, contact is within a narrow zone (usually an ellipse) in the vicinity of a point (ball bearings) or a line (roller bearings). Because of this concentration, stress is high and results in local elastic deformation. The actual area of contact is determined by the load, the geometrical shape of the contacting parts and the elasticity of the materials involved. The mathematical determination of the contact conditions was first outlined by Hertz in 1886, such contacts thereafter being described as Hertzian and accepted as "elastic".

10.2.1 Bearing materials

It is obvious that the considerable differences between sleeve and ball/roller bearings will lead to completely different materials of construction being chosen.

In the case of sleeve bearings, the journal surface is usually made of a soft material which will conform readily to the harder shaft material. It is preferable to select materials which have a considerable difference in hardness so that the permanent shape of the bearing is determined by the harder surface.

Thus in a fan bearing, where a unidirectional load is transmitted from a rotating shaft to a stationary bearing housing, the shaft would be manufactured from an alloy steel, which would retain its shape, whilst the bearing housing would be lined with "white" metal or "babitt", which would take up the shape of the shaft as shown in Figure 10.1.

In the past the bearing lining would be scraped by hand to conform to the shaft. The author, in his apprenticeship days, spent many happy hours blueing, rolling and scraping! Now, however, it is usual to machine slightly oversize. Conformity is then achieved from a light "running-in".

Assuming that the shaft is truly round, the surfaces will rapidly settle down to close conformity with negligible wear.

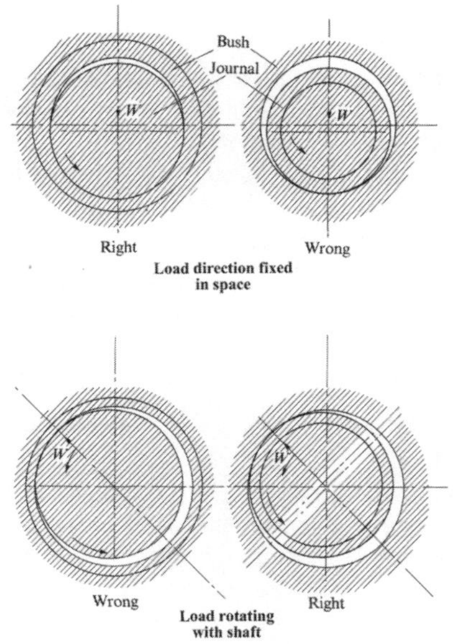

Figure 10.1 Position of bearing lining relative to direction of load

For concentrated contacts, as in anti-friction (ball/roller) bearings, high values of Hertzian stress dictate that very hard materials be used for all contacting surfaces. Either case-hardened or through-hardened steel is normally used.

10.2.2 Lubrication principles (hydrostatic and hydrodynamic)

The differences between sleeve and antifriction bearings are also most apparent when considering lubrication. When load and relative sliding velocity are low, lubrication requirements may be minimal and indeed unnecessary. The only problem is to dissipate the heat generated, there being no circulated lubricant to aid the process.

Where loads are substantial, oil, water or even gas may be forced between the surfaces at sufficient pressure to balance the external load, and to separate them. This is known as "hydrostatic" lubrication.

When the closely conforming surfaces of a lower pair are slightly modified to produce a wedge-shaped gap filled with lubricant and when the surfaces are rotated, a pumping action will be generated within the bearing. This is called "hydrodynamic" lubrication.

Although it had obviously been used within bearings for many years it was not until Tower described some experiments conducted by him in 1885, that its existence was recognised. Some journal bearings used by the London Metropolitan Railway had a plug in a hole in the loaded crown. This was repeatedly ejected during his oil bath lubrication experiments. As a result he investigated the oil pressure distribution with the results shown in Figure 10.2. To preserve the historical flavour, the original Imperial units have been retained.

10.2.3 Reynolds' equation

The theoretical basis for lubrication was derived by Reynolds in 1886. Despite its age, the equation continues to give accurate results, except at the extremes of the parameters detailed.

Thus:

$$\frac{\delta}{\delta x}\left(\frac{h^3}{\eta}\frac{\delta p}{\delta x}\right)+\frac{\delta}{\delta z}\left(\frac{h^3}{\eta}\frac{\delta p}{\delta z}\right)=6\left[\left(U_1+U_2\right)\frac{gh}{dx}+2V\right]$$

where:

p = pressure

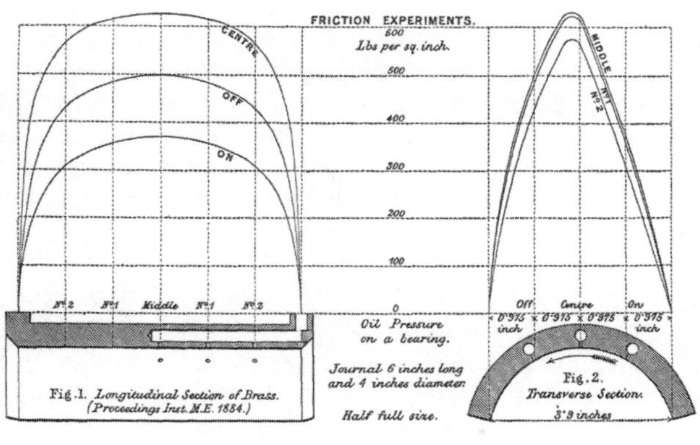

Figure 10.2 Beauchamp Tower's experimental results

U_1 & U_2 = tangential velocity of the two surfaces

v = velocity of approach

η = viscosity of the lubricant

h = distance between surfaces

x = measured in the direction of motion

z = measured at right angles to the motion

For hydrodynamic action to be complete, the fluid film must be sufficiently thick to separate the shaft and bearing journal by an amount which exceeds the sum of the peaks on the two surfaces.

The thickness h of the lubricant film is therefore of critical importance. In any particular case it is determined by the product of two factors – a hydrodynamic factor in which applied force is matched against the combined action of viscosity and velocity, and a geometrical factor dependent on the type of pad.

10.3 Plain bearings

Very small fans may have the simplest of bearings consisting of a plain sleeve in which the shaft rotates. The sleeve material may be sintered brass or phosphor bronze impregnated with a lubricant. If oil is the lubricant, a felt pad may be incorporated as an oil reservoir. Plastics materials may be used where the presence of oil is prohibited but these may not be suitable for high speed. PTFE impregnated bearings are also used on small fans and provide good performance over a wide operating range. Graphite sleeves can be used in locations where other materials are sometimes not suitable.

The shaft and bearings need to be manufactured to tight tolerances for optimum performance, the shaft usually being hardened and polished. The bearing sleeve may have a spherical seating to overcome misalignment and a flange to accommodate limited axially loading.

10.3.1 Sleeve bearings

For other than the smallest of fans the above arrangement is not an acceptable system and rolling bearings are universally used on most other small and medium size fans. On the largest fans and some ultra-quiet fans, sleeve bearings with a lubrication system may be favoured particularly as the life can be superior to that of rolling element bearings. The complexity of sleeve bearings and sometimes the need for a separate cooling system make the cost greater than that of rolling element bearings. Sleeve bearings of this type are generally only suitable for horizontal running.

On some large high speed fans, sleeve bearings may be the only viable bearing system as rolling element bearings have a short life and/or insufficient load carrying capacity. As a rough guide, a peripheral speed of about 8 m/s is required for an oil film and wedge to form for satisfactory operation. Below this speed sleeve bearings may not be viable.

A typical sleeve bearing will consist of a plain hard shaft journal and a soft metal sleeve which is often split on the horizontal centreline to aid assembly. Lubrication oil is fed into the sleeve area by means of rings running on the shaft and in grooves in the sleeve or by means of oil from an integral header tank, topped up by a disc system. In each case the oil is contained in a reservoir under the bearing and the rings or disc are immersed in the oil.

Often the exterior of the housing is provided with fins to help dissipate the heat which has been generated (see Figure 10.3).

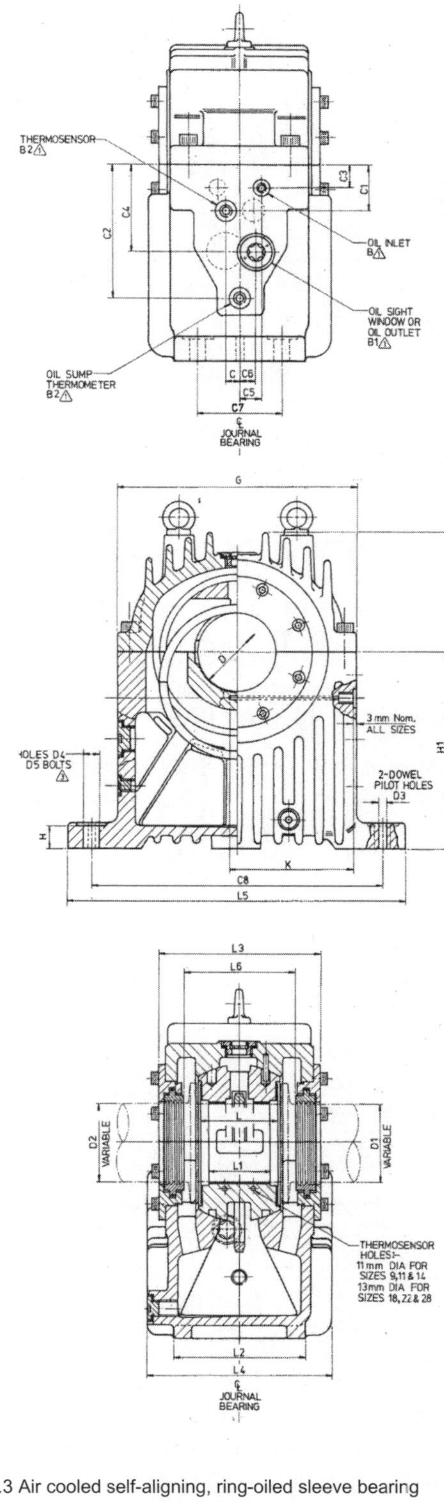

Figure 10.3 Air cooled self-aligning, ring-oiled sleeve bearing

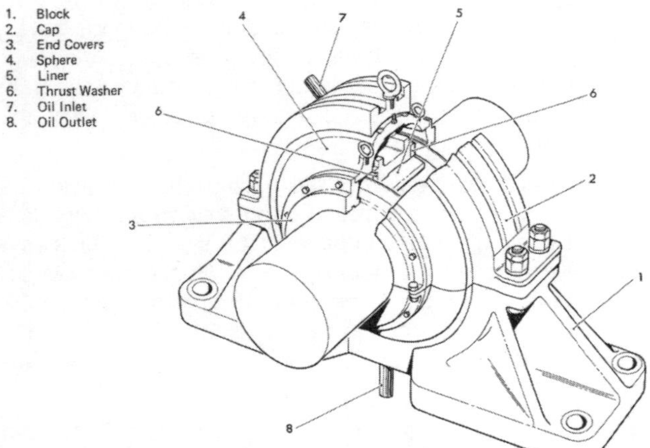

Figure 10.4 Ring-oiled sleeve bearing

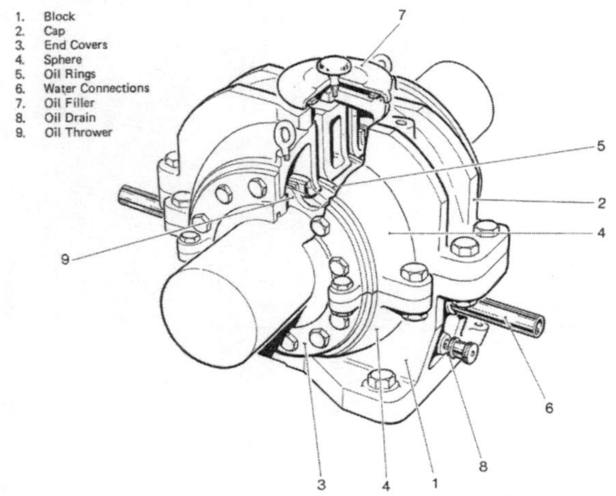

Figure 10.5 Water cooled sleeve bearing

Because of their special nature, bearings of this type are often designed and manufactured by the fan company itself. However, some transmission suppliers have also entered the field, and typical ring-oiled sleeve bearing plummer blocks are shown in Figures 10.3, 10.4 and 10.5.

A table of typical applications of sleeve bearings for large fans is shown in Table 10.1.

Fan application	Lubrication/ cooling	Bearing diameter	Fan speed rev/min	Radial load N	Thrust load N
CFB fluidising air	Oil circulation	90	3565	4000	1000
Steelworks B.O.S	Oil circulation	125	1445	22000	3000
Boiler forced draught	Ring oiled Water cooled	125	1485	7000	12000
Boiler primary air	Ring oiled Water cooled	140	1490	20000	14000
Boiler gas recirculation	Ring oiled Water cooled	180	743	37000	3000
Boiler forced draught	Ring oiled Water cooled	180	743	68000	4000
Boiler induced draught	Ring oiled Water cooled	200	990	54000	4000
Steelworks sinter waste gas	Oil circulation	250	1000	112600	5000
Boiler induced draught	Oil circulation	300	740	178000	15000

Table 10.1 Typical applications of sleeve bearings for large fans

Courtesy of Howden Group

In the case of the disc, a lip ensures that oil is picked up and contained within the outer part of the disc by centrifugal force action and then a scoop extracts oil from the lip region to top up the oil chamber above the bearing. The oil reservoir can have sufficient surface area to ensure the oil temperature is kept within limits and large bearings will usually have this outer surface provided with cooling fins. In the case of large, high-speed fans (approximately 2000 kW and above) a separate cooling fan driven off the main fan shaft and blowing air over the reservoir may be required. Alternatively, the oil is pumped through a separate cooler, or cooling water pipes are incorporated in the reservoir. On high pressure, high speed fans, even at only moderate power the bearings may be forced lubricated from a separate oil lubrication system with its own pump.

For the bearing to operate, the oil must form a wedge between the journal and the sleeve. This oil wedge is not present immediately after start-up and so rubbing between the journal and sleeve surfaces will occur until sufficient speed is reached. At start, the shaft journal will tend to climb up the side of the sleeve and draw oil in to form the wedge. At very low speeds some wear will take place, but normally a transition speed is quickly reached with partly metal-to-metal contact and some oil film present before a full, load-bearing, oil wedge is established. The wedge is formed because the journal is running eccentric with respect to the sleeve and so the shaft centreline position can vary between stationary, start-up and running conditions. The journal-to-sleeve clearance (normally referred to as "bearing clearance") is small and the different shaft positions can be accommodated by the shaft system and coupling.

Plain sleeve bearings can exhibit a whirling action within the bearing whereby the journal, in addition to the normal rotation, rotates about a centre offset slightly from the geometric centre. It arises because the journal may try to roll around the inside of the sleeve. This is often at half the shaft rotational speed, and is known as "half-speed whirl". It is particularly evident if the journal bearing is lightly loaded, as may be the case with a vertical-shaft fan - using plain sleeve bearings - this is one reason why such bearings are rarely used on vertical motors. It may also occur with narrow high speed centrifugal blowing fans. In some cases shaft whirling may give rise to unacceptable vibrations.

Whirling can be overcome by using non-circular sleeves, either in the form of lobes or wedge shapes as shown by the examples in Figure 10.6 These shapes may be confined to a limited axial length at the centre of the bearing, essentially forming shallow pockets and leading to the name "pocket bearings". Where wedge shapes are used only one direction of rotation is possible.

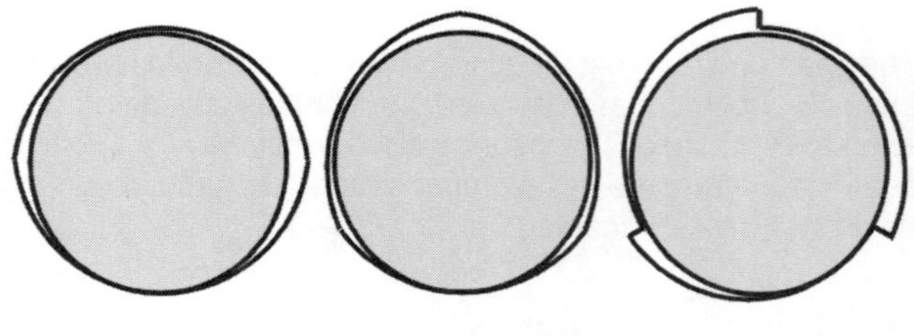

Figure 10.6 Examples of non-circular sleeve shapes

Figure 10.7 shows a schematic diagram of a plain sleeve journal bearing lubricated by means of a single ring in an oil reservoir.

The bearing sleeve is shown as fitting into a spherical seating which is the usual practice on large bearings of this type. At either end of the bearing enclosures, seals - often labyrinth seals - are embodied. The shaft can slide axially within the bearing and this end float is typically ±5 mm.

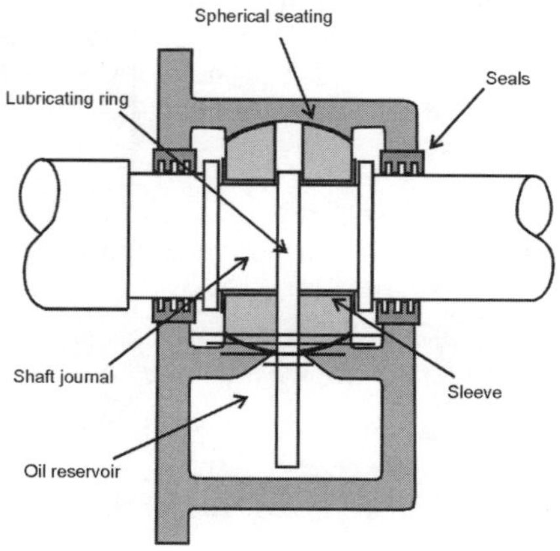

Figure 10.7 A schematic diagram of a plain sleeve journal bearing

10.3.2 Tilting pad bearings

The ultimate extension of film lubrication my be seen in the tilting pad bearing, first introduced by the British engineer A.G.M. Michell, FRS, when working in Melbourne, Australia.

10.3.2.1 General principles

When a well-lubricated journal bearing runs with normal clearance between shaft and bush, a tapered oil film is naturally formed, the thinnest portion of it being that under the load. As the shaft turns, oil is drawn in to feed this wedge (some, of course, being squeezed out at the sides) and an internal oil pressure is set up in the film exactly balancing the bearing load. The faster the shaft revolves, the more oil is drawn in and the thicker and stronger the film becomes.

Moreover the internal film pressure builds up from zero to a maximum just where it is wanted at the point where the load is greatest. But only about one-third of a journal half-brass is really effective. Obviously therefore if each redundant side can be cut out and replaced by a pad which can help to share the total load, the bearing will be much more efficient and this is what is done in the Michell Bearing.

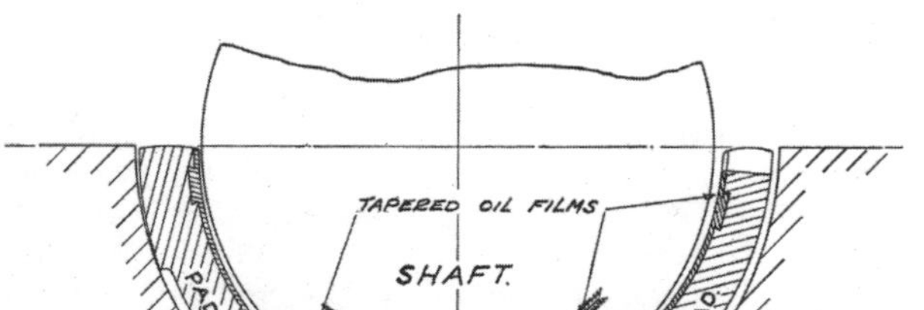

Figure 10.8 Michell thrust pad
Courtesy of Michell Bearings

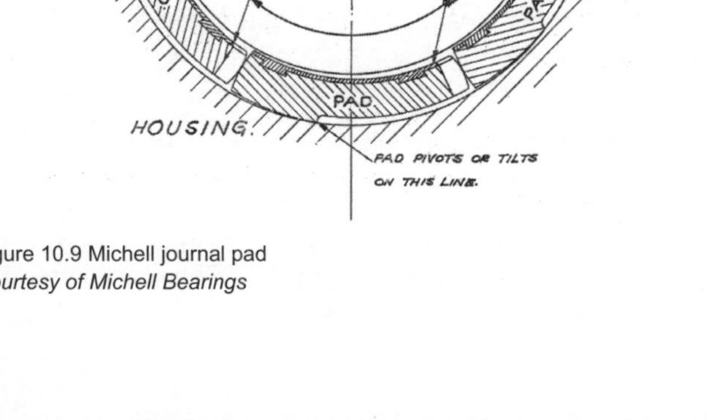

Figure 10.9 Michell journal pad
Courtesy of Michell Bearings

The manner in which persistent, positive and indestructible pressure-oil-films are produced and maintained between the bearing surfaces is clearly shown in Figures 10.8 and 10.9. Figure 10.8 illustrates the action in a Michell thrust block and Figure 10.9 shows a similar process taking place in a Michell journal bearing.

It will be observed that the tapered pressure-oil-film or wedge of lubricant is self-generated by the mere motion of the shaft or collar and is not dependent on any extraneous pressure from an oil pump.

All Michell bearing pads, whether for thrusts or journals, are so designed and proportioned that they tilt and float the load on their own oil films. The stream-like photograph in Figure 10.10 shows how some of the lubricant escapes at the sides of a Michell thrust pad leaving the remainder to feed the trailing edge.

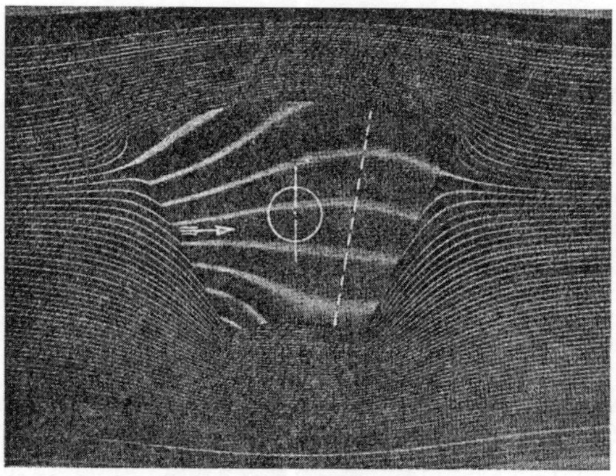

Figure 10.10 Stream-lines of oil flow in tilting thrust pad
Courtesy of Michell Bearings

It is natural to suppose that, as there is no metallic contact, it is unnecessary to white-metal the faces of Michell thrust and journal pads. The reasons for so doing are because white metal is the least liable to damage from minute particles of grit and foreign matter which occasionally find their way into the lubricating oils of even the best kept systems; and also during the boundary conditions (or partial lubrication) when starting and stopping.

10.3.2.2 Tilting pad thrust bearings

The thrust bearing functions on the lines just described. Naturally, flat thrust surfaces cannot adapt themselves (as does a journal bearing) to create any form of tapered oil film, so Michell conceived the idea of dividing the thrust carrying surface into a number of pads, each pad being supported - not by a flat abutment - but by a pivot or step which allows it to tilt slightly. As the thrust collar revolves in its oil bath, the oil adhering to its surface is carried round and lifts every pad at its leading edge to admit the tapered oil film. Thus each of the pads round the thrust collar generates a tapered pressure oil film of a thickness appropriate to the load, the speed, and the viscosity of the lubricating medium.

The position of the pivot, which is the edge of a radial step on the back of the pad, is of some importance. For maximum efficiency - in other words minimum friction - the pivot is beyond the centre of the circumferential width of the pad measured from its leading edge, and these pads are termed "off-set", being right or left-handed to suit the direction of rotation.

The Michell thrust bearing is a simple single-collar unit capable of carrying at least 20 times the load per unit area of a flat multi-collar thrust bearing, with only about one twentieth of the frictional loss. No subsequent adjustment is required when once the thrust bearing is installed and the entire absence of wear at all speeds, even when overloaded, makes it one of the most reliable pieces of machinery.

10.3.2.3 Tilting pad journal bearings

It is clear that when effective films are induced at other parts of the circumference than that just under the load, the carrying capacity of a journal bearing is correspondingly increased.

As in tilting pad bearings, the same principle of segmental pads is adopted in Michell journal bearings. The usual pair of solid brasses gives place to a series of pads, generally six in number, surrounding the shaft journal. Each pad is free to tilt slightly in its cylindrical housing and is prevented from cross-winding by suitable flanges engaging the machined ends of the housing. Oil is automatically introduced between each pair of pads from an annulus in the housing and any surplus that is not carried all the way across escapes naturally at the ends of each pad. As the shaft revolves, all the pads tilt to admit oil along their leading edges, and each one thus creates its own characteristic tapered oil film.

At speed, the shaft thus becomes surrounded by a close-fitting oil garter, constantly renewing and maintaining itself, which under the severest conditions of load and shock, has never been known to fail. Loads up to and exceeding 360 kgf/cm^2 of projected surface have been registered experimentally, and pads, after many years of hard service, have shown no signs of wear for the very good reason that metallic rubbing contact has never occurred.

The load carrying capacity of such bearings is enormously greater and the friction much less than the best solid brass types, and they can be made much shorter in consequence. This is often a matter of supreme importance where space and weight are restricted.

For ordinary conditions of bath lubrication, journal bearings are provided with a light collar secured to the shaft in halves and dipping into an oil well below. Oil is lifted over the top centre by this revolving collar and the resulting spate of oil guided to the top of the bearing and into the oil annulus feeding the pads. No packed end glands are necessary, any surplus oil being prevented from creeping out along the shaft by special oil deflectors fitted at the ends of the bearing. These bearings are entirely self-lubricating and self-contained and can be adapted for certain duties where automatic functioning for prolonged periods without attention is a requirement.

10.3.2.4 Load carrying capacity of tilting pad bearings

The load that can be safely carried on the oil films of a tilting pad bearing depend on its diameter, length, peripheral speed and oil viscosity. The load carrying capacity also increases with the revolutions, and loads exceeding 400 kgf/cm^2 have been sustained on prolonged tests. These bearings are in successful operation at all speeds ranging from five revolutions per minute, up to the highest speeds encountered in modern fan technology.

10.3.2.5 Friction losses

In the foregoing it has been impossible to ignore friction entirely – there must be friction in every type of bearing. Tilting pad bearings however are unique in that whatever friction there may be, it is never metallic friction but simply oil friction. In other words, the only resistance to relative motion between shaft and bearing pads is that required to shear the intervening layers of oil comprising the film. This resistance is a measurable quantity and can be calculated from the rotational speed, pressure and oil viscosity. Certain experiments with a bearing loaded to 40 kgf/cm^2 gave a coefficient of friction (μ) of 0.0020 against a calculated figure of 0.0022 - near enough for all practical purposes. The coefficient of friction of a good ordinary bearing is 0.036 - about eighteen times as much. The coefficient of friction in tilting pad bearings ranges from .001 to .005 and varies with the factors mentioned above. When starting under load, the friction is naturally considerably greater for the first half revolution, by which time the oil film is generated.

10.3.2.6 Cooling

The heat generated in a tilting pad bearing is affected more by speed than load and there are three methods of dissipating the heat.

1. **Air cooling by natural radiation**. This covers the majority of applications of moderate speed.
2. **Water cooling**, which becomes necessary at higher speeds.
3. **Circulated oil**, which is required for the highest speeds.

In the first case air cooling is obtained by means of suitable external ribs on the bearing casing.

In the second case the self-contained oil in the bearing casing is kept cool by means of a water jacket incorporated in the housing or by water passing through solid drawn coils or tubes in the oil well.

In the third case the oil is pumped through an external cooler in the oil circuit.

It should be noted that when circulated oil is used it is not necessary to have a high oil pressure at the pump. All that is required is sufficient to ensure a free flow through the circuit of the amount required for cooling. Forced lubrication, as usually understood, is not necessary, the oil pressure in the films being generated by the action of the tilting pads.

10.4 Anti-friction or rolling element bearings

10.4.1 Deep-groove ball bearings

The commonest form of ball bearing is the deep-groove type as shown in section in Figure 10.11. These are the most popular of the rolling element types and can operate with both radial and axial loads and at high speed. For fans where quiet running is required, deep-groove ball bearings are the first choice with special "low noise" versions available for silent running. This only applies to small fans where other sources of noise generation can also be minimized or eliminated.

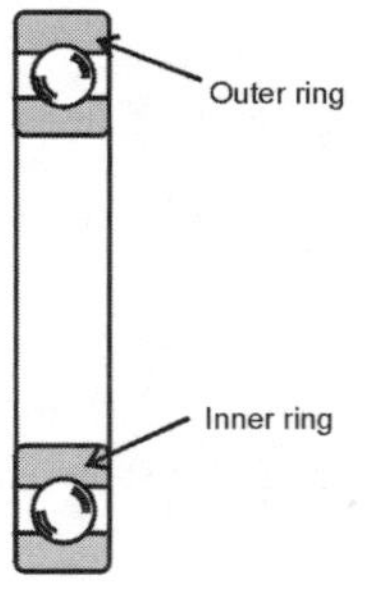

Figure 10.11 Deep-groove ball bearing

The only disadvantage of this type of bearing is its inability to accept misalignment of the inner and outer rings. At most a misalignment of 10 minutes of arc can be tolerated with some bearings only able to tolerate 2 minutes of arc. If the bearing rings are misaligned then the life is reduced and the noise level can increase appreciably.

The clearance is defined as the total distance that one ring can be moved relative to the other in either the radial direction (radial internal clearance) or axial direction (axial internal clearance). The interference fits with respect to the shaft and bearing housing, operating loads and thermal effects usually reduce

Figure 10.12 Self-aligning ball bearing

the clearance (operational clearance), and ideally this should be virtually zero, otherwise some preload may develop. The initial clearances usually conform to ISO 5753 being designated as either C1, C2, C3, C4 or C5 (the lowest numeral being the lowest clearance) with C3 being the most widely used. Many suppliers designate normal clearance CN and this is likely to be between C2 and C3.

Bearings can be supplied with two rows of balls or as matched pairs for extra load carrying capacity but these arrangements can tolerate even less misalignment and usually run with an increased noise level.

10.4.2 Self-aligning ball bearings

Self-aligning bearings have two rows of balls with the outer ring having a spherical race as shown in Figure 10.12. The two rows of balls are staggered with respect to each other. This type of bearing can be used where the shaft may suffer misalignment, either because of errors that could occur due to the method of assembly or due to shaft deflections. They can be run at high speed, but not to the same extent as deep-groove ball bearings, and are reasonably quiet in operation. As with deep-groove ball bearings they are unsuitable if axial displacement takes place with the bearing performance and life suffering as a consequence. They cannot tolerate any axial load. The permitted misalignment is generally in the range 1° to 3° depending on design and size.

10.4.3 Angular-contact ball bearings

By displacing the ball races in the two rings the bearing can be optimized to withstand a combined axial and radial load. The bearing performance is similar to that of deep-groove ball bearings except they are not able to run at quite the same high speed and the noise level is slightly higher. A section through a typical angular-contact bearing is shown in Figure 10.13. The contact angle is as shown in the Figure and this is usually about 40°. Figure 10.14 shows typical bearings with the cage details.

Angular-contact ball bearings cannot tolerate misalignment and there must be at least a small load on the bearing for satisfactory operation. A bearing with a contact angle of 40° should have an axial load greater or equal to the radial load. As with

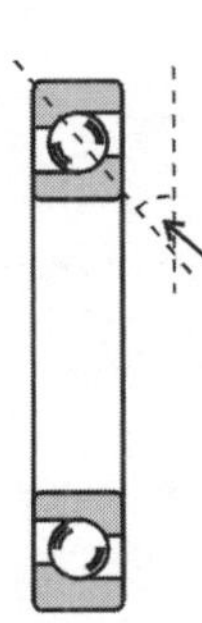

Figure 10.13 Angular-contact ball bearing

Figure 10.14 Examples of angular-contact ball bearings
Courtesy of ABB Drives

deep-groove ball bearings, angular-contact bearings can be supplied with two rows of balls to operate with the axial load in either direction or as matched pairs for increased load capacity.

A version of the angular-contact ball bearing is the four-point ball bearing which can operate well with axial loads in either direction. In this case both the outer and inner race is in the shape of a "V" as shown in Figure 10.15.

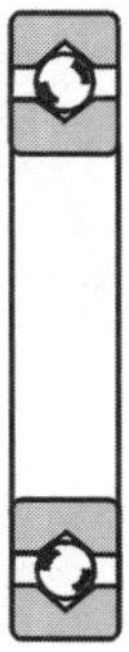

Figure 10.15 Four-point, angular-contact ball bearing

When the axial load is in excess of the radial load a modified version of the deep-groove ball bearing can be used as an angular-contact bearing. Known as a duplex bearing, either the outer or inner ring is split into two separate rings. Figure 10.16 shows an example with the outer ring split.

Figure 10.16 Duplex angular-contact ball bearing

10.4.4 Cylindrical roller bearings

For improved radial load-carrying capacity and greatest bearing stiffness, roller bearings can be used. A typical cylindrical roller bearing is shown in Figure 10.17. This may have longer rollers for enhanced load carrying or long small-diameter rollers (needle bearings) if space is limited. As shown in the figure, the inner ring has flanges to retain the rollers in position but this may equally well be on the outer ring.

This type of bearing is ideal for non-location bearings because axial displacement is possible within set limits. However misalignment is limited to about 3 minutes of arc for most bearings and 4 minutes of arc for bearings with short length rollers. They

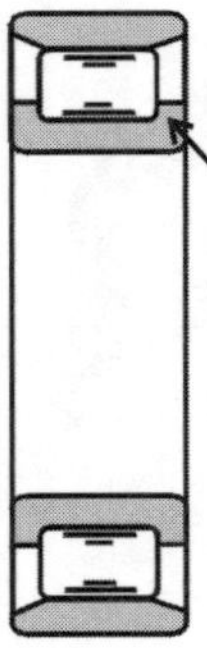

Figure 10.17 A typical cylindrical roller bearing

can be used at high speed and run reasonably quietly. The two bearing rings can be separated and this may make assembly easier in some cases.

10.4.5 Spherical roller bearings

As with ball bearings, if a spherical outer race is provided then self-aligning properties can be obtained. In this case the rollers are also required to be spherical and by using two rows - as with self-aligning ball bearings - a self-aligning bearing with good radial load carrying and some axial load carrying capability is obtained. The maximum running speeds are not quite as high as with cylindrical roller bearings and the noise level can be higher. A typical arrangement is shown in Figure 10.18.

Figure 10.18 Cylindrical roller bearing

Nevertheless, the all-round capabilities of this bearing make it a very popular choice for general purpose centrifugal fans.

10.4.6 Tapered roller bearings

The roller equivalent of the angular-contact ball bearing is the tapered roller bearing with the bearing inner and outer races tapered to a single point on the bearing axis if the surfaces are extended. This gives optimum running with the angle of the taper on the outer race determining the amount of axial load compared to the radial load that the bearing can withstand. A typical tapered bearing arrangement is shown in Figure 10.19.

If a radial load is imposed on the bearing an axial load is induced and this must be counteracted by another bearing; it is normal therefore to employ two tapered roller bearings at each end of a shaft system to balance the loads or to use a bearing

Figure 10.19 Tapered roller bearing

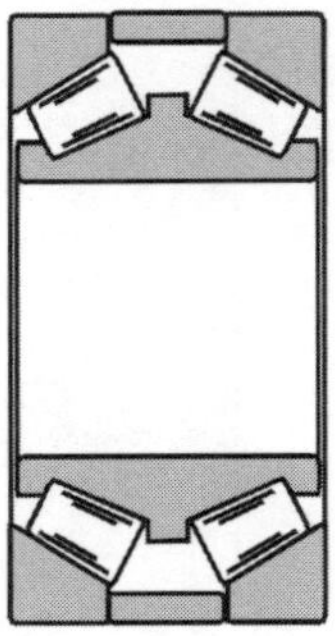

Figure 10.20 Double-row tapered roller bearing

with two rows of rollers tapered in the opposite directions, as shown in Figure 10.20.

10.4.7 Thrust bearings

Thrust bearing versions of most of the journal bearing types described above are available. Figure 10.21 shows a typical thrust ball bearing orientated to withstand a vertical thrust such as the weight of a rotor - but this type of bearing, and indeed any journal or thrust bearing, can be used in any attitude.

To withstand thrust in either direction, two rows of balls are required as shown in Figure 10.22. This shows the outer rings held by housing washers with spherical seatings to compensate for misalignment during assembly. The inner ring is attached to the shaft, embodying a suitable shoulder and collar to withstand the thrust loads.

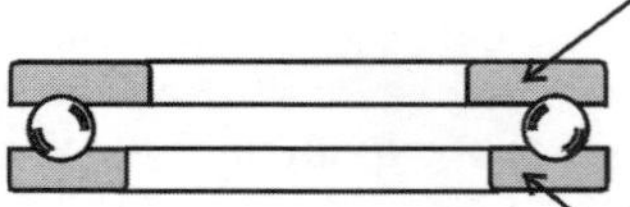

Figure 10.21 Thrust ball bearing

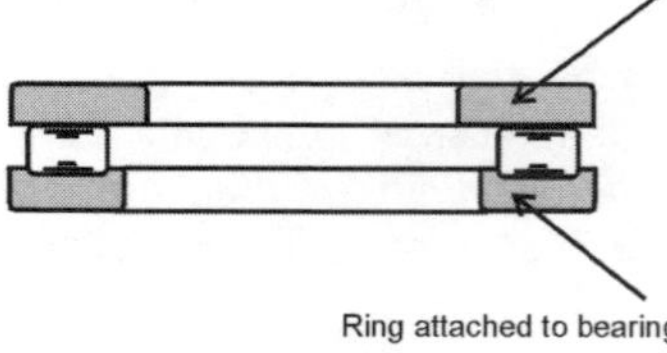

Figure 10.22 Double thrust ball bearing

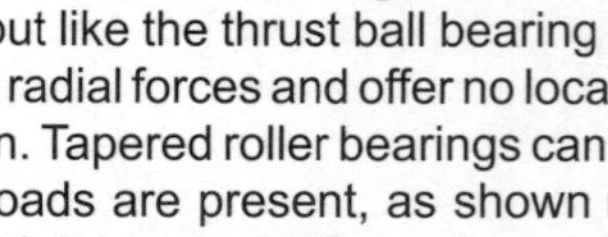

Figure 10.23 Cylindrical-roller thrust bearing

Cylindrical-roller thrust bearings can be used, as shown in Figure 10.23, but like the thrust ball bearing these cannot accommodate any radial forces and offer no location function in the radial direction. Tapered roller bearings can be used where thrust and radial loads are present, as shown in Figures 10.19 and 10.20, and high bearing stiffness is required.

For high thrust loads where radial loads are present and misalignment may be a problem, the spherical-roller thrust bearing is necessary, as shown by Figure 10.24.

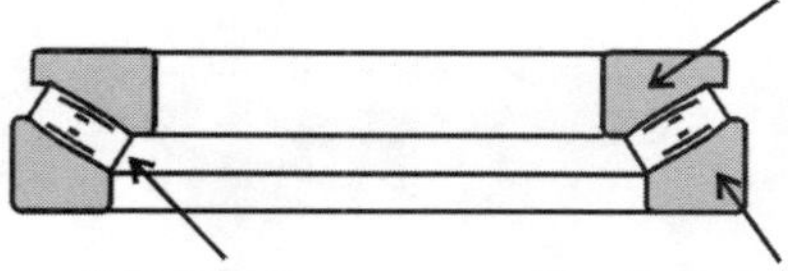

Figure 10.24 Spherical-roller thrust bearing

10.4.8 Other aspects of rolling element bearings

Rolling element bearings are available in versions with various features that are suitable for particular applications and the bearing supplier should be consulted for special applications and hazardous environments. Clearances may need to be non-standard in some applications (See Chapter 8, Section 8.6.4) and different materials are available for the ball or rollers, the rings or raceways and the bearing cage.

Carbon chromium through-hardening steel is a common material with manganese and molybdenum added on large bearings to improve the hardening. Equally common is chromium nickel and manganese-chromium alloys as case-hardening steels with little difference in performance. These materials are acceptable up to about 125°C but for higher running temperatures a special heat treatment and/or special material is required and advice should be sought from a bearing manufacturer. If corrosion resistance is required, stainless steel - typically chromium or chromium/molybdenum based - can be supplied but with a reduced bearing load capacity.

The rolling elements are held in place and with the correct spacing by means of a cage. The cage also serves to hold lubricant and, where bearing rings are separable, hold the rolling elements together. The cages must present a minimum friction, withstand the inertia forces and be acceptable in the environment (the external environment as well as the grease or oil used for lubrication). The cage must be centred on the rolling elements or one of the rings.

Cages are made of steel, brass or plastics and for a given type and size there will be a normal standard cage material. Plastic cages, for example fibre reinforced polyamide, have a temperature limit, depending upon the lubrication, of between about 80°C and 120°C and are unsuitable at very low temperatures, below about -40°C. Pressed steel cages can be used up to 300°C and are usually used on large size bearings whereas brass cages can be used up to the same temperature but are more common on medium and small size bearings. Brass cages in some environments can suffer from "season cracking" and steel cages can become corroded in the presence of water. Experience has shown that the cage design and material can affect the noise performance.

10.4.9 Other features

Other features which may be available include lubrication holes in the outer ring and circlip grooves in the outer ring to provide axial alignment.

Perhaps the most popular feature for fan manufacturers has been the provision of a tapered bore instead of a cylindrical bore. This is used with a tapered adaptor sleeve and locking nut. By this means the bearing may be clamped on to a parallel shaft without the need for shoulders or complicated fitting procedures, (see Figure 10.25).

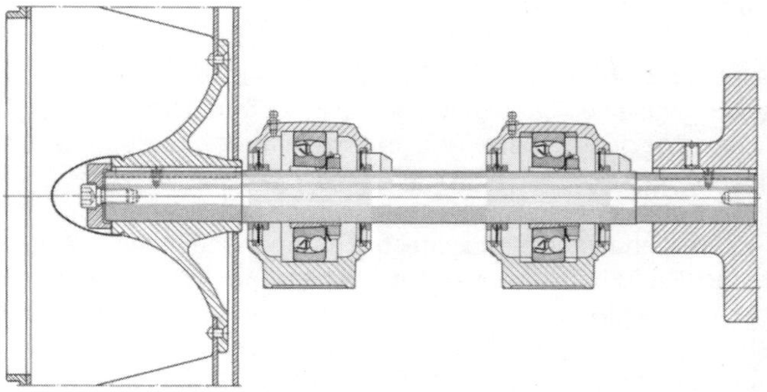

Figure 10.25 Bearings with taper sleeve adaptors fitted to parallel shaft
Courtesy of SKF (UK) Ltd

10.4.10 Bearing dimensions

The main dimension of any rolling element bearing is the bore size but for a given bore there can be numerous outer diameters and bearing widths. The International Organization for Standardisation (ISO) has published several "Dimension Plans" to cover dimensions which are followed by most bearing suppliers. Publication ISO 15 covers radial bearings, except for tapered roller bearings which are covered by ISO 355, and thrust bearings which are covered by ISO 104.

The Dimension Plans are based on a series of outer diameters for each bore diameter and for each outer diameter there is a series of widths (or heights in the case of vertical thrust bearings). Each diameter and width series is designated by a numeral. In the case of tapered roller bearings the numerals are replaced by letters and a numeral is introduced to cover the contact angle. There are numerous additional numerals and/or letters to indicate the bearing type and its features and this complicates the final form of the bearing designation.

10.5 Needle rollers

10.5.1 Introduction

Needle rollers are an extension to normal roller bearings and a basic part of some manufacturers' product range. They can be used either on their own as a bearing arrangement or in combination with components such as cages, drawn outer cups, outer and inner rings and seals to give a wide range capable of meeting the technical and economic demands of many different applications. Whilst not common for fans they are used in certain applications for high speeds and high radial loads.

10.5.2 Dimensions

Needle rollers may conform to DIN 5402-3, grade G2 or ISO 3096, type B, with flat ends. They are made as standard from through hardened rolling bearing steel in accordance with DIN 17230. They have a core hardness of at least 670 HV and a precision machined surface. Standard diameters usually range from 1 to 6 mm, and the length is generally between 3 and 11 times the diameter.

Needle rollers are grouped in sorts corresponding to tolerance groups for the diameter measured at the centre of the needle roller length. The ends of the needle rollers are of a profiled form, with a curved transition from the longitudinal surface to the end face. This has the effect of reducing the edge stresses that would occur at the ends of the roller if it were not in completely flat contact with the raceway. Needle rollers can be used for full complement needle roller arrangements, or alternatively as pins or axles.

10.5.3 Design options

A full complement needle roller arrangement is one in which the entire available space between the inner and outer raceway is filled with needle rollers. This gives a particularly compact bearing arrangement with high load carrying capacity and high rigidity. When needle rollers are used in such an arrangement, they require a shaft and a housing bore as inner and outer raceways respectively, both of which must be hardened and ground in order to provide the necessary characteristics. If the raceways are of sufficient geometrical accuracy, a full complement bearing arrangement will have high runout accuracy and adjustable radial internal clearance. Such designs are preferably used for applications involving swivel type motion and high loads.

Figure 10.26 Needle roller and cage assembly
Courtesy of INA Bearing Company Ltd

Needle rollers can be used not only in full complement arrangements but also in assemblies in which the rollers are separated and guided by a metal or plastic cage (see Figure 10.26). These are particularly suitable for applications involving high speeds, since separation of the needle rollers allows faster rotation without generating unacceptable levels of friction and heat. Due to the relatively narrow crosspieces of the cage, the cage can still accommodate a large number of needle rollers and these assemblies therefore offer high load carrying capacity. As in the case of the full complement arrangements, a hardened and ground shaft and housing bore are required as raceways and high runout accuracy can be achieved if these are of sufficient geometrical accuracy. Depending on the needle roller sorts and the shaft and housing tolerances, adjustable radial internal clearance is possible. Needle roller and cage assemblies are available in single row design for shaft diameters from 3 to 265 mm and in double row design for shaft diameters from 24 to 95 mm.

10.6 CARB® toroidal roller bearings

One of the most significant advances in fan design in recent years has been the introduction of the CARB® toroidal roller bearings. These are particularly appropriate where, as in high temperature fans, expansion of the shaft takes place.

10.6.1 Description

The CARB® bearing is a single row roller bearing with relatively long, slightly crowned roller and is used in conjunction with other types of locating bearings such as ball or spherical rollers (see Figure 10.27).

The inner and outer ring raceways are correspondingly concave and symmetrical. The outer ring raceway geometry is based on a torus, hence the term toroidal roller bearing.

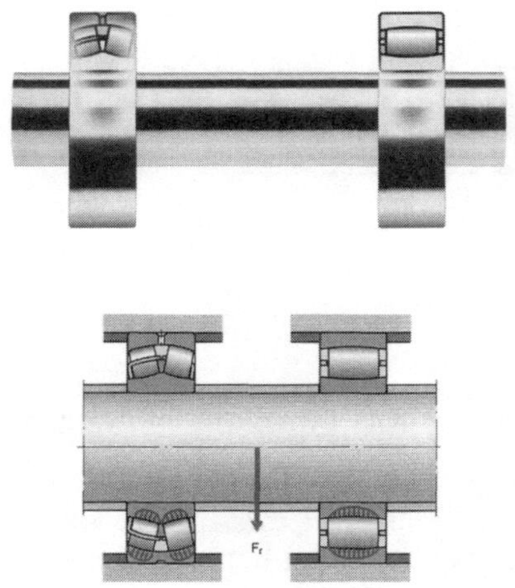

Figure 10.27 Spherical roller bearing (located) and CARB® toroidal roller bearing compensate for angular misalignment
Courtesy of SKF (UK) Ltd

This toroidal roller bearing is designed as a non-locating bearing that combines the self-aligning ability of a spherical roller bearing with the ability to accommodate axial displacement like a cylindrical or needle roller bearing. Additionally, if required, the toroidal roller bearing can be made as compact as a needle roller bearing.

10.6.2 Applicational advantages

An application incorporating a CARB® toroidal roller bearing provides the following:

Self-aligning capability

The self-aligning capability of the bearing is particularly important in applications where there is misalignment as a result of manufacturing or mounting errors or shaft deflections. To compensate for these conditions, the bearing can accommodate misalignment up to 0.5 degrees between the bearing rings without any detrimental effects on the bearing or bearing service life

Axial displacement

Previously, only cylindrical and needle roller bearings could accommodate thermal expansion of the shaft within the bearing. Now the toroidal roller bearing can be added to that list.

The inner and outer rings of the bearing can be displaced, with respect to each other, up to 10% of the bearing width. By installing the bearing so that one ring is initially displaced with respect to the other one, it is possible to extend the permissible axial displacement in one direction. In contrast to cylindrical and needle roller bearings that require accurate shaft alignment, this is not needed for toroidal roller bearings, which can also cope with shaft deflection under load. This provides a solution to many problem cases.

Long system life

The ability to accommodate misalignment plus axial displacement with virtually no friction enables this type of bearing to provide benefits to the bearing arrangement and its associated components

- Internal axial displacement is virtually without friction; there are no internally, induced axial forces, thus operating conditions are considerably improved.
- The non-locating bearing as well as the locating bearing only need to support external loads.
- The bearings run cooler, the lubricant lasts longer and maintenance intervals can be appreciably extended.

High load carrying capacity

It is claimed that this toroidal roller bearing can accommodate very high radial loads. This is due to the optimized design of the rings combined with the design and number of rollers. It is also claimed that the large number of long rollers make CARB® bearings the strongest of all aligning roller bearings.

Also, these bearings can cope with small deformations and machining errors of the bearing seating. The rings can accommodate these small imperfections without the danger of edge stresses. The high load carrying capacity plus the ability to compensate for small manufacturing or installation errors provide opportunities to increase machine productivity and uptime.

Increased performance or downsizing

For bearing arrangements incorporating this toroidal roller bearing as a non-locating bearing, internally-induced axial forces are prevented. Together with high load carrying capacity it is also claimed that:

- for the same bearing size in the arrangement, performance can be increased or the service life extended, or
- new machine designs can be made more compact to provide the same, or even higher performance.

Reduced vibration

Self-aligning ball or spherical roller bearings in the non-locating position need to be able to slide within the housing seating. This sliding, however, causes axial vibrations which can reduce bearing service life considerably. Bearing arrangements that use CARB® toroidal roller bearings as the non-locating bearing are stiff because the bearing can be radially and axially located in the housing and on the shaft. This is possible because thermal expansion of the shaft is accommodated within the bearing. The stiffness of the bearing arrangement, combined with the ability of the bearing to accommodate axial movement, substantially reduces vibrations within the application to increase service life of the bearing arrangement and related components.

Full dimensional interchangeability

The boundary dimensions of these toroidal roller bearings are in accordance with ISO 15:1998. This provides dimensional interchangeability with self-aligning ball bearings, cylindrical roller and spherical roller bearings in the same dimension series. The range also covers wide bearings with low cross-sections normally associated with needle roller bearings.

10.7 Rolling element bearing lubrication

Rolling element (or anti-friction) bearings need to be lubricated to prevent inter-metallic contact between the balls or rollers, raceways and cages. The lubricant however also has the additional function of protecting the bearing against corrosion or other sources of environmental wear.

Bearings may be lubricated with grease, oil or in rare cases with a solid. The best operating temperature for a bearing is obtained when the minimum of lubricant necessary to ensure reliable operation is provided. However the lubricants become contaminated in service and must therefore be replenished or changed from time to time. The choice of lubricant depends on the operating temperature range, environmental conditions and rotational speed.

As previously noted, rolling element bearings are used for the great majority of fan applications. Wherever possible grease is used for lubrication as it is more easily retained in the bearings no matter what the inclination. It also helps to seal the housing against outside impurities such as dust and water. Lubricating greases are thickened mineral oils or synthetic fluids. Their consistency depends on the quantity and type of thickening agents included. Consistency, miscibility, operating temperature range and rust inhibiting properties are the important properties of a good lubricant.

The lubrication interval is dependent on bearing size, rotational speed, operating temperature and grease type. Figure 10.28 is applicable to bearing temperatures around 70°C. Below this temperature the intervals are likely to increase, but above this temperature they will reduce considerably. Reference should be made to the fan and/or bearing manufacturer for further information.

For small ball bearings, especially those used in electric motors, the lubricating interval may be longer than their service life. They may then be fitted with shields or seals and are sealed for life.

The amount of grease needed for a charge can be obtained from the formula

$$G = K\,D\,L \qquad \text{Equ 10.1}$$

where:

G	=	grease quantity (g)
K	=	constant 0.005
D	=	bearing outside diameter (mm)
L	=	bearing axial length (mm)

The means of relubrication will depend on the frequency necessary. Where convenient, the housing caps can be removed and fresh grease can then be packed between the rolling elements. If more frequent relubrication is necessary, grease nipples may be fitted to the bearing housings and a grease gun used.

In all cases, too much grease will lead to overheating and maintenance staff must be encouraged not to lubricate every time they pass the fan. High-speed fans however, often require frequent greasing. There is then a danger that the used grease will collect in the bottom of the bearing housings. In this case grease escape valves should be fitted. These enable excess grease to be discharged. They permit greasing to be carried out without having to stop the fan.

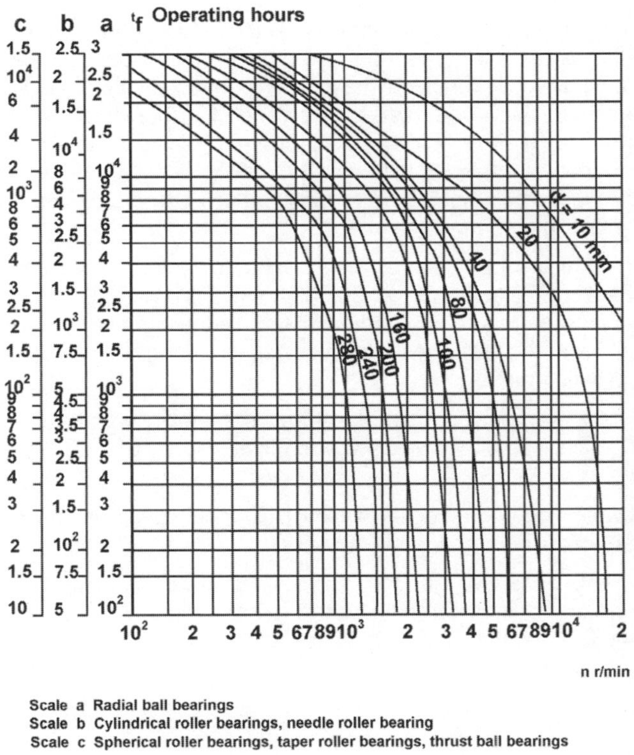

Figure 10.28 Typical lubrication intervals for rolling element bearings

Proprietary grease dispensers may also be fitted to the bearing housings which ensure that the correct amount of grease is dispensed at the appropriate time interval.

Oil lubrication is used not only for most sleeve bearings, but also for rolling element bearings when the rotational speed is above the allowable limit for grease. It may be essential where high operating temperatures make grease unsuitable.

The simplest method of oil lubrication is by use of an oil bath, but increasing speed raises the bearing temperature, and leads to oxidation of the oil. To avoid frequent lubricant changes the oil may be filtered and externally cooled before being recirculated by means of a pump. Oil jets or mist may be necessary to ensure that the lubricant reaches the parts where frictional heat is generated.

Solvent-refined mineral oils are normally used for oil-lubricated fan bearings. Additives to improve lubricant film strength or oxidation resistance are only required in extreme circumstances. Viscosity is one of the most important properties of a lubricating oil and the requisite value must be maintained at the operating temperature. It is unwise therefore to change the oil characteristics without reference to the fan and bearing manufacturers.

10.8 Bearing life

The size of a bearing to be used for a fan application is normally determined from its known load bearing capacity. This may need to be modified dependent on a minimum diameter necessary to satisfy shaft critical speed requirement.

In general the basic dynamic load ratings of the bearings will have been determined by the bearing manufacturer in accordance with the methods specified in ISO 281:1990. The life of a rolling bearing is defined as the number of revolutions which the bearing is capable of performing, before any signs of fatigue are evident on its rings or rolling elements. Such signs might be flaking or spalling of these elements.

At a constant rotational speed, it is then possible to convert the number of revolutions into an operating life for

$$\text{Life hours} = \frac{\text{revs to fatigue}}{\text{revs / min} \times 60} \qquad \text{Equ 10.2}$$

By experience we know that apparently identical bearings operating under the same load and ambient conditions will have varying lives, even if they have been correctly installed and lubricated. Usually we use the so-called L_{10} (basic rating) life, which is the life at which a sufficiently large group of these bearings can be expected to have a 10% failure rate. The L_{10} life for the application should be known and/or agreed between the parties to a contract.

In general small clean air fans will be designed with bearings rated to give an L_{10h} life of 20,000 hours rising to 40,000 hours for a medium size light industrial fan. Heavy duty public utility fans are frequently designed for an L_{10h} bearing life of 100,000 operating hours.

The average life of a sufficiently large sample of bearings under identical load and temperature conditions will be 5 times the L_{10} life.

It will be noted that an increase in rotational speed results in a reduction in operating life in hours.

The ISO Standard in fact specifies the basic rating life L_{10} in terms of millions of revolutions for a basic dynamic load rating and the formula which interconnects the various factors is:

$$L_{10} = \left(\frac{C}{P}\right)^p \quad \text{or} \quad \frac{C}{P} = L_{10}{}^{1/p} \qquad \text{Equ 10.3}$$

where:

L_{10} = basic rating life, millions of revolutions

C = basic dynamic load rating N

P = equivalent dynamic bearing load N

p = exponent of the life equation

$p = 3$ for ball bearings

$p = {}^{10}\!/_3$ for roller bearings

For fan bearings operating at constant speed it is usual to calculate with a basic rating life expressed in operating hours using the equation

$$L_{10h} = \frac{1000000}{60n}\left(\frac{C}{P}\right)^p \qquad \text{Equ 10.4}$$

or

$$L_{10h} = \frac{1000000}{60n} L_{10} \qquad \text{Equ 10.5}$$

where:

L_{10h} = basic rating life (operating hours)

n = rotational speed (r/min)

At elevated bearing temperatures dynamic load carrying capacity is reduced. This reduction is taken into account by multiplying the basic dynamic load rating C by a temperature factor as shown in Table 10.2.

Bearing temperature °C	150	200	250	300
Temperature factor	1.00	0.90	0.75	0.60

Table 10.2 Temperature factors

Satisfactory operation of the bearings at elevated temperatures also depends on whether they have adequate dimensional stability for the operating temperature, whether the chosen lubricant will retain its lubricating properties and whether the materials of the bearing seals, cages etc., are suitable.

It must be emphasised that this temperature is the temperature of the bearing race. Usually, unless the bearing is in the air stream, this is much below the air or gas temperature. Where the impeller is overhung on the shaft, there is often the possibility of introducing an auxiliary cooling fan between the casing side and the inner bearing to reduce the heat transmitted along the shaft. A "spacer" coupling or slots in the shaft can perform a similar function.

The radial loads acting on the bearings are simply calculated using the theory of moments. It is assumed that the fan shaft acts as a beam resting in rigid, moment-free supports for fixed bearings, or simple supports if the bearings are contain in self-aligning housings. (See Chapter 8, Section 8.6.3.)

Whilst the "dead" weight of the impeller, shaft and where applicable, pulleys are known, there are other loads which are variable and have to be estimated. Thus the impeller weight will be augmented by a fluctuating load due to its residual out-of-balance. This will have been allowed for at the design stage, but may increase due to erosion, corrosion, or dust build-up.

Many centrifugal and mixed flow fans are driven through vee belts, and these are also used to a lesser extent with axial flow fans. The effective belt pull is dependent on the transmitted torque and will be an important load in the determination of bearing radial loads. (See Chapter 11.)

One of the fan bearings will also be subject to an axial load due to the impeller end thrust. This is a function of the fan pressure, its distribution between the inlet and outlet ducting, the inlet area of the fan impeller and the momentum change due to the flowrate.

If the resultant load is constant in magnitude and direction, the equivalent dynamic bearing load can be obtained from the general equation.

$$P = XF_r + YF_a \qquad \text{Equ 10.6}$$

where:

P	=	equivalent dynamic bearing load (N)
F_r	=	actual radial bearing load (N)
F_a	=	actual axial bearing load (N)
X	=	radial load factor for the bearing
Y	=	axial load factor for the bearing

An additional axial load only influences the equivalent dynamic load P for a single row radial bearing if the ration F_a/F_r exceeds a certain limiting value, but with double row radial bearings even light axial loads are significant.

Equation 10.6 is also applied for thrust bearings, which can take both axial and radial loads, e.g., spherical roller thrust bearings. For thrust bearings, the equation can be simplified, provided the load acts centrally, viz.

$$P = F_a \qquad \text{Equ 10.7}$$

It will be appreciated that axial loads higher than design (due to excessive system resistance) will adversely affect bearing life. Double inlet, double width centrifugal fans have essentially balanced end thrusts and their bearings are therefore only subject to radial loads. Nevertheless a minimum axial load is necessary to ensure correct "centring" of the bearing, which often results from the blocking effect of a pulley in one inlet.

The L_{10h} life is only achieved when the bearings are correctly installed, correctly lubricated and correctly maintained. If the lubricant is unsuitable for the application and is replenished incorrectly in both quantity and frequency then premature failure will occur. Over-greasing is often more harmful than under-greasing. Corrosion and external wear may also affect the bearings, and seals must be inspected to confirm that they are preventing the ingress of contaminants.

10.9 Bearing housings and arrangements

Bearing arrangements for fans may be designed in a variety of ways dependent on the size, operating conditions and rotational speed. Cost also is a consideration together with the expected life. The comments and selections which follow are to a certain extent in ascending order of price and reliability.

10.9.1 Light duty pillow blocks

These are normally recommended for light duty fans having a shaft diameter of 50 mm or less. Such bearings have a zinc-coated bore and an extended inner ring with eccentric locking collar.

In the arrangement shown in Figure 10.29 the fan impeller is supported by Y-bearings mounted in cast iron housings. As both Y-bearings are located, the sheet steel sideplates of the fan must accommodate possible thermal elongation of the shaft.

As the bearing bore tolerances are to plus limits to permit mounting on drawn steel shafts (say tolerance h9/IT5) a clearance fit is obtained. This leads to a slightly eccentric operation with resulting vibration, therefore the use of Y-bearings should be confined to low or medium speed operation. Relubrication is not normally required as the bearings are supplied lubricated for life. However, if necessary, Y-bearings fitted in cast iron housings can be relubricated.

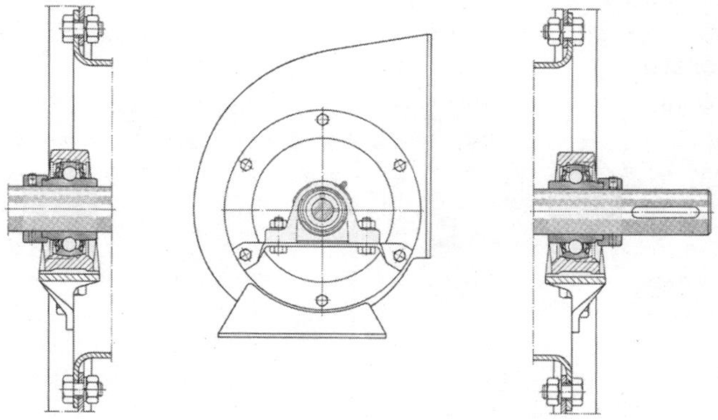

Figure 10.29 Light duty double inlet, double width (DIDW) centrifugal fan fitted with pillow blocks and ball bearings
Courtesy of SKF (UK) Ltd

10.9.2 Plummer block bearings

Where silent running is stipulated with relatively high speeds, self-aligning ball bearings mounted on adapter sleeves are recommended for light and medium duty fans with shaft diameters up to and including 110 mm. For heavier duty fans spherical roller bearings mounted on adapter sleeves, may be necessary. Normally the bearing is mounted in a cast steel plummer block housing. Various types of seal are available. Relubrication can be arranged if there is a suitable grease escape arrangement for use with the seal.

Figure 10.25 in Section 10.4.9 shows an arrangement using a self-aligning bail bearing mounted in an SNA plummer block housing with grease escape valve, type TAV. The efficiency of relubrication has been much improved by mounting an extra V-ring inboard of the V-ring seal washer at the side where grease is supplied, so that grease can only leave the housing at the opposite side after passing through the bearing. It should be noted that grease is usually supplied to these housings on the side away from the lock nut.

Tolerances

Shaft		h9/IT5
Housing	Standard plummer block	H8
	Plummer block with grease escape valve	H7

Lubrication

A high quality lithium base grease is normally recommended.

10.9.3 Plummer block bearings for oil lubrication

Spherical roller bearings with cylindrical bore and also with tapered bore plus the relevant adapter sleeve, are recommended for the larger heavy-duty fans. Appropriate housings will be found for both cylindrical and taper bores, in most bearing manufacturers' catalogues.

Where long relubrication intervals are desirable oil lubrication is recommended and specially designed plummer block housings can be used. These have an adequate space for an oil reservoir and have been developed mainly for high speed fans. They are equipped with effective labyrinth seals to eliminate oil losses.

For applications where low vibration and silent operation are required, preference is given to the use of spherical roller bearings with cylindrical bore mounted in series, see Figure 10.30.

Spherical roller bearings with tapered bore mounted on adapter sleeves are frequently used where easy mounting is required.

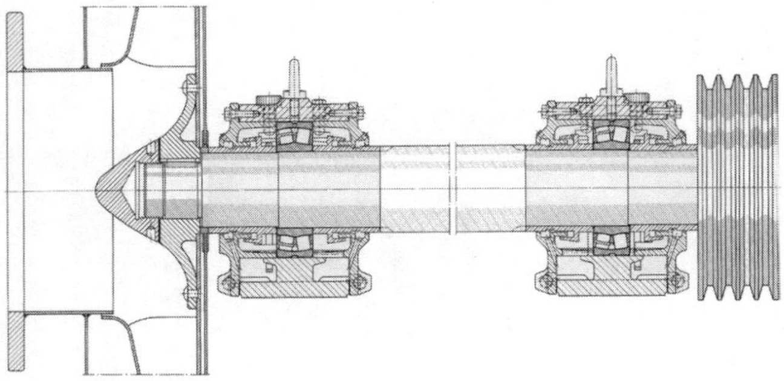

Figure 10.30 Heavy duty fan with oil lubricated plummer blocks
Courtesy of SKF (UK) Ltd

In this case a different design of housing is available in three variants:

- Shaft end, non-locating bearing - suffix AL
- Through shaft, non-locating bearing - suffix BL
- Through shaft, locating bearing - suffix BF

Tolerances

Shaft		
	Cylindrical seatings - direct mounting	m6
	Cylindrical seatings - mounting on sleeves	h9/IT5
Housing		F6

Lubrication

Oil lubrication is used. To keep the bearing temperature as low as possible with the minimum amount of oil in the bearing, the oil is lifted from the reservoir to a collecting trough, as the shaft rotates, by a pick-up ring which hangs loosely on a sleeve on the shaft and dips into the oil in the lower half of the housing. The oil then passes through the bearing on its way back to the reservoir.

10.9.4 Bearing arrangements using long housing cartridge assemblies

Deep groove ball bearings, paired angular contact ball bearings and cylindrical roller bearings have all been used in various combinations in two bearing cartridge housing assemblies. Such housings are available from the bearing manufacturers complete with their shafts, but are also manufactured by the larger fan manufacturers with special features to suit the application.

Perhaps the most common combination of races within a long housing is for a deep groove ball bearing at the impeller end and a cylindrical roller bearing at the drive end. (Figure 10.31.) The ball race "looks after" the end thrust whilst the cylindrical roller can take the radial load imposed by a vee belt drive. It t will

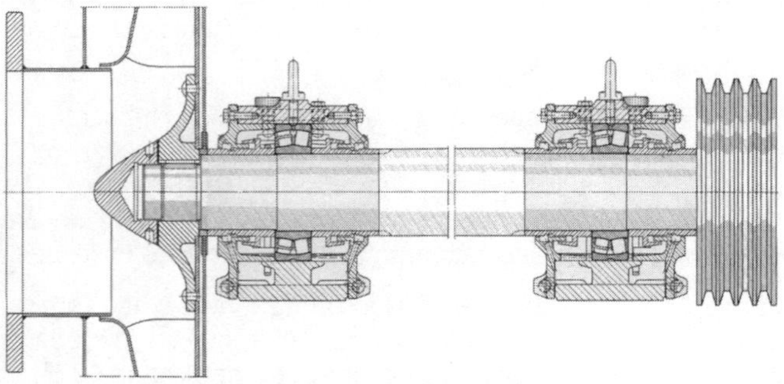

Figure 10.31 Two bearing cartridge assembly fitted with ball and roller bearings for grease or oil lubrication
Courtesy of SKF (UK) Ltd

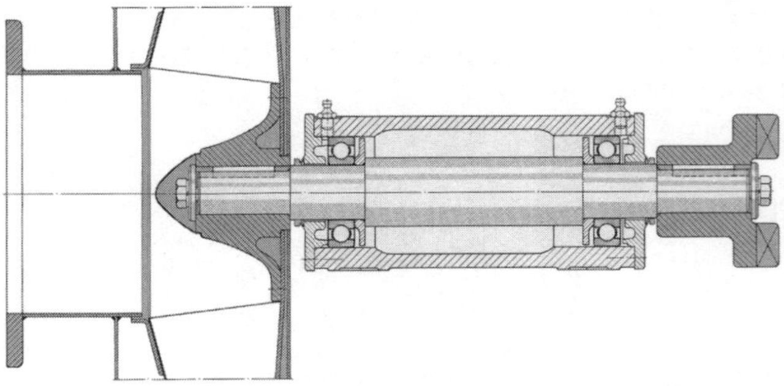

Figure 10.32 Cartridge assembly with single row deep groove ball bearings
Courtesy of SKF (UK) Ltd

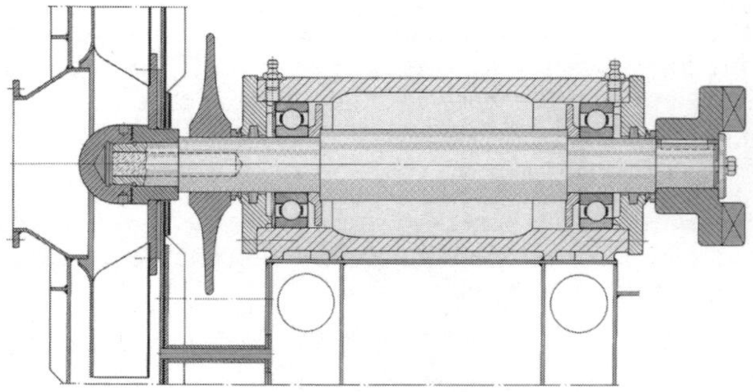

Figure 10.33 Hot gas fan fitted with cooling disc, heat shield and grease lubricated bearings
Courtesy of SKF (UK) Ltd

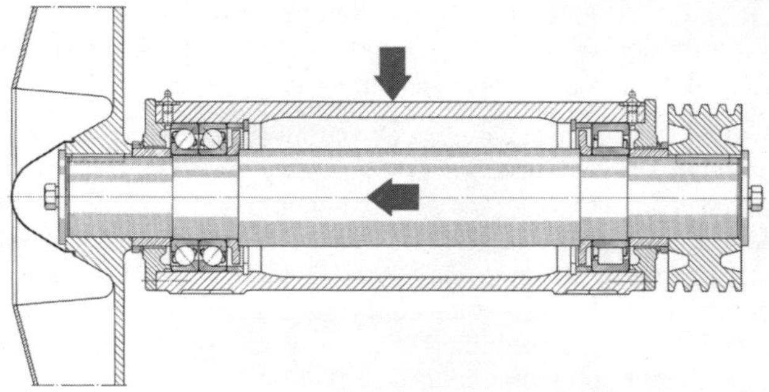

Figure 10.34 High pressure fan fitted with angular contact ball bearings and roller bearing to take vee belt drive loading
Courtesy of SKF (UK) Ltd

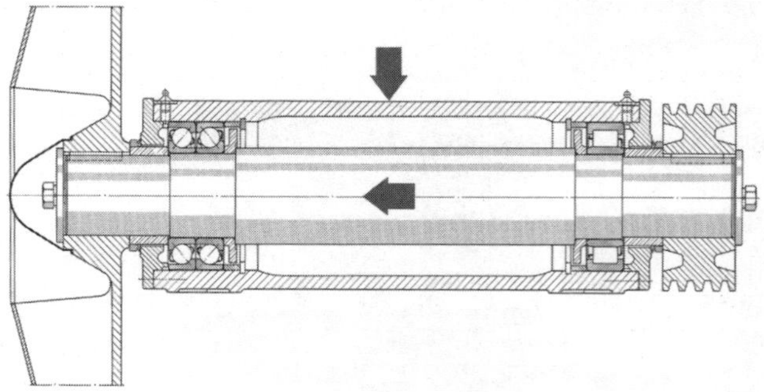

Figure 10.35 Cartridge assembly for heave radial loads (roller bearings) and ball race for location
Courtesy of SKF (UK) Ltd

be seen that grease or oil lubrication are both possible. However, many other combinations are available as shown in Figures 10.32 to 10.35.

10.9.5 Spherical roller thrust bearings

Spherical roller thrust bearings may be used in conjunction with deep groove all bearings, cylindrical roller bearings and spherical roller bearings. When high axial forces have to be accom-

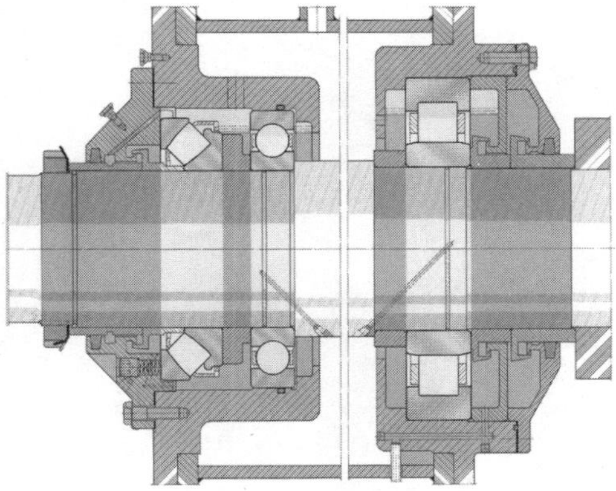

10.36 Spherical roller thrust bearing for horizontal shaft fan
Courtesy of SKF (UK) Ltd

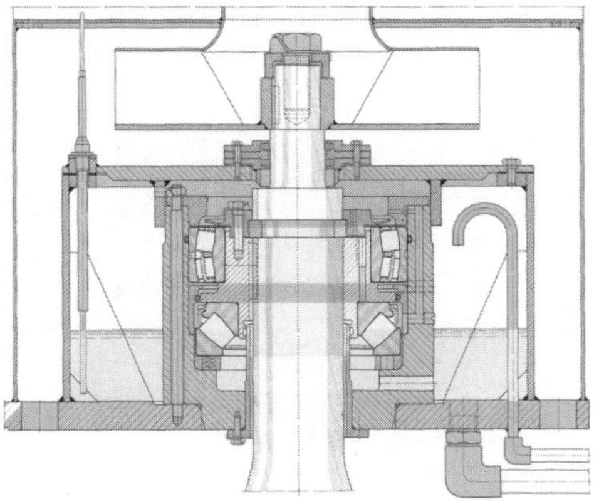

10.37 Spherical roller thrust bearing used for centrifugal fan with vertical shafts
Courtesy of SKF (UK) Ltd

modated, it is sometimes necessary to use a thrust bearing for the support.

Figures 10.36 and 10.37 show respectively a horizontal and a vertical fan, each fitted with a spherical roller thrust bearing. In each case, the spherical roller thrust bearing is radially free and therefore only axially loaded; the housing washer is loaded by using several springs, equally spaced around the periphery, to prevent the bearing from separating when the fan is started or the thrust load reversed.

Tolerances

Shaft	Deep groove ball bearings	
	d : 100 mm	k5
	d > 100 mm	k6
	Cylindrical roller bearings	
	d ' 140 mm	m5
	d > 140 mm	n6
	Spherical roller bearings	
	d : 140 mm	m6
	d > 140 mm	n6
	Spherical roller thrust bearings all diameters	j6
Housing	Deep groove ball bearings (with O-ring to prevent creeping)	H7
	Cylindrical roller bearings	M7
	Spherical roller bearings (with O-ring to prevent creeping)	H7
	Spherical roller thrust bearings	clearance

Lubrication

Circulating oil lubrication is used for the bearings in the horizontal fan. Oil bath lubrication is preferred for the bearings in the vertical fan. The pumping action of the spherical roller thrust bearing is utilised to ensure lubrication of both bearings in this arrangement.

10.10 Seals for bearings

10.10.1 Introduction

Whatever the bearing arrangement or type of bearing used, the bearings must be sealed to prevent contaminants and moisture entering the bearing in addition to retaining the lubricant. When seals are an integral part of a rolling element bearing, the bearing can be greased and sealed for life. However bearings used on medium and large motors and many small motors have to withstand load and speed conditions for a life which is outside the ability of sealed bearings. Hence the seals are generally part of the bearing housing in all but the smallest motors, because access for oil lubrication or greasing is required.

10.10.2 Shields and seals for bearing races

Shields and seals may be fitted. A shield does not form a complete seal and is fitted to the non-rotating ring with a small gap between the shield and the rotating ring, whereas seals are fixed to one ring and have a low-friction sliding face or fine clearance on to the other ring. Shields and seals may be fitted to one or both sides of a bearing and serve to keep contaminants out of the bearing and the lubricant in the bearing. Seals are usually of a synthetic rubber and thus usually have a temperature limitation of about -40°C to 120°C whereas metallic shields can be used outside this range. Shielded bearings are only suitable where water is not present and contamination is very light. It is more normal for fan bearings, except for very small sizes, to be fitted into housings with seals as part of the housing.

10.10.3 Standard sealing arrangements for bearing housings

Fan manufacturers will normally have standard bearing housings incorporating suitable seals to cover most applications and the operating conditions of the motor, but if there are particularly harsh operating conditions then special sealing arrangements may be necessary. Seals that form part of the bearing housing can be of non-rubbing or rubbing types. The non-rubbing type has the advantage of very low friction and no wear and is ideally suited to high speed and high temperature. Rubbing seals rely on a rubbing contact with a means of applying a light contact pressure and can provide a much more reliable seal than a non-rubbing type, when running and stationary. However, wear does take place and friction losses are generated, thus making them normally unsuitable for high peripheral speeds. If not fitted correctly, rubbing seals can give problems and contaminants that try to enter the seal can cause damage.

Non-rubbing seals are simply narrow gaps either axially, radially or a combination of both; the deciding factor being the likely movement of the shaft relative to the bearing housing. For example, a shaft that is likely to move axially either because of load influences or thermal expansion - but is restrained radially, would require a radial gap. Labyrinth seals are more effective than plain gaps and take many forms, examples of which are shown in Figure 10.38.

The third example of Figure 10.38 requires a split outer ring for assembly purposes. All the examples can improve the sealing properties by using a grease within the seal, a water-insoluble lithium or calcium based grease is recommended. The first example can have shallow grooves machined into the shaft adjacent to the seal and these grooves may be helical to drive lubricant back into the bearing, but this is only suitable for one direction of rotation.

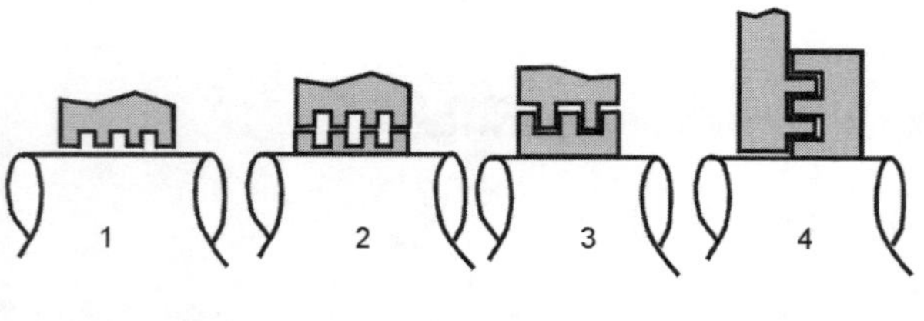

Figure 10.38 Examples of labyrinth seals

Another form of labyrinth seal involves washers with integral spacing flanges which are designed to fit either onto the shaft or into the bearing housing. By alternately placing the washers onto the shaft and into the housing a seal is created, the efficiency improving with the number of washers used.

Rubbing lip seals are generally manufactured from a synthetic type rubber, either of a form that gives a natural pressure from deflection of the seal or enhanced pressure by using a garter spring. Sections through typical rubbing seals are illustrated in Figure 10.39.

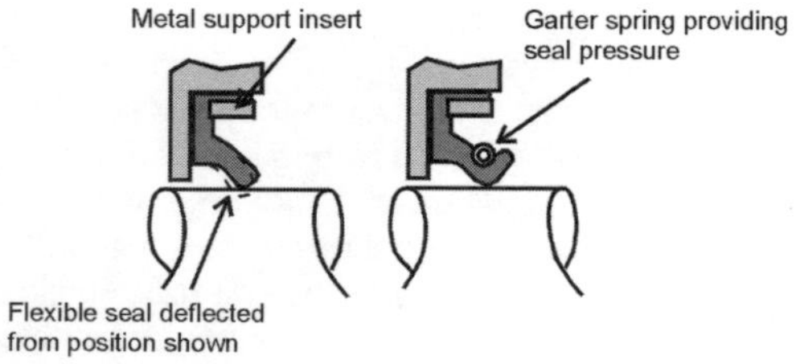

Figure 10.39 Examples of rubbing seals

The seal material type determines the operating temperature range, but generally -40°C to 200°C can be achieved without resorting to expensive special materials. The sealing surface on the shaft should be ground for best performance. At peripheral speeds in excess of about 4 m/s this is essential and at speeds higher than about 8 m/s the surface should be fine ground and hardened. As shown in Figure 10.40, the bearing is assumed to be positioned to the left of the seal and the seal is most effective at keeping contaminants from the bearing. If it is more important to keep lubricants in the bearing then the seal should be reversed.

A simple form of rubbing seal is the V-ring seal as shown in Figure 10.40. Made from synthetic rubber, it can be stretched over the shaft and provide enough grip to rotate with it, whilst the flexible lip rubs on the fixed sealing surface. Considerable misalignment can be permitted at low speeds and the sealing surface need not be exceptionally smooth. If the peripheral speed exceeds about 7 m/s, axial location is necessary and above about 12 m/s a steel support ring must be used to prevent the seal lifting from the shaft. The sealing lip is likely to lift off the sealing surface and create a small gap at above about 15 m/s peripheral speed.

An inexpensive seal, but limited to low temperatures and peripheral speeds below 4 m/s, is the felt insert. This is a simple felt ring soaked with oil within and located in a suitable retaining groove. It is an effective seal for grease lubricated bearings.

The seals described above are for the retention of grease or oil in bearing housings and to prevent moisture or contaminants entering the bearing. Seals for preventing the egress of contaminants or the ingress of air to fan casings are described in Chapter 7.

Figure 10.40 V-ring seal

10.11 Other types of bearing

There are several other types of bearing which have been developed for special applications, unsuited to the more standardised types of sleeve or rolling element bearings. Because of their unique features they are only briefly described to give an indication of what is available should the need arise.

10.11.1 Water-lubricated bearings

Where the fan/motor combination cannot be adequately sealed against the escape of oil, water has been used as a lubricant. This can mean a much lower film thickness because of the lower viscosity of water. However satisfactory bearings for certain applications have been designed.

10.11.2 Air-lubricated bearings

Air may also be used as a lubricant in sleeve bearings if supplied under pressure. It produces little friction loss but is really only suitable for small high speed bearings running in excess of about 6000 rev/min.

10.11.3 Unlubricated bearings

Sleeve bearings may be manufactured with porous bushes impregnated with substances such as PTFE. This produces a reasonably low coefficient of friction such that they can be used in small fans where the radial and thrust loads are low and the rotational speeds are not too high.

10.11.4 Magnetic bearings

Magnetic bearings have been used in large units operating at high radial loads and high rotational speeds. As there is no physical contact of lubricant, frictional power losses are virtually zero. However, power circuits, position sensors and controls are all needed to keep the shaft central within the housing. Provided that the fan duty remains fairly constant and, therefore, that the power absorbed also remains steady, successful bearings can and have been designed. At the present time development continues in an endeavour to reduce the very high cost.

10.12 Bibliography

The Friction of Lubricated Journals, carried out for the Institution of Mechanical Engineers by Beauchamp Tower, first reported in 1883 and 1884.

On the theory of lubrication and its application, to Mr. Beauchamp Tower's experiments, including an experimental determination of the viscosity of olive oil, Royal Society, Phil. Trans., Pt. 1, 1886.

Lubrication its Principles and Practice, A G M Michell, 1950, Blackie

ISO 5753:1991 *Rolling bearings — Radial internal clearance*

ISO 15:1998 *Rolling bearings — Radial bearings — Boundary dimensions, general plan*

ISO 355:1977 *Rolling bearings — Metric tapered roller bearings — Boundary dimensions and series designations*

ISO 104:2002 *Rolling bearings — Thrust bearings — Boundary dimensions, general plan*

ISO 3096:1996 *Rolling bearings — Needle rollers — Dimensions and tolerances*

DIN 17230 / ISO 683-17 *Ball and roller bearing steels*

DIN 5402-3 *Rollers for needle roller bearings*

ISO 281:1990 *Rolling bearings — Dynamic load ratings and rating life*

11 Belt, rope and chain drives

In the interest of energy efficiency, it would be preferable for all fans to be arranged for direct drive. There are however, many reasons for incorporating an indirect drive through vee belts, ropes or chains etc. A degree of flexibility can be introduced which will cater for a system resistance which has been imprecisely calculated or which may vary through the lifetime of the fan.

These drives may allow the use of standard motors and also enable the manufacturer to cover the duty envelope with a reduced number of models.

Contents:

11.1 Introduction

It might be thought desirable to arrange all fans to be directly driven, i.e. with the fan impeller mounted directly on the shaft extension of the driving motor. There are however, a number of reasons for arranging for an indirect drive through belts, ropes or chains and suitable pulleys or sprockets.

From a user viewpoint, such drives give a degree of flexibility to the fan installation, permitting easy changes in the fan speed. If the system resistance as calculated proves to be incorrect, it is a relatively simple matter to make a change to the pulleys and/or belts. Thus a new fan speed to give the required duty can be arranged. Provided that the fan is mechanically suitable for any such increases then it is also possible to upgrade the performance over time. This might be necessary with extensions to a building and its associated HVAC system. In a mine ventilating plant for example, the duty could be increased as the mine working lengthened. There are many other reasons for changing the fan duty and the reader will be able to identify these for his particular industry.

From a manufacturer's viewpoint, indirectly driven fans enable him to reduce the number of models, which he has to produce in order to provide an adequate cover of the duty range at a reasonable efficiency. Theoretically, provided it could be driven fast enough, one fan model could meet all fan duties, albeit in many cases at low energy efficiency.

11.2 Advantages and disadvantages

Apart from duty flexibility, there are many other considerations in the decision as to whether to incorporate a direct or indirect drive.

To take an extreme case, a requirement to produce a high volumetric flowrate at a low pressure will inevitably mean a large diameter fan running at a low speed, if multiple fans cannot be considered. If direct drive were to be specified, then, with an AC electric driving motor, this would require a large number of poles and a large frame size with a correspondingly high purchase price. It might also result in a somewhat lower efficiency motor with less starting torque available.

Conversely, with a belt or rope drive interposed between the fan and motor, it is possible to select a much cheaper motor at a better efficiency with improved starting characteristics. It is also possible to select fans running at greater than the two pole motor speed on an AC supply i.e. approximately 3,000 rev/min on 50Hz AC. All these advantages can more than offset the disadvantage of the transmission efficiency, which will of course be less than 100%.

There are many cases in industrial applications where the gas stream is at a temperature higher than ambient or contains corrosive/erosive/explosive constituents. Any indirect drive may then permit the driving motor to be positioned away from these dangers such that with minimal precautions, a relatively standard machine can be used. A disadvantage of rope and belt drives is the need for maintenance. Tension in the belt or rope(s) has to be correctly maintained to ensure that the power is transmitted without slip. This is especially important in multiple vee belts when each belt has to have an equal tension to ensure that it correctly transmits its share of the absorbed power. In the past matched sets of belts, in regard to length, were specified. Now, however, the manufacturers are able to guarantee, by improved manufacturing processes, that nominally identical ropes are equal in length to within very close tolerances.

11.3 Theory of belt or rope drives

In these drives, the power transmitted depends upon the friction between the rope or belt and the rim of the pulley (denoted as sheave in American parlance).

Referring to Figure 11.1 (a), let q be the angle of wrap i.e. the angle at the pulley centre made by each end of the belt or rope in contact with the pulley rim. Alternative forms of this rim are shown in(b) to (d) Figure 11.1.

The so-called vee belt or rope (c) is now by far the most popular, having benefited from standardization and the resultant mass production by a number of reputable manufacturers. Circular cross-section ropes (d) are now rarely used for fan drives, but the flat belt (a) has shown some signs of a revival. Its reduced radial thickness compared with vee ropes means that centrifugal forces tending to make the belt(s) leave the pulley are minimized and high belt speeds (and therefore power transmitted) are possible. It should be noted that whilst the belt is flat, the rim of the pulleys used with it are in practice slightly "crowned", since this has been found to help in maintaining the belt centrally on the pulley.

If the tension at one end of the belt is T2 and the tension T1 at the other end is increased gradually, then the belt will eventually start to slip bodily around the pulley rim. The value of T1 at which slip takes place will depend upon the values of T2, q and the coefficient of friction m between the belt and the rim.

Consider a short length mn of belt, which subtends and angle dq at the pulley centre.

Let T be the tension on the end m and T+ dT must be due to the friction between the length mn of the belt and the pulley rim, and it will depend upon the normal reaction between mn and the rim and the side of the groove for the sections (c) and (d). Let R be the radial reaction between the pulley rim and the length mn of

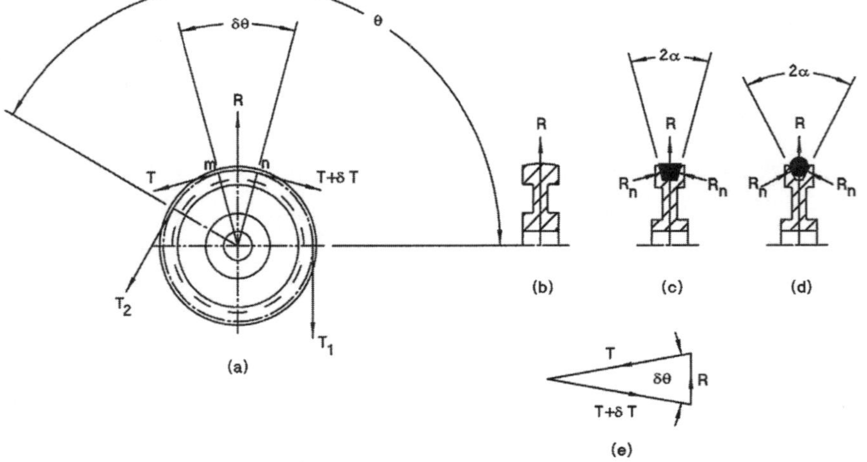

Figure 11.1 Diagrammatic view of pulley and belts or ropes

belt or rope and let Rn be the normal reaction between each side of the groove and the side of mn for the sections (c) and (d).

Then for section (b):

$$\delta T = \mu R \qquad \text{Equ 11.1}$$

and for sections (c) and (d):

$$\delta T = 2\mu R_n$$

But for these sections the radial reaction R is the resultant of the two normal reactions Rn, so that $R = 2R_n \sin\alpha$ and, substituting for R_n in terms of R,

$$\delta T = \frac{\mu R}{\sin\alpha} = \mu_1 R \qquad \text{Equ 11.2}$$

where:

$$\mu_1 = \frac{\mu}{\sin\alpha} = \cos ec\ \alpha \qquad \text{Equ 11.3}$$

It follows, therefore, that the friction between mn and the grooved rim is the same as that between mn and a flat rim, if the actual coefficient of friction μ is replaced by the virtual value $\mu_1 = \frac{\mu}{\sin\alpha}$.

In the plane or rotation of the pulley the three forces which act on mn are the tensions T and T + δT on the ends m and n and the radial reaction R. Since mn is in equilibrium under this system of forces the triangle of forces may be drawn as shown in (e) of Figure 11.1.

From this triangle, since $\delta\theta$ and δT are small, $R \cong T.\ \delta\theta$, and substituting this value of R in equation 11.1:

$$\delta T \cong \mu T \delta\theta \text{ or } \frac{\delta T}{T} \cong \mu\delta\theta$$

If both sides of this equation are integrated between corresponding limits, then :

$$\int_{T_2}^{T_1} \frac{dT}{T} = \int_0^{\theta} \mu\delta\theta$$

$$\therefore \log_e \frac{T_1}{T_2} = \mu\theta$$

$$\text{or } \frac{T_1}{T_2} = e^{\mu\theta}$$

Equ 11.4

As it stands this equation applied to the flat rim (b), but if μ_1 is substituted for μ, it will apply equally well to the grooved rims (c) and (d).

It must be emphasized that equation 11.4 gives the limiting ratio of the tensions T_1 and T_2 when the belt or rope is just about to slip bodily round the pulley rim. The actual ratio of the tensions may have a lower value, but cannot have a higher value than this limiting ratio.

The limiting ratio is very much increased, for given values of μ and θ, by using a grooved section. For instance if q is 165° and μ is 0.25, the limiting ratio for the flat rim is given by:

$$\frac{T_1}{T_2} = e^{0.25\frac{11\pi}{12}} = 2.054$$

If a vee rope or belt is used with a groove angle of 40°, then

$$\mu_1 = \frac{0.25}{\sin 20°} = 0.731 \text{ and } \frac{T_1}{T_2} e^{0.731\frac{11\pi}{12}} = 8.21$$

Similarly, if a rope of circular section is used with a groove angle of 45°, then

$$\mu_1 = \frac{0.25}{\sin 22.5°} = 0.653 \text{ and } \frac{T_1}{T_2} = e^{0.653\frac{11\pi}{12}} = 6.56$$

The maximum effective tangential pull exerted by the belt or rope on the pulley rim is, in each case, given by the difference between T_1 and T_2. It may be expressed in terms of the tension T_1 of the tight side, the magnitude of which is, of course, determined by the cross-section of the belt or rope and the allowable stress in the material.

For the flat belt under the above conditions the effective tension

$$T = T_1\left(1 - \frac{T_2}{T_1}\right) = 0.513T_1$$

for the vee belt or rope belt, $T = 0.878T_1$

and for the circular section rope, $T = 0.848T_1$.

It is clear from these figures that the use of a grooved pulley rim with a suitable vee or circular rope section enables the material to be employed more efficiently than where a flat rim is used.

So far it has been assumed that the pulley is stationary. If the pulley is mounted on a shaft, which is supported in bearings, then the effective tangential force exerted by the belt or rope on the pulley may be used to transmit power from the belt or rope to the pulley and thence to the shaft. The power transmitted may be determined when the effective tension and the speed of the belt or rope are known. But when the belt or rope is in motion, the stresses in the material are not simply those which arise form the power transmitted. There is in addition the centrifugal stress due to the inertia of the belt or rope as it passes round the pulley rim. The magnitude of this stress may be determined as shown in the following section.

11.3.1 Centrifugal stress in a belt or rope

Referring to Figure 11.2, let r be the radius of the pulley, v the speed of the belt or rope, a the cross-sectional area and w the weight of the belt or rope per unit length.

The weight of the short length mn which subtends to angle $\delta\theta$ at the pulley centre, is $w \cdot r\delta\theta$ and the centrifugal force on mn is given by:

$$F_c = \frac{wr\delta\theta}{g} \cdot \frac{v^2}{r} = \frac{wv^2}{g} \cdot \delta\theta$$

This force acts radially outwards and, if the pulley rim is flat, the only possible way in which it can be resisted is by applying two forces T_c to the ends of mn. The short length of belt is in equilib-

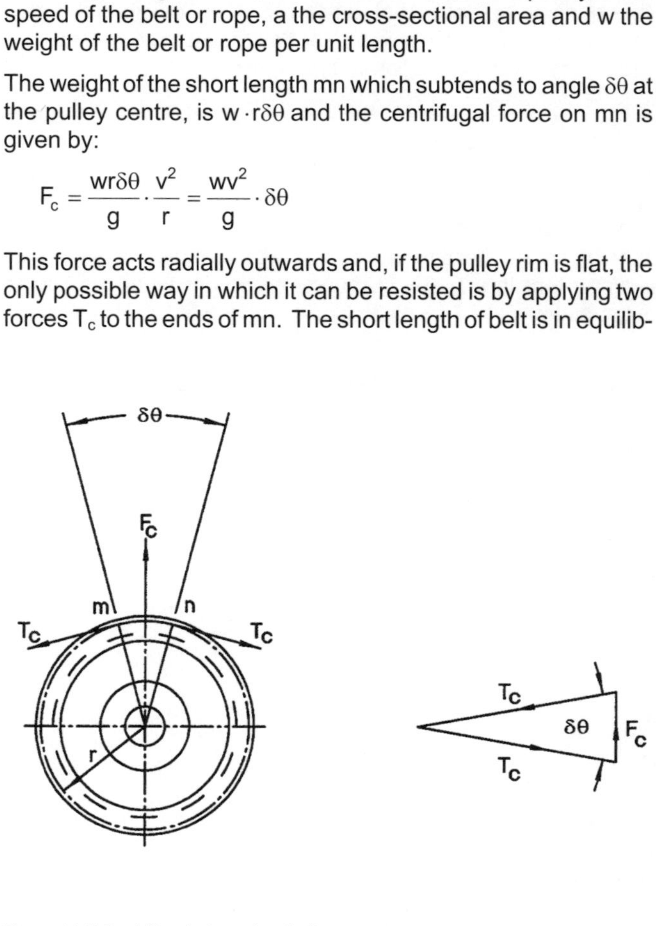

Figure 11.2 Centrifugal stress in a belt or rope

rium under these three forces and the triangle of forces may be drawn. From the triangle of forces T_c may be expressed in terms of F_c. Since $\delta\theta$ is small, $F_c \approx T_c\delta\theta$ and substituting for F_c from the above equation:

$$\frac{wv^2}{g} = \delta\theta = T_c \cdot \delta\theta$$

$$\therefore T_c = \frac{wv^2}{g} \qquad \text{Equ 11.5}$$

The stress per unit area of the belt or rope material due to the inertia is given by:

$$f_c = \frac{T_c}{a} = \frac{w}{a}\frac{v^2}{g} \qquad \text{Equ 11.6}$$

It should be particularly noticed that the centrifugal stress is independent of the radius of curvature of the path. It has been assumed so far that the rim of the pulley is flat and that the centrifugal inertia force therefore gives rise to a stress in the belt or rope material which is additional to the stresses caused by the tensions T_1 and T_2.

If, however, the pulley rim is grooved as at (c) and (d) in Figure 11.1, it would appear at first sight that the centrifugal force may be either wholly or partly balanced by the friction between the sides of the belt or rope and the sides of the groove, in which case F_c will be either zero or will have a value less than that given by equation 11.6. But there are two other factors which have to be taken into account in this connection.

First, if the power transmitted by the belt or rope is such that limiting friction exists in the tangential director i.e. if the belt or rope is just on the point of slipping bodily round the rim, there can be no friction force opposed to the centrifugal force. Since this condition of limiting friction rarely, if ever, exists in practice, there can be no doubt that the centrifugal stress in that part of the belt or rope, which is in contact wit the rim, will be less than the stress calculated from equation 11.6.

Secondly, and more importantly in any actual drive, the part of the belt or rope between the pulleys is not straight but hangs in a curve. The free parts of the belt must therefore be subjected to the centrifugal stress given by equation 11.6. Hence, there is not justification for the assumption which is sometimes made that the centrifugal stress in a belt or rope running on a grooved pulley is less than that in the same belt or rope when running on a flat pulley.

11.3.2 Power transmitted by a vee rope or belt

The power transmitted by a vee rope or belt may be calculated from the effective tension $T_e = T_1 - T_2$ and the belt speed

$$T_1 - T_2 = \frac{\text{Power P (watts) per rope}}{\text{Rope speed } v_b(\text{m/s})}$$

$$= \frac{\text{kW x 1000}}{\pi \text{ x } \frac{d_{p(nm)}}{1000} \text{ x } \frac{N(\text{rev/min})}{60}} \qquad \text{Equ 11.7}$$

and

$$\frac{T_1}{T_2} = e^{\mu\theta\text{cosec}\frac{\alpha}{2}} \qquad \text{Equ 11.8}$$

These tensions and powers are for one rope. By utilizing multiple ropes, the power transmitted is directly proportional to their number i.e. three ropes will transmit three times the power.

11.4 Vee belt drive Standards

Classical vee belts have been available since 1920 and until the 1970s were manufactured to the various editions of BS 1440.

The later, narrow wedge vee belts introduced around 1960 were covered by BS 3790:1973. More recently both types of vee belt have been manufactured to BS 3790:1995 and ISO 4184, it being recognised that as the included angle of the ropes are the same, the width of the belt or rope merely determines exactly where it sits in the groove and thus defines the effective pitch angle of the pulley.

Both types of vee belt have a trapezoidal cross-section consisting of a tension member contained within a rubber base and surrounded by a rubber-impregnated fabric cover. They are variously known as belts or ropes being a compromise between each.

To meet the wide range of speeds and powers, various rope sections have been standardised as shown in Table 11.1.

Type	Section	Pitch width (mm)	Top width (mm)	Height (mm)	Angle (degrees)
Classical	Y	5.3	6.5	4	40
	Z	8.5	10	6	40
	A	11	13	8	40
	B	14	17	11	40
	C	19	22	14	40
	D	27	32	19	40
Wedge	SPZ	8	9.5	8	40
	SPA	11	13	10	40
	SPB	14	16	14	40
	SPC	19	22	18	40

Table 11.1 Standardised vee belt sections

An indication of the likely belt section is shown in Figures 11.3 and 11.4. However it is recommended that a reputable manufacturer be consulted for the most appropriate selection.

It should also be noted that vee belts continue to be made to other standards such as the American RMA IP20 and DIN 2215 etc.

Whilst most of the requirements for classical and wedge type vee belt drives are contained in BS 3790:1995, it should be noted that the list of ISO Standards in Section 11.7 Bibliography, is extensive and also encompasses the specification of synchronous (toothed or timing) belts as well as some of the other alternatives mentioned in Section 11.5.

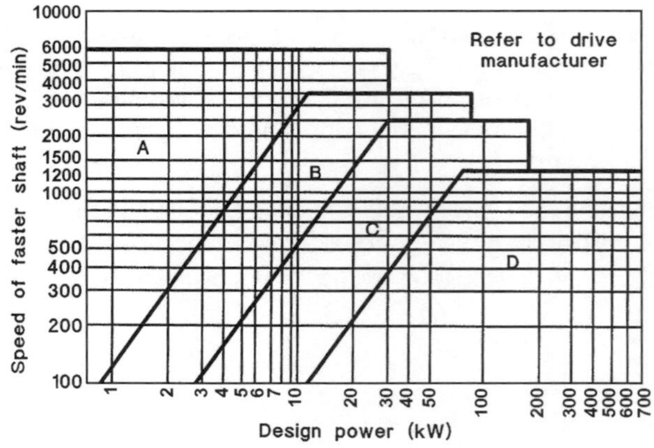

'Y' and 'Z' section belts should be used for design powers lower than those shown, or when pulley diameters are smaller that the recommended minimum for A-section belts

Fig 11.3 Selection of classical vee belt cross-section

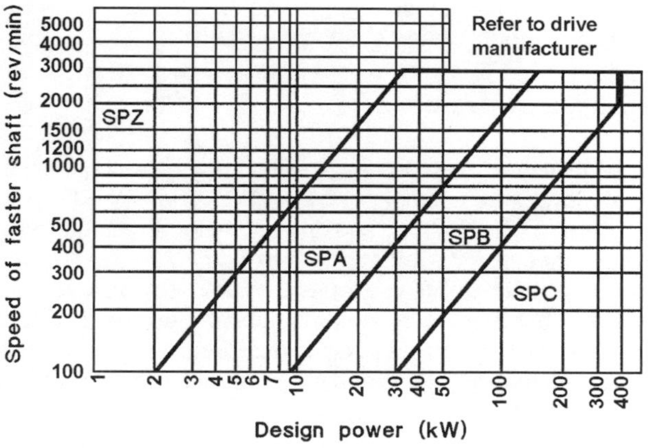

Fig 11.4 Selection of wedge belt cross-section

As previously noted, powers beyond the capacity of a single belt are covered by using multi-grooved pulleys and a matched set of belts. Since both classical and wedge belts are manufactured from the same materials and have the same included angle, it follows that the tension ratio is not influenced by belt section.

British and International Standards effectively assume that μ = 0.175, in both cases, i.e. well below the limiting coefficient of friction and thus if the angle of wrap is 180° (π radians).

$$\frac{T_1}{T_2} = 2.7183^{0.175 \times \pi \times 2.9238}$$

or

$$\frac{T_1}{T_2} = 5$$

or

$$T_2 = \frac{T_1}{5}$$

It is repeated that μ = 0.175 is a very pessimistic value and was chosen to give a margin of safety on the frictional grip between the rope and pulley.

The total running tension, which has to be resisted by the drive end fan bearing and the nose motor bearing = $T_1 + T_2$.

Thus:

$$T_1 + \frac{T_1}{5} = k_v\left(T_1 - \frac{T_1}{5}\right)$$

where k_v is constant or

$$1.2\ T_1 = k_v \times 0.8\ T_1$$

or

$$k_v = \frac{1.2}{0.8} = 1.5$$

i.e.

$$T_1 + T_2 = 1.5(T_1 - T_2)$$

The line of action of this pull will be determined by the number and section of the belts. A moment will be produced at the bearing and this will be reduced by keeping the pulleys as close as possible to the bearings.

The tension is that theoretically required for running and is usually exceeded in practice. Where the tensioning of the drive is in accordance with the manufacturer's recommendations, the figure should be multiplied by a safety factor of 1.25. Poor fitting, however, can result in considerable over tensioning when a factor of 2 is more appropriate.

When the pulleys are rotating, the belts tend to leave the pulley grooves due to the effects of centrifugal force. An additional tension is therefore given to the belts to overcome this effect. Thus the static load e on the bearings will be greater than the running load. It should be especially noted that bearing loads for correctly tensioned drives are the same for classical and wedge belts when the belt speed, pulley diameter, and power are the same.

With wedge belts, however, due to their smaller section and therefore greater flexibility, it is possible to use smaller pulleys. This then results in lower belt speeds and correspondingly increased tensions. There has therefore been reluctance by some users to employ wedge belts. Provided that the minimum pulley diameters and maximum pulley widths specified in Tables 11.2 and 11.3 are followed and that drives are correctly tensions, both classical and wedge belts will function satisfactorily and will give acceptable motor and fan bearing lives.

Motor frame size	Min pulley dia (mm)	Max pulley width (mm)
D63	50	50
D71	63	50
DD80	75	100
D90S	75	150
D90L	115	100
D100L	160	100
D112M	200	100
D132S	160	160
D132M	215	125
D160M	180	200
D160L	245	160
D180M	260	160
D180L	280	160
D200L	315	200
D225S	355	200
D225M	400	200

Table 11.2 Pulley dimensions for electric motors

Fan Shaft dia (mm)	Min pulley dia (mm)	Max pulley width (mm)
20	80	100
30	90	100
40	140	125
50	180	140
55	250	160
60	280	160
65	315	160
70	355	170
80	400	170
90	450	170
100	500	170
125	630	170

Table 11.3 Pulley dimensions for fan shafts

11.4.1 Service factors

When determining the number of ropes in a multi-vee rope drive, it is usual to apply a service factor to the calculated power thus increasing the number of ropes above that theoretically necessary. This service factor is to take account of the increased loads likely when starting and for more arduous conditions during running (see Table 11.4).

It should be noted that such factors inevitably mean that under normal running conditions the drive may be over-engineered and thus of lower efficiency. The problem may be particularly serious where low power fans may be specified with two belts when one might be sufficient. The value of low maintenance has to be weight against lowered energy efficiency. A soft start electric solution may be an alternative.

Special Cases	Types of prime mover					
	"Soft" starts			"Heavy" starts		
For speed increasing drive of: Speed ratio 1,00 – 1,24 multiply service factor by 1,00 Speed ratio 1,25 – 1,74 multiply service factor by 1,05 Speed ratio 1,75 – 2,49 multiply service factor by 1,11 Speed ratio 2,50 – 3,49 multiply service factor by 1,18 Speed ratio 3,50 and over multiply service factor by 1,25	Electric Motors: AC – Star Delta start DC – Shunt Wound Internal Combustion Engines with 4 or more cylinders All prime movers fitted with Centrifugal ???			Electric Motors: AC – Direct-on-Line start DC – Series & Compound Wound Internal Combustion Engines With less than 4 cylinders Prime movers not fitted with soft Start devices		
Types of Fan	Hours per day duty					
	< 10	> 10 to 16	> 16	< 10	> 10 to 16	> 16
Blowers, exhausters and fans of all types up to 7.5kW	1,0	1,1	1,2	1,1	1,2	1,3
Blowers, exhausters and fans of all types above 7.5kW	1,1	1,2	1,3	1,2	1,3	1,4

Table 11.4 Service factors

11.5 Other types of drive

Whilst most fan drives have for many years been of the vee rope type, it should be noted that interest has also recently been shown in other types.

11.5.1 Flat belts

These have improved tremendously and now incorporate synthetic tension members having great shock absorbing capacity, strength, suppleness, and dimensional stability.

The high coefficients of friction enable large power to be transmitted, but care must be taken in selection to minimise bearing loads. Efficiency can be as high as 98%. With the light weight, centrifugal effects are small and there is not permanent stretch so that tension adjustment is rare.

11.5.2 Toothed belts

These incorporate optimum grades of neoprene with glass fibre tension cords and nylon facings giving considerably improved lives with the new tooth profiles now used. As they do not rely on friction, tensions are lower and therefore bearing loads are lower.

Once installed they do not require re-adjustment, but must be carefully aligned to minimise wear. At start-up under conditions of rapid acceleration, high transient tensions can result due to

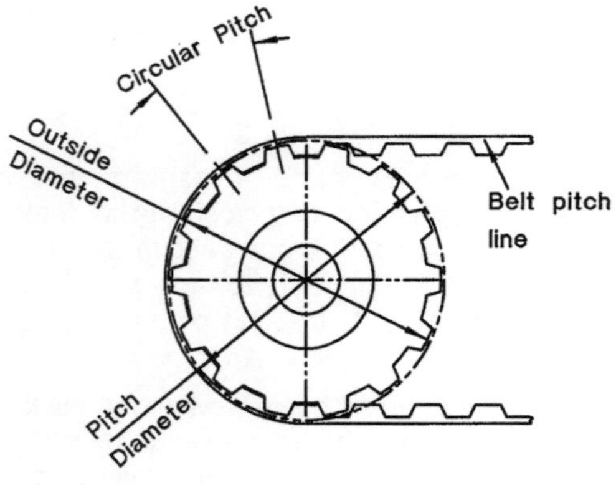

Figure 11.5 Toothed belt

the fan inertia and these must be determined to prevent belt breakage or tooth shear, see Figure 11.5.

11.5.3 Micro-vee belts

These combine the simplicity and flexibility of a single flat belt with the properties of higher power and higher speed ratios of vee belts. The belt is constructed with an uninterrupted strength member of synthetic cord extending across the whole width of the belt. Unlike vee ropes they do not operate by wedging action but there is continuous contact between the ribbed surface of the belt and pulley grooves. Being a single belt, there are no matching problems and they cannot turn over as a result of shock loads.

11.5.4 Banded belts

In applications where pulsating or shock loads can cause normal vee ropes to turn over, twist or whip, then banded belts are a solution, as shown in Figure 11.6. By joining together a number of vee ropes with a tie band and thus forming a compromise between the flat belt and vee ropes, the lateral rigidity is increased sufficiently to resist turn over etc. Also by ensuring that the underlying ropes enter the pulley grooves in a straight line, excessive jacket wear is avoided, resulting in a longer life.

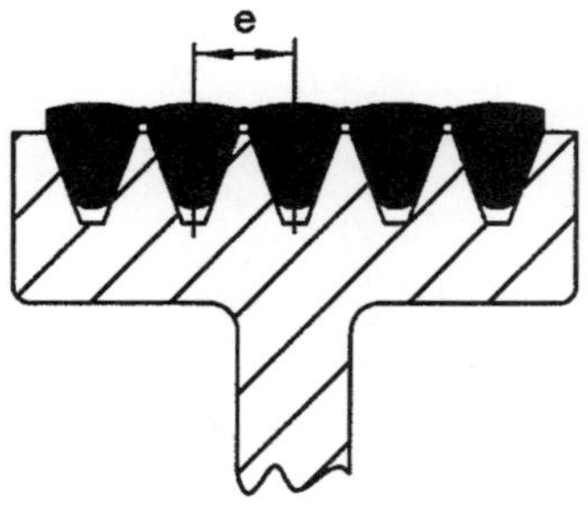

Figure 11.6 Cross-section of banded belt and pulley rim

When using banded belts it is important that the correct groove profile is selected. The groove spacing i.e. dimension "e" is given in Table 11.5.

Belt section	Groove spacing e (mm)
SPZ	12.0
SPA	15.0
SPB	19.0
SPC	25.5

Table 11.5 Spacing of grooves for different belt sections

11.5.5 Raw-edged vee belts

It was noted in Section 11.4 that both classical and wedge type vee ropes consist of a tension member contained with a rubber base and surrounded by a fabric cover. Of recent years it has come to be recognised that the fabric at the sides of the rope could be deleted without affecting the strength, particularly with the improved wear properties of modern synthetic rubbers. This gives a so-called raw edge and leads to greater flexibility in the belt. Reduced pulley sizes are possible and better wrap is achieved. Greater drive efficiencies are also attained.

This revolution in drives has led to the Standards being outdated, such that the purchaser is strongly recommended to consult a reputable manufacturer for an up-to-date selection of any drive. As with all such advances, it may take some time for the Standards to "catch up".

11.5.6 Chain drives

These are now rarely used for fan drives, due to their limitations in speed and power. There is also a need for lubrication and maintenance, beyond that required for vee ropes.

A chain may be regarded as a belt, built up of rigid links, which are hinged together in order to provide the necessary flexibility for the wrapping action round the driving and driven sprockets. These sprockets have projecting teeth, which fit into suitable recesses in the links of the chain and thus enable a positive drive to be obtained. The pitch of the chain is the distance between a hinge centre of one link and the corresponding hinge centre of the adjacent link. The pitch circle diameter of the chain sprocket is the diameter of the circle on which the hinge centres lies, when the chain is wrapped round the sprocket.

11.5.6.1 Types of chain

There are two types of chain in common use for transmitting power, namely:

- the roller chain
- the inverted tooth or silent chain.

The roller chain. The construction of this type of chain is shown in Figure 11.7. The inner plates A are held together by steel bushes B, through which pass the pins C riveted to the outer links D. A roller R surrounds each bush B and the teeth of the sprockets bear on the roller. The rollers turn freely on the bushes and the bushes turn freely on the pins. All the contact surfaces are hardened so as to resist wear and are lubricated so as to reduce friction.

Figure 11.8 (a) shows a simple roller chain, consisting of one strand only, but duplex and triplex chains, consisting of two or three strands, may be built up as shown in Figure 11.7 (b), each pin passing right through the bushes in the two or three strands.

The inverted tooth or silent chain. The construction of this type of chain is shown in Figure 11.8 (a). It is built up from a series of flat plates, each of which has two projections or teeth. The outer faces of the teeth are ground to give an included angle of 60° or, in some cases, 75°, and they bear against the working faces of the sprocket teeth. The inner faces of the link teeth take no part in the drive and are so shaped as to clear the sprocket teeth. The required width of chain is built up from a number of these plates arranged alternately and connected together by hardened steel pins which pass through hardened steel bushes inserted in the ends of the links.

The pins are riveted over the outside plates. The chain may be prevented from sliding axially across the face of the sprocket teeth by outside guide plates without teeth, or by a centre guide plate without teeth which fits into a recess turned in the sprocket.

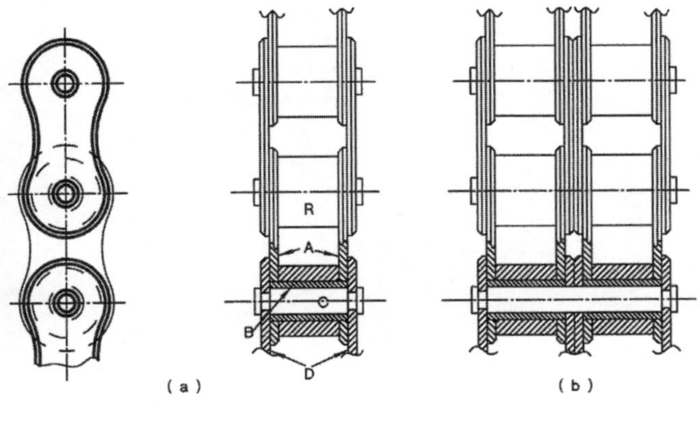

Figure 11.7 Details of roller chain

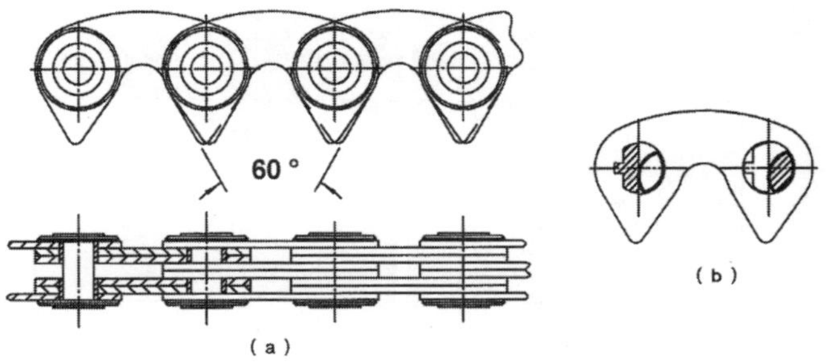

Figure 11.8 Details of inverted tooth chain

Figure 11.8(b) shows the type of hinge used in the Morse silent chain. This reduces friction by substituting a hardened steel rocker on a hardened steel flat pivot for the pin and bush.

When the chain is new, the position which it takes up on the sprocket is shown in the upper part of Figure 11.9. Each link, as it enters the sprocket, pivots about the pin on the adjacent link which is in contact with the sprocket. The working faces of the link are thus brought gradually into contact with the corresponding faces of the sprocket teeth. A similar action takes place as each link leaves the sprocket. Hence there is no relative sliding between the faces of the links and the faces of the sprocket teeth.

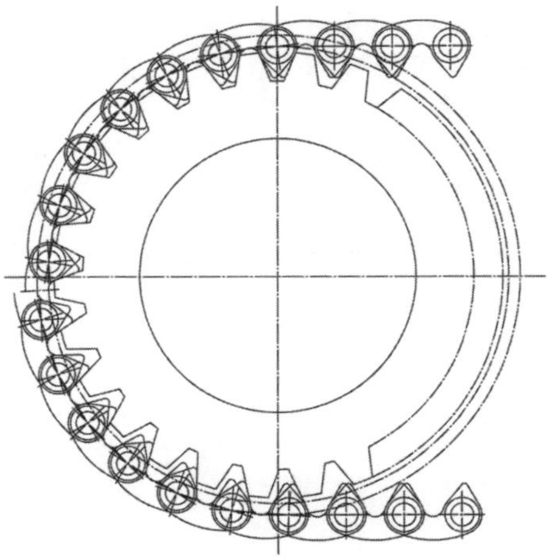

Figure 11.9 Sprocket and silent chain

As wear takes place on the pins and bushes, the smooth action of the chain is not impaired, but the chain rides higher up the sprocket teeth and the effective pitch circle diameter of the sprocket is increased, as shown in the lower part of Figure 11.9

11.5.6.2 Standards for chain drives

The Standards for chain drives are not nearly so comprehensive as those for vee belts. However, the ISO standards given in Section 11.7 Bibliography, are relevant:

11.5.7 Drive efficiency

Many of these alternative drives have been designed to overcome some of the shortcomings of the standard vee rope drive.

Normal belts suffer from tension decay, resulting in slip and loss of efficiency. They require frequent adjustment to maintain performance. Being a single member, these alternatives do not suffer from matching problems. In a multi-belt drive, where there is a variation in length, however small, the shorter belts will be under tension and transmitting the power whilst the longer belts are running slack and contributing little. Effectively the drive is under-designed and will have a short life.

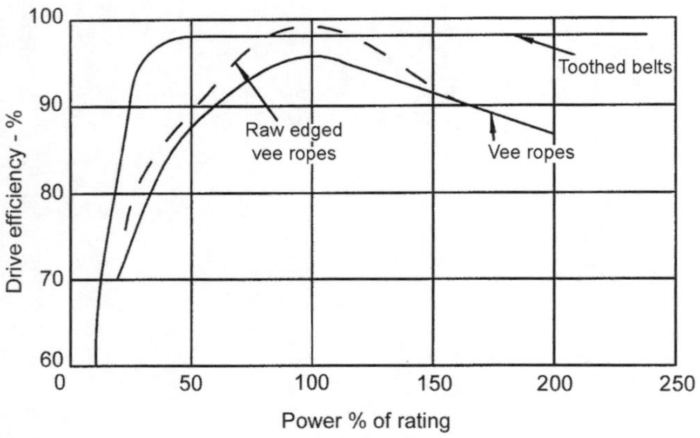

Figure 11.10 Efficiency of toothed and vee belt drives

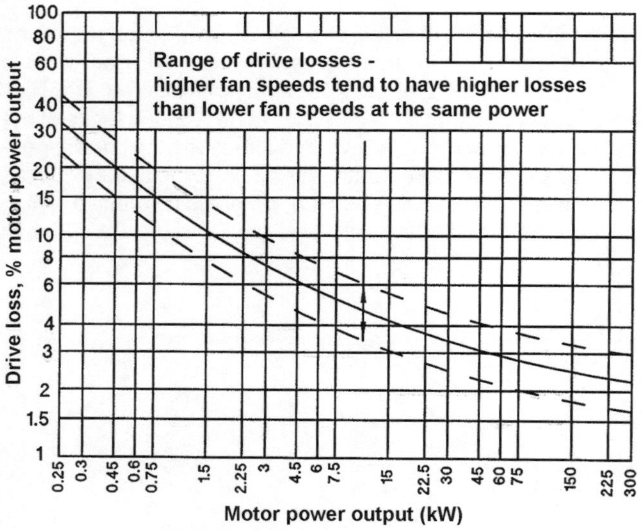

Figure 11.11 Estimated vee belt drive losses

Drive efficiencies can be maintained over a wider range of powers and can in any case exceed the 97% possible with vee ropes. It should here be noted that if a vee drive is either under- or over-engineered, efficiency will suffer as shown in Figure 11.10.

With very small drives, the difference in power transmitted between, say, one and two belts or between two and three belts, is obviously substantial. The chart in Figure 11.11 has been based on AMCA International data and may be used to estimate the losses in a standard vee belt drive. Such losses will need to be added to the fan power to determine the power required from the motor.

Example 1 Motor power output P_m is determined to be 9.9kW. From curve drive loss = 5.8%. Drive loss P_1 = 0.58 X 9.9 = 0.6kW. Fan power input P_f = 9.9-0.6 = 9.3kW.

Example 2 Fan power input P_f = 0.75 kW. In this case it is necessary to estimate motor power input. Motor power output = 0.88 kW. From curve drive loss = 15%. Drive loss P_1 = 0.15 X 0/88 = 0.13. Fan power input = 0.75 + 0.13 = 0.88 kW which is correct.

11.6 Installation notes for vee belt drives

1. Pulleys should always be fitted so that the effective centre of the belt or rope is as near as possible to the motor or fan bearing.
2. The load must not in any case be applied beyond the end of the fan or motor shaft extension.

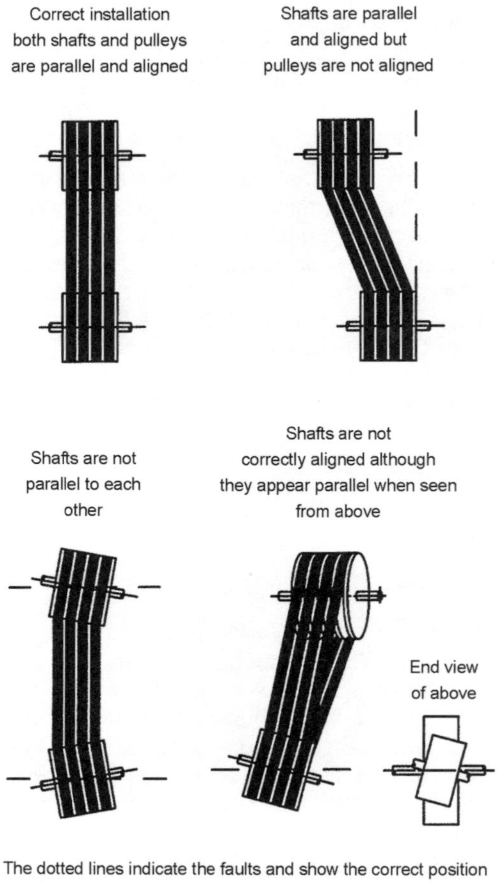

Figure 11.12 Frequent installation faults for vee rope drives

3. To avoid the danger of imposing excessive stresses, it is advisable to consult the fan and motor manufacturers for all drives on shafts above 48mm diameter.
4. It is recommended that only direct coupled drives be used for motors in sizes D160M and above at 2-pole speeds.
5. Clean all oil and grease from pulley grooves and bores.
6. Remove any burrs or rust.
7. Reduce the centre distance until belts can be placed in the pulley grooves without forcing.
8. Align the pulleys correctly using a straight edge to ensure that the pulleys are in line and the shafts parallel. (see Figure 11.12)
9. Tension the drive using the motor slide rail bolts.
10. Check that the vee belts are correctly tensioned (see Figure 11.13):
    a) Measure the span.
    b) Apply a force at right angles to the belt at the centre of the span.
    c) This force should deflect one belt 0.016 mm for every millimetre of span length.

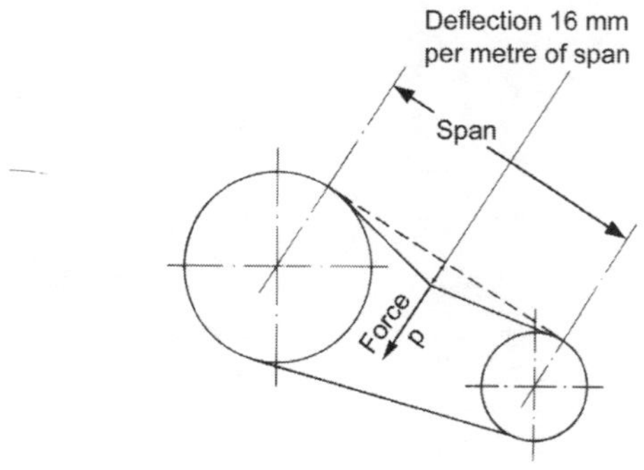

Figure 11.13 Belt deflection measurement

d) The average value of the force in each belt should be compared with Table 11.5 and should initially be tightened to the higher values. If the measured force falls within the values given in Table 11.5 the drive tension should be satisfactory. A force below the lower value indicates under-tensioning. When starting up, a new drive should be tensioned to the higher value to allow for stretch during the running in period. After the drive has been running a few hours the tension should be re-adjusted to the higher value. The drive should be re-tensioned at regular maintenance intervals. Make adequate provision for tensioning the belts during their life.

Belt section	Small pulley pcd (mm)	Belt speed		
		0 to 10 m/s	10 to 20 m/s	20 to 30 m/s
SPZ	95	12 to 18	10 to 16	8 to 14
	95	18 to 26	16 to 24	14 to 22
SPA	140	22 to 32	18 to 26	15 to 22
	140	32 to 48	26 to 40	22 to 34
SPB	250	38 to 56	32 to 50	28 to 42
	250	56 to 72	50 to 64	42 to 58
SPC	355	72 to 102	60 to 90	50 to 80
	355	102 to 132	90 to 120	80 to 110
Z	50		4 to 6	
A	75		10 to 15	
B	125		20 to 30	
C	200		40 to 60	
D	355		70 to 105	

Table 11.5 Correct vee belt tensions: required force N at centre of span for belt speed

To obtain kgf divide N by 10 to give the approximate value.

Note: These figures are reasonable for most applications but should be checked with the manufacturer for specific installations.

11.7 Bibliography

BS 1440:1971, *Endless V-belt drive sections* (withdrawn replaced by BS 3790).

BS 3790:1995, *Specification for endless wedge belt drives and endless Vee belt drives.*

Rubber Manufacturers of America, RMA IP20 *(Classical)*

DIN 2215, *Classical endless V-belts.*

ISO Standards for vee belt drives:

ISO 22:1991, *Belt drives – Flat transmission belts and corresponding pulley – Dimensions and tolerances.*

ISO 155:1998, *Belt drives – Pulleys – Limiting values for adjustment of centres.*

ISO 254:1998, *Belt drives – Pulleys – Quality, finish and balance.*

ISO 255:1990, *Belt drives – Pulleys for V-belts (system based on datum width) – Geometrical inspection of grooves.*

ISO 1081:1995, *Belt drives – V-belts and V-ribbed belts, and corresponding grooved Bilingual edition.*

ISO 1604:1989, *Belt drives – Endless wide V-belts, for industrial speed-changers and groove profiles for corresponding pulleys.*

ISO 1813:1998, Belt drives – *V-ribbed belts, joined V-belts and V-belts including wide section belts and hexagonal belts – Electrical conductivity of antistatic belts: Characteristics and methods of test.*

ISO 2790:1989, *Belt drives – Narrow V-belts for the automotive industry and corresponding pulleys – Dimensions.*

ISO 4183:1995, *Belt drives – Classical and narrow V-belts – Grooved pulleys (system based on datum width).*

ISO 4184:1992, *Belt drives – Classical and narrow V-belts – Lengths in datum system.*

ISO 5288:1982, *Synchronous belt drives – Vocabulary Trilingual edition.*

ISO 5290:1993, *Belt drives – Grooved pulleys for joined narrow V-belts – Groove section 9J, 15J, and 25J (effective system).*

ISO 5291:1993, *Belt drives – Grooved pulleys for joined classical V-belts – Groove section AJ, BJ, and DJ (Effective system).*

ISO 5292:1995, *Belt drives –V-belts and V-ribbed belts – Calculation of power ratings.*

ISO 5294:1989, *Synchronous belt drives – Pulleys.*

ISO 5295:1987, *Synchronous belts – Calculation of power rating and drive centre distance.*

ISO 5296-1:1989, *Synchronous belt drives – Belts – Part 1. Pitch codes MXL, XL, L, H, XH and XXH – Metric and inch dimensions.*

ISO 5296-2-1989, *Synchronous belt drives – Belts – Part 2: Pitch codes MXL and XXL – Metric dimensions.*

ISO 8370-1: 1993, *Belt drives – Dynamic test to determine pitch zone location – Part 1: V-belts.*

ISO 8370-2:1993, *Belt drives – Dynamic test to determine pitch zone location – Part 2: V-ribbed belts.*

ISO 8419:1994, *Belt drives – Narrow joined V-belts – Lengths in effective system.*

ISO 9563:1990, *Belt drives – Electrical conductivity of antistatic endless synchronous belts – Characteristics and test method.*

ISO 9608:1994, *V-belts – Uniformity of belts – Test method for determination of centre distance variation.*

ISO 9980:1994, *Belt drives – Grooved pulleys for V-belts (system based on effective width) – Geometrical inspection of grooves.*

ISO 9982:1998, *Belt drives – Pulleys and V-ribbed belts for industrial application – PH, PJ, PK, PL and PM profiles: dimensions.*

ISO 12046:1995, *Synchronous belt drives – Automotive belts – Determination of physical properties.*

ISO 13050:1999, *Curvilinear toothed synchronous belt drive systems.*

ISO Standards for chain drives:

ISO 487:1998, *Steel roller chain, types S and C, attachments and sprockets.*

ISO 606:1994, *Short-pitch transmission precision roller chains and chain wheels.*

ISO 1275:1995, *Double-pitch precision roller chains and sprockets for transmission and conveyors.*

ISO 1395:1977, *Short pitch transmission precision bush chains and chain wheels – Amendment 1: 1982 to ISO 1395:1977.*

ISO 3512:1992, *Heavy-duty cranked-link transmission chains.*

ISO 4347:1992, *Leaf chains, clevises and sheaves.*

ISO 6971:1982, *Welded steel type cranked link drag chains and chain wheels.*

ISO 6972:1982, *Welded steel type cranked link mill chains and chain wheels.*

ISO 10823:1996, *Guidance on the selection of roller chain drives.*

12 Shaft couplings

This Chapter sets out the factors which influence the relationship between shaft couplings and the fan unit. It includes a short review of the different types of coupling and continues with an explanation of the various types of misalignment and the forces and moments which are transmitted. Advice is given on “service factors” with special emphasis on the torque produced when starting electric motors. Several other factors are dealt with, and as shaft alignment is considered to be of importance, several different methods are explained. A check-list of important factors related to couplings is also included.

Contents:

12.1 Introduction

Chapter 9 showed that there are a considerable number of mechanical arrangements for fans, both centrifugal and axial flow. When looking at how the drive is transmitted from the prime mover to the fan impeller, it can immediately be seen that these can be resolved into three basic classifications:

- where the fan impeller is directly mounted on the motor shaft extension and thus runs at the motor speed.
- where the fan impeller is mounted on a separate shaft running in its own bearings and there is an indirect connection through belts, chains or gears to the prime mover.
- where the fan impeller is mounted on a separate shaft running in its own bearing(s) and there is a direct connection through a shaft coupling to the prime mover.

In this Chapter we are particularly interested in the last category. The coupling may be "rigid" or "flexible", transferring torque between two in-line, or nearly in-line, rotating shafts. Torque in the two shafts will of course, be equal in magnitude. If slipping or disengagement is possible however, there may be variations in speed. In its basic form the coupling is used as a simple way of joining shafts. Another requirement is to join two shafts which are not necessarily in perfect alignment with each other – indeed the author's experience is that they rarely are.

Perfection is not possible in this world and so the coupling must be capable of accommodating such misalignment. Modern couplings, between fans and their drivers, must be capable of rapid disassembly, especially in capital intensive plant where down-time can affect profitability.

It should be noted that coupling drives are invariably used on larger fans where the impeller is too heavy for the motor shaft or where vee belt drive would require lay-shafts and/or too many belts.

Shaft couplings can perform many different functions and have varying characteristics. They are usually divided into three main groups with sub-divisions, namely:

Non-disengaging couplings

- solid
- torsionally rigid
- torsionally flexible

Disengaging couplings

- clutch with manual over-ride mechanism
- free-wheeling clutches

Limited torque couplings

- non-controlled
- controlled and variable

Some of the requirements for flexible couplings, including definitions, performance and operating conditions, dimensions of bores, reference to components as well as an appendix on alignment are to be found in BS 3170. Friction clutches and power-take-off assemblies for engines, and their requirements are included in BS 3092. Process fans to API 610 Standard may have spacer couplings in accordance with API 671.

For fan applications it is common to use a coupling from the first group above, although special installations make use of disengaging clutches and limited torque couplings. Thus it is possible to incorporate centrifugal clutches to reduce starting loads when using a direct-on-line starting induction motor. Hydrodynamic clutches can be used for reducing starting loads and speed regulation. Combinations of brakes and reverse locks can be used to prevent reverse fan rotation.

Power recovery hydraulic turbines have been used in public utility and process fans when they have been coupled to the non-drive end of the fan motor so that the turbine can "unload" the motor.

The coupling used is a free-wheel type with manual over-ride so that the fan/motor can start-up before the turbine. Once the turbine runs up, as it tries to rotate faster than the motor, the clutch locks automatically and power is transmitted.

12.2 Types of coupling

Non-disengaging couplings maintain, after assembly, a more or less flexible but continuous transmission of the rotational movement. The connection is only broken for disassembly, repair, etc. Flexible couplings of one form or another, which are capable of absorbing residual misalignment, are most common; although solid couplings do have their areas of use, see Figure 12.1.

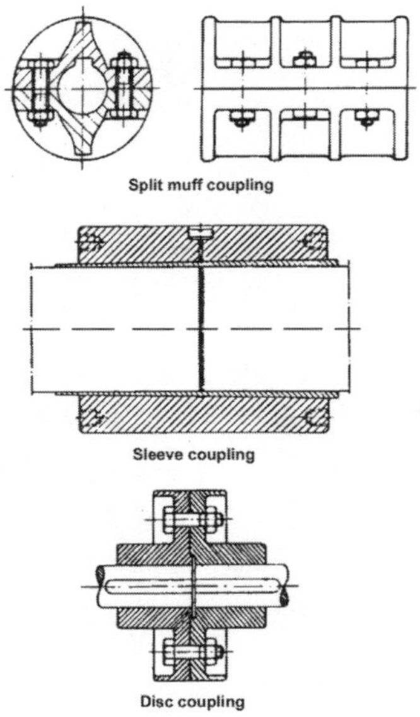

Figure 12.1 Examples of solid shaft couplings

One example is the split muff coupling, the main advantage being its ease of assembly. It is best used for low speed applications due to the difficulties in balancing. The sleeve coupling is mounted and removed by oil-injection; being almost symmetrical, balancing is easy.

In the early days of fan engineering rigid couplings were frequently used, as witness the Keith mine fan in Figure 1.21 in Chapter 1. However, extremely careful alignment was necessary if additional loads were not to be imposed on the fan or motor bearings.

It did, however, give the possibility of using only one fan bearing. Reference to Chapter 9, Figure 9.3 show that rigid couplings were used in arrangements 5 and 6 of the NAFM (USA) Bulletin 105. It is not without significance that these arrangements are now withdrawn. Fitters would nowadays have apoplexy if called upon to align three or four bearings!

Torsionally-rigid flexible couplings consist of various types of diaphragm and gear couplings, shown in Figure 12.2. Couplings with a single functional element have the ability to take up angular and axial misalignment. Couplings with two functioning elements separated by a fixed "spacer", are also able to cope with radial misalignment, whereby the magnitude of the radial misalignment is determined by the angular misalignment multiplied by the distance between the coupling elements.

Torsionally-flexible shaft couplings usually consist of flexible rubber, plastic or even steel elements, as in Figure 12.3. The first mentioned coupling elements require somewhat larger

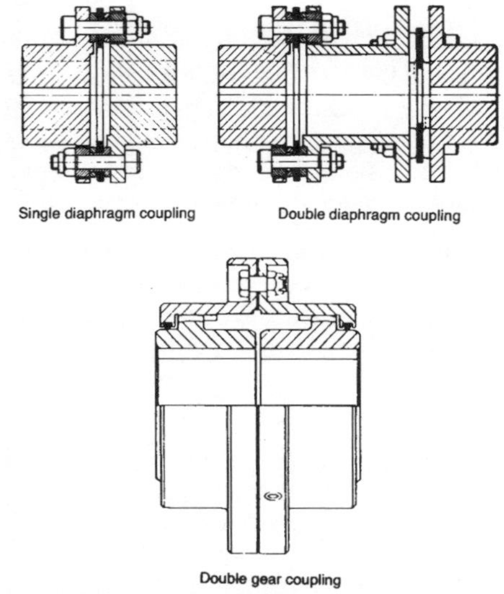

Figure 12.2 Examples of torsionally-rigid flexible couplings

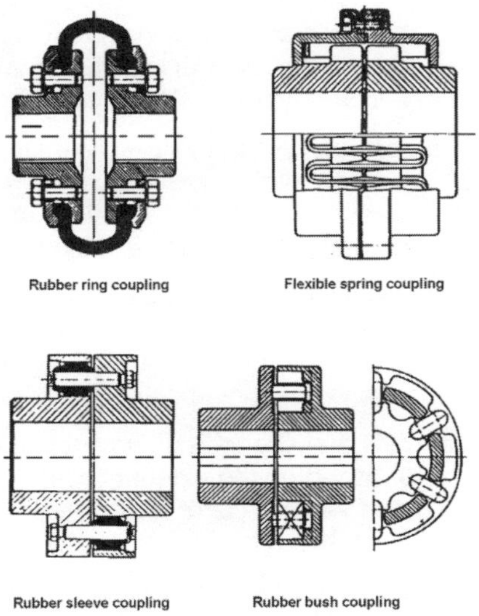

Figure 12.3 Examples of torsionally flexible couplings

coupling diameters because of their lower load carrying capacity. Single element couplings can accommodate radial misalignment as well as angular and axial. The flexible spring coupling is interesting because it is designed to have a variable torque/deflection characteristic. Together with dampening provided by the grease lubricant, the variable torque/deflection characteristic provides a powerful torsional vibration dampener.

The torsionally-flexible couplings shown can be built with two working elements and a spacer to allow additional radial misalignment. In order to simplify disassembly and service of some machines, spacer couplings can be used. An example of these is shown in Figure 12.4 b.

Removal of the spacer enables the rotating elements to be serviced without necessitating the removal of the whole machine. A limited end float feature is available for driving or driven machines not fitted with an axially located bearing as shown in Figure 12.4 a.

Cardan shaft couplings with rubber end stops as shown in Figure 12.4 c are also available.

12.3 Misalignment

Three types of movement or deviation can occur between two shafts, see Figure 12.5, namely:

- Radial misalignment, where the shafts are parallel although not lying on a common centre line.

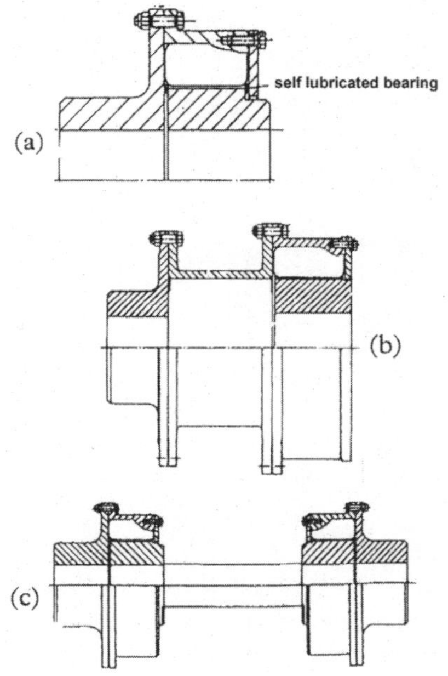

Figure 12.4 Shaft coupling examples

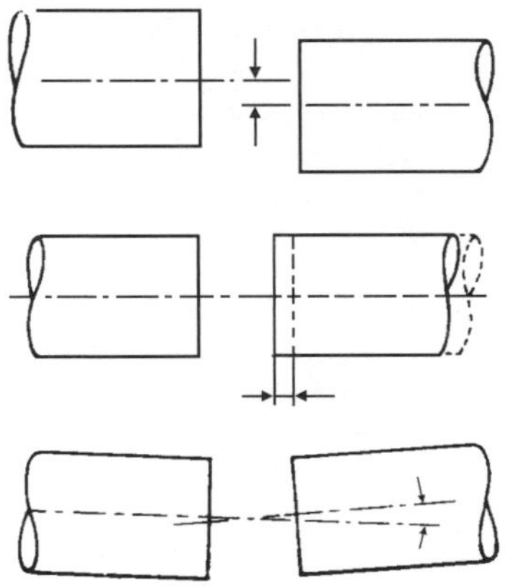

Figure 12.5 Types of misalignment

- Axial misalignment, end float, where the shaft centre lines are in alignment although the axial position is incorrect and axial movement may be possible.
- Angular misalignment, where the centre lines of the respective shafts are not parallel.

The deviations can occur singly or in combinations. Also the individual deviations can change with operating conditions. A typical changing condition is from cold to running temperature conditions. Thermal growth causes machine centre heights to increase slightly as they warm up.

High temperature fans may be centreline mounted to avoid thermal growth of the fan casing, and imposing strain on the connection ductwork. It might also lead to loss of clearance between the fan inlet cone and the impeller. However, the motor driving a centreline mounted fan is usually foot-mounted and may itself have thermal growth. In this situation motors are mounted low so that the growth expands the centreline height of the motor into near perfect alignment.

In large machines changes in ambient temperature or sunshine can affect the alignment. The thermal growth phenomenon can be further complicated when the drive and non-drive ends of a machine expand at differing rates. Not only does the radial alignment change, but also the angular alignment. Accurate on-line measurement is necessary to check for this condition.

Suppliers of couplings provide information relating to the maximum permissible deviations, usually stated for each individual

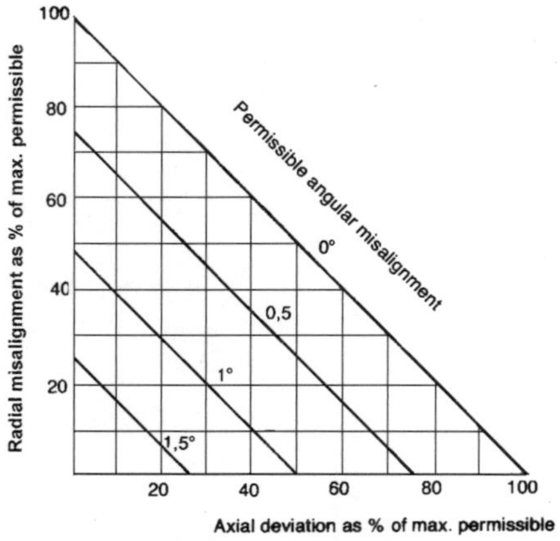

Figure 12.6 Permissible angular misalignment as function of axial deviation and radial misalignment for a particular size of double-diaphragm spacer coupling

type of deviation. It is important to know the maximum permissible values of combined misalignment, see Figure 12.6, and how the maximum permitted deviations are influenced by speed and the torque transmitted.

The service life of both couplings and machines, normally machine bearings, are influenced by misalignment. Just how much the life of the machine is affected can only be judged when information regarding the precise magnitude of the torque and forces transmitted due to misalignment is known. It is usual to refer only to the amount of misalignment permitted for a specific coupling type. But it is the amount of misalignment tolerated by the machine, Figure 12.7, which should really be investigated.

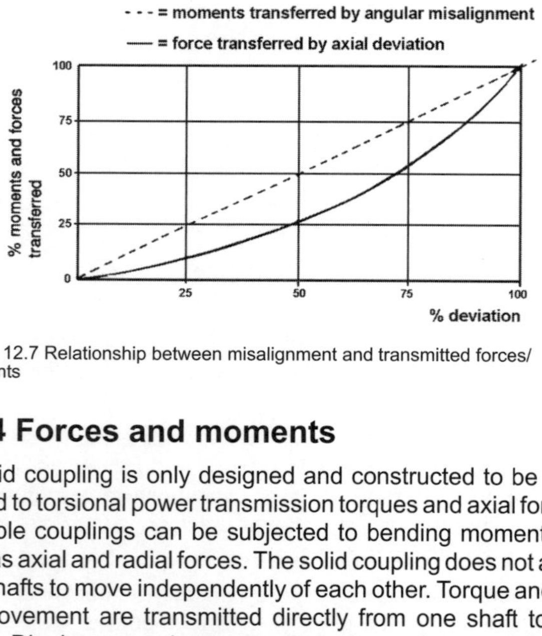

Figure 12.7 Relationship between misalignment and transmitted forces/moments

12.4 Forces and moments

A solid coupling is only designed and constructed to be subjected to torsional power transmission torques and axial forces. Flexible couplings can be subjected to bending moments as well as axial and radial forces. The solid coupling does not allow the shafts to move independently of each other. Torque and axial movement are transmitted directly from one shaft to the other. Diaphragm and gear couplings transmit torque directly but react differently to axial and radial movement.

- A diaphragm coupling allows the shafts to move axially and radially, the diaphragms are deformed, and both an axial force and a radial moment are generated.
- The double gear coupling also allows axial and radial movement. No axial force is produced, but a radial load is produced rather than a moment.
- A torsionally-flexible coupling produces radial loads rather than moments.
- The rubber ring coupling will produce an axial force when axial movement takes place, whereas the other types of coupling will slide to accommodate axial movement.

12.5 Service factors

When determining the size of flexible and solid couplings, it is usual to evaluate a so-called "service factor". The cynics amongst us would suggest that this is a euphemism and should more correctly be designated a "safety factor". It will cover our lack of knowledge of **all** the operating conditions.

Most coupling manufacturers publish nominal ratings for each of their products, together with lists of service factors for various applications. User groups also give advice and it is perhaps significant that those published by the American Petroleum Institute in its Standard No 613 are higher than those given by designers.

Drives with squirrel-cage motors and fans are usually stated by manufacturers to have a service factor of 1.0. However, it is wise to remember that where the absorbed power can vary, then this should be taking into account. Increased power can result from measurement uncertainties in the original base design manufacturing variations between nominally identical units, temperature variations in the gas/air handled, and whether the system resistance varies or has been incorrectly calculated (especially important with fans having a rising power characteristic e.g. forward curved bladed centrifugal fans).

To compare different couplings objectively a method has been developed which takes into consideration the frequency of starting, temperature, the moments of inertia of the driving and driven machine, normal torque and maximum torque.

This method has been presented in the German coupling Standard DIN 740, which, apart from the method of calculation, also contains dimensional standards. There are, however, two additional service factors which should be considered.

The first is the effect that shaft misalignment can have on the coupling. A factor based on the extent of allowable misalignment expressed as a percentage of the maximum permissible deviation, should also be given.

The second factor should take into consideration the level of vibration of both the fan and its driver.

Note that for fans, vibrational velocities above 5 mm/sec may well be permissible. In this respect the reader is referred to ISO 14694 (BS 848 Part 7) for the appropriate grades AN1 to AN4 and their corresponding balance quality grades.

The size of the various factors and their influence on coupling speed varies with different types, which is why the calculations and values given in DIN 740 must be used with a certain amount of caution and always with due regard to the suppliers' instructions, which must apply.

A very important point in this context, to which too little consideration is given, is the magnitude of the starting torque in the case of direct-on-line starting of a squirrel-cage induction motor. Measurements have shown that almost immediately after connection, approximately 0.04 s, a maximum torque is reached which is between 6-10 times the rated torque and even higher in some cases. This is a result of the electrical sequence in the actual motor and the fact that connection of the three phases does not occur absolutely simultaneously. The actual maximum torque is therefore much greater than the starting torque quoted in motor catalogues.

An important factor for coupling calculations is the relationship between the moments of inertia of the driving and driven machine. This quotient determines the percentage of torsional moment which is to be used for the acceleration of the motor and fan rotors. When starting, the torque passing through the shaft coupling is:

$$M_k = M_i\left(1 - \frac{J_{mo}}{J_{ma} + J_{mo}}\right) = M_i\left(1 - \frac{t_o}{t}\right) \qquad \text{Equ 12.1}$$

where:

M_k = coupling torque at start (Nm)

M_i = internal motor torque (air-gap torque) at start (Nm)

J_{mo} = moment of inertia of motor (kgm^2)

J_{ma} = moment of inertia of driven machine (kgm^2)

t_o = motor starting time without load (s)

t = motor starting time with load (s)

By inserting appropriate figures in equation 12.1 and assuming that M_i may be 6 to 10 times the rated torque, values for coupling torque at starting may be up to 4 times the rated torque for 4-pole motors and 8 times the rated torque for 6-pole motors. Care must therefore be taken when sizing couplings for fans which are started direct-on-line, especially when the fan has a large inertia.

12.6 Speed

Centrifugal forces increase with speed squared. The material of the coupling and the permissible peripheral velocities must be calculated. The maximum peripheral velocity for grey iron, for example, is 35 m/sec. To avoid vibrational damage it is necessary, for couplings which are not fully machined, to carry out both static and dynamic balancing at much lower speeds than those which are fully machined.

The mass of the coupling is often quite small in relation to the rotating masses in the driving and driven machines. For a fan unit the relationship of coupling/total rotor weight may be as low as 0.02. It therefore follows that out-of-balance in the coupling normally has less effect on bearings and vibration than out-of-balance in the actual main components. However the actual position of the coupling relative to the bearings may change this.

The following relationship applies:

$$F = m \cdot e \cdot \omega^2 \cdot 10^{-3} \qquad \text{Equ 12.2}$$

where:

F = out-of-balance force (N)

m = out-of-balance mass (kg)

e = distance from centre of rotation to centre of gravity of out-of-balance mass (mm)

ω = angular velocity (rad/s)

For highly resilient rubber element couplings with a spacer, the out-of-balance can be further increased by whirling. It is also important that balancing is carried out using whole keys, half keys or without keys, depending upon the method of balancing the attached component.

Example:

A fully-machined coupling can be assumed to have an inherent degree of balancing, without dynamic balancing, equivalent to G16 to G40, i.e. approximately 0.08 mm permissible centreline deviation at 3000 rev/min.

If the concentricity tolerance for the shaft bore in the hub is 0.05 mm, the maximum centreline deviation can therefore be 0.13 mm. This is not abnormal. In many cases the tolerance alone reaches this value. This centreline deviation generates an out-of-balance force of about 12 N per kg coupling weight at 3000 rev/min. A coupling for 50 kW can weigh 10 to 15 kg, which thus generates a rotational out-of-balance force of 120 to 180 N.

Most couplings have no components which can move radially to create out-of-balance forces. The gear coupling is different. The teeth on the hubs and the spacer must have clearance at the top and bottom; this allows the spacer to move radially. In theory, the angle of the teeth flanks should provide a centralising force to counteract any tendency for the spacer to run eccentrically. Problems have been experienced with gear couplings and special attention should be paid to radial clearances and spacer weight.

The flexible spring coupling has a spring which could move and run eccentrically. These couplings are usually used on fans running at speeds which are low enough not to have balance problems.

12.7 Size and weight

The importance of small size and low weight to achieve as little a moment of inertia as possible, as well as reducing the out-of-balance forces, has been mentioned previously.

In certain extreme cases light-alloy metal spacers and diaphragms are used to reduce weight. Apart from the need to maintain a small size/transmitted torque ratio, it is also important, from the cost and standardisation point of view that the coupling should be able to accommodate large variations in shaft diameter. Figure 12.8 shows the normal range of shaft diameters possible.

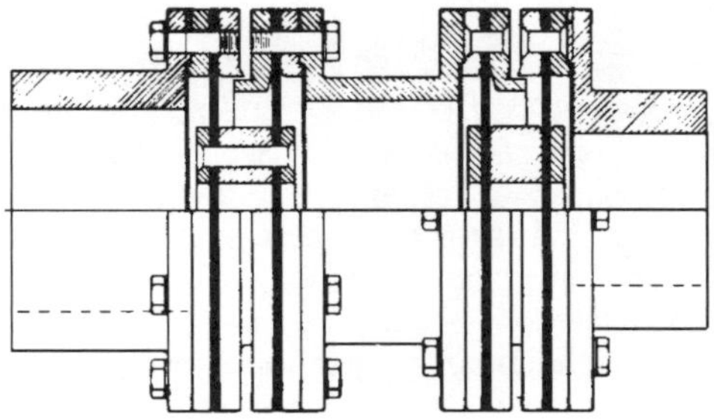

Figure 12.8 Non-sparking diaphragm coupling

12.8 Environment

Corrosive and abrasive environments affect the service life of the coupling by causing abnormal wear to the component elements. Extremes of heat and cold affect the strength and elasticity of the component materials. Oils, chemicals, sunlight and ozone can completely destroy a rubber element. A coupling made entirely of metal such as a diaphragm or flexible spring coupling, for example, is usually the only solution in such cases.

The process industries offer a very poor environment. In the petrochemical industry for instance, in refineries as well as oil and gas tankers, for example, it is necessary to use non-sparking couplings.

A non-sparking diaphragm coupling can be manufactured by making the diaphragm of Monel and the remaining components of carbon steel or bronze. Non-sparking types are usually used in conjunction with flameproof electric motors in environments where there is risk of explosion, either continuously or normally during operation. Statutory regulations must be observed, see also EN 14461.

A flexible spring coupling has the important elements housed in a seal cover and coated with lubricant, in the form of grease. Environmental changes have little effect on the coupling. Instances of spring breakage are rare, but any parts which could create a spark are fully enclosed, see Figure 12.9.

Another method of overcoming explosion risks, especially on board ship and with engine drivers, is by means of gas-tight bulkheads and bulkhead fittings consisting of two mechanical seals with barrier fluid between them, together with bellows which absorb misalignments. This type of fitting must be equipped with non-sparking shaft couplings.

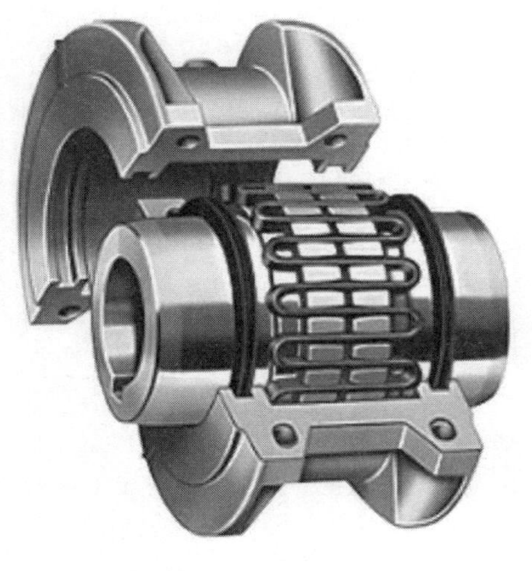

Figure 12.9 Cut-away view of a flexible spring coupling

Thread diameter mm d_1	d_2	d_3	d_4	t_1 $^{+2}_{0}$	t_2 min	t_3	t_4	Shaft journal diameter d_6
M 3	2.5	3.2	5.3	9	13	2.6	1.8	7-10
M 4	3.3	4.3	6.7	10	14	3.2	2.1	11-13
M 5	4.2	5.3	8.1	12.5	17	4	2.4	14-16
M 6	5	6.4	9.6	16	21	5	2.8	17-21
M 8	6.8	8.4	12.2	19	25	6	3.3	22-24
M 10	8.5	10.5	14.9	22	30	7.5	3.8	25-30
M 12	10.2	13	18.1	28	37.5	9.5	4.4	31-38
M 16	14	17	23	36	45	12	5.2	39-50
M 20	17.5	21	28.4	42	53	15	6.4	51-85
M 24	21	25	34.2	50	63	18	8	86-130

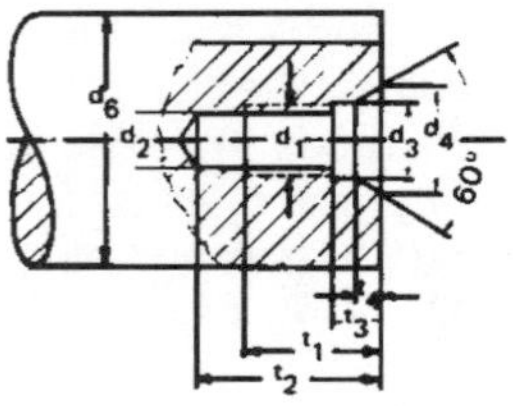

Figure 12.10 Tapped assembly hole in electric motor shaft

12.9 Installation and disassembly

To maintain maximum operational reliability and to simplify assembly and service it is important that the machines connected are securely mounted, preferably on a common foundation and baseplate. Guards must be fitted to rotating parts according to safety requirements, see Section 12.13.

Alignment of couplings or, more correctly, alignment of the shafts which the coupling is to connect, should be carried out as accurately as possible. For fans packaged on baseplates with their driver and other equipment, provisional alignment should be achieved by "chocking" the baseplate during levelling. After grouting, the alignment should be set correctly by adjusting the shims. A perfect alignment should be considered as an economic possibility, since alignment can considerably affect both service life and maintenance costs. See Section 12.11 with regard to methods of shaft alignment.

It is normal practice to bolt the fan directly to the baseplate. Other drive train equipment is shimmed to achieve correct alignment. In the case of cardan shafts the angular deviation should be equally distributed between the two joints to avoid unequal rotational velocities. Furthermore, a universal coupling should always rotate with a slight amount of angular misalignment to promote lubrication.

The attachment of a coupling half to a shaft usually presents a dilemma. The hub should be securely attached and preferably absorb part of the torque, to reduce the load on the key, as well as being easy to detach. The practice of hub attachment is similar to that for motor shafts where the fit is usually H7/k6, light push fit up to 48 mm diameter. A push fit H7/m6 is preferred for diameters above 55 mm. Some fan manufacturers prefer a positive interference fit, typically 0.001 mm per mm of shaft diameter. These couplings are heated for mounting and dismounting. Large couplings become unwieldy. Oil injection on shallow taper shafts, without keys, can be very successful.

The tighter fit is brought about by the fact that the height of the key is reduced from 12.5% of the diameter at 24 mm diameter to only 6% at 100 mm shaft diameter. This reduction should also be compensated for by increasing the length of the hub. In the case of electric motors the key does not normally extend right to the end of the shaft, which also increases the strain on the key. This must also be compensated for by increased hub length.

Assembly and disassembly of the coupling halves must be carried out carefully to avoid damage to the shaft ends and bearings. This operation could be simplified considerably if motor, fan and coupling suppliers fitted their equipment with suitable lugs, etc., to assist the attachment of pullers. For electric motors a tapped hole in the end of the shaft, as shown in Figure 12.10 can be supplied at extra cost, and ought to be standardised on all equipment.

Other methods of attaching the coupling halves are shrink fits, bolted joints or some form of clamping sleeve, Figure 12.11. Taper bushes are used primarily for vee belt pulleys, but can be a useful alternative for couplings where space permits. Some manufacturers offer taper bushes as an alternative to parallel bores. The hydraulically loaded clamping sleeve shown is a relatively new innovation and is not used extensively in fans.

The resilient elements in the shaft coupling must be easy to purchase, replace or repair. That it must be possible to replace without disturbing the machines or coupling hubs, goes without saying.

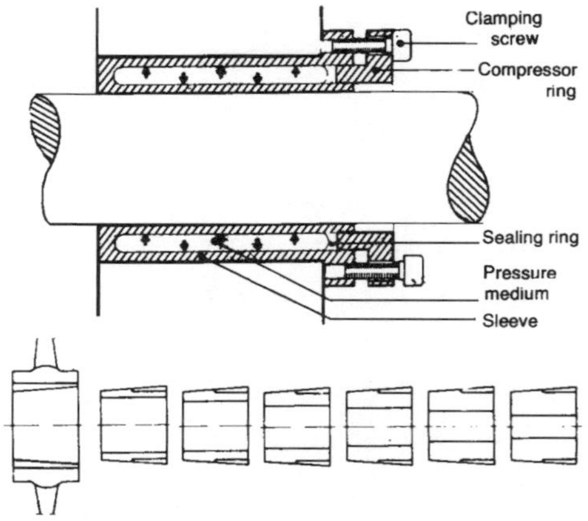

Figure 12.11 Examples of clamping sleeves

12.10 Service life

The life of the coupling is influenced by many factors, which vary according to the style of construction. One which above all affects couplings with rubber elements is the surrounding environment. The service life of a gear coupling is largely dependent upon regular lubrication using the correct type of lubricant according to the ambient temperature, etc. Flexible spring couplings are available with special grease which lasts five years, require almost no routine maintenance, and have no effects on the environment. Alignment affects the service life of all couplings irrespective of type or manufacturer.

For certain types of installations it can be desirable to use a coupling that allows a certain amount of emergency drive even in the event of failure of the flexible element. For other installa-

Method	Shaft coupling type	Measuring device and location	Zero setting and notation rules**	Parallel misalignment mm	Inclination* mm per 100 mm measured length	Remarks
I	Short shaft coupling. Machined outer diameter. Machined on insides	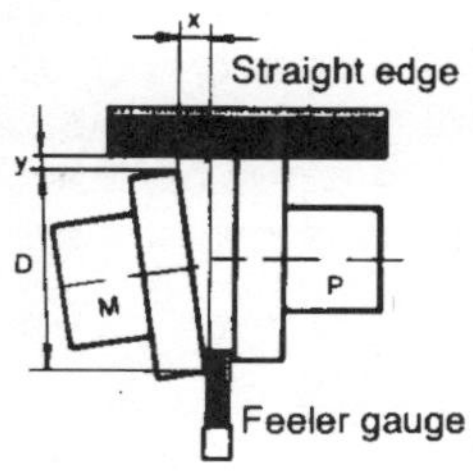	Misalignment according to the figure is positive i.e. the difference is measured above on the motor side.	Measured directly as dimension y	$L=\frac{100 \cdot x}{D}$	Make due allowance for bearing end float in the machines.
II	Short shaft coupling Requires at least a good surface at measuring pointer. Machined on insides	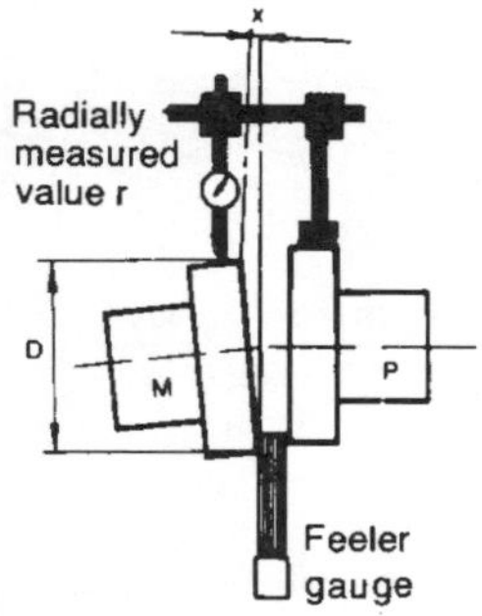	For vertical location zero set the dial above. Measured value is read after rotating one half turn.	$y=\frac{r}{2}$	$L=\frac{100 \cdot x}{D}$	Make due allowance for bearing end float in the machines. (Zero set the dial indicator underneath if the pointer is resting on the pump half.)
III	Short shaft coupling. Good surfaces at the measuring pointers	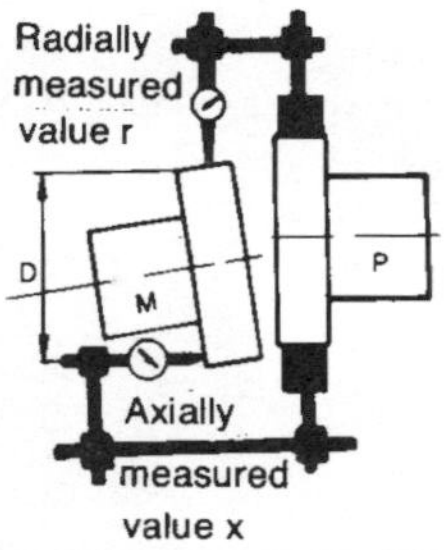	For vertical location zero set both dial indicators in the position shown, i.e. for radial deviation above and axial deviation underneath. The dials are read after rotating one half turn	$y=\frac{r}{2}$	$L=\frac{100 \cdot x}{D}$	Make due allowance for bearing end float in the machines. If both dial gauges are placed with their pointers on the pump half, then zero setting should be carried out from underneath.
IV	Long shaft couplings, i.e. couplings with a distance between the coupling halves. Good surfaces at the measuring pointers	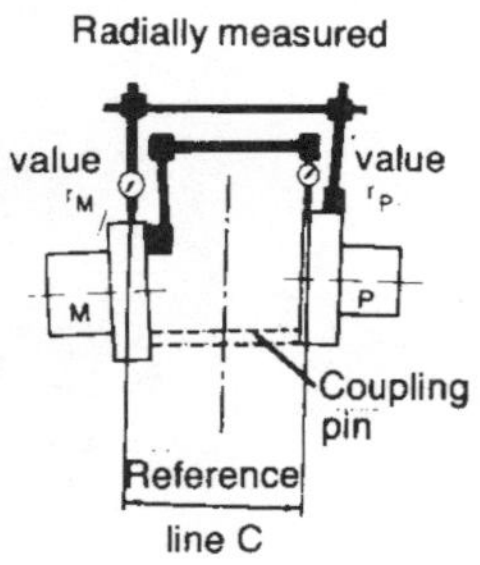	Zero set both dial gauges from above. The measurements r_M and r_P are read after rotating one half turn	$y=\frac{r_M-r_P}{4}$ ***	$L=\frac{r_M+r_P}{2 \cdot C} \cdot 100$	N.B. Measurement can also be carried out on "smooth" shaft ends.
V	Long shaft couplings, i.e. couplings with distance between the coupling halves. Good surfaces at the measuring pointers	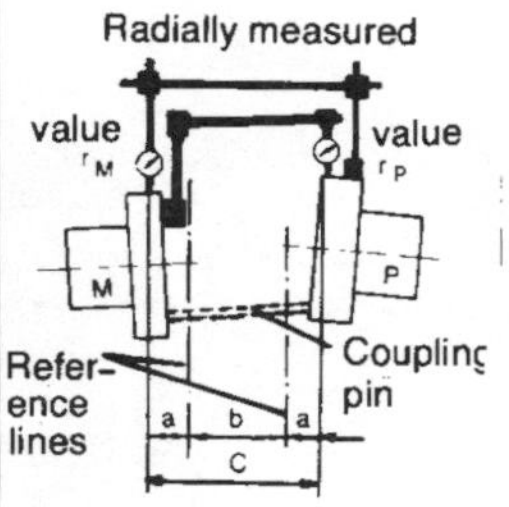	Zero set both dial gauges from above. The measurements r_M and r_P are read after rotating one half turn.	$y_M=\frac{r_M}{2}$ *** $y_P=\frac{r_P}{2}$	$L=\frac{r_M+r_P}{2 \cdot C} \cdot 100$	Similar to method IV. Notice the position of the reference lines for calculating angular misalignment.

Figure 12.12 Shaft alignment methods

tions it may be necessary to use a limited torque coupling with overload protection.

It is important to carry out regular service and alignment checks according to the manufacturer's instructions, and equally important that these instructions are placed in the hands of the personnel concerned. Unfortunately methods or regulations for assessing the degree of wear are often lacking.

12.11 Shaft alignment

12.11.1 General

Flexible shaft couplings are normally used to transfer torque between rotating shafts where the shafts are not necessarily in perfect alignment. It should be noted that a flexible coupling is not an excuse for poor alignment. Careful alignment is important for the purpose of achieving maximum operational reliability whilst reducing service and maintenance.

When carrying out alignment, consideration must be given to relative movements of the respective machines due to thermal expansion and deformation caused by pipe forces/moments and setting of baseplates on foundations, etc. In certain cases, such as electric motors with plain bearings, notice must be taken of the electric motor's magnetic centre. Alignment should be carried out at various stages during installation.

When alignment is carried out at cold temperatures, it is necessary to make allowances to compensate for the thermal expansion caused by the difference in temperature to that of the normal operating temperature of a fan and driver. When possible, a final check should be made at operating conditions after a few weeks in service. Alignment checks should then be carried out at regular intervals. Misalignment, apart from being caused by any of the previously mentioned loads and deformations, can depend upon worn bearings and loose holding down bolts. An increase in vibration levels can often be caused by a change in alignment.

Within the petrochemical industry and refineries, reports are frequently made with respect to alignment. The reports note the alignment prior to and after operation, before removing the fan or dismantling for repairs. The same procedure is carried out to check alignment of hot gas fans after warm running.

Correct alignment can be achieved in many ways depending upon the type of equipment and degree of accuracy required. Information regarding alignment requirements is usually to be found in the fan manufacturer's instructions.

Never use the limiting values for the coupling as given by the coupling manufacturer since they greatly exceed the values for machines if smooth running and long service life are to be achieved.

As a guide, a final alignment check should not produce greater parallel misalignment than 0.05 to 0.1 mm or an angular misalignment exceeding 0.05 to 0.1 mm per 100 mm measured length. For the definition of misalignment see Section 12.11.2.

Alignment is adjusted by means of brass or stainless steel shims, usually placed beneath the machine supports. Baseplates are generally machined so that a minimum number of shims are always required under the motor. Horizontal adjustment is performed by moving the machine sideways on its mountings. Large machines must have horizontal jacking screws fitted. Sometimes the fan and driver are fixed after final adjustment by means of parallel or tapered dowels.

12.11.2 Methods of alignment

In principle, alignment is based upon the determination of the position of two shafts at two points. Measurement or assessment can be made by straight edges, feeler gauges and dial indicators for the various radial and axial distances or run-out, see Figure 12.12. Adjustment is continued until these deviations are zero, or nearly zero.

12.11.2.1 Misalignment and reference lines

Two shafts in a vertical plane, for example, can display two deviations from their common centreline, namely parallel misalignment and angular misalignment, see Figure 12.13. The amount of misalignment at the flexible section of the coupling is that which is of interest. It is therefore appropriate to use a reference line which passes through the flexible section. Parallel and angular misalignments are then referred to this reference line, Figure 12.14.

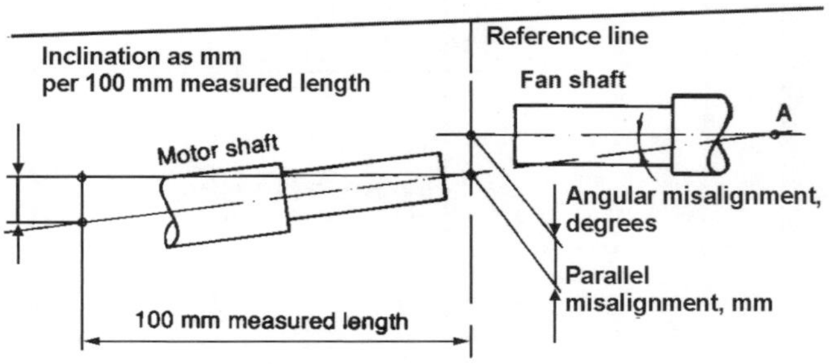

Figure 12.13 Misalignment of two shafts in a common plane

In Figure 12.13 it is important to note that if the reference line were to be chosen at the intersection point of the two centre lines of the shafts, point A, then only angular misalignment would exist. From a practical point of view angular misalignment is best measured as an inclination expressed as mm per 100 mm measured length rather than as an angular measurement in degrees.

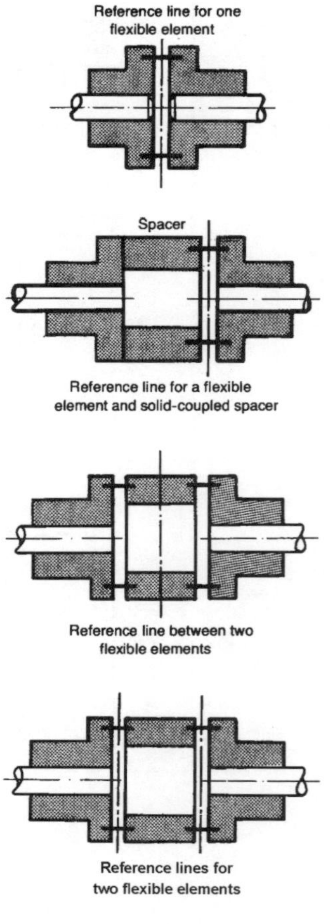

Figure 12.14 Location of reference lines for various types of coupling

The position of the reference line depends upon the type of coupling and naturally should always be located in relation to the flexible section of the coupling. For couplings with spacers and one or two flexible elements the position of the reference line is shown in Figure 12.14. Unless otherwise stated by the coupling manufacturer the permitted misalignment is considered to be that which is measured from the reference line.

12.11.2.2 Alignment procedure

In the case of a horizontal unit, alignment is best carried out by first aligning in the vertical plane, followed by transverse alignment. For vertical units alignment is measured in two directions at 90° to each other. For a horizontal unit, alignment is carried out in the following steps:

1. Align the machines visually and check that the coupling is not crushed in any way.
2. Attach the measuring device(s) and check that the dial indicator(s) moves freely within the area to be measured.
3. Check possible distortion of the motor mounting or baseplate by tightening and loosening each, holding down bolt individually. Shim the motor feet if distortion is present.
4. Set the dial indicator(s) to zero in the position shown in Figure 12.12.
5. For methods II, III, and IV in Figure 12.12, rotate both shafts simultaneously through 180°, half revolution, thus eliminating the influence of run-out between shaft bores and the outer diameter of a coupling half. The coupling halves need not then be cylindrical. Determine the measured values according to Figure 12.12. Note the measured values with plus or minus signs, see Figure 12.12 for notation. Determine parallel and angular misalignments.
6. Determine shim thickness according to Section 12.11.3 or 12.11.4 and adjust.
7. Carry out checks according to steps 4 and 5.
8. Carry out transverse alignment in the same manner as in the vertical plane.
9. Perform final alignment checks in both vertical and transverse directions and record for future reference remaining parallel or angular misalignments in both vertical and transverse directions. Also make note of operational conditions at the time of alignment, for example, cold motor with warm fan.

12.11.2.3 Choice of measuring method

Figure 12.12 shows the five most common measuring methods. From the point of view of accuracy it is difficult to compensate for manufacturing tolerances between the two halves of the coupling by using a straight edge and feeler gauge, method I.

The difference in accuracy between method III and method IV is determined by the differences in the dimensions D and C respectively. Accuracy increases in both cases as each respective dimension increases, whereby method III is chosen if D is larger than C and method IV or V is chosen if C is larger than D. The choice of method is also determined, apart from accuracy, by the available measuring surface and by attachment facilities and space requirements of the measuring devices.

The difference between methods IV and V lies in the location of the reference lines. Method IV is universally applicable and suitable for smooth shafts or where it is sufficient to measure the total parallel misalignment and inclination. In the case of a coupling with two flexible elements, method V is suitable if the angular misalignment for each element is first calculated individually.

Optical methods are also available. Light sources and mirrors are attached to each coupling half. The units are connected to a small dedicated portable computer which, when supplied with information regarding the feet position, will calculate the respective feet adjustments. Similar optical devices can be attached to machine casings to detect differential expansion when warming up.

12.11.3 Determination of shim thickness

Using the measured parallel and angular misalignment, the necessary shim thickness can be calculated directly. The misalignment is expressed as positive or negative, + or -, according to Figure 12.15, which shows positive misalignment.

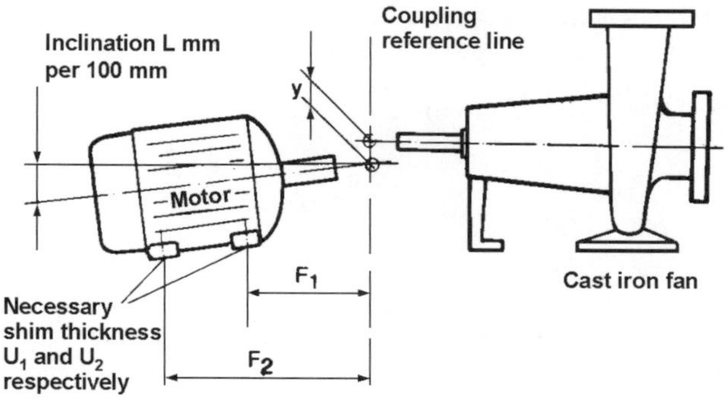

Figure 12.15 Positive misalignments y and L

The shim thicknesses are calculated from the simple relationship:

$$U_1 = y + L \cdot \frac{F_1}{100} \qquad \text{Equ 12.3}$$

$$U_2 = y + L \frac{F_2}{100} \qquad \text{Equ 12.4}$$

where:

U_1	=	shim thickness at foot 1 (mm)
U_2	=	shim thickness at foot 2 (mm)
y	=	signed parallel misalignment (mm)
L	=	inclination expressed as mm per 100 mm measured length
F_1 & F_2	=	distance in mm from coupling reference line to each respective foot, see Figure 12.15. The coupling reference line usually passes through the middle of the coupling.

Example:

Indicator reading shows parallel misalignment y = +0.28 mm and inclination L = -0.06 mm/100 mm.

The distances to the feet are F_1 = 300 mm and F_2 = 500 mm.

The shim thicknesses required are:

$$U_1 = 0.28 = -0.06 \cdot \frac{300}{100} = 0.28 - 0.18 = 0.10 \text{ mm}$$

$$U_2 = +0.28 - 0.06 \cdot \frac{500}{100} = 0.28 - 0.30 = 0.02 \text{ mm}$$

Shims of thickness 0.1 mm are placed under foot 1. The calculated value of U_2 = -0.02 mm means that 0.02 mm should be removed from foot 2, but can probably be accepted as permissible misalignment.

Equations 12.3 and 12.4 can also be combined so that parallel and angular misalignments can be determined in cases where it is not possible to fit the calculated shim thickness. In which case:

$$y = \frac{U_1 + U_2}{2} \qquad \text{Equ 12.5}$$

$$L = \frac{U_2 - U_1}{\frac{F_2}{100} - \frac{F_1}{100}} \qquad \text{Equ 12.6}$$

where:

y and L are residual misalignments

U_1 and U_2 respectively (with sign notation) are shim thickness deviations.

For the previous example, when the proposed correction has been carried out, the residual misalignment is:

$$y = \frac{0 - 0.02}{2} = -0.01 \text{ mm}$$

$$L = \frac{-0.02 - 0}{\frac{500}{100} - \frac{300}{100}} = -0.01 \text{ mm} / 100 \text{ mm}$$

12.11.4 Graphical method of determining shim thickness

The required shim thickness can also be determined graphically by drawing the position of the shaft in respect of the measured values using a greatly enlarged vertical scale, 100:1 for example, and a reduced horizontal scale, 1:5 or 1:10 for example.

The method is illustrated by the following example carried out according to measuring method IV or V in Figure 12.12 with the various stages:

1. Fit the measuring device according to method IV or V and take readings r_P and r_M on the dial gauge.

Example:

dial reading at fan half gives r_P = -1.40 mm

dial reading at motor half gives r_M = +1.20 mm

2. Determine the dimensions C, F1 and F2. Note that the reference line in this example has been chosen to pass through the measuring pointer as shown in Figure 12.16.

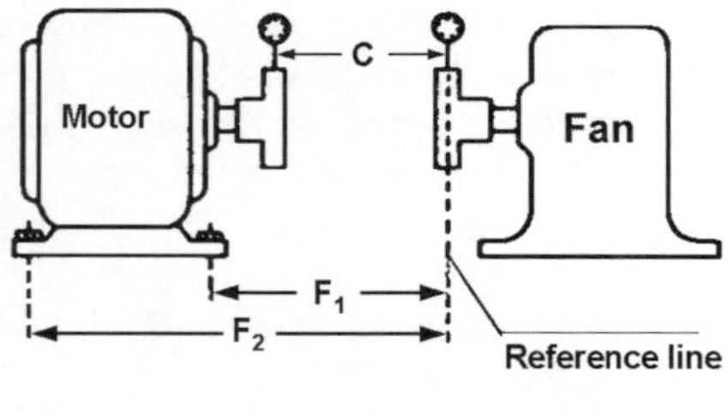

Figure 12.16 Length measurements and location of reference line

Example:

Measured results

C = 180 mm

F_1 = 470 mm

F_2 = 890 mm

3. Draw up a diagram on squared paper as shown in Figure 12.17. Mark in the dimensions C, F_1 and F_2 on the horizontal scale.

4. Mark half the measured value at the fan half, 0.5 r_P, on the vertical axis furthest to the right. The positive sign for r_P means that the motor shaft lies above the fan shaft and is marked upwards, whilst a minus sign is marked downwards. The reading r_P = -1.4 mm should thus be marked as -0.7 mm, i.e. downwards.

5. Mark half the measured value at the motor half, 0.5 r_M, at distance C. The reading's positive value means that the motor shaft lies below the fan shaft and should be marked as a minus value and vice versa for negative readings. The reading r_M = + 1.2 mm should thus be marked as - 0.6 mm, i.e. downwards.

6. Join both points and extend the line to the motor feet locations F_1 and F_2 respectively. The motor shaft shown in the example lies 0.44 and 0.21 mm too low at the respective foot locations and should be raised by shims of corresponding thickness, after which transverse alignment is carried out in the same manner.

7. The alignment can be checked simply by using the two measured values and r_M and the distance "b" between the two flexible elements. In the case of couplings with two flexible elements, only the total angular misalignment of each element should be calculated. Parallel misalignments are experienced as angular misalignments by the coupling.

To calculate angular misalignment, the parallel misalignment at the flexible element must be calculated first, i.e. calculated at both reference lines. These misalignments are:

$$h_p = \frac{r_p}{2} - \frac{a \cdot L}{100} \qquad \text{Equ 12.7}$$

$$h_M = \frac{r_M}{2} - \frac{a \cdot L}{100} \qquad \text{Equ 12.8}$$

The angular misalignment in the vertical plane is then determined from the relationship:

$$\propto_M = \frac{h_P}{b}(\text{radians}) = 57.3 \cdot \frac{h_P}{b}(\text{degrees}) \qquad \text{Equ 12.9}$$

$$\propto_p = \frac{h_M}{b}(\text{radians}) = 57.3 \cdot \frac{h_M}{b}(\text{degrees}) \qquad \text{Equ 12.10}$$

The angular misalignments in the horizontal plane β_M and β_P are calculated in the same way.

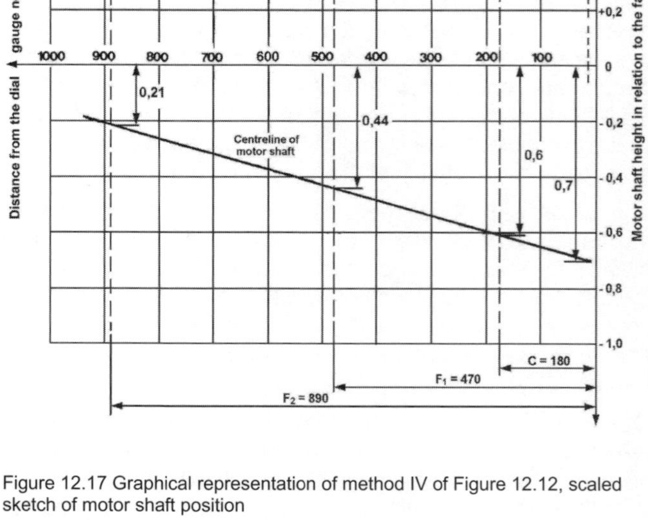

Figure 12.17 Graphical representation of method IV of Figure 12.12, scaled sketch of motor shaft position

Thereafter, the total angular misalignment, θ, per flexible element is calculated from the relationships:

$$\theta_M{}^2 = \propto_M{}^2 + \beta_M{}^2 \qquad \text{Equ 12.11}$$

or

$$\theta_P{}^2 = \propto_P{}^2 + \beta_P{}^2 \qquad \text{Equ 12.12}$$

12.11.5 Optical alignment

Recent advances in micro-electronics and laser technology have allowed optical alignment techniques to become portable and cost effective. A laser source is mounted on one shaft and a mirror is mounted on the other. The source module includes a detector which measures the position of the returned beam. The shafts are rotated incrementally through 90 and readings stored. A small control unit, sometimes small enough to be hand held, which is programmed with the drive geometry calculates the shim adjustment necessary to achieve good alignment. Figure 12.18 shows a typical set up for a small cast iron fan. Laser alignment can be used for shafts which are 10 m apart.

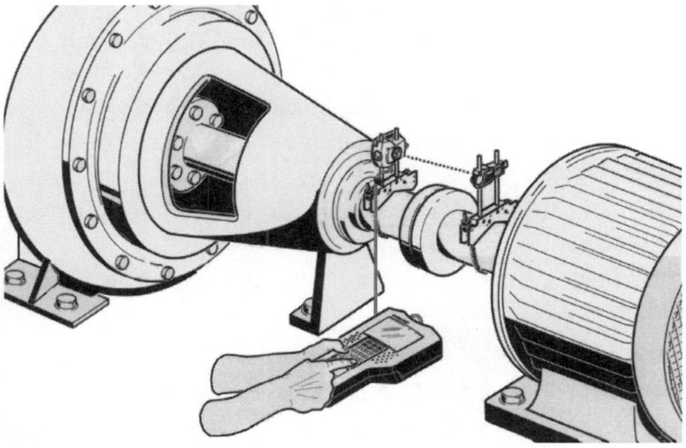

Figure 12.18 A typical laser alignment set up
Courtesy of Pruftechnic Ltd

Similar equipment can be attached to fan casings, gearboxes, motor stators or baseplate pads to monitor movement or deflection under operating conditions.

12.12 Choice of coupling

12.12.1 Costs

In general the cost per kW of a coupling is only a fraction of that of a fan or motor, a fan usually costing at least 30 times that of a coupling and a 4-pole electric motor at least 20 times. The cost varies according to the size and the type. The market for couplings is very competitive; the cost difference between manufacturers is usually small.

Gear couplings are the most costly. If a spray oil lubrication system is required this obviously increases the total cost considerably. Diaphragm and flexible spring couplings, together with the rubber buffer couplings, are about the same cost. Some of the rubber ring couplings are surprisingly expensive.

A good way to compare the cost of couplings is to set the price in relation to the torque and range of shaft end sizes to which the coupling can be fitted. The same fan shaft can, for example, be used for a torque range of 1:20 which occasionally means that the shaft end dimension and not the torque is used when selecting the size of a coupling.

Furthermore, the motor shaft may be larger than the corresponding fan shaft. The motor shaft may be dimensioned for bending stress to a greater degree than the fan shaft; for example a motor is often used for belt drive. This can also mean using a larger size coupling.

12.12.2 Factors influencing choice

It is important, not least of all from an initial cost point of view but also cost and space required for spare parts, to establish a viable internal standard by which a small number of type or style variations can cover the majority of coupling requirements within a company or plant. The factors reviewed in the check-list, Table 12.1 should be considered.

Factor	Influencing parameters
Type of coupling	Non-disengaging Disengaging Torque limitations Torsionally rigid Torsionally flexible
Type of movement	Radial and axial deviation Angular deviation
Forces and moments	Torsional moment Bending moment Axial and radial forces
Operational factors	Frequency of starting Connection frequency Operating time Ambient temperature Moment of inertia Method of calculation
Speed	Balancing Strength Throw protection (safety flange)
Size, weight	Shaft bore Space requirements Spacer for disassembly
Environment	Corrosive Abrasive Temperature Explosive (spark-free, flameproof)
Installation and disassembly	Horizontal and vertical shafts Alignment Fit
Others	Attachment facilities etc. for alignment measuring device. Replaceable wear elements Service life Routine maintenance Internal standard Costs Coupling safeguards

Table 12.1 Check-list for shaft coupling selection

For many centrifugal fans, the diaphragm spacer coupling has become the standard. These couplings are very reliable and can easily cope with the loads and speeds encountered in most situations. For higher speed applications, e.g. fans driven by steam turbines, the gear coupling is preferred by some users. Smaller fans operate better with a torsionally flexible coupling; flexible spring and couplings with rubber cushioning are favourites.

Users who have a large number of fans usually choose a single coupling manufacturer whenever possible. This philosophy increases the purchasing power of the user while reducing inventory requirements for spares.

12.13 Guards

The fan manufacturer is normally responsible for machine guards. In the case of standard fans, a distributor may package the fan with its driver and other equipment and it would become the distributor's responsibility to supply and fit guards.

Standard guards are generally made of painted steel. Sometimes aluminium is used because it is easier to bend and may not need painting. When fans are to be installed in a potentially hazardous environment special motors are used to reduce the chances of the motor igniting any gas present. A steel coupling rubbing on a steel guard could cause a spark and is not appropriate. Onshore, in these situations, an aluminium or bronze guard would be fitted. Offshore fans in potentially hazardous atmospheres have brass guards; the salt laden atmosphere offshore is not compatible with most aluminium alloys. Aluminium and brass guards would be described as "non-sparking" guards.

With high speed couplings the distinction between high and low speed is subjective. There is a remote chance that the coupling may fail physically and explode due to the centrifugal force acting on the pieces. It is generally thought that bolting is the weakness link and may be sheared due to an unforeseen overload. If the coupling is not "burst-proof", see Figure 12.19, then the guard must be capable of retaining any scattered material.

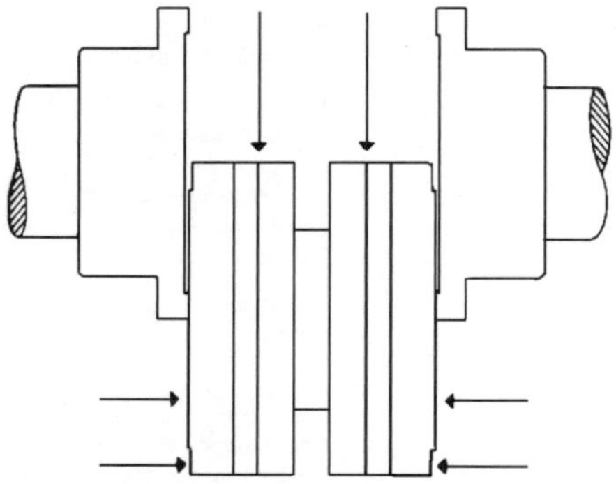

Figure 12.19 Burst-proof diaphragm coupling with spigotted spacer

Within Europe, the safety of machinery in general is covered by the Machinery Directive which is implemented by EN 292, Safety of Machinery. The safety of fans is covered by prEN 14461. Guards are specifically regulated by EN 953, Safety of machinery; general requirements for the design and construction of guards (fixed, movable).

Other interesting safety Standards worth reviewing include BS 5304, DIN 31001, ANSI B15.1 and OSHA coupling guard requirements.

12.14 Bibliography

ISO 8821:1989, *Mechanical vibration - Balancing - Shaft and fitment key convention.*

ISO 5406:1980, *The mechanical balancing of flexible rotors.*

BS 6861-1:1987, ISO 1940-1:1986, *Mechanical vibration. Balance quality requirements of rigid rotors. Method for determination of permissible residual unbalance.*

VDI 2060 Q40, *Dynamic balance of rotating bodies which include propshafts (for shafts with slight wear).*

NFE 90600 (France), *Balance Class, Flexible couplings.*

ANSI/AGMA 9000-C90 (R2001), *Flexible Couplings - Potential Unbalance Classification.*

ANSI/API 671, *Special-Purpose Couplings for Petroleum, Chemical, and Gas Industry Services*

ANSI/API 613, *Special Purpose Gear Units for Petroleum, Chemical and Gas Industry Services*

DIN 740, *Power transmission engineering; flexible shaft couplings; technical delivery conditions*

EN292, *Safety of Machinery - Principles of Design*

Mekanresultat 72003, *Shaft couplings*, Product information issued by the Swedish Association for Metal Transforming, Mechanical and Electro-mechanical Engineering Industries.

Couplings and Shaft Alignment, M Neale, P Needham, R Horrell, - Professional Engineering Publishing, ISBN 1860581706

ISO12499:1999, *Industrial fans - Mechanical safety of fans -Guarding.*

ISO14694:2003, *Industrial fans - Specifications for balance quality and vibration levels.*

prEN 14461, *Industrial Fans - Safety requirements.*

AMCA 202-1998, *Trouble-shooting.*

AMCA 240-1996, *Laboratory Method of Testing Positive Pressure Ventilators for Rating.*

BS EN 953:1998, *Safety of machinery. Guards. General requirements for the design and construction of fixed and movable guards.*

DIN 31001-1, *Safety design of technical products; Safety devices.*

OSHA 1910.211, *Occupational Safety and Health Standards - Machinery and Machine Guarding.*

13 Prime movers for fans

The majority of fans are driven by an electric motor, the squirrel cage induction type being the most popular, except in the smaller sizes. This Chapter points the user to the selection of appropriate types of prime movers for fans, and also describes starting and running characteristics.

Just as important to the selection of the correct motor type is a knowledge of how the power absorbed by the fan varies with time, temperature and barometric pressure. The inertia of the impeller may be significant and will affect both the motor type and its control.

Contents:

13.1 Introduction

The majority of fans are driven by a separate electric motor. There are some exceptions to this general statement e.g. so called "inside out" electric motors may incorporate the fan impeller within their overall construction. It would then be difficult to separate the fan impeller from the (rotating) motor stator without a major de-construction.

Furthermore, fans driven by internal combustion engines are not unknown in the agricultural and marine industries. The public utilities, especially, use fans driven by steam turbines.

The type of fan and the energy sources available can have an important influence on the choice of driver. Fans can vary from very slow speeds (e.g. forward curved centrifugals) to very high speeds (e.g. narrow backward bladed high pressure fans). To develop any worthwhile pressure, axial fans also need to run at high peripheral speeds.

The most efficient fan and control systems will be directly driven, obviating any transmission losses, but this assumes that the operating conditions can be correctly calculated. As the demand for energy saving increases, variable speed transmissions become ever more popular in a successful fan system.

For mains-fed motor applications, induction motors and electronically commutated (EC) motors mainly are used. Switched reluctance motors have not been used in the past because of their poor noise behaviour. However, significant improvements are now being made.

Universal motors are series commutator motors able to work from AC and DC supply. The commutator and the carbon brushes produce electrical interference, acoustic noise and limit motor life expectancy significantly. Therefore, this type of motor has not been used in a large number of applications.

Squirrel-cage induction motors, as well as EC motors, have only the bearings as a wearing part. They therefore have a high lifetime expectancy.

EC motors have some important technical advantages: wide speed range, easy speed controllability and high efficiency. However, because of the higher price of mains-fed EC motors, AC induction motors will remain a considerable part of the market, where low cost positioning is important.

For higher power, 3-phase induction motors are often used. For single phase supply, shaded-pole motors and capacitor-run motors can be utilized. An induction motor with only one phase winding does not have a rotating magnetic field. The single winding, fed with AC, simply produces a pulsating flux in the air gap.

The motor will not start from rest. The start can be achieved by using the principle of shaded-pole motor or with an auxiliary winding. The stator of a shaded-pole motor is slotted to receive the shaded ring which is a single short-circuited turn if thick copper or aluminium. The time variant stator flux induces a voltage which causes a current in the ring. The phase-lagged magnetic field of this current produces together with the main flux of the motor a starting torque.

Capacitor-run or also called permanent split capacitor (PSC) type induction motors are squirrel-cage induction motors with two windings. The current in the second ("auxiliary") winding is supplied from the same single-phase source as the main winding, but a series capacitor caused to have a phase-lag. In that way, a rotating magnetic field is generated which makes possible an adequate starting torque and a higher efficiency.

Single-phase induction motors are robust and reliable; especially shaded-pole motors are very inexpensive. However, shaded-pole motors tend to have low power density and poor efficiency because part of the active pole is permanently short-circuited. For example, a shaded-pole motor with 10 W nominal output power only has an efficiency of 24%. Capacitor-run motors are more efficient (35-40% at the same output power). Further advantages are favourable acoustic behaviour and a power factor (cos φ) approaching unity (1.0).

13.2 General comments

Fans may be driven by a varied range of machines, as indicated in Section 13.1.

The most common are:

- fixed speed electric motors of the synchronous and induction types
- variable speed electric motors
- steam turbines
- internal combustion engines of the petrol, diesel oil or gas types

If a suitable supply of steam is available, for example where steam is produced in a power station or is a by-product of an industrial process, a steam turbine driver may well be the most appropriate choice. It has the advantage of being easily adjusted to a variable speed, resulting in a more efficient method of providing an output matched to demand.

If a suitable steam supply is not available e.g. domestic or commercial buildings, agriculture etc., etc., then the most reliable and economical form of driver is invariably an electric motor, provided of course that an adequate and sufficiently robust source of electricity is present.

The most reliable type of electric motor is generally accepted to be the induction design. This rotates at a little below synchronous speed which for a two pole machine running on a 50Hz AC supply limits the maximum speed to something just less than 3000 rev/min or 3600 rev/min on a 60Hz AC supply.

Some fans may need to operate at speeds in excess of this, in which case a speed increase belt drive or a step-up gearbox may be necessary. An alternative is to convert the supply to a much higher frequency e.g. 400 Hz when much higher speeds are possible.

The driving motor should in all cases be sized to provide the power demanded by the fan impeller plus any losses in bearings, vee belt drives etc. As far as the power supply is concerned, it will be necessary to provided for additional losses in the electric motor itself together with losses in the control gear.

The driver should also be sized to provided the power required by the fan, its bearings and transmission under all expected operating conditions with a suitable margin to cover:

- uncertainty or inaccuracy in the definition of the fan duty
- variation in the fan duty due to changes in air/gas density
- deterioration in the fan performance due to erosion, corrosion or dust build-up
- uncertainty in the measured performance
- variation between a prototype and a production machine due to manufacturing tolerances
- deterioration in performance of the driver such as gradual breakdown of electric motor insulation or fouling and erosion of a steam turbine
- variations in the energy source e.g. power supply voltage or steam pressure

The likely magnitude of this margin may need to be considered in detail. A minimum recommendation, which is a reasonable approximation for most centrifugal fans cases, is given in Table 13.1.

Impeller type	Width		
	Narrow	Medium	Wide
Backward inclined	14%	10%	7%
Backward curved	8%	7%	5%
Aerofoil	8%	6%	5%
Forward curved	20%	17%	15%
Shrouded radial	14%	12%	12%
Radial tipped	16%	14%	12%
Open paddle	14%	12%	12%
Backplate paddle	14%	12%	12%

Table 13.1 Approximate margins to be added to absorbed power

13.3 Power absorbed by the fan

This will be obtained from the duty requirements of air/gas volume flow, pressure to be developed, and known air/gas conditions at fan inlet. It is also necessary to consider how all these factors may vary during fan operation.

For example, it is usually difficult to assess accurately the fan pressure. The system designer often therefore adds a "safety margin" to his calculated pressure to ensure that he achieves the design flow. If he can subsequently add in additional resistance by orifice plates or similar to bring the flow back to specification then there will be no problem. Alternatively he may be able to partially close a damper in the system to dissipate the unwanted pressure. If this is not possible, and the speed cannot be changed, then the fan will handle more air and this may affect the power consumption.

With "non-overloading" fans fitted with backward inclined backward curved, or aerofoil, the volume flow against power curve is relatively flat over the working range, i.e. an increase in capacity with reduced pressure has only a small effect, if any, on the power absorbed. With impellers having blades radial at the outlet, i.e. shrouded radial, open paddle, backplate paddle, and radial tipped, the power increases uniformly with capacity.

The forward curved impeller has a flow versus power curve, which increases ever more rapidly towards the "free air" or zero pressure condition. For this reason it is suggested that the margins given in Table 13.1 be added to the fan absorbed power, simply to cater for the normal inevitable errors in system resistance calculations.

Where the system resistance is accurately known, or where a small loss of capacity is acceptable, then it may be possible to reduce these margins.

It is also important to know if the power absorbed can vary with time. In a ventilating system with a fan handling "outside" air the only variation will be that due to a variation of air density with changing barometric pressure or ambient temperature. Calculations of both fan duty and system resistance are normally made under "standard" conditions i.e. with air having a density of 1.2 kg/m^3. Typically this would correspond to dry air at a temperature of 20°C and a barometric pressure of 101.325 kPa. Alternatively air at 16°C temperature, 100 kPa barometric pressure, and 62% relative humidity also has the same density. Between summer and winter there will be variations in both temperature and barometric pressure, and these will affect the air density. Typically temperature could fall to -3°C (270° K) and barometric pressure could rise to 105kPa. The effect on air density would then be:

$$1.2 \times \frac{105}{101.325} \times \frac{273+20}{273-3} = 1.35 \text{ kg/m}^3$$

i.e. an increase of 12%.

If such variations in conditions do occur, then the necessary margin must be allowed. A possible alternative is for the motor to be "overloaded" for short periods of time. This is not necessarily a danger, as motor performance (usually determined by winding temperature) can improve at low temperatures.

A more important case of varying temperature would be for hot gas fans where the starting condition could be with ambient air, but the normal condition is at a reduced gas density. The motor may have to be rated to cover the higher horsepower, although where the working temperature is rapidly achieved, the margin can be minimal. Often in such cases a damper is incorporated in the system. This is closed either fully or partially on start-up and opened when the temperature is achieved. The fan motor need then only be rated to cover the hot gas conditions, provided the power with damper closure is materially lower. An example will illustrate the problem.

13.3.1 Example of a hot gas fan starting "cold"

A fan has an absorbed power of 75 kW when handling gas at a temperature of 325°C. It is started on air at 20°C with the gas-tight damper in the system fully closed. Reference to the fan characteristic curve shows that the power at zero flow is 35% of that at the rated flow.

$$\text{Power at start up} = 75 \times \frac{273+325}{273+20} \times \frac{35}{100} = 53.4 \text{ kW}$$

If the fan had been started on air at 20°C with a fully open damper, the power would have been:

$$75 \times \frac{273+325}{273+20} = 153.1 \text{ kW}$$

The power at zero flow is a function of the fan design. Generally the narrower the fan, the lower will be the percentage of maximum. Backward bladed fans have a higher zero flow power than forward curved, with radial intermediate. If the percentage was 50% then the power at zero flow would be:

$$75 \times \frac{273+325}{273+20} \times \frac{50}{100} = 76.5 \text{ kW}$$

This is higher than the duty power. At intermediate flows, the power being a greater percentage of maximum, care will need to be taken to ensure that the temperature has risen sufficiently. If not, the power absorbed could rise significantly above the start-up and duty conditions. The motor will need to be rated for the highest power consumption.

It should be noted especially that many dampers are not completely gas tight and allow a flow even when fully dosed. This may typically be of the order of 5% to 10% of the rated flow. The power under these conditions can be significantly higher than at zero flow, dependent on the shape of the fan/power characteristic. Reference to the curves is therefore recommended. There is also an additional power loss in the transmission, be it a belt drive or coupling. This is discussed in Chapter 11.

13.4 Types of electric motor

It is not the intention of this Chapter to be a comprehensive guide to the various types of electric motor. *Guide to European Electric Motors, Drives and Controls* gives a detailed description of the whole electric motor market and the variants available. Performance characteristics, design features and accessories such as starters are all described.

However a brief resumé of the most popular types used with fans is included for completeness.

13.4.1 Alternating current (AC) motors

Motors for alternating current fall into two main groups:

- induction motors
- all other types

From the point of view of characteristics, induction motors are similar to direct current (DC) shunt wound motors and are said to possess shunt characteristics. They are inherently constant speed machines, which run at just a little lower than synchronous speed for the supply frequency and the number of poles on the field of the machine. The difference between the actual running speed and synchronous speed is known as the "slip". A further rather important point about induction motors is that although poly-phase machines will start without assistance, single-phase induction motors are inherently non-self-starting. This is the reason for the many different types of single-phase motor.

The relationship between poles and speeds of alternating current motors is given in Table 13.2.

Frequency	40 cycles		50 cycles		60 cycles	
No. of Poles	Speed – r.p.m.		Speed – r.p.m.		Speed – r.p.m.	
	Synchronous	Nominal approx.	Synchronous	Nominal approx.	Synchronous	Nominal approx.
2	2400	2240	3000	2800	3600	3350
4	1200	1120	1500	1400	1800	1670
6	800	720	1000	900	1200	1080
8	600	560	750	700	900	830
10	480	455	600	570	720	685
12	400	375	500	470	600	565
14	343	320	430	400	514	480
16	300	290	375	360	450	430

Table 13.2 Relationship between poles and speeds of alternating current motors

Apart from synchronous motors (which run exactly at synchronous speed) and induction motors, all other types of AC machines may be said to possess series characteristics and are not limited to speeds dependent on the supply frequency.

However, the majority of AC fan drives are performed by induction motors, as they are more reliable and generally require less attention than other types of AC machines. Invariably they are also less expensive. Any speed tolerances quoted in this section for induction motors assume exact maintenance of supply frequencies, and since supply systems are often heavily loaded an additional tolerance of plus or minus 4% may easily arise from this cause.

13.4.2 3-phase motors

13.4.2.1 Squirrel cage induction motors

These consist of a stator wound normally for 3-phase supply and with a rotor of squirrel cage construction, (see Figure 13.1). They are essentially a constant speed drive, but motors specially designed for fan drives may be arranged to give speed regulation of up to about 50% of normal speed by means of voltage reduction. Pole-changing motors are available giving two speeds in the ratio of 2 to 1 by re-connection of the stator windings. Alternatively, multiple-wound stators provide two or occasionally more speeds in any ratio.

This type may be purchased in sizes up to quite large powers. Low kilowatt machines, up to about 4 kW may generally be

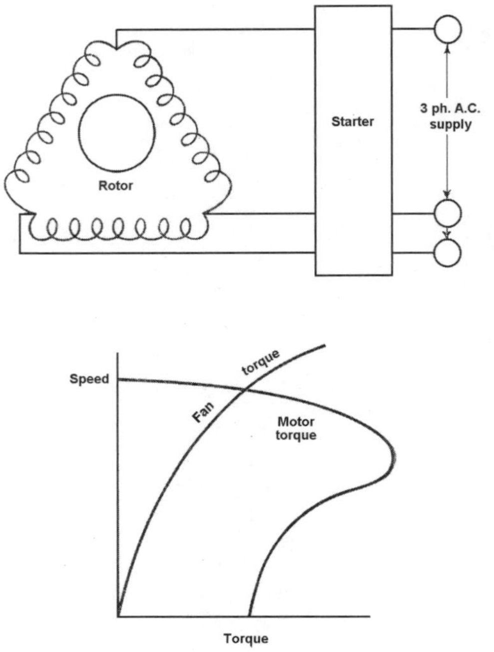

Figure 13.1 3-phase AC squirrel-cage induction motor

started direct-on-line. For greater powers the following two main methods are used for starting:

1. The voltage is reduced by means of a resistance or auto-transformer (usually wound in open delta for economy). The machine is generally started on light load, as the starting torque is reduced when the voltage is reduced.
2. Star-delta starting is used quite often on moderate power. This is achieved by arranging that the motor has the end connections of each winding brought out to six terminals. The machine is designed to run normally with its winding connected in delta, that is, with each winding connected to the full supply voltage. During the starting period, however, the windings are connected in star by means of a special switch, which in effect reduces the voltage across each winding to about 57% of the supply voltage and consequently reduces the starting current drawn from the mains to one-third of that for direct starting. When the machine is running close to full speed the switch is operated and the machine is delta-connected for running, thus putting full voltage on each of the windings. There is no radio interference from this type of machine.

Important note: Induction motors may also be used as variable speed machines by altering the frequency of the AC supply. This is best achieved by the use of an inverter, a method which has now received universal acceptance. The method is discussed more fully in Chapter 5.

Typical characteristics of squirrel-cage induction motors:

kW range	0.25 to 100
Starting torque	150% to 250% of full load torque
Starting current	6 to 8 times full load current
Power factor	0.8 to 0.9
Speed tolerance	± 5% for small sizes and low speeds
	± 2% for larger sizes

13.4.2.2 Wound-rotor induction motors

These machines are different from the squirrel cage induction motor in that the rotor is wound, and the end of the windings brought out to slip rings. (See Figure 13.2.) They are inherently speed regulating machines, this being achieved by adding resistance to the rotor circuit via the slip rings. They make excellent fan drives, particularly when volume regulation is required, the range of speeds obtainable being virtually from standstill to

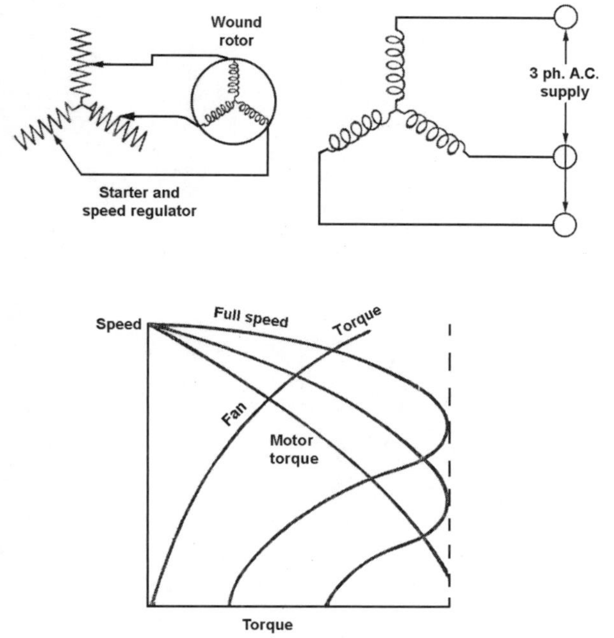

Figure 13.2 3-phase AC wound-rotor induction motor

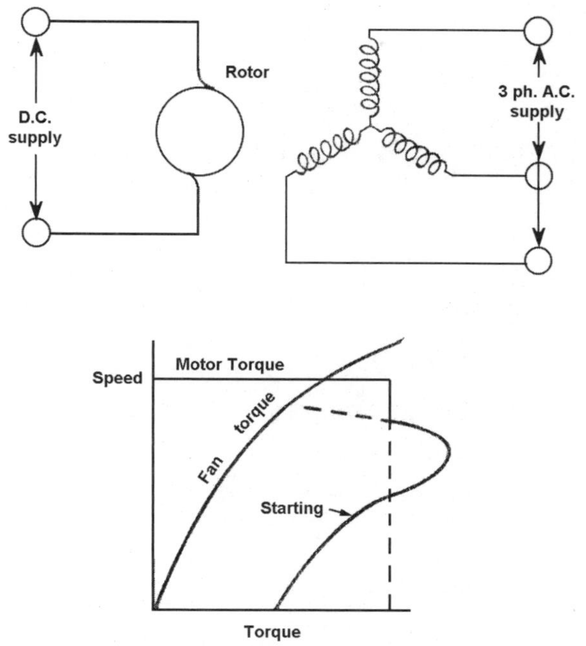

Figure 13.3 3-phase AC synchronous induction motor

full speed of the machine. However, in order to keep the speed regulator to economical proportions, it is usual to regulate from full speed down to about 50% of full speed. They are available in any size, though machines of larger powers are more common because of the comparatively high expense of the lower power machine compared with other types of AC motor of similar horsepower.

In order to limit the current on starting, the machines are usually arranged to start at the lowest speed position of the speed regulator and interlocks are normally fitted to ensure that this occurs. Starting currents may be kept down to 1.5 times full load current. There is generally no radio interference from these machines, but some may be experienced if the slip rings and collectors are allowed to get into poor condition.

Typical characteristics of wound-rotor induction motors:

kW range	5 to 1000 and over
Starting torque *	150% to 300% of full load torque
Starting current *	1.5 to 3 times full load current
Power factor	0.7 to 0.9 according to degree of speed regulation
Speed tolerance	± 2% at full speed

*at lowest speed

13.4.2.3 Synchronous induction motors

Synchronous motors are rarely used for fan drives, except where power factor correction is necessary for a large continuous-running fan installation. The leading power factor current drawn by the synchronous motor compensates for the low power factor of other installed electrical equipment. Synchronous motors usually have field supplied by AC, while the rotor is supplied by DC generated by an excitor mounted on the same shaft, (see Figure 13.3).

They are inherently non-self-starting and must be run up to speed on light load either by means of an auxiliary motor or, as is more common by means of a squirrel cage or other windings constructed in the pole faces of the rotor. In the latter case the machines are started up under light load as induction motors , after which the rotor DC supply is switched on and the machines have sufficient torque to pull themselves into synchronous speed. The windings in the pole faces of the rotor then act as damping windings to prevent hunting with load fluctuations.

Synchronous motors are also made in very small sizes with permanent magnetic rotors, and these are becoming popular for fan applications.

The DC excitor emits continuous radio interference and provision for suppression should always be installed.

Typical characteristics of synchronous induction motors

kW range	15 to 100 and over
Starting torque	50% to 150% of full load torque
Starting current	2 to 5 times full load current
Power factor	1.0 to almost anything leading

13.4.2.4 Polyphase AC commutator motors

It is probable that the majority of polyphase commutator motors are built for specific purposes rather than for general industrial drives. A well-known type of commutator motor, which has been used as a fan drive where speed regulation with minimum loss is required, is the Schrage motor. It comprises a rotor with a primary winding, connected to the supply by slip rings, and a low voltage commutator winding in the same slots. The secondary windings on the stator (one for each phase) are fed from the commutator by means of brushes whose positions may be varied simultaneously, giving speed variation above and below synchronous speed. It has two main advantages. At a given brush setting it possesses shunt characteristics, i.e. speed varies very little with torque variation. Also, losses due to speed regulation are low.

Provision should be made for suppression of radio interference.

Typical characteristics of the Schrage motor:

kW range	3 to 2000
Starting torque*	150% of full load torque
Starting current*	1.5 times full load current
Power factor	0.8 to 1.0
Speed tolerance	± 5%

* When started at lowest speed

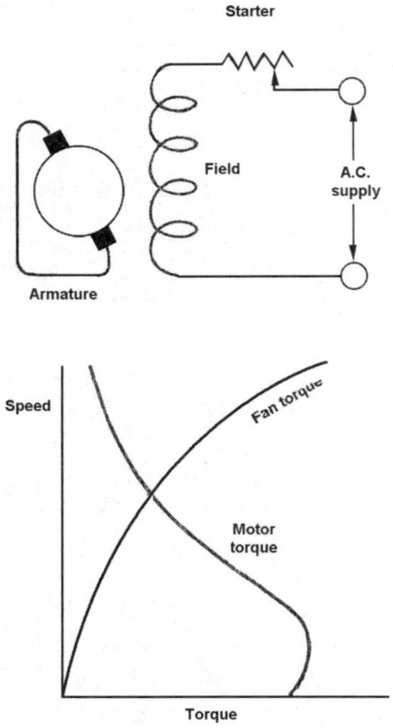

Figure 13.4 Single-phase AC motor

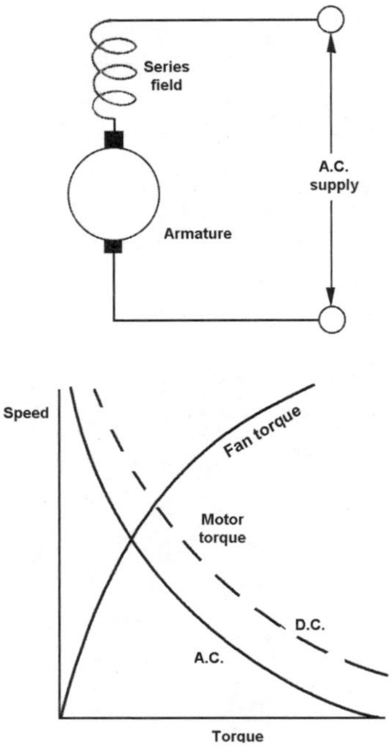

Figure 13.5 Single phase AC (or AC/DC) series motor

13.4.3 Single-phase AC motors

These machines have a single field winding and a wound rotor with short-circuited brushes. (See Figure 13.4.) The speed and direction of rotation are dependent on the position of the brush axis. They are sometimes used for fan drives and are available in low power sizes. Low-power machines may be started direct on to the supply, whilst higher-powered machines are arranged to have the voltage reduced on starting by means of auto-transformer, series choke, or series resistance.

In some machines starting and speed regulation are obtained by moving the position of the brushes. The starting torque is quite high. The machines emit continuous radio interference, which should be suppressed.

Typical characteristics of AC range motors:

kW range	0.33 to 7.5
Starting torque	300% to 400% of full load torque
Starting current	3 to 4 times full load current
Power factor	0.7 to 0.8
Speed tolerance	below 0.33 h.p. per 1000 r.p.m ± 20%
	above 0.33 h.p. per 1000 r.p.m ± 15%

13.4.3.1 AC series motors

In fractional kW sizes these machines are invariably known as universal motors, as they may be run on both alternating or direct current. Their speed torque characteristics are generally similar to those of DC series motors, but the same machine will run at a higher speed on DC than on the same voltage AC (see Figure 13.5).

They are sometimes used for fan drives where speeds in excess of maximum AC synchronous speeds are required, and for AC/ DC supplies where it is not essential to have the same speed on both supplies. Alternatively they run on a different voltage on either supply. Speed regulation on fan loads may be obtained by means of a series resistance.

At speeds below about 5000 r.p.m. commutation is generally poor on AC For this reason these machines are usually made only in fractional power sizes and high speeds. They are invariably short-time time rated. Starting is usually direct on line, and the starting torque is high.

Continuous radio interference is emitted and suppression devices should therefore be fitted.

Typical characteristics of AC series motors:

kW range	0.01 to 0.4
Starting torque	300% to 500% of full load torque
Starting current	5 to 9 times full load current
Power factor	0.5 to 0.7
Speed tolerance	below 0.25 kW per 1000 r.p.m. ± 20%
	above 0.25 kW per 1000 r.p.m. ± 15%

13.4.3.2 Single-phase AC capacitor-start, capacitor-run motors

These motors have a stator with two windings, the phase of one of them being practically 90° (electrical) different from the

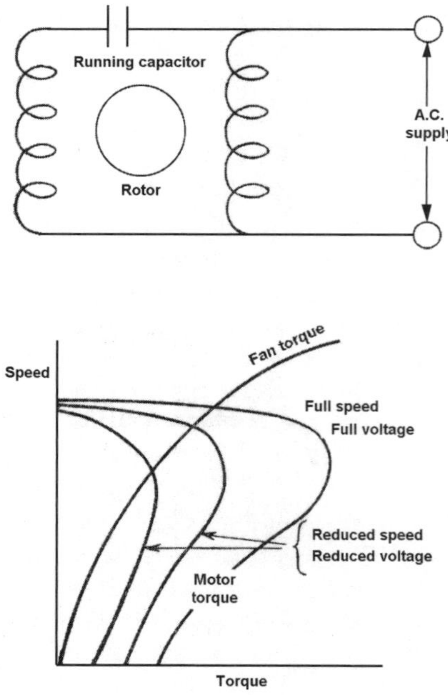

Figure 13.6 Single phase AC capacitor-start, capacitor-run motor

phase of the other. This is achieved by the insertion of a capacitor (condenser) permanently in series with one of the windings. The rotor is of squirrel cage construction, (see Figure 13.6). The performance of these machines can be quite high, approaching that of a true 2-phase motor. The power factor is high and the motor forms an excellent fan drive.

A limited range of speed variation on fan loads only may be obtained with a specially designed machine of this type. By regulating the voltage to the stator by means of an auto-transformer or series choke, speed reductions of about 50% of nominal speed may be achieved. Two speeds may be obtained by means of double winding or pole changing. The machine is normally made in fractional and low power sizes, although machines up to 7.5 kW have been produced.

Reversal is quite easily obtained by reversing the connections of one of the stator windings. In low power sizes the machine is usually started direct on to the supply. A compromise must be made by the designer in the choice of capacitor to permit both starting and running of the machine on a single capacitor, which gives a lower starting torque than is ideally obtainable. Higher power machines are usually fitted with an extra capacitor, which is used during the starting period only, giving additional starting torque. When the machine is up to running speed this capacitor is switched out and the machine runs on the remaining capacitor, which has been chosen for optimum performance at running speed. The machine with two capacitors is not suitable for speed regulation. Capacitors must be extremely reliable and are usually of a high quality paper insulated type. In the case of high power machines it may also be necessary to reduce the voltage on starting by means of an auto-transformer, series choke, or series resistance. 'There is no radio interference from this type of machine.

Typical characteristics of capacitor-start, capacitor-run motors:

kW range 0.33 to 7.5

Starting torque 200% to 300% of full load torque (some special permanent capacitor types for fan drives have only 75%)

Starting current 1.5 to 2.5 times full load current

Power factor 0.95

Speed tolerance ± 5% for small sizes and low speeds ± 2% for larger sizes

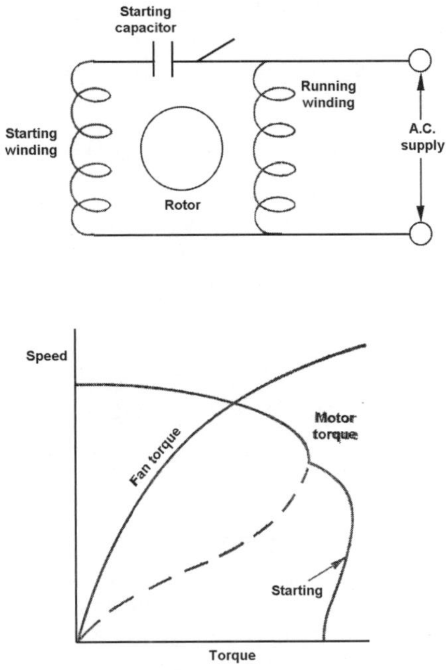

Figure 13.7 Single phase AC capacitor-start induction motor

13.4.3.3 Single-phase AC capacitor-start, induction-run motors

These are generally similar to capacitor-run motors, but the capacitor and additional winding are used only for starting, after which they are cut out at speed by means of a relay or switch, usually a centrifugal type mounted on the motor shaft, (see Figure 13.7). They then run, as single-phase induction motors.

The capacitor is usually a short-time-rated electrolytic type. The motor is normally a constant speed machine. Reversal may be achieved by reversing the connections of the starting winding. The starting torque is quite high with correspondingly high starting current. These motors are less suitable for fan drives than the capacitor-start, capacitor-run type. They cannot be regulated, since speed reduction would cause the re-connection of the starting condenser and rapid burn-out of the machine. They have an inferior efficiency and power factor, while the high starting torque provided is unnecessary for fan drives. No continuous radio interference is emitted, but clicking will be heard when the centrifugal switch operates.

Typical characteristics of capacitor-start, induction-run motors:

kW range	0.1 to 1
Starting torque	200% to 300% of full load torque
Starting current	3 to 5 times full load current
Power factor	0.65 to 0.75
Speed tolerance	± 5% for small sizes and low speeds ± 2% for larger sizes

13.4.3.4 Single-phase AC split phase motors

In the case of the two types of motor, just described, a capacitor is employed to achieve electrical angular displacement between the magnetic fields of the two windings, producing approximately two-phase conditions. In the split phase machine there are again two windings, but the displacement is achieved either by inserting resistance in series with the starting winding, or by so constructing the starting winding to give a higher ratio of resistance to reactance than the main winding, (see Figure 13.8).

Either method creates a displacement of phases between the fields of each winding sufficient to start the machine. When the motors have attained normal speed, the starting winding is cut out by a switch which may be operated manually, by a relay con-

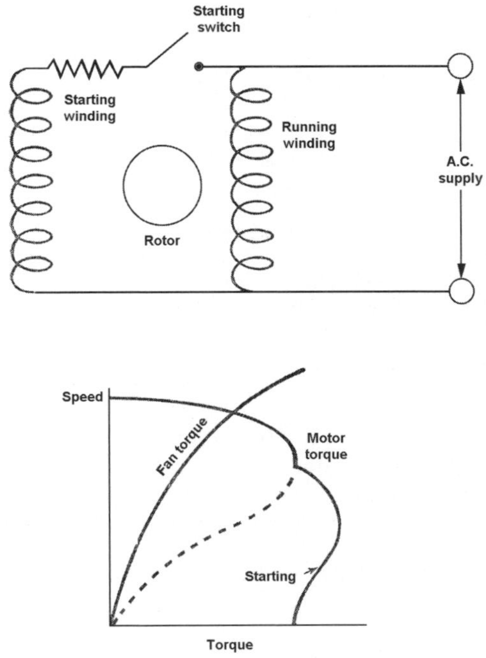

Figure 13.8 Single phase AC split phase motor

trolled by the main winding current, or more commonly by a centrifugal switch mounted on the shaft. The motors then run as single-phase induction motors. These machines have the same disadvantage for fan drives as the capacitor-start, induction-run type.

Two speeds may be obtained by either double winding or pole changing. Reversal is possible by reversing the connections the starting winding. They are made only in fractional sizes and are suitable for low power fan drives. They are started direct on supply. There will be no continuous radio interference, but clicks will be heard when the centrifugal switches operates.

Typical characteristics of split phase induction motors:

kW range	0.03 to 0.25
Starting torque	100% to 200% of full load torque
Starting current	4 to 6 times full load current
Power factor	0.5 to 0.7
Speed tolerance	± 5% for small sizes and low speeds
	± 2% for larger sizes

13.4.3.2 Single-phase shaded pole motors

These are the simplest form of self-starting, single-phase induction motors. They have a squirrel cage rotor and the field is so constructed as to have an offset short-circuited coil producing a magnetic field displaced electrically from the main field, (see Figure 13.9). Compared with other types of single-phase motor the performance is poor and power factor very low, but this is counter balanced by cheapness and robustness. As losses are normally quite high it is generally impossible to damage the machine by overload.

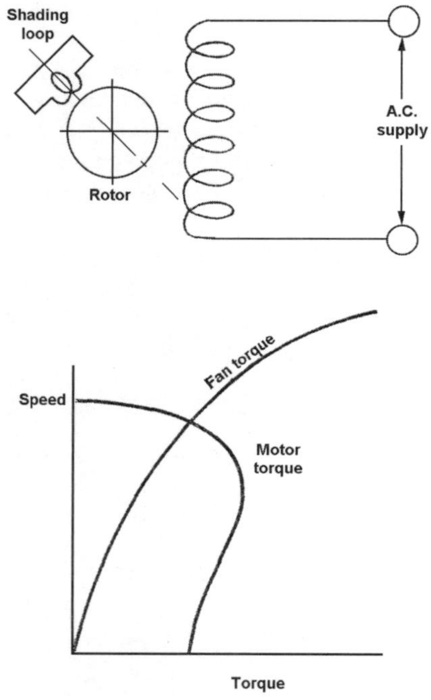

Figure 13.9 Single phase AC shaded pole motor

The speed may be regulated on fan loads only from full speed to 50% of full speed by voltage reduction. The machines are essentially non-reversing. Their starting torque is very low. They are a very popular drive for small fans requiring powers not exceeding 1/50 horsepower and may be started direct on the supply. There is no radio interference from these motors.

Typical characteristics of shaded pole induction motors:

kW range	0002 to 0.15
Starting torque	50% to 150% of full load torque
Starting current	10.5 to 2 times full load current
Power factor	0.4 to 0.6
Speed tolerance	± 5% for small sizes and low speeds

13.4.4 Single-phase repulsion-start induction motors

These machines have a single field winding and are similar to the repulsion motor in that they have a wound rotor and commutator, (see Figure 13.10). They are started as a repulsion motor, that is, the brushes are short circuited. When running speed has been attained a centrifugal switch operates a short-circuiting ring making contact with all of the commutator segments. The machines then run as single-phase induction motors. They may be reversed at rest by altering the brush position.

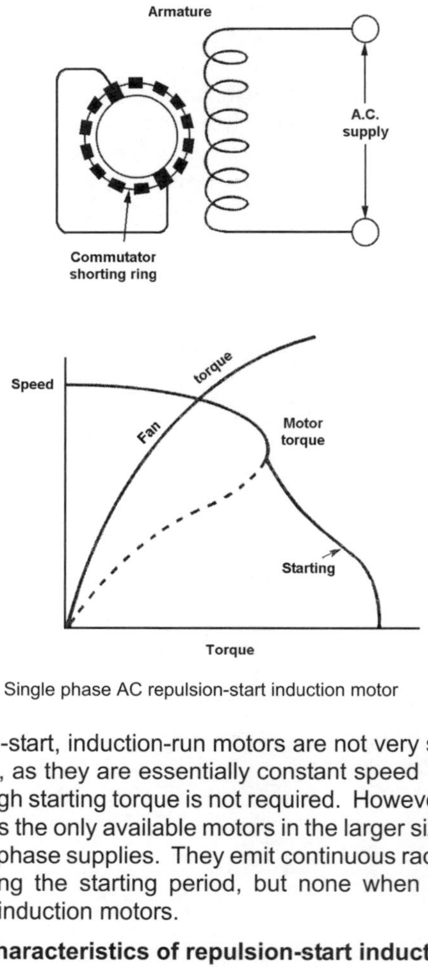

Figure 13.10 Single phase AC repulsion-start induction motor

Repulsion-start, induction-run motors are not very suitable for fan drives, as they are essentially constant speed machines, and the high starting torque is not required. However, they are sometimes the only available motors in the larger sizes for use on single-phase supplies. They emit continuous radio interference during the starting period, but none when running at speed as induction motors.

Typical characteristics of repulsion-start induction motors:

kW range	0.2 to 3.5
Starting torque	300% to 500% of full load torque
Starting current	4 to 6 times fun load current
Power factor	0.7 to 0.8
Speed tolerance	± 5% for small sizes and low speeds
	± 2% for larger sizes

13.4.5 Direct current (DC) motors

13.4.5.1 Series wound motors

These motors are eminently suitable for use as direct fan drives as the speed of the motor will adjust itself until the motor output balances the fan load, (see Figure 13.11). They are quite simple to speed regulate, but where the full speed power exceeds 1 kW, the regulators tend to be rather bulky and the electrical losses in the regulator rather high when the fan is being regu-

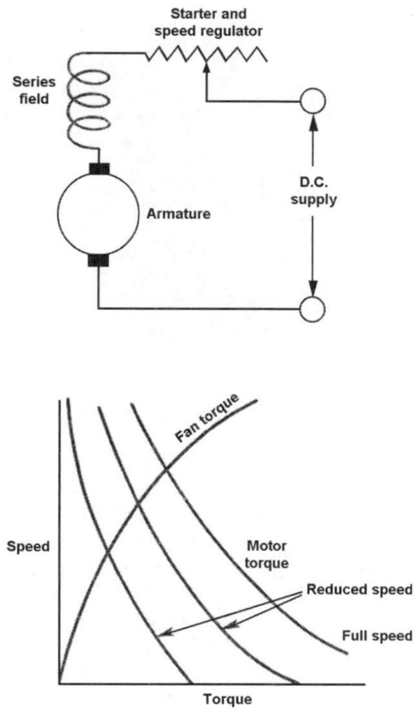

Figure 13.11 DC series wound motor

lated. Series motors should not be used on indirect fan drives because if the load is disconnected, for example through belt failure, the speed will rise to a dangerous level. Reversal may be obtained by reversing the connections of the armature.

The starting torque of these motors is high. When the machine is connected directly to the supply the starting current is of the order of 5 to 8 times full load current. With large motors this may be higher than the permissible current allowed by the authorities; in that case a controller is used whose function is to limit the normally high starting current. As the same current passes through the field and armature, a series resistance will serve to reduce the rating of the motors on starting and so reduce the current consumed. This resistance is made variable so that it can be gradually reduced as the machines gather speed. Control is generally by hand, but automatic controllers are produced. The starting current with a controller is usually limited to 1.5 times full load current.

These machines emit continuous radio interference and provision should always be made for suppression. A tolerance of plus or minus 10% on speed is normally to be expected from series wound fan motors, rising to plus or minus 20% for the fractional powered versions.

13.4.5.2 Shunt wound motors

Shunt wound motors are essentially for constant speed, although speed regulation is possible by adjusting the strength of the field. In this case the frame would be larger than would be necessary with a constant speed machine of the same power. These motors are suitable for a constant speed drive of any horsepower and may be reversed, if suitably designed, by reversing the connections to the armature. The starting torque of these motors is not as high as that of a series wound motor.

A starter is usually necessary to avoid instability during the starting period, (see Figure 13.12). This starter is arranged to limit the starting current to about 1.5 times full load current and to ensure starting on full field if the motor is of the shunt field regulating type. The starting resistance in this case is in series with the armature only while the field receives full supply voltage. Starting is usually carried out manually, although automatic starters are available.

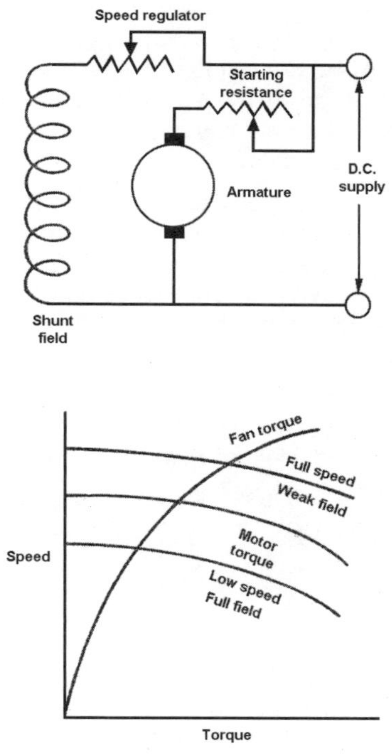

Figure 13.12 DC shunt motor

Radio interference is continuous and provision should always be made for suppression. Normal tolerances on speed to be expected in the manufacture of these machines are as follows:

Below 2 kW	per 1000 r.p.m.	plus or minus 10%
Between 2 & 7.5 kW	per 1000 r.p.m.	plus or minus 7.5%
Over 7.5 kW	per 1000 r.p.m.	plus or minus 5%

13.4.5.3 DC compound wound motors

Compound wound motors may be designed to exhibit characteristics ranging from those of the series machine to those of the shunt machine. When used for fan drives the best type is probably one, which, whilst exhibiting characteristics similar to those of a series machine, is sufficiently compounded to prevent dangerously high speeds on light load. Although suitable for power, they are normally used where drives of 1.5kW or above are required.

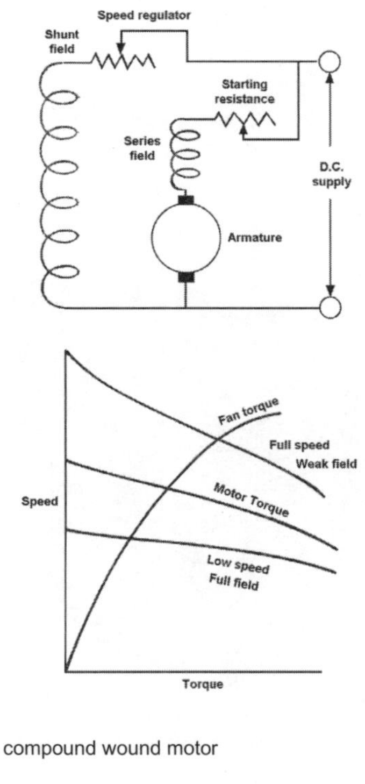

Figure 13.13 DC compound wound motor

Speed regulation is usually achieved by reducing the strength of the shunt field, (see Figure 13.13). As in the case of shunt motors, the frame for a regulating machine would be larger than for that of a constant speed machine of the same power. If suitably designed the machine may be reversed by reversing the connections to the armature.

The method used to start a compound wound DC motor is to use a variable resistance in series with the armature and series field. The shunt field is given the full supply voltage and exercises a retarding influence on both speed and current. Starting gear is generally designed to limit the starting current to about 1½ times the full load current.

Radio interference is continuous and provision should always be made for suppression. Normal tolerances on speed to be expected in the manufacture of these machines are as follows:

Below 2kW	per 1000 r.p.m.	plus or minus 10%
Between 2 & 7.5 kW	per 1000 r.p.m.	plus or minus 7.5%
Over 7.5kW	per 1000 r.p.m.	plus or minus 5%

13.4.6 "Inside-out" motors

Single-phase as well as 3-phase induction motors can be built as conventional inner rotor motors or as "inside-out" external rotor motors. For fan application, an external rotor motor, in which the cowl-shaped rotor revolves around the inner stator wound with copper wire, is especially advantageous. The short length of the winding head enables space-saving design and reduced copper losses. In addition, such motors are very compact because of the bearing system (sintered sleeve bearings or precision ball bearings) integrated into the stator's interior. The motor installed inside an impeller results in a fan unit requiring minimum space. The unique integration of the motor

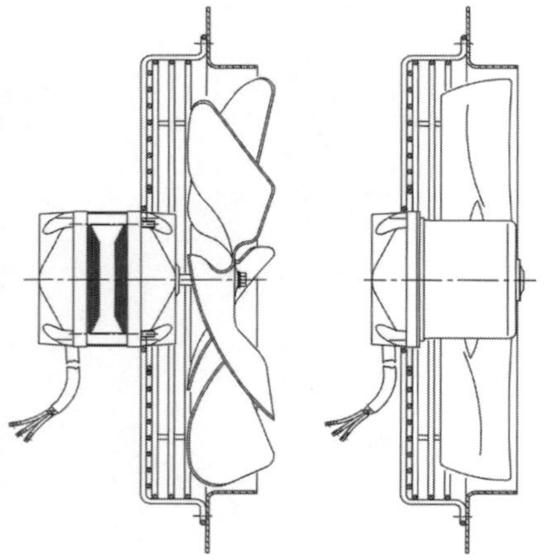

Figure 13.14 Comparison of space required for an axial flow fan fitted with an "inside-out" and conventional motor respectively

Figure 13.15 View of forward curved centrifugal fan fitted with "inside-out" motor
Courtesy of PM°DM Precision Motors Deutsche Minebea GmbH

Figure 13.16 Cross-sectional view of "inside-out" motor fitted to forward curved bladed impeller
Courtesy of PM°DM Precision Motors Deutsche Minebea GmbH

and the impeller permits precise balancing which guarantees low loads to the bearing system. The motor is positioned directly in the air stream, so the very efficient cooling extends lifetime expectancy.

Figure 13.14 shows the space saving possible for an axial flow fan, whilst Figures 13.15 and 13.16 show this motor variant applied to a small forward curved bladed centrifugal fan.

13.5 Starting the fan and motor

During start up, the motor has to accelerate from zero to full speed. If there were no resistance this would be achieved rapidly, but with a fan the "inertia" of the rotating parts resists this acceleration. Fans, perhaps more than any other application, have high inertia relative to the power requirements. The power absorbed by a fan impeller varies as its speed cubed (see Chapter 4, Section 4.6 on fan laws) i.e.

$$\frac{P_i}{P_{100}} = \left(\frac{N_i}{N_{100}}\right)^3 \qquad \text{Equ 13.1}$$

where:

P_i = power at any instant

P_{100} = power at full speed

N_i = speed at any instant

N_{100} = full speed

For vee belt-driven fans there will be additional small power losses in the bearings and belt (varying directly as the speed), but for the following analysis, these are ignored.

It is usual for electric motor manufacturers to produce torque speed curves. It is therefore necessary to calculate the torque required by the fan.

Now $P_i = N_i T_i$ and $P_{100} = N_{100} T_{100}$

$$\therefore \frac{T_i}{T_{100}} = \left(\frac{N_i}{N_{100}}\right)^2 \qquad \text{Equ 13.2}$$

It will be seen that this is a square relationship. We may therefore draw a curve of torque versus speed. This starts at the origin, for when $N_i = 0$ then $T_i = 0$. N_{100} and T_{100} will be the full speed and corresponding torque taken by the fan under the stated conditions of gas air density and point of operation (damper closure etc). In fact, with a fan impeller mounted on a shaft running in bearings, there will be a small amount of torque at the instant of starting. This is due to the "stiction" in the bearings and is known as the "break away torque". It is only of any significance with sleeve bearings and again will be ignored in the present analysis.

If the torque developed by the motor were the same as that required by the fan, then they would be in balance, and the fan would neither accelerate nor slow down. During the run up period, therefore, the excess of motor torque over torque required is available for accelerating the fan to full speed.

The relationship is:

$$T_{im} - T_{if} = T_{ia} + I\alpha_i \qquad \text{Equ 13.3}$$

where:

T_{ia} = torque available for acceleration

T_{im} = torque developed by motor

T_{if} = torque required by fan

I = inertia of rotating parts

αi = acceleration all at any instant

We may determine the run up time from the following further analysis:

m = mass of rotating parts (kg)

r = radius of gyration (m)

I = inertia of rotating parts (kg.m^2) = mr^2

N = rotational speed (rev/min)

t = run up time (S)

T = torque (Nm)

α = angular acceleration (rad/s^2)

P = power (kW)

w = angular velocity (rad/sec)

$\varpi = \frac{2\pi N}{60}$

Suffix

f = fan

m = motor

t = total

I = instantaneous

100 = full speed

Now generally:

$$\alpha = \frac{\omega}{t} = \frac{2\pi N}{60t} = \frac{T}{I} \qquad \text{Equ 13.4}$$

also

$$T = \frac{P}{T} = \frac{60}{2\pi t} \times 1000 \qquad \text{Equ 13.5}$$

Inertia referred to motor shaft:

$$= I_t = I_m + I_f \left(\frac{N_f}{N_m}\right)^2 \qquad \text{Equ 13.6}$$

Torque referred to motor shaft:

$$T_r = T_f \times \frac{N_f}{N_m}$$

$$t = \frac{2\pi N}{60} \times \frac{I_t}{T_m} \text{ or } t = \left(\frac{2\pi N}{60}\right)^2 \times \frac{I}{P \times 1000} \qquad \text{Equ 13.7}$$

This analysis assumes that 100% of the full load motor torque is available during the run up period.

In fact the torque for acceleration is varying all the time from zero rev/min to full speed. Figure 13.17 shows this. The formula must therefore be amended by a factor "f" which gives the average torque available for acceleration (average of all ordinates taken over very small Increments of speed).

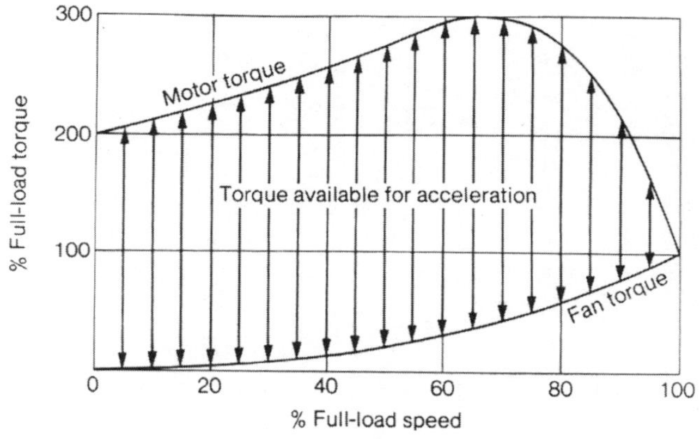

Figure 13.17 Torque available for acceleration

In the examples which follow "f" is approximated for some of the most popular types of motor and starter. However, there is no substitute for a detailed analysis when actual fan and motor torque/speed curves are drawn to scale on the same base. This will enable "f" to be accurately assessed.

The time allowable for starting is dependent on a number of factors. Acceleration produces additional stresses in the fan impeller and shaft but these are not usually of significance. More important are the effects of higher motor winding temperatures, suitability of starter overload relays, and the ability of power lines to accept the additional current. Usually a time of around 18 seconds is therefore recommended, but this may not be achieved with very large units. The whole installation must then be discussed between fan, motor, and starter manufacturers to achieve the best solution.

To assist in the calculation of these times, it is necessary to have accurate values of the inertia of both motors and fans. However, typical values are given in Tables 13.2, 13.3 and 13.4, which may be used for initial calculations at the project stage. They should be replaced by actual values, once the fan and motor manufacturers have been selected.

In most cases the power absorbed by the fan will be within a small percentage of the motor installed power. Assuming them to be equal, at this stage of the analysis, we may then plot curves for the motor and fan. The various types of motor and starter may now be considered and factors "f" determined to give approximate run up times:

Direct-on-line (DOL) induction motor

This method of starting is usually employed up to about 7.5 kW, and for motors of this size the torque/speed characteristic is generally as shown in Figure. 13.15. As may be seen the available torque varies from 200% to 0% of the motor full-load

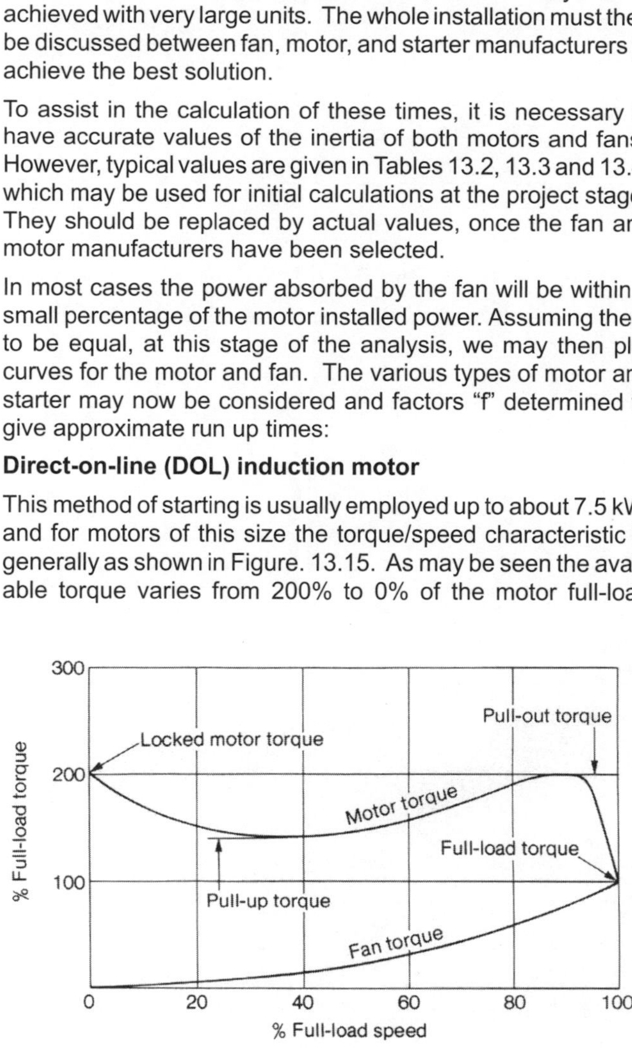

Figure 13.18 Direct-on-line starting

Frame size	Moment of inertia mr^2 kgm^2			
	2-Pole	4-Pole	6-Pole	8-Pole
D63	3.63×10^{-4}	3.65×10^{-4}	—	—
D71	5.33×10^{-4}	5.43×10^{-4}	—	—
D80	1.14×10^{-3}	1.56×10^{-3}	1.61×10^{-3}	—
D90S	1.61×10^{-3}	3.43×10^{-3}	3.40×10^{-3}	3.40×10^{-3}
D90L	1.99×10^{-3}	3.93×10^{-3}	3.88×10^{-3}	3.88×10^{-3}
D100L	6.43×10^{-3}	1.15×10^{-2}	1.16×10^{-2}	1.16×10^{-2}
D112M	7.35×10^{-3}	1.35×10^{-2}	1.38×10^{-2}	1.38×10^{-2}
D132S	1.90×10^{-2}	3.10×10^{-2}	3.35×10^{-2}	3.35×10^{-2}
D132M	—	3.38×10^{-2}	4.15×10^{-2}	4.15×10^{-2}
D160M	4.63×10^{-2}	7.18×10^{-2}	1.02×10^{-1}	1.02×10^{-1}
D160L	5.20×10^{-2}	8.53×10^{-2}	1.20×10^{-1}	1.20×10^{-1}
D180M	6.00×10^{-2}	9.83×10^{-2}	—	—
D180L	—	1.52×10^{-1}	1.99×10^{-1}	1.99×10^{-1}
D200L	1.87×10^{-1}	1.88×10^{-1}	3.59×10^{-1}	2.49×10^{-1}
D225S	—	3.43×10^{-1}	—	4.16×10^{-1}
D225M	2.04×10^{-1}	3.78×10^{-1}	4.71×10^{-1}	4.71×10^{-1}

Table 13.3 Typical moments of inertia for TEFV induction motors

Fan size D mm (impeller diameter)	Moment of inertia mr^2 kgm^2				
Width	Extra narrow	Narrow	Medium	Wide	Extra wide
160				7.19×10^{-3}	1.10×10^{-2}
180				9.61×10^{-3}	1.44×10^{-2}
200				1.26×10^{-2}	2.04×10^{-2}
224				1.73×10^{-2}	2.88×10^{-2}
250			2.29×10^{-2}	2.41×10^{-2}	3.38×10^{-2}
280			2.47×10^{-2}	2.74×10^{-2}	3.83×10^{-2}
315	4.15×10^{-2}	4.28×10^{-2}	4.76×10^{-2}	5.01×10^{-2}	7.26×10^{-2}
355	6.10×10^{-2}	6.35×10^{-2}	7.06×10^{-2}	7.43×10^{-2}	1.17×10^{-1}
400	8.89×10^{-2}	9.26×10^{-2}	1.03×10^{-1}	1.07×10^{-1}	1.69×10^{-1}
450	1.35×10^{-1}	1.41×10^{-1}	1.57×10^{-1}	1.74×10^{-1}	2.69×10^{-1}
500	2.31×10^{-1}	2.43×10^{-1}	2.70×10^{-1}	3.00×10^{-1}	4.81×10^{-1}
560	4.32×10^{-1}	4.55×10^{-1}	5.04×10^{-1}	5.60×10^{-1}	9.53×10^{-1}
630	7.18×10^{-1}	7.64×10^{-1}	8.49×10^{-1}	1.01	1.53
710	1.21	1.29	1.43	1.83	2.78
800	2.49	2.68	2.98	3.21	5.12
900	4.31	4.63	4.67	5.19	7.65
1000	1.39×10	1.49×10	1.66×10	1.74×10	2.82×10
1120	2.1×10	2.28×10	2.53×10	2.66×10	4.71×10
1250			3.58×10	4.01×10	7.53×10
1400			5.93×10	6.43×10	1.10×10^2
1600				1.05×10^2	1.98×10^2
1800				1.58×10^2	2.91×10^2
2000				2.69×10^2	4.74×10^2

Note:
1. These figures are for a range of light duty centrifugal impellers. They are of the backward inclined typed, spot/plug welded up to size 1900 mm diameter and fully welded above.
2. For other blade types refer to Table 13.5
3. Units are "engineers" i.e. mass kg x radius of gyration^2 m^2

Table 13.4 Typical moments of inertia for a range of centrifugal fans

Impeller type	Sizes 160 to 900	Sizes 1120 to 2000
Backward curved	1.00	1.05
Forward curved	1.09	1.18
Shrouded radial	1.05	1.10
Open paddle	1.12	1.12
Aerofoil	1.21	1.16

Table 13.5 Typical multiplier for other blade forms

torque over the run-up period and for this reason it is usual to assume an average 100% full-load torque available for the whole period. No correction is therefore necessary to the general formula. See Figure 13.18.

Star-delta starting induction motor

Normally used for motors between 7.5 kW and 45 kW this method reduces the line voltage (and hence current) on starting to prevent large surge currents. Unfortunately, it also reduces available torque as may be seen in Figure 13.19. An average value of torque available is 30% of the full-load value and therefore a correction factor of 3.33 may be used.

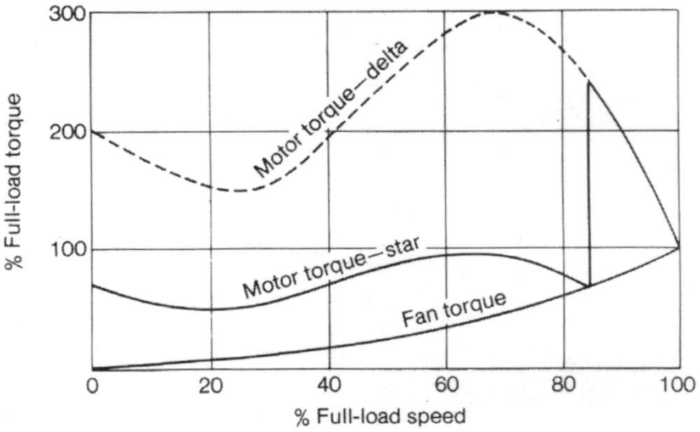

Figure 13.19 Induction motor characteristics, star-delta starting

Note: Some motors, particularly between 15 kW and 30 kW, have a torque characteristic with a pronounced "dip" limiting the speed that may be attained in star. This is shown in Figure 13.20. Here the fan torque characteristic cuts the motor torque characteristic at a low speed and the motor will not accelerate beyond this point. Changing to delta connection at this speed will mean the line carrying a very high current for which the cables, fuses, and overloads must be adequately sized.

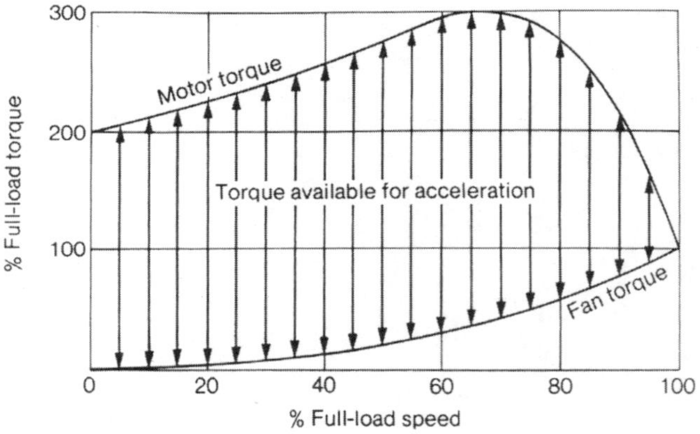

Figure 13.20 Induction motor characteristics, unsatisfactory torque

It is difficult to generalize in this case, but it may be assumed that the lowest value of the motor torque occurs at 30% full-load speed and is approximately 40% full load torque in star. Should the fan torque at this speed exceed this low value of motor torque, alternative starting methods should be used.

$$T_r = \frac{P_f}{N_m} \times \frac{60}{2\pi} \times 1000 \times 0.3^2 \qquad \text{Equ 13.8}$$

$$\cong 860 \times \frac{P_f}{N_m}$$

The torque absorbed by the fan at 30% motor speed referred to motor shaft.

Auto-transformer starting

Autotransformer starting again reduces voltage current and torque, but in a greater number of stages (usually three, but can be two or four) thereby giving a higher average available torque. Tappings may be at 40%, 60%, 80% voltage and a correction factor of two is then used. Figure 13.21 gives typical characteristics.

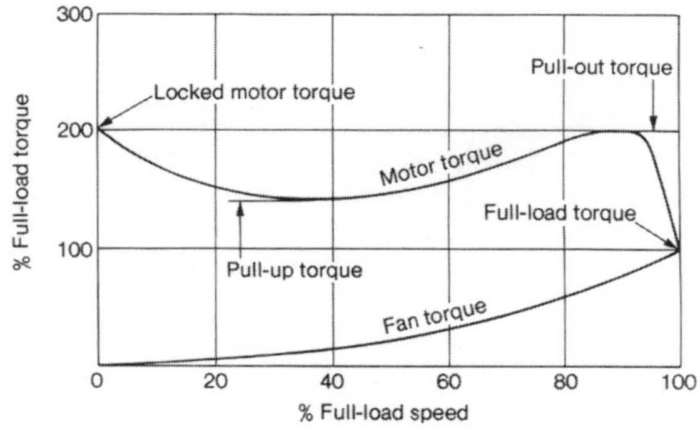

Figure 13.21 Induction motor characteristics, auto-transformer starting

Slip-ring motors/stator-rotor starting

This is one of the most satisfactory methods of fan starting since by inserting resistance in the rotor circuit, the torque characteristic is arranged such that maximum is available when required. Figure 13.22 shows a higher torque is available than in most other cases. The correction factor may be as low as 0.4 although 0.5 is a reasonable figure to use.

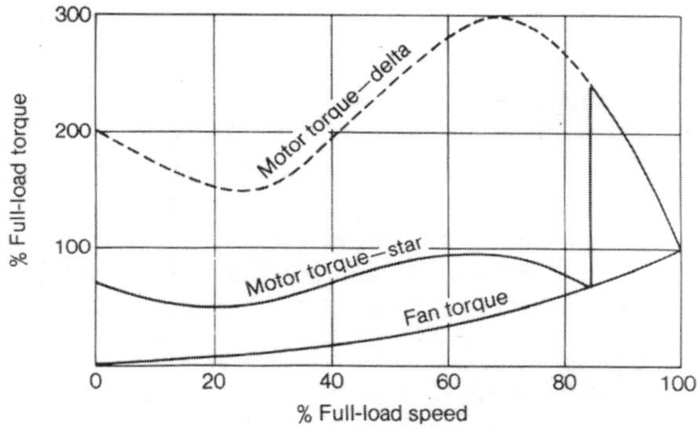

Figure 13.22 Slip-ring motor characteristics, stator-rotor starting

Correct voltage selection is also important, and care should be taken to ensure that the motor is rated at the line voltage.

For example, a motor wound for 440 volts connected to a 380 volt supply will develop only $\left(\frac{380}{440}\right)^2 \times 100$, i.e. 75% of normal torque, but more important, in star connection, the torque available for starting the fan may be as low as 20% of the direct on-line value.

Summary:

From the above remarks it can be seen that a general formula may be derived to calculate the run-up time of any AC motor, i.e.:

$$t = \left[I_m + I_f \left(\frac{N_f}{N_m}\right)^2 \right] \times \frac{N_m^2}{P_f \times 1000} \times \left(\frac{2\pi N}{60}\right) 2 \times f\, R_E^{\ 2} \qquad \text{Equ 13.9}$$

or

$$t = \left[I_m + I_f \left(\frac{N_f}{N_m}\right)^2 \right] \times \frac{N_m^2}{Pf} \times \frac{f}{105} R_E^{\ 2} \times 1.097$$

where:

R_E = ratio of the applied voltage to the motor rated voltage

f = correction factor referred to in the text

Hence, assuming the correct voltage is applied, the approximate formula for each method of starting may be simplified to:

DOL induction

$$t = \left[I_m + I_f \left(\frac{N_f}{N_m}\right) 2 \right] \times \frac{N_{m^2}}{P_f} \times \frac{1.1}{10^5} \qquad \text{Equ 13.10}$$

Star-delta Induction

$$t = \left[I_m + I_f \left(\frac{N_f}{N_m}\right) 2 \right] \times \frac{N_{m^2}}{P_f} \times \frac{3.7}{10^5} \qquad \text{Equ 13.11}$$

Auto-transformer

$$t = \left[I_m + I_f \left(\frac{N_f}{N_m}\right) 2 \right] \times \frac{N_{m^2}}{P_f} \times \frac{2.2}{10^5} \qquad \text{Equ 13.12}$$

Slip ring stator rotor

$$t = \left[I_m + I_f \left(\frac{N_f}{N_m}\right) 2 \right] \times \frac{N_{m^2}}{P_f} \times \frac{0.55}{10^5} \qquad \text{Equ 13.13}$$

In all cases it is good practice to limit the value of t to about 18 seconds. The value of P_f to insert in the formula is that relating to the conditions of start up.

It is important to note that these approximate formulae make the assumption that the fan absorbed power and the motor rating are almost equal and certainly within 10% of each other. If a larger motor is installed then this will reduce the starting time. Strictly speaking a new correction factor should be assessed. However, an indication of the starting time, likely to result, may be obtained by the use of the graph in Figure 13.23. Simply by multiplying the time calculated by the use of equations 13.10 to 13.13 by the factor k_T, the reduced time may be calculated.

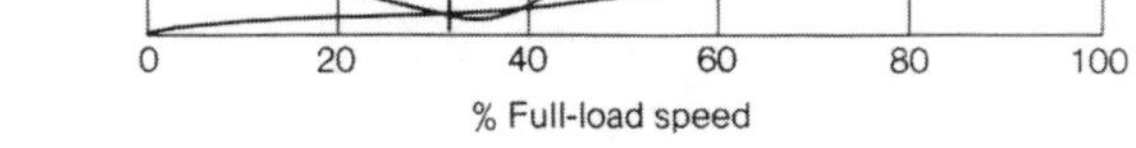

Figure 13.23 Indication of reduction in starting time

Example:

A fan is driven by an induction motor and controlled by a direct on-line starter. It absorbs 5 kW and is fitted with a 5½ kW motor. The run up time calculated from Equation 13.10 is 18 seconds. If the motor power is increased to 7½ kW what will be the new starting time?

Thus:

$$\frac{P_m}{P_f} = \frac{7.5}{5} = 1.5$$

$$\therefore k_T = 0.61$$

$$\therefore t_{revised} = 18 \times 0.61 = 11 \text{ seconds}$$

Note: k_T has been calculated for a range of typical TEFV squirrel cage induction motors with direct-on-line starting. The factors are expected to be somewhat smaller, and the starting times shorter, for induction motors with autotransformer starting or slip ring motors with stator–rotor starters.

13.6 Motor insulation

Insulation is an essential part of all motors. Sufficient insulation must be provided to ensure live conductors within the motor are insulated from each other and from the motor frame, which is normally earthed. Different materials combine to form an insulation system, which varies according to the nature and condition of the component to be insulated. Components include motor windings, leads, terminals, slip rings, commutators, brushes and numerous auxiliary devices.

By their nature, insulation materials cannot withstand temperatures as high as most other parts within motors and consequently most performance aspects are usually limited by the insulation system. As elevated temperatures also degrade the materials used, the life of most motors is determined by the insulation system. Most motor failures occur because of an insulation related problem, whether this is due to excessive temperatures, vibration damage, supply voltage transients, contamination or simply expiry of the expected insulation life.

This Section gives background information on the classification of insulation systems. Manufacturers normally decide the system materials and how they are combined and processed to give a reliable insulation system. However, in some cases there are alternative generic systems, which may be specified by the purchaser.

It is also important for the purchaser to understand the supply system and whether there could be any abnormal conditions that could affect the insulation integrity. The higher the supply system voltage, the more important it becomes that the insulation system and its manufacturer's testing programme are properly specified.

13.6.1 Temperature classification

Insulation materials and insulation systems are classified according to the maximum temperature at which they can satisfactorily operate. Insulation has been progressively improved to Class E such that modern motors operate at higher temperatures then those manufactured 50 years ago.

The lettering does not follow an alphabetical progression due to the insertion of additional improved grades with the passing years.

13.7 Motor standards

13.7.1 Introduction

There has been a gradual process of change from countries using their own Standards to the adoption of European and International Standards to ensure uniformity in the widest international meaning. This process is continuing, in particular with the advent of the European Union and associated legislation.

There are already established standards that are recognized throughout Europe and beyond. The basis of most Standards originates with the International Electro technical Commission (IEC), which are then adopted either as National Standards or as European Standards. National committees throughout Europe play a large part in drafting and agreeing the contents of the standards either through the IEC or the European Committee for Electro technical Standardization (CENELEC). Countries worldwide recognize the work of the IEC and IEC Publications often form the basis of national standards.

Because of the involvement within Europe of IEC, CENELEC, CEN and national standard bodies, for example the British Standards Institution (BSI) in the United Kingdom, there tend to be standards published with three types of identification systems (International Standard - IEC, European Standard – CENELEC and National Standard for what are often the same basic standard.

The IEC Publication IEC 60034 is a good example of the variety of designations that can arise from the publication of the many parts that make up this Standard. The main motor Standard within Europe is IEC and after national agreement parts of this standard have become European Standards under CENELEC. Some parts before agreed by CENELEC were used as the basis for national standards. In addition parts of IEC 60034 appeared as Harmonization Documents (HD) under CENELEC control.

The British equivalent of IEC 60034 is British Standard BS 4999 and this itself had many parts when first issued. When re-issued from 1987 onwards, some parts were combined and the part numbers were adjusted to line up with IEC 60034 part numbers where appropriate. But to avoid confusion with the original part numbers the new part numbers commenced at Part 101 with 100 added to the IEC part number where it applied.

Standards are used wherever possible for the principle motor dimensions to ensure interchangeability. This applies particularly to the main fixing dimensions and the shaft end. Standard dimensions are covered by IEC Publications IEC 60072-1 (small and medium size motors) and IEC 60072-2 (medium and large size motors). These also give standard symbols for each significant dimension.

British Standard BS 4999:Parts 103 and 141 are related to these IEC publications and have some additional symbols and standard dimensions which are included in the figures below where appropriate. Dimensions are generally based on preferred numbers but there are some dimensions that are a carry-over from imperial measurements. The Standards include tolerances for all dimensions that affect interchangeability.

For frame sizes from 56 up to 400 inclusive, standard dimensions uniquely define the motor, but for larger motors this is impractical because of a number of design constraints. Standard dimensions are primarily intended for low voltage induction motors. For motors of 355 size and above there is a set of preferred dimensions - the overlap of the 355 and 400 sizes with standard dimensions allows for special designs and motors other than induction motors.

There is international agreement on the nomenclature of small motors from 56 to 400 sizes inclusive. This is extended to cover larger motors in a modified form with the 355 and 400 sizes included when these are not to standard dimensions.

It is still possible to obtain some small motors to imperial dimensions, as specified in British Standard BS 2048:Part 1. The frame size is based on the shaft centre height multiplied by 16. For example, a motor with a shaft centre height of 3 in is a 48 frame size. Frame sizes 36, 42, 48, 56 and 66 are available and should be prefixed with the letter B - this should avoid confusing the imperial and metric 56 sizes.

For motors below the metric 56 frame size there are no universal standard dimensions. This covers the majority of small DC and AC motors. Consequently manufacturers of these motors have their own frame size conventions and dimensions to suit their products. However, most base the frame size on the frame diameter, and where motors are fitted with a square flange, this is often the flange main dimension.

13.7.2 Frame nomenclature system

Small motors, particularly of the induction motor type, are internationally recognized by the frame nomenclature which gives the basic enclosure type, the size and method of mounting. This does not replace the IP, IC and IM codes which give a more detailed description of the motor, but serves to readily identify the common types by means of a simple nomenclature.

The system described in IEC Publication IEC 60072-1 consists of number/letter combinations to denote the centre height for motors with feet, the shaft diameter and/or the flange size. A motor with normal feet is designated by the centre height of the shaft above the base of the feet in millimetres followed by a letter denoting the frame length as either "S" for short, "M" for medium or "L" for long followed by the shaft diameter in millimetres, for example 112 M 28.

Flange-mounted motors can be of three basic types denoted by the letters FF for flange with clearance holes on a pitch circle diameter greater than the spigot diameter, FT for flange with tapped holes but otherwise as FF flanges and FI for flanges with tapped holes but the pitch circle of the holes inside the spigot diameter.

These letters follow the shaft diameter and are themselves followed by the flange fixing-holes pitch-circle diameter in millimetres, for example 28 FF 215. In cases where a motor has both feet and a flange the designation appears as 112 M 28 FF 215, for example.

The basic system outlined in British Standard BS 4999 : Part 103 differs from the IEC Standard and consists of a letter, number, letter combination of which the meanings are as follows:

a) First letter to indicate the basic enclosure either as "C" for enclosed ventilated or "D" for totally enclosed. (It should be noted that the letter "E" has been used to indicate flameproof enclosures but this is not covered by the standards. When the system is extended to large motors an extra letter is often added to indicate a particular variant, for example "DW" for totally enclosed, water cooled or as a range identifier, for example "GD" for the manufacturer's G range of totally enclosed motors.)

b) Number of two or more digits indicating the centre height of the shaft above the base of the feet of horizontal motors in millimetres. For flange-mounted motors or others without feet, the same basic frame size retains the same number. The numbers are from the R20 preferred number series except for the 132, which is approximately half way between 125 and 140.

c) First suffix letter to characterize the longitudinal dimension where more than one length is used, specified as either "S" for short, "M" for medium or "L" for long. (Some large motors using the same basic system have had additional letters added by some manufacturers to indicate a further length step, for example "MX" as a length between "M" and "L".).

d) For other than foot-mounted motors an additional letter to indicate the type of mounting as either "D" for flange, "V" for skirt, "C" for face flange, "P" for pad or "R" for rod. (The "P" mounting can usually be used for rod mounting.)

As an example a motor of the 180 size, of an enclosed ventilated type, with a medium length and for flange mounting would be called a C180MD.

13.8 Standard motors and ratings

13.8.1 Standard motor features

There is no IEC publication covering standard ratings associated with frame sizes, but British Standard BS 5000 : Part 10 does give ratings against frame size and shaft number generally from 56 up to 315 sizes depending upon the type of motor. Although this standard was first published during 1978, and has been amended more recently, it is still current and forms the basis for standard ratings for motors within this range.

The motors covered by the Standard are described as "general purpose induction motors" and meet various parts of British Standard BS 4999 (this generally therefore meets the IEC publications on which BS 4999 is based where appropriate). The motors are suitable for connecting to 3-phase, 415 V, 50 Hz supplies but by agreement may be wound for any voltage not exceeding 660 V. Class E, Class B or Class F insulation may be used with the ambient conditions not exceeding 40°C or 1000 m altitude. BS 5000 : Part 10 should be consulted for full details.

13.8.2 Standard motor ratings

The standard ratings are specified for single-speed motors with synchronous speeds of 3000, 1500, 1000 or 750 r/min. In most cases the shaft sizes are the same for all speeds, except for 3000 r/min on some of the larger standard frame sizes.

Table 13.6 gives standard outputs and shaft sizes for totally enclosed fan-ventilated (TEFC) cage motors where the cooling system is defined as IC411 and the degree of protection as IP44. These motors are fitted with either feet or flanges. The standard allows the same ratings for airstream rated motors with feet or flanges without specifying the air velocity.

Frame No.	Output (kW)				Shaft No.	
	Synchronous speed (r/min)					
	3000	1500	1000	750	3000	1500 or less
D56	0.09 & 0.12	0.06 & 0.09	-	-	9	9
D63	0.18 & 0.25	0.12 & 0.18	-	-	11	11
D71	0.37 & 0.55	0.25 & 0.37	-	-	14	14
D80	1.1	0.75	0.55	-	19	19
D90S	1.5	1.1	0.75	0.37	24	24
D90L	2.2	1.5	1.1	0.55	24	24
D100L	3	2.2 & 3	1.5	0.75 & 1.1	28	28
D112M	4	4	2.2	1.5	28	28
D132S	5.5 & 7.5	5.5	3	2.2	38	38
D132M	-	7.5	4 & 5.5	3	38	38
D160M	11 & 15	11	7.5	4 & 5.5	42	42
D160L	18.5	15	11	7.5	42	42
D180M	22	18.5	-	-	48	48
D180L	-	22	15	11	48	48
D200L	30 & 37	30	18.5 & 22	15	55	55
D225S	-	37	-	18.5	55	60
D225M	45	45	30	22	55	60
D250M	55	55	37	30	60	65
D280S	75	75	45	37	65	75
D28CM	90	90	55	45	65	75
D315S	110	110	75	55	65	80
D315M	132	132	90	75	65	80

Table 13.6 Standard outputs and shaft numbers for totally enclosed fan-ventilated (TEFC) cage motors

In the case of airstream rated motors with pad or mountings classified as IC418, the ratings are as given in Table 13.7 with the average air velocity at least the value given by Table 13.8 when measured 50mm radially from mounting pads.

Frame No.	Output (kW)				Shaft No.	
	Synchronous speed (r/min)					
	3000	1500	1000	750	3000	1500 or less
D80	1.1	0.75	0.55	-	19	19
D90L	1.5 & 2.2	1.01 & 1.5	0.75 & 1.1	0.37 & 0.55	24	24
D100L	3	2.2 & 3	1.5	0.75 & 1.1	28	28
D112M	4	4	2.2	1.5	28	28
D132M	5.5 & 7.5	55 & 75	3, 4 & 5.5	2.2 & 3	38	38
D160L	11, 15 & 18.5	11 & 15	7.5 & 11	4, 5.5 & 7.5	42	42
D180L	22	18.5 & 22	15	11	48	48
D200L	30 & 37	30	18.5 & 22	15	55	55
D225M	45	37 & 45	30	18.5 & 22	55	60
D250M	55	55	37	30	60	65

Table 13.7 Standard outputs and shaft numbers for pad or rod mounted cage motors

The standard ratings for enclosed ventilated cage motors are given in Table 13.9. These motors have a cooling system classified as IC01 and a degree of protection classified as IP22.

Frame No.	Average air velocity (m/s)			
	Synchronous speed (r/min)			
	3000	1500	1000	750
D80	10	7.5	6.5	5
D90	12.5	9	7.5	6
D100	15	10	8	7
D112	16.5	11	9	7.5
D132	18	12	9.5	8
D160	19	12.5	10.5	8.5
D180	20	13.5	11	9
D200	21	14	11.5	9.5
D225	22	14.5	12	10
D250	23	15	12.5	10.5

Table 13.8 Average air velocity for cooling totally enclosed airstream rated motors

Frame No.	Output (kW)				Shaft No.	
	Synchronous speed (r/min)					
	3000	1500	1000	750	3000	< 1500
C160M	11, 15	11	7.5	5.5	48	48
C160L	18.5 & 22	15 & 18.5	11	7.5		
C180M	30	22	15	11	55	55
C180L	47	30	18.5	15		
C200M	45	37	22	18.5	60	60
C200L	55	45	30	22		
C225M	75	55	37	30	60	65
C250S	90	75	45	37	65	75
C250M	110	90	55	45		
C280S	-	110	75	55	65	80
C280M	132	132	90	75		
C315S	160	160	110	90	70	90
C315M	200	200	132	110		

Table 13.9 Standard outputs and shaft numbers for enclosed ventilated cage motors

It should be noted that the air velocities specified in Table 13.8 are in many cases extremely low for low hub-to-tip ratio axial flow fans which have a high flowrate. In consequence the air velocities flowing over the motor will be considerable greater than those given in the Table. The power produced can therefore be appreciably greater, without exceeding safe temperature rises in the windings or the motor surfaces. Fan motors may therefore take advantage of this situation provided that the nose motor bearing can accommodate both the increased torque requirement and also the radial and thrust loads imposed by the fan impeller.

This has lead the major fan manufacturers, some of whom manufacture their own electric motors, to develop machines specifically appropriate to the application. Such solutions are especially the case in the smaller frame sizes where quantity requirements make such motors economically viable.

13.9 Protective devices

When electric fan motors are connected to the public supply, protective devices are required for two main purposes. In the first place it is necessary to ensure that a breakdown in the insulation of the motor, its control gear or connecting wiring, shall not cause overheating of the supply cables or interruption of the supply to the whole premises. Fuses perform this function effectively and economically for small and moderate power circuits, while circuit breakers are employed for high power applications. These devices must be kept for their proper function of interrupting instantaneously the heavy rush of current which flows into an earth or short-circuit before it has time to open the main breakers further back; otherwise the power interruption will spread beyond the particular motor or controller which is faulty.

In the second place it is desirable to limit the amount of damage, which may be done to a fan motor by accidental overloads or minor faults. This is largely an economic matter, and it would be clearly unsound to load a small fan motor of low first cost with the comparatively heavy cost of fully protective control gear, when the chance of breakdown is in any case small. Moreover, fan motors are inherently unlikely to encounter overloads, except with the forward curved centrifugal fan. Nevertheless it is sound practice to instal starters with overload protection when the power exceeds about 0.33 kW.

13.10 Bibliography

Guide to European Electric Motors, Drives and Controls, Dr. David Searle, ISBN 860583393.

IEC 60034-1 Ed. 11.0 b:2004, *Rotating electrical machines - Part 1: Rating and performance.*

IEC 60072-1 Ed. 6.0 b:1991, *Dimensions and output series for rotating electrical machines - Part 1: Frame numbers 56 to 400 and flange numbers 55 to 1080.*

IEC 60072-2 Ed. 1.0 b:1990, *Dimensions and output series for rotating electrical machines - Part 2: Frame numbers 355 to 1000 and flange numbers 1180 to 2360.*

BS 4999-103:2004, *General requirements for rotating electrical machines. Specification for symbols.*

BS 4999-141:2004, *General requirements for rotating electrical machines. Specification for standard dimensions.*

BS 2048-1:1961, *Specification for dimensions of fractional horse-power motors. Dimensions of motors for general use.*

BS 5000-10:1978, *Rotating electrical machines of particular types or for particular applications. General purpose induction motors.*

14 Fan noise

The principle source of noise in any air moving system is the main fan. Rules for determining fan noise and noise-producing mechanisms are covered as well as a review of the sound laws. If the ducting resistance has been incorrectly assessed, the fan noise can be significantly affected.

This Chapter points out some of the pitfalls in the selection of ductwork of the ventilation system which contribute to the addition of unforeseen noise.

Contents:

14.1 Introduction

A prime source of noise in any air moving system is the main fan. It has the ability to direct its duct-borne noise to the farthest corners of any occupied space and can be a major irritant. The problem can, of course, be magnified by the addition of system generated noise. To the humble fan engineer, it seems remarkable from a noise point-of-view, therefore, that so little apparent attention is given, in the design of a ventilation system, to the correct selection of the fan. To this must be added the often less than ideal ductwork connections to the fan, which can result in an additional unforeseen noise.

It is the intention of this Chapter to point out some of the pitfalls and to suggest that the requisite information be obtained from a reputable manufacturer at the earliest possible time. Unfortunately this is not always possible, as the fan supplier will only be chosen late in the building programme when much of the design has been "frozen". It would be beneficial, however, to conduct a feasibility study using results obtained from experiments beforehand.

The user's primary aim is to ensure that the fan will satisfactorily perform its duty. That is to say, it will handle the required volume flowrate at the system pressure and for the stated power. Even more important, however, is what nuisance will be caused, by its noise, to operators of the plant, to neighbours, or to inhabitants of the conditioned area. So many misconceptions, half-truths, and errors have been propagated in the field of acoustics, that one might imagine it had replaced alchemy as the "black art" of 20th century man.

This Chapter is not intended to be a textbook of noise measurement, and those who wish to know more are referred to the references in Section 14.15. However, in order to give meaningful information, it is worth reminding the user of some of the terms employed and their values and underlying concepts.

14.1.1 What is noise?

Noise may simply be defined as:

Sound undesired by the recipient.

14.1.2 What is sound?

Sound may be defined as any pressure variation in a medium - usually air - that can be converted into vibrations by the human eardrum, causing signals to be sent to the brain. As with all other sensations, the result can be pleasant or unpleasant.

14.1.3 Frequency

To vibrate the eardrum it is necessary for the pressure variations in the medium to occur rapidly. The number of variations per second is called the frequency of the sound, measured in cycles per second or Hertz. The human ear can detect sounds from about 20 Hz to 20,000 Hz - the lowest and highest sounds respectively. As a guide, the lowest note on a piano has a frequency of 27.5 Hz, whilst the highest note is at 4186 Hz.

14.1.4 Sound power level (SWL)

The noisiness of a fan can be expressed in terms of its sound power (the number of watts of power it converts into noise). It is unusual to do this, however, as the range of values found in practice would be very large. Fan noise can be measured by its sound power level, a ratio which logarithmically compares its sound power with a reference power, the Pico Watt (10^{-12} watts). The unit of sound power level is the decibel.

Sound power level may be defined as:

$$SWL = 10 \log \frac{W}{W_o} \qquad \text{Equ 14.1}$$

where:

SWL = sound power level in decibels (re 10^{-12} watts)

W = sound power of the noise generating equipment (watts)

W_o = reference power (re 10^{-12} watts)

Table 14.1 shows how the logarithmic scale compresses the wide range of possible sound powers to sound power levels having a practical range of 30 dBW to 200 dBW.

Sound Power (Watts)	Sound power level dBW	Source
40 000 000	196	Saturn rocket
100 000	170	Ramjet
10 000	160	Turbo jet engine 3200 kg thrust
1 000	150	4 propeller airliner
100	140	
10	130	Full orchestra
1	120	Large chipping hammer
0.1	110	Blaring radio
0.01	100	Car on motorway
0.001	90	10 kW ventilating fan
0.0001	80	Voice – shouting
0.00001	70	Voice – conversational level
0.000001	60	
0.0000001	50	
0.00000001	40	
0.000000001	30	Voice – very soft whisper

Table 14.1 Sound powers expressed as sound power levels

14.1.5 Sound pressure level (SPL)

The sound power level of a fan is comparable to the power output of a heater. Both measure the energy (in one case — noise energy, the other — heat energy) fed into the environment surrounding them. However, neither the sound power level nor the power output will tell us the effect on a human being in the surrounding space.

In the case of a heater, the engineer, by considering the volume of the surroundings, the materials of the room, and what other heat sources are present, can determine the resulting temperature at any point. In a similar way, the acoustic engineer, by considering very similar criteria, can calculate the sound pressure level at any point. (Remember, it is sound pressure that vibrates the eardrum membrane and determines how we hear a noise.)

Sound pressure levels are also measured on a logarithmic scale but the unit is the decibel re 2 x 10^{-5} Fa. There is another advantage in using the decibel scale. Because the ear is sensitive to noise in a logarithmic fashion, the decibel scale more nearly represents how we respond to a noise.

$$SPL = 20 \log \frac{p}{p_o} \qquad \text{Equ 14.2}$$

where:

SPL = sound pressure level in decibels (re 2 x 10^{-5} Fa)

p = sound pressure of the noise (Pa)

p_o = reference pressure (= 2 x10^{-5} Pa)

It should be realised that in specifying a sound pressure level, the distance from a noise source is implied or stated. In Table 14.2 the position of the observer relative to the source is indicated.

Sound pressure Pa	Sound pressure level dB	Typical environment
200.0	140	30 m from military aircraft at take-off
63.0	130	Pneumatic chipping and riveting (operator's position)
20.0	120	Boiler shop (maximum levels)
6.3	110	Automatic punch press (operator's position)
2.0	100	Automatic lathe shop
0.63	90	Construction site – pneumatic drilling
0.2	80	Kerbside of busy street
0.063	70	Loud radio (in average domestic room)
0.02	60	Restaurant
0.0063	50	Conversational speech at 1 m
0.002	40	Whispered conversation at 2 m
0.00063	30	
0.0002	20	Background in TV and recording studios
0.00002	0	Normal threshold of hearing

Table 14.2 The position of the observer relative to the source

Note: The engineer must clearly distinguish and understand the difference between sound power level and sound pressure level. He must also appreciate that dB re 10^{-12} watts and dB re 2 x 10^{-5} Pa are different units.

It is impossible to measure directly the sound power level of a fan. However, the manufacturer can calculate this level after measuring the sound pressure levels in each octave band with the fan working in an accepted standard acoustic test rig.

What he cannot do is unequivocally state what sound pressure levels will result from the use of the fan. This can only be done if details of the way the fan is to be used, together with details of the environment it is serving, are known and a detailed acoustic analysis is carried out.

14.1.6 Octave bands

Noise usually consists of a mixture of notes of different frequencies, and because these different frequencies have different characteristics a single sound power level is not sufficient in itself to describe the intensity and quality of a noise.

Noise is therefore split up into octave bands (bands of frequency in which the upper frequency is twice that of the lowest) and a sound pressure level is quoted for each of the bands. The octave band frequencies universally recommended have mid-frequencies of 63, 125, 250, 500, 1000, 2000, 4000, and 8000 Hz.

It is now becoming an increasing requirement for data at 31.5 Hz and 16000 Hz to also be included, although for a number of reasons the former is exceedingly difficult to measure with any degree of certainty.

The noisiness of a fan is specified by a number of sound power levels (in decibels re 10^{-12} watts), each corresponding to an octave band of frequencies. For research and other purposes it is also possible to measure the noise in more precise bands e.g. ⅓ octave or at so-called discrete frequencies.

As with sound power levels, sound pressure levels must be quoted for each octave band if a complete picture of the effect of the noise on the human ear is required.

14.1.7 How does sound spread?

The effect of a sound source such as a fan on its environment can be likened to dropping a pebble into a pond. Ripples will spread out uniformly in all directions and will decrease in height as they move from the point where the pebble was dropped. Normally the ripples will be circular in shape unless affected by some barrier. See Figure 14.1

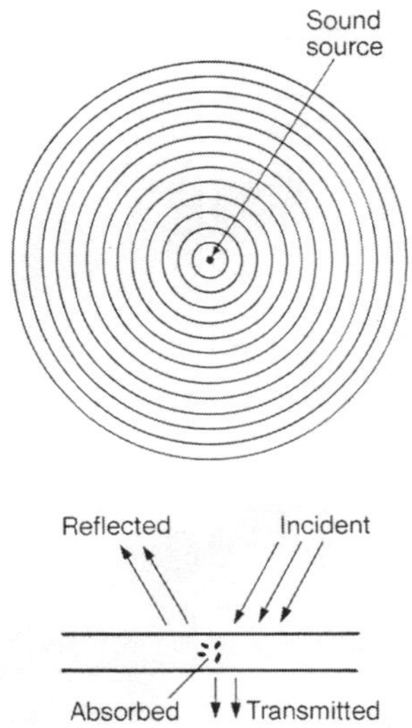

Figure 14.1 Sound in a free field (above) and sound incident on a surface (below)

It is just the same with a sound source in air. When the distance doubles, the amplitude of the sound halves, and this is a reduction of 6 dB, for using equation 14.2:

$$\text{Reduction} = 20 \log \frac{p_2}{p_1} = 20 \log 2 = 6 \text{ dB}$$

But the power of the sound source and therefore the SWL is unchanged.

To summarise, if you move from one metre from the source to two metres, the SPL will drop by 6 dB. If you move to four metres it will drop by 12 dB, eight metres by 18 dB, and so on. But this is only true if there are no objects in the path of the sound, which can reflect, or block.

Ideal conditions where the sound can spread unhindered are termed "free field". If there is an object in the way, some of the sound will be reflected, some absorbed, and some transmitted right through. How much is reflected, absorbed, or transmitted depends on the properties of the object, its size, and the particular wavelength of the sound. Generally speaking an object must be larger than one wavelength to have an effect.

$$\text{Wavelength} = \frac{\text{Speed of sound} \cong 340\,/\,\text{s}}{\text{Frequency Hz}}$$

For example

$$\text{Sound of 8K Hz : wavelength } 340 = \frac{340}{8 \times 1000} = 0.425 \text{ m}$$

$$\text{Sound of 63 Hz: wavelength} = \frac{340}{63} = 5.4 \text{ m}$$

Hence for a high frequency noise even a very small object will disturb the sound field and absorb or isolate it. But low frequency noise, whilst less objectionable, is more difficult to block.

14.1.8 Sound absorbing or anechoic chambers

If we wished to make measurements in a free field without any reflections, then the top of a very tall but small cross-section flagpole in the middle of the Sahara desert (after it had been raked flat) would probably be ideal. Obviously there are difficulties and an anechoic room is a reasonable alternative. Here the walls, ceiling and floor are covered in a highly sound absorptive material to eliminate any reflections. Thus the SPL in any direction may be measured. See Figure 14.2.

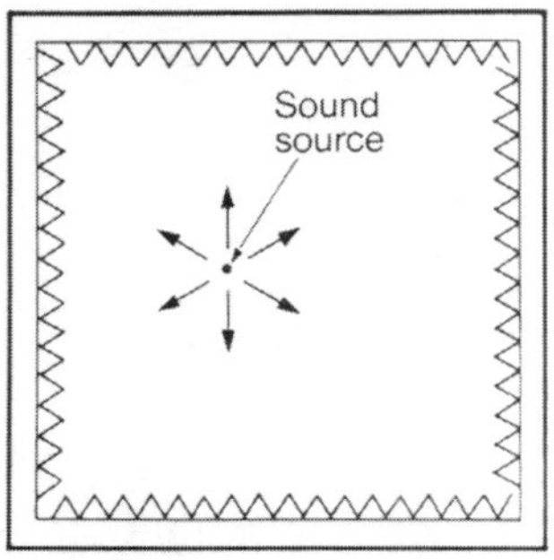

Figure 14.2 Sound in an anechoic chamber

14.1.9 Sound reflecting or reverberation chambers

This is the opposite of the anechoic chamber. All surfaces are made as hard as possible to reflect the noise and all the walls are made at an angle to each other so that there are no parallel surfaces. Thus the sound energy is uniform throughout the room and a "diffuse field" exists. It is therefore possible to measure the SWL, but the SPL measurements in any direction will be meaningless due to the many reflections. Such rooms, see Figure 14.3, are cheaper to build than anechoic chambers and are therefore very popular.

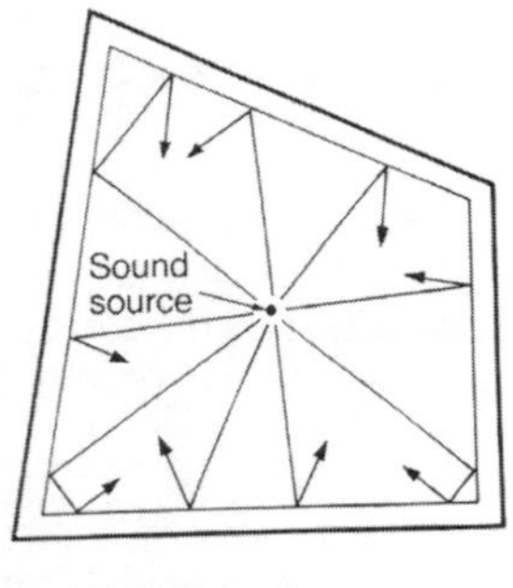

Figure 14.3 Sound in reverberation chamber

14.1.10 The "real room"

In practice we usually wish to make measurements in a room that is neither anechoic nor reverberant, but somewhere in between. It is then difficult to find a suitable position for measuring the noise from a particular source.

When determining noise from a single fan, several errors are possible. If you measure too closely, the SPL may vary considerably with a small change in position when the distance is less than the wavelength of the lowest frequency emitted or less than twice the greatest dimension of the fan, whichever is the greater. This is termed the "near field" and should be avoided.

Other errors arise if measurements are made too far from the fan. Reflections from walls and other objects may be as strong as the direct sound. Readings will be impossible in this reverberant field. A free field may exist between the reverberant and near field and can be found by seeing if the level drops 6 dB for a doubling in distance from the fan. It is here that measurements should be made. Sometimes, however, conditions are so reverberant or the room so small, that a free field will not be present. A fan in a "real room" is shown diagrammatically in Figure 14.4.

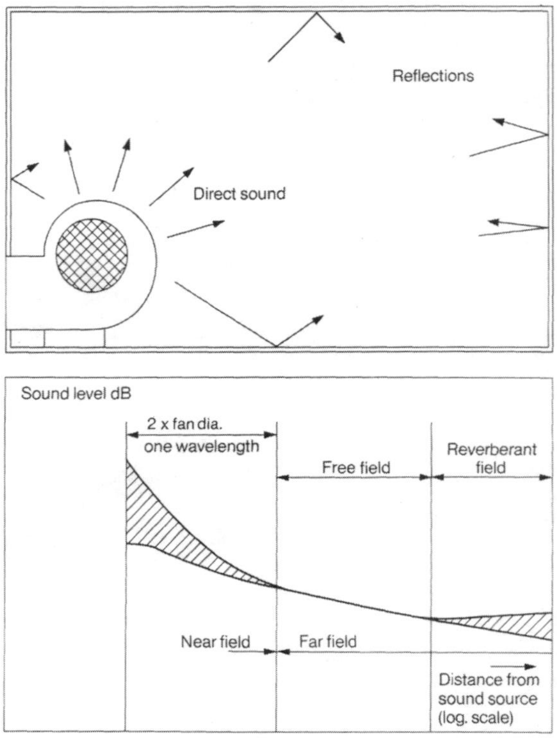

Figure 14.4 Fan in a "real room"

14.1.11 Relationship between sound pressure and sound power levels

The relationship between SPL and SWL is given as:

$$SPL = SWL + 10\log\left[\frac{Q_\theta}{4\pi r^2} + \frac{4}{R_c}\right] \qquad \text{Equ 14.3}$$

where:

SPL	=	sound pressure level dB (re 2 x 10^{-5} Pa)
SWL	=	sound power level dBW (re 10^{-12} W)
r	=	distance from the source (m)
Q_θ	=	directivity factor of the source in the direction of r
Rc	=	room constant = $\frac{S\alpha_{av}}{1-\alpha_{av}}$ (m^2)
S	=	total surface are of the room (m^2)
α_{av}	=	average absorption coefficient in the room

The first term, within, the brackets is the "direct" sound, whilst the second term is "reflected" sound.

The value of the average absorption coefficient α_{av} can be calculated.

If we have an area S, of material in the room having an absorption coefficient α_1, and area S_2 with absorption coefficient α_2, and so on, $\alpha_{av} = \frac{1}{S}(S_1\alpha_1 + S_2\alpha_2 + S_3\alpha_3 + \text{etc})$

α not only varies with the material, but also differs according to the frequency of the noise. It is therefore necessary to calculate the SPL from the SWL in each frequency. Some typical values of absorption coefficient α can be found in Table 14.3.

For special proprietary acoustic materials and all other surface finishes, refer to the manufacturers.

Material	Hertz							
	63	125	250	500	1000	2000	4000	8000
Brickwork	.05	.05	.04	.02	.04	.05	.05	.05
Breezeblock	.1	.2	.45	.6	.4	.45	.4	.4
Concrete	.01	.01	.01	.02	.02	.02	.03	.03
Glazed tiles	0.05	0.05	0.05	0.05	0.05	0.05	0.05	0.05
Plaster	.04	.04	.05	.06	.08	.04	.06	.05
Rubber floor tiles	.05	.05	.05	.1	.1	.05	.05	.05

Table 14.3 Typical values of absorption coefficient α

The surface area of a sphere equals $4\pi r^2$. Thus if the fan is in the geometric centre of the room, its sound will be equally dispersed over a sphere. If the fan is at the centre of the floor, the sound will be radiated over a half sphere for which the surface are is $2\pi r^2$. This is half the previous surface area and thus inverse of the proportion of the sphere's surface area. This is known as the directivity factor Q_θ.

The directivity factor can thus be assessed for all likely fan positions. See Figure 14.5 and Table 14.4.

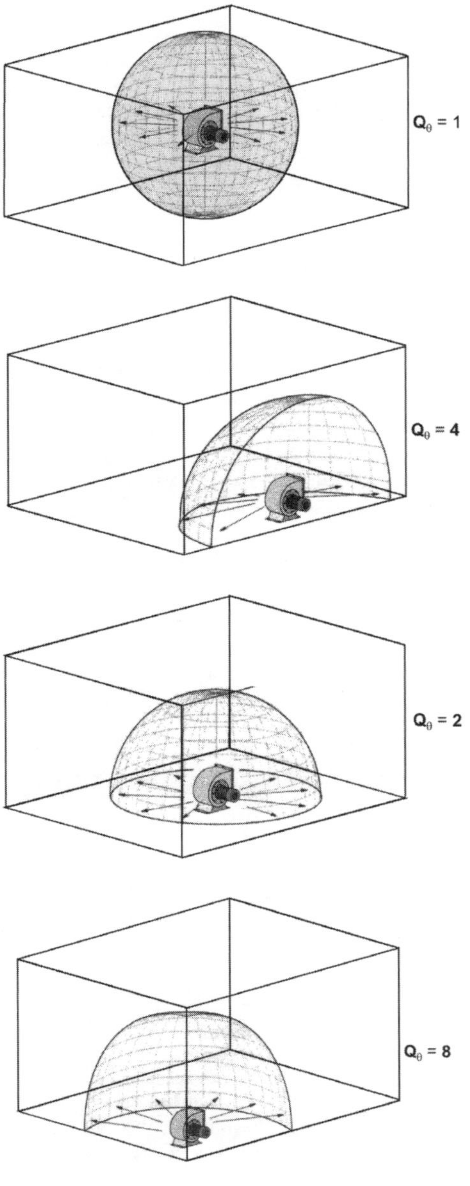

Figure 14.5 Fan source at different positions in a "real room"

Position of source	Directivity factor Q_θ
Near centre of room	1
At centre of floor	2
Centre of edge between floor and wall	4
Corner between two walls and floor	8

Table 14.4 Values of the directivity factor, assuming fan source in a large room

Certain fan manufacturers will quote the sound pressure level of their units at a specified distance – usually 1.5 m or 3 impeller diameters under "free field conditions" and assuming spherical propagation. These would exist if the fan was suspended in space and there were no adjacent floor or walls to either absorb or reflect the noise.

Using the formula in equation 14.3

$$Q_\theta = 1 \text{ and } R_c \rightarrow \infty$$

Thus:

$$SPL = SWL + 10\log\frac{4}{4\pi r^2} = SWL - 10\log 4\pi r^2$$

and

$$\text{if } r = 1.5 \text{ then } SPL = SWL - 14.5 \text{ dB}$$

Other manufacturers calculate for "hemispherical" propagation under the same free field conditions, i.e. it is assumed that the fan is mounted on a hard reflecting floor. Q_θ then equals 2.

Thus:

$$SPL = SWL + 10\log\frac{2}{4\pi r^2} = SWL - 10\log 2\pi r^2$$

and

$$\text{if } r = 1.5 \text{ then } SPL = SWL - 11.5 \text{ dB}$$

For three diameters, knowing the impeller diameter in metres, the difference in both cases may be calculated. See Figure 14.6.

Whilst these figures may be used as a basis for comparison between different units calculated in the same manner, it must be realised that the SPLs measured on site with a meter may be either above or below these values. The actual result is as much a function of the room as of the fan characteristics. The analogy of an electric fire in a room with or without heat losses should be remembered.

The internal areas of modern commercial and industrial buildings have hard boundary surfaces, which cause a high proportion of sound energy incident upon them to be reflected and a

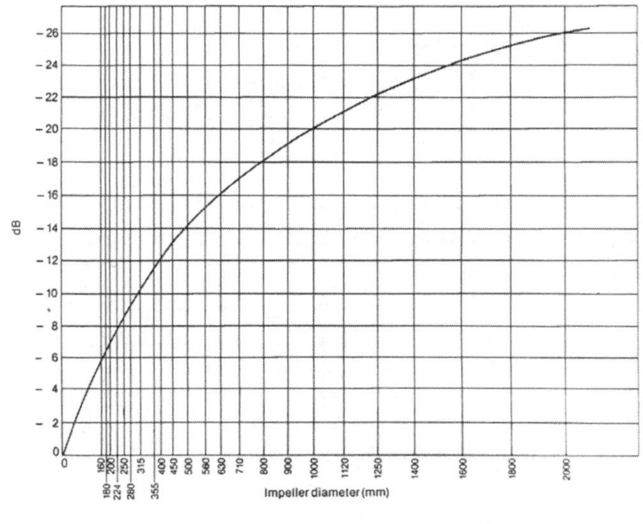

Figure 14.6 Conversion from sound power level to sound pressure level

high reverberant sound pressure level to be built up. When this occurs, the sound pressure level readings indicated on a sound meter are independent of the distance from the noise source.

Understanding the difference between sound power level and sound pressure level is important, but the engineer must also know how acceptable levels of sound pressure can be specified.

It is inconvenient to quote a series of sound values for each application. Efforts therefore have been made to express noise intensity and quality in one single number. The ear reacts differently according to frequency. All these single figure indices mathematically weight the sound pressure level values at each octave band according to the ear's response at that frequency.

To obtain basic sound pressure level, re 2 x 10^{-5} Pa under free field conditions, assuming spherical propagation, measured at 3 fan diameters distance or 1.5 m (whichever is the greater) from impeller centre, deduct the value indicated by fan diameter from the sound power level (re 10^{-12} watts).

14.1.12 Weighted sound pressure levels

A, B, C, and D noise levels are an attempt to produce single number and sound pressure indices. To obtain them, different values are subtracted from the sound pressure levels in each of the frequency bands, subtracting most from those bands which affect the ear least. The results are then added logarithmically to produce an overall single number sound level. The graphs (see Figures 14.7 to 14.10), show the different weightings employed. The resulting noise levels are known respectively as dBA, dBB, dBC, and dBD.

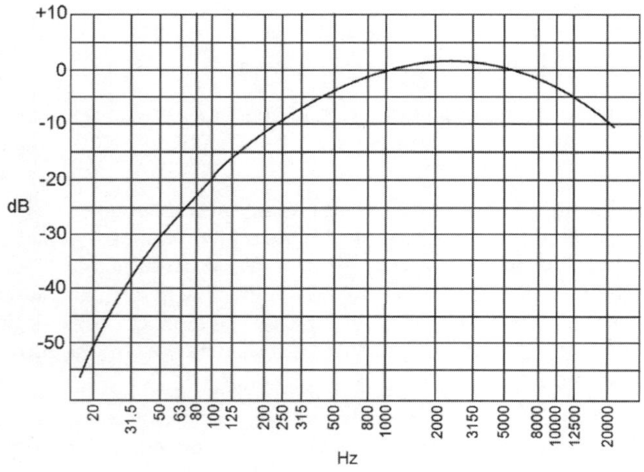

Figure 14.7 Weighted sound pressure curve A

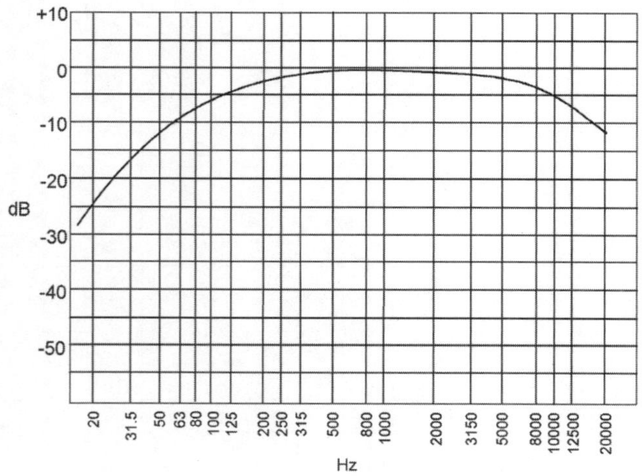

Figure 14.8 Weighted sound pressure curve B

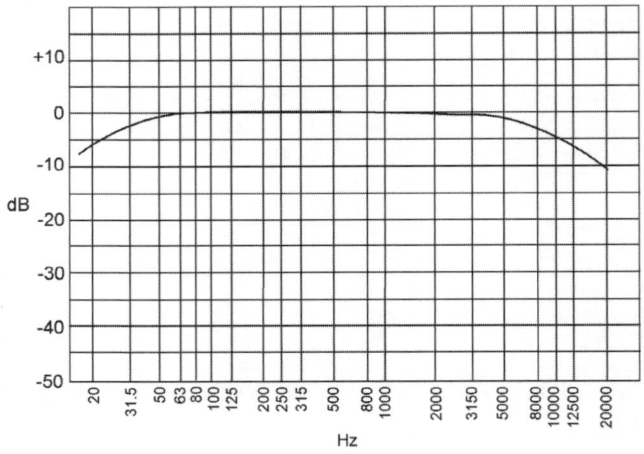

Figure 14.9 Weighted sound pressure curve C

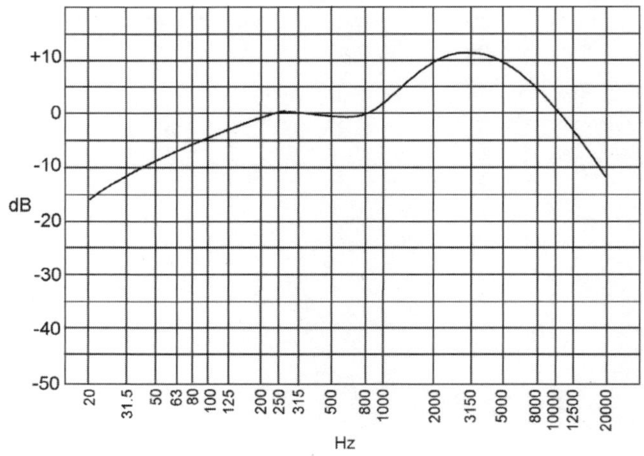

Figure 14.10 Weighted sound pressure curve D

Theoretically dBA values apply up to levels of 55 dB only, dBB for levels between 55-85 dB only and dBC for higher levels only. dBD is reserved for special noise, e.g., aircraft. However dBA is now used almost exclusively whatever the level. Engineers should check what weighting curves have been used by manufacturers and, if necessary convert them to a common base before comparisons are made.

A, B, C and D weightings are useful for making initial assessments (inexpensive sound level meters are available which measure directly on these scales). Unfortunately too much information is lost in combining all the data into one figure for it to be of use for calculation and design work. Most noise control depends on frequency analysis.

14.2 Empirical rules for determining fan noise

The desire to have a simple rule by which the noise output of a fan could be deduced from its operational duty is apparent. An early attempt was made by Beranek, Kamperman and Alien, when the following relationship was proposed:

$$PWL = 100 + 10\log \ HP \text{ dB re } 10^{-13} \text{ W} \qquad \text{Equ 14.4}$$

where:

PWL = overall acoustic power level of noise transmitted along ducts fitted to the inlet and outlet of fan operating at or near its peak efficiency

HP = nameplate horsepower of the driving motor

At that time the Americans were using a different base reference level and if updated for present day units, the above formula becomes:

$$\text{PWL} = 91.3 + 10\log \text{kW dB re } 10^{-12}\text{ W} \qquad \text{Equ 14.5}$$

which looks far less attractive and could well have been a deterrent to its use!

It will be appreciated that this formula was of necessity approximate only, and was based on a series of fans tested at pressures up to about 500 Pa. Subsequently, with the steady increase in system pressures up to 2500 Pa in many cases, a revised formula was suggested:

$$\text{PWL} = 100 + 10\log \text{HP} + 10\log p \text{ dB re}10^{-13}\text{ W} \qquad \text{Equ 14.7}$$

where:

p = pressure (ins. w.g.)

Again in modern units this becomes:

$$\text{PWL} = 67.3 - 10\log \text{kW} + 10\log p \text{ dB re}10^{-12}\text{ W} \qquad \text{Equ 14.8}$$

where:

p = pressure (Pa)

Bearing in mind that there can be a considerable difference between absorbed and nameplate power (especially in the case of forward curved centrifugal fans), it was also suggested that the former be inserted in the formula.

A further manipulation of the power term is possible for:

$$\frac{Q \times p}{10 \times \eta\%} = \text{kW}$$

where:

Q = m^3/s

p = Pa

η = fan efficiency %

then

$$\text{PWL} = 57.3 + \log Q + 20\log p - 10\log \eta\% \text{ dB re } 10^{-12}\text{ W} \qquad \text{Equ 14.9}$$

This formula gives the total noise. Assuming that inlet and outlet noise are equal, then these would each, of course, be 3 dB less.

And there the exercise should end, for one has to say that for very large fans and for fans at pressures above 1000 Pa, the uncertainty when compared with actual noise tests can be as much ±15 dB using any of these formulae, even when the fan has been selected at its peak efficiency. This is hardly surprising for whilst some fan ranges which were current in 1955 are still available, research over the past thirty years or so has meant that we now have a very much better idea of the noise generating mechanisms within fans.

Research into the cut-off and volute design of centrifugal units has, in itself, led to improvements of over 10 dB whilst in axial fans, the importance of tip clearance, impeller-casing concentricity, rotor-stator gap, and rotor-stator vane numbers, have all been the subject of important work.

It might be said that use of empirical formulae, such as those above, has by experience given results similar to manufacturers' claims. This does not necessarily confirm their correctness — indeed it may simply show that that particular manufacturer does not have noise measuring facilities, and therefore, uses the self-same formulae.

Noise measuring equipment and laboratories are extremely expensive. It is a matter of regret that only a few of the major manufacturers have invested in such facilities and that many of the others continue to use such empirical formulae. The alternative is to sub-contract such sound testing to one of the many independent laboratories now capable of this.

14.3 Noise-producing mechanisms in fans

There are three principal noise generating agencies at work in the production of a fan's total acoustic output. These may be summarised as follows:

- Aerodynamic
- Electromagnetic
- Mechanical

In most industrial fans, the order given is indicative of their relative importance, although for units at the extremities of the size range, mechanical noise becomes an increasing hazard. Electromagnetic noise, as would emanate from an electric motor, is often masked by the aerodynamic noise, especially where, as with a direct driven axial flow fan, this driving unit is contained within the casing and, therefore, the moving airstream. It can, however, be of great importance in slow speed machines driven, for example, by 6 to 12 pole motors which are inherently more noisy. In these cases, the electromagnetic contribution may be of a higher magnitude than the aerodynamic signature, especially in the lower frequency domain.

For centrifugal fans, where the motor is usually outside the airstream, electromagnetic noise will not contribute to the in-duct sound power level. It may, however, mask the breakout noise from the fan casing and ducting system. Many electric motors used with such fans are of the totally enclosed fan ventilated type, and in these the cooling fan may itself be the dominant noise source in the free field around the unit.

14.3.1 Aerodynamic

There are three recognised ways in which acoustic energy may be derived from the kinetic energy produced by a fan impeller in its action on the airstream (Figure 14.11). They are, in descending order of radiation efficiency:

Monopole source: The most efficient generating mechanism in which the conversion from kinetic to acoustic energy is achieved by forcing the gas within a fixed region of space to fluctuate. This may be visualized as a uniformly radially pulsating sphere surrounded by a perfectly homogeneous material of infinite extent, such that no end reflections occur.

Dipole source: This is thought to be the predominant sound generating mechanism in low speed turbo machinery such as

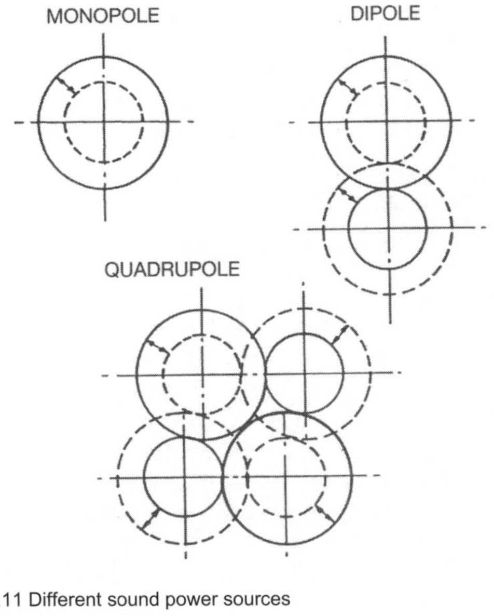

Figure 14.11 Different sound power sources

fans. Energy conversion requires the momentum within a fixed region of space to fluctuate, the process being equivalent to a uniformly pulsating sphere oscillating in the x-direction as a rigid body. Alternatively, it may be thought of as two adjacent monopoles where one is at its maximum dimension, when the other is at a minimum. Thus the dipole is vibrating along one axis. This accounts for the directional nature of the sound generated, the normal particle velocity on the sphere surface being a function of its polar location.

Quadrupole source: This is the least efficient energy conversion mechanism in which sound is generated aerodynamically, with no motion of solid boundaries, as in the mixing region of a jet exhaust. Within a fixed region of space, there is no change of either mass or momentum.

Energy conversion is achieved by forcing the rates of momentum flux across fixed surfaces to vary. Momentum flux is the rate at which momentum in the x_i direction is being transported in the x_j direction, with corresponding velocities v_i, v_j. A quadrupole source may be modelled as a double dipole, both oscillating along the same axis. It exhibits complex directionality.

The acoustic pressure generated by these different sources may be deduced as follows:

$$\tilde{\Delta}\,\text{monopole} \propto \frac{1}{r}\frac{\delta M(t)}{\delta t}$$

$$\tilde{\Delta}\,\text{dipole} \propto \frac{\delta}{\delta x}\left[\frac{r_{sp}}{r}, \frac{\delta M(t)}{\delta t}\right]$$

$$\tilde{\Delta}\,\text{quadrupole} \propto \frac{\delta}{\delta x_i}\frac{\delta}{\delta x_j}\left[\frac{1}{r}, v_i, v_j, \rho, D^3\right], F(\nu, \Delta t)$$

where:

$M(t)$ = rate of addition of mass from the neighbourhood of the source to its surroundings

r = polar distance to the observer

r_{sp} = radius of the sphere

x = direction of oscillation

v = momentum flux velocity

D = characteristic dimension

ρ = ambient density of the air or gas

ν = air or gas viscosity

Δt = temperature change across the region

Generally the dissipation of acoustic energy into heat by viscosity and heat conduction, is negligible over distances of less than say 100m, in which case the viscosity and temperature defect terms in the quadrupole equation may be neglected.

The equations detailed above may be applied to single sources, but within the acoustic field of a fan, the degree of radiation will depend also on the level of phase cancellation between adjacent sources. Indeed, this whole question of phase difference is seen as the way forward in the reduction of fan noise. It is leading to the introduction of scimitar-shaped blades, angular cut-off pieces and other devices.

It was Lighthill who first applied dimensional analysis to the acoustic power radiated by the different sources of sound pressure and derived the proportionality relationships with respect to velocity. The writer has, however, extended these identities in the final column by recognising that, in a homologous series of fans, all velocities will be proportional to the impeller tip velocity, i.e.

$$v \propto \pi DN \qquad \text{Equ 14.10}$$

where:

D = characteristic dimension, is recognised as the impeller tip diameter (m)

N = impeller rotational speed (rev/sec)

Thus:

for a fluctuating mass or monopole the generated sound power

$$\propto \frac{\rho D^2 v^4}{C} \quad \propto \frac{\rho D^6 N^4}{C}$$

for a fluctuating force or dipole the generated sound power

$$\propto \frac{\rho D^2 v^6}{C^3} \quad \propto \frac{\rho D^8 N^6}{C^3}$$

for turbulent mixing or quadrupole the generated sound power

$$\propto \frac{\rho D^2 v^8}{C^5} \quad \propto \frac{\rho D^{10} N^8}{C^5}$$

Now we know that the air power P, i.e. the power absorbed by the fan impeller

$$P \propto \rho D^5 N^3 \times fn\,(Re_F)$$

We may therefore state that:

Sound Power W_n

$$\propto P\frac{DN}{C} \propto PMa_F \times f_n\,(Re_F) \text{ for a Monopole} \qquad \text{Equ 14. 11}$$

$$\propto P\left[\frac{DN}{C}\right]^3 \propto PMa_F{}^3 \times f_n\,(Re_F) \text{ for a Dipole} \qquad \text{Equ 14.12}$$

$$\propto P\left[\frac{DN}{C}\right]^5 \propto PMa_F{}^5 \times f_n\,(Re_F) \text{ for a Quadrupole} \qquad \text{Equ 14.13}$$

as, by the re-introduction of π, we can recognise that $\frac{\pi DN}{c}$ is the Mach number related to the impeller tip speed, i.e. Ma_F.

In high speed fans this can approach 0.3. The Reynolds number function has the effect of reducing these indices.

We can also see that in a homologous series of fans, the generated sound power $W_n \propto D^{\kappa}N^{\iota}$ where κ must lie between 6 and 10 whilst ι is some number between 4 and 8.

Overall sound power radiation for any homologous series of fans will have a sound power/rotational velocity relationship, which depends on the relative contributions of the three sources. However, it is not simply a matter of how an acoustic mechanism varies with a typical speed, but rather how the flow conditions related to that acoustic mechanism vary with speed.

Whilst a considerable amount of work has been done in attempting to define a consistent relationship between fan rotational speed and the generated sound power, unless strict similarity is ensured, or design variations accounted for, the empirically derived equations may give rise to considerable error. Consequently, results from various researchers differ and the exponents have been variously quoted between 6 to 8 for k and 4 to 6 for ι. It should be noted here that the Beranek formula and its extrapolations assume $\iota = 5$ as power absorbed $\propto$ Qp and $Q \propto v$, $p \propto v^2$ and the pressure term has a coefficient of 20.

The first theoretical study of noise from rotating machinery was probably that of Gutin in 1936. His basic equation assumed a steady state where the blade loading distribution was independent of time. Here an element of gas within the area swept by the rotor was considered to receive an impulse periodically with the passing of a blade. The impulses were treated as a series of dipole sources distributed throughout the swept area, and of constant strength at any radius. The dipole source amplitudes were obtained from the thrust and torque loading conditions, the fundamental frequency of the noise generated being zN, where z is the blade number and N is the rotational frequency (revs/sec). The resultant sound field can be analysed into a series containing the fundamental frequency and its integer harmonics. It is assumed that the acoustic pressure satisfies the homogeneous wave equation:

$$\frac{\delta^2 p}{\delta t^2} - C^2 \frac{\delta^2 p}{\delta x^2} = 0 \qquad \text{Equ 14.14}$$

The fluid surrounding the blade surfaces must, therefore, have velocities which are low compared to the speed of sound, such that acoustic waves can travel radially from their source, this may not be the case and it is then necessary to consider the fluid as a perfect acoustic medium containing quadrupole sound sources of $T_{ij} = \rho v_i, v_j + p_{ij} - C^2 \delta_{ij}$.

As previously stated, the last two terms in this stress tensor may usually be ignored as the quadrupole strength density becomes equal to the "fluctuating Reynolds Stress" of the gas around the blades.

It is, therefore, possible to itemise the source components of the whole radiation field such that sound produced by a fan may be regarded as generated by monopole sources related to volume displacement, dipoles distributed over the machine surfaces and quadrupoles of strength density T_{ij} distributed throughout the surrounding gas.

Lighthill's acoustic analogy was to regard density variations within the gas as being driven by a source distribution

$$K = \frac{\delta^2 P}{\delta t^2} - C^2 v^2 \rho$$

for the general case of an unbounded fluid, but in the real world, solid boundaries are present. Modifications to the theory are, therefore, necessary to take account of reflections at these surfaces and also for an uneven quadrupole distribution as these may only exist external to the blades. These have been considered by Curle and John E Ffowcs Williams who have taken into account surface force distributions and moving boundaries.

Practically, sources of aerodynamic noise within a fan may be grouped under the following headings:

- thickness noise due to the passage of blades through the air - a quadrupole source
- torque and thrust noise - quadrupole sources
- rotation noise due to the blades passing a fixed point e.g. cut-off - a dipole source
- vortex shedding due to flow separation from the blades - a dipole source with some Reynolds number dependence
- air turbulence noise due to shear forces when the blades are stalled - a quadrupole source
- interference noise due to contact between turbulent wakes and obstructions
- pulsation noise - where at high system pressures the flowrate regularly varies and a pitched tone is produced a the frequency of the pulses - a monopole source.

An overall assessment of the aerodynamic generating mechanisms has been made by Neise and these are shown in Figure 14.12.

It will be noted that both pure tones (discrete frequencies) and broadband (random) noise is produced. Rotating blades displace a mass of gas periodically and generate sinusoidal pressure fluctuations in the adjacent field so that thickness noise is found in all but the very highest pressure fans, the acoustic radiation efficiency is low and thickness noise is not, therefore, of great importance.

Often a fan will operate in a duct system where the approaching airstream is not fully developed. The velocity profile may be "peaky", contain swirl, or indeed be axially distorted. Thus its impeller will be subjected to unsteady fluid forces, since both the magnitude of these velocities and their angle of attach will change with angular position.

Tyler and Sofrim have shown that the phase velocity of these unsteady blade forces may be much higher than the relevant impeller peripheral speed, and even be greater than the speed of sound. Their acoustic radiation efficiency will thus be very high and tonal noise will be produced at blade passing frequency and its harmonics. The usual cause of such noise will be the presence of bends or transformation pieces adjacent to the fan inlet. Even sagging flexible connections can be a problem. In the fan design itself, upstream guide vanes or motor supports can cause wakes before the impeller and again result in unsteady blade forces.

The most important source of noise in a well-designed fan and duct system is due to vortex shedding from the backs of the impeller blades. This is a dipole source and is usually broadband, although instances of discrete frequency have also been noted. Thus the noise generated in such fans $W_n \propto v^6 \propto D^8 N^6$.

The spectral shape of the noise from a fan varies according to its design. In very general terms, an axial flow fan may have

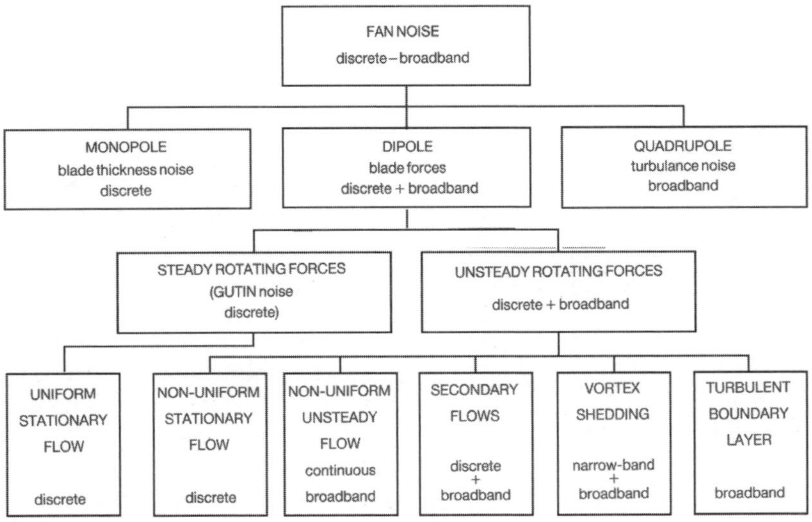

Fig. 14.12 Summary of aerodynamic fan noise generation mechanisms

high noise in the octave band containing the blade passing frequency, zN (Blade number x rev/sec) with a declination of around 2 dB per octave on either side. The peak at blade passing frequency can exceed the general spectral level by 4 to 10 dB, being especially severe where the impeller is eccentric in its casing. There may also be additional tones generated at interactive frequencies determined by (blades + vanes), (blades-vanes) etc., the strength of these being dependent on the gap between them, and the ratio $\frac{\text{Blade No.}}{\text{Vane No.}}$.

Furthermore, much recent testing of axial flow fans has shown high noise levels in the 31.5 Hz and 63 Hz bands.

Perhaps there has been too much extrapolation of idealised spectra in the past. It should be remembered that in the 1950s and 1960s, measurement of noise below the 125 Hz octave was next to impossible with the state of instrumentation and knowledge at that time.

A centrifugal fan will have a spectrum with its peak towards the lower frequencies. The declination is of the order of 3 to 7dB per octave band dependent on blade shape, but this general statement requires a host of provisos. In backward-bladed fans, the blade passing tone and its harmonics may be of especial importance. With the flat inclined type, they are easily identified above the general broadband background. With backward-curved blades, they are not so pronounced, and are lowest with backward aerofoil designs.

Sound waves produced by a source within a duct will also undergo reflection, interference and decay according to the frequency of the emitted wave. Centrifugal fans usually run at lower Mach numbers than axial fans and the predominant tones have wavelengths larger than characteristic impeller or duct dimensions. The overall radiated sound power may be greatly affected by reflection properties of the casing and ductwork. This can lead to some distortion of the sound power and directivity pattern, especially at low frequencies.

Whilst an uncased centrifugal impeller usually gives a flat frequency spectrum, the addition of a case leads to enhancement of the noise at well defined frequencies, related to the casing geometry. Flowrate variations do not significantly affect the overall shape of the cased spectra, although the magnitude, in particular frequency bands, can vary.

It is clear, therefore, that the overall radiated sound power can be quite different from the generated power. The casing may act as a Helmholtz resonator and a major casing dimension may relate to the wavelength of some important frequency. Overall, this can mean a reduction in the speed and size indices

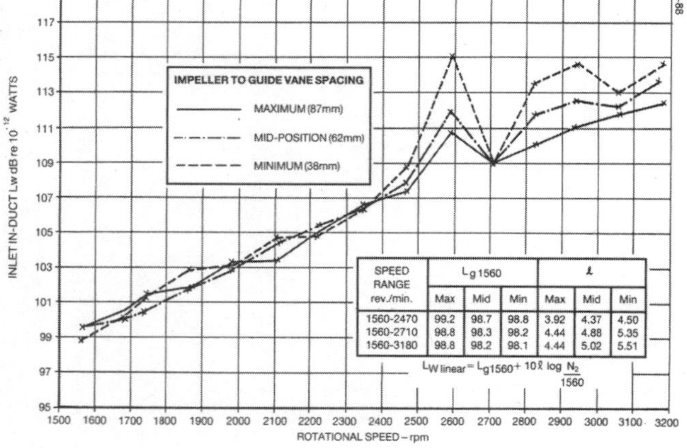

Figure 14.13 Sound power levels for a mixed flow fan at a range of rotational speed

over most of the fan performance envelope with sudden increases at identifiable speeds (Figure 14.13).

In the example shown, the first peak was seen to be where the blade passing frequency coincided with the duct cut off frequency (change from plane wave propagation to more complex modes). The second peak occurred where impeller resonance coincided with the second harmonic of blade passing frequency.

14.3.2 Electromagnetic

Whilst a very small number of fans may be driven by prime movers such as steam turbines or petrol engines, the vast majority — in excess of 98% — are driven by electric motors. With axial flow fans, it is common for the fan impeller to be mounted directly on the motor shaft extension. Centrifugal fans, may, of course, be vee belt drive or directly driven either through a flexible coupling with or without an intermediate gearbox (this is common in the UK on large mine ventilation fans).

Again with the majority of fans, electric motors are of the totally enclosed squirrel cage induction type suitable for a three phase supply. Single phase motors are usually limited to fractional horsepower outputs.

The induction motor is extremely reliable and robust. In nearly all cases it may be considered symmetrical both mechanically and electrically. The windings are balanced between phases and slots. Care is taken to ensure that the rotor runs in the correct position axially within the stator field, and that the airgap between the rotor and stator is the same at all axial and radial positions. However, especially with direct driven fans, there will be an end thrust due to the impeller action and this will "try" to take the rotor out of the magnetic field, being resisted by the magnetic forces and also such devices as wave washers in the bearing housings.

Skewing of rotor slots is often resorted to, to improve starting performance, and has also been considered as a means of reducing magnetic noise. This, however, has been the subject of much debate. Certainly an axial thrust is generated which may lead to increased noise emission.

Many fans are driven by 2 pole motors running at approximately 49 rev/sec on a 50 Hz or 59 rev/sec on a 60 Hz AC supply. If the rotor does not run in the centre of the stator, or if the stator core presents an unequal reluctance path, then a homopolar flux is generated which tries to circulate through the core, along the shaft returning via the end cover plates and frame. This causes noise and vibration at twice line frequency.

The heart of an induction motor is its laminated iron core and the stator and rotor windings. As the core is in no way connected to the power supply nor is power directly removed from it, it can be considered as passive. It is, however, the path of minimum resistance for the flux generated by the magneto motive force (mmf) set up by the stator winding, which itself is the path of least resistance for the input current.

Magnetic noise is produced by vibration of the laminations, its form being complex and taking place about all axes.

The problems of producing a low noise electric motor are severe. Yang has "de-mystified" the subject to a very large extent and shown that the noise emitted by a motor depends not only on the electromagnetic forces but also on the response to those forces by the motor carcase, and end- shields and to their radiating characteristics. He has also shown the value of parallel path winding.

The rotor must be concentric with the stator bore, and this requires that the bearing and end-shield location and stator pack tolerances all be closely controlled during manufacture. Bearing housings and end-shields need to be sufficiently rigid to avoid distortion during assembly. If the motor casing is of fabricated construction, stress-relieving is desirable before the final machining operation. In general terms, the greater the size of iron core per kilowatt of output at a given speed, the lower will be the level of magnetic vibration and noise.

Other features that have an effect are:

- core material, size and geometry,
- natural frequency of the core, core-to-frame fit and core-pack axial pressure,
- lamination insulation and burr height, number of stator slots, type and fit of stator coils,
- type and fit of slot wedges, pitch of coils, connection of coils and coil groups,
- impregnation, number of rotor slots, air-gap length and frame stiffness.

In summary, the power supplied to a three-phase stator winding sets up a rotating magnetic field. This induces an opposing current in the rotor winding and thus another magnetic field. Interaction of these two fields produces a tangential force. As the rotor shaft is only restrained by its bearings, it has to rotate.

Viewed from a fixed point on the rotor, the air-gap performance around a rotor with R slots will have R cycles of variation. Similarly, a stator with S slots will produce S cycles of variation. As the power to the stator has a frequency f_1 Hz, and as the winding is distributed around the stator in slots, the stator will produce vibrations, and therefore noise, proportional to field strength squared, related to the supply frequency, winding pitch and number of slots per pole-pitch.

Harmonics will also be present and, together with all the interactive frequencies, a very complex situation results. The rotating magnetic field of the stator produces low frequency vibration and noise, whereas rotor slot performance variation and its reactions with supply frequency lead to higher frequencies.

These may be calculated from:

$$(R \times f_1) - 2f_L,\ \text{Hz} \qquad \text{Equ 14.15}$$

$$R \times f_1,\ \text{Hz} \qquad \text{Equ 14.16}$$

$$(S \times f_1) + 2f_L \qquad \text{Equ 14.17}$$

where

f_L = line frequency

f_1 = rotational frequency

When R > S, equation 14.15 is usually of more importance. If S > R, equation 14.17 predominates. Again, many harmonics will be present.

At the design stage, the stator-rotor slot combination can be chosen to minimise vibration. To achieve this, the number of vibration nodes should be as high as possible:

$$\text{number of nodes} = (2R - 2S) \pm 2P \qquad \text{Equ 14.18}$$

where:

P = number of poles

Nevertheless, the "magic" combinations of stator/rotor slot numbers should be viewed with suspicion at the very least.

Forces in the air-gap between rotor and stator tend to pull these together and produce vibration at double the line frequency. Normally, this vibration is small, except in 2-pole motors, and if the air-gap varies, or if the tightness of stator laminations or winding in the stator varies. The second and third harmonics may also be important.

In general, slip frequency (= f_L - f_1 Hz) will not in itself be important, as it will be of very low frequency. Its interaction with higher frequencies can, however, produce pulsations.

If the rotor is severely unbalanced, the high spot will come closer to the stator than other points. As it passes the stator poles, a greater pull is exerted and the vibration occurs at double the slip frequency on a 2-pole motor. The magnitude of the readings in this frequency can indicate whether the problem is simply due to the lack of balance, a change in the air-gap, worn journals, broken rotor bars, etc.

If a resonance condition exists within the motor at the line frequency, large vibrations can be produced. More often this is a result of an unbalanced magnetic pull and can be overcome by changing stator connections.

With suspected electrical sources of noise and/or vibration, a simple check is to switch off the motor, when they should "die", This is the opposite to mechanical sources, which will gradually decay with decreasing fan speed. The translation of vibration into noise will depend on the constructional stability of the motor and, therefore, the "radiation efficiency" of vibrating surfaces.

From all the above, it will be appreciated that the prediction of motor noise at the design stage is nearly impossible and that similarity rules to interpolate/extrapolate the measured noise from one frame-size to another, do not exist. It is fortunate for the motor designer (and unfortunate for the fan engineer) that except in the case of low synchronous speed motors (6 or more poles) the fan noise often masks the motor noise.

Care must, however, be exercised with all motors subject to variable speed control through inverters. The electrical waveform may be distorted sufficiently from the ideal sinusoidal shape, that the motor noise may increase with reduced speed such that it dominates the fan noise.

14.3.3 Mechanical

Sources of noise under this heading are legion. Those of most importance to the fan designer are, however, restricted to a small number and may be categorised as follows:

- Bearings
- Couplings
- Gearboxes
- Vee belt drives
- Component vibration

Bearings

Bearings used in fans are of two main types:

- plain
- rolling element.

Plain bearings, whilst used to a great degree in the past on slow speed centrifugal fans, are not now nearly so popular in ventilation applications. Of recent years, therefore, their use has been confined to the larger, special purpose fans where their ability to handle high journal and thrust loads is desirable. This may require tilting load pads and/or forced lubrication.

Except for the very lightest loads when porous lead impregnated or PTFE bushes may be used, plain bearings are oil lubricated to minimise sliding friction. The performance of the bearing, in fact, depends on maintaining an oil film between the shaft and journal under the load and temperature conditions imposed. Where the fan is handling hot gases, a water jacket may be included within the housing to take away the heat transmitted along the shaft and in turn, to the oil (which would otherwise lose its lubricant properties).

Variations in the surface finish of shaft and journal and the means of circulating the lubricant are, therefore, the only cause of any noise emitted and these bearings do not usually contribute to the fan noise signature, being effectively masked by other components.

The more popular bearings in fan use are rolling element, or "antifriction" types, as they require considerably less maintenance, have reduced "stiction" at start-up, and are less restricted in the attitude at which they can operate. Grease lubrication is particularly favoured and in many cases, the race can be sealed-for- life.

A rolling element bearing consists of four sets of working components as compared with the one in a plain bearing, these being:

- outer race
- inner race
- elements (balls or rollers - cylindrical, taper or spherical)
- cage for maintaining the relative positions of the elements.

The operation is a combination of rolling and sliding contact. Rolling element bearings are considered to have point (ball) or line (roller) contact between the raceways and the elements. In reality, these conditions cannot exist where a load is applied, since the smallest force would induce an infinite stress. Deformation, therefore, takes place and this leads to the emission of noise. The contact is over an area sufficiently large to result in a stress value that can be accepted by the bearing materials. To ensure that the stress is within the elastic limit, and to keep the contact area to a minimum, the steels used are hardened. High stresses, nevertheless, result so that under normal use, the major cause of failure is fatigue, which leads to flaking of the raceway and elements and a marked increase in noise.

It has been shown by Glew that the noise emitted by a rolling element bearing is a direct function of its internal clearances. Unfortunately, many users are now requesting C3 increased clearance bearings as these are less susceptible to misalignment and, therefore, require lower skill levels by maintenance staff during replacement.

Where loading and application permit, ball bearings should be preferred to roller. An initial preload on the outer race of the bearing by a spring waved washer will also control bearing clearances (Figure 14.14).

Tapered roller thrust bearings in vertical motors have been shown to increase the noise of a 450 kW 2 pole machine by 10dB in the 2 kHz octave band, compared to the same machine running horizontally and fitted with ball/cylindrical roller bearings.

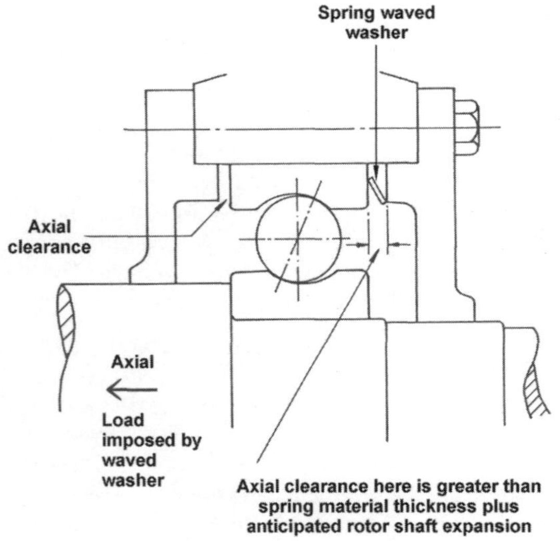

Figure 14.14 Control of bearing clearances

Bearings are often incorrectly installed which can lead to an increase in their noise emission. Even the very smallest misalignment (well within acceptable manufacturing tolerances) can be detected. Less frequently, flaws may be present on the elements and these usually result in an increase in the high frequency noise. These faults may be detected from a vibrational frequency analysis:

Flaw in outer raceway or variation in stiffness around housing

$$f_2 = f_1 \times \frac{n}{2}\left[1 - \frac{d}{D}\cos A\right], \text{Hz} \qquad \text{Equ 14.19}$$

Flaw in inner raceway

$$f_3 = f_1 \times \frac{n}{2}\left[1 + \frac{d}{D}\cos A\right], \text{Hz} \qquad \text{Equ 14.20}$$

Flaw in ball or roller

$$f_4 = f_1 \text{ x } \frac{D}{d}\left[1 - \frac{d^2}{D^2}\cos^2 A\right], \text{Hz} \qquad \text{Equ 14.21}$$

Irregularity in cage or rough spot on ball/roller

$$f_5 = f_1 \times \frac{1}{2}\left[1 - \frac{d}{D}\cos A\right], \text{Hz} \qquad \text{Equ 14.22}$$

where

n	=	number of balls or rollers
d	=	diameter of balls or rollers
D	=	pitch circle diameter of race
A	=	angle of contact of ball/rollers
f_1	=	fundamental frequency (equivalent to N rev/sec)

It should be noted that such vibrations are attenuated before being transmitted to the rest of the fan and emitted as noise. They are therefore best recognised by vibrational velocity readings on the bearing housing. Severe misalignment of a race will sometimes result in vibration at a frequency of n x f_1 Hz, even when the bearing itself is satisfactory.

In summary, modern ball and roller bearings are manufactured to a high standard and with correct installation/lubrication they are unlikely to increase the fan noise. Where the noise does increase, it is more often the fault of vibration due to imbalance, misalignment or use at speeds/loads/temperatures in excess of those recommended by the manufacturers. When faults are present, noise levels at the relevant frequencies may be as much as 7 dB greater than the readings of a good bearing.

Great care should be taken in the selection of shaft and housing limits. An interference fit of the bearing to the shaft and a small clearance between the outer raceway and the bearing housing are preferable. Bearing end caps should be of a substantial design, incorporating a sufficient number of setscrews or bolts, but differing from the number of balls or rollers.

The demand for high quality and low price necessitates quantity production of all anti-friction bearings. Machine designers are required to select from a standard range, the items that most closely meet their requirements covering: dimensional and speed properties, frictional drag and heat generated, noise output, deflection under load, rate of wear and lubrication and life in relation to load.

Of these, the life is probably of most importance, especially at the normal speeds and loads of these fans. Correct selection for life usually ensures that performance under the other headings is also acceptable.

Couplings

Couplings are not a dominant source of noise. Where misalignment is severe, they can lead to the vibration of adjacent parts and this, in turn, leads to an emission of noise dependent on the radiation efficiency of the material and its geometry.

Where torsional oscillation is present, the interaction of the coupling elements may also lead to noise dependent on the materials involved and the amount of deformation which takes place.

Gearboxes

Gearboxes are only incorporated in special purpose units such as the fans for the main ventilation of coalmines. By this means, cheaper, higher speed motors can be used to direct drive the fan at the relatively low speed required. Pinion changes can also be made where development of the mine tunnels dictates an increase in duty. Vee belt drives are usually impractical due to the very high powers involved up to about 2500 kW.

Even gears with a perfect involute form emit noise due to deflection of the teeth under load, and more importantly, the sudden changes in deflection as the load is shared and changed between differing numbers of teeth. Noise is, therefore, emitted at the meshing frequency and its harmonics. Where the gearbox contains more than two pinions to give the necessary speed reduction, side band frequency noise will be generated at the sums, differences and products of the fundamental frequencies.

Vee belt drives

These are not a source of noise except in so far as windage may be a problem with the larger spoked pulleys. Unless there are faults such as unbalance or misalignment, they can, therefore, be ignored in this analysis.

They are, of course, an extremely popular form of power transmission with centrifugal fans up to about 300kW as they enable the fan speed to be matched to the duty requirements and also have a good resistance to shock and vibration. Sometimes, because they may be seen to whip and flutter, especially when the belts are unmatched for length, they are incorrectly identified as a source of noise.

Vibration from faults in pulleys and belts may be transmitted by adjacent flat metal surfaces where these are of sufficient size. Belt faults are identified at multiples of belt speed. The relevant frequencies are:

$$1,\ 2,\ 3 \text{ or } 4 \times \frac{\text{pulley diameter}}{\text{belt length}} \times \pi \times f_p \text{ Hz} \qquad \text{Equ 14.23}$$

where:

f_p = pulley rev/sec

Likely faults are pieces broken off, hard or soft spots etc.

Faults in pulleys, such as chipped grooves etc., will be identified at the speed of the relevant pulley f_pHz.

Component vibration

Vibration can be a source of noise subject to certain conditions. Usually such vibration is itself due to some fault within the fan such as imbalance, misalignment, looseness, increased clearances, etc. Every component will have a natural frequency at which it likes to resonate. This will be "resisted" by the effects of inertia, stiffness and damping. Aerodynamic forces can excite casing panels.

Basically, any semi-rigid flat sheet surface in the fan such as the casing side plates or bearing pedestals, can act as a noise radiator where its size is equal to or greater than the wavelength of the vibration frequency transmitted to it. Its efficiency as a noise producer will be inversely proportional to the self-damping properties of the material used.

As wavelength $\lambda = \frac{C}{f}$

where:

C = the speed of sound (m/s)

f = frequency (Hz)

It follows that sound at 100 Hz could be transmitted by an unsupported panel of 3.4 m width, this reducing to 340 mm at 1 kHz. The need for stiffening and adequate metal thicknesses is, therefore, apparent.

Airborne noise will be emitted from any resonant point, whether an efficient radiator or not, where the excitation frequency coincides with the natural frequency of the element at that point, or with one of its modes as defined by its resonant frequencies.

It will be seen that component noise should not be a problem in a well-manufactured and designed fan. Where a fan has to operate at a range of speeds, however, it may be subject to resonance in some component. Often the energy in this resonance will be insufficient to cause failure, but may lead to an unforeseen increase in noise.

It might be thought that mechanical and electrical sources of noise would be masked in all cases by those of aerodynamic origin. There are, however, a number of examples where this is not the case.

To isolate non-aerodynamic sources is difficult. The usual method is to replace the fan impeller by a solid disc of the same weight, so that bearing loads and drive losses are the same. This method does not, however, reproduce any end thrust effects, nor is the electric motor under load. End thrust may be re-introduced by tilting the assembly as shown in Figure 14.15. With the addition of a belt dynamometer, the motor will be loaded, when its noise level will increase by up to 5 dB. Incidentally, we have found with electric motors that a change in core length has reduced overall fan noise by 3 dB linear.

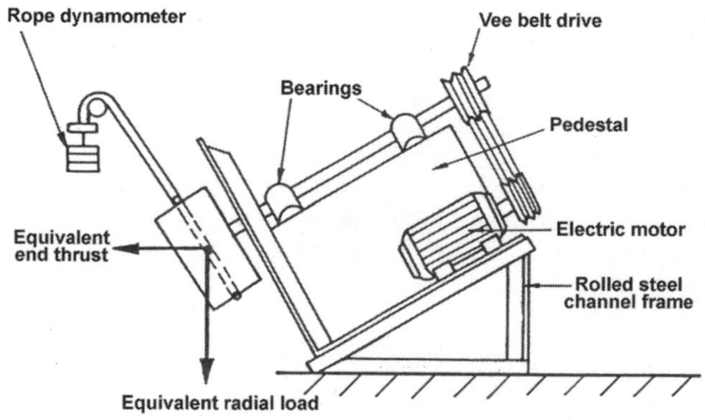

Figure 14.15 Assembly for measuring mechanical and electrical noise

14.4 Fan noise measurement

For many years it has been known that the aerodynamic performance of a fan is dependent on the ductwork connections attached to the fan inlet and/or outlet. If the fan is to develop its maximum pressure capability, then air must be presented to its inlet as a symmetrical and substantially fully developed velocity profile. In like manner, outlet ducting should permit the recovery of excess kinetic energy in the uneven velocity pressure at the discharge plane and its conversion to useful static pressure further along this duct.

The form of the inlet connection can have a significant bearing on the aerodynamic and acoustic performance, according to how the fan is ducted. Thus, a spigot may be ideal for a unit attached to its system via a flexible connection. If, however, the

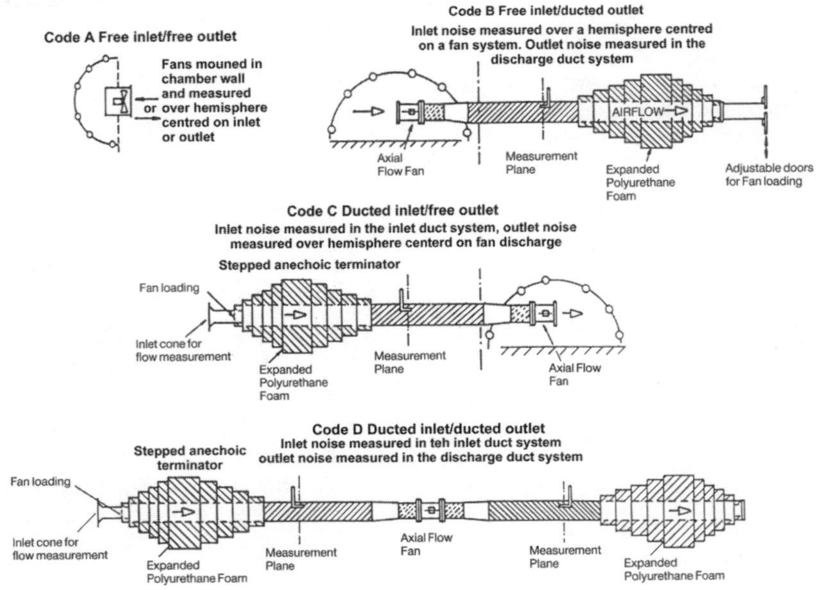

Figure 14.16 Arrangement of test ducting for measurement on in-duct and free field sound power levels

fan is unducted, and drawing its air from free space, the spigot will lead to the formation of a "vena contracta" with corresponding reduction in fan pressure and flow and an increase in noise. In such a case a bellmouth at entry will render any losses negligible.

It is only of recent years that these performance differences have been recognised in test standards and four installation categories defined in ISO 5801.

Code A: free inlet - free outlet

Code B: free inlet - ducted outlet

Code C: ducted inlet - free outlet

Code D: ducted inlet - ducted outlet

ISO 13347 and ISO 5136 have determined parallel test methods for noise, the ducting arrangements being shown in Figure 14.16. In similar manner, fan sound levels used to be considered a fixed quantity (Figure 14.17) dependent only on the position of the operating point on the fan's aerodynamic characteristic. Inlet and outlet sound power levels in open spaces around the fan inlet/outlet were calculated according to classical formulae using end reflection corrections. Research in the 1970s

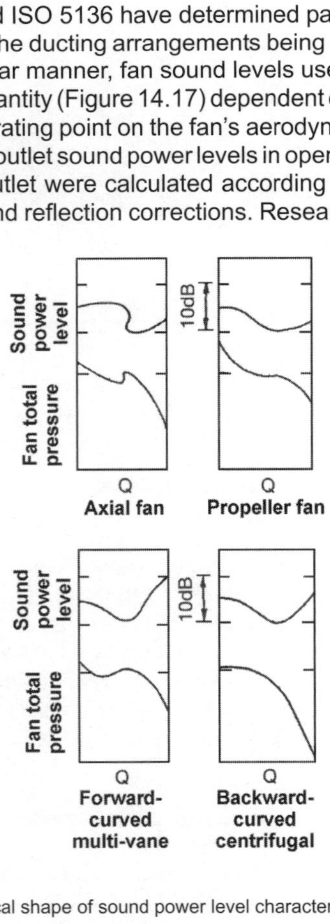

Figure 14.17 Typical shape of sound power level characteristics

by Baade suggested that this approach was no longer valid but it is only recently that differences in fan sound power levels, according to how a unit is ducted, have even begun to be recognised by industry.

We now have considerable experimental evidence to support the theory that the sound generated and radiated or transmitted by a fan, is dependent on the acoustic loading at its inlet or outlet. Hence the cross-sectional area, length and geometry of any ducting will all have an effect on the sound power levels measured.

For each of the installation categories specified above, there will, therefore, be a definitive inlet and outlet sound power. An example of these differences is shown in Figure 14.18 for a mixed flow fan.

Additionally, noise will be radiated from the fan casing, to which will be added the noise from any external motor and transmission. It will thus be seen that there are a number of noise levels that may be specified for any particular flow and rotational speed.

But even this is not the end of the story, for Bolton in 1986 also showed that outlet in-duct sound power levels measured in an anechoically terminated duct, changed when the open ended inlet duct was altered in length (Figure 14.19).

Not all researchers (see Bibliography, Section 14.15) in the field are convinced that the differences in these various levels are incapable of resolution. Whilst sound power spectra in the plan wave mode, determined by in-duct tests are invariably higher than those obtained under free field or reverberant room conditions, it is claimed that the differences can be attributed to the reflection of the sound waves at the fan inlet/outlet when the duct is removed.

Tests, however, have produced results where the differences cannot be explained by end reflections, alone. The change in acoustic loading on the inlet side due to removal of the anechoic duct leads to a reduced total (i.e. logarithmic addition of inlet and outlet) sound power output of the fan. Such an effect is not thought to be present on the outlet side.

Conversely, in the frequency range of higher order modes, in-duct sound power levels have been shown to be lower than those measured under free field conditions. It is thought by some that this may be explained by inaccuracies in the terms for "modal correction" and "flow velocity correction" contained in the standards.

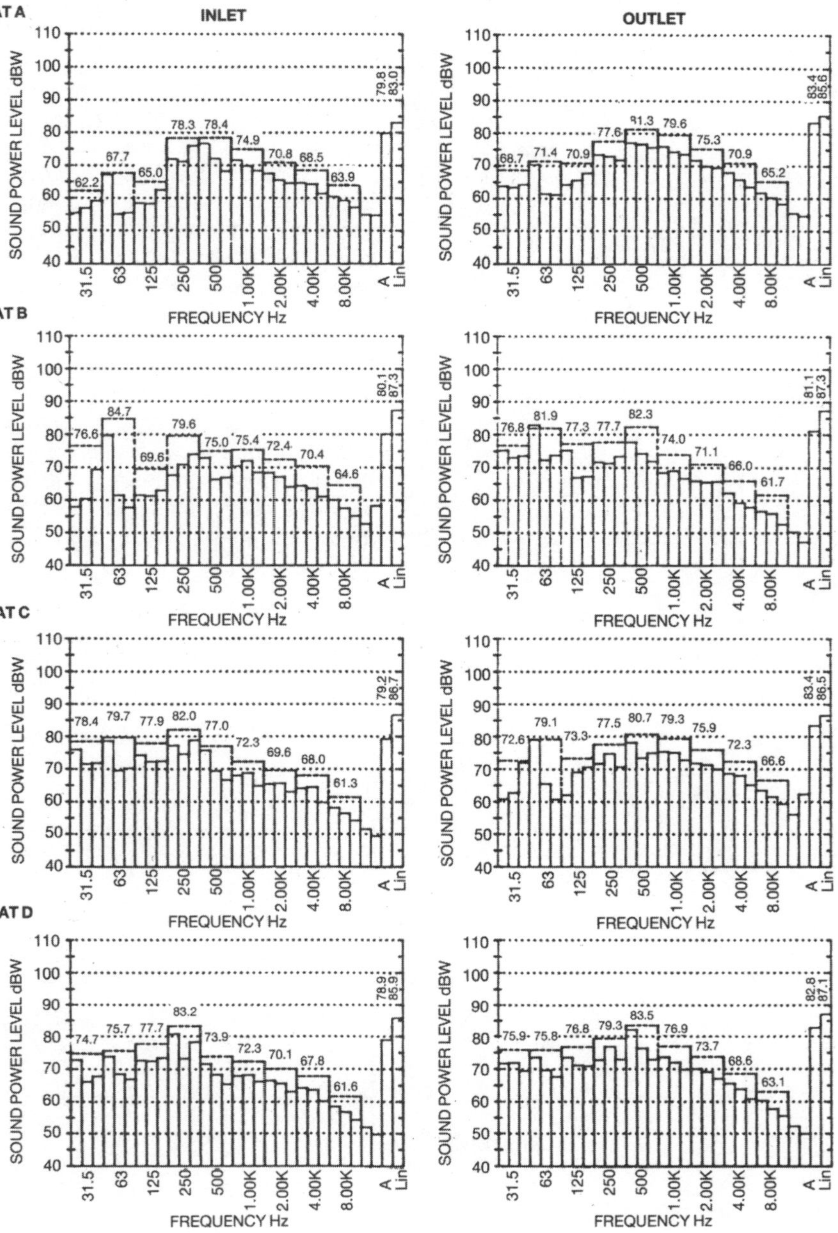

Figure 14.18 Inlet and outlet sound power differences for a 315 mm mixed flow fan at 2850 rev/min and max efficiency (0.53 to 0.54 m^3/s)

14.5 Acoustic impedance effects

An alternative and/or parallel explanation for some of the differences in sound level which have been noted, is the acoustic impedance of the ductwork configuration. Until recently, there were severe practical difficulties in making impedance measurements but these have been reduced with recent advances in digital frequency analysis and correlation techniques.

Whereas it was previously necessary to investigate the standing wave patterns by a microphone traverse along the duct for each discrete frequency of interest, it is now possible to use phase-matched condenser microphones for simultaneous measurement of sound pressure levels at a known separation. The signals may then be processed through a Fast Fourier Transform (FFT) twin channel frequency analyzer to derive impedances from the cross-spectral density function (see Bibliography, Section 14.15) or by a transfer function method.

The specific acoustic impedance i may be defined as the ratio of acoustic pressure p to acoustic particle velocity u and in air is equal to pC.

In a duct, however, this is not a particularly helpful concept and the acoustic impedance I is used, defined as the ratio of acoustic pressure p to the acoustic volume velocity q.

With plane wave propagation along a duct of cross-sectional area A and with no reflected waves, then

$$I = \frac{p}{q} = \frac{p}{Au} = \frac{i}{A} \qquad \text{Equ 14.24}$$

Where reflected waves are present, the pressure and volume velocities are the sum of incident and reflected pressures and the difference between forward and reflected velocities respectively so that the ratio of $\frac{p}{u}$ is generally complex. Knowing the impedance at a point together with either the acoustic pressure or volume velocity, it is possible to calculate the unknown parameters.

Whilst the main applications of these acoustic impedance concepts have been in reactive silencer design, an impedance model of a ducted fan as been given by Baade where it is considered as a dipole source of noise with internal impedance I_F.

Acoustic loads of impedance IL_i and IL_o are coupled to the end of straight inlet and outlet ducting respectively. Acoustic impedances seen by the fan impeller are I_i and I_o. The volume velocities q_i and q_o are equal in magnitude but of opposite sign and are related to the dipole source strength by equation:

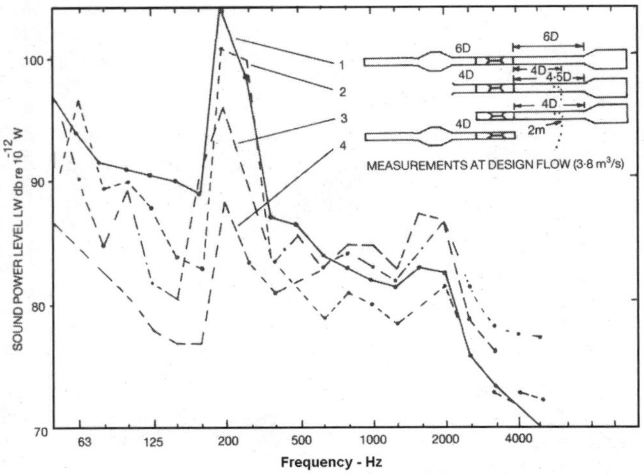

Inlet

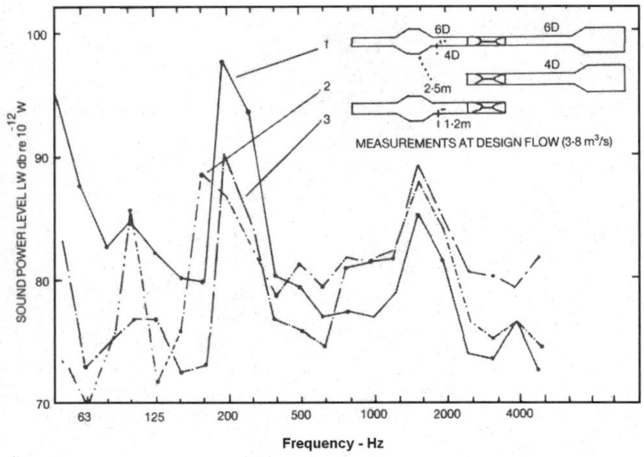

Outlet

Figure 14.19 Mixed flow fan noise at inlet and outlet under various test configurations

$$q_i = \frac{\Delta p}{l_i + l_F + l_o} = -q_o \qquad \text{Equ 14.25}$$

By manipulation of these terms and noting that the acoustic power flow $W_o = q_o{}^2 R|l_o|$

$$= \frac{\Delta p^2 R_o}{(l_i + l_F + l_o)^2}$$

Baade deduced that:

$$W_o = \frac{p^2 R\left[\frac{l_{Lo} + j\tan kl_i}{1 + j l_{Lo}\tan kl_o}\right]}{\left[\frac{l_{Li} + j\tan kl_i}{1 + j\, l_{Li}\tan k_{li}} + l_F + \frac{l_{Lo} + j\tan kl_o}{1 + j\, l_{Lo}\tan kl_o}\right]} \qquad \text{Equ 14.26}$$

It will be noted that the sound power in the discharge duct is a function not only of the outlet duct length and outlet terminating load, but also of the inlet duct length and inlet terminating load.

Bolton and Margetts have also looked at the influence of changing duct configurations on the noise generated and concluded that, there was no way of estimating the inlet or outlet sound power for one particular installation category from tests carried out on another. Tests are, therefore, necessary in all four categories from which it may be possible to identify those fan designs that are installation sensitive.

It will also be noted that it should be possible in a fully ducted situation (Installation category D) to position the fan for the minimum noise at a desired observer location. Figure 14.20 shows the differences for a bifurcated for the same aerodynamic duty but with vary distribution of the resistance on the fan inlet and outlet.

For any meaningful comparisons to be made between noise tests and fans in a homologous range, and also to compare sound power levels of fans of different types, it must be accurate and repeatable. They must provide information that can be used by a system designer for noise management and, where necessary, attenuation. To do this, it is necessary that they are conducted under a similar ducting configuration and if at all possible, under a similar distribution of inlet to outlet ducting resistance.

To repeat, the ducting acts as an acoustical impedance. The noise output at inlet and outlet not only varies according to the point on the fan characteristic. It also varies according to how it is ducted and the distribution of this ducting. We thus have at least eight different noise levels (four installation categories to be measured for inlet and outlet noise). If we add to these the "breakout" noise levels, then a further four levels can be expected.

Frequency Hz	Mixed flow	Mixed flow	In-line radial	Bifurcated axial	Axial	Backward curved centrifugal
	Duct configuration					
	Type D	Type B - Type C	Type B - Type C	Type B - Type C	Type B - Type C	Type B - Type C
	dBW	dBW	dBW	dBW	dBW	dBW
50	5.0	17.0	-1.9	7.1	-0.4	-4.0
63	8.5	9.9	-2.1	1.9	1.8	0.8
80	6.0	12.3	-2.9	-5.3	-1.1	2.1
100	8.0	5.2	-7.8	-5.2	-1.7	4.4
125	9.5	8.0	-10.7	-9.3	-11.3	4.7
160	9.0	16.3	-1.6	-16.2	-10.7	4.1
200	5.5	5.3	-4.8	-4.0	-3.4	-2.7
250	4.0	4.8	-4.2	0.7	-2.0	0.5
315	6.5	6.6	-4.2	-6.1	-0.5	3.9
400	7.0	9.6	0.1	-5.0	-0.2	3.0
500	7.0	8.4	3.5	-3.3	0	5.1
630	5.5	3.1	4.8	-2.6	-2.1	5.0
800	5.0	1.5	1.5	-1.8	-2.2	4.9
1000	2.5	-0.7	-0.7	-3.2	-1.8	5.3
1250	-2.5	-3.7	-0.6	-4.3	-2.3	3.6
1600	1.0	2.4	-0.8	-3.3	-2.5	3.3
2000	2.0	4.7	-1.7	-3.7	-2.7	3.0
2500	0	3.0	-2.6	-3.7	-3.0	1.8
3150	-4.5	1.2	-3.3	-4.7	-3.7	2.2
4000	-2.5	3.2	-3.6	-5.7	-3.4	2.5
Total	4.6	6.8	-2.1	-2.9	-3.7	-0.9

Table 14.5 Difference between outlet and inlet sound power levels for various fan types each at their design flowrate expressed as (outlet - inlet) in sound power level re 10^{-12} watts

Some representative differences for different fan types are shown in Table 14.5. And still our misery is not ended! The actual type of microphone head used can affect the results (unless correction factors are included in the measurement code) see Figure 14.21.

ISO 5136 gives these correction factors, for the different types of shield identified, according to the flow velocity and modal effects. The turbulence screen is recommended for the highest velocities, but a foam ball is adequate for the velocities experienced in normal HVAC applications.

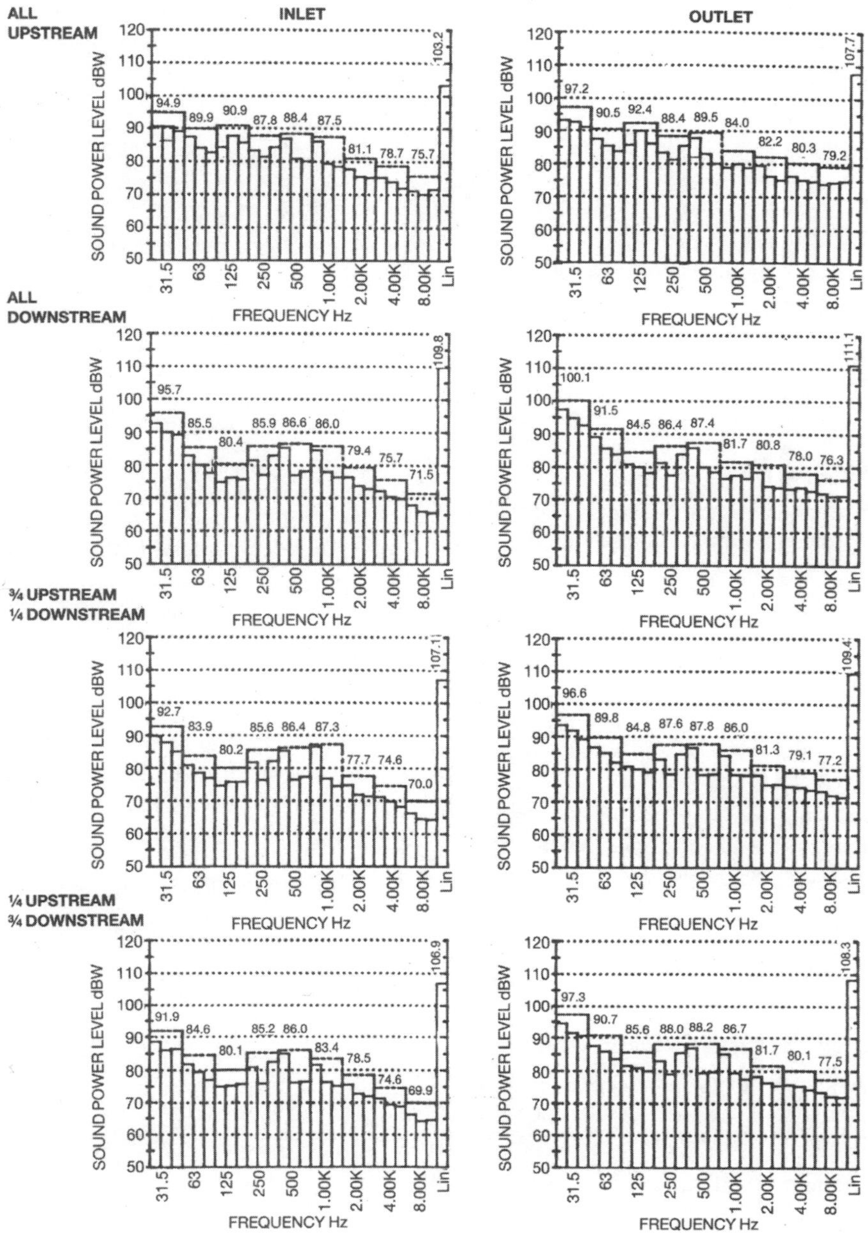

Figure 14.20 Variations in sound power levels for 610 mm bifurcated axial flow fan

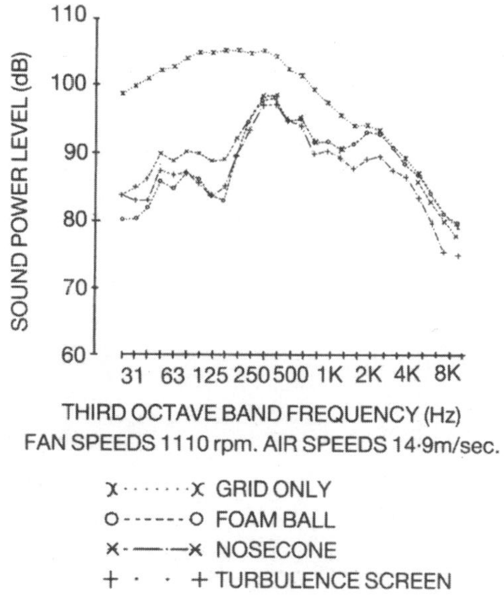

Figure 14.21 Variations in measured sound power levels for a mixed flow fan according to microphone shield used

14.6 Fan sound laws

Not withstanding the analysis of sound sources, in Section 14.3, and how they vary with rotational speed and diameter, it is felt that a simplified approach may prove useful when carrying out predictive calculations.

Just as we can calculate the air performance of a given size of fan at a given speed, from tests on another size at another speed, in a range of similar units, so it is desirable to be able to establish similar scaling laws for the acoustic performance. It is understood that such laws would be subject to the same limitations as those for air flow, i.e. strict geometric proportionality with respect to all air passages and impellers, applicable only to corresponding points of operation (equal fan efficiencies) and valid only for a specified range of fan Reynolds numbers.

The great majority of fans operate in the turbulent gas flow regime and thus generate fluctuating forces which are received by the ear as noise. If the fluctuation is regular, a fixed pitch "note" is produced, but if the process is random, then broad band noise results.

As the noise output of a fan dBW is expressed in watts, i.e. a power, it can be expected that the noise will bear some fixed relationship to the impeller power. There will also probably be Mach number and Reynolds number effects.

We therefore say that the fan sound power level dBW $\propto$ fan impeller power x f_1 (Ma) x f_2 (Re).

Now for standard air conditions:

Impeller power	$\propto N^3D^5$
Mach number	$\propto$ Tip speed $\propto$ ND
Reynolds number	$\propto$ Tip speed x diameter $\propto ND^2$

Therefore in the general case:

$$\text{Fan SWL} \propto N^3D^5 \times (ND)^a \times (ND^2)^b \qquad \text{Equ 14.27}$$

Whilst we may call this a "fan sound law" it must be appreciated that it can vary widely and is not nearly so accurate as the fan aerodynamic laws given in Chapter 4.

Considering firstly Mach number effects, the noise produced will increase with the velocities involved and according to the source of noise:

For a monopole source	a = 1
For a dipole source	a = 3
For a quadrupole source	a = 5

Reynolds number effects are more difficult to identify but for straightforward boundary layer separation one would expect a negative and fractional index, i.e. $0 > b > -1$.

Such effects will depend on absolute velocities through the impeller, thicknesses, clearances, number of blades, blade angle, etc. It seems reasonable to suggest therefore that there will be some interdependence with fan specific speed $N_S = \frac{NQ^{0.5}}{p^{0.75}}$ and that the lower N_S the nearer b will approach -1

14.7 Generalised fan sound power formula

From Section 14.6, assuming that dipole sources predominate, and that the Reynolds number exponent has an average value of -0.5, we may state fan sound power level:

$\propto$ fan power x (Mach number)^3 x (Reynolds number)$^{-0.5}$

$\propto N^3D^5 \times N^3D^3 \times -0.5\ D\text{-}1$,

or fan sound power level SWL $\propto N^{5.5}D^7$.

This of course is an average of the extreme variations which could occur but is unlikely to lead to drastic errors for small variations in speed and diameter.

Converting to a logarithmic decibel scale we may say in the general case:

fan sound power level SWL (dBW re 10^{-12} watts)

$$= X + 55 \log \frac{N_2}{N_1} + 70 \log \frac{D_2}{D_1} \qquad \text{Equ 14.28}$$

where:

X	=	constant for a particular design of fan operating at a particular point on its characteristic
N_1	=	original fan speed (rev/min)
N_2	=	final fan speed (rev/min)
D_1	=	model fan diameter (m or mm)
D_2	=	final fan diameter (m or mm)

This holds good over reasonable ranges of speed and size if the inherent inaccuracies and tolerances of the British Standard are recognised.

Example 1:

Suppose a fan 500 mm diameter operating at 1100 rev/min has a sound power level of 76 dBW. What will this increase to at 1350 rev/min?

$$\text{Increase} = 55 \log \frac{1350}{1100} = 4.89 \text{ dBW}$$

i.e. the new sound power level will be 81 dBW approximately (no greater accuracy is justified).

In like manner we may calculate the sound power level of a 630 mm diameter fan at 1100 rev/min in the same geometric series.

$$\text{Increase} = 70 \log \frac{630}{500} = 7.03 \text{ dBW}$$

i.e. the new sound power level will be 83 dBW to nearest 1 dBW.

To assist the fan and system designer, these calculations may be plotted on a Nomogram (Figure 14.22). For ease in the manipulation of the figures they can be related to a datum for a 1000 mm diameter fan at 1000 rev/min.

Thus if we have the base figure X for a 1000 mm diameter fan at 1000 rev/min at a certain point on its characteristic, Δ the increment to be added for any other size or speed in the same fan range can be obtained.

Connect the point on rev/min scale to point on size scale. Where this intercepts Δ scale is the amount to be added to the X value. (The effects of variation in air density may usually be ignored, being of the order of 1 dB or less.)

Example 2:

For the given design of fan, X has a value of 95.5 dBW. What will be the fan sound power level of a 500 mm fan at 1100 rev/min in the same series?

Joining 500 on the size scale to 1100 the increment is -19.5 dB. Thus fan sound power level = 95.5 -19.5 = 76 dBW

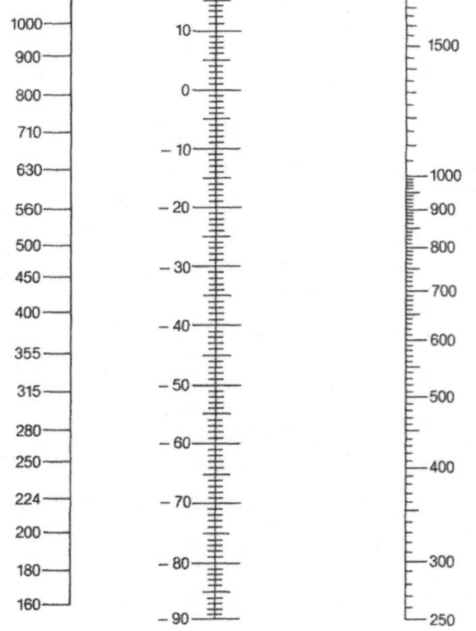

Figure 14.22 Sound level nomogram

In reverse manner, knowing the sound power level of a particular fan at a particular speed, we can calculate X.

Example 3:

A 710 mm fan at 1800 rev/min has a sound power level of 99 dBW.

Increment Δ = +3.5 dB

$\therefore$ X + Δ = 99 dBW or X = 95.5 dBW.

14.8 Disturbed flow conditions

To achieve catalogue performance in respect of flowrate and developed pressure, it is essential that the velocity profile of the air presented to the fan is fully developed, symmetrical and free from swirl. It is also essential to have a sufficiently straight duct on the discharge to permit the recovery of velocity pressure and its conversion into useful static pressure.

Similar considerations are also applicable to any noise ratings. One of the main noise sources has been observed to be due to turbulence and non-uniform inflow. Figure 14.23 shows outlet sound spectra for a specific fan fitted with different bends at varying distances from its inlet. Objects downstream of a fan will affect the noise generated by causing a non-uniform flow. This will in turn result in non-steady loading of the blades. The effect, however, is usually less severe than that for obstructions upstream.

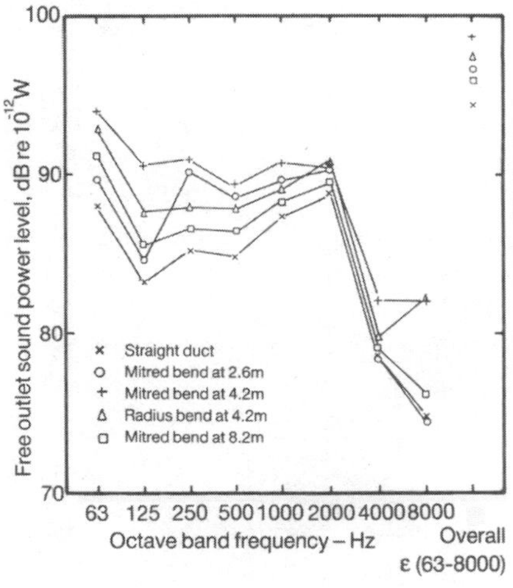

Figure 14.23 Effect of upstream bend on axial fan noise at outlet

14.9 Variation in sound power with flowrate

At a constant fan speed, the sound power generated will be dependent on the system resistance against which the fan has to operate. It is, therefore, of importance to ensure that this has been correctly calculated.

The change in noise at constant fan speed for some typical designs was shown in Figure 14.17. Differences up to 15 dB are common and will occur quite sharply if the characteristic contains a marked stall point.

In recent years, variable air volume systems have become of great importance and are recognised as an energy efficient solution to the whole question of air conditioning. It is rare for a building to continuously require the design flowrate determined by temperature, occupancy, solar heat gain, relative humidity or other criteria. Some percentage of the maximum flowrate must, therefore, be delivered by the fan.

A distribution curve (Figure 14.24) can be constructed and this indicates the percentage running time against percentage flow.

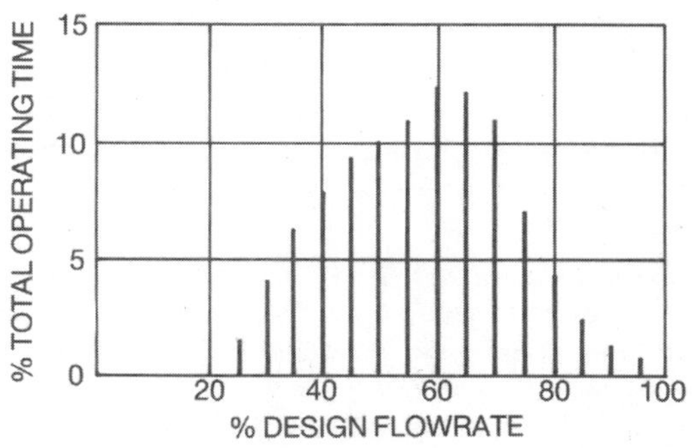

Figure 14.24 Typical fan operating load profile

How this affects the noise produced, depends on the method used.

a) **Simple damper control:** in this case the fan simply works along its characteristic. Noise will generally increase according to fan design as previously stated.

b) **Speed control:** noise may be expected to reduce with decreasing fan speed according to the relationship

$$PWL_2 - PWL_1 = 10a\frac{\log N_2}{\log N_1}$$

where:

a = exponent between 3.5 and 7.5 with an average value of say 5.5.

Note: Because of resonances and phenomena still the subject of analysis, the variation may not be continuous. It is also likely to vary at different points along the fan characteristic.

There can be "peaks" on the graph (see Figure 14.13). It should also be remembered that this curve does not take account of motor noise. Where the motor is contained within the fan duct, as with a typical direct driven axial flow fan, the reduction in noise may be less. With certain types of inverter control the electrical waveform may be sufficiently distorted to increase the motor noise at reduced speed. Figure 14.25 shows the overall effect.

Caution should still be used when controlling by inverter on lightweight fans, or where fans are mounted on flimsy structures and in any installation where the fan is run in an open environment. In these situations the torque characteristics suited to fans should always be utilised, ($V \propto f^2$). The fan type will also affect the amount of noise radiated into the system, and if possible indirect drive should be considered for critical applications.

It is likely that with careful application of damping materials and the design of fan hardware to suit the problems of general inverter drives, a reduction in the resultant noise level could be

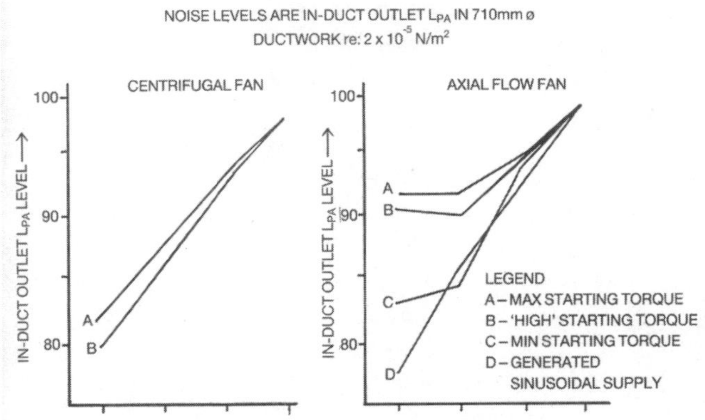

Figure 14.25 Variation in fan noise levels with speed according to motor type and control method

expected. However, this can result in considerable increase in initial fan cost, and may make the option of inverter control less attractive.

Some inverters are available that have a fundamental switching frequency in the ultrasonic range, and these noise problems can then be eliminated.

c) **Inlet vane control:** this type of control may be used with mixed flow fans, with a noise penalty of up to 10dB at small opening angles (see Table 14.6). It should not be used with axial fans where the noise penalties are severe (Table 14.7) With centrifugal fans, the effect on noise down to about 50% design flow is small, but below this figure instability can be a problem with the wider high flow designs, such that noise will increase (Table 14.8).

Vane Angle	Flowrate m^3/s	Fan pressure Pa	In duct PWL dB re 10^{-12} W								
			Tot	63	125	250	500	1k	2k	4k	8k
Full open	2.4	410	91	84	79	83	86	83	80	75	67
80°	2.37	405	92	85	80	84	87	84	80	75	67
70°	2.3	382	94	88	82	86	88	85	80	75	67
60°	2.17	356	96	91	85	87	89	86	81	76	67
50°	2.05	320	97	94	87	89	90	87	81	76	67
40°	1.89	277	99	96	89	90	90	87	81	75	67
30°	1.67	221	100	98	91	90	90	86	80	75	67
20°	1.39	154	98	96	90	89	88	84	79	74	67
10°	0.76	56	100	98	92	89	87	85	80	76	69
Closed	0	6	101	98	96	92	91	88	84	81	74

Table 14.6 Typical noise levels of mixed flow fan with inlet vane control

Vane Angle	Flow rate m^3/s	Fan pressure Pa	In duct PWL dB re 10^{-12} W								
			Tot	63	125	250	500	1k	2k	4k	8k
Fan only	24.1	Free inlet and delivery	105	93	90	96	94	94	92	98	95
Full open	23.5		122	98	100	122	110	108	101	97	96
79°	22.8		122	98	100	122	115	111	103	96	94
67°	21.8		123	97	101	119	117	111	101	96	92
56°	20.6		122	98	102	118	116	109	100	95	90
45°	19.2		120	100	103	118	112	106	100	94	87
34°	17.4		118	101	104	118	109	106	100	93	86
23°	14.9		117	102	104	116	107	105	99	93	85
11°	13.1		117	102	102	116	107	105	98	91	84
Closed	0		116	101	101	115	106	104	97	90	83

Table 14.7 Typical noise levels of axial flow fan with inlet vane control

Vane Angle	Flow rate m^3/s	Fan pressure Pa	In duct PWL dB re 10^{-12} W								
			Tot	63	125	250	500	1k	2k	4k	8k
Fan only	4.85	790	107	104	100	96	92	93	87	82	74
Full open	4.73	751	107	105	101	97	93	94	88	83	75
77°	4.37	641	106	103	100	97	93	94	88	83	75
60°	4.12	570	106	102	100	97	93	94	88	83	75
54°	3.90	520	106	102	100	97	94	95	89	84	76
24°	2.30	178	107	103	101	98	95	96	90	86	78
Closed	0.34	4	108	104	102	99	96	97	91	88	80

Table 14.8 Typical noise levels of centrifugal fan with inlet vane control

$\frac{b_E}{b}$	Flow rate m^3/s	Fan pressure Pa	In duct PWL dB re 10^{-12} W								
			Tot	63	125	250	500	1k	2k	4k	8k
Fan only	4.85	790	107	104	100	96	92	93	87	82	74
Full open	4.85	790	107	104	100	96	92	93	87	82	74
80%	4.51	683	105	103	99	95	92	93	87	82	74
60%	4.2	592	103	100	96	92	90	91	85	80	72
40%	3.90	520	100	95	95	92	88	89	83	78	70
20%	2.30	178	100	91	97	92	86	88	82	77	69
Closed	0	0	99	91	97	91	87	88	82	77	69

Table 14.9 Typical noise levels of centrifugal fan with disc throttle control

d) **Disc throttle control:** this patented control for centrifugal fans (UK number 2, 119, 44OB) varies the flow by narrowing the effective blade width and a monotonic reduction in noise with decreasing flowrate is achieved (see Table 14.9). The reductions are especially noteworthy at low frequencies where other controls are ineffective.

e) **Variable pitch in motion axial fans:** noise reduces with decreasing flow throughout the whole range of performance and no discontinuities or distortions are apparent (Figure 14.26). This graph also shows the differences in

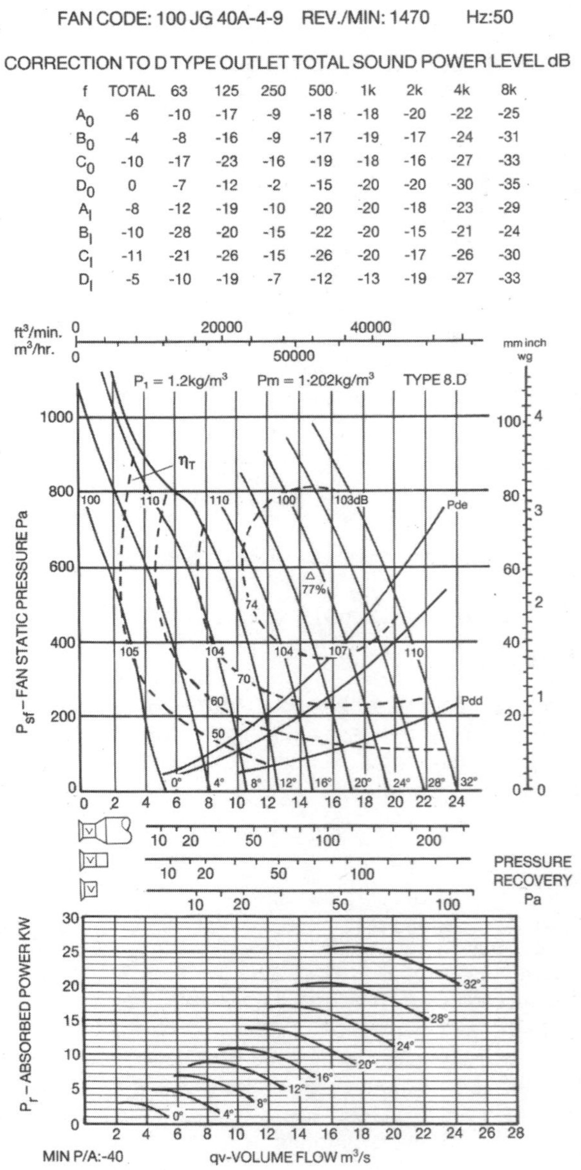

Figure 14.26 Variations in sound power levels according to installation category for an axial flow fan

spectra for both inlet and outlet noise according to ducting configuration.

14.10 Typical sound ratings

From the Section above, it will be seen that it is virtually impossible to determine the sound power of a fan for a specific duty, without knowing the characteristics of the particular design to be used.

Nevertheless, it is appreciated that a demand will still exist for some predictive measurement. In an attempt to meet this demand, Figure 14.27 has therefore been produced. Again, it is assumed that the fan has been selected to operate at its best efficiency point and is handling air of standard density.

PWL is the level of sound power transmitted along a duct attached to the fan inlet or outlet (this in itself may be ±3 dB).

LP is derived from the fan total pressure and L_p from the volumetric flowrate.

$$PWL = L_p + L_q \text{ dBW re } 10^{-12} \text{ W} \qquad \text{Equ 14.29}$$

The air duty has been used rather than the size, speed or mechanical power input so that fans of differing type or efficiency may be compared. On the diagram in Figure 14.27, straight lines have been drawn through 84 dBW, 250 Pa at slopes corresponding to PWL $\propto N^{3.5}$ to $N^{7.5}$ where N is the fan speed.

The area bounded by the dashed lines covers the range within which L_P may be expected to lie. Axial and forward curved centrifugal fans will be located around the middle of the band, whilst backward curved and mixed flow designs will be in the lower half. The lowest values will be found from aerofoil bladed centrifugal fans. At very high pressures radial tipped blades often have to be used for strength considerations. These are not so quiet and hence the power limit line has been curved upwards.

It will be noted that this graph shows a much wider spread between the best and worse fans than previously thought.

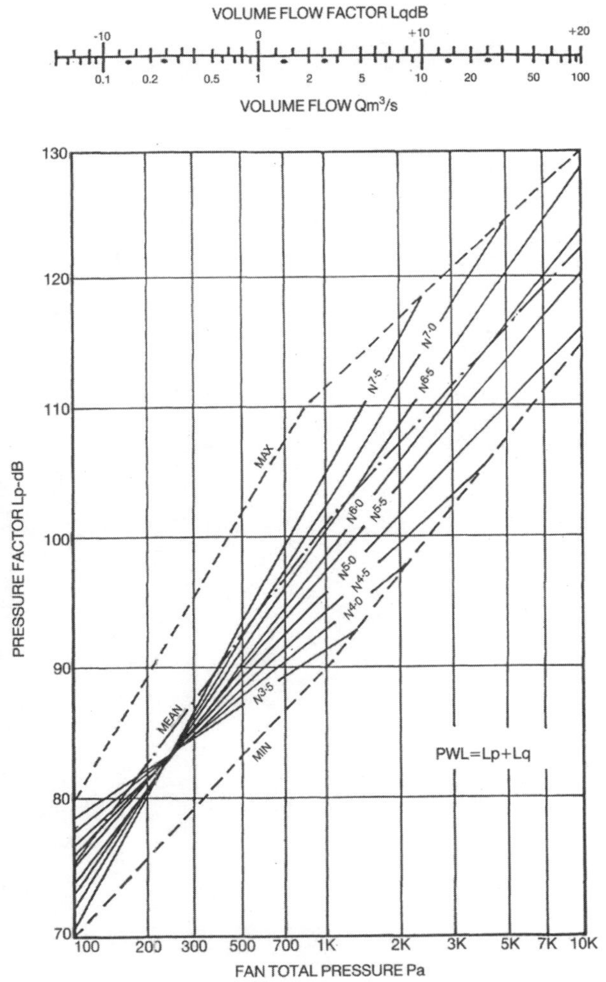

Figure 14.27 Sound power level and fan duty

14.11 Installation comments

1. In an installation where the fan and system are entirely within the space being considered (Figure 14.28), such as might be encountered in process work, local dust control systems, furnace cooling cycles, etc, the total sound is radiated to the space and the SWL values represent the total noise. If duct systems are installed on the fan discharge and inlet and the separate terminations are considered far apart, they should be calculated as separate sources. Each source can be considered as approximately 3 dB less than the total SWL of the fan, if separate data for inlet and outlet noise is not available.

   The amount of sound radiated from the fan casing is generally well below the fan inlet and outlet sound levels. It may however be required when calculating noise levels in plantrooms. If there is considerable absorption in the ductwork or system, the radiation could be a factor in the near field which is absorbed by insulating the fan.

2. Where the fan discharge (or inlet) has been ducted or connects directly to the outside space and the sound radiated through this section of the ductwork is not a factor in the determination of the sound pressure level in the space (Figure 14.29), the sound radiating from the non-ducted opening of the fan is one-half of the total sound. For ducted fans, this is the total sound power level SWL minus 3 dBW.

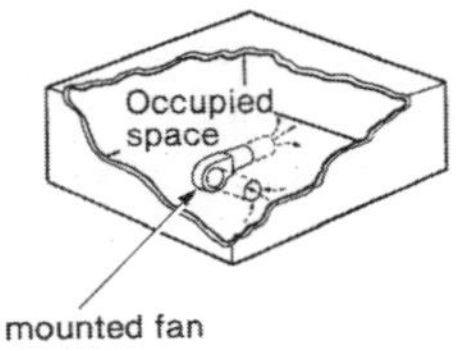

Figure 14.28 Installation — fan and system within space

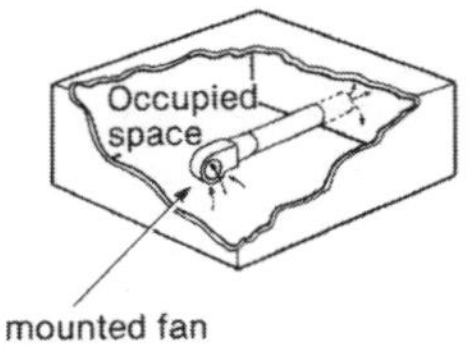

Figure 14.29 Installation — fan discharge directed to outside space

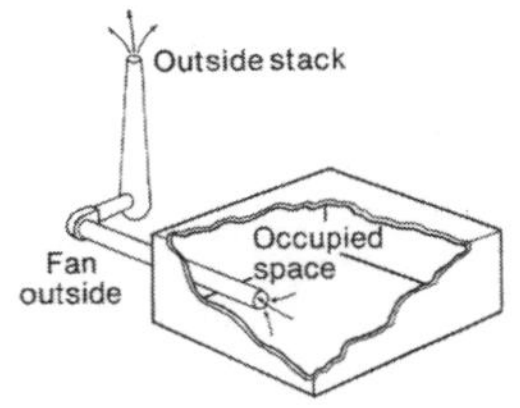

Figure 14.30 Installation — fan connected to adjacent space

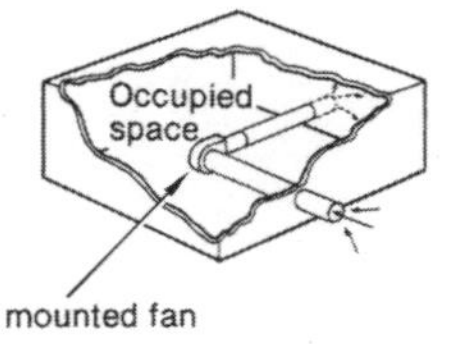

Figure 14.31 Installation — inlet and outlet ducted from room

3. Where the fan is connected to an adjacent space and sound is transmitted through the ductwork to the occupied space (Figure 14.30), the sound power level radiated from the inlet is used to calculate the resulting sound pressure level in the occupied space and is approximately equal to the total SWL values minus 3 dBW.
4. Where both inlet and outlet are ducted from the room, (Figure 14.31), it should be noted that SWL values may not specifically cover sound radiated from the fan housing. This is not a serious shortcoming since the housing radiation will not be the primary source of sound.

In most cases there are two other sources of sound that will predominate. One is the flexible connection used in most fan installations to isolate the fan vibration from the ductwork. Usually this is relatively light flexible material and becomes a source of sound far more important than that radiated from the fan. Secondly, the ductwork is, in most cases, of lighter construction than the fan housing and more sound will be transmitted through the duct walls than through the fan casing.

Depending on the fan size and casing thickness, and based on experience with installations of this kind, it is recommended that the total sound power level be reduced by up to 20 dBW to estimate the sound level in the fan house. In installations where special isolation points (special flexible connections) and heavy ductwork are used, there can be a reduction of up to 35 dBW in the occupied space.

14.12 Addition of sound levels

If the noise levels of two machines, such as a fan and its driving motor or two fans, have been measured individually and you want to know how much noise the machines will make when operating together, the two sound levels must be added.

However, when using dBW one cannot add the sound levels directly as the scale is logarithmic and:

$$dBW_{Total} = 10 \log \left[10^{\frac{SWL_1}{10}} + 10^{\frac{SWL_2}{10}} \right] \qquad \text{Equ 14.30}$$

Figure 14.32 will assist in this calculation, the procedure being as follows:

1. Measure the levels of machine 1 and machine 2.
2. Find the difference between these levels.
3. From the bottom of the chart with this difference, intersect the curve, obtaining increment on the vertical axis.
4. Add the value indicated at the vertical axis to the level of the noisiest machine. This gives the sum of the noise levels of the two machines.

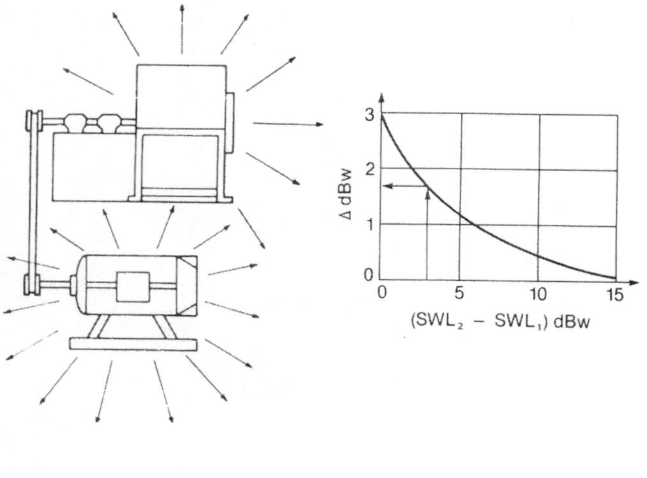

Figure 14.32 Calculation of combined sound level for fan and motor

Example:

1. Fan = 95 dBW
   Motor = 92 dBW
2. Difference = 3 dB
3. Correction (from chart) = 1.7 dBW
4. Total noise = 95 + 1.7 = 96.7 dB

14.13 Noise rating (NR) curves

It is apparent that the combination of a single figure index such as dBA, with more information on the shape of the frequency content would be useful. Noise rating curves (NR) were evolved by ISO to meet this need, largely replacing the very similar NC curves which did not follow mathematical laws and were therefore more difficult to handle on a computer. Nevertheless, such curves continue to proliferate and we now have PNC curves and who knows what else.

NR curves consist of a family of octave band spectra, with each curve marked with its own NR rating number. The octave band spectrum of the noise being analysed is plotted on the same grid and the NR rating of that noise corresponds to the highest NR curve touched by the noise spectrum.

Figure 14.33 shows a set of NR curves and Table 14.10 gives recommended levels for various environments. The spectrum of a noise with an NR rating of 35 is also shown on the grid.

NR ratings are particularly suitable for selecting and assessing suitable background noise levels for ventilating and air conditioning systems.

Warning: NR curves assume SPLs in the environment and are not directly applicable to fans without knowing the room charac-

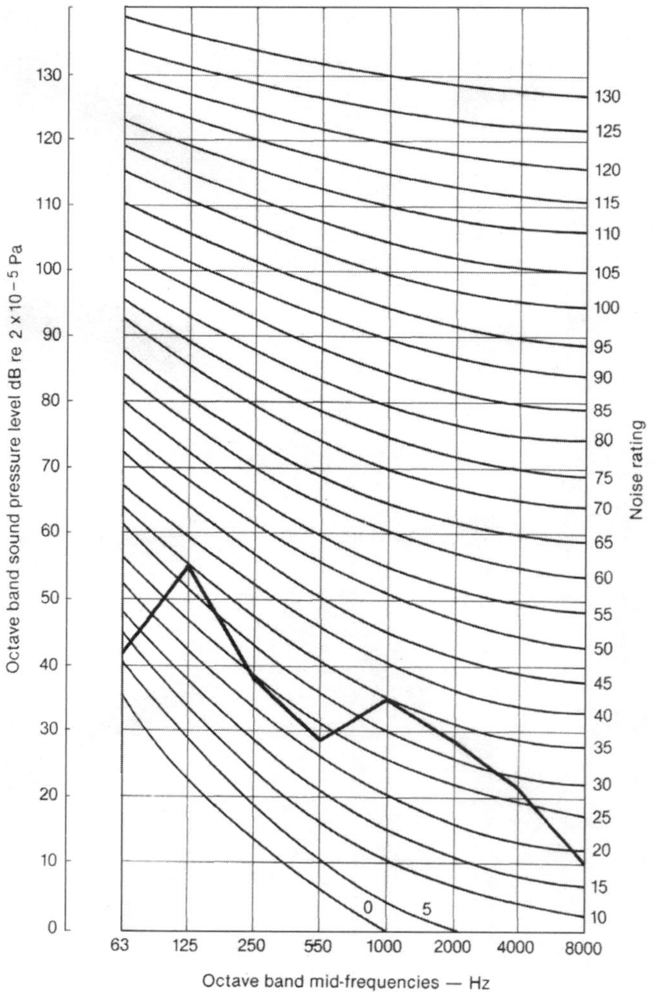

Figure 14.33 Noise rating (NR) curves

teristics, distances from sound sources to point of measurement, etc. They are best calculated by acoustic specialists knowing the fan SWL levels.

Environment	NR criterion
Concert halls, opera halls, studios for sound reproduction, live theatres (> 500 seats)	20
Bedrooms in private homes, live theatres (< 500 seats), cathedrals and large churches, television studios, large conference and lecture rooms (> 50 people)	25
Living rooms in private homes, board rooms, top management offices, conference and lecture rooms (20-50 people), multi-purpose halls, churches (medium and small), libraries, bedrooms in hotels etc., banqueting rooms, operating theatres, cinemas, hospital private rooms, large courtrooms	30
Public rooms in hotels, etc., ballrooms, hospital open wards, middle management and small offices, small conference and lecture rooms, (< 20 people), school classrooms, small courtrooms, museums, libraries, banking halls, small restaurants, cocktail bars, quality shops	35
Toilets and washrooms, large open offices, drawing offices, reception areas (offices), halls, corridors, lobbies in hotels, hospitals, etc., laboratories, recreation rooms, post offices, large restaurants, bars and night clubs, department stores, shops, gymnasia	40
Kitchens in hotels, hospitals, etc., laundry rooms, computer rooms, accounting machine rooms, cafeteria, canteens, supermarkets, swimming pools, covered garages in hotels, offices, etc., bowling alleys	45
NR50 and above NR50 will generally be regarded as very noisy by sedentary workers. Higher noise levels than NR50 will be justified in certain manufacturing areas.	

Table 14.10 Recommended noise rating (NR) levels

14.14 Conclusions

The use of empirical "laws" to determine fan noise can be fraught with danger. Even the use of so-called "fan sound laws", when applied to test data can lead to serious error. In all possible cases, reference should be made to actual tests, and results taken from as near as possible to the same size, speed and installation category.

If the flowrate varies, care should be taken in selecting an appropriate method. The sound output may increase if the ducting resistance has been incorrectly assessed and the fan does not operate at the correct point on its characteristic. Ductwork impedance can determine the fan noise, particularly at low frequencies. The need for good inlet and outlet connections cannot be understated.

14.15 Bibliography

Journal of the Acoustical Society of America, 1955. Beranek, Kamperman and Alien.

On Sound Generated Aerodynamically, M J Lighthill, I. Proceeding of the Royal Society of London, A211: 564-587, 1952.

Über das Scholifield einer Rotierenden Luftschraube, L Gutin, Physik. Zeits. SowjetUnion, 9:57-71, 1936.

The Influence of Solid Boundaries upon Aerodynamic Sound, N Curie, Proceedings of the Royal Society,1955.

Theory relating to noise of rotating machinery, J E Ffowcs Williams and D L Hawkings, Journal of Sound and Vibration, Volume 10, Issue 1, July 1969.

Fan Noise - Generation Mechanisms and Control Methods, W Neise, Proceedings Inter-noise, 1988.

Axial Flow Compressor Noise Studies, J M Tyler and TG Sofrim, Society of Automotive Engineers Transactions, 1962.

Low Noise Electric Motors, S J Yang, Oxford University Press, 1981.

The Origins and Control of Induction Motor Noise, C N Glew, Paper 3 Industrial Motors Symposium, GEC Ltd, 1977.

Effects of Acoustic Loading on Axial Flow Fan Noise Generation, P Baade, Noise Control Engineering, 1977.

A New Fan Noise Measurement Standard BS848: Part 2: 1985, A N Bolton, Proceedings of the Air Movement and Distribution Conference, Purdue University, Indiana, 1986.

Experimental Comparison of Standardised Sound Power Measurement Procedures for Fans, W Neise, F Holste and G Hoppe, Proceedings Inter-noise, 1988.

Experimental Determination of Acoustic Properties using a-two-microphone random excitation technique, A F Seybert and D F Ross, Journal of Acoustics Society of America, 1977.

Transfer Function Method of Measuring Induct Acoustic Properties, J Y Chung and D A Blaser, Journal of Acoustics Society of America, 1980.

Transfer Function Method of Measuring Induct Acoustics Properties, A N Bolton and E J Margetts, Paper C124184 Conference on Installation Effects in Ducted Fan Systems, (I.Mech.E.), 1984.

ISO 5136:2003, *Acoustics - Determination of sound power radiated into a duct by fans and other air moving devices – In-duct method.*

ISO 13347-1:2004 , Parts 1 to 4, *Industrial fans - Determination of fan sound power levels under standardized laboratory conditions.*

Woods Practical Guide to Noise Control, I. Charland, Woods of Colchester Ltd.

15 Fan vibration

Noise may be regarded as the transmission of pressure waves through a fluid, usually air (and less usually through some other gas) on its way to the human (or some other animal) ear. Vibration may be seen as a similar phenomenon, but transmitted through a solid to some other part of the recipient's anatomy.

This is a fast moving subject in which the electronics industry has become much involved. There have been numerous amalgamations of the companies concerned, whilst new ones have started up. There is however, one certainty for the author – all his descriptive material will be long out of date by the time this book is published! Modern instruments are remarkable in their versatility and ability to capture data for analysis and diagnosis. They are very much in the "black box" category, but the earlier instruments did have the capacity for displaying everything – so you thought you understood what was going on!

Contents:

15.1 Introduction

When describing the performance of a fan, the customer is accustomed to specifying the volumetric flowrate, the fan pressure and even the noise. These are met with the supplier's response of a fan size and model, a fan speed and motor requirements.

Just as fan noise has been added to the specification over the past 20 years, so vibration is now recognised as an important parameter. It gives an indication of how well the fan has been designed and manufactured and can also provide advanced warning of possible operational problems. The measured results may be useful in determining the adequacy or otherwise of concrete foundations, or the necessary stiffness of supporting structures.

It will be realised that this chapter follows on logically from Chapter 14. Noise may be regarded as the transmission of pressure waves through a fluid, usually air (and less usually through some other gas) on its way to the human (or some other animal) ear. It can however be transmitted through a liquid, such as water, and this is used in submarine detection and for communication between whales and other sea mammals. In this progression, Vibration may be seen as a similar phenomenon, but transmitted through a solid.

Vibration measurements may be required for a number of reasons of which the following are but examples:

- design/development evaluations
- in-situ testing
- as baseline information for condition monitoring programmes
- to inform the designers of foundations, supporting structures, ducting systems etc., of the residual vibration which will be transmitted into their part of the system
- as a quality assessment at the final inspection stage.

15.1.1 Identification

Perhaps the most important cause of vibration is unbalance. Reference is made to the relevant Standards and recommendations made as to an acceptable grade. Fan unbalance manifests itself as a periodic vibration characterised by a sine wave. The so-called simple harmonic motion.

With the necessary instruments three properties can be directly measured:

- displacement,
- velocity,
- acceleration.

The importance of each is discussed and the relationship between them shown. The keys to the identification of the cause of a vibration are in its frequency and velocity - NOT necessarily its amplitude except below about 10Hz. It is therefore of value to obtain a vibration signature and the analysis of this will lead to possible sources of trouble being identified. Unbalance, misalignment, eccentricity, looseness, aerodynamic forces, bearing and electric motor problems are all discussed and the troublesome frequencies identified. Particular attention is devoted to bearing defects and the concepts of shock pulse, spike energy and Kurtosis factor are introduced and the meters for their measurement described.

15.1.2 History

From the very early years of fan manufacture the problems of vibration and its reduction or isolation have given engineers many happy (?) hours of listening and analysing. The absence of vibration came to be seen as a sign of a fan's health. Perhaps this was why the old-timers used a stethoscope to hear the odd rumblings coming from the bearings!

Over the last decade or so a completely new science has emerged for accurately measuring and identifying the causes of vibration in our modern highly stressed, high speed fans. Using transducers to convert the vibrations into electric signals, these could be amplified, integrated, filtered and metered.

15.1.3 Sources of vibration

It is virtually impossible to avoid all vibration as this arises from the dynamic effects of out-of-balance, misalignment, clearances, rubbing or rolling contacts, the additive effects of tolerances etc. Sometimes the vibrations from these sources may be small, but excite the resonant frequencies of the stationary parts such as casings or bearing pedestals. Where the fan is directly driven by an electric motor, electromagnetic disturbances will also exist, these producing further vibrations.

15.1.4 Definitions of vibration

Vibration may be defined as the periodic motion in alternately opposite directions about a reference equilibrium position. The number of complete motion cycles which take place during unit time is called the frequency. This frequency may also be measured in cycles/minute which is useful for a direct comparison with the fan revs/minute. In recent years, however, the SI unit has come into prominence and frequency is usually now given in Hertz (Hz) equivalent to cycles/second.

The motion could consist of a single frequency as with a tuning fork. With a fan however there are likely to be several motions taking place simultaneously at different frequencies. These various motions can be identified by frequency analysis - or the plotting of a graph showing vibration level against frequency.

15.1.5 Vibration measuring parameters

There are three properties of a vibrating element which can be measured. Each is of value and may be recorded according to the application:

a) **Displacement**, or the size of the movement is of importance where running clearances have to be maintained for efficient performance or where contact between stationary and rotating surfaces could take place. Most weight is given to low frequency components.

b) **Velocity**, which is directly proportional to a given energy level and therefore where low and high frequencies are equally weighted. The disturbing effects on people and other equipment are by experience related to velocity.

c) **Acceleration**, which is a measure of the forces and stresses set up within the fan and motor, or between these and the foundations. Weighted towards the higher frequencies and therefore should be used where such components exist.

15.2 Mathematical relationships

15.2.1 Simple harmonic motion

The three parameters described above are mathematically connected in the case of a simple harmonic or sinusoidal vibration such as that produced by out-of-balance.

The displacement "e" is proportional to sin ωt where ωt is an angle which goes through 360° in one vibratory cycle. Angular velocity (or circular frequency) ω is equal to $2\pi f$ where f is the frequency in Hertz, or $\frac{2\pi N}{60}$ for balance problems where N equals r/min.

The other properties are also sine waves, the velocity "v" having a 90° phase lead (one quarter or a cycle with respect to time) whilst acceleration "a" is advanced by half a cycle i.e. a 180° phase lead. This is shown in the equations below:

Displacement $e = e_{peak} \sin \omega t$

Velocity $v = \omega e_{peak} \sin\left(\omega t + \frac{\pi}{2}\right)$

Acceleration $a = \omega^2 e_{peak} \sin(\omega t + \pi)$

These three parameters are illustrated in Figure 15. 1 whist Table 15.1 gives their values with respect to e_{peak}.

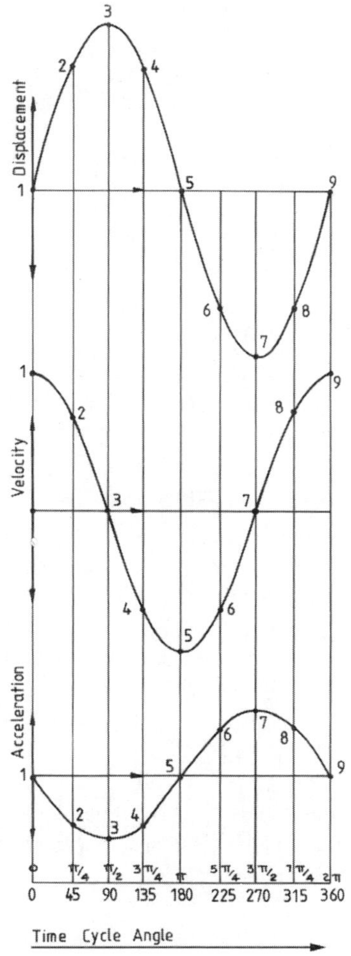

Figure 15.1 Sinusoidal vibration

Point in cycle			Displacement	Velocity	Acceleration
No.	Radians	Degrees			
1	0	0	0	ωe_{peak}	0
2	0.25π	45	1.71 x e_{peak}	0.71 x ωe_{peak}	-0.71 x $\omega^2 e_{peak}$
3	0.5π	90	e_{peak}	0	$-\omega^2 e_{peak}$
4	0.75π	135	0.71 x e_{peak}	-0.71 x ωe_{peak}	-0.71 x $\omega^2 e_{peak}$
5	π	180	0	$-\omega e_{peak}$	0
6	1.25π	225	-0.71 x e_{peak}	-0.71 x ωe_{peak}	$\omega^2 e_{peak}$
7	1.5π	270	$-e_{peak}$	0	-0.71 x $\omega^2 e_{peak}$
8	1.75π	315	-0.71 x e_{peak}	0.71 x ωe_{peak}	-0.71 x $\omega^2 e_{peak}$
9	2π	360	0	ωe_{peak}	0

Table 15.1 Values of parameters expressed as function of peak displacement

15.2.2 Which vibration level to measure

It will be seen that all these quantities vary with time. For analytical purposes it is desirable to reduce them to single figures and those for displacement are shown in Figure 15.2.

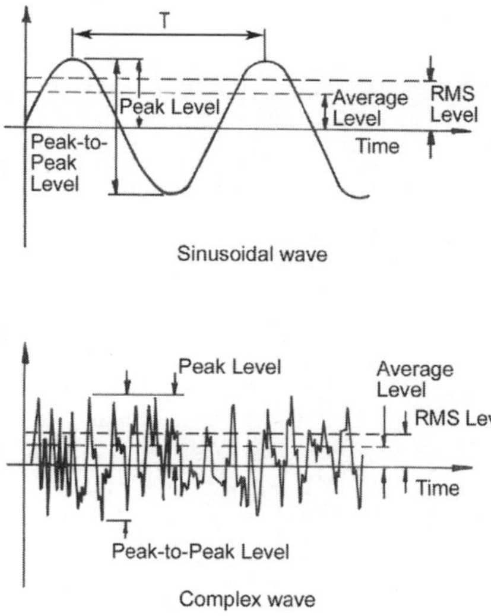

Figure 15.2 Relationship between various vibration levels

The **peak-to-peak** value indicates the total excursion of the wave and is useful in calculating maximum stress values or determining mechanical clearances.

The **root-mean-square** value is probably the most important measure because it takes account of the cycle time and gives an amplitude value which is directly related to the energy content and therefore the destructive capabilities of the vibration.

For sine wave vibrations e.g. out of balance $e_{rms} \times \sqrt{2} = e_{peak}$.

Peak and average values may also be calculated but have a limited value.

$$e\sqrt{\frac{1}{T}\int_0^T e^2(t)dt}$$

$$e_{av} = \frac{1}{T}\int_0^T |e|\, dt$$

Velocities and accelerations are given in similar terms and the root-mean-square velocity is especially important as it is used in ISO 2954-1975 as the measure of vibration severity in the range 600 to 12000 r/min (10 to 200 Hz).

Again for a sinusoidal vibration:

$$v_{mas} \times \sqrt{2} = v_{peak}$$

It must be emphasized that the relationships connecting root-mean-square and peak values only apply to sine waves. Vibrations arising from certain other sources e.g. rough rolling element bearings or air turbulence may not follow this form. Consequently the equivalents in Table 15.1 will not hold and the acceleration values especially may be much higher.

Where sine wave conditions do exist, by taking time-average measurements the effects of phase may be ignored and:

Displacement $e = \frac{a}{4\pi^2 f^2} = \frac{v}{2\pi f} = \int v dt$

Velocity $v = \frac{a}{2\pi f} = \int a dt$

Acceleration $a = 2\pi f v$

The values of e, v or a may be either root-mean-square or peak as applicable.

15.3 Units of measurement

15.3.1 Absolute units

All these parameters may be measured in either metric or Imperial units. The latter are still used in the USA and hence are commonly available in this country because of the wide availability of American instrumentation. Those commonly used are shown in Table 15.2. Reference may also be made to Chapter 22 Units and Conversions, for further guidance.

Property	SI = Metric	Imperial = US
Displacement	μm = 0.001mm	thous = mils = 0.001 in
Velocity	mm/s	In/s
Acceleration	m/s^2	g's (1g = $32.17ft/s^2$)
Frequency	Hz = cyc/sec	cyc/min

Table 15.2 Units used in vibration measurement

15.3.2 Decibels and logarithmic scales

Frequency is almost invariably plotted logarithmically to keep the scale length down to a reasonable size. It results in the lower frequency part being expanded whilst the high frequency part is compressed. A constant percentage resolution is obtained over the whole chart.

In like manner logarithmic scales may be used for plotting vibration velocities and accelerations. As the absolute values can vary enormously, and to enable vibration levels to be easily compared, decibel scales are often used. From our knowledge of noise levels it is appreciated that the decibel (dB) is the ratio of one level with respect to a reference level. It therefore has no dimensions. To obtain absolute values the reference level must be known.

It is an unfortunate fact that there are two commonly used sets of reference levels – marine/defence and those recommended in ISO 1683. These are set out in Table 15.3.

For the same absolute values ISO levels will therefore be 20dB higher than marine/defence levels. In the fan industry it is believed that the latter are almost universal perhaps because the values for a fan's vibration closely align with the figures obtained for the Noise Power Level in dBW ref 10^{-12} watt. We all get a little worried using values above 120 AdB!

Property	Definition	ISO	Marine/Defence
Acceleration	$L_a = 20 \log \left[\frac{a}{a_O}\right]^{AdB}$	$A_O = 10^{-6}\ m/s^2$	$A_O = 10^{-5}\ m/s^2$
Velocity	$L_v = 20 \log \left[\frac{v}{v_O}\right]^{VdB}$	$V_O = 10^{-9}\ m/s$	$v_O = 10^{-8}\ m/s$

Table 15.3 Vibration definitions for decibel scales

15.3.3 Inter-relationship of units

From Section 15.2.1, it can be seen that there is a relationship between any measured quantity such as displacement, velocity or acceleration for a single frequency event.

This can also be extended to the logarithmic scales noting the appropriate reference levels.

Again, it should strictly be for a single frequency simple harmonic motion. However, where one property such as unbalance, dominates all others, it can be applied to more complex wave forms, without undue error.

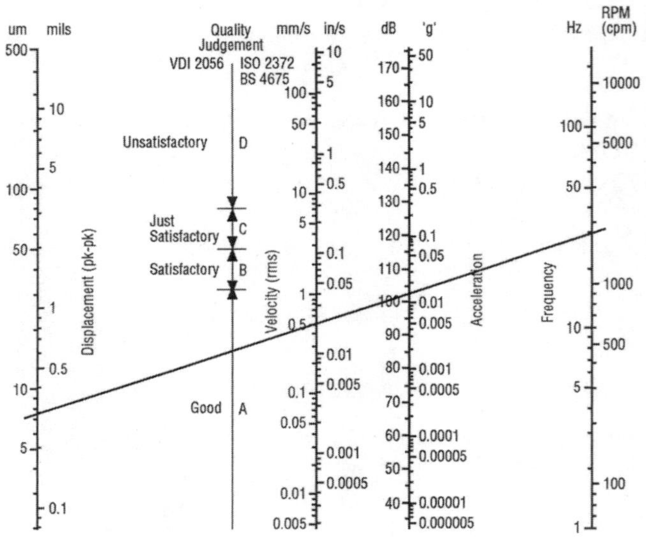

Figure 15.3 Machine vibration nomogram for converting absolute parameters into decibel values

The nomogram in Figure 15.3 is a simple way of carrying out these conversions.

15.4 Fan response

The fan and its parts may be likened to a spring-mass system. An understanding of this fact is useful in resolving many vibrational problems. It is also of importance in revealing the causes of resonance.

Every fan will have three basic properties:

a) Mass "m" measured in kg or $lbf.sec^2/in$. The force due to the mass of the system is an inertia force or a measure of the tendency of the body to remain at rest.

b) Damping "C" is the damping force per unit velocity of a system. It is a measure of the slowing down of vibrations and is given in N.sec/mm or lbf.sec/in.

c) Stiffness "k" is a measure of the force required to deflect part of the fan through unit distance. Measured in N/mm or lbf./in.

The combined effects of these restraining forces determine how a fan will respond to a given vibratory force e.g. unbalance. Thus we may state that:

$$\frac{md^2e_p}{dt} + \frac{Cde_p}{dt} + ke_p = M_u\omega^2 r \sin(\omega t - \phi) = M\omega^2 e \sin(\omega t - \phi) \qquad \text{Equ 15.1}$$

or

$$me_p\omega^2 \sin \omega t + Ce_p\omega \sin\left(\omega t + \frac{\pi}{2}\right) + ke_p = M_u\omega^2 r \sin(\omega t - \phi) = M\omega^2 e \sin(\omega t - \phi) \qquad \text{Equ 15.2}$$

where:

e	=	displacement of centre of gravity from centre of rotation
ep	=	displacement of part due to vibratory force
M	=	mass of rotating parts
M_u	=	mass of residual unbalance
r	=	distance of unbalance from rotating centre

φ = phase angle between exciting force and actual vibration

or

Inertia force + Damping force + Stiffness force = Vibratory force

It will be seen that the three restraining forces are not working together and that the inertia and stiffness forces are 180° out of phase and tending to cancel each other out. At the frequency where they are equal "resonance" occurs, and there is only the damping (which is 90° out of phase) to keep the system vibrations down.

All fans together with their supporting bases consist of a number of different spring-mass systems each having its own natural frequency possible with various degrees of freedom and a different resonant frequency for each. So far we have only considered unbalance as the exciting force, but there will be numerous other sources such that resonance can be a common problem.

15.5 Balancing

Balancing is the process of improving the distribution of mass in an impeller so that it can rotate in its bearings without producing unbalanced centrifugal forces. Perfection is impossible and even after balancing there will be residual unbalance, its magnitude being dependent on the machinery available and the quality necessary for the application.

Fan application category	Balance quality grade for rigid rotors/impeller
BV -1	G 16
BV -2	G 16
BV -3	G 6.3
BV -4	G 2.5
BV -5	G 1

Table 15.4 Balance quality grades

Application	Examples	Driver power kW limits	Fan application category BV
Residential	Ceiling fans, attic fans, window AC	< 0.15 > 0.15	BV-1 BV-2
HVAC & agricultural	Building ventilation and air conditioning; commercial systems	< 3.7 > 3.7	BV-2 BV-3
Industrial process & power penetration etc.	Baghouse, scrubber, mine, conveying, boilers, combustion air, pollution control, wind tunnels	< 300 > 300	BV-3 See ISO 10816-3
Transportation & marine	Locomotive, trucks, automobiles	< 15 > 15	BV-3 BV-4
Transit/tunnel	Subway emergency ventilation, tunnel fans, garage ventilation	< 75 > 75	BV-3 BV-4
	Tunnel jet fans	none	BV-4
Petrochemical process	Hazardous gases, process fans	< 37 > 37	BV-3 BV-4
Computer chip manufacture	Clean rooms	none	BV-5

Note 1 This standard is limited to fans below approximately 300kW. For fans above this power refer to ISO 10816-3. However, commercially available standard electric motor may be rated at up to 355 kW (following an R20 series as specified in ISO 10816-1). Such fans will be accepted in accordance with this standard.

Note 2 This table does not apply to the large diameter (typically 2.8 m to 12.5 m diameter) lightweight low speed axial flow fans used in air cooled heat exchanges, cooling towers, etc. The balance quality requirements for these fans shall be G16 and the fan application category shall be BV-3.

Table 15.5 Fan application categories

The relevant grades are specified in ISO 14694 : 2003. Recommendations are given for various types of fan impeller to avoid gross deficiencies or unattainable requirements. If the balance quality grades shown in Table 15.4 are adopted according to the fan application categories shown in Table 15.5 then satisfactory running due to this cause should result. There may however be vibration resulting from other faults. Large fans for public utilities are included with ISO 10816-3.

An unbalanced impeller will create forces at its bearings and foundations and the complete fan will vibrate. At any given speed the effects depend on the proportions and mass distribution of the impeller as well as the stiffness of the bearing supports. In the past residual unbalance has been resisted by massive supports. Now, it is recognised that a preferable solution is to reduce this unbalance so that unnecessary weight need not be added to the bearing pedestal.

For narrow impellers (width less then 20% of diameter) the static unbalance is of primary importance. Two unbalances (in different planes) in the same direction usually cause a greater disturbance than two equal unbalances in opposite directions. With wider impellers (width up to 50% of diameter) couple effects become of importance.

Static unbalance, sometimes called force or kinetic unbalance, can be detected by placing the impeller on parallel knife edges. The heavy side will swing to the bottom. Correction weight can be added or removed as required and the part is considered statically balanced when it does not rotate on knife edges regardless of the position in which it is placed (see Figure 15.4).

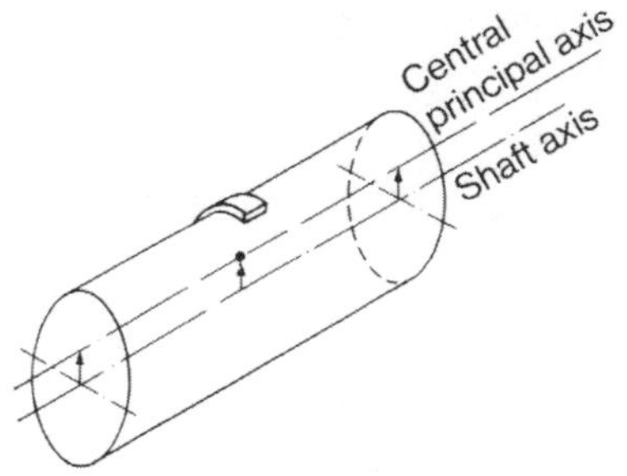

Figure 15.4 Static unbalance

Dynamic unbalance is a condition created by a heavy spot at each end of the impeller but on opposite sides of the centreline. Unlike static unbalance, dynamic unbalance cannot be detected by placing on knife edges. It becomes apparent when the impeller is rotated and can only be corrected by making balance corrections in two planes (see Figure 15.5).

An impeller which is dynamically balanced is also in static balance. Thus there is no need for the two operations where a dynamic balancer is used, despite the many specifications calling for both.

In general, the greater the impeller mass, the greater the permissible unbalance. It is therefore possible to relate the residual unbalance U to the impeller mass m. The specific unbal-

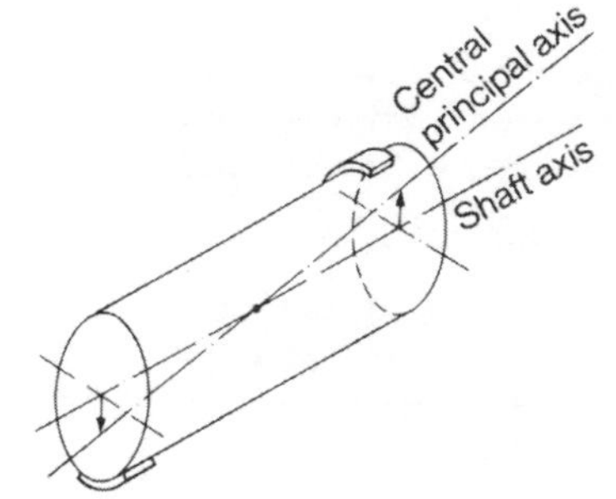

Figure 15.5 Dynamic unbalance

ance $e = \frac{U}{m}$ is equivalent to the displacement of the centre of gravity where this coincides with the plane of the static unbalance.

Practical experience shows that e varies inversely with the speed N over the range 100 to 30000 rev/m in for a given balance quality.

It has also been found experimentally that eN = constant (see Figure 15.6).

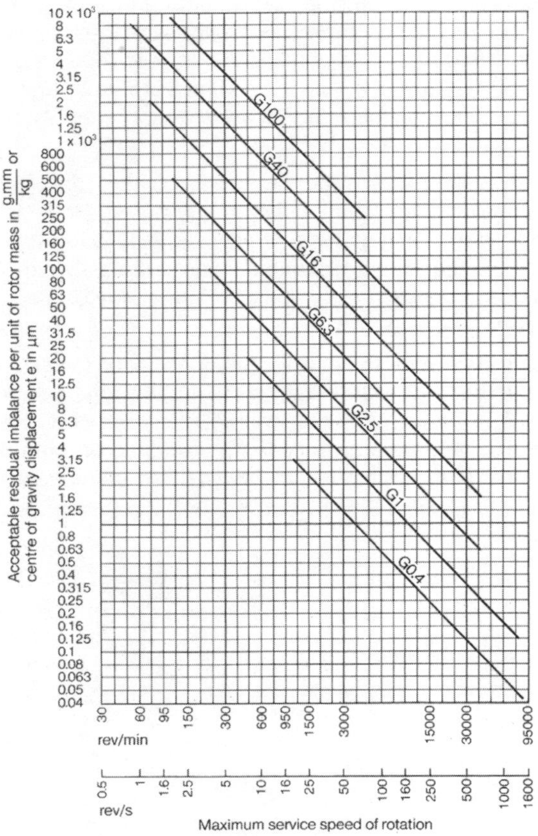

Figure 15.6 Balance quality grades to ISO 14694 and ISO 1940

Example: For an impeller of 40 kg mass the recommended value e = 20 µm is found from the graph for a maximum service speed of 3000 rev/min. If this is of the DIDW pattern and the centre of gravity is located within the mid third of the distance between the bearings, then one half the recommended permissible residual unbalance should be taken for each correction plane, i.e., 400 g.mm.

The balancing machine used must be capable of determining the magnitude of the unbalance forces: in other words, it must be objective. It is insufficient for the machine to be subjective in approach, relying on the centring of a "spot" on, a screen.

15.6 Vibration pickups

From the information given so far it will be appreciated that exactly how the vibration is measured and the equipment used becomes of prime importance. The actual "pickup" or transducer is a sensing device which converts the mechanical vibration into electrical energy. Several types exist as follows:

Seismic velocity pickup

This consists of a coil of wire supported by springs in a magnetic field created by a permanent magnet which is part of the case. For details of the construction see Figure 15.7.

When it is held against or attached to a vibrating machine, the permanent magnet, being attached to the case follows the vibratory motion. The coil of wire or conductor being supported

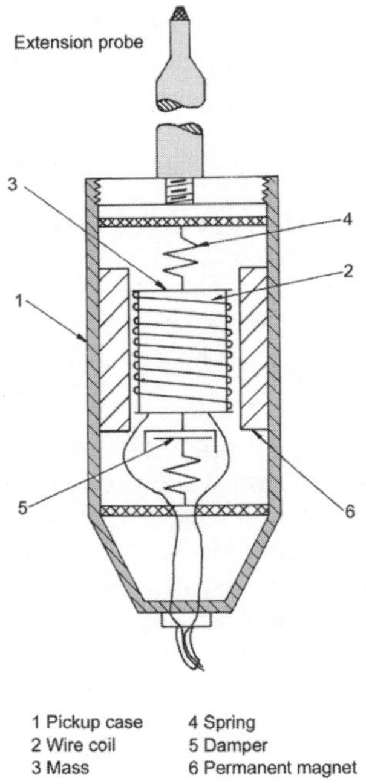

Figure 15.7 Cross-section of a velocity pickup

by springs of low stiffness remains stationary in space. Thus the conductor is moving through a magnetic field and a voltage is therefore induced. The voltage generated is directly proportional to the velocity.

Piezoelectric accelerometer

This consists of a mass rigidly attached to certain crystal or ceramic elements which when compressed or extended produce an electrical charge (see Figure 15.8).

The voltage generated by the element is proportional to the force applied and since the mass of the accelerometer is a constant, is proportional to the acceleration. As acceleration is a

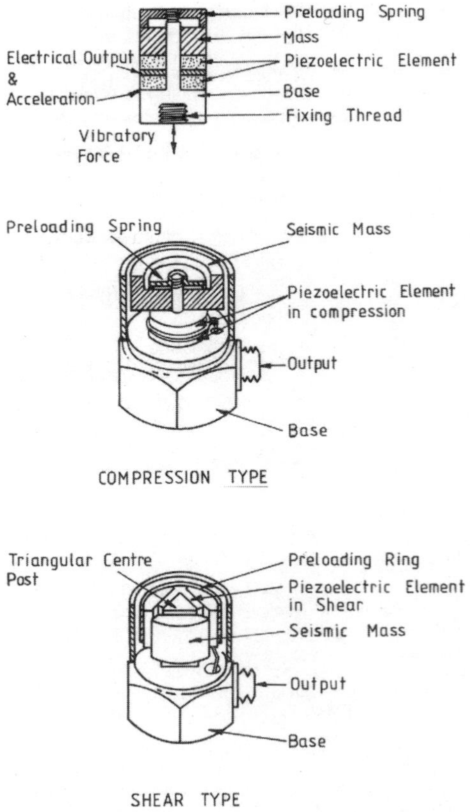

Figure 15.8 Construction of accelerometers

function of frequency squared they are most sensitive to high frequency vibration.

15.7 Vibration analysers

It is not the intention of this Chapter to be a technical manual of vibration measuring equipment. Suffice it to say that just as the voltages generated are a function of the property being measured, so the analyser to which they are attached by cable, can reconvert the signals backs to velocity or acceleration. Furthermore, due to the mathematical relationships which exist, the addition of an integrator in the circuitry allows the other vibration properties to be obtained.

Low and high pass filters are included, and these can be adjusted to limit the frequency range to that of interest for examination, whilst linear to logarithmic converters enable the signal to be displayed correctly. Output sockets are also provided so that a complete vibration signature over the full frequency spectrum can be obtained and displayed on a chart recorder, oscilloscope or tape recorder.

ISO 14695 gives full information on the mounting of fans, measuring equipment and the positioning of transducers.

It will have been realized from Section 15.4, that vibrations measured at the fan bearings may only provide an indication of vibratory stresses or motions within the fan. They do not necessarily give evidence of the actual vibratory stresses or motions of critical parts, nor do they ensure that excessive local vibratory stresses may not occur within the fan itself due to some internal resonance.

15.8 Vibration limits

15.8.1 For tests in a manufacturers works

The acceptable vibration limits for complete and assembled fans in accordance with ISO 14694 are given in Table 15.6. These are r.m.s. velocity values filtered to the fan rotational frequency and to be taken at the design duty.

Fan application category	r.m.s. velocity mm/s	
	Rigidly mounted	Flexibly mounted
BV-1	9.0	11.2
BV-2	3.5	5.6
BV-3	2.8	3.5
BV-4	1.8	2.8
BV-5	1.4	1.8

Table 15.6 Vibration limits for the manufacturer's works tests

15.8.2 For tests on site

The in-situ vibration level of any fan is not the sole responsibility of the manufacturer. Apart from the design and balance quality, it also depends on installation factors such as the mass and stiffness of foundations for supporting structures.

application Category	Rigidly mounted r.m.s velocity mm/s			Flexibly mounted r.m.s velocity mm/s		
	Start-up	Alarm	Shut-down	Start-up	Alarm	Shut-down
BV-1	10	10.6	*	11.2	14.0	*
BV-2	5.6	9.0	*	9.0	14.0	*
BV-3	4.5	7.1	9.0	6.3	11.8	12.5
BV-4	2.8	4.5	7.1	4.5	7.1	11.2
BV-5	1.8	4.0	5.6	2.8	5.6	7.1

* To be determined from historical data

Table 15.7 Vibration limits for in-situ tests

The vibration levels give in Table 15.7 are guidelines for acceptable operation and are for filter-out measurements taken on the bearing housings. Newly commissioned fans should be at or below the start-up level increasing with time, as wear and tear take place, until it reaches the "alarm" level. Remedial action should then take place.

15.8.3 Vibration testing for product development and quality assessment

Just as measurement of displacements will give most weight to low frequencies, so acceleration measurements will weight the level towards the higher frequency components. Velocity measurements are intermediate and most fans have a reasonably flat velocity spectrum.

Fans produced for higher pressures and flowrates - greater speeds and stresses – may be required for more critical applications. With direct drive units, especially at 2-pole speeds, high frequency vibrations will be generated by the bearings and also by the many electromagnetic forces.

Nevertheless, a quick method of vibration testing for production purposes is considered essential. It may be that for initial acceptance/rejection, acceleration decibel readings in the usual frequency octave bands can be a quality tool. The method of mounting the accelerometer to the measuring point is of paramount importance in obtaining accurate and repeatable results.

Bad mounting can drastically reduce the frequency range of the accelerometer. Whilst a threaded stud onto a flat machined surface is an ideal fixing, this is very seldom possible. An intermediate holding block for adhesive fixing may therefore be used, this being stuck in position with Araldite® or Loctite®. The design of such a block is shown in Figure 15.9. It will be seen that the tapped holes for the accelerometer are in three planes. Thus it is possible to obtain. readings in the horizontal, vertical and axial directions.

Accelerometer positions may be standardised as shown in Figure 15.10. As the absolute readings may be very low, it is essential for the fan to be soft-mounted and an "A" frame assem-

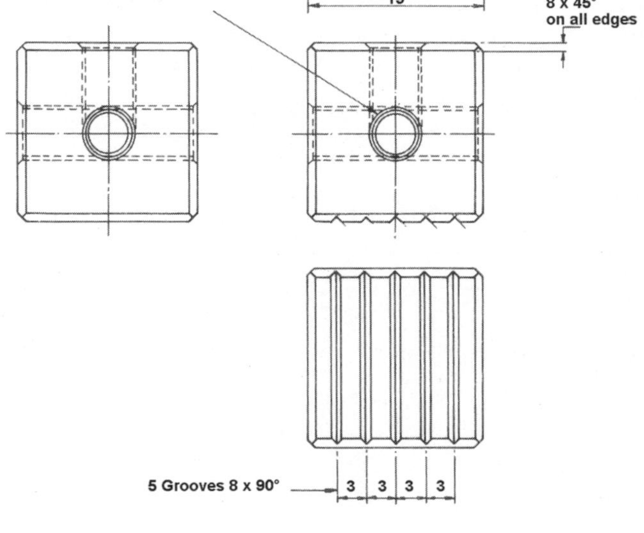

15.9 Vibration limits

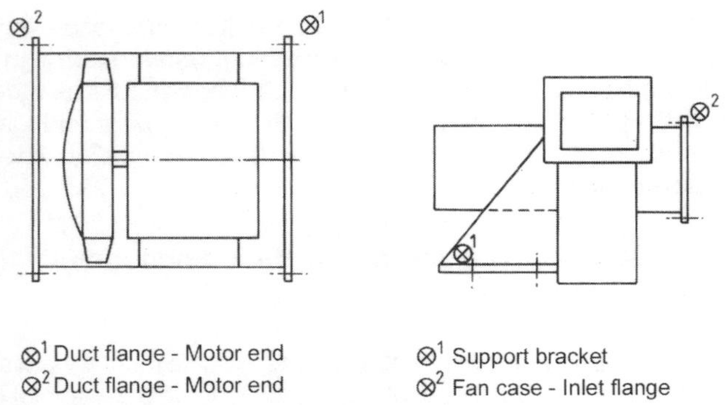

Figure 15.10 Accelerometer position for sling testing

bly with rubber cords and nylon slings in accordance with ISO 14695 should be used as shown in Figure 15.11 and Figure 15.12 . Whilst not absolute in its accuracy, it would enable consistent readings to be taken and comparability to be established. The low natural frequency of the ropes ensures that the fan is completely isolated from any outside influences.

The first 20 fans of a given type should be tested and readings taken at the prescribed accelerometer positions. All these fans have to be assessed as satisfactory according to the normal subjective inspection then current. In this case the acceptance level AdB in each octave band may be set at 85% pass i.e. the acceptance level is set at the fourth highest reading obtained for all units in all directions. It must be appreciated that these levels will be unique to a particular design of fan at a particular speed. Those for some typical small machines are shown in Table 15.8.

Readings must be taken in all three directions and be within the acceptance level set down. No differentiation is made between horizontal, vertical or axial measurements. Such acceptance levels are constantly under review. Each fan should be logged and trends noted. The intention should be to gradually reduce the acceptance levels.

Recent improvements in balancing procedures (compound balancing of fan impeller and motor rotor/shaft to quality grade G 1) has indicated that levels in the appropriate band - 63 Hz for 2-pole speeds, < Hz for all others, can be significantly reduced. For large Category 2 and 3 fans about 1250 mm diameter, it is

Figure 15.12 Vibration test on 34" axial flow fan

Fan size mm and type	Speed r/min	Power kW	AdB re 10^{-5} m/s^2 in each Octave band Hz								
			<45	63	125	250	500	1k	2k	4k	>5.6k
180 Axial	3500	0.3	79	86	90	102	102	117	108	107	104
800 Axial	1180	22.5	82	91	98	109	113	111	110	108	98
315 Centrifugal	3500	1.5	86	97	102	105	109	111	109	111	112
900 Centrifugal	1180	10.0	91	95	100	98	98	104	103	96	91

Table 15.8 Typical vibration acceptance standards for small Category 3 or marine fans

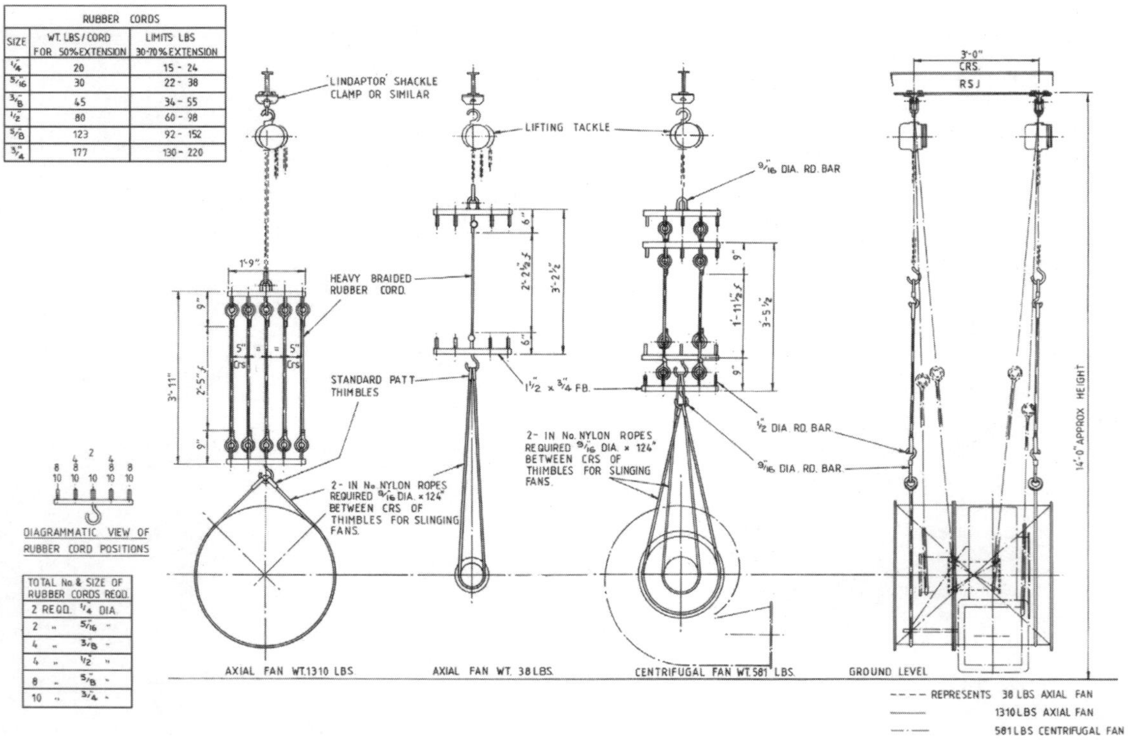

Figure 15.11 Assembly for vibration testing of axial and centrifugal fans

impractical to suspend the unit due to the weight and physical dimensions involved.

Furthermore, it is desirable to obtain as much information as possible, with a view to determining the source of all vibrations. Such fans therefore should be bolted down on a rigid foundation block. Complete discrete frequency analyses of displacement, velocity and acceleration should be taken at each fan bearing together with motor bearings where applicable. It may also be necessary to take readings at other particular points of interest e.g. shaft seals, fan feet etc.

To enable objective assessments to be finalized and for acceptance standards to be set, a manufacturer will need to make routine tests. The combinations of fan size, speed, blade form, duty, specific width etc., lead to many permutations. Repeatable tests will take a long time. Nevertheless the velocity standards set in Table 15.5 can be followed and fans must meet these before despatch.

15.9 Condition diagnosis

Units which fail to meet acceptable criteria should be given a complete frequency analysis. This applies to all weights and speeds of machine. Again readings should be taken at the various positions. A typical frequency analysis is shown in Figure 15.13.

So far we have discussed vibration in a general sense and indicated permissible overall limits. For important and/or arduous applications however, we need to be able to identify the causes of vibration and their likely effects on the machine which could be catastrophic in the event of a total breakdown.

The keys to the identification of the cause of a vibration are in its magnitude and frequency over most of the spectrum. Below about 10Hz displacement will be of primary importance, whilst above about 1kHz, acceleration is paramount. Over the remainder of the range, velocity measurements will be sufficient.

Different causes of vibration occur at different frequencies. For example, a faulty ball bearing would cause high frequency vibration at many times the rotational frequency whilst unbalance or misalignment produce vibration at the rotating speed frequency. To rely on an overall or linear response reading of vibration velocity could lead one to ignore a developing problem.

To obtain the installation's "vibration signature", a pick-up is used, which can either be handheld or more rigidly attached, feeding a meter giving a visual display. For analysis some type of tunable frequency analyser is necessary together with a stroboscopic light. The strobe permits rapid tuning to rotational speed when it "views" the rotating element and apparently sees a stationary image. From the analyser an X-Y recorder can be fed to show the magnitude of the vibration in narrow frequency bands right across the spectrum which the analyser can identify. By fixing the probe in either the vertical, horizontal or axial modes, different traces can be obtained which will inform the operator as to possible sources of vibration. Any "peaks" in the reading at specified frequencies will indicate the onset of certain troubles.

It is to these discrete frequencies we now look. Remember vibration severity is a quality judgement whilst frequency will indicate the cause. An increase in the severity of a particular frequency during inspection, commissioning or operation may indicate the onset of a particular problem. By referring to the initial signature and specifying that the particular frequency level should not increase by more than a predetermined amount, it is possible to construct planned maintenance procedures.

In the comments which follow f_1, is defined as rotational frequency and equals (rev/min ÷ 60) Hz.

15.9.1 The machine in general

Unbalance As the heavy spot would give a "pulse" to the pick-up once every revolution, unbalance will be identified by

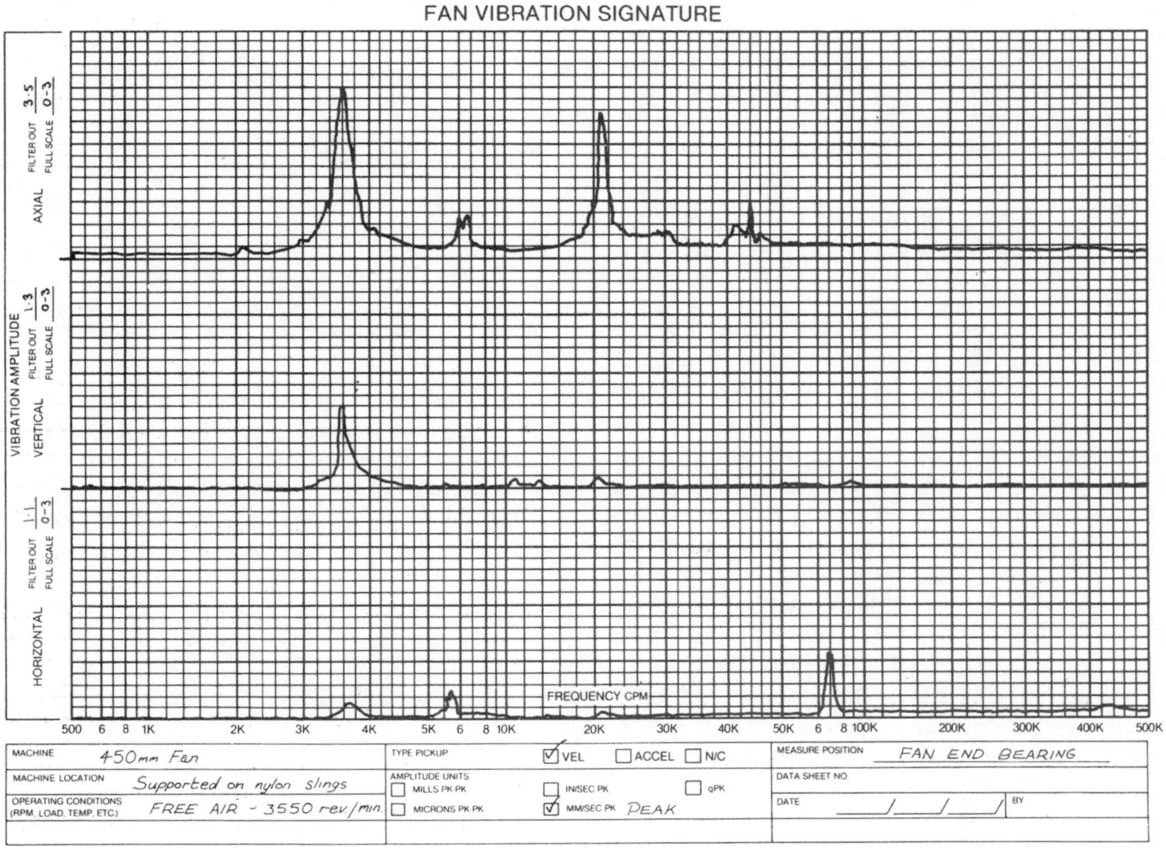

Figure 15.13 Typical vibration signature for fan fitted with 2-pole 440 volt 3-phase 60 Hz AC motor and running at 3550 rev/min

high readings in the horizontal and vertical directions at the rotational frequency, i.e. f_1Hz. This is the most common cause of vibration.

High readings at commissioning can indicate residual unbalance in manufacture or "sag" if the rotating element has not been turned over regularly during storage on site. The build up of dust on a rotor, erosion or corrosion will also lead to increasing figures. Scaling at high temperature (above 400°C) or soaking in heat whilst stationary are other possible causes.

Misalignment This is almost as common as unbalance despite the use of self-aligning bearings, which should still be lined up as well as possible. Flexible couplings can be out of line both by height and angle. A bent shaft produces angular misalignment. Radial and axial forces are always produced, the size of such forces, and therefore the vibration which results being proportional to this misalignment.

As previously stated, the axial readings are usually 50% or more of the radial readings and again the frequency is normally f_1 Hz. When the misalignment is severe however vibration at $2f_1$Hz and even $3f_1$Hz may be experienced. Misalignment can also occur where a machine has been distorted by tightening down onto foundations which themselves are not level. With sleeve bearings this will produce vibration according to the amount of residual unbalance, but with ball or roller bearings an axial vibration would be produced even if the unit were "perfectly" balanced, which is physically impossible.

Another very common fault is when pulleys and ropes of vee belt drives are not correctly aligned. This results not only in destructive vibration, but also leads to rapid wear of belts and production of frictional heat through to shafts and bearings.

Eccentricity An example of this could be where an impeller centre with excessive bore is "pushed over" by a taper key. The centre of rotation does not then coincide with the geometric centre. As far as the impeller is concerned, this leads to more mass being on one side of the rotational centre than the other, i.e. unbalance.

It can therefore be corrected by rebalancing provided that the rebalancing takes place in its own shaft and bearings and that with ball/roller bearings the position of the inner race on the shaft also does not change. The predominant frequency is of course f_1Hz. Where a fan is gear driven, eccentricity can produce reaction forces between pinions because of the cam-line action.

The largest vibration will occur along a line joining the centres of the two pinions at a frequency equal to (pinion rev/min ÷ 60) Hz of the one which is eccentric. It will look as if it is unbalanced but cannot be corrected by re-balancing. A similar situation can arise with vee belt pulleys which are eccentric. The largest vibration will be in the direction of belt tension at a frequency of (pulley rev/min ÷ 60) Hz of the eccentric pulley. Again re-balancing cannot cure.

Looseness Common forms are loose foundation bolts and excessive bearing clearances. It will not be manifest unless there is some exciting force such as unbalance or misalignment to encourage it. Only small forces are necessary however to excite the looseness and produce large vibrations. Although rebalancing or realignment may therefore assist, extreme accuracy would be necessary which may be impossible to achieve. It is essential to tackle the problem at its source.

To determine the characteristic frequency of looseness, let us consider an unbalanced rotor fitted to a shaft running in a bearing with loose holding down bolts. When the heavy spot is downward, the bearing will be forced against its pedestal. When the heavy spot is upward it will lift the bearing whilst at positions 90° away, the force will neither lift nor hold down, and the bearing will drop against the pedestal due to weight alone. Thus there are two applied forces each revolution of the shaft and the vibration frequency is $2f_1$ Hz. This is the characteristic frequency of looseness.

Resonance The section on fan response (Section 15.4), showed that every object has a natural frequency at which it "likes" to vibrate. Should the forcing frequency coincide with the natural frequency of a part, then resonance will occur. To overcome the problem with a vee belt-driven unit is simple, a small change in rev/min will normally suffice. With a direct drive unit stiffening or a change in the design may be indicated, although unlikely.

Many of these problems can only be identified at the commissioning stage or during service. Figure 15.14 shows a technician taking readings on site.

Figure 15.14 Site testing with hand-held vibration analyser
Courtesy of Schenck RoTec GmbH

15.9.2 Specific vee belt drive problems

Many of the problems found in impellers will also be present in vee belt drives. Often the balancing of pulleys has been overlooked and must be specified when ordering. Misalignment has already been mentioned.

In Chapter 14 a number of problems were identified, which could result in additional noise. However as these problems were essentially mechanical in origin, they are also manifest as vibration.

Such drives have good resistance to shock and vibration but may be blamed for causing trouble as they can be readily seen to whip and flutter especially when the belts are unmatched. Belts are often changed unnecessarily when the fault is really that of unbalance, misalignment etc. Nevertheless, the importance of using matched sets of belts cannot be emphasised enough.

Vibration from faults in the belts themselves occur at multiples of belt speed. The relevant frequencies are:

$$1,\ 2,\ 3 \text{ or } 4 \times \frac{\text{pulley diameter}}{\text{belt length}} \times f_p \text{ Hz} \qquad \text{Equ 15.3}$$

where

$$fp = \frac{\text{pulley rev/min}}{60}$$

Likely faults are pieces broken off, hard or soft spots etc.

Faults in pulleys, such as chipped grooves etc., will be identified at the speed of the relevant pulley f_p Hz.

15.9.3 Electric motor problems

Most electric motor vibrational problems are mechanical in origin e.g. unbalance, misalignment, bolting down to foundations which are not level, loose foundation bolts, faulty bearings etc. Previously described frequencies are therefore applicable using, of course, the motor rev/min where this differs from the machine rev/min. Again the noise sources identified in Chapter 14 will also be identified as vibration.

With induction motors, forces act in the air gap between rotor and stator tending to pull these together and produce vibration at 2 x line frequency Hz. Normally such vibration is small except in 2-pole motors, but if the air gap varies, or if the tightness of stator laminations or winding in the stator vary, then this vibration will increase considerably. The second and third harmonics may also be important.

Generally,

$$\text{slip frequency} = 2 \times \frac{\text{line frequency}}{\text{no. of poles}} - f_m \text{ Hz} \qquad \text{Equ 15.4}$$

where

$$f_m = \frac{\text{motor rev/min}}{60}$$

This will not in itself be important as it will be of very low frequency. However, its interaction with higher frequencies can produce pulsations.

If the rotor is severely unbalanced, the high spot will come closer to the stator than other points. As it passes the stator poles more pull is exerted. Thus vibration occurs at 2 x slip frequency on a 2-pole motor, 4 x on a 4-pole motor and so on. The magnitude of those readings in these frequencies can indicate whether the problem is simply due to the lack of balance, change in the air gap, worn journals, broken rotor bars etc.

Vibrations may be produced at a frequency equal to no. of rotor bars x f_1 Hz and at no. of stator slots x f_1 Hz. Vibrations at interactive frequencies may also be important.

If a resonance condition exists within the motor at line frequency, then large vibrations can be produced. More often however this is the fault of an unbalanced magnetic pull and can be cured by changing stator connections.

With all suspected electrical sources of vibration, the simple check is to switch off the motor when they should "die".

15.9.4 The specific problems of bearings

Sleeve bearings Problems with these generally result from excessive clearance, wiping, erosion of the journal surfaces (e.g. builders' dust on site entering the bearings before start up), looseness of the white metal, inadequate lubrication (poor maintenance), lubrication with an incorrect grade of oil, or chemical corrosion.

Characteristic frequencies are f_1 Hz, $2f_1$ Hz or random, for the reasons already given. It will be appreciated that some of these problems are prevalent under modern conditions, becoming especially important on high speed fans, and have encouraged the trend to ball and roller bearings.

Ball and roller bearings Races which have flaws on the balls, rollers or raceways will not only cause additional noise but also high frequency vibration identified in Chapter 14 but repeated here as follows:

Flaw in outer raceway or variation in stiffness around the housing:

$$f_2 = f_1 \times \frac{n}{2}\left[1 - \frac{d}{D}\cos A\right] \text{ Hz} \qquad \text{Equ 15.5}$$

Flaw in inner raceway:

$$f_3 = f_1 \times \left[1 + \frac{d}{D}\cos A\right] \text{ Hz} \qquad \text{Equ 15.6}$$

Flaw in ball or roller:

$$f_4 = f_1 \times \frac{D}{d}\left[1 - \frac{d^2}{D^2}\cos^2 A\right] \text{ Hz} \qquad \text{Equ 15.7}$$

Irregularity in cage or rough spot on ball/roller:

$$f_5 = f_1 \times \frac{1}{2}\left[1 - \frac{d}{D}\cos A\right] \text{ Hz} \qquad \text{Equ 15.8}$$

where:

n = number of balls or rollers

d = diameter of balls or rollers

D = pitch circle diameter of race

A = angle of contact of ball/roller

Such vibrations are not easily transmitted to the rest of the fan (except where there are large flat mounting surfaces) and will therefore be recognised by velocity readings on the bearing housing.

Severe misalignment of a race will sometimes result in a frequency at n x f_1 Hz, even when the bearing itself is satisfactory.

15.9.5 Selection and life of rolling element bearings

15.9.5.1 Bearing parameters

Modern ball and roller bearings are a precision made item. With correct selection, installation and lubrication, premature failure is unlikely. When this does happen it has usually been caused by machine out-of-balance, misalignment or use at speeds/loads/temperatures in excess of those recommended by the manufacturers.

The demand for high quality and low price, necessitates quantity production of all anti-friction bearings. Machine designers then have to select from a standard range the items which most closely meet their requirements as to:

- Dimensional and speed properties
- Frictional drag and heat generated
- Noise output
- Deflection under load
- Rate of wear and lubrication
- Life in relation to load

Of these, the life is of most importance, especially at moderate speeds and loads. Correct selection for life will usually ensure that performance under the other headings is acceptable.

15.9.5.2 Fatigue life

Earlier in *Fans & Ventilation*, we considered a rolling element bearing to have point or line contact between the raceways and the ball or roller. In reality these conditions cannot exist where a load is applied since the smallest force would induce an infinite stress. Deformation therefore takes place and the contact is over an area sufficiently large to result in a stress value which can be accepted by the bearing materials.

To ensure that the stress is within the elastic limit, and to keep the contact area to a minimum, the steels used are through hardened. Accordingly high stresses still result, and the major cause of failure becomes metal fatigue.

The time at which the first fatigue crater appears cannot be measured precisely. A batch of apparently identical bearings run under the same conditions of speed, load, lubrication and temperature will fail at different times. By using a statistical approach and analysing data accumulated over the years, it is possible for the manufacturer to quote the probability that a bearing running under the specified conditions will last for a given period of time, but this cannot be predicted with certainty.

The calculation of bearing life has now been made the subject of a Standard, namely ISO 281. The L_{10} life in hours or running is defined as that at which 10% of a group of apparently identical bearings can be expected to have failed by rolling fatigue. Being a statistical forecast, the result will be more accurate, the greater the numbers tried. Conversely 90% of all bearings can be expected to exceed their L_{10} lives, whilst the average life should be five times as great.

15.9.5.3 The need for early warning techniques

With such a wide spread of hours to failure, it has become desirable to monitor bearings, so that we may predict their lives much more accurately. When a bearing does fail, the damage to associated machine parts, and the production losses, are often far in excess of the actual cost of replacement.

The characteristic of fatigue failure is to increase the very high frequency vibrations at between 2.5 kHz and 80 kHz. Whilst these will always be present at a low level in nominally perfect bearings, they may be expected to increase by a factor of many hundreds before the onset of complete failure. During this time, the vibrations at the lower frequencies related to rotation and its multiples may not increase very much or may be attributed to other causes. Many vibration analyzers have a maximum cut-off frequency of about 1 kHz and by the time they are able to detect a significant increase in this vibration level, failure due to fatigue may be imminent.

Other techniques have therefore been developed, and these all monitor the high frequency vibrations in some form or another. It is to these methods which we now turn, describing the features and advantages of each. It is left to you to detect by inference or omission their respective disadvantages!

15.10 Equipment for predicting bearing failure

15.10.1 Spike energy detection

What is spike energy?

Normal vibration analysers which measure displacement, velocity and acceleration over a frequency range of 10Hz to 1 kHz have now become available with an additional readout of "spike energy". This is defined as the ultrasonic microsecond-range pulses caused by impacts between bearing elements which have microscopic flaws.

Special circuits have been designed to detect this pulse amplitude, the rate of occurrence of the pulses and the amplitude of the high frequency broad band vibratory energy associated with bearing defects. These three parameters — pulse amplitude, pulse rate and high frequency random vibratory energy — are electronically combined in the single quantity g-SE. This is recognised as a measurement of bearing condition and has the units of acceleration but in the ultrasonic frequency range.

Early spike energy meters

An example of an early spike energy meter is shown in Figure 15.15. The meter had a cable input from a transducer with hand-held probe or a more permanent magnetic pick-up, this being applied to the bearing housing with a light, steady pressure so that it did not chatter. To establish a programme for checking the condition of anti-friction bearings, a comparison

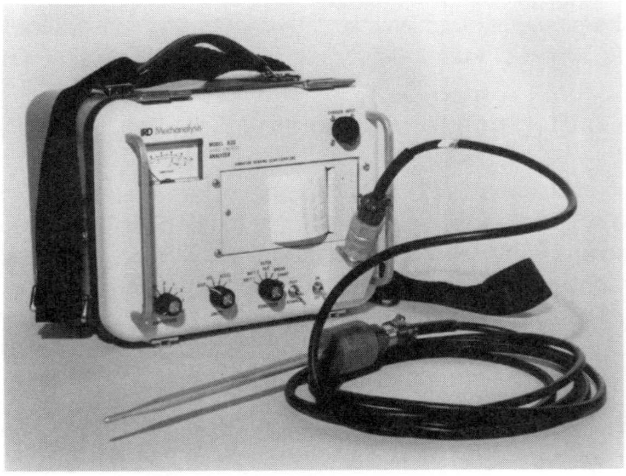

Figure 15.15 Early spike energy meter

technique was used. Thus the g-SE levels of similar machines were measured and those which diverged from the average were identified. A close watch was kept on any such bearings as being a source of potential trouble. The method led to the quick establishment of criteria for determining whether a bearing was good or bad.

It should be noted that g-SE is dependent on rotational speed rev/min. A doubling in speed would result in the spike energy measurement doubling for the same bearing condition. From a vibration severity standpoint it should, however, be remembered that low speed bearings can usually tolerate more damage than high speed bearings since the former will tend to deteriorate more slowly. With single machines, measurements had to be taken periodically and any trends noted.

An unchanged level of g-SE over a period of time would indicate a good bearing, but any significant upward trend would signal imminent failure. General experience over the range of 600 to 3600 rev/min and using the 9 inch hand-held probe shown, g-SE value of over 0.5 usually indicated a defective bearing. This value was used with caution as it might have been dependent on bearing type and mounting.

Apart from bearings, there are other sources of spike energy. Incipient gear defects, rubbing of seals or guards are all possible causes. Where these elements are close to the bearings,

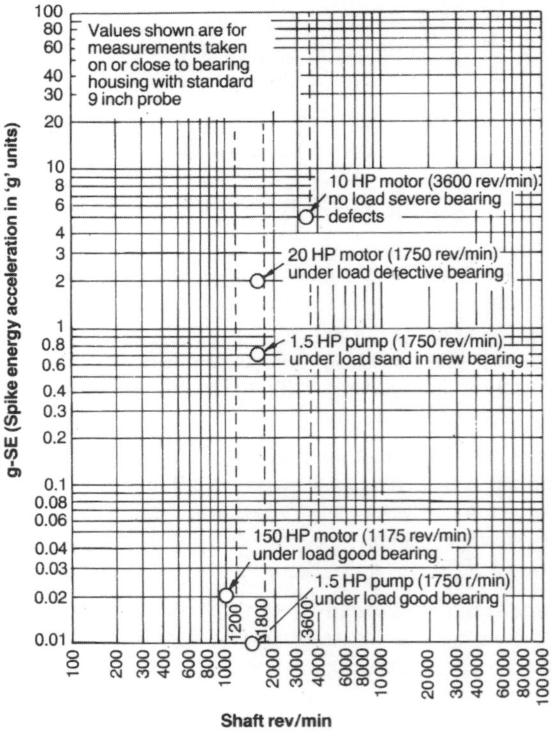

Figure 15.16 Rolling element g-SE severity chart

additional readings should be taken to avoid misinterpreting the data.

The meter was best used with spike energy severity charts, as shown in Figure 15.16. These led to the establishment of g-SE severity criteria for a given machine and its bearings. No specific severity levels such as smooth, good etc. were given, since they were dependent on the machine, its bearing type(s), speed and loads. Some case histories have nevertheless been plotted to indicate the resultant range.

Present day spike energy meters

These are now produced by Rockwell Automation Ltd who offers products for the fan site engineer. Figure 15.17 shows these instruments, remarkable for their reduced size when compared with Figure 15.15.

One is a small lightweight portable data collector/analyser that monitors the condition of equipment found in many process industries. It is easy to use and has features normally associated with bulky real time analysers. It also uses the latest advances in analogue and digital electronics and screen technology to provide speedy and accurate and both unlimited and reliable data collection.

Another is a Windows-based, 2-channel data collector and signal analyser. It enables easy condition monitoring of equipment including vibration information. Bearing assessment is also available which integrates with other information systems and software.

Figure 15.17 Data collector and data collector/analyser
Courtesy of Rockwell Automation Ltd

15.10.2 Shock pulse measurements

Theory

This method detects development of a mechanical shockwave caused by the impact between two bodies. As an example, consider a ball dropping onto a bar as shown in Figure 15.18.

At the moment of impact, or initial phase 1, no detectable deformation of the material has yet taken place. An infinitely large particle acceleration therefore results, its magnitude being solely dependent on the impact velocity v. The result is unaffected by the sizes of the two bodies or by any mechanical vibration present. Two compression waves are set up, that in the bar propagating ultrasonically in all directions, whilst the other travels through the ball. The magnitude of the wavefront is an indirect measure of the impact velocity v.

During the second phase of impact 2, the ball and bar surfaces deform, the energy deflecting the bar and setting up vibrations. This is the motion detected during normal vibration analysis.

It must be emphasized that the shock pulse method is concerned solely with phase 1 by detecting and measuring the magnitude of a mechanical impact from the resultant compression wavefront. A piezoelectric accelerometer is used, which is

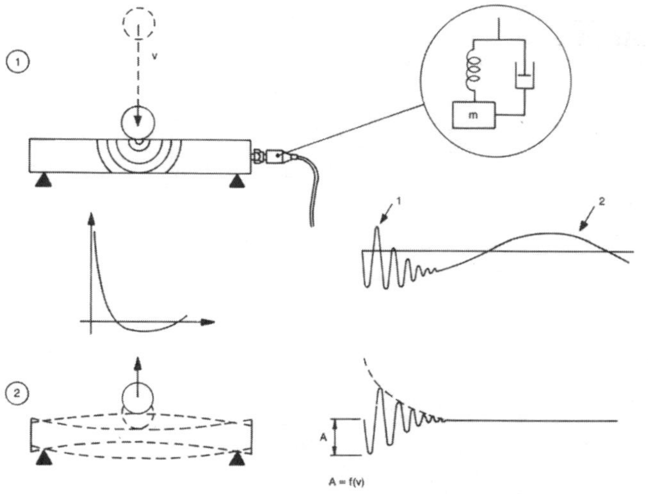

Figure 15.18 Illustration of shock pulse mechanism during impact

not influenced by background vibration or noise. This transducer is tuned mechanically and electrically to have a resonance of 32 kHz. The compression wavefront or shock pulse sets up a dampened oscillation in the transducer at its resonant frequency. This also is shown in Figure 15.18 as the dampened transient electrical output caused by the impact. The peak amplitude of this oscillation (A) is therefore directly proportional to the impact velocity v.

As the transient is well defined, and decays at a constant rate, it is possible to filter out electronically all the normal vibration signals. The measurement and analysis of its maximum value is the basis for determining the condition of rolling element bearings.

Testing anti-friction bearings

As previously stated, the running surface of a bearing will always have a degree of roughness, from microscopic flaws or indentations which will increase as it approaches failure. When the bearing rotates these surface irregularities or fatigue craters will cause mechanical impacts between the rolling elements and thus become a shock pulse generator. The magnitude of the shock pulses is dependent on the surface condition and the peripheral velocity of the bearing ($\propto$rev/min x size). As the shock pulses increase with age it is possible to follow the progress of a bearing's condition from installation, through the various stages of deterioration to ultimate failure.

Shock pulses generated by a typical bearing will increase by a factor of up to 1000 times from when it is new to when it is replaced. To simplify the readout of such a large range, figures in decibels (dB) are used. It should be remembered that the decibel is by definition a ratio on a logarithmic scale. Apart from noise, it can and is used for a number of other purposes e.g. acceleration values. In the present case the intensity of the shock pulses generated by the bearing is measured in dBsv (decibel shock value) and the scale thus compressed to 60 dB sv, i.e. $20 \log \frac{1000}{1}$.

Readings expressed in dB sv refer to the total or absolute value of the shock pulses.

Empirical testing has shown, as expected, that even a new, properly installed and properly lubricated bearing will generate shock pulses. This initial value or dB_i is primarily dependent on rotational speed rev/min and bore diameter mm (see Figure 15.19).

As the bearing ages and deteriorates the dB_{sv} total shock pulse value increases. This increase is defined as its dB_N or normalized value i.e. $dB_N = dB_{sv} - dB_i$.

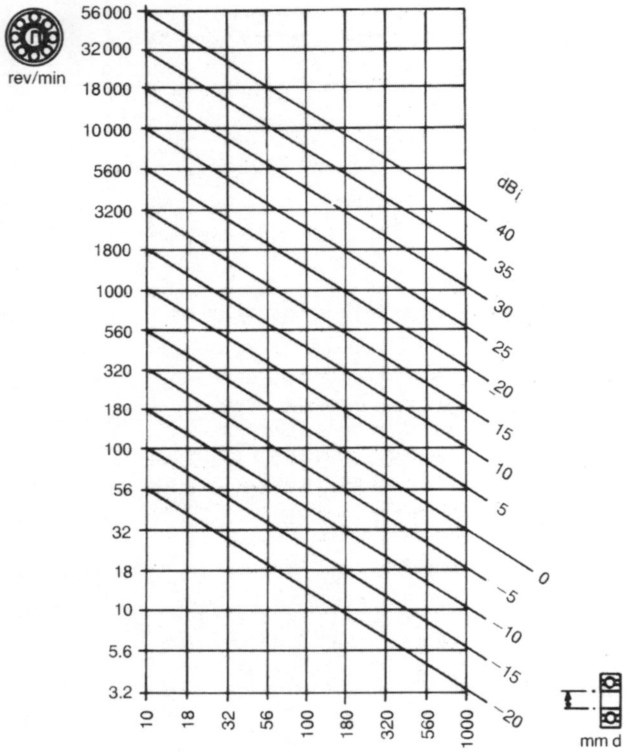

Figure 15.19 Initial value of dB_i - Relationship with bore and speed

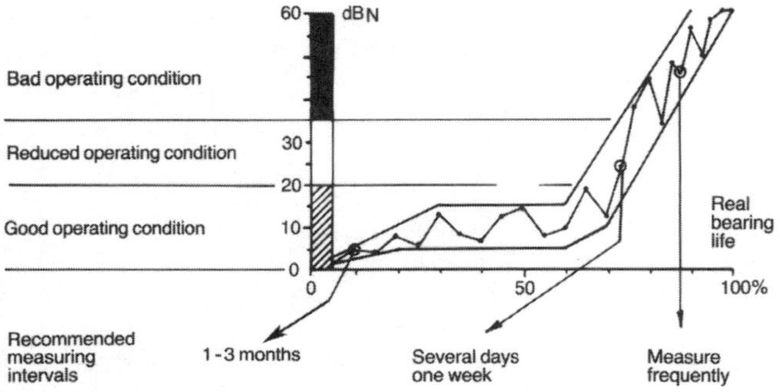

Figure 15.20 Relationship between bearing condition and percentage life

Figure 15.20 shows the relationship between bearing condition and percentage bearing life.

Zone	dBN value	Bearing condition
Green	Less than 20	Good operation
Yellow	20 to 35	Caution

Table 15.9 Bearing operating zones

By experience the dB_N scale has been divided into three zones as shown in Table 15.9.

Periodic measurements should be taken and, in the early days were plotted on the chart shown in Figure 15.21. Decisions can then be made as to when bearings should be changed.

It is worthy of note that over the years, since the author bought his first shock pulse meter, with the increasing miniaturization, many of the functions and calculations are now performed within the instrument itself in the most recent versions. However this explanation of the earlier versions is given as it most readily describes the theory and workings of shock pulse.

An early shock pulse meter

The early portable meter was hand-held and battery-powered as shown in Figure 15.22. Before any readings were taken the bore diameter mm and speed rev/min were dialled into the meter by aligning their values on the respective scales. The dB_i of the bearing was then automatically subtracted from the trans-

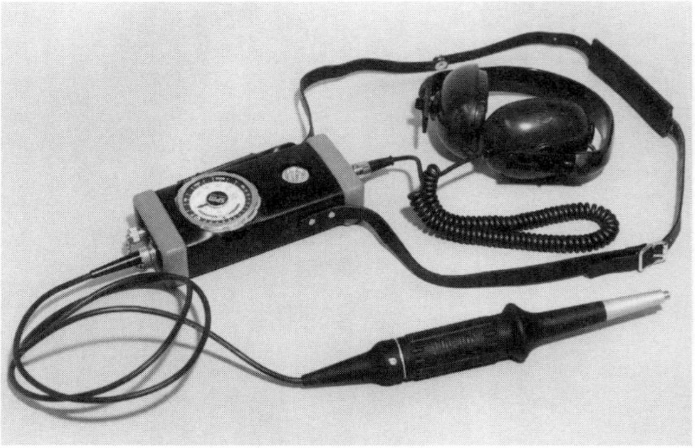

Figure 15.22 Early shock pulse meter

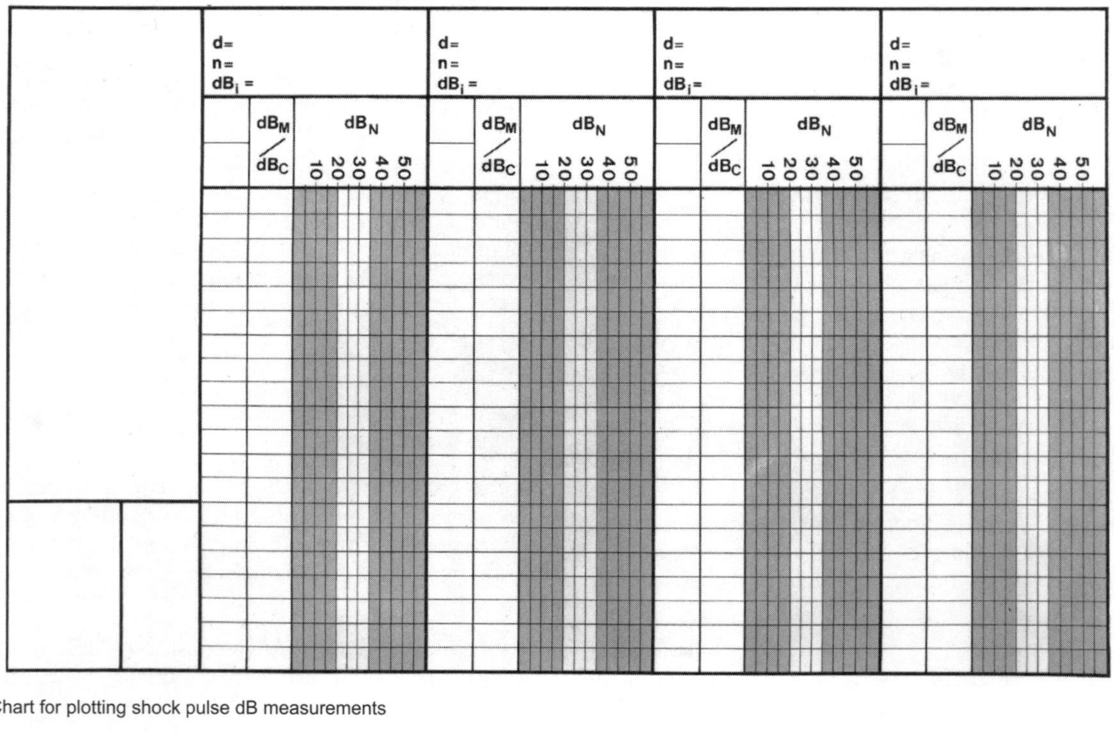

Figure 15.21 Chart for plotting shock pulse dB measurements

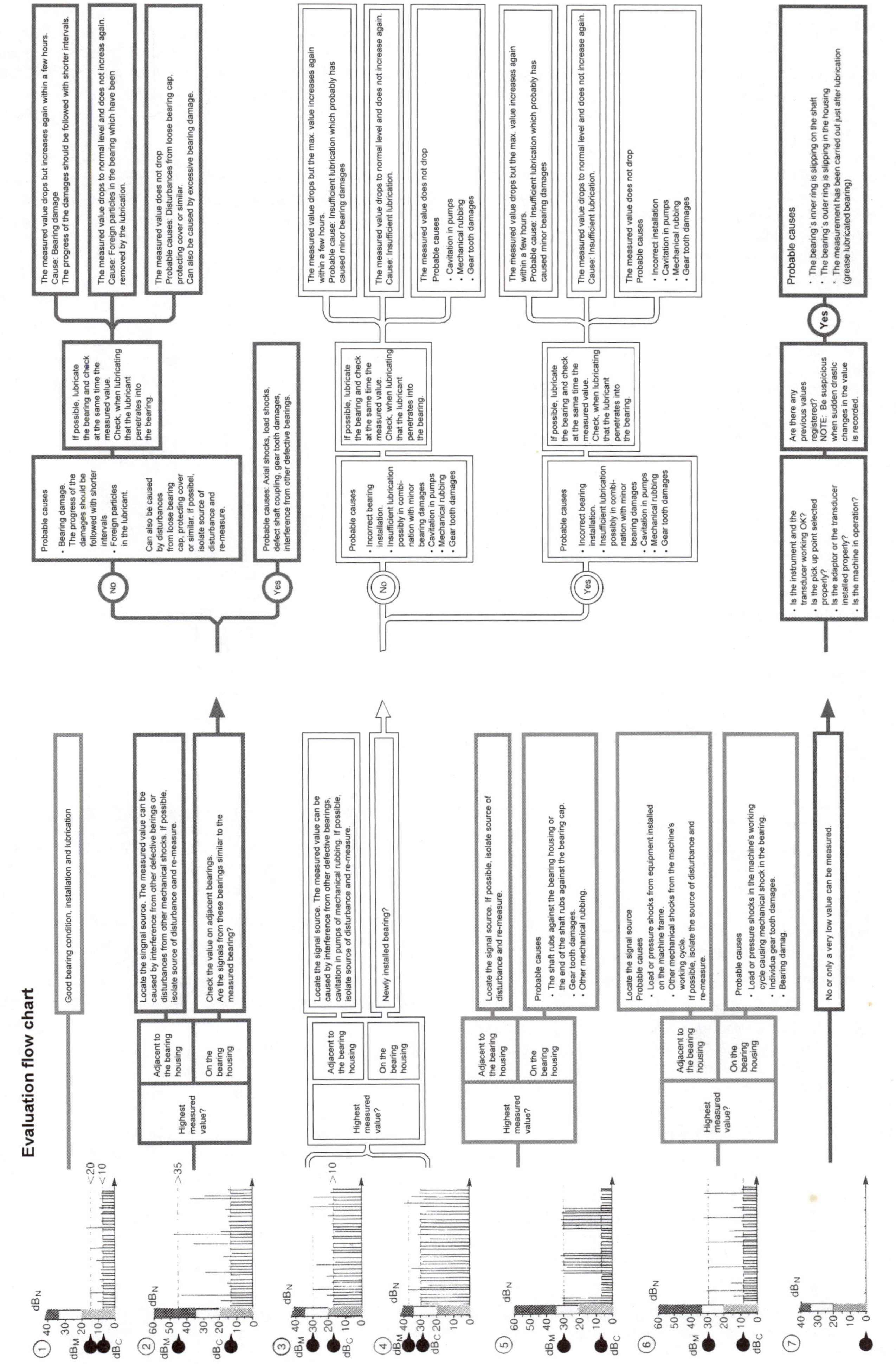

Figure 15.23 Evaluation flowchart showing examples of individual shock pulse measurements

ducer output which measured dB_{sv}. This additional amount was, of course, the dB_N and a direct indicator of bearing condition.

The transducer signals were compared within the meter to a manually set threshold level, which could be adjusted by rotating the large outer dial relative to the large black stationary arrow. Starting with a dial setting of 0 dB_{N1} a continuous tone, generated by the instrument, was heard from the built in speaker and external earphones. As the dial was turned to higher scale values, a point would be located where the tone was intermittent. This dB_N reading was defined as the bearing's carpet value dB_C. By continuing to turn the scale to higher readings, the tones became more and more intermittent, until they finally disappeared. This value of dB_N was defined as dB_M maximum or peak, and indicated the bearing condition.

Amplitude distribution

During bearing operation, not only peak shocks appeared, but a number of differing amplitudes and rates of occurrence. The relationship between shock amplitude read on the dB_N scale and rate or number per unit time gave the amplitude distribution of the bearing shocks. Again the distribution was assessed by listening to the built in speaker on the meter or the external earphones.

Figure 15.23 is an evaluation flow chart where every individual shock pulse measured at the meter was represented by a vertical line whose height corresponded to the shock amplitude dB_N. Bearing condition, installation, fit, alignment and lubrication were all assessed by measurements of maximum and carpet values. Additional comments in explanation of some of the items in the flowchart are:

a) Good bearing, properly installed, properly lubricated

In a good bearing, the shocks are mainly caused by the rolling contact on normal surface roughness, which means that there will be a low shock noise carpet and random shocks with slightly higher value. The carpet value should be under $10dB_N$ and the peak value under $20dB_N$.

b) Damaged bearing

When the bearing raceways or rolling elements are damaged, high peak amplitude shocks will appear.

Through coincidence between different damages in different running surfaces, these shocks will appear randomly. Often, the carpet value will be below $20dB_N$. However if the bearing is badly damaged, the overall surface roughness will increase and so will the carpet value. Usually however there is a large difference between the peak and carpet values.

c) Improper installation or lack of lubricant

These are operating condition problems. If the bearing is improperly installed (out-of-round or pinched housing, too tight or loose a fit) the internal load in the bearing will increase locally and thereby the shocks caused by the rolling motion will also increase even if the bearing is not yet damaged on its running surfaces. It is characteristic of this type of problem that the peak and carpet values are relatively close together.

A bearing running with insufficient lubricant has a shock pattern similar to an improperly installed bearing. The lack of lubricant will increase the carpet value. Lack of lubricant will normally only appear in greased bearings. Therefore, greasing the bearing is recommended when an increase in carpet value is noticed. The carpet value should decrease after lubrication.

d) Mechanical rubbing

Mechanical rubbing near the bearing between a rotating and stationary part (for example, rubbing between the bearing seal and shaft) will cause rhythmic shock bursts at a certain dB_N level. They are easy to identify because of their repetitive nature.

e) Machine cycle load shocks

If a bearing is exposed to a cyclic shock load, a measurable shock signal may appear in the bearing. These shocks will appear with a rhythm related to the machine working cycle and are therefore simple to determine and isolate. They will be very repetitive but the peak and carpet values of the bearing can usually. be determined.

Pinion damage in a gear box can also generate a shock pattern similar to the above load shocks. These shocks will appear with a rhythm related to the speed of the shaft involved. Moreover, it is typical for pinion damage to generate the same repetitive shock pattern on all the bearings involved.

Present day shock pulse meters

These too have changed considerably from the early meters.

One of the meters is produced by SPM Instruments AB and is a portable, multi-functional instrument for bearing and lubrication condition monitoring, vibration analysis. It includes corrective maintenance features such as balancing and alignment, see Figure 15.24.

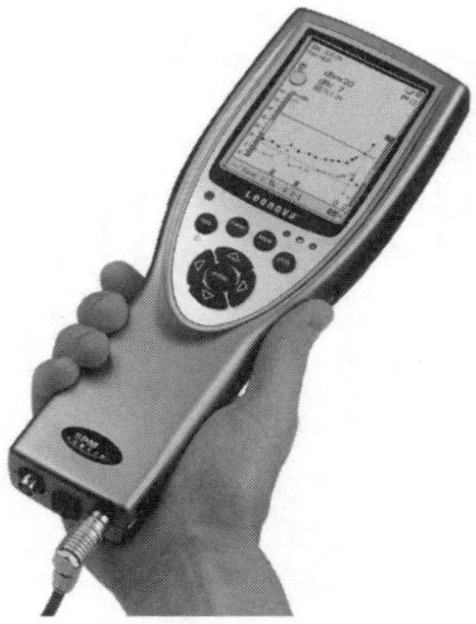

Figure 15.24 Portable machine condition analyser
Courtesy of SPM Instrument AB

15.11 Kurtosis monitoring

The Kurtosis meter as applied to vibration measurement was originally manufactured by CML Systems under licence from the then British Steel Corporation. Both companies have long been subsumed within larger industrial enterprises – CML by Rockwell Automation and British Steel by the Corus Group.

At the present time therefore the Kurtosis meter is not available. Because of its potential for producing a result which was not wholly dependent on trending, this is of some regret to the author. He therefore felt that the following descriptive material deserved a permanent record:

15.11.1 What is Kurtosis?

It should firstly be recognised that Kurtosis is a statistical parameter widely used in the analysis of distribution curves. If we have a number of measurements to plot, the value which occurs most frequently is called the mode. In a normal distribution, a symmetrical bell-shaped curve can be drawn having its peak at the mode. Originally derived by Gauss, it is often called the Gaussian curve. The Kurtosis value β_2 is defined in the equations below:

$$\beta_2 = \frac{1}{\sigma_4} \times \int_{-\infty}^{+\infty} (x-\bar{x})4P(x)dx \qquad \text{Equ 15.9}$$

where:

x = measurement

$\bar{x}$ = mean value of x

P(x) = probability of x

σ = standard deviation or Root Mean Square for a zero mean signal

As an alternative we may say:

$$\beta_2 = \frac{\mu_4}{\mu_2} \qquad \text{Equ 15.10}$$

where:

μ_4 = the fourth moment of the measurement distribution density function

μ_2^2 = the second moment (variance) of the measurement distribution density function

The Kurtosis value of the normal or Gaussian distribution is 3. This level is used as a reference to judge the "peakiness" of the distribution curve. Greater than 3 would be more peaky than Gaussian whilst less than 3 would indicate a flatter curve.

As mentioned before this work was introduced by ISVR (Institute of Sound and Vibration Research) at Southampton University whilst carrying out an investigative contract for the former British Steel Corporation. Kurtosis, when applied to the monitoring of bearing condition, is protected by Patent Specification 1536 306 owned by the former British Steel Corporation and its successors.

Using the statistical theory outlined above, it was decided that peak acceleration values of vibration should be obtained over a frequency spectrum of 2.5 kHz to 80 kHz. Inserting these measurements in the formulae, it could be anticipated that the Kurtosis factor for a good bearing would equal 3. A deviation of more than ± 8% from this figure would indicate the presence of damage.

Further research showed that if Kurtosis measurements were taken in discrete frequency bands and used in conjunction with overall velocity and/or acceleration measurements of vibration, then a more detailed assessment could be made, together with a trend analysis.

It should be remembered that the system does not rely on obtaining an absolute vibration measurement. The process of obtaining a Kurtosis reading is a statistical one based upon acceleration distribution. Thus although the variation in transmissibility of the vibration signals over the frequency band will produce a wide dynamic range of signals, the Kurtosis value will hardly be affected.

15.11.2 The Kurtosis meter

In its commercial form the instrument was known as the K meter. It consisted of a battery powered portable instrument with its own inbuilt microcomputer, together with a transducer (accelerometer) and input cable. The batteries could be re-charged from the mains. A carrying case was also provided and the whole equipment is as shown in Figure 15.25.

Vibration signals were monitored either by using a probe fitted into the end of the transducer, or preferably by mounting the transducer using its hand nut to secure it to a stud fitted to the bearing housing under investigation. If the probe was used, then it had to be firmly held, and applied to a point on the machine adjacent to the bearing race. The position selected should preferably have given the highest acceleration values of vibration in g RMS. The location should have been marked for future repeatability.

Grips could also be used where a hand probe might be dangerous but sensitivity could have been reduced. Nevertheless, the meter adjusted itself to suit the strength of the vibration signal available and the operator did not have to range the instrument.

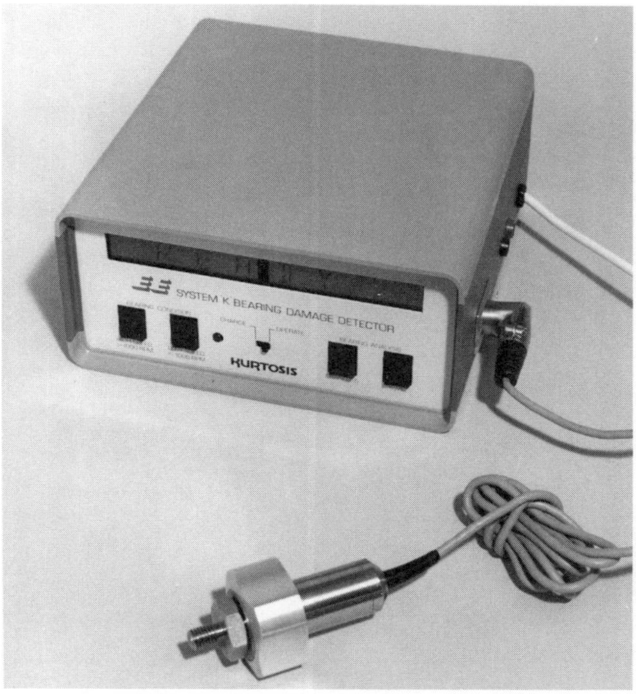

Figure 15.25 Early bearing damage detector for Kurtosis measurements

The method was also virtually unaffected by bearing size, a speed change or increase in bearing load.

15.11.3 Kurtosis values relative to frequency

The various stages of damage to a bearing are shown in Figure 15.26 together with the effect on the acceleration and Kurtosis value in each frequency band. It will be seen that these change significantly. The relative shape of the graphs will be true for a given amount of damage no matter where the bearing is installed. These curve shapes can be recognised by the microcomputer within the meter and thus the degree of damage can be indicated on the display.

The meter was operated in three different modes:

- Assessment
- Analysis
- Enveloping

Assessment

This was the most simple, and for many cases did suffice. Having fixed the transducer to the bearing housing and switched on the instrument, a battery check took place. The display panel I indicated if this was satisfactory or not, and whether re-charging was necessary. When the panel indicated "READY" the bearing condition - LOW SPEED (less than 1000 rev/min) or HIGH SPEED (greater than 1000 rev/min) was pressed. Even this was not critical, as selection of the wrong button simply extended the time taken to analyse the data and display the results.

The meter in the meantime responded with "BUSY LS" (low speed) or "BUSY HS" (high speed) whilst the data signals from the transducer were gathered and the calculations carried out. If the data was unstable, or if the accelerometer was detached from the machine then "ERROR" appeared on the display, and the bearing condition button had to be pressed again. Once the instrument had accepted the data and carried out its internal calculations, it indicated bearing condition directly as "GOOD", "LODAMAGE" (indicating early damage of the bearing) or "HIDAMAGE" (indicating a serious condition and imminent failure).

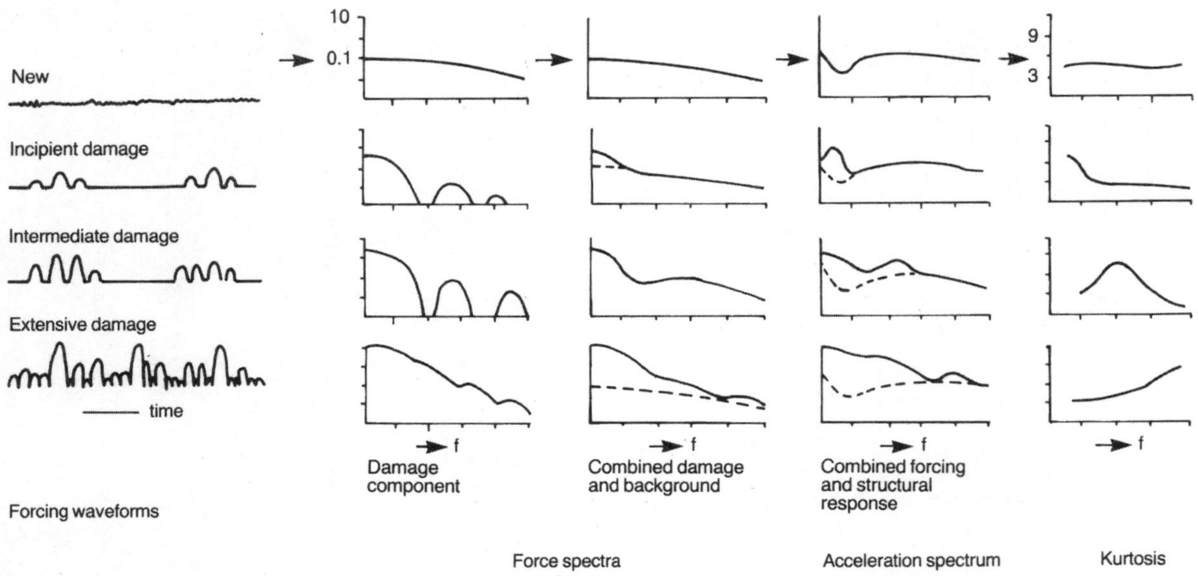

Figure 15.26 Diagram showing value changes with increased bearing damage

Bearing details		K1	K2	K3	K4	K5	G1	G2	G3	G4	G5	V mm/sec RMS
1 Pump bearing No. 4	Speed rev/min 1500, Date 7/10 Assessment Good	√	√	√	√	√	-1 5.12	-1 6.24	-2 7.92	-2 2.98	-3 3.1	2.1
2	Speed rev/min 1500, Date 10/11 Assessment	√	√	√	√	√	—	—	—	—	—	2.3
3	Speed rev/min 1500, Date 9/12 Assessment Early	5.0	4.7	4.8	4.6	4.6	-1 8.7	-1 9.1	-1 2.1	-2 3.8	-3 4.6	2.2
4	Speed rev/min 1500, Date 15/1 Assessment Advanced	3.7	3.8	4.2	4.8	5.2	11.2	1.5	-1 7.1	-2 6.1	-2 2.7	2.5
5	Speed rev/min 1500, Date 12/2 Assessment Advanced	√	√	3.7	4.2	5.1	4.3	6.1	1.1	-1 2.1	-2 4.2	3.2
		For further interpretation of results and operational details see K meter model 4100 handbook				If no data mark* If K = < 3.5 mark √						

Figure 15.27 Typical Kurtosis result sheet

Analysis

In the assessment mode, whilst data was collected in five discrete frequency bands, the evaluation was automatically carried out to arrive at the final assessment. For analysis, the more proficient operator could use the switches at the right of the meter.

Switch "f BAND 1-5", which selected the required frequency band:

Band	Frequency range kHz		
1	2-5	to	5
2	5	to	10
3	10	to	20
4	20	to	40
5	40	to	80

and the "KgVE" switch, which selected either:

K - Kurtosis value

g - RMS acceleration

V - Velocity RMS mm/sec

E - enveloping function

Both switches had a stepping function, for example the display might have shown:

03.87 KB3 which indicated a Kurtosis value of 3.87 in frequency band 3. By pressing the KgVE button the display could change to:

12.67 g B3 showing an acceleration level of 12.67g in frequency band 3.

If a bearing was in a "GOOD" state, it suggested that gRMS values were recorded for all frequency bands. When the bearing entered the "LODAMAGE" condition both gRMS and K should have been taken. A trend in the readings then showed the progress of damage, see Figure 15.27. g values increased whilst K factors will probably "peaked" at higher frequencies. By using this technique an experienced operator could predict the time to failure and thus the number of useful hours left in the bearing.

Enveloping

This was a facility used with an external analyser and provided an operator with the ability to identify damage repetition rate and thus that relating to machine speed. The resulting spectrum analysis showed whether the vibration signal was random in phase and amplitude or whether there was a repetitive waveform present.

The meter, once it had provided an assessment, stored indefinitely all the readings in its memory, and the last assessment, until power was switched off, the batteries run down or the speed buttons were pressed.

15.11 Conclusions

The intelligent use of condition monitoring techniques can assist greatly in the determination of necessary maintenance and the replacement of rolling element bearings. Systems are now available which have proved successful in giving warning of impending fatigue failure. Whilst often viewed with suspicion by the more conservative amongst us, it is believed that they will become widely accepted in the future. Only where there is the danger of imminent damage or malfunction should it be necessary to stop machinery.

15.12 Bibliography

Mechanical Vibration and Shock Measurements and *Frequency Analysis,* Brüel & Kjaer Ltd.

Preventative Maintenance Programme Handbook and *Vibration Measurement/Vibration Analysis Instruction Manual* – IRD Mechanalysis.

The Shock Pulse Method for Determining the Condition of Anti-Friction Bearings – SPM Instrument AB.

The Kurtosis Method of Bearing Damage Detection – Environmental Equipments Ltd.

ISO 10816-1:*1995 Mechanical vibration - Evaluation of machine vibration by measurements on non-rotating parts - Part 1: General guidelines.*

ISO 10816-3:1998, *Mechanical vibration - Evaluation of machine vibration by measurements on non-rotating parts - Part 3: Industrial machines with nominal power above 15 kW and nominal speeds between 120 r/min and 15 000 r/min when measured in situ.*

ISO 14694:2003, *Industrial fans - Specifications for balance quality and vibration levels.*

ISO 14695:2003, *Industrial fans - Method of measurement of fan vibration.*

ISO 1940-1:2003, *Mechanical vibration - Balance quality requirements for rotors in a constant (rigid) state - Part 1: Specification and verification of balance tolerances.*

ISO 281:1990 *Rolling bearings - Dynamic load ratings and rating life.*

ISO 2954-1975, *Mechanical vibration in rotating machinery. Requirements for instruments for measuring vibration severity.*

ISVR (Institute of Sound and Vibration Research), University Road, Highfield, Southampton S017 1BJ UK Tel: +44 (0)23 8059 2294 Fax: +44 (0)23 8059 3190 www.isvr.soton.ac.uk

16 Ancillary equipment

A number of ancillaries are available for fans and some of these are described in this Chapter. Whilst flexible connections, matching flanges and guards are obvious additions, the list is virtually endless and, indeed, seems to be growing by the day. There is also some competition between those manufacturers who provide at least some of these "appurtenances" and specialist suppliers for items such as dampers.

With the increasing importance of issues such as noise and vibration, the demand for attenuators and anti vibration mountings has increased. The problems of adequate maintenance have also become important leading to continuous monitoring of bearings and to automatic greasing systems, etc.

In HVACR, tunnel ventilation and grain drying applications, automation proceeds apace. Instruments are now being added to ensure that the fan is only activated when it can do useful work.

The moral is obvious – don't just read this Chapter for information on ancillaries. You may well find the information for a particular instrument in Chapters 8, 14, 15 or even 21.

Contents:

16.1 Introduction

In addition to the special features detailed in Chapter 8, fans may also be furnished with ancillaries, which enable a working fan set to be self-sufficient. In American parlance, these ancillary pieces are known as "appurtenances". Exactly what differentiates a special feature from an ancillary may be the subject of debate e.g. bolted-on upstream guide vanes on an axial flow fan are designed to provide contra-rotation to the airstream and thus increase pressure development. In like manner diffusers fitted to the discharge side of all types of fan convert high velocity pressures into useful static pressure.

There are, however, a number of bolted-on ancillaries for which there can be no doubt. Many of these such as:

- flexible connections
- matching flanges
- guards
- dampers (back draught and controllable)
- noise attenuators

can be fitted to both the inlet and outlet and are shown in Figure 16.1 for centrifugal fans, but similar "extras" are also available for axial and mixed flow fans.

16.2 Making the fan system safe

Improper installation, use or maintenance can make fan units a danger. The following Sections are intended to assist in the safe installation and use of fans and to inform operating and maintenance personnel of the dangers inherent in all rotating machinery and especially those used in air or gas movement. Often only the fan is supplied by a manufacturer.

The customer/user must therefore consider how the rest of the system - motors, drives, starters, etc, may affect fan operation.

Installation and maintenance must be carried out by experienced and trained persons, as discussed in Chapter 18. As well as the manufacturer's own instructions, it is important that all national and local government requirements are complied with. In the United Kingdom, the Health and Safety at Work Act 1974 should be followed.

16.2.1 Guards

All fans have moving parts which may require guarding. It is a fact of life that two danger areas are the fan inlet and outlet. Perfect guarding would require these to be blanked off completely - but then there would be no air/gas flow. Fan guards have to be designed to reduce the fan's performance as little as possible whilst giving a good measure of safety. This requires that they do not deflect when leant against.

In areas accessible only to experienced and trained personnel, a standard industrial-type guard may be adequate. This will prevent the entry of thrown or dropped objects with the minimum restriction of airflow.

Where the fan is accessible to untrained personnel or the general public maximum safety guards should be used, even for DIDW fans, at the cost of some loss of performance. Fans located less than 2 metres above the floor require special consideration. Even roof-mounted equipment will require guards when access is possible, for example, by climbing children.

For full information on this subject the customer/user should refer to ISO 12499 and AMCA 410.

16.2.1.1 Inlet and outlet guards

These are not necessary for an Installation Category D fan, provided that access to the ducting cannot be made whilst the fan is in operation (Figure 16.2). With the same proviso, an inlet

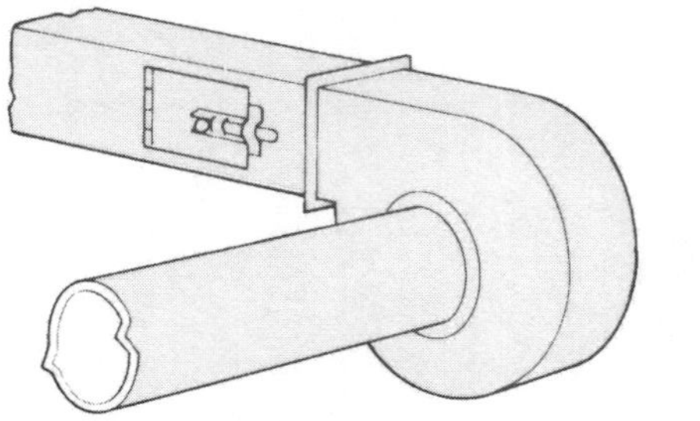

Figure 16.2 Fan protected by ductwork

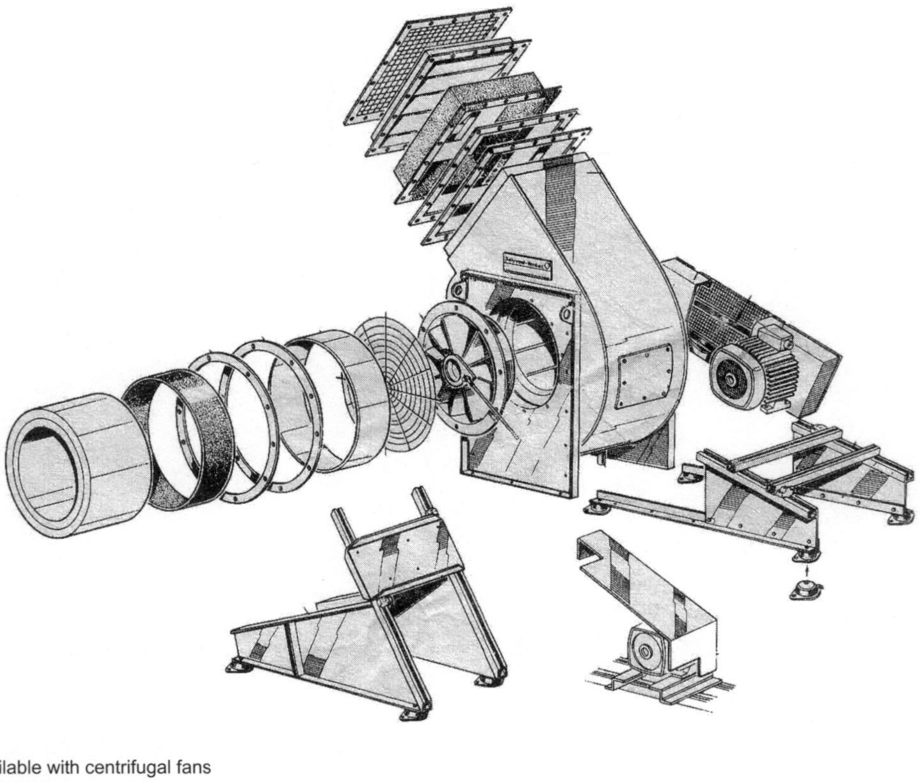

Figure 16.1 Ancillaries available with centrifugal fans

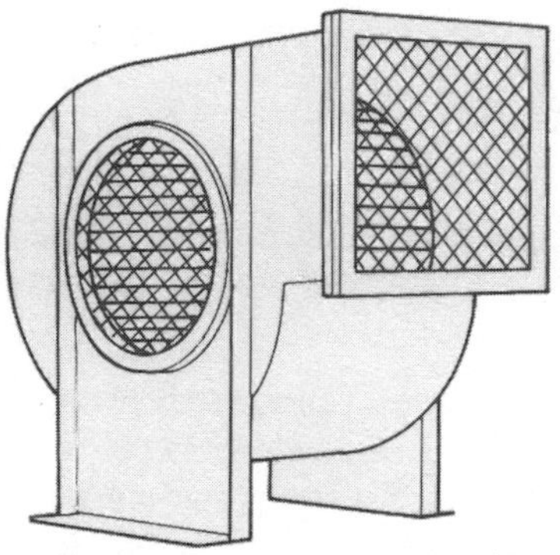

Figure 16.3 Inlet and outlet guards (Installation Category A)

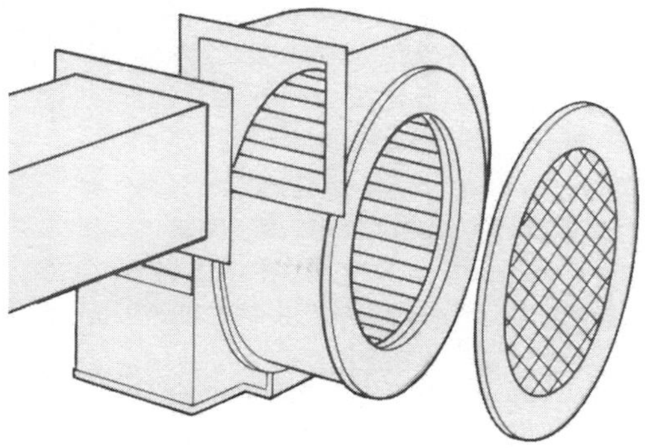

Figure 16.4 Ducting at outlet, guard at inlet (Installation Category C)

guard must be provided for a Category A or B fan whilst an outlet guard must be provided for a Category A or C fan.

The intention with all inlet or outlet guards is to prevent finger or arm contact with the internal moving parts. The distance from the guard to the moving part will determine the mesh size. Thus a backward bladed centrifugal fan, which has a relatively long inlet cone, can have an inlet guard with a more open mesh than say an axial flow fan, where the guard is closer to the impeller.

The illustrations Figures 16.2 to 16.4 are self-explanatory. The customer should advise the manufacturer how the fan is to be installed and the guards which he requires.

16.2.2.2 Drive guards

Fans may be driven directly from the motor shaft or through a belt drive. In every case where the bearing assembly, rotating shaft, sheaves, or belts are exposed, a suitable guard should be provided, (see Figures 16.5 and 16.6). Most centrifugal fan manufacturers include a combined shaft (and cooling disc if fitted) guard as standard, but it is as well to check.

Customers often prefer to provide their own motors, drives, and drive guards on indirect driven fans. They should in all cases follow the recommendations of BS 5304:1975 and BS 3042:1992, or other relevant local standards.

In restricted access areas, one-sided guards of expanded metal may be acceptable. Readily accessible locations will require maximum protection guards, and in many cases a fully enclosed sheet metal guard. The loss of fan performance on DIDW fans must be weighed against the degree of safety provided. Where the customer/user is in any doubt, he should purchase the complete assembly of fan, drive, motor, guarding, and combination baseplate from the fan manufacturer who can provide a fully engineered system to meet any specified standards. For indoor applications a wire mesh drive guard will be perfectly satisfactory, but for outdoor applications, a totally enclosed weatherproof driveguard will be necessary, probably manufactured from sheet steel.

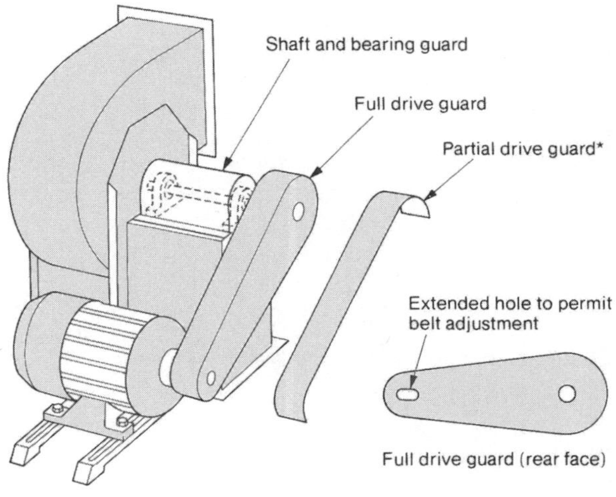

Figure 16.5 Typical example of an Arrangement 1 (belt driven) fan. (A combined guard covering the bearings and shaft, and cooling disc if fitted, should be provided.)

* **Important:** Partial guards should only be used where restricted access makes the use of a full guard impossible, and never unless the partial guard can be combined with existing stationary structure to form a complete guard.

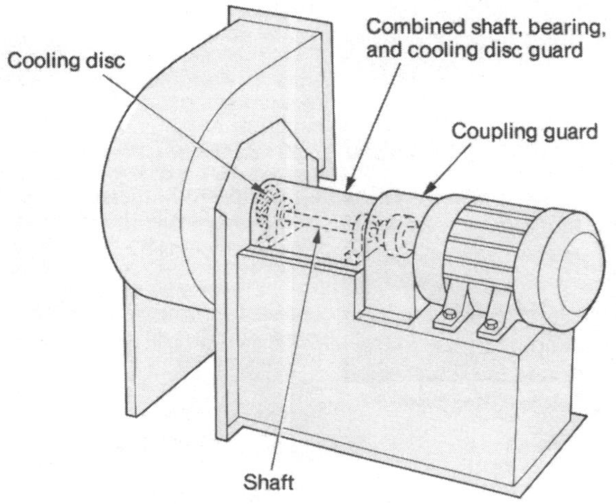

Figure 16.6 Typical example of an Arrangement 8 (coupling drive) fan. (A combined guard covering the cooling disc, bearings and shaft should be provided.)

16.3 The hidden danger

Whilst not strictly part of the fan supply, and therefore not suggesting any specific ancillaries, there are what might be termed the "hidden dangers" of fan systems. The following features which may be necessary in the ductwork system are suggested:

- As well as the normal dangers of rotating machinery, some fans (e.g. paddle-bladed), present an additional hazard in their ability to suck in loose material as well as air. Solid objects can pass through the fan and be discharged by the impeller as potentially dangerous projectiles. They can cause serious damage to the fan itself, if not allowed for in the design.

- Intakes to ductwork should whenever possible be screened to prevent the accidental or deliberate entrance of solid objects. For example, on a sawdust handling system an intake screen should be provided which will allow the entry of sawdust but prevent the entry of large pieces of wood.

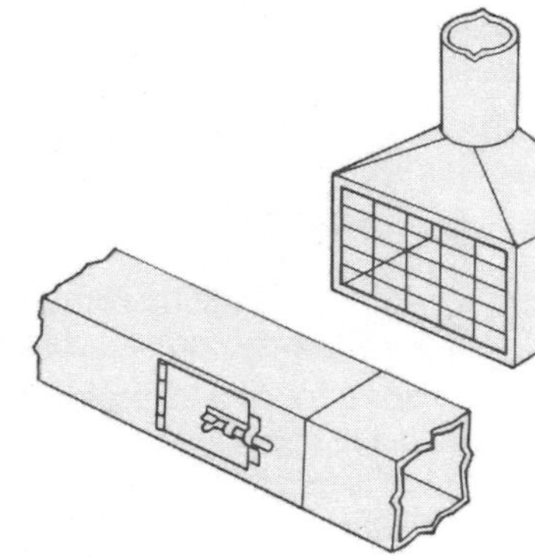

Figure 16.7 Access door in duct and specimen intake screen

- Access doors to a duct system should never be opened with the fan running.
- On the downstream (or pressure) side of the system, releasing the door with the system in operation could result in explosive opening. On the upstream (or suction) side the inflow may be sufficient to suck in tools and clothing, etc, and even cause a man to lose his balance. Where quick-release handles are provided on access doors, at least one positive bolt should be installed to prevent accidental opening.
- When a fan is being started for the first time, a complete inspection should be made of all the ductwork and the fan interior to make certain that no foreign materials have been left, which could be sucked into or blown through the ductwork, (see Figure 16.7).

16.4 Combination baseframes

A rigid base which allows the fan, motor and drives to be transported and installed as a complete unit is often desirable. It ensures that the various items are correctly aligned and that vee belt drives are correctly tensioned. When anti-vibration mountings are fitted below the baseframe it becomes essential to ensure that both fan and motor vibrate as one and that belt tension is maintained by fixing their relative positions. (It would be helpful to refer to Chapter 8, Figure 8.8.)

Baseframes are of many varieties but the following are the most popular:

a) Fabricated from rolled steel channels welded or bolted together as appropriate. A heavy duty construction, which gives the desired mass when used with anti-vibration mountings, (see Figure 16.8).

Figure 16.8 Fan set on rolled steel channel baseframe

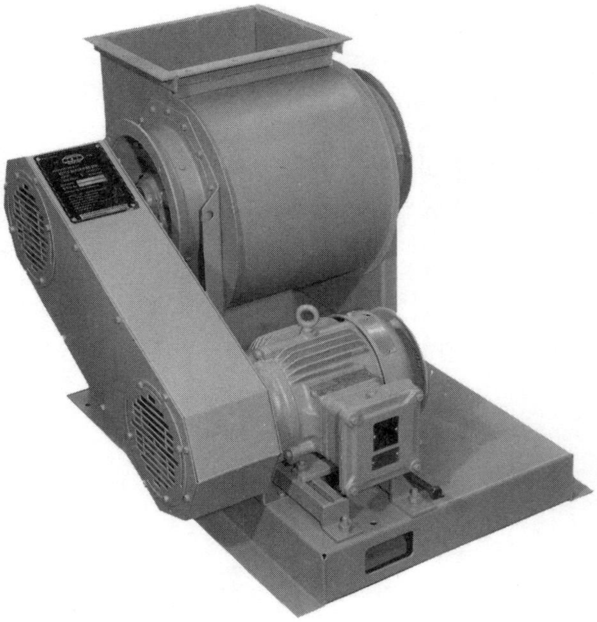

Figure 16.9 Fan set on sheet steel fabricated baseframe

b) Sheet steel fabrication of appropriate depth to give rigidity, usually constructed from parts produced by a turret punch or a laser (see Figure 16.9).

c) Angled fabrication from slotted square section (see again Figure 16.1) designed to give reduced drive centres and an overall reduced "footprint".

16.5 Anti-vibration mountings

Anti-vibration mounts come in a number of different forms. These most commonly used to reduce the vibration of a fan unit to its foundations are:

- Rubber or neoprene in shear
- High deflection steel springs

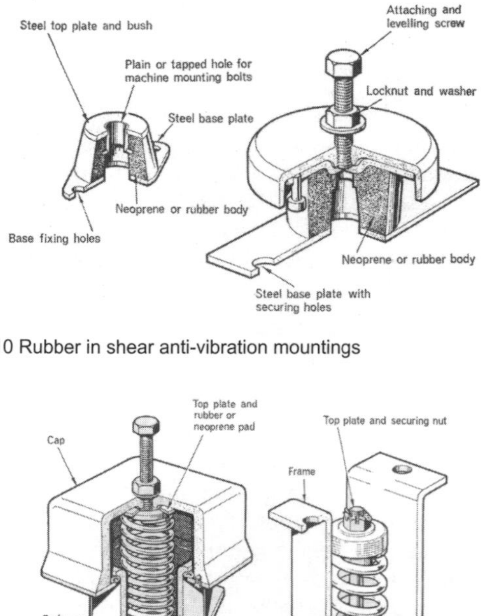

Figure 16.10 Rubber in shear anti-vibration mountings

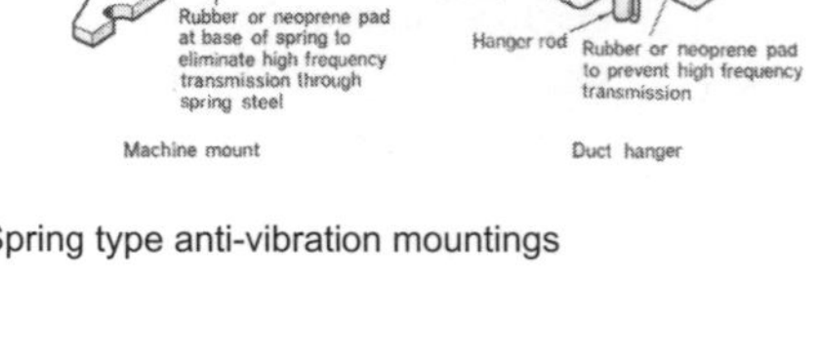

Figure 16.11 Spring type anti-vibration mountings

Rubber in shear mounts are generally used for deflections up to about 12.5 mm. This means that their natural frequency is higher then spring mounts, which are to be preferred for units operating at relatively low rotational speeds. The latter also have the advantage of maintaining their linear stiffness over a wide range of operating conditions and are impervious to humid or oily environments.

Examples of the two types are shown in Figures 16.10 and 16.11.

16.6 Bibliography

ISO 12499:1999, *Industrial fans - Mechanical safety of fans - Guarding.*

Health and Safety at Work -Act 1974, .London HMSO, reprinted 1991, ISSN: ISBN 0105437743.

AMCA 410-96, *Recommended Safety Practices For Users and Installers of Industrial and Commercial Fans.*

BS 5304:1988, *Code of practice for safety of machinery*

PD 5304:2000, *Safe use of machinery*

BS 3042:1992, IEC 61032:1990, *Test probes to verify protection by enclosures.*

17 Quality assurance, inspection and performance certification

This Chapter details the requirements for the inspection of the base materials and components used in the manufacture of fans.

Knowledge of the particular Standards to which they are produced is particularly important especially at a time when national standards are being replaced by European standards.

Apart from aluminium alloys, where relevant grades are well defined, many non-ferrous metals and non-metal materials are purchased against trade names. It then becomes important to assess the quality reputation of the supplier and to determine whether he possesses a Quality Assessment such as ISO 9001.

This Chapter also describes the various inspection functions and the tests possible to confirm functionality. It defines what documentation the purchaser may expect to receive with the fan and also what additional information he may request if he wishes to carry out his own inspection. Guidelines are given for purchaser quality requirements.

When fans are "designed to order" and have to meet the purchaser's specification, the whole question of quality becomes of major importance. In these instances it is recommended that the quality plan is produced and agreed before the signing of any contract. Non-compliance procedures should be included and tolerances on performance, dimensions, manufacturing defects etc., should all be defined.

It should be especially recognised that a balance has to be maintained between the customer's aspirations and the price he is prepared to pay. The extent of control that is required ultimately depends upon the confidence which the customer has in a particular supplier, the trust which he is prepared to give, any legal requirements which may exist and any requirements laid down by insurance companies.

Contents:

17.1 Introduction

The fan industry has a claim to being "mature". Its products have been around for up to 150 years. During this time much expertise has been built up by the long surviving companies. However, over the last 30 years, we have witnessed major upheaval and old established companies with excellent products have succumbed to the machinations of administrators without the same level of financial expertise. Small companies have been formed to fill the gap, but whilst their prices may be attractive, this is often because they do not have the necessary supporting structure to validate the design of their products.

Inspection of all components should be carried out by the manufacturer as a matter of course. The degree of inspection will be dependant upon the criticality of the component, its nature and its function. It may also be dependant on the batch size and whether "sampling" is appropriate. Mass-produced parts do not normally require the same degree of inspection as a small number of specially made parts. With batch production it may be sufficient to check the "first off" to ascertain that production has been correctly set up and then sample occasionally to check adherence.

The basic fan assembly stage having been completed, a final dimensional check should be carried out. Components not having the necessary fit or clearance should be readily apparent.

17.2 Physical properties of raw materials

The majority of fans have their components manufactured from materials supplied by others. Thus many casings and impellers will be fabricated from sheet steel, but have cast iron or steel hubs, die cast aluminium blades, plastic casings and/or impellers, etc. Even stainless steel or nickel chrome alloys may be appropriate in applications where the air or gas contains corrosive elements or is at high temperature.

Most fans have their major components manufactured from sheet steel whilst other components may be of cast iron or machined from an alloy steel. Iron ore is the basis of all these materials and can be converted into iron by these methods:

- blast furnace,
- sintering or pelletised/blast furnace,
- direct reduction.

In a blast furnace, iron ore reacts with hot coke to produce pig iron. The sintering or pelletising process prior to the blast furnace operation is added to allow blending of iron ores and also to control the size of the blast furnace feed. Sintering or pelletising improves the blast furnace operation and reduces energy consumption. Direct reduction produces sponge iron from iron ore pellets by using natural gas. Most iron is produced from sintered iron ore and coke. The steel maker controls the sintering process to produce a consistent iron quality.

Modern blast furnaces are fitted with many instruments and, together with computer modelling, enable in-process control. Iron is taken from the blast furnace as finished material for iron foundries. Iron is transferred to the oxygen steel process for conversion to various grades of steel. Iron from direct reduction plants is mixed with scrap steel in an electric arc furnace to produce various grades of steel.

Standard tests are applied, solely to assess compliance with the published specifications. Some materials are characterized only by their physical properties or chemical composition, others by both. Grey cast iron is specified by its physical properties. Some low grades of carbon steel are specified by their chemical composition, no physical properties are necessary. Most materials are described by both.

For the physical properties defined in Section 17.2, standard test pieces are stretched in a machine which simultaneously measures the increase in length and the applied load. There are several different test piece sizes which give slightly different results. One standard test piece is very small, this fits a machine called a Hounsfield Tensometer. Very small test pieces are useful when samples must be taken from castings or finished parts.

The various tests undertaken are now outlined.

17.2.1 Ultimate tensile strength

The strength of the material when it fractures. See Chapter 7, for typical values.

17.2.2 Limit of proportionality

The strength of the material when the relationship between stress and strain ceases to be linear. In low carbon steel this is classified as the yield point, the onset of plastic deformation, the material does not return to its original length when the load is removed. Most designs do not stress materials beyond the limit of proportionality.

17.2.3 Elongation

How much the material has increased in length when it fractured. Different test pieces have different gauge lengths, each gauge length gives a slightly different result. Good elongation properties, 15 to 20%, are required for complex components which are highly stressed. Good elongation indicates ductility. Ductility is necessary so that components can deform very slightly to spread the load. A good cast iron may be 4%.

17.2.4 Reduction in area

Ductile materials "thin" slightly as they are stretched. When the material fractures, the cross-sectional area of the fracture is less than the original test piece. Reduction in area is reported in most American standards but not used very much in Europe.

17.2.5 Hardness

The ability of the material to withstand surface indentation. No special test piece is required, raw material and finished parts can be tested. Several scales of hardness are used; Brinell Hardness Number, Vickers Pyramid Hardness and Rockwell. Approximate conversions are available between scales (see Chapter 23). In carbon steels, the hardness is directly related to the strength.

17.2.6 Impact strength

The ability of the material to withstand shock or impact. A special test piece is required to fit the test machine. Most materials lose impact strength as the temperature reduces. Depending upon the material, impact properties should be checked when operating below 0 °C. Two different tests are used which give different results, very approximate conversions are available. Charpy and Izod are the most popular. A benchmark for offshore equipment is 27 J at the design temperature. It is normal to check three test pieces.

17.2.7 Fatigue strength

All the tests defined so far can be performed fairly quickly; "test the pieces today, get the results tomorrow". Fatigue is very different. A special test piece is either subjected to repeated tensile loads or repeated bending loads.

For repeated tensile loads, the test piece experiences cyclic loads from 0 to + value. A bending test piece is loaded from -value to +value. To find the endurance limit the test piece must not fail. A test piece may appear satisfactory if it lasts five million cycles. If the machine runs at 3000 r/min this will take 1667 minutes, i.e. 28 hours. Of course, it will not be possible to guess the correct stress so several tests must be run.

Testing for fatigue in clean air is the most simple. However, these results may not be applicable to the actual fan environment. Valid conclusions may only be drawn by conducting the tests in air/gas containing the actual contaminants.

It s not common for fatigue strength of materials to be checked on a contractual basis. Such tests would take too long to reach any valid conclusions. It should be especially noted that the fatigue strength of aluminium products continues to fall with the number of stress reversals. The asymptotic curve assumed in may specifications just does not exist.

Most centrifugal fan designs are not based on fatigue but axial fan blades are cantilevered. An important factor in their design is therefore due to fluctuating stresses and hence fatigue failure. The manufacturer should state if the life of the blades, or any other component is limited by running at the rated conditions, or indeed any other likely situation, such as running in reverse.

17.2.8 Creep resistance

Creep is the permanent distortion of the material after being subjected to a stress for a long period of time. This is not many a problem in fans, although those built of GRP, PVC, PTFE or other engineering plastic, may suffer at any temperature. It must however be considered for fans operating at gas/air temperatures above about 400 °C. Creep testing is similar to fatigue testing but creep tests can last for years. Published research data is therefore often used when necessary.

17.2.9 Limitations

Many mechanical properties of a material are dependent on its grain direction. Unless specified otherwise all these values relate to the longitudinal direction. Properties in the transverse direction or the through direction may well be lower, dependent on the physical treatment of the material and its grain structure.

17.3 Heat treatment

Many materials require heat treatment to achieve the correct condition or strength. Carbon steels are hardened and tempered to achieve high strength, usually at the expense of ductility. Austenitic stainless steels are stress-relieved, softened or solution-annealed to modify the physical or chemical properties. The final condition is usually confirmed by taking hardness readings. When components are heat treated to achieve specific physical properties a test piece is heat treated as well. The necessary physical tests are conducted on the test piece.

The customer may request a certificate detailing the duration of the heat treatment and the temperature achieved at specified intervals. If necessary a continuous trace of the temperatures may be provided.

17.4 Chemical composition

When a metallic material is produced as a raw material, its chemical composition is checked. When cast iron is converted to carbon steel in the oxygen process, all the relevant elements are weighed before being put into the converter. Before the steel is poured, the chemical composition is checked. When the steel is poured a sample is cast. The sample is analysed and its chemical properties are the properties of the melt. Certificates will show the name of the steelmaker and the melt, cast or heat number.

The chemical composition may show elements which are not required by the specification. Low carbon steels may show traces of nickel, chromium and molybdenum. The trace elements are a welcome addition because they tend to enhance the physical properties of the material. Impurities such as sulphur and phosphorous, will be shown very accurately. The chemical composition of specific components, when necessary, can be traced back to the original melt.

On rare occasions, a sample will be taken from a component and analysed. Modem techniques only require very small samples. It is possible to analyse material without destroying it. Two devices are available which can analyse material without removal from the component. Neither method can detect carbon. However sufficient accuracy is present to differentiate between 304 and 316 stainless steel.

17.5 Corrosion resistance

Corrosion resistance of materials is judged from published research. A few manufacturers carry out long term research on corrosion to develop materials to cope with specific problems. If a fan user wishes to handle a new gas of which previously no fan manufacturer has had experience, the user should conduct basic corrosion testing.

17.6 Non-destructive testing

Raw material, raw castings and completely finished components can be examined physically to determine the quality of certain aspects of the material. This type of examination falls into two categories:

- surface inspection,
- interior inspection.

Surface inspection looks for discontinuities in the surface which could be detrimental to the service life of the component. Cracks in the surface create stress raisers which can lead to fatigue failures. Pinholes in the surface may indicate porosity.

Internal examinations can show the integrity of the material and identify any impurities, inclusions or voids in critical locations. Impurities, inclusions and voids detract from the cross-sectional area available for stressing and create stress raisers. Porosity can lead to problems of leakage.

When flaws are detected it has to be decided whether the flaw is serious, if it can be repaired or whether it should be repaired. Some national standards, particularly pressure vessel standards, have categories for defects. The manufacturer's requirements may be more or less stringent than published standards. If the flaw is in raw material, a casting or piece of plate, it may be more cost-effective to scrap it rather than expend more time and money on repairs. If the flaw is in a semi-finished piece there may be more incentive to repair. If the flaw is in a finished component there may be compelling financial reasons for a repair.

17.6.1 Visual inspection

Sand cast axial impeller blades and hubs for all types of fan may be made by pouring molten metal into a prepared mould and allowing it to solidify. Following shake-out from the mould and clean-up, many such casting will be heat treated and machined. During this process certain surface and subsurface imperfections may become evident.

Surface imperfections in sand castings can vary in the level of importance from significant to superficial. Surface imperfections in their approximate order of importance based on the imperfection type and its effect on casting serviceability are now discussed:

a) Cracks in castings appear as tight, linear separations in the material that are continuous or intermittent. Cracks may be jagged or straight. Cracks are not acceptable.

b) Surface hot tears are likely to be found at tight curvatures in the casting or where there is an abrupt change in casting thickness. Hot tears are not acceptable.

c) Surface shrinkage is occasionally visible on the cast surface where a riser has been removed or on a machined surface.

d) Surface and subsurface porosity or pin-holes in castings are formed as a result of gas formation during solidification. Sub-surface gas inclusions or porosity are evaluated by the radiographer if the casting is radiographic quality. Surface porosity is often the result of moisture in a sand mould which has not been pre-heated properly.

   Generally, surface porosity in castings is not considered harmful if it is 0.8 mm diameter or less and not concentrated. In such cases, it is customary to explore grind 10% of the indications and accept the condition if the porosity is shallow and no subsurface pockets are opened. Porosity in castings is considered unacceptable when it is concentrated in a specific area.

e) The surface of a casting may show evidence of trapped sand which was dislodged from the sand mould during a pour and which floated to the upper part of the casting (cope side). Rough surface indentations are evidence of sand inclusions where the sand has been removed during abrasive clean-up. It should be kept in mind that such criteria are for guidance.

f) In order to accelerate the solidification process in specific areas of a casting, the foundry mould designer places metal chills in those areas of the mould. Such chills are expected to be melted by the molten metal and fused completely into the solidified casting. Occasionally, the chills do not completely fuse with the cast metal.

g) Veins are irregular, linear ridges on the surfaces of castings which are produced by cracks in the sand mould. For most applications, moderate veining is considered acceptable.

h) Rat tails are irregular, linear depressions on the surface of castings which result from ridges in the mould surface. Such depressions are relatively shallow. For most applications rat tails are considered acceptable.

I) Wrinkles, laps and coldshuts are surface irregularities caused by incomplete fusion or by folding of molten metal surfaces. Where such imperfections are of negligible depth the condition is considered acceptable. Where a lack of fusion exists, the imperfection should be explored to sound metal.

j) Scabs are raised imperfections adhering to the cast surface. They are usually sand crusted over a thin porous layer of metal and are often considered unacceptable. Removal by grinding usually verifies that scabs are surface imperfections only.

17.6.2 Radiographic inspection

Radiography, X-ray, is accepted as the highest grade of internal inspection and the most costly. Radiography is good because a permanent record of the inspection is available at any time for review. It is used mostly for welds and also for critical areas in castings. Components can be taken to fixed X-ray equipment. Large components or assemblies are radiographed using radioactive isotopes. Strict safety precautions must be enforced. In view of its importance for the integrity of high performance fans, this method is described in the greatest detail.

Figure 17.1 Room based real-time radiographic system

The most technically advanced companies in the fan industry have facilities which by using real time techniques cut the time defects by about two-thirds, enabling production and delivery improved. The system (Figure 17.1) is more sensitive, more t also provide a far more comprehensive and easily accessible s: previous X-ray units. Each moving part is stamped with a all X-ray images are automatically archived onto 50 mm laser. 30 years, providing full component traceability.

Inspection by X-rays is carried out by irradiating one surface of the specimen with X-rays whilst a radiation sensitive electronic imaging sensor is held against the opposite surface. The radiation, in passing through the specimen, is differentially absorbed by discontinuities caused by flaws, voids, changes in thickness or material density, and an image of the variations integrated throughout the sample thickness, is produced on the surface of the electronic sensing screen.

After the electronic image has been noise reduced, it is displayed on screen where variations within the specimen appear as shadow objects of differing half tones, from which information may be obtained about the presence of flaws. The record produced in this way is known as a real time radiograph. Real time because the image display is live and if the specimen is moved the X-ray radiograph changes to show the corresponding incident shadow on the image display. The use of X-rays to produce a radiograph is called X-Radiography. Figures 17.2 and 17.3 show two examples of impeller blade radiographs.

X-rays are a form of electromagnetic radiation which may be generated by causing a stream of fast-moving high energy electrons to strike a metal target. The sudden deceleration of electrons gives rise to radiation of photons (X-rays) with a continuous energy spectrum.

X-rays possess great penetrating power which increases with increasing energy of the waves (increasing frequency or shorter wavelengths). X-ray equipment is defined by the energising voltage, which can typically range from 25 kV to 15 m V. X-rays can be used to examine items varying from layers of paper to steel of thickness up to 0.5 metre. All materials are penetrated by X-rays, but the greater the density, the less the penetration.

Radiation of short wavelength produced by high target potential is said to be of high energy and is described as a hard X-ray, having greater penetrating power. Longer wave radiation produced by lower target potential is said to be low energy and is described as soft X-ray, having lower penetrating power.

The penetration ability of X-radiation may be expressed in terms of a given material thickness (e.g. steel or aluminium) that can be adequately inspected.

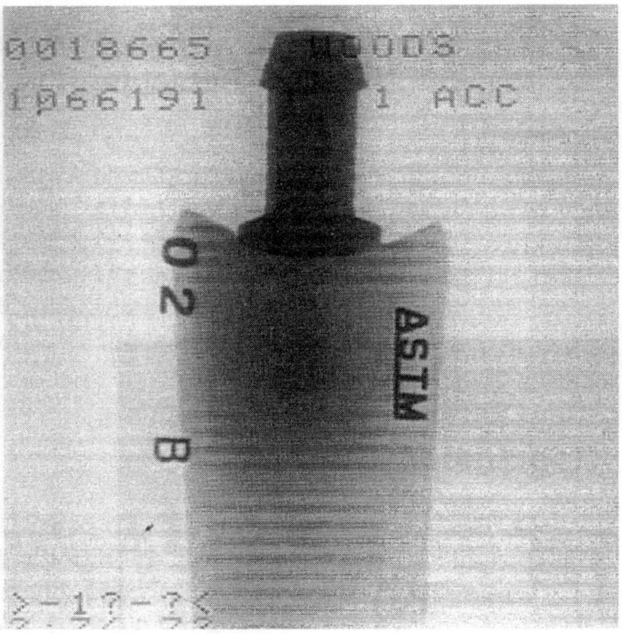

Figure 17.2 Example of acceptable blade radiography

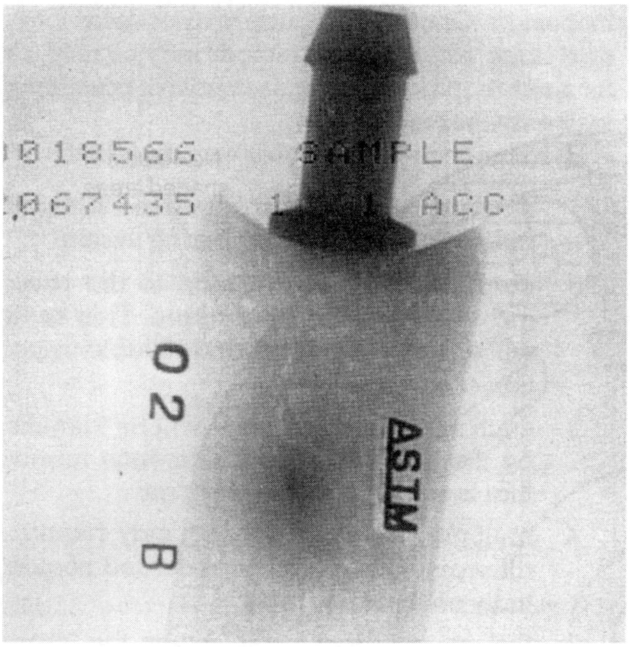

Figure 17.3 Example of unacceptable blade radiography

For low energy, constant potential X-ray generators, the beam intensity produced by the X-ray tube is determined mainly by the magnitude of the filament current, and to a lesser extent by the target potential. A near linear relationship exists between filament current and beam current so it is customary to express the output capability of such a tube in terms of filament current.

The quality of a real time X-ray radiograph is nearly always quoted in terms of the amount of detail discernible on the image of an image quality indicator (IQI) of the same material as the specimen placed on the surface of the specimen. This IQI sensitivity depends upon the radiographic technique used, the type of IQI and specimen thickness. When radiographing other materials other than steel it is customary to use conversion tables related to the material and radiation energy to obtain approximate equivalent thickness factors.

In the UK two different patterns of IQI are recommended, known as the "wire" type and the "step hole" type and one or the other is commonly used in most European countries. In the USA an ASTM-plaque is generally used.

IQI sensitivities are expressed as percentage values, i.e. the size of the minimum discernible IQI details is expressed as a percentage of the specimen thickness, thus a smaller numerical value implies a better sensitivity. Typical radiographic sensitivities range between 0.5 and 2.5 percent depending upon inspection variables.

A recommended procedure for reporting weld and castings defects in a radiograph is to use a three-part code:

(1) A number to denote the horizontal or vertical distance in inches between "the reference mark or the lowest number on the radiograph and the start of the defect.

(2) A code letter or letters to denote the type of defect (see abbreviations in Table 17.1).

(3) A number to denote the approximate length in inches over which the particular defect extends.

Surface imperfections	
Code	**Description**
SXP	Excessive penetration
SRC	Root concavity
SGI	Incompletely-filled groove
SGS	Shrinkage groove
SUC	Undercut
SSP	Weld spatter
SED	Underflushing (excessive dressing)
SMG	Grinding mark
SMC	Chipping mark
SMH	Hammer mark
STS	Torn surface
SPT	Surface pitting

Internal defects	
Code	**Description**
K	Crack
KL	Longitudinal crack
KT	Transverse crack
KE	Edge crack
KC	Crater crack
L	Lack of fusion
LS	Lack of side fusion
LR	Lack of root fusion
LI	Lack of inter-run fusion
RP	Incomplete root penetration
I	Inclusion
IL	Linear inclusion
IT	Tungsten inclusion
IC	Copper inclusion
PG	Gas pore
P	Porosity
PU	Uniform porosity
PL	Localised porosity
PP	Linear porosity
EC	Elongated cavities
WH	Worm hole (pipe)
CP	Crater pipe
BT	Burn-through
DM	Diffraction mottling

Table 17.1 Type of defect-coding abbreviations

For example, an X-ray radiograph image showing the existence of lack of fusion commencing 50 mm (2 inches) from the reference mark over a length of 25 mm (1 inch) and the defect repeated 150 mm (6 inches from) the reference mark over a length of 25 mm (1 inch), and also localized porosity for 19 mm (0.75 inch) at a distance of 150 mm (6 inches) from the reference mark, the code would be 2-L-1: 6-PL-0.75: 8.5-L-0.5.

Real time radiographic facilities are used in any of the following modes:

a) Intermediate stage product inspection or intermediate radiography. As a general rule when the items are cast, an inspection at that stage segregates the good castings and rejects before any value is added to the casting. This minimises wastage of time before the casting is handled and cleaned for burrs and excessive materials.

   Radiographic technique is usually defined for different products. This inspection stage can be carried out any number of times before the finished product stage. In this mode product inspection records are not normally required. Good castings are transferred to the next production stage and rejects dealt with as appropriate.

b) Intermediate radiography with image storage. Quality demands on products may stipulate a minimum acceptance quality for defect sizes and type. Proof of acceptance based on records may be required by independent inspectors. Once the products are accepted for the next stage, radiographic records may be required for short term storage requirements, possibly 6 to 24 months.

c) Real time radiography with record keeping and digital or analogue long term image storage. Safety critical and sensitive application products normally require inspection records to be held in archives for the duration of the product life. A complete history of the product has to be maintained. Stringent quality control inspection specifications are stipulated and adherence to the specification is mandatory.

Recommended procedures demand that:

1. Each product item to be inspected is identifiable with a unique numbering system.
2. Each product type will adhere to radiographic inspection techniques. This setup will ensure repeatable and reliable inspection of castings.
3. Each radiographic image will be identified by the unique product reference number that is stamped on the castings.
4. Multiple views of a product may require different set up and recorded sequentially on video.
5. The operator manually logs on the inspection records the verdict of the image and total acceptability of the items.

It should be noted that subsurface imperfections can only be determined by methods such as radiography and ultrasonic. There have been many instances where an apparently good casting has failed, only to reveal quite massive internal faults.

Subsurface imperfections include shrinkage, hot tears and inclusions as follows:

a) **Shrinkage - subsurface**: this is often referred to as centreline shrinkage since it occurs near the mid point of the casting wall which is the last to solidify. Since shrinkage is a subsurface condition, it should be evaluated by radiography.

b) **Hot tears**: Casting tears usually appear at points of thickness transition and are attributable to contraction stresses during cooling and the low hoop strength of the casting.

c) **Inclusions**: Subsurface non-metallic inclusions such as sand, slag and trapped gas pockets or porosity are readily identifiable during radiographic inspection.

17.6.2.1 Acceptance criteria for X-ray examination

It will be apparent that strength and integrity are closely related to the quality of a component as cast. The criteria for acceptance are generally those described in ASTM Standard E155 together with its reference radiographs. A procedure should be adopted which defines the inspection process. Before X-ray or fluoroscopic examinations are carried out, the following checks should be made:

- In aluminium die castings, there should be no visible signs of surface porosity or cracks.
- In aluminium sand castings, there should be no visible signs of blowholes or mould misalignment
- Aluminium heat treated castings should be checked for hardness.
- Blade carriers in malleable cast iron should have no blowholes or surface scabs.

Additional criteria are given for blades and hubs according to the stresses imposed on them during operation.

Blades may be divided into 3 basic categories as shown in Figure 17.4 where areas required to be of high integrity are shown cross-hatched, whilst areas having a lower integrity requirement are shown plain.

Areas of high integrity should be substantially free of any defects, the maximum allowable being:

- An area of porosity not greater than 5 mm diameter as sample micrograph No. 1, as defined in ASTM E155.
- A single isolated defect not greater than 2 mm diameter.

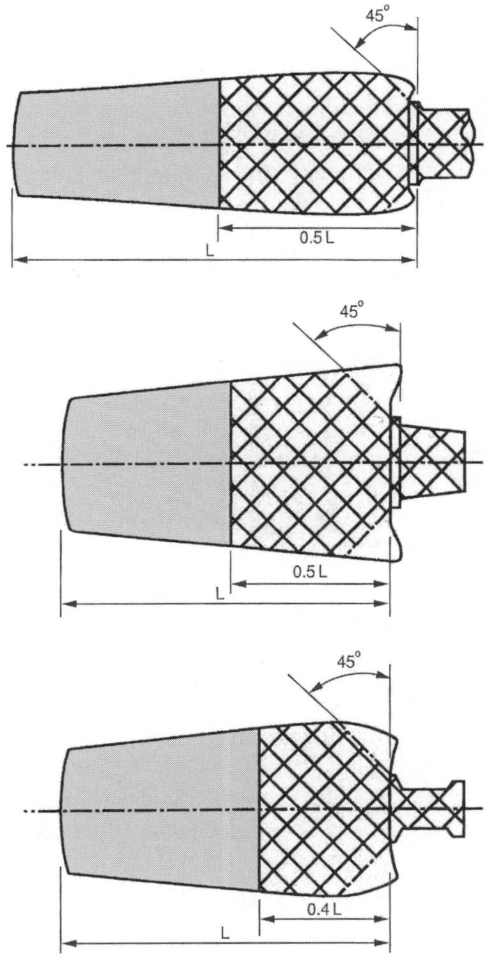

Figure 17.4 Acceptance criteria for blades (High-Low integrity areas)

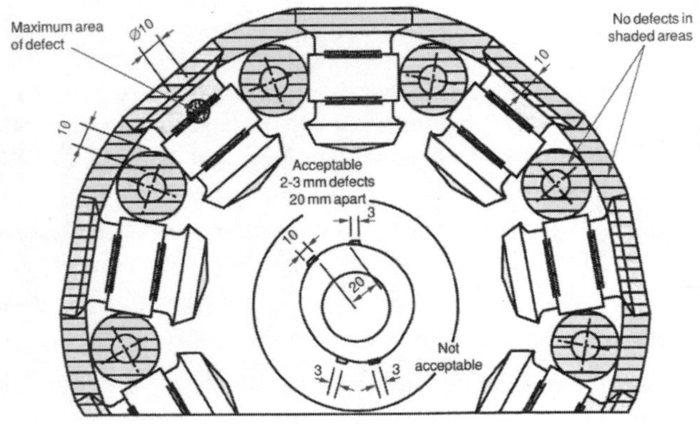

Figure 17.5 Acceptance criteria for hubs

In areas of low integrity the maximum defect allowable should be:

- An area of porosity not greater than 10 mm diameter.
- A single isolated defect not greater than 5 mm diameter.

The defects should not be within 5 mm of the boundary of the casting and only one defect should be allowed in each component.

Hubs for axial flow impellers are as shown in Figure17.5.

The acceptance criteria for cast aluminium hubs and clamp plates should be:

1. There shall be no surface porosity visible on the casting.
2. There shall be no porosity breaking through into cored holes.
3. There shall be no porosity within 10 mm of any boundary surface looking in the axial direction.
4. There shall be no inclusions greater than 1 mm in diameter.
5. There shall be no group of inclusions (each less than l mm diameter) greater than l0 mm diameter in total and shall comply with 3. above.
6. There shall not be more than two such groups in note 5. and they shall not be adjacent to one another.
7. Gas holes or porosity shall be acceptable so long as they comply with 1. to 6. above. Shrinkage cavities and porosity, foreign materials, micro shrinkage etc. shall not be accepted.
8. The level of porosity shall be no greater than plate 4 aluminium - gas porosity (round) as per ASTM E155.
9. There shall be no continuous defect line during X-ray between the insert and aluminium casting. The defects shall not be longer than 3 mm. The total defects shall not be greater than 6 mm and shall not be adjacent to one another.
10. On the machining of the casting, a fine "Witness" line will be tolerated as long as a fine pointer scriber will not penetrate more than 0.5 mm deep.
11. Flaking or pitting of the area between the insert and aluminium is not acceptable.

The techniques described will act as a powerful tool to identify areas for improvement or for process variables to be tailored to improve overall quality of product. It is essential that a design and testing procedure is adopted which recognizes that a major cause of failure especially in axial flow impellers is due to insufficient knowledge of the fatigue criteria and how they are affected by casting quality.

Close co-operation between design and production departments is necessary to ensure that the stated operating life is achieved. Constant vigilance is, nevertheless, indicated with continual research to improve knowledge. By this vigilance, product integrity can be assured.

17.6.3 Ultrasonic inspection

Ultrasonic inspection is popular because it can be conducted with portable equipment. The display on a cathode ray tube indicates the position of flaws in respect to the thickness of the material. The size of the flaw must be assessed by an experienced operator. This technique is very good for inspecting large flat plates prior to fabrication or forgings. It is also of great value in assessing the porosity of cast aluminium motor rotors.

17.6.4 Dye penetrant inspection

Dye penetration, or "dye pen" is a surface inspection method which is ideal for finding pinholes and hairline cracks.

The surface is first sprayed with dye which is allowed to soak into any surface defects. After a specified time, the dye is washed off and the component cleaned. Chalk is finally sprayed onto the surface. If surface defects exist, the dye trapped in the defect is drawn into the chalk by capillary action. Defects are outlined by dye indications in the chalk. The technique is generally used on finished machined surfaces. A skilled operator can judge the depth of the defect by the size of the "bleed-out".

"Dye pen" is particularly useful on welds and surface coatings. Weld integrity is degraded significantly by the presence of surface imperfections. "Dye pen" can indicate if a surface coating has achieved complete coverage without porosity, pinholes or cracks. It works on non-metallic coatings as well as metallic.

17.6.5 Magnetic particle inspection

Magnetic particle, or "mag particle" is a surface inspection method which is popular for cast materials. This inspection can only be conducted on materials which can be magnetised by an electric current. The surface is coated with a liquid bearing small magnetic particles. If the surface contains flaws, the magnetic flux is concentrated around them, drawing the magnetic particles towards the flaw.

17.7 Repair of castings

When repairs are undertaken, usually by welding, the repair must be inspected. Initially the faulty material must be removed by any suitable means and then checked to ensure that all faulty material has been removed; this is normally by dye penetrant or magnetic particle. The repair is then carried out. The component must then be inspected by the same method which found the original flaw. Fans built specifically for a purchaser may have repairs categorised into "major" and "minor".

17.8 Welding

Welding is a skilled occupation. There are many types of welding and the welder undertaking any particular job must be proficient at that type of welding. Welders are graded by the types of components they weld, the positions of the welds and by the types of equipment used. The welding of pipework is thought to be the most difficult type of welding.

Welders for pressure vessels have to be qualified and this is usually assessed by an insurance company. Designing weld methods for exotic materials requires very special materials knowledge. All welding for process parts should be supported by certification.

These are a number of Standards applicable to particular grades EN287 and EN288 are most relevant to the general welding of centrifugal fan casings and impellers.

17.9 Performance testing

Implicit in the use of ISO 9001 is the need for agreement between the supplier and customer as to the product specification and performance. Essentially the customer is buying a given airflow against a pressure. He will have calculated these according to known formulae or experience. Much data is available on the pressure loss through duct elements and terminations. Without knowledge of the interaction of fan and system, however, problems can arise.

17.9.1 Aerodynamic testing

In the early days of fan manufacture, each individual company would have its own methods of test. These might vary greatly according to the whims of individual engineers. Whilst some, but not all, would be grounded in acceptable theory and practice, the results achieved could not be easily compared. Definitions and measurement of pressure were a particular difficulty - as witness the terms used such as total pressure, effective pressure, resistance head, total static pressure. How these could be correlated with calculated system pressures was a mystery.

Into this morass stepped the then Institution of Heating and Ventilating Engineers who set up a Fan Standardisation Committee. Around 1924 it made recommendations as to standardised methods of testing a fan's performance. It is believed that this was the first attempt anywhere to settle what is still a challenge to the specialist in flow measurement.

The conclusions were made available to the British Standards Institution (BSI) from which grew BS 848:1939, authorised by the Electrical and Mechanical Industry Committees as an extension of work originally asked for by the India Store Depot. It was part of a series of test Codes covering various fan types. Others produced around the same time were BS 367:1932 for Ceiling Fans, BS 38 :1930 for Desk Fans and BS 707:1936 for Mines Fans.

By 1952 these Codes were generally found to be unsatisfactory. The Fan Manufacturers' Association (FMA), noting the urgency and unwilling to wait for a new British Standard, produced FMA Code 3:1952. This introduced new methods primarily for the testing of axial flow fans which were becoming popular, and extending the range of pressures which could be measured by taking into account the compressibility of air.

This work, in turn, was made available to BSI who updated its own Standard to produce BS 848:1963 Part 1 which largely followed the FMA Code but was brought into line with BS 1042 Code for Flow Measurement. The subject of tolerances was also introduced.

Concurrent with all the above, similar activity was being carried out in other major industrialised countries. Test codes were produced in the USA, France, Germany and Italy which were characterised by certain significant differences. Further information is given in Chapter 4 from which it will be noted that international agreement has been achieved after more than 30 years of effort.

Customers are strongly advised to specify that all performance data shall be obtained from tests carried out in accordance with ISO 5801. Only if this is the case can valid comparisons be made between competing nations' products. The tolerances applied to this data are detailed in the shortly to be published ISO 13349.

17.9.2 Sound testing

BS 848 Part 2 Fan Noise Testing was published in 1966 as a first attempt in the UK to satisfy the need for meaningful results which could be used by acousticians.

An earlier attempt had been made in the USA with the NAFM test Code of 1942 (NAFM was a forerunner of AMCA, the Air Movement & Control Association International). Whilst this enabled comparisons to be made between competing products, the information produced was not in a form which could be used by acousticians in the design of systems.

This need had to await the production of BS 848 Part 2:1985, which gave a comprehensive series of tests designed to give information of open inlet /open outlet noise, noise break-out and in-duct noise levels.

It must be recognised that there are at least 12 noise levels associated with a fan for a particular duty. Both the manufacturer and the customer must agree on which particular level is appropriate for a particular installation.

Further information on Fan Noise is given in Chapter 14 from which it will be noted that six new international noise Standards have been produced which update the methods described in BS 848 Part 2:1985 and additionally incorporate French, German and American methods.

Again customers are strongly advised to specify that all sound data shall be carried out in accordance with ISO 13347 Parts 1 to 4, ISO 5136 or ISO 10302 as appropriate.

17.9.3 Balance and vibration testing

Again, the world has changed considerably from the days when balancing of a fan impeller was carried out with a piece of chalk and a length of string. This is not to decry the efforts of the skilled craftsmen of those years. Their results were often as good as those obtained from the early balancing machines.

However, the need for a quantitative Standard was apparent which resulted in BS 848 Part 6:1989. This specified in detail how such tests were to be carried out and was largely a copy of a UK MOD Defence Standard. It did not, however, lay down standards for specific types of fan and the first attempt at this was incorporated in AMCA 204.

Again international work has largely mirrored that carried out for sound testing and we now have ISO 14695 which is largely a rewrite of BS 848 Part 6 and ISO 14694 which is a specification for balancing and vibration. This latter is very similar to AMCA 204 but with vibration levels quoted in r.m.s. (root mean square) values as internationally agreed rather than the peak-to-peak levels favoured in North America.

Again it is recommended that these two international Standards be specified so that valid comparisons can be made. It should be noted that out-of-balance forces are not the sole reason for fan vibration. Further information is given in Chapter 15.

17.9.4 Run tests

For many years, after final assembly, it was customary to give every fan manufactured a run test. This would typically be for an hour or more, during which time a check would be made on bearing temperatures, vibration, impeller integrity, tightness of fastenings etc.

With the advent of batch and even mass production of small fans, especially axial flow units, the need for these tests was questioned. It was pointed out that tests at other parts of the production cycle could obviate the need for run tests whilst the adequacy of the customer's foundations (a more likely cause of poor running) was never questioned.

The author's belief is that, for the present, this must continue to be the subject of agreement between the manufacturer and the customer. He would however strongly recommend the inclusion of a run test in the specification of any fan above about 18.5 kW.

It should be noted that every fan supplied for the Sydney Harbour Tunnel (see Chapter 1, Section 1.2.8) was subjected to a 24 hour run test; the electricity consumption, and therefore the cost, being considerable.

17.10 Quality Assurance Standards and registration

17.10.1 Introduction

Whenever anyone mentions Quality Assurance, a picture is created of myriads of white-coated inspectors, running around with sheaves of paper. That they are actually contributing to a company's efficiency would probably produce a horse laugh. That they are giving a benefit to the customer would create his disbelief.

17.10.2 History of the early Certificate of Air Moving Equipment (CAME) Scheme

In November 1979 a report to the UK National Economic Development Council was made by the Heating, Ventilating, Air Conditioning and Refrigeration Equipment Sector Working Party. Whilst this was not for publication, some important comments were made which had a galvanising effect on some of the larger UK fan manufacturers. These implied that there would be growing pressure at home from European manufacturers who, attracted by the abnormal buoyancy in the market at that time, were increasing their share. They would be expected to consolidate this hold when the world moved into recession, and as UK firms could comfortably supply normal home demands, competitive pressures would be fierce. How right they were!

At the same time, conditions in other world markets could not by any means be described as similar. UK manufacturers had been achieving a growing reputation in Europe for reliable deliveries and competitive price, albeit from an earlier period where they could claim neither. However, they were coming up against less obvious, but just as effective, barriers to trade. Many countries had certification schemes and demanded that UK fans be passed through their test houses before they would be allowed into their country. This could take 12 months and lose the order. No prizes are given for guessing who used these sorts of subterfuge in protecting their home industry! More annoyingly, some third world countries specified non-British certification schemes, placing UK products at a considerable disadvantage.

The UK Government had separately been expressing overall concern at the generally perceived low quality of all British goods and was asking how this impression could be corrected. (It later resulted in a National Quality Year and the publication in 1983 of a White Paper - "Standards, Quality and International Competitiveness".) Industry had in mind that Government would be able to trade off a British scheme with those of the country into which it hoped to export i.e. reciprocal recognition of each other's registration. This would at best unlock the unseen barriers to trade, or at lease ensure a "tit-for-tat" relationship.

The scene was therefore set, and a demand for registration was voiced on many sides. The first of many meetings took place between the then Department of Trade and Industry (DTI), Department of the Environment (DOE), users and fan manufacturers. Following the accord between the government and BSI, the latter were chosen to develop and administer the resultant scheme. The essential elements of the Certification of Air Movement Equipment (CAME) Scheme were eventually decided and were:

I) A Quality Assessment Schedule ref 3284/37 relating to the design, manufacture and testing of air moving equipment and ancillaries coming within the scope of BS 848 - Fans for General Purposes.

ii) Registration of the applicant company as a "firm of assessed capability" when measured against the requirements of BS 5750 Part 1 - Quality Systems Specification for design, manufacture and installation.

iii) All performance data to be published in accordance with BS 848 and the manufacturer to be able to substantiate such data by the necessary tests. BSI inspectors to visit the company periodically to audit its quality procedures.

The scheme offered manufacturers the ability to demonstrate a commitment to quality, thereby building customer confidence and eliminating the need for multiple assessments. Having been formulated by BSI with the assistance of both users and manufacturers it provided, by means of assessment and subsequent surveillance, an independent assurance of a firm's capability of working to specification. This, before the event, assurance is distinct from redress under contract warranty or the law, which can only operate after the event.

It conformed to the guidelines of the National Quality Campaign launched by the UK Government in 1983 and had received the approval of the Department of Trade and major purchasing organisations. The first four companies to be registered received their certificates at the House of Commons on Wednesday 14 December 1983. By the 11 March 1985, 10 companies had been registered with further additions "in the pipeline". It was suggested therefore that the Property Services Agency (PSA) commitment to make the scheme mandatory when there was sufficient choice of suppliers had now arrived. As a certain Frenchman said when defending the use of the guillotine, "implementation of this requirement encourages the others".

It must be said, however, that at this stage there was a feeling within the UK industry that whilst it was endeavouring to play a good game of cricket (European companies were quickly registered under CAME), some overseas opposition indulged in rule-less all-in wrestling. The hoped-for reciprocal agreements were not implemented and the delays in achieving other countries' approvals continued. In consequence the CAME Scheme "died", leaving a lot of disillusioned fan manufacturers. Only the Quality Assurance element survived and the industry continued its search for a certification scheme (see Section 17.11).

17.10.3 What is quality?

If I went to my local car showroom to buy a Ford and proffered my £8000, I would not expect to receive a Rolls-Royce. Each, however, could be said to be a quality product.

Before we discuss quality assurance, therefore, we must agree on what is meant by quality. Amongst the many definitions which have been proposed, perhaps the best and simplest is "fitness for purpose". But this really begs the question. What does being fit for a particular purpose actually mean? For how long should it be fit and who is to decide?

Within the building services industry there are many different people who have to be satisfied - consulting engineer, contractor, user company - and right down to its maintenance engineer. Ultimately it is the customer who has to be satisfied. With some temerity it is suggested that his expectations could be unreasonably high (he may only have paid for a "banger" and he may have wanted a quality product). If the customer (whoever he may be) is not to feel disappointed then there must be some agreement right from the initial enquiry as to what those reasonable requirements really are.

They may be obvious from custom and practice in the industry, or they may have to be worked out in detail before any order can be placed. The implications, where technology is rapidly changing, where the time to produce specifications is limited, where new companies (home or overseas) are entering the market place, and where contracts are the subject of intense competition is that:

- there is a need for some independent assessment of quality
- minimum requirements need to be supported by industry standards which must not however inhibit innovation
- like must be compared with like when determining the successful tenderer
- when coming back for "more of the same", the user must be assured that the product will indeed be repeatable.

With all the above in mind, the following revised definition of quality is therefore suggested:

"The ability to meet a customer's reasonable needs and expectations, bearing in mind the technical constraints, time constraints and price charged."

And here our problems begin!

17.10.4 Quality Assurance

In a company of any size or complexity desirable things don't just happen of their own accord; they have to be made to happen. Devotees of Northcott Parkinson and Murphy will know, however, that undesirable things will happen with spontaneity, inevitability and rapidity. We must therefore have a quality system which ensures that the customer's needs are first identified and then implemented.

The 1982 UK Government White Paper on quality standards and international competitiveness explained the needs and laid down the objectives in a form which for succinctness, the author believes, has not been improved upon:

"Quality assurance, in the form of sound technical and administrative procedures for ensuring quality, offers more scope for reducing cost and enhancing competitiveness and profitability than many other management controls. It does this by reducing materials wastage, lost production times, re-work, extra handling and rejections. Improved quality and reliability, by improving customer satisfaction, lead to increased sales competitiveness, reduced warranty claims and premium pricing."

The elements that a quality system must cover were worked out, codified, and made available in a UK Ministry of Defence Specification 05-21 and out of this grew BS 5750 (which was to be expected as the Chairman of the committee was an Admiral!). Out of this in turn grew ISO 9001, (= EN 29001). There are now other similar Standards in other parts of the world but in general, if a company meets the requirements of the latest edition of ISO 9001, the others can usually be met without difficulty.

All quality systems require procedures to be documented, work to be done against written instructions, the results of inspections and tests to be recorded, and so on. In a sentence - everything must be written down. The system must be regularly reviewed to ensure it remains effective and can successfully pass any audits carried out either in house or by the customer/assessment authority. Instances of error or poor quality should be identified and corrected. Where this does not happen, or where there are delays in correction, then senior management must be advised.

It should be especially noted that Quality Assurance is not the same as Quality Control. Whilst Control comprises inspection and other activities designed to ensure that defective goods do not leave the factory gate, Assurance is aimed at minimising the chance of error before it actually happens. It may well use inspection and other similar techniques to provide a feedback on how the company is performing, but it is much more far-reaching.

The introduction suggested that there would not be lots of paperwork. Why then all this writing down? It must be emphasised that once a quality system is in being, there is then less need for individual written instructions relating to a particular contract. They will all exist and should cover the totality of the firm's work. The initial effort in producing a Quality Manual will be amply repaid many times over.

Other advantages may be summarised as follows:

- written procedures capture know-how or experience, and the company becomes less vulnerable to the absence or loss of staff
- the training of new staff is made easier
- written instructions reduce the possibility of misunderstandings
- documentation provides the objective evidence against which assurance can be given that the company is working well.

The paperwork is there simply to ensure that the right materials and information arrive at the right time at the place where the work is to be done. It also ensures that those doing the work know what is required of them, and that once the work is done, the results are passed on to the right place in the correct form.

17.10.5 The Quality Department

In any organisation the scope of each director or manager has to be defined to ensure harmonious working. So it is with the Quality Assurance Manager, referred to in most Standards as the "management representative". He and his department are responsible for the implementation, development and maintenance of the documented procedures. Ideally the Quality Manager should report directly to the Managing Director, or failing this, have direct access to him in case of dispute. Only in this way can the quality versus quantity conflict be resolved and the company retain its quality reputation.

By carrying out these tasks, the quality staff will gain expertise. It may therefore be convenient, although not essential, for them to be responsible for other activities inspection, standards or the issue of quality related documents. Even if there is no specialist department, such documents would still have to be produced. It is totally illogical to assume that they are part of some mysterious additional activity invented by the Quality Manager.

The basic responsibilities of the Quality Manager may therefore be summarised as follows:

- to develop and maintain the quality system
- to negotiate with the customer's quality representatives
- to negotiate with the quality representatives of his own firm's suppliers
- to ensure that quality problems with the company are resolved
- to carry out audits of internal and suppliers' quality systems.

Generally speaking, the Quality Assurance Manager does not solve the quality problems, but rather seeks to ensure that the system highlights any difficulties. Those that stem from faulty design have to be tackled by the engineering department, whilst those which result from faulty manufacture are the responsibility of the production department.

17.10.6 Quality performance

If one wishes to measure the quality performance of a company, there are perhaps three important questions to be asked:

I) is there a system for reporting the costs involved in rework, scrap, trouble-shooting, customer complaints and warranty claims?

ii) is there a complete system for detecting quality problems, finding the real causes, providing solutions and reporting delays in their resolution?

iii) does the quality system cover all activities in the company, including design, contracts, purchasing, manufacture, installation and service?

Unless affirmative answers can be given to all these questions, senior management is unlikely to know the size of the quality problem, or to have ensured that the quality assurance effort is sufficient. However good the system may appear, it is unlikely to be effective. It must be remembered that all parts of a company must operate efficiently to achieve a quality product.

17.10.7 Quality assessment

The major step for any company wishing to be assessed to ISO 9001 is to produce a quality manual. This document provides the basis for the subsequent independent audit. Other essential elements of the manual are detailed below, with some brief comments. When assessing the company, the inspector will ascertain whether instructions are being followed and whether any non-compliance seriously affects the company's ability to manufacture products which adhere to specification. Remedial action, and a timescale for completion, may be recommended. In serious cases Certification may be withheld.

1) **Organisation**: The company structure should be identified and the responsibilities of each director or manager defined. A key post and focal point within the system organisation is that of Quality Assurance Manager, referred to as the "Management Representative".

2) **Review of quality system**: The Quality Assurance Manager is responsible for reviewing the Quality System, normally by means of an audit. This should be carried out at regular intervals to suit the size of company and type of work.

3) **Planning**: The progress of a contract through design, development, manufacture, packing and despatch, requires careful co-ordination. The Quality system must have a plan for assessment prior to work commencing. This planning must allow for everyday items, such as installation, operating and maintenance instructions, as well as the "special" where, for example, the material specification might change, or special testing may be required.

4) **Work instructions**: Clear and completely documented instructions, referring to a method of work or a procedure, are required to ensure a consistent standard and quality of the product. When training is required, the Work Instructions maintain this consistency and can be supplemented with "on the job" training.

5 **Records**: For a quality system to work records must be maintained. This provides documented evidence for the customer and demonstrates the performance of the quality system.

6) **Corrective action**: All faults should be reported to the Quality Assurance Manager, who can instigate the necessary actions to correct the fault. Such corrective actions should require a feedback so that he can monitor their effectiveness. Where different interests conflict, he should seek a suitable solution with the persons involved. If a dispute should arise with the Quality Assurance Manager, then he must have access to the Chief Executive of the company for a final decision.

7) **Design control**: The company has to show that it has control of this function, however performed. A manual setting out all procedures and methods of calculation usually satisfies the requirement. Whenever changes are made, it is essential that all drawings are updated, the issue status changed, and some means of identifying at which particular unit, the change was incorporated. Drawing records are essential, and these should detail the reason for any amendments.

8) **Documentation and change**: All paperwork relating to quality must be maintained up to date. This requires that records be kept of all holders of documents, that there is an acceptable method of updating and that all those out of date are destroyed.

9) **Control of inspection, measuring and test equipment**: All equipment used shall be numbered and its details recorded in a central register. The dates when last calibrated and the interval between calibrations shall be known and adhered to. Any in-house calibration shall be by an experienced person and traceable to an independent, approved, external test house.

10) **Control of purchased material and services**: The systems employed by the company should ensure that all outside materials conform to specification and are uniquely identified.

11) **Manufacturing control**: All operations must be carried out under controlled conditions and the criteria for workmanship shall be prescribed to the greatest practicable extent. Inspection shall be carried out after each operation that affects quality.

12) **Purchaser supplied material**: Free issue material should be subject to the same sort of control to ensure that the correct quantities are received and that it is suitable for use.

13) **Completed Item inspection and test**: These are necessary to confirm that the goods fully comply with the specified requirements.

14) **Sampling procedures**: These should be to a statistically proven method or as agreed with the customer.

15) **Control of non-conforming material**: All such items should be segregated and recorded so that the success rate of the system can be measured and updated as necessary.

16) **Indication of inspection status**: Means shall exist for identifying whether material is acceptable or not at each stage of manufacture.

17) **Protection and preservation of product quality**: Procedures shall exist for clean and secure storage, stock rotation where appropriate and adequate packing and transport.

18) **Training**: A company policy is required to ensure that functions are carried out by persons with suitable training.

17.11 Performance certification and Standards

17.11.1 Introduction

Diligent readers of Section 17.10.2 will have perceived a certain disappointment of the author that the CAME Scheme did not come to full fruition. Some colleagues in the industry were even more damning in their condemnation of all things governmental. They felt that they had gone along with the requirement for quality assurance but had not been given Performance Certification in return. All the good work completed by the National Engineering Laboratory (NEL) in the production of the rules for the CAME Scheme had been a wasted effort.

The situation in the United States could not have been more different. Unlike the scheme developed in the UK, they had a private programme produced by the manufacturers without government support or interference. The trade body, AMCA (Air Movement & Control Association International) had developed its own approach and widely publicised it. Indeed the author's company joined in 1976. We had been told in no uncertain terms by the customer in the USA that the order was ours provided we achieved AMCA Certification for the relevant products. We did, and we got it!

In more recent times, the European Committee of Air Handling and Refrigeration Equipment Manufacturers (Eurovent/ Cecomaf), decided that the time was ripe for a similar scheme in Europe. Many of its manufacturers however, had in the meantime joined the AMCA Scheme and insisted that reciprocity between the two programs was essential. A memorandum of understanding between AMCA and Eurovent Certification Company (ECC) was signed by the author (then AMCA President) and Alan Duttine (ECC Chairman) in late 2003.

It is hoped that there soon will be global performance certification scheme for fans to assure customers that what they are buying (airflow and pressure) is what they will get. In the meantime customers should look for the AMCA Seal if they wish to have performance assured to close tolerances. Failing this they should consider a witnessed performance test.

The AMCA programme, in the absence of any other scheme at present, is considered so important that it is detailed fully in the following Section.

17.11.2 AMCA International Certified Ratings Programme

17.11.2.1 Purpose

- To provide the buyer, user and specifier assurance that the manufacturer's published performance ratings of air moving equipment are reliable and accurate, and further, to provide these parties with information on how the product was tested, what appurtenances were included, and other pertinent information so that they may be able to select a fan that will provide the performance required.
- To provide a procedure for verification of the manufacturer's performance ratings on a regular schedule by check-testing of the certified product line in the AMCA Laboratory.
- To provide assurance that competitors' ratings are based on standard test methods and ratings procedures.

17.11.2.2 Scope

Products that can be licensed by AMCA to bear the AMCA Certified Ratings Seal are centrifugal fans, axial fans, power roof ventilators, air curtains, agricultural fans and other air moving devices within the product scope of AMCA.

The programme applies only to complete air moving devices, and is not applicable to component parts such as fan impellers or impellers and housings.

The programme applies to fans within the scope of AMCA for which performance rating catalogues are published and made available to the public. When only a portion of a catalogued series of sizes are licensed, at least a majority of the sizes catalogued may be licensed. It does not apply to special units for which performance ratings are not published. When performance ratings for both licensed and unlicensed products are contained in the same catalogue, there must be a clear distinction made between licensed and unlicensed products.

When one or more licensed products are used as component parts of a larger unit, the AMCA Certified Ratings Seal may not be applied to the complete unit unless the complete unit has been itself licensed in accordance with this programme.

17.11.2.3 Administration

The administration of the Certified Ratings Programme is the responsibility of the Executive Director.

17.11.2.4 Responsibilities

The AMCA staff is responsible for administering the Certified Ratings Programme in accordance with the requirements set out. They will verify that the air performance ratings developed by the manufacturer were developed in accordance with the requirements of the programme, conduct pre-certification test(s) to confirm the licensee's test results, and conduct check tests of the licensee's products as required by this programme.

The AMCA staff is also responsible for verifying that the catalogue published by the licensee conform to the requirements of the programme.

17.11.2.5 Definitions

1) AMCA is the Air Movement and Control Association International Inc., a non-profit corporation organized under the laws of the state of Michigan, USA, having its principal office at 30 West University Drive, Arlington Heights, Illinois 60004, USA.
2) The AMCA Certified Ratings Programme Air Performance is a programme "for certifying a product's aerodynamic performance rating", as defined in AMCA Publication 211.
3) An AMCA Registered Laboratory is a laboratory that has been registered in accordance with AMCA Publication 111. A registered laboratory may be owned and operated by certified ratings licensees or by an independent individual or organization.
4) The AMCA Laboratory is the Association's laboratory in Arlington Heights, Illinois.
5) Other independent registered laboratories may be designated as an AMCA Laboratory by the Board of Directors for the purposes of performing pre-certification tests, check tests and challenge tests.
6) A Certified Rating is a published performance rating of a product that the manufacturer or seller of the product certifies has been tested and rated in accordance with the requirements of the AMCA Publication 21 and that has been licensed by AMCA to bear the AMCA Certified Rating Seal.
7) The license is a legal contract between AMCA and a person, firm or corporation which authorizes participation in the AMCA Certified Ratings Programme and the affixing of the AMCA Certified Rating Seal to products listed in the appendices of the license. A separate Appendix to the license will be issued for each certified product line.
8) A Licensee is a person, firm, or corporation that entered into a License Agreement to certify the ratings of a product

or products with AMCA in accordance with the requirements of this programme.

9) The AMCA Certified Ratings Seal (the Seal) is a registered trademark of the Association which can be applied to products to indicate that the product is licensed by AMCA to bear this seal. The seals are the property of AMCA and may be used only on the particular product of the Licensee listed in the appendices of the License Agreement. Reproductions of the AMCA Certified Ratings Seal may be used in catalogues and other publications only as permitted by Section 10 of this Certified Ratings Programme.

10) A product line is a series of product sizes, or an individual product where only one size is offered for sale, with a common design purpose and generally similar aerodynamic features, but not necessarily aerodynamically similar, that are catalogued under the same product description or name and/or identifying reference and that are within the scope of this programme.

11) The performance rating of a product is a statement of the airflow versus pressure performance of the product at a given speed and fan shaft power or motor input power at standard inlet air density for a printed catalogue, or at a specified density for an electronic catalogue. The rating may be published in tabular and/or graphical format. Specific performance rating requirements are given in the Product Rating Requirement Subsections A through H of AMCA 211.

12) A catalogue for the purpose of this programme is defined as a printed document that contains the dimensional data and performance ratings of a product line, and that meets the other requirements of the Certified Ratings Programme. It is also defined as an electronic media (computer disk) that provides performance data for a licensed product.

13) Catalogue data is the performance rating of each model of the product line in the specific format that is to be published.

14) Test data is a record of airflow, pressure, speed, density and power as measured for each test point of operation, as well as the data reduced to standard density and constant speed, where applicable.

15) A licensed product is a product that has met the requirements of this programme and for which an appendix to the License Agreement listing the product identification has been issued.

16) The Directory of Licensed Products is AMCA Publication 261 which lists all products that are currently licensed by AMCA to bear the Certified Ratings Seal date of issue. Agricultural products are listed in AMCA Publication 262.

17.11.2.6 Procedure for participation

Eligible parties participate in the AMCA Certified Ratings Programme by first entering into a Certified Ratings Programme License Agreement with AMCA. Thereafter, any individual product line may be licensed to bear the AMCA Certified Ratings Seal after the requirements of this programme are met, and an appendix to the License Agreement has been issued by AMCA for the product line.

In general, the steps shown in Table 17.2 will need to be completed in order for a product line to become licensed to bear the AMCA Certified Ratings Seal.

Steps	Description	Responsibility
A	Test the product(s) to be licensed	Applicant
B	Apply for a product line license	Applicant
C	Submit the required data	Applicant
D	Staff review of submitted data	AMCA
E	Conduct the pre certification check test	AMCA
F	Submit proposed catalogue data	Applicant
G	Staff review of proposed catalogue data	AMCA
H	Issue of notice of acceptability for license	AMCA
I	Submit proof copy of catalogue	Applicant
J	Staff review of catalogue proof copy	AMCA
K	Submit finished catalogue	Applicant
L	Staff review of catalogue and issuing of license	AMCA

Table 17.2 AMCA licence procedure

17.11.2.8 Licence procedure for similar products

Manufacturers of a product line or lines that are aerodynamically similar to a product already licensed by the manufacturer may apply for an AMCA Certified Ratings Seal by submitting two copies of form CRP-A and sufficient dimensional data to verify the products are in fact aerodynamically similar.

The AMCA staff will review the application and notify the applicant if any changes or further required information is required to comply with this programme. Once aerodynamic similarity has been verified by AMCA, the applicant should follow the procedure in Table 17.2.

17.11.2.9 Requirements for maintaining the certified ratings license

The licensee is responsible for maintaining conformance to the requirements of the Certified Ratings Programme by submitting new or revised catalogues to AMCA for review prior to publication, and by control of manufacturing procedures to ensure that the product as manufactured will perform in accordance with its published ratings, within the tolerances allowed for the product as defined. Products should be submitted for check tests.

The purpose of a check test is to ensure that the air performance of the as manufactured unit remains consistent with the ratings originally certified.

Each licensed product line will be subject to a check test within 36 months of license issue and within 36 month periods thereafter under a continuing license. In the event that the data submitted with the original application was obtained on a preproduction sample (prototype), the first check test is required within 12 months of license issue, and within 36 month periods thereafter under a continuing license.

AMCA staff will be responsible for selecting a suitable product sample. The sample may be obtained from the Licensee or by purchase on the open market, when appropriate.

17.11.2.10 AMCA Certified Ratings Seal

The Seal is a registered trademark of AMCA which may be affixed to a licensed product or reproduced by the Licensee in catalogues and other publications, as permitted by the document.

The AMCA Certified Ratings Seal applies to the AMCA Certified Ratings for air performance only.

In order to certify the sound performance of any device, all the requirements of AMCA Publication 311 must be met. A separate Seal for air and sound is available for products that also meet the requirements of Publication 311.

17.11.2.11 Catalogues and publications

In order to comply with the programme, the Licensee should publish a catalogue which identifies the product and may include descriptions, photographs or renderings of the product, and contains the AMCA Certified Ratings Seal, qualifying statement(s) and performance ratings. Performance ratings should be developed in accordance with the applicable product rating requirements.

17.11.2.12 Challenge test procedure

Any person, firm, or corporation, whether a member of AMCA or not, may challenge the performance of a licensed product by requesting that a check test be performed in the AMCA Laboratory under the following conditions:

- The test unit to be provided by the challenging party at no cost to AMCA.
- Test data obtained in an AMCA Registered Laboratory or the AMCA Laboratory, showing that the test unit does not perform within the specified performance tolerances for the product being tested, to accompany the request for a challenge test.
- The request to be made in writing to the AMCA Executive Director with a copy to the Licensee.
- The challenging party, by making a request for a challenge test, agrees to all applicable provisions of the Certified Ratings Programme.

17.11.2.13 Directory of licensed products

AMCA Publication 261 and AMCA Publication 262 are published each year or at such other times as the Executive Committee authorizes. They will list all products that are licensed to use the AMCA Certified Ratings Seal at the date of publication and identify the latest approved catalogue containing the ratings of each licensed product.

17.11.2.14 Appeals and settlements of disputes

If the applicant or Licensee does not agree with the administration or interpretation of the requirements of the scheme by the AMCA staff, or if they want to request additional time beyond what is allowed to correct a violation, they may appeal to the Executive Committee of the AMCA Board by written request directed to the Executive Director.

In the event that agreement cannot be reached between the Executive Committee and the appealing party, the appeal will be put before the AMCA Board of Directors and the decision of the Board will be considered final.

17.11.2.15 Other comments

These paragraphs are of necessity only a brief resumée of the AMCA Scheme. They contain those parts of the programme considered of importance to manufacturers and users.

The Scheme is industry based – but there are no more vigilant interrogators than one's peers in the industry, i.e. one's competitors!

17.12 AMCA Laboratory Registration Programme

17.12.1 Purpose

The purpose of the AMCA Laboratory Registration Programme is to provide a means for laboratories to be qualified for testing of products in accordance with appropriate AMCA standard test methods, or other test methods recognized by AMCA. Only data obtained in an AMCA Registered Laboratory or the AMCA Testing Laboratory will be accepted for use in AMCA's Certified Ratings Programme.

17.12.2 Scope

The AMCA Laboratory Registration Programme applies to laboratories capable of testing in accordance with the standard test methods. The Laboratory Registration License will specifically specify for which standard test method, or methods, the laboratory is registered.

17.12.3 Definitions

17.12.3.1 The Licence

1) The License is a legal contract which specifies the terms under which AMCA authorizes a person, firm, or corporation to participate in the AMCA Registered Laboratory Programme.
2) Amca Registered Laboratory is a laboratory equipped and staffed to conduct tests according to the appropriate AMCA test method listed in Appendix A and which has been inspected by a staff engineer and duly approved by the Executive Director.
3) Amca Testing Laboratory is the association's laboratory in Arlington Heights, Illinois.

17.12.4 Procedure

17.12.4.1 Application

Prospective licensees may have their laboratories registered by making application to the AMCA office and completing the registration requirements. Applications for registration (form LRP 2) shall be obtained from the AMCA office or the AMCA Testing Laboratory, completed and returned with the information specified therein. The application will be reviewed by the AMCA staff and the applicant notified of any changes or further information necessary to comply with the requirements of the standard test method or methods for which registration is requested.

17.12.4.2 Witness test

When the staff is satisfied with the information provided, a witness test will be arranged on a product, or products, selected by AMCA within the limitations of the laboratory to be registered. The products used for the witness test may be furnished by the laboratory or by AMCA.

The object of the witness test is to determine whether or not the laboratory is equipped and staffed to carry out tests in accordance with all of the requirements of the appropriate standard test method. One or more tests will be witnessed by an AMCA staff engineer. Laboratory personnel normally assigned to testing products shall conduct the test. The AMCA staff engineer will take no part in the test but will observe calibration methods, instrument readings and check critical dimensions on the test setup and the test device.

Immediately following the completion of the witness test, the AMCA staff engineer shall be provided with copies of log sheets and calibration curves.

17.12.4.3 Check test

The results of the witness test will be sent to the AMCA office along with the product(s) tested. The products will then be tested in the AMCA laboratory. The results of the witness test and the test in the AMCA laboratory will be compared by the AMCA staff. The results shall be as allowed by the requirements of the appropriate Certified Ratings Programme in Appendix B.

17.12.4.4 License agreement

If, in the opinion of the AMCA staff, the laboratory is in compliance with the specified AMCA test method, the laboratory will be registered upon the signing of a License Agreement and the payment of the scheduled fees. At this time a Certificate of Registration will be issued by the AMCA office.

17.12.5 Reference to AMCA registered laboratory

17.12.5.1 Literature or advertisement

The term "AMCA Registered Laboratory" or any variation thereof shall not be used in any catalogue product literature, or advertisement that contains product performance, except when all products are licensed to bear the AMCA Certified Ratings Seal for all data published.

In catalogues, product literature or advertisements that are general in nature and do not contain product performance data, reference may be made to the fact that the company maintains an "AMCA Registered Laboratory".

The term "AMCA Registered Laboratory" shall be marked with an asterisk referenced to the following statement which must be included on the same page.

Product performance data based on tests in an AMCA Registered Laboratory are not to be construed as being licensed to bear the AMCA seal.

17.12.5.2 Individual test data

Individual test results may indicate that the data was obtained in an AMCA Registered Laboratory by use of the following statement:

This test data obtained in a laboratory registered by AMCA for (insert appropriate test method: AMCA Publication 210 Air Performance, AMCA Publication 500 Leakage, etc) testing. Data is not certified by AMCA.

17.12.5.3 Other statements

No statements shall be made which implies that the results performed in one AMCA Registered Laboratory are better, more accurate, or more reliable, than those from any other registered laboratory.

17.12.6 Settlement of disputes

Any disagreement concerning administration of the Laboratory Registration Programme between the Executive Director and the licensee that cannot be resolved will be submitted by the Executive Vice President to the Executive Committee. If the disagreement still cannot be resolved, it shall be put before the Board of Directors. In the event that resolution is not achieved by the Board of Directors, the AMCA President will obtain the services of three impartial persons, acceptable to the licensee and not directly associated with the air movement and control industry, to form a Final Appeals Board.

The decision of the Final Appeals Board shall be binding on both parties and the costs of convening the Board, together with the cost of laboratory tests to resolve the dispute, will be borne by the party in error.

17.12.7 Other comments

These paragraphs are of necessity a brief outline of the Laboratory Accreditation programme. They confirm those parts of AMCA Publication 111, which are considered of importance to manufacturers and users.

17.13 Bibliography

ASTM E155, *Standard Reference Radiographs for Inspection of Aluminum and Magnesium Castings.*

EN 287, *Approval Testing of Welders Fusion Welding, Part 1- Steels; Part 2 - Aluminium and Aluminium Alloys.*

EN 288, *Specification and Approval of Welding Procedures for Metallic Materials, Part 1- General Rules for Fusion Welding; Part 2 - Welding Procedure Specification for Arc Welding.*

BS 1042, *Measurement of fluid flow in closed conduits.*

ISO 5136:2003, *Acoustics -Determination of sound power radiated into a duct by fans and other air-moving devices - In-duct method.*

ISO 10302:1996, *Acoustics - Method for the measurement of airborne noise emitted by small air-moving devices.*

ISO 5801:1997, *Industrial fans - Performance testing using standardized airways.*

ISO 13347:2004, *Industrial fans - Determination of fan sound power levels under standardized laboratory conditions.*

ISO/DIS 13348, *Industrial fans - Tolerances, methods of conversion and technical data presentation.*

ISO 13349:1999, *Industrial fans - Vocabulary and definitions of categories.*

ISO 9001:2000, *Quality management systems - Requirements.*

BS 848-6:1989, *Fans for general purposes. Method of measurement of fan vibration.*

AMCA 111 *Laboratory Registration Programme.*

ANSI/AMCA 204-96, *Balance Quality and Vibration Levels For Fans.*

AMCA 210, *Laboratory Methods of Testing Fans for Aerodynamic Performance Rating.*

AMCA 211, C*ertified Ratings Programme - Air Performance.*

AMCA 261, *Directory of Licensed Products. A listing of all the products licensed by AMCA to bear the Certified Rating Seal.*

AMCA 262-03, *Directory of Agricultural Products With Certified Ratings.*

AMCA 311-05, *Certified Ratings Programme - Product Rating Manual For Fan Sound Performance*

18 Installation, operation and maintenance

This Chapter gives advice on the correct installation of industrial fans. It also gives details on fan and ducting installations and in particular the need for ancillary equipment to ensure a safe system.

The Sections on the care and maintenance of fans include commissioning and trouble-shooting and will provide guidance for operating staff at fan installations. Maintenance staff unfamiliar with fans should also find this Chapter of use.

Contents:

18.1 General

It is a fact of life that a fan installation may be the responsibility of a number of different parties each with their own interests. These persons may have a view of only one aspect of a particular plant. It is, however, incumbent on them to at least talk to each other so that they are aware of the problems which may arise. They should also agree on the extent of each party's responsibility and the dividing lines so that problems do not disappear "down the cracks".

The installation of fans can be carried out by the fan manufacturer, the driver manufacturer or the contractor building the installation. Large fans are usually installed as early as possible while access at site is at its best. Small fans can usually be fitted in at any time. The fan should be ordered so that its delivery occurs at the correct time in the overall site programme.

The certified fan drawing must be available to allow foundations or structural steelwork to be designed and manufactured before the fan is delivered.

The space allocated for the fan must be sufficient to permit installation, care, disassembly and maintenance. Transport routes and lifting facilities must be available. Sufficient space must be allocated around the installed fan for the access of personnel. Operators may need to make frequent running adjustments. Maintenance personnel will require space to remove components and assemblies. Some fan designs require withdrawal space at the non-drive end. The motor may need space for the withdrawal of the rotor. The building or plot must have drainage facilities for water if used for bearing cooling. The risks of flooding the fan unit, in addition to the electrical and control systems, which can be very costly and time consuming to dry out, must also be taken into account. The fan should be sited in a place permitting the shortest possible inlet duct with the minimum number of bends.

The ventilation of the fan site is very important. The electric motor must receive the necessary cooling. If hazardous gases, harmful to the environment or highly flammable, are to be extracted, special ventilation requirements must be observed. It should be noted that the heat generated by the fan's electric motor can be considerable. The electrical and control systems for the fan must also be protected against damage.

Consideration must also be given to the noise caused by the fan, its motor and drive train. More and more attention is being devoted to noise from fan units and their ducting systems, and special measures may have to be taken. To reduce vibration transmission, which may be further conveyed through the building structure, it may be necessary for the fan foundations and parts of the duct system to use isolation mountings.

Mounting the fan unit on isolation mountings can create ducting and bearing problems. Ducting close to the fan can fracture if it vibrates. This problem can be solved by using flexible connections adjacent to the fan. Short bearing life can be a problem if the fan does move appreciably on its isolation mountings. It is better to use an isolated foundation block.

Acoustic enclosures may have to be erected around the fan unit. Permissible noise levels are often controlled by legal, safety and environmental requirements at places of work. In the case of noise exceeding 75 to 85 dB(A), specific measures must usually be taken. Staff can be protected by declaring an "Ear protection zone".

It is the user's responsibility to ensure the fan unit is stored in suitable conditions and that the factory preservation instructions have been followed.

On unpacking the fan unit it must be checked thoroughly for any damage which may have occurred in transit. Also the packing list must be checked to ensure all parts are present. Some delicate instruments may have been removed and packed separately. If there is damage or shortages, the manufacturer must be informed immediately. If delicate parts have been packed separately, re-pack them temporarily. Remove any temporary bracing or locking clamps; replace parts as instructed by the manufacturer.

Before placing the fan on its foundation, thoroughly clean the top of the foundation, remove any thin ridges and roughen the top to provide a good key for the grout. Prepare enough shims to level the baseplate, two sets for each foundation bolt. The shims should be longer than the width of the bottom flange. A packer should be placed either side of each foundation bolt, about 20 to 35 mm thick. Remove coupling spacers or driving pins. Lift the fan over the foundation block and fit the foundation bolts into the baseplate holes before lowering onto the packers. Level the baseplate, by adding shims, using an accurate spirit level on the machined pads. Long baseplates may be fitted with targets for laser or optical alignment.

18.1.1 Receiving

Most fans have been correctly aligned, checked and inspected prior to delivery. They have also been subjected to a run test before leaving the works. When you receive the equipment, examine it carefully for damage caused in transit. If anything is wrong, both the manufacturer and the carriers should be contacted within three days of receipt. The goods will be accompanied by a delivery note, which should be signed and returned to the manufacturers. An advice note is also sent by post. Check that all items indicated have been received, and, if not, contact the manufacturer immediately. If the fans are to be put into store by non-engineering personnel, it is recommended that inspection on receipt be carried out by a skilled person. Fitters provided by the makers can usually be hired on a day rate basis.

18.1.2 Handling

The equipment should be handled carefully to prevent damage. Always use all the lifting points provided. Extra care should be given to the impeller or the dynamic balance may be affected. The shaft and bearings are also very important for on them depends the vibration-free running. Where the fan has been despatched complete (up to about size 1250 mm) never "sling" under the shaft.

18.1.3 Storage

When fans are to be stored or installed for any length of time before running, special care should be taken as follows:

a) Where not specifically designed for outdoor use, they should be protected against the elements, special care being given to bearings, motors, and rotating parts.

b) Slowly turn rotating parts at regular intervals to re-distribute the bearing lubricant, making sure that the shaft finishes at 180° to its former position. Never leave the fans stationary for any length of time adjacent to other vibrating machinery. These precautions will diminish the possibility of brinelling of the bearings and/or serious shaft damage.

c) It is recommended that fans stored for any length of time should be inspected again by a the manufacturer's fitter before installation and start up.

18.2 Installation

18.2.1 Introduction

With the exception of axial or mixed flow fans and some light duty "in-duct" centrifugal fans directly incorporated into the ducting, a fan requires a solid foundation. This is especially important where the plant is handling corrosive or erosive fumes or solid particles such as wood refuse or coal dust. Foundations may be of concrete or a steel structure.

Standard fan baseplates are usually designed to be grouted to a continuous concrete foundation block. If the fan is to be installed on a steel structure, the manufacturer should be informed of the size and position of the steelwork. Baseplate modifications may be necessary.

It is the user's responsibility to provide adequate support for the fan unit. The size of a foundation block will depend upon the nature of the sub-soil and the magnitude of the vibrations produced by the fan. Increasing the mass of the foundation block will reduce the amplitude of radiated vibration. A block isolated by proprietary elastomer mats may be necessary. Structural steelwork will be much more flexible than a concrete foundation block. Consideration must be given to the natural frequency of the support structure.

Further, the foundation should be sufficiently high to facilitate the connection of ducts and to ensure adequate space for drainage. The foundation block should be longer and wider than the baseplate to provide extra physical protection, from such items as forklift trucks and barrows.

Every fan set should therefore be erected on firm foundations of adequate depth, taking particular care with the levelling and alignment. Reinforced concrete is recommended, the minimum weight being four times the combined weight of all the rotating parts, or twice the dead weight of the whole unit, whichever is the greater.

Special care is necessary where sets are mounted on steel supporting structures. These can be used but must be level and well braced in all directions to ensure adequate rigidity. Such foundations are vital for trouble-free quiet operation. The minimum natural frequency of any part of the structure must be 50% higher than the running speed of the equipment. Before erection, foundations should always be checked against the fan arrangement drawings.

18.2.2 Concrete foundations

Check the height and ensure that there is the required grouting allowance between the foundation and the fan base. Suitable pockets should be provided for the bolts to be grouted in after levelling and aligning. For vee belt indirect drives, the foundation bolts should be of adequate length and well embedded in the concrete. The location of plinths should be checked in relation to the fan layout drawing. Use steel packers to obtain the correct height of the fan; the packers should be approximately the same width as the base plate and placed each side of and close to the H.D. bolts.

Note: It is the upper surfaces of the machined pads on the baseplate which are important for levelling. The underside of the baseplate is usually not machined and will not be exactly straight.

Horizontal fans and motors are usually supplied complete and aligned on a common baseplate. Whether the fan has been tested or not, the fan and motor will have been accurately aligned prior to dispatch from the manufacturer's works. Check the coupling alignment, see Chapter 12, Section ??. Adjust the shimming until the coupling alignment is close, and consult the manual for thermal growth correction. Protect the tops of the foundation bolts and fit shuttering around the block in preparation for grouting. A good mixture for grout is one part of cement to two parts of sharp sand. The final consistency should be able to flow easily.

Pour the grout and ensure the top of the foundation is covered evenly. If full depth cross members are fitted, ensure complete support. If a cast iron baseplate is used, fill it completely. Remove the shuttering after five days, but allow the grout to harden for at least 10 days. Do not allow the grout to dry out too quickly, and protect it from direct sunlight. In hot environments it may be necessary to cover the grout with damp sacks. Also, protect the grout from frost if the temperature is low. Sacks, covered with polythene sheets will be adequate unless the temperature is very low. Tighten up the foundation bolts and smear with grease.

Check the coupling alignment, remember thermal growth corrections and adjust the shimming under the motor feet if necessary. Record all the alignment settings. Spacer couplings are not an excuse for poor alignment. Diaphragm coupling life will be short with poor alignment. Fans with Cardan shafts should have some radial misalignment, and the manufacturer's instructions should be followed.

The suction and discharge ducting should be finalised after the grout has hardened and final alignment has been completed. The ductwork must line up naturally with the fan connections. Do not force ductwork into alignment. Remove the ducting sections adjacent to the fan and clean out the ductwork, as best as possible. Recheck coupling alignment with recorded figures and modify the ductwork if necessary. Jog the driver to check for correct rotation direction before replacing coupling spacers or drive pins. It is important that coupling bolts are tightened with the correct torque and that the appropriate quality of bolt is used. Information on the subject should be obtained from the manufacturer's instruction book. Fit any loose equipment which was removed for transportation and wire up the unit, locking-off all local isolators.

Very large fans driven by electric motors over 1 MW, or steam or gas turbines, will probably be delivered in at least three sections, fan casing, rotating assembly and driver. Starting with the heaviest section, level and grout in as described above. After the grout has hardened, align, level and grout in the second section. When the second section grout has hardened, then proceed with final alignment and assembly.

18.2.3 Supporting steelwork

Steelwork levels should be checked (including holes in beams). The steelwork should be level and rigid. Make certain that all bolts are tight: welded supporting structures are preferable.

18.2.4 Erection of complete units

Unless too large for transit (size 1400 mm and above), fans are assembled in the manufacturer's works. It is therefore only necessary to mount them on a level foundation and fix in position using all the foundation points provided. It is important however to ensure that after the foundation bolts have been initially tightened, the fan pedestal is not twisted, as this may affect the bearing alignment. In extreme cases it will strain the casing which in turn may cause the inlet cone/venturi to foul the impeller. A spirit level should be used for levelling the unit and if necessary the support points should be shimmed.

The following should also be noted:

a) Correct positioning of a fan to the drawing should be based on the discharge flange.

b) The foundation/H.D. bolts, when set in concrete, should be left after initial grouting for 5 days to allow them to fully harden. If the foundation bolts are into steel, they can be tightened down evenly immediately the fan is considered to be level.

c) Connecting ductwork must not be tightened to a fan unit until it is fully and securely bolted down.

d) Inferior concrete foundations or grouting can be a cause of fan vibration, and if this is considered to be so, the only satisfactory solution is to renew concrete or grouting with a stronger mixture of good quality material.

e) The plinth should be feathered into a concrete floor.

Where ductwork, pipe connections, or other ancillary equipment are connected to the inlet or outlet of the fan, it is essential that they should be supported entirely independently of the machine. Conversely, apart from the small fans noted in Section 18.2.1, fans must be independently supported and not suspended from ductwork, etc. When handling air or gas at high temperatures suitable expansion joints should be provided between the inlet or outlet and the connecting ductwork.

18.2.5 Erection of CKD (Complete Knock Down) units

As the title of this section implies, these fans will be despatched in a completely knocked down state. They will, however, have usually been trial-assembled at the manufacturer's works and the parts identified with part numbers and match points. Nevertheless, it is strongly recommended that assembly on site should be carried out by the manufacturer's own erectors. Failing this the manufacturer could supply a supervisor, with the contractor providing the unskilled labour. Incorrect methods of lifting and assembly could result in damage and/or the balance quality of the rotating assembly being affected.

18.3 Making the system safe

18.3.1 Introduction

Fans are made to many different arrangements and cover a variety of sizes and impeller types. Properly installed, run and maintained, they assist in the creation of better living conditions, cool other equipment, provide essential air for combustion, convey materials, and efficiently carry out many other functions.

All fans have moving parts which may require guarding and access doors to duct systems will also be needed. For further details of guards and doors, refer to Chapter 16, Sections 16.2 and 16.3.

18.3.2 Noise hazards

Excessive noise can be a health hazard. The sound pressure level at any given location is dependent on the effect of all noise generating equipment and the acoustic environment within the vicinity of the reference point, the fan being only one of the contributing sources. It is, therefore, difficult to predict the sound level without a complete survey of all equipment, orientation of each sound source, acoustical characteristics of the structure, and distances involved to each noise source.

Acoustical engineering services should be employed to determine compliance with noise regulations and to make recommendations on any necessary attenuation devices.

18.3.3 Start-up check list

Before putting any fan into operation the following operations should be completed.

1. Cut out primary and secondary power source.
2. Make sure the foundation or mounting arrangement and the duct connections are adequately designed in accordance with recognised acceptable engineering practices.
3. Check and tighten all holding down bolts.
4. Spin impeller to see if rotation is free and does not bind or rub.
5. Inspect impeller to see if it is the proper handing for the fan design.
6. Check all set screws and tighten if necessary.
7. Check vee drive or coupling for alignment — use recommended belt tension.
8. Check vee drive for proper pulley selection and make sure they are not reversed or the fan could run to excessive speeds as well as overloading the motor.
9. Make certain there is no foreign loose material in ductwork leading to and from fan or in fan itself.
10. Properly secure all safety guards.
11. Check security, correct alignment, and fixings of flexible connectors.
12. Secure all access doors to fan and ductwork.
13. Close any inlet or outlet fan dampers.
14. Switch on electrical supply and allow fan to reach full speed.
15. Progressively open dampers, making certain the system continues to function satisfactorily and that motor does not overload.
16. Check carefully for:
    - Correct impeller rotation (shown by rotation arrow on fan casing).
    - Excessive vibration.
    - Unusual noise.

If any problem is indicated switch off immediately. Cut out the electrical supply, check carefully for the cause of the trouble, and correct as necessary.

Even if the fan appears to be operating satisfactorily, shut down after a brief period and recheck items 3 to 12 as the initial start-up may have relieved tightness of bolts and set screws, again ensuring electrical supply is cut out before attempting other checks.

The fan may now be put into operation, but during the first eight hours of running it should be periodically observed and checked for excessive vibration and noise. At the same time checks should be made on motor input current and motor temperatures to ensure that they do not exceed manufacturer's recommendations.

After eight hours of satisfactory operation, the fan should be shut down and the power cut out to check the following items and adjust if necessary.

1. All set screws and hold down bolts including guard fixings.
2. Drive coupling alignment.
3. Vee drive alignment.
4. Vee drive belt tensions should be readjusted to recommended tension.
5. Security of flexible connections.

18.3.4 Electrical isolation

Every fan must be provided with a disconnect switch which will allow it to be isolated completely from the electrical supply.

Most roof-mounted fans and many others are started by remote switches or push-buttons, by interlocks with other equipment, or by automatic controls. In these cases a disconnect switch must be provided close to the fan so that maintenance personnel can "positively" cut off the power when working on the fan. See Figure 18.1.

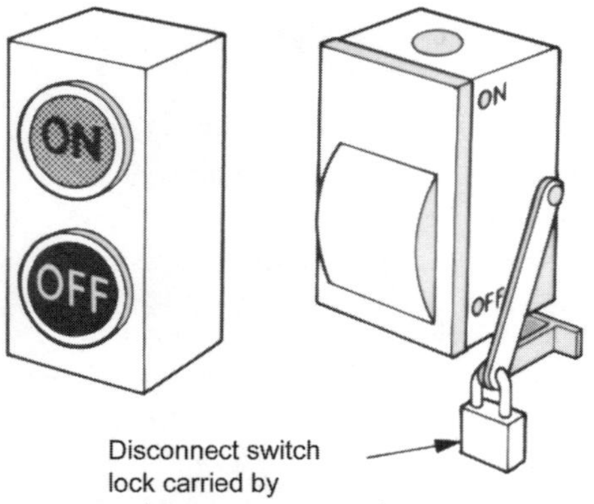

Figure 18.1 Remote switch - left; disconnect switch - right

In some installations other equipment, such as gas burners, may be interlocked with the fans so that disconnecting the fan will automatically shut off the burner or other device. Maintenance on systems of this type should be performed only under the supervision of competent technical staff.

18.3.5 Special purpose systems

Fans which are used to move anything other than clean air at normal temperatures (-40°C to +75°C) may require special precautions to ensure safe operation. Explosive or toxic fumes or gases, transported solids, high temperatures, and corrosive contaminants will present special hazards which must be carefully considered. All national and local codes should be reviewed together with any applicable industry standards. The manufacturer's recommendations for the specific application should be closely followed.

When the system will handle explosive or inflammable fumes or gases, fans of spark-resistant construction should be used. The manufacturer should be consulted when specifying fans for this use. There is now a draft European Standard prEN 14986 to meet the requirements of the ATEX Directive.

If the fan is handling toxic or explosive fumes — even in traces — care must be taken to ensure that fumes have not collected in areas which require access by workmen. Concentrations of fumes can collect in "air trap" areas, particularly when a system is shut down.

Fans with radial or paddle impellers are specially designed to allow the fan to handle a specific type of material without excessive accumulation of material on the fan wheel. To ensure satisfactory operation it is essential to observe the manufacturer's limits concerning the type of material to be handled by the fan.

Fan ratings and maximum speed limits are based on the use of air at 20°C. At temperatures above the normal range (over say 150°C) a reduction must be made in the maximum speed limit. Information on this and on other precautions to be taken for high temperature applications may be obtained from the manufacturer.

Corrosive contaminants can be formed when moisture combines with an active airborne chemical. Unprotected fans subjected to corrosive attack will eventually fail, but suitable protective coatings or material used in the fan construction will resist corrosion. Even protected fans must be regularly inspected to ensure that the protection remains effective.

18.4 Commissioning and start-up

18.4.1 General

Many fan units are supplied with large quantities of operating and maintenance manuals and other paperwork such as Certificates of Conformity and the like. It is a particular gripe of the author that they are rarely available to the process operators or site maintenance staff. These people need such information on a daily basis, but instead the manuals reside in a project file, gathering dust.

Whilst they may be of interest when the plant is duplicated, this is **not** their purpose. Even worse is the situation where the contractor takes out what **he** thinks is the essential information to give the operators!

The transition from installation to operation is commissioning and start-up. Commissioning can often be performed by site staff but if they are unfamiliar with any features of the fan then they should seek information from the manufacturer. It is often then possible to combine commissioning with staff training.

18.4.2 Start-up

When erection is complete, shafts and impellers should be checked for freedom of rotation. As erection proceeds care should be taken to ascertain that no tools, pieces of packing etc, are left in the fan or ducts to cause obstruction or damage when the machine is started up.

When direct motor-driven, the direction of rotation of the motor should be checked with the fan, preferably before connecting up to impeller, coupling, or vee belt drive. This is important where special thrust bearings may be employed. The correct impeller rotation is shown by an arrow on the fan casing.

When the machine is driven through a coupling it is particularly important that correct alignment of the two shafts be achieved by checking the coupling by recognised methods for the particular type; levelling up for this purpose is obtained by interposing steel packing between the underside of the bedplate and the top of the foundations. See Chapter 12, Section 12.11.

Alignment should be re-checked at the running temperature of the machine after the set has been run sufficiently long to allow for possible foundation settlement.

When the machine is indirect-driven the pulleys should be checked for correct alignment and spacing, and the driving and driven shafts must be parallel to each other to ensure that the vee belts run truly in the grooves and with recommended tension of the vee belts. Excessive tension will overload bearings and cause possible damage to shafts and vee belts; conversely slack tension can result in slip leading to excessive wear on ropes and pulleys.

18.4.3 Precautions and warnings

Make sure that the guarding requirements outlined in Section 18.3 have been met and that the recommendations detailed in ISO 12499 have been complied with.

Before starting up check especially that the storage, installation and lubrication instructions have been followed. Then close dampers, on the inlet and/or outlet of the fan. Start and run up to full speed with dampers closed, then open gradually until the required duty is obtained. These precautions will minimise starting time and avoid excessive load on the driving motor.

Note: An electric motor may be overloaded if a fan is allowed to run up to speed with inlet and outlet fully open. The set should be connected to the ductwork system, or the inlet or outlet dampered. Where shaft seals are provided, adjustment may be necessary. If circumstances necessitate the fan being on site for a considerable time before erection and starting up, the lubricant it the bearing should be replaced. If it has been subjected to vibration from other machinery while standing, the bearings themselves may have suffered from indentation of the races by the roller and they should be examined and if necessary replaced before putting to work.

Remember:

1 Ensure that no tools, pieces of wood, etc, are left in the fan casing or associated ductwork.

2 Check freedom of rotation of shaft assembly.

3 Check vee belt drive alignment and tension (or coupling gap and radial alignment).

4 Check lubrication of bearings (and coupling if fitted).

5 Check that air connections are correctly made.

6 Check that shaft and cooling fan guards are correctly fitted and adequate clearances are maintained with rotating parts.

7 Fully close inlet and/or outlet damper.

8 Check motor and fan rotation by a flick of the starter on first start. Rotation should agree with rotation arrow on fan casing. If incorrect consult the motor manufacturer's manual.

9 Start motor and allow to run up to speed.

10 Gradually open any damper(s) ensuring that motor amperes do not exceed duty figures.

18.5 Maintenance

18.5.1 Introduction

The care of a fan unit should be carried out by routine inspection performed according to a definite schedule. The timing intervals required are dependant upon the working conditions and environment of the fan and the demands of operational reliability. The latter should have been a major requirement in the fan specification and in the criteria for evaluation of competing products.

Good maintenance practices cannot reverse deficiencies of fan selection. Some fans may require attention every day; topping up oil reservoirs for example. Periods exceeding one week are inadvisable, especially in the case of fans in distant fanhouses and lacking alarm systems. Observations such as listening and feeling for vibrations, and checks on pressure, flow and power consumption should be performed with every inspection in addition to the checking of gaskets and shaft seals. More detailed investigation should, of course, be undertaken if deviations from normal operation are noted.

If several fans are included in the system, the starting sequence should be adjusted with the lag-lead switch so that the running time is divided between all the units. Automatic and alarm devices should also be regularly tested.

In the absence of any specific instructions, the recommendations given in Section 18.5.3 should be followed.

18.5.2 Routine inspection

In the selection of fans and the planning of their systems, it is necessary to take into account all aspects of maintenance. The assessment of a particular fan should be based on technical grounds connected with maintenance such as ease of dismantling, availability of spare parts, trade skills required and above all, a record of reliability.

The difference between routine inspection on the one hand and routine maintenance on the other is often difficult to decide. In general, it is recommended that inspection, as well as maintenance, should be carried out on a regular basis. Some form of database of faults noted and measures taken should be initiated and up-dated using PCs to allow the storage and retrieval of considerable quantities of information. Software is available which enables machinery data to be stored and individual fan service histories to be retrieved.

Under normal circumstances handling clean air, the system will need cleaning about once a year. However, with the service history of the fan and with regular routine inspections it should be possible to detect any unusual accumulation and modify these instructions accordingly.

The fan impeller should be specially checked for build up of material or dirt which may cause an imbalance, with resulting undue wear on bearings and vee belt drives. A regular maintenance programme must be established to prevent this build-up.

The rotating assembly should be inspected regularly to detect any weakening of the impeller shaft and bearings resulting from corrosion, erosion, or metal fatigue.

Do not attempt any detailed inspection of a fan unless the electrical supply has been completely disconnected. If a disconnect switch has not been provided, remove all fuses from the circuit and lock the fuse panel so that they cannot be accidentally replaced.

18.5.3 Routine maintenance

Maintenance in this context refers to preventive maintenance intended to reduce the number of breakdowns and the resulting unscheduled shut-downs. Attempts have been made to calculate optimum maintenance statistics for fans, but the results are uncertain and the systems are difficult to handle.

Generalisation of fan maintenance is a waste of time. Fans operate in widely differing circumstances with a vast array of materials. The best policy is to initiate a strict, very regular, routine inspection of the equipment and continue this until a pattern of equipment behaviour is apparent. At this stage, it may be possible to relax the inspection routine in some areas.

Running and maintenance instructions should be issued with every fan, if the requirements of the Machinery Directive are to be met. It is surprising how often this instruction is ignored. In the absence of any specific instructions, the following are suggested:

Every shift

When taking over plant at beginning of shift, operators should check that all bearings are cool.

Every week

Check for undue vibration. If present, stop fan at earliest opportunity, check impeller for any dirt build-up on the blades, and clean as necessary.

Every 6 months

(a) Consult motor manufacturer's manual and carry out instructions.

(b) Examine vee belt pulleys for any chipping, tension of ropes, or

c) Check coupling alignment and condition.

Every 12 months

a) Examine impellers, fan bearings, inlet spigots/venturi. Check vee belts and pulleys or coupling element(s). If any wear, replace as necessary.

b) Check clearance at impeller, level of shaft, and general alignment. Adjust as necessary.

c) Check all H.D. bolts for tightness.

d) Refer to motor and control gear manufacturer's maintenance instructions and act accordingly.

e) Grease-lubricated bearings should be cleaned out and grease renewed.

Refer also to all proprietary item literature and act as instructed. Lubrication should be carried out regularly according to operating conditions.

18.5.4 Bearing lubrication

The whole question of bearing lubrication is addressed in Section 10.6. It will have been noted that the greasing intervals for ball bearings are invariably higher than for roller bearings. Nevertheless, spherical roller bearings are used particularly in many indirectly driven fans (e.g. vee belt drives) as they can resist the high radial loadings due to belt pull.

The instructions which follow do not apply to arrangement 4 and 5 fans (see Chapter 9) which should follow the motor manufacturer's recommendations.

Bearing size (mm)	Fan speed rev/min					
	125	250	500	1000	2000	4000
20	8000	8000	6300	3150	1600	710
30	8000	8000	5000	2500	1250	560
40	8000	8000	4500	2240	1000	400
50	8000	8000	4000	1800	800	315
55	8000	7500	3550	1700	750	280
60	8000	7100	3150	1600	710	224
65	8000	6300	2800	1400	630	180

Table 18.1 Relubrication interval (operating hours) for spherical roller bearings

18.5.4.1 Split roller bearings

Lubrication — Grease lubrication is usually satisfactory up to the likely fan maximum speeds, subject to consideration for temperature and axial loads. Greases are grouped according to maximum working temperature at the bearing. They are often water-absorbent and many contain moisture and oxidation inhibitors. Extreme pressure additives can be advantageous especially for high axial loads.

For high temperatures and speeds it is always advisable to consult the fan manufacturer.

Procedure — Apply grease as follows:

For speeds up to dn* = 20 000 the roller cage should be coated with grease and the other parts lightly covered for protection, grease weights are as shown in Table 18.2. The remaining space in the lower half cartridge should be filled with grease or the whole cartridge may be completely filled to aid sealing in wet or dirty conditions.
(*d = bearing bore (mm), n = rev/min)

For speeds over dn = 20 000, the cage and parts should be coated as above plus 25% of listed grease weight in the cartridge. Cartridges fitted with thrust bearings, which are used only up to dn = 20 000, should be completely filled with grease on assembly, including the bore of the thrust bearing.

All cartridge end bore seals should be well lubricated on assembly including the bores of the revolving triple labyrinth seals. Blanking plates should be sealed with grease or jointing compound. Swivel seatings should be lubricated: anti-scuffing compounds such as Molycote are useful.

Note: Never assemble the bearings dry or inject the grease after closing the cartridge during assembly.

Bearing bore (mm)	Series		
	01	02	03
50	0.03	0.05	
60	0.05	0.07	
65	0.05	0.07	
70	0.06	0.10	
80	0.10	0.15	
90	0.12	0.20	0.40
100	0.17	0.30	0.46
125	0.20	0.37	0.46

Table 18.2 Recommended grease weights (kilograms) for split roller bearings

Courtesy of Cooper Roller Bearings Ltd

18.5.4.2 Routine greasing

Expansion EX bearings: One or two shots from a grease gun two or three times a year (say every 1000 hours) is usually sufficient.

Fixed GR bearings for thrust: One or two shots from a grease gun every two weeks (say every 100 hours) or longer according to duty and experience.

18.5.4.3 Recommended lubricants

All the major oil companies have greases suitable for anti-friction (ball and roller) bearings. Competition is fierce and continued improvements are being made. The bearing manufacturer should always be consulted, but those shown in Table 18.3 have been found to be satisfactory in many types of fan.

Manufacturer	Normal temp (up to 80 ° C at brg)	High temp (up to 120 °C at brg)
Shell	Alvania RA	Darina R2
Mobil	Mobilplex 47 or 48 Mobilux 2 or 3	Mobiltemp 1
Esso	Beacon 2 or 3	
	Spheerol AP2 or 3	
	LS 2 or 3	
	Regal AFB 2	Ultra Temp
	Admax L2 or L3	Admax B3
	Multi-purpose No 2	Hi temp
	Lupus A2	Bellatric 2
Rocol	BRB 1200	BRB 1200

Table 18.3 Recommended lubricants for many types of fan

Fixed GR bearings used for location only: Treat as expansion bearings.

Clean out and replace the grease yearly or as determined by the conditions.

Lubrication points — Cartridges are tapped $\frac{1}{8}$ or $\frac{1}{4}$ pipe according to size and series. Lubricating nipples or temporary plugs are fitted as standard. The lubricant is injected through the outer race directly to the rolling surfaces.

Grease weights — The weights given are sufficient to coat the roller bearing as described. A similar amount fills the remaining space in the lower half of the cartridge and thus three times the values given will completely fill the bearing and cartridge. All weights are approximate.

Extreme pressure greases are usually normal range and suffixed EP. High temperature greases should be checked for speeds over dn = 100 000, and replenishment intervals may be reduced.

Deviation from these standard recommendations is notified separately when required. It is essential not to overgrease as this will raise the running temperature of the bearing and may shorten its life.

18.5.5 Excessive vibration

Check for build-up of material on the wheel. Generally this will show up as material flaking off the fan wheel and causing an imbalance which may lead to fatigue failure of the wheel. Never allow a fan to operate if the amplitude of vibration is above the maximum safe limit. Contact the manufacturer for this information, if it is not included in maintenance instructions.

18.5.6 High motor temperature

Check that cooling air to the motor has not been diverted or blocked by dirty guards or similar. Check input power: an increase may indicate that some major change has been made in the system. For other motor problems refer to motor manufacturers' instructions.

18.5.7 High fan bearing temperature

Usually caused by improper lubrication (either "over" or "under"). In every case if the cause of the trouble is not easily seen, experienced personnel should examine the equipment before it is put back into operation. (See Section 18.5.4.)

18.6 Major maintenance

18.6.1 Introduction

Fan construction varies enormously from one manufacturer to another. There are however, two types which are characterised by their popularity throughout the industry:

a) Semi-universal construction with bolted on sideplates.

b) Fixed discharge construction with split casing.

Readers are referred to Chapter 8 for fuller details. Suffice it to say that where major maintenance is anticipated, this should be given in the operation and maintenance manuals. It should be noted that axial flow fans usually only have motor maintenance to contend with, when the appropriate instructions should be followed. Many lightweight centrifugal fans (Category 1 according to ISO 13349, are assumed to be a commodity purchase when replacement may be more appropriate.

Before removing any pieces of equipment, the relative positions of mating parts should be marked to simplify erection.

1 Remove any coupling or drive guards.

2 Remove any shaft guard by unfastening the set screws with spring washers.

3 Remove any vee belts or coupling elements.

4 Disconnect inlet ducting, supporting as necessary.

The removal of the rotating assembly from the casing is detailed for semi-universal and fixed discharge fans with taper bushed hubs. Many other types are manufactured, and these are mere

18.6.2 Semi-universal fans

Alternative maintenance methods are possible, dependent on the disposition of the ducting.

1 The complete inlet sideplate with spigot and shaped inlet may be removed giving access to the impeller, which may then be removed through the inlet side. To prevent the volute "dropping", it will need to be chocked up underneath.

2 The complete pedestal and rotating assembly may be removed by undoing the fasteners on the drive side plate. Sufficient space will be required behind the fan and it will also be necessary to lift the pedestal over the H.D. bolts. Alternatively the assembly may be slid back if mounted on extended steelwork. Again the casing volute must be supported by chocks.

18.6.3 Fixed discharge fans

1 Disconnect inlet spigot from inlet sideplate unbolting the setscrews. Slide shaped inlet away from impeller until it is clear of the impeller shroud. Support the loose venturi (shaped inlet).

2 Remove both halves of any shaft washer, unfastening the screws.

3 Unfasten the bolts along the horizontal join in the drive sideplate.

4 Unfasten the bolts along the horizontal join in the inlet sideplate.

5 Secure slings through the lifting eyes on top of the casing sideplates.

6 Unfasten the bolts along the horizontal join in the scroll.

7 Disconnect any outlet ducting from the casing, unfastening the bolts on each sideplate and the bolts on top and below.

8 Disconnect the tongue from its half of the scroll.

9 The top half of the casing can now be carefully hoisted above the unit and removed.

10 Place slings around shaft adjacent to impeller, on the main bearing side, and adjacent to tail bearing on the impeller side. The impeller must be propped in the casing as it will tend to tip the shaft upwards when the pedestal top halves are removed.

11 Remove top halves of bearing blocks, unfastening the bolts on each pedestal. Now carefully lift the shaft, impeller, and roller bearings out of the unit, taking care not to damage any of them. The drive end of the shaft should be held down to avoid tipping.

Note: The instructions given above do not apply fully to fans in Arrangements 4 and 5 where the impellers are mounted directly on the motor shaft extension and there are no separate shafts and bearings.

18.6.4 Removal of impeller from shaft

1 Mount the shaft on suitable supports terminating in hardwood vee blocks. The coupling /pulley end of the shaft must be restrained from moving upwards under the tipping action of the impeller. Ensure also that the impeller is clear of the ground.

2 Mark positions of centre bush on shaft and make a note of impeller blade angle relative to rotation and bearings.

3 Slacken off all 3 screws in tapered bush and remove them. Replace 2 of the screws in the "jacking off" holes and screw in alternately, after oiling the thread and point of grub screw (or thread and under cap of a cap screw), continuing screwing in until the bush is loose in the hub and both bush and hub are loose on shaft. Oil shaft and slide impeller and bush towards inlet end of shaft and remove, taking care not to damage the shaft. Remove key from keyway. When removing impeller it should be supported on nylon or padded slings, between backplate and shroud, and around 4 blades.

18.6.5 Removal of bearings from shaft

This is done after the impeller and any coupling element or vee belts have been removed, as the bearings help to balance the assembly.

18.6.5.1 Spherical roller adapter sleeve bearings

1 Remove the bearing cap setscrews and take off bearing caps, being careful to note which way round they are fitted, and to which plummer block. Note also which is located before removing rings.

2 Lift out shaft and bearings being careful to use rope slings and not to damage anything.

3 Position shaft in wooden vee blocks on a suitable bench.

4 Carefully prise out tab of washer. Loosen off lock-nut and remove.

5 Bearings may now be removed by carefully tapping down taper of the adapter sleeve.

6 Adapter sleeves may now be removed.

18.6.5.2 Split roller bearings

1 Remove socket head cap screws and take away both halves of cartridge complete with both halves of the outer race.

2 Remove jointing clips of the roller cage and take away both halves of roller cage complete with rollers.

3 To remove outer race from cartridge housing on expansion bearing, remove radial socket head cap screws from each half of housing and slip out race.

4 To remove outer race from cartridge housing on located bearing loosen side screws and their side rods in addition to the radial socket head cap screws.

5 The seals (triple labyrinth type) on both bearings are split in one place and secured by half circlips. These circlips are removed and the seal unwrapped off the shaft.

6 Undo socket head cap screws on each clamping ring and remove both halves of each clamping ring.

7 Finally remove both halves of inner race.

18.6.6 Refitting of new bearings on to shaft

18.6.6.1 Spherical roller adapter sleeve bearings

1 The bearing should not be taken from its packing until required. Do not remove the protective grease except from the bore which should be wiped with a clean cloth dampened with white spirit.

2 Saturate the felt seals in a mixture of two-thirds lubricating oil and one-third tallow at 80-85°C.

3 Caps and bases are not interchangeable or reversible — **do not mix them**.

4 Clean the inside of the housing thoroughly. Press the felt seal into the groove in the base, and cut off the ends flush with the machined face. Pack the remaining felt in the cap groove, and trim off the surplus. Make sure there are no frayed edges to prevent the cap fitting correctly; but do not cut off too much as a gap between the ends will allow foreign matter to enter.

5 Secure the housing base to a support and check for alignment. Lightly smear the outer ring seating of the housing with a thick oil containing a rust inhibitor.

6 Remove any sharp edges or burrs from the shaft, adapter sleeve, tab washer and nut, then wash them clean and wipe dry. Lubricate the thread and lightly oil the outer surface of the sleeve. Mount the adapter sleeve roller bearing, tab washer and nut loosely on the shaft as previously. Make sure the concave side of the tab washer faces the bevelled side of the nut. Tighten the sleeve nut, checking the clearance frequently until it is reduced by approximately 50%. When mounting bearings with C3 or C4 fit, the reduction in clearance should be less than 50%.

7 Position the shaft in the lower half of each housing. Tighten the nut of the locating bearing sufficiently to position the sleeve on the shaft, which should be supported so that there is no load on the bearing when finally tightening up.

8 Check the radial clearance between the rollers and the outer ring with a feeler gauge (in C type bearings it is easier to check the clearance at the lower part of the bearing). After tightening the nut, swivel the outer ring of the bearing to see that the inner ring is not expanded too much; over-tightening may cause premature failure.

9 Use special "D" spanners to tighten the nut. **Do not use** drifts or punches as these mutilate the nut and may damage the tab washer and bearing cage. In the case of large roller bearings, the locking washer may be damaged if placed between the nut and the sleeve during the driving up process. It is preferable to fit the washer after driving up, but if this not possible, the friction can be reduced by smearing the contact surfaces and threads with oil to lubricate them.

10 After the sleeve has been tightened, bend one tab of the washer into a convenient slot in the nut. To line up a tab and a slot tighten the nut slightly rather than slacken off. The bearing outer ring must not rest on its seating during this operation.

11 Only one fixed or located bearing is used on each shaft, the bearing being positioned axially in the housing by one or two locating rings, depending on its type. If two rings are fitted, position one on each side of the bearing; if one ring is used it is fitted on the same side as the nut. Make sure that the same bearing is located as previously.

12 The outer ring of a free bearing should be in the centre of the housing seating.

13 When fitting the impeller pulley or coupling, support the shaft so that blows cannot be transmitted to the bearings.

14 Thumb grease into both sides of the bearing and, in addition, fill the bottom half of the housing. For low speeds the bearing housing may be filled completely. Smear a little grease around the shaft adjacent to the felt seals to assist in lubrication and sealing.

15 After fitting the housing cap check that the shaft does not foul at any point.

18.6.6.2 Split roller bearings

1 Check that bearing parts are thoroughly clean and that existing parts are suitable for further service; look for example for signs of wear and tear in inner and outer races.

2 Remove all old grease from bearing housings and thoroughly clean using a clean cloth.

3 Ensure that nitrile seal is in good condition.

4 The instructions given previously (Section 18.6.5.2) for removal of split roller bearings should be reversed noting the following points.

5 Clean cartridge bore and lightly oil. Clean the outer race and place in the half cartridge so that all the pairing marks coincide and lubrication hole is in top half of cartridge. On the grooved race GR NTL located bearing, just enter the radial socket head cap screws with washers and very lightly tighten. It is important to fit the washers. Fit the side rods and side screws and very lightly tighten.

6 The expansion bearing EX NTL has no side rods. Check the shaft diameter.

7 Place the two halves of the inner race at the correct position on the cleaned shaft. A tap may be required to spring the races over the shaft.

8 Fit the clamping rings with the joints at about 90° to the inner race joint. This overlap is particularly important for side fitting clamping rings of the grooved inner races to ensure alignment of the lips. There should be approximately equal gaps at both parts of the clamping rings and races.

9 Tighten all socket head cap clamping screws equally using the correct hexagon key and tube extension.

10 Tap down each half of the inner race and clamping rings all round the shaft, interposing a fibre or hardwood block between hammer and bearing parts, and re-tighten screws. Repeat until screws are fully tightened.

11 Check that there is a gap at both joints of the inner race, coat the roller cage with grease, fill the cartridge $\frac{3}{4}$ full in each half, and grease all seals including the bores of the revolving labyrinth seals. Place the roller cage around the inner race and engage jointing clips.

12 Close cartridge and tighten socket head cap joint screws. Lubricate the spherical seating.

It is strongly recommended that ball or roller bearing replacements should be of the same type, diametrical clearance, and fit as originally supplied.

18.6.7 Refitting of impeller on to shaft

1 Ensure that the mating tapered surfaces on hub and bush are completely clean and free from oil and dirt. Insert bush into hub so that holes line up.

2 Oil thread and point of grub screw or thread, and under head of cap screws. Place screws loosely in holes thread in hub.

3 Clean shaft and fit side fitting key with top clearance in position in keyway. Fit hub and bush to shaft as one unit and locate in correct position by lining up original marks. Remember that the bush will nip the shaft first and then the hub will be slightly drawn onto the bush.

4 Tighten screws finger-tight.

5 Using a hexagon torque wrench tighten screws gradually and alternately until all are pulled up correctly. Take care not to exceed recommended screw torques given in Table 4 or damage may result.

6 After the fan has been running under load for a short time, stop and check tightness of screws.

7 Fill empty holes with grease to exclude dirt.

Bush size	Screw size (in)	Wrench torque (Nm)
1210	$\frac{3}{8}$	20
1615	$\frac{3}{8}$	20
2517	$\frac{1}{2}$	48
3020	$\frac{5}{8}$	90
4545	$\frac{3}{4}$	192

Table 18.4 Recommended screw torques for fitting impeller

18.6.8 Refitting rotating assembly into unit

18.6.8.1 Semi-universal fans

The reverse procedure to that used for removal should be adopted.

18.6.8.2 Fixed discharge fans

1 Support shaft on nylon or padded slings with one sling as close to impeller as possible on the bearings side and one sling outside tail bearing.

2 Lift shaft assembly carefully, holding coupling or pulley end of shaft down to avoid tipping.

3 Move assembly over unit and lower carefully onto bearing pedestals horizontally. Care must be taken not to damage impeller or shaft or bearings. Ensure that the triple seals slide into their corresponding grooves.

4 Replace top half of pedestal housings and fit socket head cap screws on each pedestal (run shaft for a short period before finally tightening pedestal cap screws to ensure swivel alignment). Remove slings.

5 Fasten tongue to outlet half of scroll using set screws.

6 Bolt up horizontal join in inlet sideplate.

7 Bolt up horizontal join in drive side plate.

8 Bolt horizontal join in scroll: attach both halves of shaft washer to drive side plates.

9 Bolt up outlet connections to casing on each sideplate, scroll and tongue.

10 Slide shaped inlet into casing until inlet spigot is flush with casing. Bolt spigot to casing set screws. Check impeller overlap and diametrical clearances. Replace inlet connections.

11 Rotate shaft to check for any obstructions, replace coupling element or vee belts.

12 Check that pulley or coupling is in original position relative to shafts (check alignment first and correct as necessary).

13 Replace coupling or drive guard.

14 Replace shaft guard using set screws with spring washers.

Again the instructions given above are simplified for fans to Arrangements 4 and 5 where the impellers are mounted directly on motor shaft extensions.

Note: Impeller/inlet venturi alignment is most important for the airflow performance of high efficiency fans. The method outlined below is recommended for determining the optimum position.

Place template (shown shaded in Figure 18.2) against inside face of impeller shroud and adjust impeller or venturi to achieve correct alignment. Repeat at four cardinal points.

18.6.9 Vee belt drives — installation

1 Clean all oil and grease from pulley grooves and bores.

2 Remove any burrs or rust.

3 Reduce the centre distance until the belts can be placed in the pulley grooves without forcing.

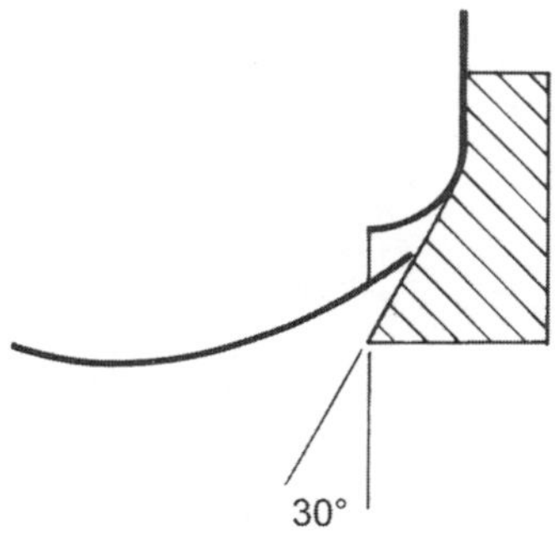

Figure 18.2 Recommended method for aligning impeller and inlet venturi

4 Align the pulleys correctly using a straight edge to ensure that the pulleys are in line and the shafts parallel.

5 Tension the drive using the motor slide rail bolts.

6 Check that the vee belts are correctly tensioned:

a) Measure the span.

b) Apply a force at right angles to the belt at the centre of the span.

c) This force should deflect one belt 16 mm for every metre of span length. See Figure 18.3.

d) The average value of the force in each belt should be compared with Table 18.5 (in accordance with BS 1440:1971 and BS 3790:1985). Belts should initially be tightened to the higher values.

Belt section	Small pulley pcd (mm)	Belt speed		
		0 to 10 m/s	10 to 20 m/s	20 to 30 m/s
SPZ	95	12 to 18	10 to 16	8 to 14
	95	18 to 26	16 to 24	14 to 22
SPA	140	22 to 32	18 to 26	15 to 22
	140	32 to 48	26 to 40	22 to 34
SPB	250	38 to 56	32 to 50	28 to 42
	250	56 to 72	50 to 64	42 to 58
SPC	355	72 to 102	60 to 90	50 to 80
	355	102 to 132	90 to 120	80 to 10
Z	50		4 to 6	
A	75		10 to 15	
B	125		20 to 30	
C	200		40 to 60	
D	355		70 to 105	

Table 18.5 Correct vee belt tensions: required force N at centre of span for belt speed (To obtain kgf divide N by 10 to give approximate value)

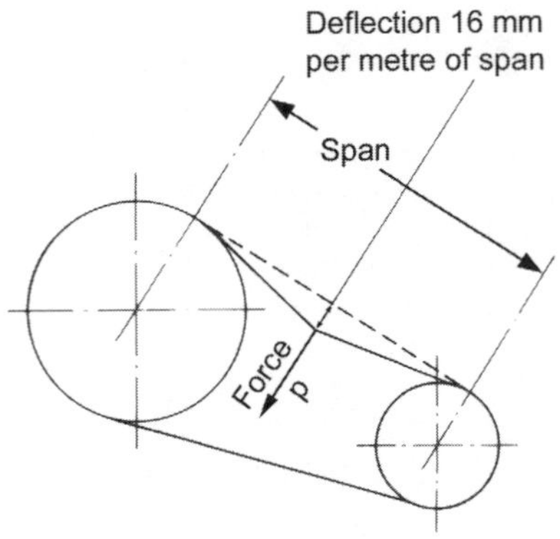

Figure 18.3 Belt deflection measurement

If the measured force falls within the values given in Table 18.5 the drive tension should be satisfactory. A force below the lower value indicates undertensioning. When starting up, a new drive should be tensioned to the higher value to allow for stretch during the running in period. After the drive has been running a few hours the tension should be re-adjusted to the higher value. The drive should be re-tensioned at regular maintenance intervals. Make adequate provision for tensioning the belts during their life.

18.6.10 Couplings and shaft seals

Couplings of the grease-filled type will require to be fully charged with suitable grease after alignment and before starting up, and replenished at monthly intervals unless otherwise advised.

Shaft seals of the packed stuffing box type generally need no lubrication except when lead wool packed, when the box should be completely filled with suitable grease and replenished periodically during operation by means of the greaser cap provided. This procedure also applies to labyrinth type glands.

Symptom	Possible cause	Remedial action
Fan won't start	Power failure Blown fuses Motor single phasing Control switch contacts open	Check all power supplies Check all fuses Check motor and wiring Check all instrument switches in motor contactor circuit
Low flow	Inlet damper closed Discharge damper closed Excessive system resistance Fan speed very low Impeller partially blocked Wrong direction of rotation	Check all inlet dampers in ducting for correct position Check all outlet dampers in ducting for correct position Check calculations, gas temperature etc. Check drive, motor frequency and voltage Fan vibrates badly - clean impeller and re-balance Reverse two phases in 3-phase motor
High flow	System resistance lower than specification Fan speed high	Check calculations, throttle discharge damper, reduce fan speed Check fan speed, check motor frequency, check pulley ratios in vee belt drives
High power consumption	Fan power high Electrical faults	Check air/gas temperature and specific gravity System resistance low, flow high, check operating conditions Fan speed too high Voltage and frequency incorrect, check supply Motor problem, check motor
Fan vibration	Poor alignment Damaged impeller Bearing problems	Check alignment Check coupling or vee drive assembly Check baseplate grouting Check ducting Check vibration harmonics, clean and/or repair impeller, re-balance Check bearing clearances Check vibration signature for bearing faults
Motor vibration	Poor alignment Bearing problems	Check alignment and coupling/vee drive assembly Check baseplate grouting Check bearing clearances Check vibration signature for bearing faults Check motor for loose rotor bars

Table 18.6 Trouble-shooting guide for fans

18.6.11 General notes

The speed of the machine should **not** be increased beyond that specified for the particular duty for which it was installed, without first referring the matter to the manufacturer.

All impellers should be accurately balanced during assembly at the works, but on certain applications, unbalance and consequent vibration may develop as a result of a build up of deposits on the impellers. This calls for periodic and thorough cleaning of the impellers to restore the original balance. Corrosion, wear, or damage on certain applications may also cause unbalance.

Where a number of machines are installed with one or more acting as standby it is good practice to use them in rotation or to give the rotors a few turns at regular intervals.

18.7 Trouble-shooting

Trouble-shooting of fans is often taken for granted. Indeed, it is frequently limited to making the fan work, when it should be making the fan do what it is specified to do.

If the resistance of the system is more or less than specified, or if other properties are not as specified, rectification of the system must be carried out first. If the necessary changes to the system cannot be carried out then this should be acknowledged. The fan may require "up-grading", a condition which is not a fault of the fan but rather the fault of the system.

Readers are referred to AMCA 202, which gives symptoms and solutions in a logical progression. Table 18.6 is a shortened resume of the information contained therein.

18.8 Spare parts

When buying fans, it is important to check the manufacturer's ability to maintain a stock of spare parts. The supplier should be able to provide a quotation for

- commissioning spares,
- spares for 1 year,
- spares for 2 years,
- insurance spares.

Spares quotations will include delivery times. The spares quoted will be the parts which the supplier knows from experience will wear or become damaged. Insurance spares may be required if the fan is on a critical service and space is not available for stand-bys. Spares are costly. Some manufacturers may offer a discount if the spares are purchased with the complete fan. If the quotation does not mention discounts, ask.

The assessment of spares requirement is dependent upon many factors; the process gas, the operating conditions, the number of hours of operation per year, delivery times for spare parts from the manufacturer, storage costs, etc. Table 18.7 shows suggested spares which should be held according to the number of identical fans.

Spare parts	Number of fans					
	2	3	4	5	6	8
	Number of spare parts					
Impellers	1	1	1	2	2	3
Shafts	1	1	2	2	2	3
Seals	2	2	2	3	3	4
Inlet cones	1	1	1	2	2	3
Vee belts (set)	1	2	2	3	3	4
Bearing races (set)	1	1	2	2	3	4
Gaskets	1	1	1	2	2	3

Table 18.7 Proposal for stocking spare parts for industrial fans

The fan supplier should also be asked about the supply of spare parts if a fan model is made obsolete and current production ceases.

When failure of the fan involves risk of damage, for example the possibilities of flooding or vital cooling systems being put out of action, a stand-by unit with an automatic starting system should always be installed if space allows. When a number of fans of the same size share a service, it is a prudent to have one complete unit in reserve for change-over in the event of a breakdown or in the case of a planned overhaul. Naturally the duct connections, electrical installation, etc. should permit really rapid change-over with the minimum of inconvenience and effort.

If overhauls and repairs are to be performed by the user's staff, it is also important that the tools required for the purpose should be obtained. Fan suppliers can make the necessary recommendations. If special tools are necessary, the fan supplier should quote costs for the tools when quoting for the fan. The use of unsuitable tools frequently results in more serious damage than that for which measures were to be taken.

18.9 Bibliography

Machinery Directive 89/392/EEC (Amended 98/37/EEC). This Directive is fully implemented into UK law by means of the Supply of Machinery (Safety) Regulations 1992 (SI 1992/3073) as amended by The Supply of Machinery Safety) (Amendment) Regulations 1994 (SI 1994/2063).

ATEX Directive 94/9/EC, *Equipment intended for use in Potentially Explosive Atmospheres.*

prEN 14986:2005, *Design of fans working in potentially explosive atmospheres.*

ISO 12499:1999 BS 848-5:1999, *Fans for general purposes. Special for mechanical safety (guarding).*

BS 1440:1971, *Endless Vee belt drive sections (withdrawn replaced by BS 3790*

BS 3790:1981, *Specification for endless wedge belt drive sections and endless Vee belt drives (technically equivalent to ISO 155, 254, 1813, 4183, 4184, 5292).*

AMCA 202-98, *Troubleshooting.*

19 Fan economics

Economic optimisation, both for new and existing fan ventilation systems is discussed in this Chapter. Emphasis is given to the need for considering the complete ventilation system, including the secondary costs of carrying out such an optimisation. Usually it is the ducting system cost and the energy cost which are the two most important factors in fan system life cycle costs. Energy consumption is greatly dependent on the flow variation required and upon the choice and methods of control and regulation.

Economic assessment criteria of various types are used to determine the viability of alternative ventilation system proposals. When calculating the profitability of an investment, it is important to consider the rate of increase of energy costs when compared with general inflation, together with the returns on the investment which can be achieved by introducing energy conservation improvements.

Contents:

19.1 Economic optimisation

19.1.1 Introduction

Present and impending legislation, as well as a greener approach from us all, dictate that fans and their systems shall become more efficient. This Chapter is intended to remind fan engineers and their customers of the fundamentals. Without consideration for the fan system and the way in which the flowrate must be controlled (see Chapter 6), it is possible to select fans, transmission motors and controls which will not give the expected power savings. The matching of all these terms is detailed and warrants more than the cursory attention often given.

Despite extensive research alternative fuels and refrigerants we have to agree that global warming is now largely accepted as a probable future threat to mankind. Not all nations (e.g. the USA) or scientists (e.g. in *The Skeptical Environmentalist*) are convinced of the validity of the case being made, but a "safety first" policy suggests that we should all adopt the Kyoto Protocol.

The six strategies outlined in the introduction to Chapter 6, Section 6.1, are of the utmost importance and should be followed.

19.1.2 The efficiency factor

An important characteristic of a fan system, from an economic point of view, is its efficiency factor, i.e. the efficiency with which it converts the absorbed energy, usually electrical energy, into air power. This is especially important, since efficiency factors normally have extremely low values, much lower than usually quoted. Reasons for low efficiency factors are discussed and suggestions for improvement are presented.

The secondary costs for the main process, of which the fan unit is a part, can be considerable if factors such as process adaptation and available capacity are not considered when designing the plant. The most precious form of energy in ventilation is air energy, which warrants attention because of the parameters which not only determine the air power requirements but also the energy consumption.

The basic requirement of every fan installation is to transport air or some other gas. The object of economic optimisation is to enable the air conveyance to be carried out at the lowest possible cost. The actual liquid transportation, however, often constitutes a small part of a much larger process. An evaluation of the air transport costs must, therefore, also include the costs of adaptation to the main process, or if adaptation is not carried out, the resultant costs of a less effective main process.

Since the required air movement cannot be accomplished without the contribution of all components in the ventilation plant, the complete system should be considered when carrying out economic optimisation. Only when the remaining costs are not affected, can partial optimisations give completely accurate results. Partial optimisations do, however, have the advantage that they are easier to evaluate and are therefore applied extensively, despite the risk of certain errors.

The cost which should be minimised is the plant whole-life cost, i.e. the summation of all the costs which occur during the total economic life span of the plant. This begins with planning and drafting and ends when the plant is written off or otherwise disposed of. Whole-life costs are traditionally divided into investment and operating costs.

The investment costs consists of the sum of the following:

- Planning
- Design and draughting
- Equipment
- Installation and commissioning
- Process adaptation
- Staff training
- Disposal, writing off

Operating costs consist of the sum of:

- Maintenance, parts and labour
- Energy consumed.

If the fan and ventilation plant is critical to a process and space is not available for stand-by capacity, the cost of lost production can be considered as a running cost.

The investment costs are basically fixed price in character, i.e. in principle they are independent of the extent to which the plant is used, whereas the operating costs increase with the number of hours of the operation. It should, however, be noted that the investment on a production plant will be very much affected by the size of the "run" and the ability to amortize the cost over a large number.

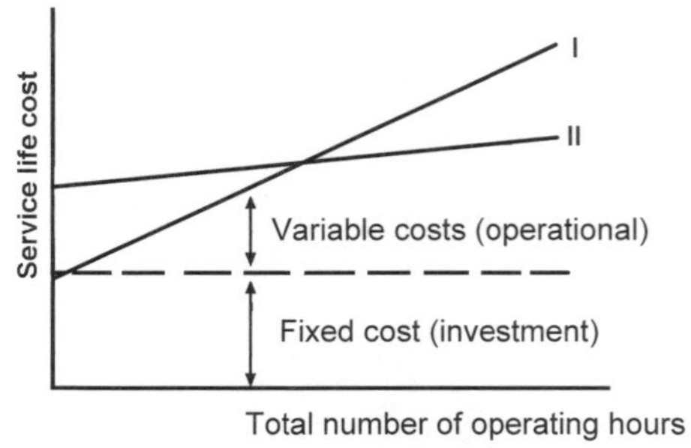

Figure 19.1 Graphical representation of whole-life costs

The two alternatives in Figure 19.1 show the influence which the number of operating hours has in respect of plant whole-life costs. Plant I, which is characterised by its simple construction with low investment costs, gives the lowest total cost despite its higher operating costs, low system efficiency, when the total number of operating hours is low.

For longer operating times, however, the relatively expensive plant II, with its high efficiency, shows the lowest whole-life cost. In practice, the operating costs will not be linear. As the plant ages, efficiency may reduce and maintenance costs and spares usage will increase.

19.1.3 New and existing plant

The design and layout of new plant offers a greater freedom of choice for economic optimisation than is the case for existing plant. For new plant the conditions are more favourable for total optimisation, since all the plant components can be chosen freely.

Normally, economic optimisation is carried out for a number of technically feasible alternatives. The costs for the various items are determined and summate as in Section 19.1.2. The alternative selected is that which fulfils the gas transportation requirements whilst incurring the lowest whole-life costs. The opportunities presented by a new plant should be seized upon to achieve good adaptation to the process. Since the major costs are often associated with the main process, of which gas transportation is only a part, considerable cost reductions can be achieved by suitably designing the ventilation plant.

The following examples are given to explain what is meant by the term "process adaptation".

- The ventilation of hazardous gases; careful selection of the fan size and design of the shaft seals reduces the costs of environmental protection.
- The gentle pneumatic conveyance of food products; reduces product damage and costs.
- The correct choice of low speed fans for abrasive gas-solid mixtures; reduces wear, costs of spare parts and labour costs.
- The scheduled adjustment of fan configuration, adding a stage, changing impeller diameter, changing impeller pitch angle; matches flow/pressure requirements to system changes reducing energy consumption.

Optimisation of existing plant should also consider the costs of process adaptation. Here, however, many parameters are already established which imposes limitations. Optimisation of existing plant assumes therefore, a partial optimisation function. The question then, is whether an extra investment can reduce the operating costs sufficiently to make the improvement profitable.

Examples of modifications proposed for existing plant are:

- Replacement of fan, replacement where possible of impeller with larger diameter, reduce impeller diameter, change pitch angle.
- Change method of regulation or control.
- Ductwork or system modifications to reduce pressure losses.
- The introduction of improved shaft sealing arrangements.

All modifications are introduced with the intention of reducing future maintenance and energy costs. Another type of optimisation occurs when the air movement requirements can be performed by alternative existing fan units.

Such situations appear, for example, in large duct networks, which are supplied by several fan stations. Here, since the fixed costs for the plant cannot be changed, the rule is to choose the fan unit which results in the lowest operational costs. The operational costs in these instances are complicated by differing fan efficiencies and varying system pressure losses due to distance and duct size.

19.2 Economic assessment

19.2.1 Investment calculation - new plant

19.2.1.1 Present capitalised value method

The ventilation plant whole-life cost K_L can be calculated as the sum of the investment cost K_I and the capitalised operational costs K_{Dcap}. The operational costs must be paid annually and therefore be summated to a capitalised present value. This is carried out with the aid of a present capitalised value factor F.

The whole-life cost becomes:

$$K_L = K_I + K_{Dcap} = K_I + F \cdot K_D \qquad \text{Equ 19.1}$$

where

K_L = whole-life cost (currency)

K_I = investment cost (currency)

K_{Dcap} = capitalised operational cost (currency)

F = present capitalised value factor

K_D = annual operating costs (currency/annum)

The present capitalised value factor is calculated from the relationship:

$$F = \frac{1 - (1 + r)^{-n}}{r} \qquad \text{Equ 19.2}$$

where

F = present capitalised value factor

r = estimated annual profits (decimal)

n = number of years

Example:

Two alternative designs of ventilation plant have been proposed. Their cost review is as follows:

Investment I = £10000 annual operational costs = £1000

Investment II= £7500 annual operational costs = £1500

Which plant is more suitable from an economic point of view, if the estimated profits are 15% and the service life is 10 years?

If r = 15% and n = 10 years, the present capitalised value factor F = 5, from Figure 19.2.

Thus:

I K_L = £10000 + 5 × 1000 = £15000

II K_L = £7500 + 5 × 1500 = £15000

Both plants are similar in this particular instance. If it is probable that the service life will not exceed 10 years, then alternative II would normally be preferred since it requires a smaller initial capital investment.

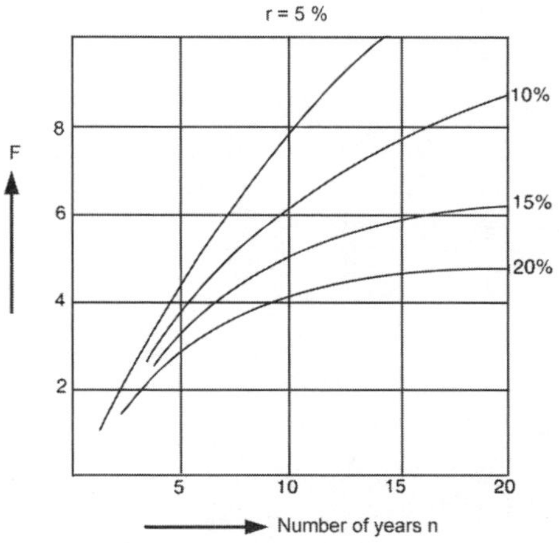

Figure 19.2 Present capitalised value factor

19.2.1.2 Annuity method

For the annuity method the annual plant costs are calculated and minimised. The annual costs consist of the sum of the annuity interest and repayments for a proposed loan which covers the investment cost and the annual operational costs.

$$K = A + K_D = a_F \cdot K_I + K_D \qquad \text{Equ 19.3}$$

where

K = annual cost (currency/annum)

A = annuity (currency/annum)

K_D = annual operating costs (currency/annum)

a_F = annuity factor (decimal)

K_I = investment (currency)

The annuity factor is identical to the reciprocal of the present capitalised value factor

$$a_F = \frac{I}{F} \qquad \text{Equ 19.4}$$

and is determined by service life and estimated profit. Applying this method, for the two alternatives in the previous example:

$$\text{I} \quad K = \frac{10000}{5} + 1000 = £3000$$

$$\text{II} \quad K = \frac{7500}{5} + 1500 = £3000$$

The two alternatives are equivalent, as before.

19.2.2 Investment calculation - existing plant

19.2.2.1 Present capitalised value method

For existing plant the question is often asked if an improvement of the plant can reduce the operational costs in such a way as to reduce the whole-life costs. According to the present capitalised value method the summation of the savings during the service life will be expressed at today's current value.

$$B_L = K_{Dcap} - K_I = F \cdot K_D - K_I \qquad \text{Equ 19.5}$$

where

- B_L = total savings during working life (currency)
- K_{Dcap} = capitalised reduction of operating costs (currency)
- F = present capitalised value factor
- K_D = annual operating cost reduction (currency/annum)
- K_I = investment cost (currency)

Example:

By investing £3,000 it is possible to reduce the annual costs for air movement by £1,000. The plant is designed to operate for 15 years and the estimated profits to be a minimum of 15%. How great are the savings during the service life of the plant at today's currency value?

According to Figure 19.2 the present capitalised value factor F = 5.9 for n = 15 years and r = 15%.

$$B_L = 5.9 \cdot 1000 - 3000 = £2900$$

Over and above the repayment of investment, including estimated profits, an extra £2900 is obtained.

If several alternatives are available to achieve a similar technical improvement, then it is normal to choose the alternative which produces the greatest savings.

19.2.2.2 Annuity method

The question as to how great the first year's saving will be at today's currency value is answered by using the annuity method.

The power used depends upon the useful air power required to maintain the specified flowrate in the ducting system and upon the efficiency of converting the electrical power (or other equivalent) into this useful air power.

This efficiency is usually called the overall efficiency and is defined as:

$$\eta_O = \frac{P_A}{P} = \eta_F \cdot \eta_m \cdot \eta_T \cdot \eta_c \cdot \eta_A \qquad \text{Equ 19.6}$$

where

- η_O = overall fan efficiency factor (decimal)
- P_A = required air power (kW)
- P = power absorbed (kW)
- η_F = fan efficiency (decimal)
- η_T = transmission efficiency (decimal)
- η_m = motor efficiency (decimal)
- η_C = control efficiency (decimal)
- η_A = other ancillary efficiencies

The flow control illustrated in Figure 19.3 takes into account the power losses which are caused as a direct result of the method of flow regulation.

$$P = q_v \times p_F \quad \cdot \eta_O$$

$$= \frac{m^3/s \times Pa}{\cdot \eta_O \times 1000} \qquad \text{Equ 19.7}$$

where

- q_v = flowrate (m^3/s)
- p_F = system resistance (Pa)

or

$$P = \frac{m^3/s \ x \ kPa}{\eta_F \cdot \eta_m \cdot \eta_T \cdot \eta_c \cdot \eta_A} = \frac{m^3/s \ x \ kPa}{\cdot \eta_O} \qquad \text{Equ 19.8}$$

It will be noted that if P is quoted in Watts then resistance p_F should be quoted in Pascals Pa. If resistance is given in kilo-Pascals kPa then the power will be given in kiloWatts kW.

Flow regulation efficiency illustrated in Figure 19.3, takes into account the power losses which result directly from the method chosen to regulate the flow. Such power losses are caused by damper regulation in the discharge duct, by-pass duct or on-off control.

The expression for regulation efficiency for on-off control consists of the quotient of energy requirements. Whereas the others are represented by the quotient of power requirements.

The fan efficiency will obviously be determined for the actual operating point and not the rated duty point efficiency, maximum flow point or any other arbitrary operational point. The

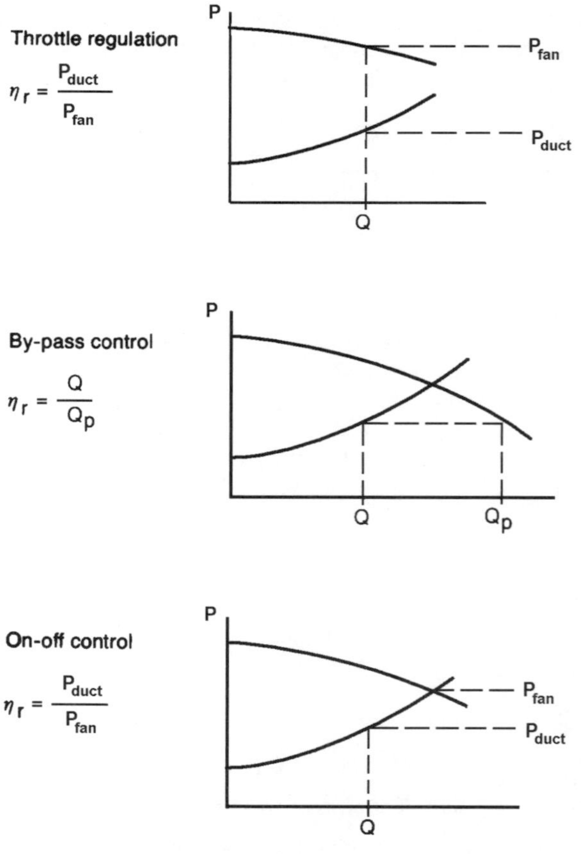

Figure 19.3 Illustration of flow regulation efficiency

power absorbed by any auxiliary systems, such as lubrication oil, seal oil or cooling water must be included in the overall fan efficiency.

The transmission efficiency represents losses in gears, couplings, vee belt drives and speed variators. It is important to consider the transmission efficiency when speed regulation is used. Losses in rectifiers, frequency inverters, regulation resistance, additional motor losses, etc., i.e. losses when utilising electrical methods of speed regulation, are traditionally calculated as transmission losses.

The motor efficiency should be taken at the actual operating point. A common misconception is to assume that the motor efficiency is still high even at part load. At part load the percentage of reactive power is increased and this may be costed as a separate item. It is important to remember about the power absorbed by separately driven cooling fans; this is an integral part of the motor efficiency.

The efficiency of other components takes into consideration losses for phase compensation and protection against supply disturbances, power requirements for supplementary ventilation and other environmental power consuming arrangements which can be directly associated with energy conversion in the ventilation plant when working.

It is extremely important to remember that it is the overall fan efficiency factor and not the efficiency of isolated components which is the characteristic unit for energy conversion efficiency. If all the component efficiencies are high at the actual operating point then a high overall fan efficiency factor will obviously be obtained. However, it only requires one of the component efficiencies to be low to cause a poor overall efficiency.

Example:

Determine the overall fan efficiency factor for a flow of 70% of maximum flow for the installation shown in Figure 19.4, for (I) speed regulation and (II) throttle regulation.

Using estimated component efficiency values:

Case I		**Case II**	
η_F	= 0.75 (estimated)	η_F	= 0.75 (estimated)
η_T	= 0.75 (estimated)	η_T	= 1 (estimated)
η_m	= 0.90 (estimated)	η_m	= 0.90 (estimated)
η_c	= 1 (not included)	η_c	= 1 (not included)
η_O	= 0.51	η_O	= 0.30

The efficiency factor for on-off control becomes evident, if it is assumed, for example, that the fan operates at full flow, 100% flow, for 75 % of the time. During 30% of the time the fan is then at rest. In this case higher duct losses are obtained during the time the fan operates and the efficiency factor becomes:

$$\eta_O = 0.75 \times 1 \times 0.90 \times 0.60 = 0.41$$

For multi-fan systems the efficiency factor can be determined in a similar way. Here again, it is only the total efficiency for the actual flow and not the maximum efficiency of individual components which is characteristic for energy conversion efficiency. For on-off control of multi-fan systems there are a number of significant operating points. The efficiencies for these flows are greatly influenced by the sizes and designs of the fans selected.

An example of this is shown in Figure 19.5. In the case of three identical fans having maximum efficiency at Q_{max}, poor part load efficiencies, shown as circles on the graph, are obtained. By choosing larger fans, with η_{max} at a somewhat higher Q value, better part load efficiencies can be obtained, shown as crosses on the graph. Only by choosing three different sizes of fan can the highest efficiency be achieved for all three operating points. Three different fans, however, involve the stocking of additional spare parts and more expensive maintenance, and the fans cannot be used as stand-bys for each other.

The basic fan efficiency is very important. Fan efficiency is greatly influenced by size, design and choice of fan. Information on fan efficiencies is given in Chapter 1, but it is also worth again noting that efficiency increases with fan size in a homogenous series.

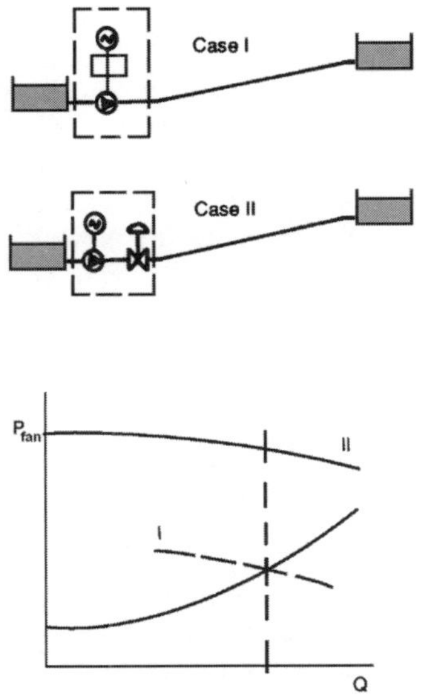

Figure 19.4 Illustration of overall fan efficiency factor

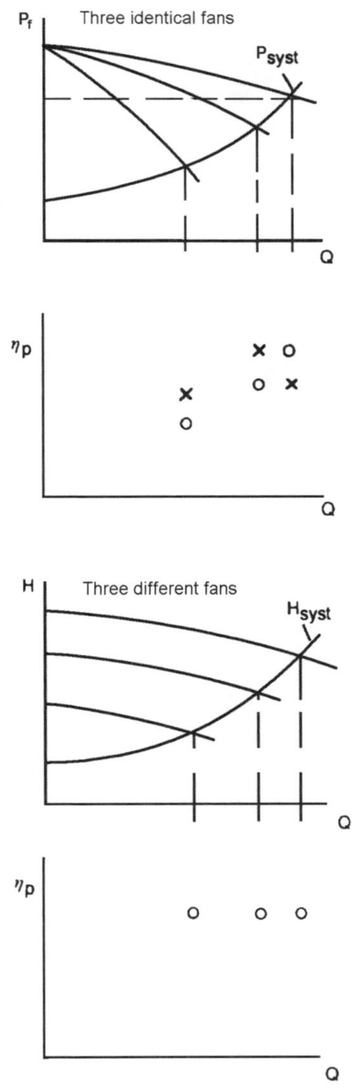

Figure 19.5 Example of overall fan efficiency for three fans connected in parallel

19.2.2.3 Pay-off method

The pay-off method uses an imaginary "repayment time" defined by the relationship

$$T_p = \frac{K_I}{K_D} \qquad \text{Equ 19.9}$$

where:

T_p = pay-off time (decimal)

K_I = investment cost (currency)

K_D = annual operating cost reduction (currency/annum)

The shorter the pay-off time the more profitable the investment. By comparing with equation 19.5 it is found that for $B_L = 0$ then $T_p = F$, i.e. the pay-off time and present capitalised value factor have the same numerical value when the total saving is equal to zero.

19.2.2.4 Investment grant

Grants for energy saving investment may apply. For up to date information it is best to seek the advice of the relevant local or central government authority.

Whilst writing this book, the most relevant schemes would seem to be:

Enhanced Capital Allowances (ECA). ECAs enable a business to claim 100% first-year capital allowances on their spending on qualifying plant and machinery. There are three schemes for ECAs:

- Energy-saving plant and machinery
- Low carbon dioxide emission cars and natural gas and hydrogen refuelling infrastructure
- Water conservation plant and machinery

Businesses can write off the whole of the capital cost of their investment in these technologies against their taxable profits of the period during which they make the investment.

This can deliver a helpful cash flow boost and a shortened payback period.

The Market Transformation Programme (MTP) is a DEFRA initiative that develops policy strategies for improving the resource efficiency of traded goods and services in the UK.

The MTP quantifies current thinking on how the daily use of products, systems and services impacts on the environment. MTP uses market projections and policy scenarios to explore alternative future developments.

The Energy Saving Trust (EST) was set up by the UK Government following the 1992 Rio Earth Summit and is one of the UK's leading organisations addressing the damaging effects of climate change. The Energy Saving Trust's goal is to achieve the sustainable and efficient use of energy, and to cut carbon dioxide emissions, one of the key contributors to climate change. The Energy Saving Trust is a non-profit organisation funded by the Government and the private sector.

19.2.3 Estimated profits and service life

19.2.3.1 Estimated profits

Profit estimates should, in principle, correspond to the interest on capital which would otherwise be realised from an alternative investment. It is a measure of a company's profitability and is higher than the current bank rate. The profit estimate increases with reduced capital resources, since it is necessary to be more particular when investment capital is limited. Profit estimates are rarely considered at less than 15%.

The methods of calculation previously reviewed assume that the annual operating cost reductions are of the same magnitude from year to year. The majority and largest savings on the operational side are achieved by reducing energy consumption. Energy costs, currency per kWh, can also be expected to rise more quickly than other costs, which mean that energy savings will become more profitable with time.

One way of considering energy cost increase when making economic calculations is to use corrected profit estimation.

$$r_k = r - e + i \qquad \text{Equ 19.10}$$

where:

r_k = corrected profit estimate (percentage)

r = uncorrected profit estimate (percentage)

e = rate of increase of energy (percentage)

i = general inflation rate (percentage)

A more rapid rate of increase of energy costs can in this way be transferred to a reduced profit estimation requirement for energy saving investment.

19.2.3.2 Service life

The economic service life is determined by factors such as write-off rules, the technical service life of components and the planned period of use of the plant. As with other parameters for fans, the economic service life is dependent upon the size and type of industry.

As an approximation with the exception of small plant the following applies:

Buildings	40 years
Ducting, underground	50 years
Other ducting	20 years
Machines	15 years
Control equipment	10 years
Instrumentation	10 years

The service life of control equipment and instrumentation may be taken as 10 years for financial planning. The rate of change within the electronics industry is very rapid. Control systems and instrumentation should be reviewed every 5 years using the "Present capitalised value" method, to see if improved equipment or improved control strategies could reduce existing operating costs and hence reduce whole-life costs.

19.5.4 Energy costs

Energy costs depend upon the amount of energy consumed the prevailing energy price scale and fixed costs for the supply installation. Premiums may be levied if the "maximum demand" is exceeded. In the most usual cases of electric motor operation, energy prices are determined by the relevant electricity tariff.

Tariffs

The basis of most forms of tariff is a fixed charge dependent upon the "maximum demand" taken by the consumer and designed to cover:

- the costs dependent upon maximum demand, e.g. interest and depreciation of generating plant, rates, taxes, insurance, salaries,
- the costs incurred for each consumer, e.g. transformers, meters, meter reading labour, service cabling. And a running charge depending on the energy supplied e.g. fuel, losses and maintenance of the supply plant, equipment, etc.

For industrial consumption two-part tariffs are usual with the fixed charge proportional to maximum kW or maximum kVA demand; and a running cost kWh, which may be dependent upon the time of day and/or year, i.e. peak and off-peak periods, and also include for example, a fuel cost variation clause. A kVA maximum demand is preferable since it takes into account the effect of low power factor. It involves, however, more expensive metering equipment. The cost of metering maximum demand makes it uneconomic, in any case, for loads of less that 20 to 50 kW.

If supplies are taken at a high voltage instead of the usual 415V/380V for distribution, the maximum demand charge may be less since the consumer then has the option to provide his own transformer.

Annual energy cost

The energy costs are a product of energy consumption and cost per unit. The annual energy cost will therefore be:

$$K_E = k_e \cdot E \qquad \text{Equ 19.11}$$

where:

K_E	=	annual energy cost (currency)
k_e	=	energy cost (currency per unit)
E	=	annual energy consumption (units)

Or capitalised for the service life of the fan station.

$$K_{Ecap} = F \cdot k_e \cdot E \qquad \text{Equ 19.12}$$

19.3 Important system characteristics

19.3.1 Introduction

It is important to remember that the fan and its system have to be in balance. Thus if the system resistance has been over or under estimated, the absorbed fan power will vary according to its type. This in turn will affect the losses in the transmission motor and controls.

19.3.2 Overall fan efficiency

The absorbed fan power is only a start to understanding how to calculate the energy consumption of a fan. An important step is to calculate the input power kW at a specific duty. Multiplying this input power by the hours run at the duty will determine the kWh.

Thus input

$$kW = \frac{Q \times p_F}{\eta_F \times \eta_R \times \eta_M \times \eta_C} \qquad \text{Equ 19.13}$$

where:

Q	=	flowrate (m^3/s)
p_F	=	fan pressure (kPa)
η_F	=	fan efficiency expressed as a decimal
η_T	=	transmission efficiency at duty expressed as a decimal
η_M	=	motor efficiency at absorbed power expressed as a decimal
η_C	=	control efficiency at absorbed power expressed as a decimal

p_F and η_F may be either both total or both static.

19.3.3 Demand variations

The primary cause of a low overall fan efficiency factor is due to the variation, with respect to time, of the desired flow through the ducts, due to insufficient consideration having been given to this aspect at the design stage. Normally the mean flow per annum — total supplied volume of air divided by a calendar year — is about 10% to 15% of the installed flow capacity Q_{max}. The efficiency factor calculated as a mean efficiency over the year — theoretical air energy consumption/actual electrical energy consumption — usually gives a value in the region 5% to 40%, which should then be compared with the normal maximum momentary value of 55% to 75%.

The reasons for production or demand variations and the low mean flow are many and typical examples are:

- Seasonal variations in industry production
- Variations in ambient temperature
- Variations in ambient humidity
- Variations in building occupancy
- Variations in heating loads
- Component and calculation tolerances
- Margins of safety over and above the normal tolerances

A survey of flow variations with respect to time is a necessary basic requirement for all economic optimisations of air transportation. It is desirable that both progressive curves and constancy diagrams are produced.

Modifications for the improvement of the efficiency factor and hence the costs can be:

- To improve the part load efficiency by means of suitable flow control and regulation equipment.
- To reduce extreme flow peaks.

19.3.4 Availability

Availability is generally defined by the relationship:

$$A = \frac{MTBF}{MTBF + MTTR} \qquad \text{Equ 19.14}$$

where

A	=	availability (decimal)
MTBF	=	Mean Time Between Failures (hours)
MTTR	=	Mean Time To Restore (hours)

The MTBF should be the summation of operating hours per annum. The MTTR should be the summation of all repair times per annum. MTTR should include all those hours of routine maintenance or adjustment when the fan must be stopped.

Another concept frequently overlooked, MTBA, Mean Time Between Adjustments, includes maintenance work performed while the fan is running. MTBA does not affect availability but it does affect running costs.

Centrifugal fans used within the process industry, for well-tried, normal operating conditions, achieve values of availability in the order of A = 0.99 to 0.9999, i.e. shut-down due to malfunction is between 1 and 80 hours per annum for continuous operation. These values also include the squirrel cage induction motor but not other electrical equipment. At the other extreme some high pressure conveying fans, have an availability of 0.88 on a daily basis. Some fans must have the seals replaced frequently.

Availability calculations are applied mainly to multi-fan systems. The percentage of operating time which a given number of fans in a fan station can be expected to be available for operation, is calculated from the formula:

$$P = \frac{N!}{K!(N-K)!}(1-A)^K \cdot A^{(N-K)}$$ Equ 19.15

where

P = system availability (decimal)

N = number of fans installed (integer)

K = number of fans unavailable (integer)

A = availability of individual fans (decimal)

! = factorial (4!=24, 0!=1)

All the fans are thus assumed to have the same availability values. By using equation 19.15 the resultant availability of a fan station can be calculated and the effect of a stand-by fan can be demonstrated.

Without reserve fan		With reserve fan	
Function requirement	Resultant availability	Function requirement	Resultant availability
1 of 1	0.9900	1 of 2	0.9999
2 of 2	0.9801	2 of 3	0.9997
3 of 3	0.9703	3 of 4	0.9994
4 of 4	0.9606	4 of 5	0.9990
5 of 5	0.9510	5 of 6	0.9985

Table 19.4 Availability of fan station with and without stand-by fan

From Table 19.4 it can be seen that a fan station's availability is considerably improved by the installation of a stand-by fan

Function requirement 3 of 4 means that 3 out of 4 will be available for operation. The individual fan availability has been taken as 0.99.

The user must define the availability required at the inquiry stage, as this parameter has a crucial effect on fan selection. If the user wishes to operate for 50 weeks every year the availability would have to be 0.959. If the user wishes to operate for 154 weeks and allow two weeks for repair the availability would have to be 0.987.

For this type of availability calculation to apply in practice it is necessary to maintain the stand-by fan to ensure that it will operate when required. In cases where it is possible to calculate the cost of a non-functioning fan, for example, in the form of lost production, availability calculations offer a direct economic optimisation possibility. Stand-by fans usually run a limited number of hours regularly, one week in four say, to ensure the fan is functioning correctly.

19.3.5 Air power

19.3.5.1 General

The useful air power from the fan is used to satisfy the ducting system (and terminal loads) resistance at the actual flow.

$$P_f = q_v \cdot p_f$$ Equ 19.16

where

P_f = power absorbed by fan impeller W or kW

q_v = volumetric flowrate m^3/s

p_f = fan pressure Pa or kPa

If fan pressure is given in Pa (pascals) then power will be given in W (Watts).

If fan pressure is given in kPa (kilopascals) then power will be in kW (kilowatts).

To achieve an optimal economic result , it is important the air power is kept as low as possible. Limiting the required air power is well worthwhile, especially with overall fan efficiency factors of as low as 5% to 40%. A reduction of P_f by 1 kW can reduce the electrical power consumption by as much as 2.5 kW to 20 kW because of the low overall fan system efficiency in the conversion of electricity into air power.

19.3.5.2 Duct pressure losses

The approximate loss of pressure in straight ducting, assuming fully turbulent flow and smooth materials, can be defined as:

$$p_L = \frac{fL}{m} \cdot \frac{1}{2}\rho v^2$$ Equ 19.17

where

f = friction factor

L = length of duct (m)

m = mean hydraulic depth (m)

$= \frac{\text{area of duct cross section } (m^2)}{\text{periphery of duct (m)}}$

ρ = air density ($kg.m^{-3}$)

v = mean duct air velocity (m^2)

Recognising that

q_v volumetric flowrate = mean velocity v x cross-section area,

then for a circular cross section duct

$$m = \frac{\pi d^2}{4} \cdot \frac{1}{\pi d} = \frac{d}{4}$$

and

$$v = \frac{q_v}{Area} = \frac{q_v}{\pi d^2} \cdot 4$$

or

$$p_L = \frac{4fL}{d} \cdot \frac{1}{2}\rho \cdot \frac{q_v^2}{\pi^2 d^4} \cdot 16$$

i.e.

$$p_L = \frac{32 fL\rho q_v^2}{\pi^2 d^5}$$ Equ 19.18

If American values of f are used (i.e. based on duct diameter) then the constant becomes 8 instead of 32, but the value is still proportional to d^{-5}.

Thus for an increase in duct diameter of only 10% then the resistance, and therefore the air power absorbed, has been reduced to $\frac{1}{1.1^5}$ x previous value i.e. 62%.

Considering now a rectangular duct, whose sides have a 2:1 ration, handling the same air flow and having the same metal content and periphery (when the price may be approximately the same) then:

$$p_L = \frac{fL}{m} \cdot \frac{1}{2}\rho v^2$$

$$= \frac{fL}{2a^2} \cdot 6a \cdot \frac{1}{2}\rho \frac{q_v^2}{4a^4}$$

$$= \frac{3fL\rho q_v^2}{32a^5}$$

but

$$\pi d = 6a \text{ or } a = \frac{\pi d}{6}$$

and

$$p_L = \frac{3fL\rho}{32\pi^5 d^5} \cdot q_v^{\ 2} \cdot 6^5$$

$$= \frac{fL\rho}{d^5} \times q_v^{\ 2} \times 2.3822$$

This should be compared with the result for circular ducting where:

$$p_L = \frac{8\ fL\rho q_v^{\ 2}}{\pi^2 d^5}$$

$$= \frac{fL\rho}{d^5} \times q_v^{\ 2} \times 0.8106$$

i.e. the pressure loss and power absorbed in 2:1 rectangular ducting of the same metal content and the same flowrate will be 2.94 times as high.

This emphasises the need for ensuring the maximum diameter of ducting consistent with space and money available, and the desirability of using circular cross-sections at all times.

Duct pressure losses are not only due to the friction in straight ducting, but also due to the effects of bends, transitions, take-offs, diffusers, etc. For further information see Chapter 3.

It should especially noted that the quality of manufacture has an important effect on pressure losses due to gross turbulence at poorly aligned connections.

Many German researchers have noted a Reynolds number dependence on the factor k where pressure loss

$$p_{L\cdot fitting} = k \cdot \frac{1}{2}\rho v^2 \qquad \text{Equ 19.19}$$

Thus, despite the assumed constancy in many textbooks k may be expected to rise at low velocities. Nevertheless, it is safe to say that:

$$p_{L\cdot fitting} = \frac{constant}{d^4} \qquad \text{Equ 19.20}$$

Where there are large flow variations down almost to zero, we will inevitably enter a laminar flow region when

$p_L \cdot$straight varies as d^{-4}

and

$p_L \cdot$ fitting varies as d^{-3}

For fan installations we can say that

$$P_{fan} = constant \times q_v^{\ 3}$$

The installed air power, which is required in a system with mean flow q_{vm} and demand/production variation of $\pm\ \Delta q_{vm}$ is:

$$P_{fan} = constant\ (q_{vm} + \Delta q_v)^3$$

$$= constant \cdot q_{vm}^{\ 3}\left(1 + \frac{\Delta q_v}{q_{vm}}\right)^3$$

or

$$P_{fan} = P_{fan\cdot mean}\left(+ \frac{\Delta q_v}{q_{vm}}\right)^3 \qquad \text{Equ 19.21}$$

The energy consumption due to the fan and duct system alone becomes dependent upon the shape of the flow demand/supply curve with time and is illustrated in Figure 19.6 for three typical cases.

The energy consumption due to the fan and ducting system, but excluding the effects of motor, transmission and control losses becomes:

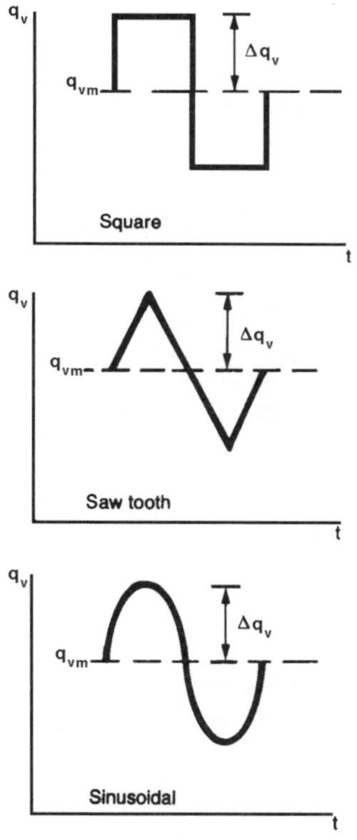

Figure 19.6 Three special cases of demand/supply variations

$$E_f = \left[1 + 3\left(\frac{\Delta q_v}{q_{vm}}\right)^2\right] P_{fm} \cdot t \qquad \text{Square}$$

$$E_f = \left[+ \left(\frac{\Delta q_v}{q_{vm}}\right)^2\right] P_{fm} \cdot t \qquad \text{Saw tooth}$$

$$E_f = \left[+ \frac{3}{2}\left(\frac{\Delta q_v}{q_{vm}}\right)\right] P_{fm} \cdot t \qquad \text{Sinusoidal} \qquad \text{Equ 19.22}$$

ΔQ/Qm	P_f/P_{fm}	E_f/P_{fm}		
		Square	Saw tooth	Sinusoidal
0	1	1	1	1
0.1	1.33	1.03	1.01	1.02
0.2	1.73	1.12	1.04	1.08
0.5	3.37	1.75	1.25	1.38

Table 19.5 Installed air power and energy consumption for different types of demand/supply variation

Table 19.5 shows that the demand/supply variations have a considerable influence upon both the installed air power and the energy consumption. From the economic point of view, reducing the flow peaks and troughs can give good returns.

19.4 Partial optimisation

19.4.1 Economic duct diameter

The two major costs for many fan plants are the costs of ducting on the investment side and the energy costs on the operational side. The investment costs increase with increased duct diameter whilst at the same time operational costs decrease, see Figure 19.7.

The total costs curve represents a minimum for a certain duct diameter. As indicated, the shape of the total costs curve is

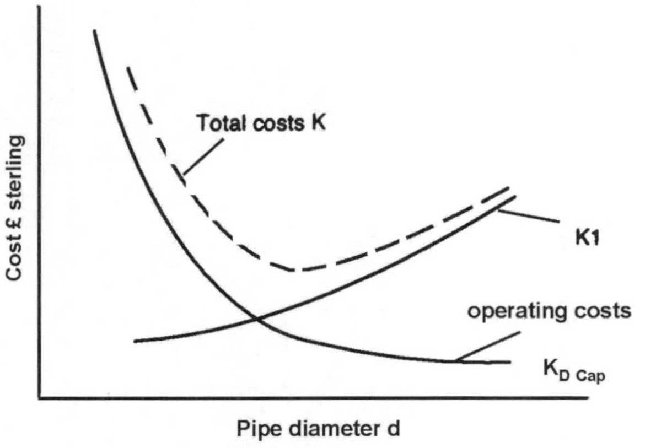

Figure 19.7 Effects on costs of duct diameter

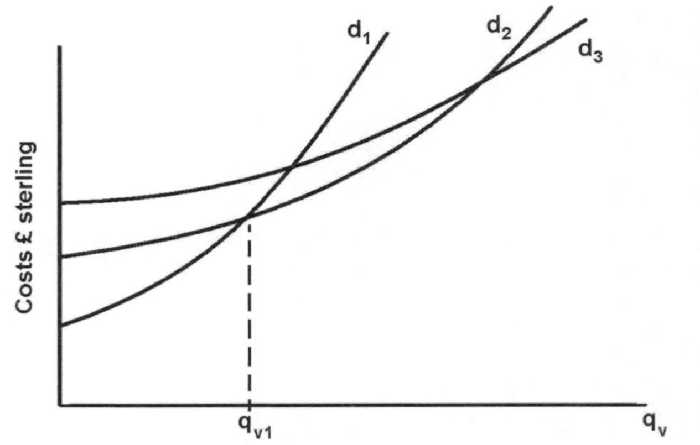

Figure 19.8 Annual costs for ventilation plant using various duct diameters

such that it rises more rapidly with reduced diameter, to the left of the economic diameter, than in cases of increased diameter, to the right of the optimum value. In cases of doubt choose the one having the largest duct diameter from the feasible alternatives.

By calculating the total costs for a number of different duct diameters it is possible to establish the most economic duct diameter. The economic diameter varies for different situations. Short operating periods and costly duct sections, for example, stainless steel, tend to reduce the economic duct diameter and to increase the economic flow velocity.

By carrying out a partial optimisation, i.e. if it is assumed that all costs are independent of the duct diameter except for the costs of the ducting and energy, the following relationship is obtained:

$$q_{v1} = \sqrt[3]{\frac{k_2 - k_1}{1/d_1^5 - 1/d_2^5} \cdot \frac{\eta_O \cdot \pi^2 \cdot a_F \cdot 10^3}{32 \cdot \rho \cdot f \cdot t \cdot k_e}} \qquad \text{Equ 19.23}$$

where

q_{v1} = the mean flow for which duct diameters d_1 and d_2 give the same annual cost (m^3/s)

k = cost/m of ducting (currency/m)

d = duct diameter (m)

η_O = fan system overall efficiency factor (decimal)

a_F = annuity factor (decimal)

ρ = liquid density (kg/m^3)

f = pipeline loss coefficient

t = operating time per year(hour/annum)

k_e = cost of energy (currency/kWh)

For this q_{v1} value the total energy costs are the same. If the mean annual flow for the plant is greater than q_v then it is economically feasible to choose duct diameter d_2. If q_{v1} is much greater than q_v then the procedure must be repeated using diameters d_2 and d_3.

The procedure for dampers and other fittings is carried out in a similar manner.

$$q_{v1} = \sqrt[3]{\frac{k_2 - k_1}{1/d_1^4 - 1/d_2^4} \cdot \frac{\eta_O \cdot \pi^2 \cdot a_F \cdot 10^3}{32 \cdot \rho \cdot P_m \cdot t \cdot k_e}} \qquad \text{Equ 19.24}$$

where

k = damper cost including connectors for duct diameter d.

It should be noted that an improved fan efficiency factor tends to reduce the economic duct diameter. (See Figure 19.8.)

19.4.2 Component efficiency

When purchasing new equipment, it is often possible to choose between a cheap fan having low efficiency and an expensive one with high efficiency. The comparative costs for the two alternatives are:

$$P_{1m} \cdot t \cdot k_e \cdot F + K_{I1} \Leftrightarrow P_{2m} \cdot t \cdot k_e \cdot F + K_{I2} \qquad \text{Equ 19.25}$$

where

P_m = consumed mean electrical power (kW)

t = operating hours per year (hours/annum)

k_e = energy cost (currency/kWh)

F = present capitalised value factor

K_I = investment cost (currency)

It is important to remember that the energy costs are debited according to readings taken from the electricity meter and not from the fan shaft. Because of cable losses, the electric power used is always greater than the fan shaft power. The capitalised value of 1 kW mean electric power is $t \cdot k_e \cdot F$. (See Figure 19.9.)

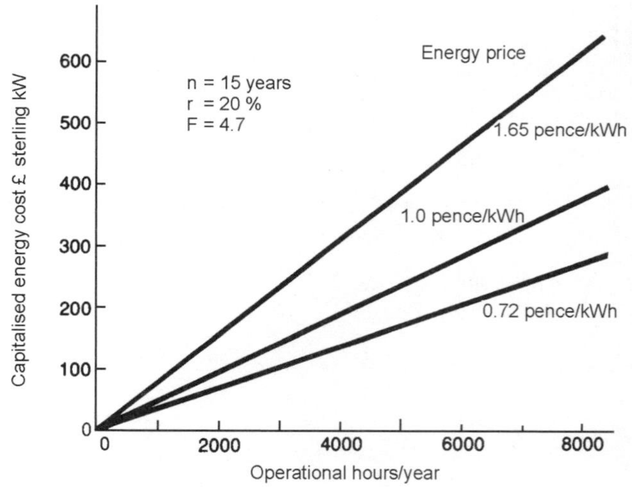

Figure 19.9 Capitalised value in £s sterling/kWh for a reduction of the mean annual electric power used

The efficiency of other components can be tested in the same way. Conversion to electrical power used must always be made, however. The closer the power consumption lies to the useful air power in the conversion chain, the greater the energy saving and the greater the motivation for additional investment. This is illustrated typically in Figure 19.10.

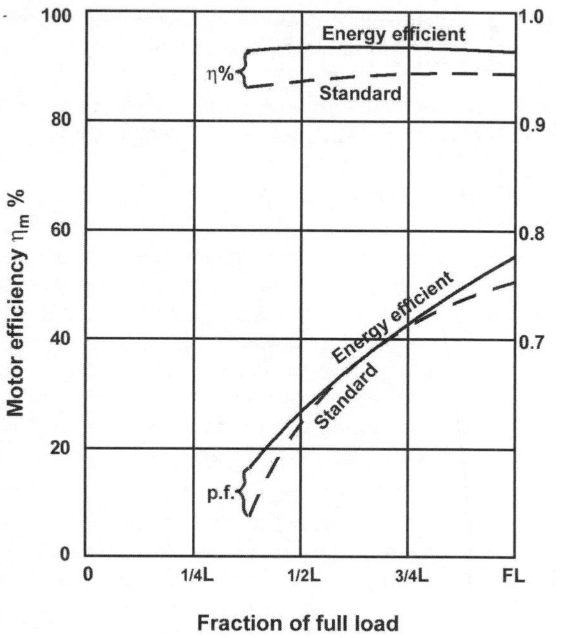

Figure 19.10 Typical efficiency and power factor values for a 37 kW 4 pole motor

19.4.3 Existing plant

In the case of existing plant there are certain limitations to energy saving possibilities. Changes to the layout and ducts are often difficult to make. Possible improvements are replacements or the addition of supplementary components which do not require too great a disturbance to the plant.

All reductions of the air power, by means of improvements which reduce the system resistance in the case of a regulated plant, only result in a greater pressure loss across the damper and are thus worthless from an energy conservation point of view. To realise a saving of energy some form of improvement must also be carried out on the fan side. For speed-regulated plant all pressure reductions are automatically used.

When planning a project at the time of determining fan specifications, it is often the case that not all of the plant details are known. It is not therefore unnatural to choose fan sizes with a certain safety margin "to be on the safe side". For this reason many fans are oversized and result, if adjustments are not made, in unnecessarily high energy costs.

In the case of basic oversizing, the following alterations to the fan may be considered:

For centrifugal fans:

- a new impeller with reduced diameter
- a new impeller with reduced width
- a new impeller with reduce blade angles

For axial flow fans:

- a new or existing impeller with reduced pitch angle
- a new or existing impeller with a reduced number of blades
- fitting a slower running motor

For indirect drive fans:

- fitting a smaller motor
- fitting a slower vee rope drive
- changing the gear ratio of a gearbox

It may also be advantageous to review the reasons for the ventilation system and what it is supposed to do. In a drying plant, for example, it may be possible to change operating procedures to reduce the demand for flow variation.

Attention should always be given to energy efficient methods of control such as speed regulation for all types of fan, variable pitch in motion for axial flow fans and radial inlet vane control for centrifugal fans. The advantages and disadvantages of each are given in Chapter 6.

19.5 Other considerations in fixed output systems

19.5.1 General

Almost without exception, both fans and motors can be made more efficient in large sizes. Where a number of units have to run continuously, consideration should be given to providing a manifold or plenum to couple up the duct systems and use one common fan/motor unit. Even allowing for the additional duct resistance, power savings can often be made. Tables 19.6 and 19.7 are typical but show the improvements possible. They show how motor efficiency increases with increasing motor frame size whilst fan efficiency also increases with fan size due to "scale" effects.

Frame size	Motor kW	Rev/min	Efficiency %	Power factor
63	0.25	2800	67	0.83
71	0.55	2800	71	0.85
80	1.1	2800	76	0.85
90L	2.2	2820	78	0.88
132S	5.5	1440	85	0.83
160M	11.0	1440	88	0.85
200L	22.0	970	90	0.80

Table 19.6 Typical motor efficiencies at rated output

Impeller diameter mm	Rev/min	Peak fan static efficiency %	Relative tip speed factor
630	2880	73.8	1.000
800	2270	74	0.995
1000	1815	75	0.992
1250	1440	77	0.989
1600	1125	78	0.987

Table 19.7 Typical peak fan efficiencies of varying fan sizes

19.5.2 Fixed speed motors

These drives are almost invariably from three phase induction motors. They should be matched to the fan to absorb as near nameplate load as possible, for both efficiency and power factor are likely to fall at reduced load. Multi-pole slow speed motors can often be used to eliminate mechanical reductions. Lower power factor can be compensated with power factor correction devices.

Wherever continuous running is a requirement, the increased efficiency of both larger and slower units is apparent, see Table 19.8. Energy efficient designs can repay their increased first cost many times over. When allied with the power savings from high efficiency fans (often of a much higher magnitude — there are many fans with a 60% peak fan efficiency which could be replaced by 80% efficient units), the effects on overall electrical demand could be enormous.

Perhaps of greater importance, especially where a fan works for considerable periods at reduced load, is the "flatter" motor efficiency curve (Figure 19.11).

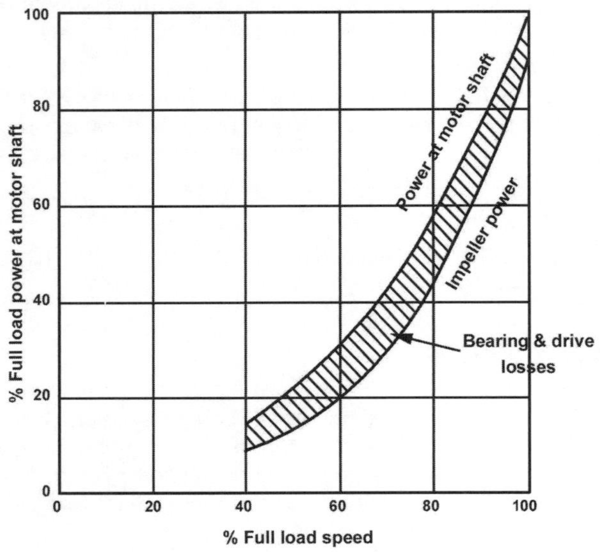

Figure 19.11 Power required at the motor shaft by a vee belt driven fan operating on a constant orifice system

Table 19.8 emphasises how motor efficiency and power factor.at full load may vary with speed for a typical manufacturer's 45 kW motor of standard or "energy efficient design". The increase in fan efficiency with reduced speed could well be larger than that for the motor.

Frame size	Poles	Rev/min	Full load efficiency %		Full load p.f.	
			Std	Energy eff	Std	Energy eff
225M	2	2955	90.0	93.9	0.85	0.91
225M	4	1470	92.0	93.8	0.86	0.87
250M	6	980	92.5	94.0	0.84	0.83
280S	8	735	92.0	94.0	0.79	0.75
280M	10	589	91.0	94.3	0.73	0.75
315S	12	490	90.5	94.8	0.73	0.75

Table 19.8 Efficiency at rated power of typical slower speed motor

19.5.3 Vee belt drives

The major fan manufacturers have families of designs covering a range of specific speeds and diameters. These ensure that all duties can be met by a fan running at a direct drive speed often with the impeller mounted on the motor shaft, thus eliminating the losses in vee rope drives. Greater effort should be made in the accurate determination of system resistance (how many enquiries for fans given an obviously rounded pressure of 500 Pa?). Thus the need for pulley changing will be obviated, a worthwhile reduction in first cost results, and the transmission losses will be eliminated. Where belt drives are inevitable, greater thought should be given to their selection. It should here be noted that if a vee drive is either under or over-engineered, efficiency will reduce as seen in Chapter 11, Section 11.5.7.

Flat and toothed belts do not suffer to the same degree, whilst their peak efficiency is usually higher. With very small belt drives, the difference in power transmitted between say one and two vee belts or between two or three is obviously substantial.

The use of larger single units is indicated, if transmission losses are to be reduced. Especial attention is drawn to the increased losses in vee belt drives when used at lower than their maximum rating on variable speed units. Whilst the fan impeller power on a constant orifice or fixed system may vary as rotational speed, (if we assume fully turbulent flow with a constant duct friction factor – which is unlikely), the losses in the bearings and shaft seals will vary approximately as the speed. Drive efficiency will fall from about 94% at rated full load to around 70% at half load so that the full "cubic" saving is not achieved where vee belts are fitted (Figure 19.11). Again the preference for direct drive fans with variable speed is indicated.

19.5.4 Electric motor design

Whilst a very small number of fans may be driven by prime movers such as steam turbines or petrol engines, the vast majority — in excess of 98 % — are driven by electric motors. With axial flow fans, it is common for the fan impeller to be mounted directly on the motor shaft extension. Centrifugal fans, may, of course, be vee belt drive, directly driven or through a flexible coupling with or without an intermediate gearbox (common in the UK on large mine ventilation fans).

Again, with the majority of fans, electric motors are of the totally enclosed squirrel cage induction form suitable for a three-phase supply. Single-phase motors are usually limited to fractional horsepower outputs. See Chapter 13 for a fuller description.

The induction motor is extremely reliable and robust. In nearly all cases it may be considered symmetrical both mechanically and electrically. The windings are balanced between phases and slots. Care is taken to ensure that the rotor runs in the correct position axially within the stator field, and that the air gap between the rotor and stator is the same at all axial and radial positions. However, especially with direct driven fans, there will be an end thrust due to the impeller action and this will "try" to take the rotor out of the magnetic field being resisted by the magnetic forces and also such devices as wave washers in the bearing housings.

The heart of an induction motor is its laminated iron core and the stator and rotor windings. As the core is in no way connected to the power supply nor is power directly removed from it, it can be considered as passive. It is, however, the path of minimum resistance for the flux generated by the m.m.f set up by the stator winding, which itself is the path of least resistance for the input current.

In summary, the power supplied to a three-phase stator winding sets up a rotating magnetic field. This induces an opposing current in the rotor winding and thus another magnetic field. Interaction of these two fields produces a tangential force. As the rotor shaft is only restrained by its bearings, it has to rotate.

The AC induction motor can be easily adapted to variable speed output by inverter control and this is perhaps the most popular form of flowrate variation. According to the type of control and its effect on the electrical wave form, this may be distorted sufficiently from the ideal sinusoidal shape that motor noise may increase and efficiency reduce as the speed is decreased.

Selection of the relationship between voltage and frequency is of special importance particularly where the fan is of lightweight design, or where it is built onto a flimsy structure. It is usual to select the so-called fan torque variant where $v \propto f^2$. If the transmission efficiency also varies with speed (as in a vee belt driven fan) this may not, however, be ideal.

Effects of the European Directive 89/336/EEC on Electromagnetic Compatibility (EMC) are still being resolved. It has been subject to three amendments and at present it is due to be repealed by the new 2004/108/EC Directive as from 20th July 2007.

Certainly an AC induction motor on a sinusoidal supply is unaffected, but inverters can present a problem where screening is inadequate or cable lengths excessive. There are, however, other types of motor currently being developed for fan drives, such as the permanent magnet type and the switched reluctance drive. It remains to be seen whether, in addition to increased efficiencies, these drives will overcome their disadvantages from an EMC viewpoint.

19.5.5 Selection of correct motor speed and type

During the early postwar years, when energy was cheap, it became accepted for motors to be designed with ever smaller frames, higher temperature rises, and consequent reduced amounts of active material. Fans also were subjected to the same thinking, and they were designed for lowest capital cost. This led to reduced diameters and increased blade widths with an ability, where direct driven, to run at 2 and 4 pole motor speeds.

In the 21st century, it is suggested that this trend ought to be reversed. A re-emergence of larger fans, driven at slower speeds, and with a high efficiency across a greater portion of the characteristic, should be encouraged. The decrease in lifetime costs, where energy consumption and maintenance costs are added to the capital cost will be readily apparent. The matching of motors and fans, and the choice of control system concepts to permit optimisation, should become a design priority.

19.6 Whose responsibility?

A measure which has been introduced in legislation by a number of countries is to limit the value of Watts/litre/sec for the fans in a particular building. It is usual to take the total flow of air — supply or extract, whichever is the greater — and divide this into the total installed motor power. The value is usually between 1 and 1.5. It should be noted that motors are invariably rated in kW whilst fans have flowrates above fractional defined in m^3/s.

However:

$$kW/m^3/s = w/l/s$$

as 1000 litres = $1m^3$

and 1000 Watts 1kW

but:

$$kW = \frac{m^3/s \times kPa}{\cdot\eta}$$

or

$$kW/m^3/s = \frac{kPa}{\cdot\eta}$$

Whilst many system designers, when confronted with such specifications, attempt to pass it on to the fan supplier, they must realise that they are the most important party in complying. It is primarily a question of reducing system pressures by lowering duct velocities, decreasing the number of bends, lengths of straight, etc. Increasing fan and motor efficiency is of lesser importance.

19.7 The integrity of fan data

Where comparisons of fan efficiency are to be made, it is essential that all data is the result of tests. These should be conducted to the same standard and test method. Individual witnessed tests are usually only possible after manufacture is complete and are very expensive. The purchaser should seriously consider if certified ratings in accordance with schemes such as AMCA 211 will give the desired assurance (see Chapter 17, Section 17.11.2).

19.8 Bibliography

The Skeptical Environmentalist, Bjorn Lomborg, Cambridge University Press 2001, ISBN 0521010683.

The Kyoto Protocol, (Kyoto Protocol to the United Nations Framework Convention on Climate Change (UNFCCC)), an international treaty on global warming. Negotiated in Kyoto, Japan in December 1997, The treaty came into force on February 16, 2005.

EMC Directive 89/336/EEC, (due to be repealed by the new 2004/108/EC Directive as from 20 July, 2007.)

AMCA 211, *Certified Ratings Programme - Product Rating Manual for Fan Air Performance.*

Enhanced Capital Allowances (ECA), DEFRA, 6/H15, Ashdown House, 123 Victoria Street, London SW1E 6DE. UK, Tel 020 7082 8709, Fax 020 7082 8708.

The Market Transformation Programme (MTP), Future Energy Solutions, PO Box 222, Didcot, OX11 0WZ UK, Tel 0845 600 895, www.mtprog.com.

The Energy Saving Trust (EST), 21 Dartmouth Street, London SW1H 9BP UK Tel: 020 7222 0101, Fax: 020 7654 2460, www.est.org.uk

The Earth Summit, (United Nations Conference on Environment and Development (UNCED)), Rio de Janeiro, June 1992.

20 Fan selection

This Chapter is concerned with the general choice of fan and ancillaries and is aimed primarily at the user. It must however, be emphasised that the choice should be finalised after detailed discussions with a reputable supplier. In the following Sections, the fan is chosen on the basis of the desired flowrate, fan pressure, air/gas temperature, humidity and density. Where solids are handled such as wood waste, textile remnants or dust particles, then these also can affect the fan selection.

Whilst centrifugal fans dominate the world market, they are not always the best choice. There are many applications where axial flow fans can be a valid alternative, especially when space is at a premium. Outside the pressure and noise limitations of the axial, the mixed flow fan should also be considered.

The correct fan selection may be complicated by the attitudes of some manufacturers. It is as well to remember that not all of them have a complete range in their armouries. Discussions with several companies may be necessary before becoming aware of what exactly is available to suit a particular application.

Fan choice is further complicated by the widely varying abilities of fan salesmen! Some companies are diligent in the training of their representatives – others are not. It is often necessary to by-pass the salesman to have sensible direct discussions with an applicational specialist.

Globalisation of fan companies has not necessarily improved the service to prospective customers. Competition between factories within a group is often just as intense as that between companies – especially where long term survival is at stake. Corporate finance arrangements may not reward the salesman or the factory for selling a fan made elsewhere in their "empire". Background information, as given in the Classification guide to manufacturers and suppliers, Chapter 24, assists in this respect.

Of ever increasing importance is the total Life Cycle Cost of the fan and its drive. In an age where global warming is seen by some as a threat to the future of mankind, it is no longer ethical to purchase on first cost considerations alone. We must consider the energy costs over the fan lifetime together with routine and breakdown maintenance, spares costs and disposal costs.

Finally, to keep the fan manufacturers "on-side" a section is devoted to the mathematics of fan selection. With the advent of computers, curve methods may be unknown to the younger generation of engineers, but can still be of value in indicating exactly where a fan is operating on its characteristic, and can also show possible alternative selections.

Contents:

20.1 General operating conditions

20.1.1 Introduction

Everything about a fan application is important. Potential purchasers should not make value judgments on what information is, or is not important. It may well be that the missing data will devalue the manufacturer's warranty. Guarantees are given for specified operating conditions. If there are operating conditions which have not been included in the specification, then the warranty does not cover them.

Whilst the majority of fans handle "clean" ambient air, others will be subject to hazards which may even be experienced when the fan is stationary — the fan can be a "heat sink" or may still be subject to corrosion.

Engineers completing or filling in fan data sheets, where these are part of a manufacturer's questionnaire, may be over cautious. They may think that they could be "caught out" and thus read too much into the questions. Data sheets often request the "discharge pressure"; the data sheet does not say "all pressures up to and including". The range of operating variables is critically important. Concentrating on maxima is a mistake; the full range of values is important.

20.1.2 Air/gas properties and operating conditions

To consider any type of fan for an application, the following information must be given:

- flowrate
- temperature
- barometric pressure
- fan pressure
- air/gas composition
- solids or other contaminants.

For many applications, such as ventilation or comfort air conditioning, the fan may be handling normal ambient air with a minimum of foreign matter — in which case say so! Fans completely suitable for such clean air applications may last less than 20% of the time predicted if abrasive or corrosive elements are present.

The properties of the air/gas are of critical importance for selecting the correct type of fan. The density of the air/gas varies with temperature and barometric pressure, whilst the pressure developed by the fan and the power absorbed is directly proportional to this density. Sometimes the required fan pressure has been calculated under "standard air conditions" which do not exist. It is necessary to be specific and to know whether a constant mass flowrate of a constant volumetric flowrate is required.

The corrosive or erosive properties of the mixture being handled will be important for determining the correct type of fan. If the most suitable material for the manufacture of the fan cannot be cast or welded, then this may restrict the choice of fan types.

Any variations in the operating conditions must be quantified. For any specific variation, the other conditions must be stated. The duration of expected running should also be specified. Changes in operating conditions can take place relatively slowly or very rapidly. It is important to know which is the case. If very slow then instability or surging is less likely to occur. Rapid temperature changes can create distortion or cracking. It is possible to crack even thick casings due to thermal shock. Any rapid changes in operating conditions must therefore be identified.

20.1.3 The duty cycle

For fans running continuously, it is desirable to specify the Mean Time Between Failures, MTBF.

Chemical processes, oil refineries, offshore platforms, power utilities and even many comfort air conditioning applications, all have fans which are required to operate for 25 or 50 weeks without a shut-down. Other fans, however, may only be required to operate for shorter periods.

The following operating descriptions can then be useful:

- Continuous — over 8 hours running in any 24-hour period
- Light — 3 to 8 hours running in any 24-hour period
- Intermittent — up to 3 hours running in any 24-hour period
- Irregular — the fan operates for differing times with various periods of extended rest between operations
- Cyclic — the fan operates with a set pattern of rest or unloaded operation followed by a period on-load

The definitions given above are based on experience. They are not standardised but do appear from time to time in specifications, especially in the oil industry. The important factor is whether the equipment warms up fully. This is critical for fan and motor bearings and for motor windings. Where parts "grow" under the influence of gas temperature, this may also be important. If equipment runs long enough for all temperatures to stabilise then it is considered to run "continuously". Small equipment can warm up continuously but large equipment may take much longer. In extreme cases even 8 hours running may be insufficient for all temperatures and dimensions to stabilise. More costly, slower equipment can, through higher availability, reliability and maintainability (the well-known "ARM" in military terminology) pay for itself very quickly from increased plant output.

20.1.4 Flow variations

Variations in the required flowrate determine whether the application should be shared by several fans. The system curve, (see Chapter 5) of flow versus pressure then determines whether fans in series or parallel are preferable. Whilst identical fans are usually desirable in the latter case, it may indicate where unequal-sized units are possible.

Variations in flowrate demand and the consequential change or otherwise, of the system pressure, determine the type of flow regulation to be used (see Chapter 6). In this context, the working time of the fan is of considerable importance. The efficiency of the fan and its driver etc., at all the operating conditions, must be used to calculate the total energy requirement kWh/annum. As stated many times in this book, the energy costs of running a fan can be as great as its purchase price after only a few months of operation. The cost of the energy supply must, therefore, be closely calculated. It may be that alternative energy sources are indicated.

Flow should be considered from both long term and short term points of view. Different regulation methods, with different initial costs and running efficiencies can be applied, depending on the frequency of the changes.

The effects of system pressure changes on the fan flow may be important. A fan with a steep pressure/flowrate characteristic (such as a narrow backward bladed centrifugal fan) may have only a small variation in flowrate over a wide range of pressures. This can simplify overall flow control methods.

Operating conditions can alter as the installation wears, corrodes or fouls. Changes to the fan unit, from "as new" to "progressive ageing" may be used to optimise performance at different duty points.

The size, nature and concentration of any particles are also of course, important factors. Hard abrasive solids will generally have a much more serious effect on fan life and efficiency than soft deformable solids such as wax. The abrasive properties of hard solids can be quantified by testing such as the Miller Number Test.

20.1.5 Fans handling solids

When handling solids either as dusts or larger pieces, the difference between abrasive and non-abrasive materials, should be recognised. The Miller Number Test, developed initially for reciprocating pumps, has sometimes been used, for classification purposes, of the dusts handled by fans. Stauffer of Escher Wyss conducted modified tests using a sand/water mixture with various metals, The relative abrasion resistance, taking a case hardened carbon steel as unity, were as found in Table 20.1

Material	Relative abrasion resistance
Cast iron	0.09
Ni Al bronze	0.12
Al bronze	0.13
Carbon steel 195 BHN	0.22
316 L stainless steel	0.26
13 Cr steel 441 BHN	0.32
27 Cr cast iron	1.00
Ni-Hard	0.89 to 1.11
Stellite 6	0.83 to 3.31
Tungsten carbide	1.39 to 4.12
Hard chrome plate	2.06 to 2.28

Table 20.1 Relative abrasion resistance to a sand/water mixture

Materials which can be considered as abrasive when transported uy air or some other gas are:

- Al_2O_3 alundum
- coal dust (pulverised fuel)
- copper concentrate
- fly ash
- Fe_2O_8 limonite
- microsphorite
- magnetile
- phosphate
- FeS_2, $CuFeS_2$ pyrites
- sand
- serpentine (magnesium silicates)
- sintered materials
- tailings
- tar sand

The particle size and concentration is of course of great importance in determining whether a particular fan will be suitable for a given application. In a homologous series of fans it is possible to build up a database of impeller lives. By the use of the formula given in Equation 20.1, it is then possible to predict the life of another fan at another speed. For worthwhile results, many fans, many applications and many speeds are essential to have confidence in the predicted lives.

$$\text{Impeller life hours} = \frac{k \times 10^3}{d} \qquad \text{Equ 20.1}$$

where:

$k = 50 \times \left(\frac{v}{u}\right)^n$

d = gm dust / kg gas

v = 25

u = impeller peripheral velocity (m/s) where $100 > u > 40$

n = an index dependent on the type of dust

e.g:

n = 1.0 for a non abrasive soft dust such as zinc oxide

n = 1.75 for limestone and cement

n = 2.5 for sinter dust

By a series of experiments it is possible for the manufacturer to arrive at similar equations (other values for k, v, etc.) for his own particular product range. The safe life would typically be set at 75% of the fan hours to catastrophic failure.

20.2 Mathematical tools

20.2.1 Introduction

Most fan duties can be met by selecting an appropriate fan size from a standard range. It is just a case of choosing the best size and type from those which are available or pre-designed. Theoretically, if it could be run fast enough, a single fan could meet all duties. Whilst much to be desired by managing directors, this would be at the expense of less than optimum efficiency, noise and outlet velocity. It might also be required to be constructed in exotic materials with high strength and ductility.

Fan selection begins with the customer specifying what he wants and ends with the supplier putting forward a solution, having evaluated many alternatives. Of the many types and sizes of fan that may be capable of producing the required flowrate and pressure, the best selection is the one that is the most economic in terms of capital, energy, maintenance and disposal costs (see Chapter 19). Low noise levels may also be of importance.

20.2.2 Specifying requirements

A customer's fan specification should give as much information as possible regarding the required performance, design life, mechanical arrangement etc, so that the best possible solution can be put forward.

If the supplier received all the information given in Table 20.2 he should be highly delighted, but undoubtedly dumbfounded! Invariably bad selections are the result of insufficient information being available for the best selection.

It is also desirable to have a schematic drawing of the ducting to and from the fan with their sizes. This will enable the system effect factors to be evaluated. The customer needs to be aware how these have been included in his assessment. Inclusion of the duct sizes will enable any diffuser/reducer regains to be assessed.

20.2.3 Fan "apparent" pressure

Users should remember that virtually all information published by manufacturers is based on the fans handling standard air. Whilst this has varied by a small amount over the years it is currently specified as air having a density of 1.2 kg/m^3. Reference to Chapter 4, Section 4.6 and the Fan Laws, showed that whilst

Information		Units
General (see Chapter 1)	No. of fans Aerodynamic type Size Installation category (see Chapter 4)	
Physical data (see Chapter 9)	No. of inlets Type of drive Arrangement No. Direction of entry for inlet boxes Rotation, discharge angle and motor position	
Gas compositions and conditions (see Chapter 2)	Gas analysis and /or name Molecular weight or specific gravity referred to out Ambient barometric pressure Temperature at fan inlet Relative humidity Dust loading	 kPa °C % kg dust/kg air
Flowrate per fan at inlet conditions (see Chapter 4)	Design MCR (maximum continuous rating) NER (normal economic rating) Minimum	m^3/s
Fan pressure at each flowrate (see Chapter 5)	Fan (total) pressure Total pressure at outlet Total pressure at inlet Fan velocity pressure Fan static pressure	Pa or kPa
Constructional features	Ancillaries (see Chapter 16) Special materials (see Chapter 7) Bearing type (see Chapter 10)	
Power evaluation (see Chapter 19)	Expected (design) life Expected operation at each rating Power tariff Demand charges	years hr/year pence/kWh £/kW
Motor data (see Chapter 13)	Electrical characteristics	volts phases frequency

Table 20.2 Information required for optimizing fan selection

the flowrate is virtually unaffected by air/gas density ($q_v \propto ND^3$), the fan pressure is directly proportional to this density. Thus if a fan is handling a hot gas with a density of 0.6 kg/m^3, then the actual pressure developed will be halved.

It is a relief to know that the resistance of most systems will also be nearly halved, apart from the second order effect of a change in air/gas viscosity (see Chapter 2). Thus the volumetric flowrate in a given system will be almost unchanged, although the mass flowrate q_m will of course be halved along with the power absorbed by the impeller. For selection purposes, therefore, where non ambient air is being handled, it is customary to calculate the fan "apparent" pressure (static or total) so that standard multi-rating tables or the various types of curve can then be used.

Thus:

Fan "apparent" pressure:

$$p_{f.app} = p_{f.act} \times \frac{\rho_{amb}}{\rho_{act}} \qquad \text{Equ 20.2}$$

20.2.4 The early history of fan catalogues

For many years the dissemination by companies of information concerning the performance of their products was shrouded in mystery. When the author entered the industry in the early 1950s, this was still very much the case. Baseline characteristic curves were kept under lock and key, with only senior engineers having access to them.

As in so many subjects, the break with the past, started in the USA. A renowned fan engineer H.F. Hagen, then with the B.F.

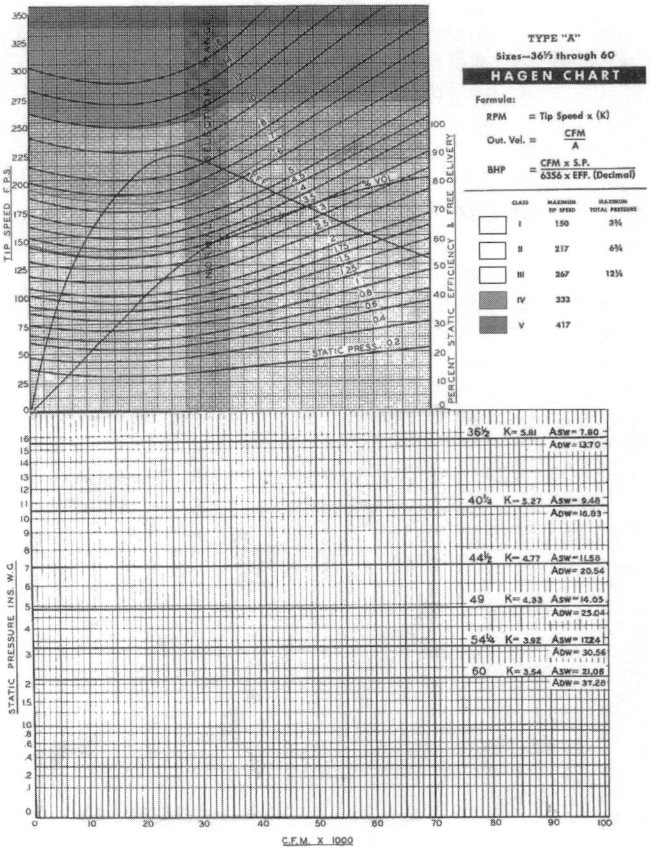

Figure 20.1 Hagen Chart for selection of aerofoil bladed centrifugal fans

Sturtevant Company of Massachusetts, was the first to devise an ingenious graphical method under US Patent No 1358107, (see Chapter 1). An example of such a "Hagen Chart" is shown in Figure 20.1. It will be noted that a number of such graphs were necessary to cover all the sizes in even one range.

20.2.5 Multi-rating tables

In the UK however, Sturtevant's associated company, Sturtevant Engineering Company Ltd, continued with the use of multi-rating tables. Again, an example is shown as Table 20.3, albeit updated for modern SI units.

The actual data from which these tables were produced was in fact surprisingly small. A small number of original test points were simply magnified by the use of the Fan Laws. (See Chapter 4.) Thus the figures at one particular pressure could be calculated for another pressure.

$$p_F \propto N_2 D_2 \text{ at constant gas density} \qquad \text{Equ 20.3}$$

or $p_F \propto (\pi ND)^2$ introducing constant π

or $p_F \propto$ peripheral speed^2

In like manner:

$$q_v \propto ND^3$$

or $q_v \propto \pi ND \times D^2$

or $q_v \propto$ peripheral speed $\times D^2$

Thus at constant pressure (i.e. peripheral speed), flowrate

$$q_v \propto D^2$$

For constant peripheral speed

$$N \propto 1/D$$

Fan Size No.	Inlet diameter mm	Inlet velocity pressure Pa	Volumetric flowrate m^3/s	Fan static pressure Pa													
				250		375		500		750		1000		1250		1500	
				rpm	kW	rpm	kW	rpm	kW	rpm	kW	rpm	kW	rpm	kW	rpm	kW
0	229	92	0.51	1410	0.41	1525	0.50	1650	0.58	1900	0.78	2140	0.99	2350	1.20	2520	1.40
		130	0.60	1575	0.60	1680	0.70	1780	0.80	2000	1.01	2220	1.25	2410	1.49	2590	1.72
		172	0.70	-	-	1870	0.93	1950	1.08	2130	1.30	2320	1.57	2500	1.85	2680	2.10
		222	0.79	-	-	-	-	2140	1.48	2275	1.70	2430	1.95	2610	2.24	-	-
1	279	100	0.79	1165	0.58	1260	0.70	1360	0.81	1570	1.10	1765	1.39	1935	1.69	2080	1.97
		142	0.94	1320	0.89	1385	0.98	1470	1.13	1652	1.42	1830	1.76	1990	2.09	2140	2.42
		187	1.08	-	-	1550	1.42	1605	1.52	1755	1.83	1910	2.20	2060	2.60	2210	2.95
		239	1.23	-	-	-	-	1765	2.09	1880	2.39	2000	2.74	2150	3.17	-	-
2	326	97	1.09	990	0.81	1070	0.96	1155	1.13	1335	1.52	1500	1.92	1645	2.33	1765	2.72
		137	1.29	1120	1.23	1178	1.36	1250	1.55	1410	1.96	1555	2.43	1690	2.89	1815	3.34
		182	1.49	-	-	1315	1.95	1365	2.11	1490	2.54	1625	3.06	1755	3.58	1875	4.10
		237	1.69	-	-	-	-	1500	2.89	1595	3.21	1705	3.80	1830	4.36	-	-
3	381	97	1.44	860	1.06	928	1.29	1000	1.49	1158	2.01	1300	2.54	1425	3.08	1530	3.60
		135	1.71	970	1.63	1020	1.80	1080	2.05	1220	2.59	1348	3.21	1469	3.80	1572	4.40
		182	1.98	-	-	1140	2.50	1180	2.59	1292	3.36	1410	4.03	1520	4.74	1628	5.41
		234	2.24	-	-	-	-	1300	3.82	1380	4.36	1429	5.02	1585	5.76	-	-
4	429	97	1.84	760	1.36	820	1.63	889	1.90	1022	2.57	1150	3.24	1260	3.92	1352	4.59
		135	2.18	860	2.07	905	2.29	960	2.61	1075	3.30	1190	4.10	1295	4.86	1390	5.61
		182	2.52	-	-	1010	3.30	1045	3.54	1140	4.29	1248	5.13	1345	6.04	1440	6.90
		234	2.86	-	-	-	-	1150	4.86	1220	5.56	1308	6.41	1400	7.35	-	-
5	483	110	2.45	660	1.80	714	2.16	771	2.54	890	3.42	1000	4.33	1097	5.23	1178	6.11
		154	2.91	748	2.76	785	3.06	834	3.49	936	4.40	1037	5.46	1128	6.49	1210	7.46
		204	3.36	-	-	878	4.40	910	4.74	995	5.70	1083	6.85	1170	8.05	1250	9.17
		259	3.81	-	-	-	-	1000	6.49	1061	7.42	1138	8.54	1220	9.77	-	-
6	533	107	2.97	600	2.18	650	2.62	702	3.06	810	4.14	910	5.22	996	6.34	1070	7.38
		149	3.52	680	3.34	715	3.69	758	4.21	852	5.33	945	6.60	1025	7.83	1100	9.06
		199	4.06	-	-	800	5.33	828	5.73	905	6.90	989	8.28	1065	9.77	1140	11.11
		249	4.60	-	-	-	-	910	7.83	965	9.02	1032	10.33	1110	11.86	-	-
7	584	107	3.54	550	2.60	595	3.12	645	3.65	744	4.92	835	6.23	915	7.53	982	8.80
		149	4.20	625	3.99	655	4.40	695	5.03	781	6.34	865	7.83	940	9.32	1010	10.74
		199	4.84	-	-	732	6.34	760	6.82	830	8.20	905	9.84	977	11.63	1045	13.20
		249	5.48	-	-	-	-	835	9.32	888	10.66	950	12.30	1020	14.09	-	-
8	635	102	4.15	508	3.04	550	3.65	595	4.29	685	5.78	770	7.31	844	8.87	907	10.37
		144	4.91	575	4.66	605	5.16	641	5.89	721	7.46	798	9.25	869	10.96	934	12.68
		192	5.66	-	-	675	7.46	700	8.02	766	9.66	835	11.56	901	13.61	965	15.51
		249	6.44	-	-	-	-	770	10.96	818	12.53	875	14.39	940	16.55	-	-

Table 20.3 Typical multi-rating table for paddle bladed centrifugal fans

All these relationships assume constant efficiency so that absorbed power can also be calculated.

It soon became apparent that these tables in their original form were inadequate, so that by the 1960s they had been expanded considerably. The manufacturers simply expanded the number of flowrates and pressures into virtual books, stating the size of fan involved to achieve the duty, together with its speed and absorbed power.

Matthews & Yates Ltd together with Keith Blackman Ltd were prime exponents of this approach.

20.2.6 Performance coefficients

It will have been noted that use of the Fan Laws is central to all methods of mathematical fan selection. By the use of these laws, fan performance may be reduced to some convenient dimensionless or standard performance.

To repeat:

$$q_v \propto ND^3$$

$$p_F \propto \rho N^2 D^2$$

or:

$$q_v = \kappa N D^3$$

$$p_F = \kappa N^2 D^2 \rho$$

In order to give reasonable values for these constants, it became the "norm" (in the pre SI era within the USA and UK) for these constants to be calculated taking the speed of rotation in thousands of revolutions per minute (per 1000 rev/min) and the impeller diameter in feet. Thus the constant became κ_Q for volumetric flowrate ie:

$$q_v = \kappa_Q \times \left(\frac{N}{1000}\right) \times \left(\frac{D}{12}\right)^3 \qquad \text{Equ 20.4}$$

where:

N = rotational speed (rev/min)

D = impeller diameter (in)

q = flowrate (cu.ft/min)

Similarly, pressure developed:

$$p_F = \kappa N^2 D^2 \rho$$

Here again it became customary to take the speed of rotation in thousand of rev/min and d in feet. Also relative density ρ_s was used for convenience instead of air density ρ. The constant was known as κ_S or κ_T according to whether comparison was being made for fan static pressure or fan (total) pressure.

Thus:

$$p_F = \kappa_T \times \left(\frac{N}{1000}\right)^2 \times \left(\frac{D}{12}\right)^2 \times \rho_S \qquad \text{Equ 20.5}$$

$$p_{Fs} = \kappa_S \times \left(\frac{N}{1000}\right)^2 \times \left(\frac{D}{12}\right)^2 \times \rho_S \qquad \text{Equ 20.6}$$

where:

$P_{F \text{ or } S}$ = pressure (total or static) ins. water gauge

In like manner, the power absorbed (then measured in b.h.p.):

$$P = \kappa_p \times \left(\frac{N}{1000}\right)^3 \times \left(\frac{D}{12}\right)^5 \times \rho_S \qquad \text{Equ 20.7}$$

If fan performance was now plotted in terms of κ_a, κ_s, κ_T and κ_P instead of volumetric flowrate (then ft^3/min) fan static or total pressure (then ins w.g.) and absorbed power (then b.h.p.) a basis of comparison between fans of different series or designs was readily available; the shape of the "standard" characteristic being in every way identical with any other fan of the same series. For axial fans this required that the curve was for the same blade pitch angle. Figures 20.2 and 20.3 show the characteristics of a Woods J Type axial flow fan, one of 24 inch diameter (2 feet or 610 mm) running at 1440 rev/min and the other in terms of coefficients κ_Q, κ_S, κ_T and κ_P.

With the change to SI units, the use of these coefficients has largely died out. There were, however, distinct advantages in being able to recognise exactly where a fan was operating on its characteristic.

Some attempt at reviving their use has therefore been noted, with certain modifications. Calculation becomes much easier if the coefficients are re-jigged on the basis of a 1000 mm, i.e. 1 m fan at 1000 rev/min. They are designated with a "c" to differentiate.

Thus:

$$q_v = c_Q \times \left(\frac{N}{1000}\right) \times \left(\frac{D}{1000}\right)^3 \qquad \text{Equ 20.8}$$

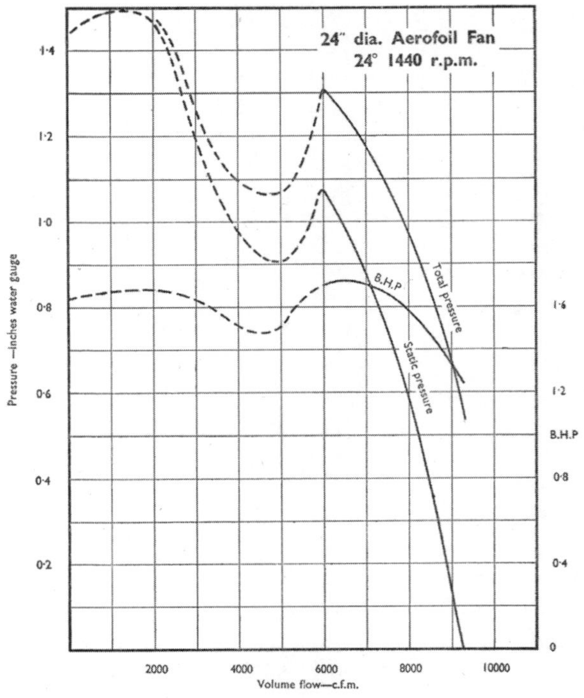

Figure 20.2 Fan characteristic in terms of cfm, ins.w.g and b/h/p

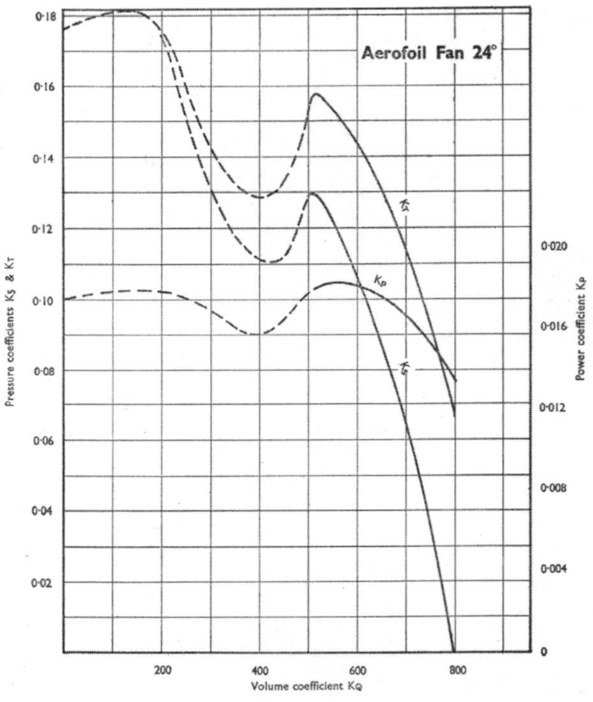

Figure 20.3 Fan characteristic in terms of flow, pressure and power coefficients

$$p_{F \text{ or } FS} = c_{r \text{ or } s} \times \left(\frac{N}{1000}\right) \times \left(\frac{D}{1000}\right)^3 \times \rho_S \qquad \text{Equ 20.9}$$

$$P = c_p \times \left(\frac{N}{1000}\right)^3 \times \left(\frac{D}{1000}\right)^5 \times \rho_S \qquad \text{Equ 20.10}$$

where:

q_v = flowrate (m^3/s)

$p_{F \text{ or } FS}$ = fan (total) or static pressure (Pa)

P = power absorbed (W)

20.2.7 R, C and E curves

When the author was a young apprentice at the Sturtevant Engineering Company, specialist engineers in the fan design department used Master curves. These were jealously guarded as knowledge is of course power! Performance curves were considered to be important sources of knowledge. They were rarely let out of their sight. Certainly, the customer was very privileged to get such information. Nowadays of course everybody asks for a performance curve. Even buyers request them – whether they understand them, or just lock them in their files is another matter.

The Master curves were based upon what are known as R, C and E values. From these curves the characteristics of any fan in the series at any speed can be calculated. They were also useful to bring into the general knowledge base any spot tests taken on installations from which the speed, flowrate, pressure and absorbed power were returned.

To prepare a master curve a full test was made on a fan of reasonable size in the series, and from the results the values of R, C and E were calculated.

The velocity pressure was the velocity equivalent of fan static pressure i.e. in Imperial units:

$$VSP = 3970\sqrt{p_s}$$

$$R = \frac{\text{tip speed ft/m}}{VSP}$$

$$C = \frac{\text{flowrate ft}^3/\text{min}}{\text{rotational speed rev/min} \times D^3}$$

$$E = \text{fan static efficiency \%}$$

where D is in feet.

It will be noted that all these parameters are dimensionless so that they can just as easily be converted to SI units provided consistent units are used.

Thus:

$$VSP = \sqrt{\frac{2p_s}{\rho}} = 1.29\sqrt{p_s} \quad \text{for standard air}$$

$$R = \frac{\text{tip speed m/s}}{VSP}$$

$$C = \frac{\text{flowrate m}^3/\text{s}}{\text{rotational speed rev/sec} \times D^3}$$

$$E = \text{fan static efficiency \%}$$

Where D is in metres

Values of R, C and E were usually taken at 8 to 10 points over the complete characteristic. These included values at closed conditions (static non delivery) or SND and fully open (free inlet and outlet) or FIO. Curves of the nature of Figure 20.4 resulted. Efficiency was assumed constant over a number of sizes without serious error.

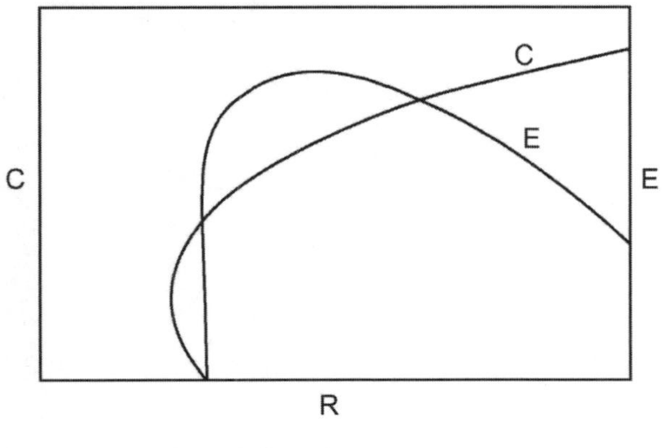

Figure 20.4 Typical R,C and E Master curve

20.2.8 Background charts and cursors

By using logarithmic graph paper (both x and y ordinates) a considerable simplification in the mathematics of the selection procedure can be made.

Consider the drawing in Figure 20.5.

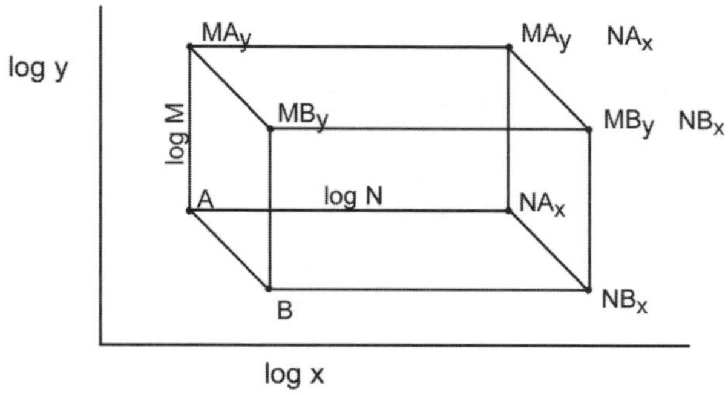

Figure 20.5 Basis for a background chart of y = f(x)

To multiply A_x by N we add log N

A_y by M we add log M

The two operations may be done together and the same operation may be carried out on point B. The line AB will remain a constant length.

If we now plot fan pressures against flowrates, an identical situation is apparent. See Figure 20.6.

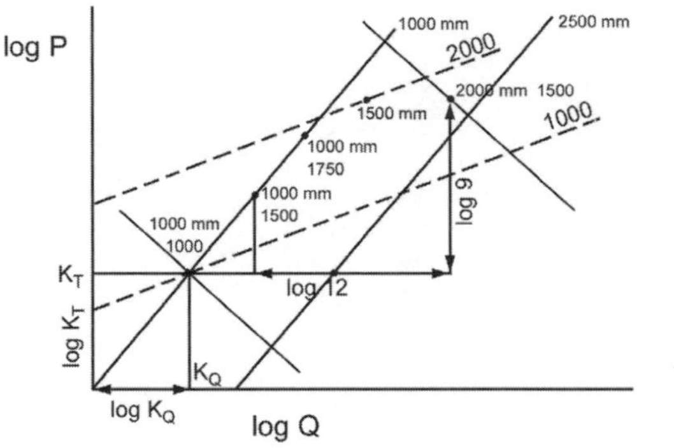

Figure 20.6 Background chart of flowrates, pressures, speeds and sizes

Thus the mathematics become:

c_Q = flowrate (m^3/s) of 1000 mm fan at 1000 rev/min

c_T = fan (total) pressure (Pa) at 1000 mm fan at 1000 rev/min

To find the q_v and p_F of a 2000 mm fan at 1500 rev/min

$$q_v = c_Q\left(\frac{1500}{1000}\right)\left(\frac{2000}{1000}\right)^3 = 12\ c_Q$$

$$p_F = c_T\left(\frac{1500}{1000}\right)^2\left(\frac{2000}{1000}\right)^2 \times 1 = 9\ c_T$$

On the diagram which has logarithmic scales, the values of q_{v2000} and p_{F2000} are found by adding lengths of log12 and log 9 respectively. This would apply to any number of points and to a curve such as the fan characteristic.

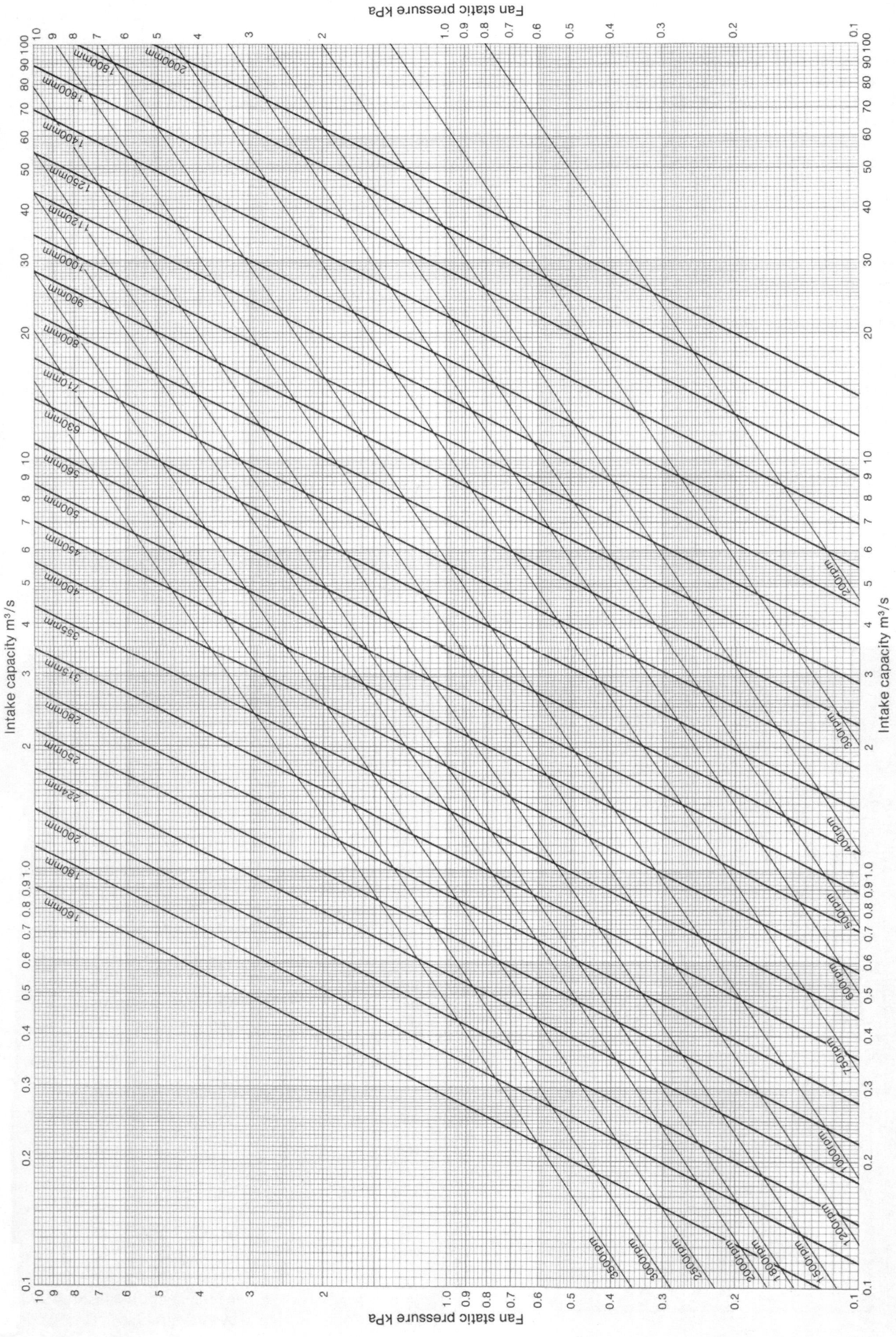

Figure 20.7 Universal background chart for a range of SI sized fans in accordance with ISO 13351:1996

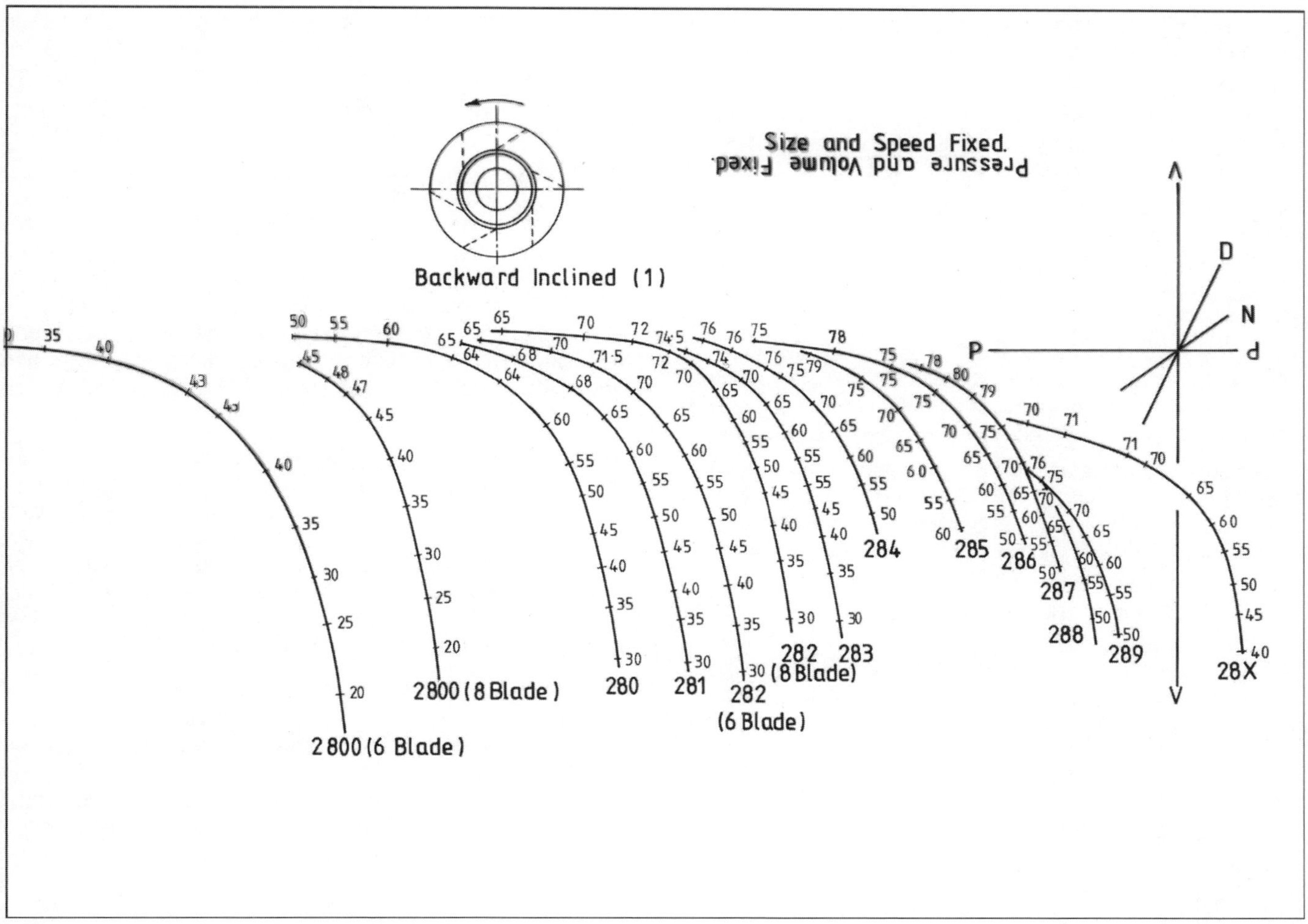

Figure 20.8 The cursor — to be used in conjunction with Figure 20.7

In like manner for a 1000 mm fan at 2000 rev/min:

$$q_v = c_Q\left(\frac{2000}{1000}\right) = 2c_Q$$

$$p_F = c_T\left(\frac{2000}{1000}\right)^2 = 4c_T$$

Whilst for a 1000 mm fan at 3000 rev/min:

$$q_v = c_Q\left(\frac{3000}{1000}\right) = 3c_Q$$

$$p_F = c_T\left(\frac{3000}{1000}\right)^2 = 9c_T$$

Now:

$$q_v \propto ND^3$$

$$p_F \propto N^2D^2 \text{ at constant gas density}$$

or:

$$D \propto \left(\frac{q_v}{N}\right)^{1/3}$$

$$D \propto \frac{p_F^{1/2}}{N}$$

Thus:

$$\left(\frac{q_v}{n}\right)^{1/3} \propto \frac{p_F^{1/2}}{N}$$

$$N^{2/3} \propto \frac{p_F^{1/2}}{q_v^{1/3}} \qquad \text{Equ 20.11}$$

Taking logarithms throughout we may say that:

$$\frac{2}{3}\log N = \frac{1}{2}\log p_F - \frac{1}{3}\log q_v \qquad \text{Equ 20.12}$$

which is a straight line relationship.

In like manner, for a 2000 mm fan at 1000 rev/min:

$$q_v = c_Q\left(\frac{2000}{1000}\right)^3 = 8c_Q$$

$$p_F = c_T\left(\frac{2000}{1000}\right)^2 = 4c_T$$

Similar points may be plotted for the value of q_v and p_F corresponding to c_Q and c_T for any size and speed of fan by using the Fan Laws. At any diameter a straight line may be drawn through the points of differing speed resulting in straight lines of constant diameter. In the same way a straight line may be drawn through the points of differing diameter resulting in straight lines of constant speed.

Since more than one pressure and flowrate results from a fan, the whole range of values of c_Q and c_T (or p_F and q_v) may be drawn through each point. To prevent this becoming tedious, the characteristic may be drawn on a separate transparent cursor to the same scale.

A reference point of a single value of c_Q and c_T is selected from which the lines of constant diameter and speed are drawn. This reference point need not lie on the characteristic, but need only be marked on the cursor. The characteristic of the same type of fan, but of another size and speed may be found by placing the cursor so that the reference point lies on the intersection of the straight lines of the desired size and speed.

The background chart of pressure and flowrate may be used for any type of fan – only the cursor being re-plotted to the new characteristic. The same reference point must be used, since the background chart is plotted to this point.

Where a manufacturer has a number of different widths of centrifugal fan (or different pitch angles of axial fan), it is possible to plot these all on the same cursor. The efficiencies may be marked on the cursor characteristic since their relative positions remain constant.

If the cursor is turned through 180°, it is possible to place the reference point over a desired duty of flowrate and fan pressure when the point where the cursor characteristic intersects a fan size will give the speed and efficiency to achieve that duty.

Figure 20.7 is a universal background chart for a range of SI sized fans in accordance with ISO 13351:1996, whilst Figure 20.8 is the cursor for a range of backward inclined bladed centrifugal fans, of varying widths, to be used with it.

20.2.9 Electronic catalogues

All that has been previously said about fan selection has been rendered superfluous by the introduction of electronic catalogues. A CD is now often provided by the fan manufacturer. Contained within the CD are all the fan laws together with the known range of sizes, maximum speeds and powers, temperature de-rating factors etc, etc. Even noise data and dimensions, together with prices can be included.

When a required duty is entered, the computer can list all possible selections in order of price, efficiency, noise level or some other parameter. It can produce a dimensioned drawing, quotation and specific performance curve. Unfortunately, or fortunately, according to your viewpoint, specialist companies have now entered this field to produce such programmes. Are we to see a situation develop where even the manufacturers' representatives do not really know what their products do, but are reliant on a CD? But then, they do save time – and time is money!

20.3 Purchasing

Many of the chapters in this book deal with the criteria and standards which are of importance in correctly specifying a fan. Chapter 17 deals with the implementation of testing and quality control. These are concerned, among other topics, with guarantees of performance and quality. Criteria for service life and reliability have not yet been standardised, however. Over and above the technical requirements, suitable general terms and conditions of delivery should be agreed.

The purchase documents should contain references to legislative regulations and requirements affecting design, to the properties of the air or gas being handled and to the operating conditions and function of the fan. Only in this way can the manufacturer ascertain if a standard unit will be satisfactory or if special materials or designs should be recommended. If particular local legislation must be observed, than a copy should be enclosed by the purchaser with his enquiry.

It is unlikely that the manufacturer will have access to obscure regulations and legislation. This is not said to encourage those prospective purchasers who enclose everything including "the kitchen sink" with their enquiries to ensure that any omissions are never their fault. Review of the enquiry should ensure that it includes all relevant material but no more! "Bumf" is the scourge of our industry sector!

20.4 Bibliography

ISO13351:1996, *Industrial fans - Dimensions.*

21 Some fan applications

This Chapter gives some indication of the many uses for fans. *Fans & Ventilation* is not intended to be a text book on ventilation and or air conditioning. Rather it seeks to address the problems that exist at the interface between fan manufacturers and users. Furthermore it hopes to show that there are many uses for fans outside these traditional areas. Chapter 1 gave a historical perspective to some of these applications, but this Chapter endeavours to demonstrate some of the sizing rules and features which should be included.

There are many manufacturers who will refuse to give a quotation for a fan unless the enquiry contains a flowrate and a pressure. Equally there are users, and even consultants, who do not know how to calculate these basic parameters. The examples which follow are just a small selection of the myriad applications of which fans are a part. They emphasize that there are a number of uses beyond the normal. There are also many other applications which could have been added such as marine and defence ventilation, bulk tanker ventilation, cooling towers, chillers, crop drying and storage, animal house ventilation, etcetera, etcetera.

This however would have presented enormous space problems and would perhaps be better left to a separate book. It is also necessary to point out that even those applications which have been included, have of necessity, been treated by an overview without all the detail which could have been included. The selection is therefore arbitrary and idiosyncratic, reflecting yet again the author's personal interests and preferences.

Finally, it should be noted that this Chapter has been titled "Some fan applications" rather than "Case studies". The latter title has been used by other product guides, but most fan companies have an applications department and there are even computer programmes which glory in the title "FANAPPS".

Contents:

21.1 Fresh air requirements for human comfort

21.1.1 Indoor air quality

Over the last few years considerable research has been conducted both here and in the USA to determine what pollutants exist in the ambient atmosphere and within buildings. These studies have principally looked at the prevalence of volatile organic compounds, pesticides, carbon monoxide and particulates.

Whilst we are all aware of reports of increased numbers of persons suffering, for example, from asthma, it is extremely difficult to correlate these with an increase in particular pollutants. A number of scientists and engineers have concluded that most people are likely to have the greatest contact with potentially toxic pollutants not outside but within buildings.

If we consider the gross amounts of one particular carcinogen i.e., benzene released into the atmosphere, the greater portion comes from automobile fuel (82%). The next highest sources are industry (14%) and domestic usage (3%). Cigarette smoke only contributes around 0.1% of the total. However, within the confines of a building, approximately 45% of the total exposure may come from smoking, 36% from inhaling petrol fumes and other common products and only 3% from industrial processes.

The corollary of these figures is to conclude that improved ventilation with an increase in fresh air levels and in association with improved filtration is a must. It could greatly improve the indoor air quality and hence reduce the risk of disease. Buildings could then claim to be more friendly to health.

Similar conclusions can be reached for many other chemicals found at quite high concentrations inside buildings - even the "perc" (perchloroethylene) used by dry cleaners or the deodorisers used by ever increasing numbers of the population, have been reported to cause cancer at high concentrations. If we add the potential risks due to carbon monoxide (from incomplete combustion in kitchens and elsewhere) and radon (a natural radioactive gas seeping from foundations and brickwork), one wonders why we should spend up to 90% of our lives within buildings. Is it just a coincidence that the author's long living grandfathers (a fisherman and a farmer) spent most of their waking hours outdoors?

21.1.2 Improving ventilation

The need for improved indoor air quality has persuaded many ventilation engineers that the way forward is to increase the amounts of outside fresh air circulated within buildings. But if that outside air is far from fresh there is a definite problem. It's more than a possibility that I can open the windows and door of my Suffolk pub for a cooling breeze. Provided I am not too disturbed by the smell of silage or fertilisers, this is the most environmentally friendly way to ensure that I don't breathe in Joe's Old Holborn tobacco fumes or whatever. It doesn't require any fan power and doesn't therefore increase the CO_2 emissions from the nearby power station at Eye.

In a big city building there are a different set of problems. The air outside is often worse than that inside. If it's not the dust, it's the petrol fumes. And if it is not the petrol fumes, then there is that other pollutant, noise.

There are, therefore, four techniques which can be used to achieve an acceptable indoor environment:

a) dispersion

b) dilution

c) filtration

d) absorption

These techniques result in the following strategies:

a) Smoking areas should be separated from non-smoking areas. Fresh air inlets should be adjacent to the non-smoking areas whilst extract should be adjacent to smoking areas.

b) Fresh air should be well above the minimum specified for preventing the build up of carbon dioxide and should be used to dilute the smoke produced from cigarettes.

c) Filtration can be incorporated to reduce the amounts of fresh air necessary by allowing air to be recirculated back to the areas of occupancy. This will also ensure that air which has been heated is not completely rejected to atmosphere, thus saving on heating bills.

d) Whilst not practised to date, it is not impossible to include gas absorbers in a ducted ventilation system so that pollutants such as carbon monoxide and petrol fumes can be removed with or without the aid of catalysts.

21.1.3 A little science!

Human beings may be considered as heat engines, taking in food (fuel) and air to produce energy and waste. To sustain life, oxygen is necessary for the metabolism of food. Human beings breathe in air (with its oxygen content) and exhale air (with a significant amount of its oxygen converted to carbon dioxide). Trees on the other hand take in carbon dioxide and change it back into oxygen. Hence the animal and vegetable worlds are in balance and rely on each other. The carbon and hydrogen in foods are "burnt" to produce carbon dioxide and water and these are rejected by the body either by exhaling or as waste. Foods can be principally classified as:

- carbohydrates
- fats
- proteins

The ratio of carbon to hydrogen in each is different. We can measure the amount of carbon dioxide produced by a person to the oxygen consumed, and this is defined as the respiratory quotient (RQ). RQ varies according to diet and is shown for typical foods below:

0.71 for a diet of 100% fat

0.8 for a diet of 100% protein

1.0 for a diet of 100% carbohydrates

A value of 0.83 is taken as a reasonable average for a normal dietary mix.

The rate at which oxygen is consumed and carbon dioxide is generated depends on physical activity. A simple equation gives the outdoor air flow rate needed to maintain carbon monoxide levels at a constant level:

$$V_o = \frac{N}{60(C_s - C_o)} \qquad \text{Equ 21.1}$$

where:

V_o = outdoor airflow rate per person (l/s)

N = CO_2 generation rate per person (l/min)

C_s = CO_2 concentration in the space (ppm)

C_o = CO_2 concentration in outside air

The CO_2 generation rate will depend on the amount of physical activity (see Figure 21.1).

For a recreational activity such as a pub, exertion will be in talking animatedly, whilst raising one's drinking arm, this will be

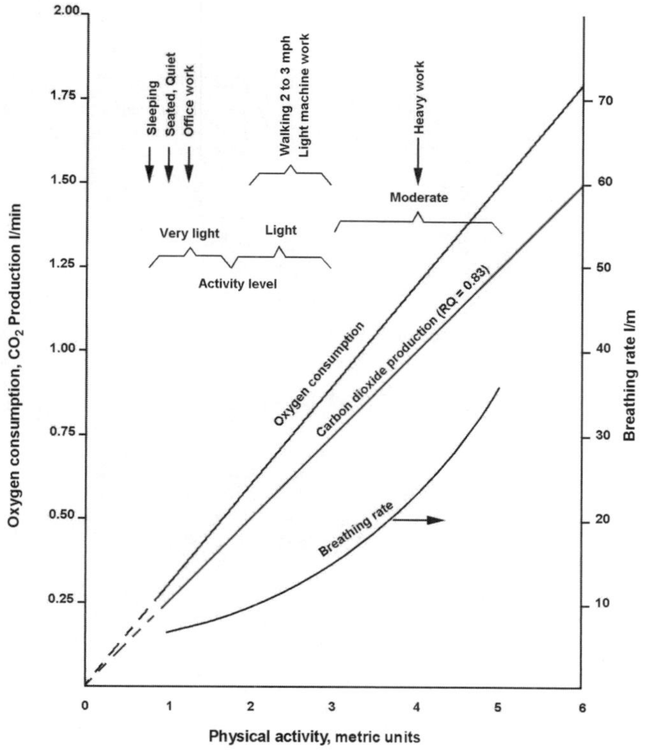

Figure 21.1 CO_2 generation with physical activity

about 0.3 l/min. If there is the chance of dancing or other forms of organised hooliganism as in a discotheque, then this could increase by a factor of 4. If the maximum space concentration is to be held at 100 ppm, and the outdoor concentration is 30 ppm, then the amount of air needed per person:

$$V_o = \frac{0.3}{60(0.001-0.0003)} = 7 \ l/s \qquad \text{Equ 21.2}$$

It should be noted that the percentage change in carbon dioxide is more significant than the reduction in oxygen level.

The calculated figure is only just below the currently recommended 8 l/s as given in many Codes of Practice. It takes no account of the additional air required for cigarette combustion, dilution of pollutants etc.

How then, are we to know what is an acceptable indoor air quality? The American Society of Heating Refrigeration and Air-Conditioning Engineers (ASHRAE) defines this as air in which "there are no known contaminants at harmful concentrations as determined by cognisant authorities and with which a substantial majority (80% or more) of the people exposed do not express dissatisfaction". That is indeed a tough target, but one to which we can all surely agree. How to calculate it without a substantial statistical survey is the real problem.

If this definition is accepted, and the air quality within a building is as good as that experienced in the external ambient atmosphere, then no one can expect more. This is not an impossible target and no doubt surprise will be expressed that tobacco smoke is not necessarily the dominant criteria. However, it must again be emphasised that the ventilation rates are higher than those which are currently specified. Let us therefore ensure that all public places have at least 8 l/s of fresh air per person but recognise that higher rates are necessary for any further improvement of indoor air quality towards that enjoyed outside in the fresh air. Unless air cleaning is also practised it is suggested that we should think in terms of:

- 30 litres/second of fresh air per smoking person and
- 12 litres/second of fresh air per non-smoker.

Suffice it to say, such a level will take care of all the other indoor pollutants.

Further, our systems should be designed for the true maximum occupancies of a particular area and these can well be above:

1 person per 1 m^2.

To repeat, ventilation at these rates would achieve an acceptable level of air quality not only in respect of tobacco annoyance, but also in other particulates, odours, carbon dioxide levels, biological aerosols, formaldehydes, radon etc.

The fan manufacturers will of course be more than happy to supply the increased numbers of fans suggested, but if larger more efficient units are selected, energy consumption will not rise disproportionately and, per air change, may even be less. The air cleaner manufacturers should also be happy, as with their equipment installed, the quantities of fresh air can be reduced to non-smoker levels and heating costs correspondingly contained.

21.1.4 Air filtration

One of the axioms in ventilation is that it is always best to deal with a problem as near to its source as possible. Thus we have noted that extract should be close to the smokers.

An alternative approach is to pass the ventilation air through some sort of filter. This will result in a reduction in the amounts of fresh air necessary and also save on the heating bills. Do not however assume that ventilation can always be reduced to 8 l/s. That would only exchange one set of problems for another — Environmental Tobacco Smoke (ETS) for an increase in CO_2 levels and possible problems from the heat generated by closely packed bodies. The author's conclusion is that one should still consider a minimum fresh air level of 24 l/s.

There are a number of different types of filter available, each with its own respective merits. The restaurant in the author's "local" uses an electrostatic precipitator — it's brilliant. The combination of high ceilings and filter means that smokers and non-smokers are mixed without problem, and the air is perfectly clear. Precipitators consist of an initial bank of positively charged ionising wires between co-planar grounded electrodes followed by a bank of grounded collection plates. Between each pair of collection plates is a positively charged repulsion plate. Using relatively low voltages and currents of positive polarity, ozone formation is reduced to a minimum.

Gas absorbers which burn off carbon monoxide is the presence of a catalyst can also be installed - most easily in a ducted system.

Other types of filter which are available include absolute paper High Efficiency Particulate Air (HEPA) filters and those using other types of media.

21.1.5 Conclusions

Efficient ventilation is a definite plus in the promotion of all public places. It can provide an atmosphere which should satisfy the most fastidious. ETS is unpleasant, even if some of one's friends are smokers, and one wishes to continue to have the opportunity to socialise with them. It is more than possible that ventilation can be installed that will allow one to do this in perfect safety even if ETS is proved to be a danger.

In the light of the current debate, it should be emphasised that hot and cold tobacco smoke are very different both physically and chemically. Certainly, a well-designed and maintained ventilation system can ensure that the atmosphere within a public space need be no worse than that in the surrounding streets. The work of Dr Andrew Geens at the University of Glamorgan show this to be the case. The proviso is that such systems must be well maintained. Many ventilation systems are not kept in good working order. A step change in the mechanical ability of publicans, hotel management, office management etc is most desirable.

In any case, ventilation is essential to provide oxygen for breathing and to remove the heat generated by the occupants. If the fans used are of the variable speed type with appropriate controls, then the rate of ventilation can be adjusted to meet the demand, according to occupancy and activity. More and more systems are in fact being provided with sensors and the control of both heating and ventilating systems can then be integrated. This is especially the case with the larger ducted system where in-duct cleaning can also be incorporated.

It should be noted that the mathematics in this Section has been concerned solely with the amounts of fresh air required. Much larger quantities of air may be re-circulated in air conditioning systems for the removal of heat.

To repeat, all such systems however, must be cleaned and maintained regularly if they are to continue to function satisfactorily. As a connoisseur of fine bitters, the author knows the value of keeping the ducts clean! It's no different with a ventilation fan or a filter!

21.2 Extract ventilation

21.2.1 Introduction

Wall mounted propeller fans have been used in the extraction mode since the very earliest days of mechanical fans. Reference to Chapter 1 shows that these were popular from the 1880s. Usually positioned in the gable ends of buildings, they are still a feature of many older industrial buildings.

Following the 2nd World War, special roof extract units were developed. These largely superseded the "Heath Robinson" arrangements designed by ventilation specialists who incorporated standard propeller fans into fabricated upstands and roof cowls.

Over the years, the application of roof extract fans has extended and they are now also used for air supply. Various options have become available and these include:

- Direct or belt drive
- Side or vertical discharge
- Propeller, mixed flow or centrifugal impeller options

Examples of these are shown in Figures 21.2 to 21.10 in cross-section.

In the first generation of out-of-town stores, roof extract fans were the most important method of ventilation. Air curtains or rotary doors were used to restrict the ingress of cold air. The high customer occupancy gave rise to considerable heat gains such that the natural buoyancy of the air assisted the extract process.

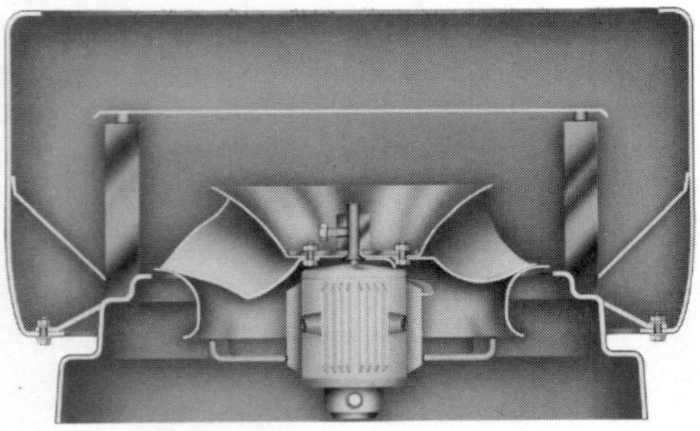

Figure 21.2 Direct drive mixed flow unit with vertical discharge

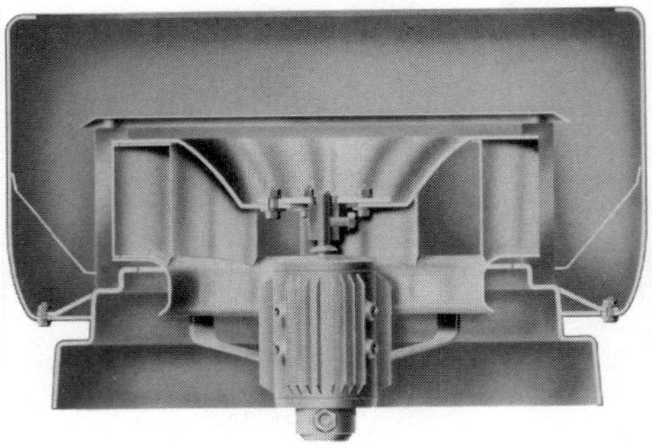

Figure 21.3 Direct drive centrifugal unit with vertical discharge

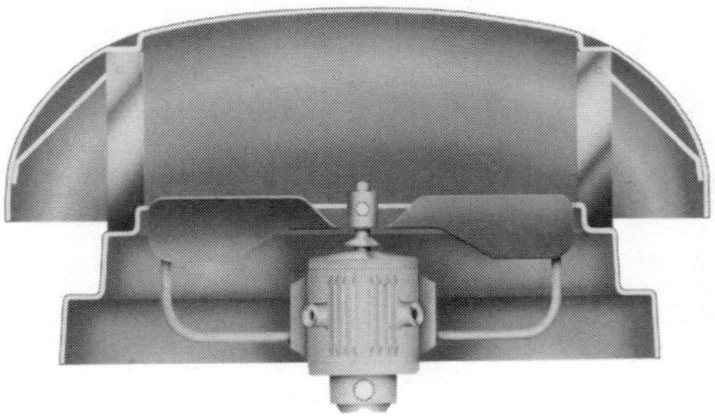

Figure 21.4 Direct drive propeller unit with side discharge

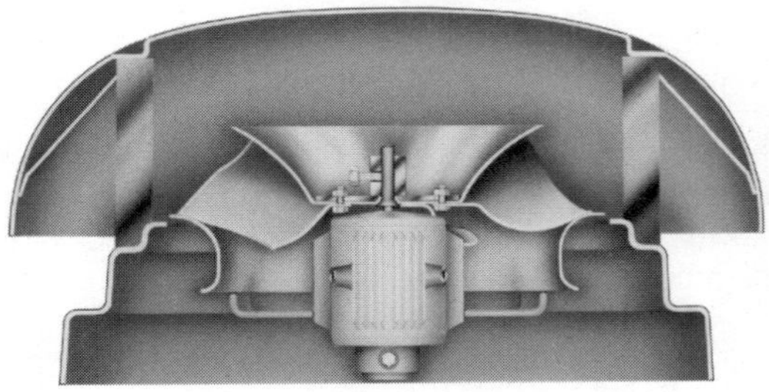

Figure 21.5 Direct drive mixed flow unit with side discharge

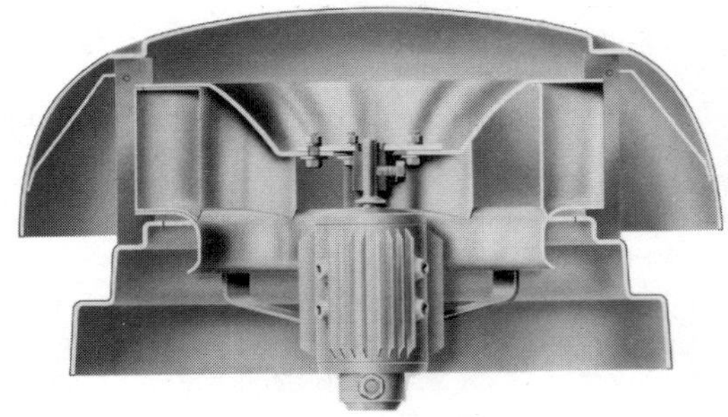

Figure 21.6 Direct drive centrifugal unit with side discharge

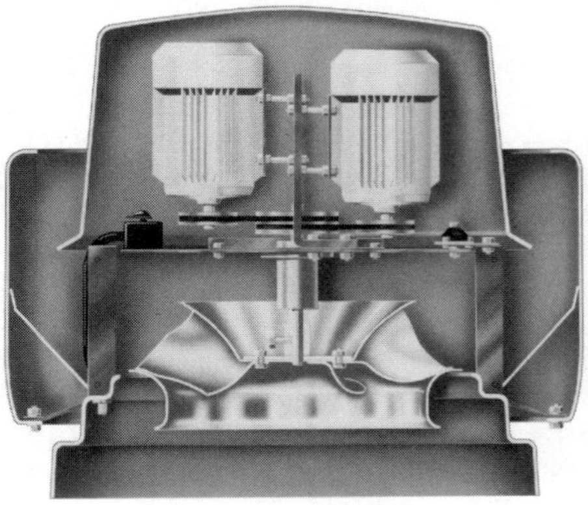

Figure 21.7 Vee belt driven mixed flow unit with vertical discharge, running and standby motors

Figure 21.8 Vee belt driven centrifugal unit with vertical discharge, running and standby motors

Figure 21.9 Vee belt driven mixed flow unit with side discharge, running and standby motors

Figure 21.10 Vee belt driven centrifugal unit with side discharge, running and standby motors

Single storey factories are also ideal for this kind of ventilation and they are widely used in agricultural buildings such a poultry and pig houses.

Indoor air quality has received increasing attention over the last few years (see Section 21.1). Indications are that this will require greater quantities of air to be extracted, contrary to the obligations of a green energy policy.

21.2.2 Powered versus "natural" ventilation

The difference between the performance of fan-powered roof ventilators and "natural" roof ventilation is not always recognised. (Perhaps the latter should be more correctly described as passive ventilators.)

The effectiveness of passive units is dependent on weather conditions and is contrary to ventilation requirements. Air movement through a natural cowl occurs because warm interior air rises and will pass through any roof outlet. The rate of air movement depends on the temperature difference between the inside air and the air immediately above the roof. It is also affected by the velocity of the wind which may increase its movement. (Figure 21.11).

Figure 21.11 The effect of natural ventilation in cold windy weather

In hot weather, when the sun raises the roof temperature to a point where the temperature is as great or greater than that of the interior air - there will be little or no air movement through the natural cowl. Thus a natural roof ventilator is most effective in cold windy weather, but is least effective in hot windless weather when ventilation is most required. (Figure 21.12).

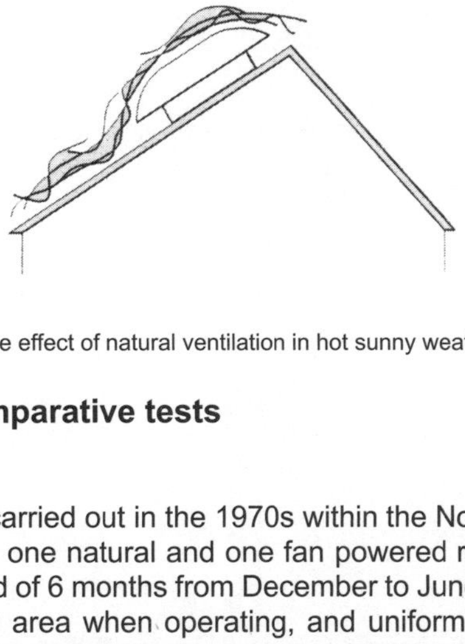

Figure 21.12 The effect of natural ventilation in hot sunny weather

21.2.3 Comparative tests

Tests were carried out in the 1970s within the Northern Hemisphere, with one natural and one fan powered roof ventilator over a period of 6 months from December to June. Both had a similar open area when operating, and uniformly, both were mounted within an identical open area whilst the tests were carried out. The units were mounted at 8.8 m above the floor level on the south facing slope of a north light roof covering a factory building of 991 m^3 with an external doorway measuring 4.25 m x 4 m wide.

Measurements of the air movement through the units were taken daily. Over the 6 month period the average air extraction rate of the natural ventilator was only 10% of the fan-powered unit. Moreover, the extraction rate of the natural unit varied considerably with the weather conditions and was very much influenced by the opening and shutting of doors. When the extraction rate of the fan-powered unit with the doors closed was taken as 100%, the following extract rates were measured, an average being taken over a number of tests under varying weather conditions.

	Door Closed	Door Open
Fan powered roof unit	100%	110%
Natural roof ventilator	8%	15%

In the case of natural ventilators that open vertically to permit maximum release, these units will admit rain unless closed in wet weather. A "rain detector" apparatus may be needed to give warning for the closure of the ventilators when the rain begins. No such requirement is needed for fan-powered roof units.

21.2.4 The justification for mechanical ventilation

The fan-powered roof extract unit gives:

- A specified rate of extraction at all times irrespective of weather changes or conditions.
- It ensures the required number of air changes
- The ventilation can always be controlled
- The actual aperture is of greater size for natural ventilation than that for a powered system.

21.2.5 Fan pressure development

Mixed flow and centrifugal roof ventilators develop static pressures ranging from 100 Pa to 1000 Pa according to size and fan speed. This fact makes them suitable not only for free-inlet extraction from areas immediately below the roof but also, for use with ducted extract systems. (Figure 21.13 and Figure 21.14).

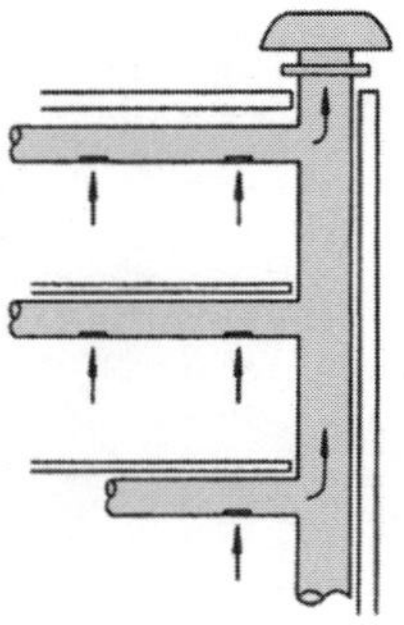

Figure 21.13 Extraction from more that one floor of a two or multi-storey building

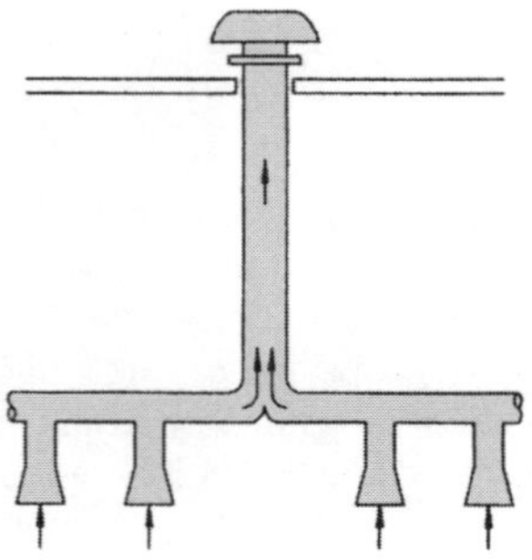

Figure 21.14 Local extraction e.g. from over vats

Two important advantages are to be gained by using them as the fan unit for ducted systems:

- The location of fans on the roof eliminates the need for plant room space.
- Flexibility is gained in designing extract systems. Varying extract requirements in different parts of the building can be handled expeditiously by roof ventilators of different sizes and speed.

21.2.6 The affordable alternative

Over the years, the word ventilation has been veiled with a cloak of secrecy usually associated with black arts. However, it is in the reach of everyone to design simple systems without requiring specialist skills. In this Section are a few simple guidelines and a calculation example is now given:

a) Usually for general ventilation requirements, such things as heat dissipation from lathes, milling machines and drills need not be considered. However, furnaces, ovens or any other large powered machine should have such values considered.

b) The first criteria to establish is the volume of the building. It can be seen from the example in Figure 21.15, the only dimension which may cause difficulty is the height. We can see that we only have a height of 4 metres, to the underside of the eaves. To save some geometric calculation, the best idea is to add an extra metre to the height as a contingency and call it 5 metres.

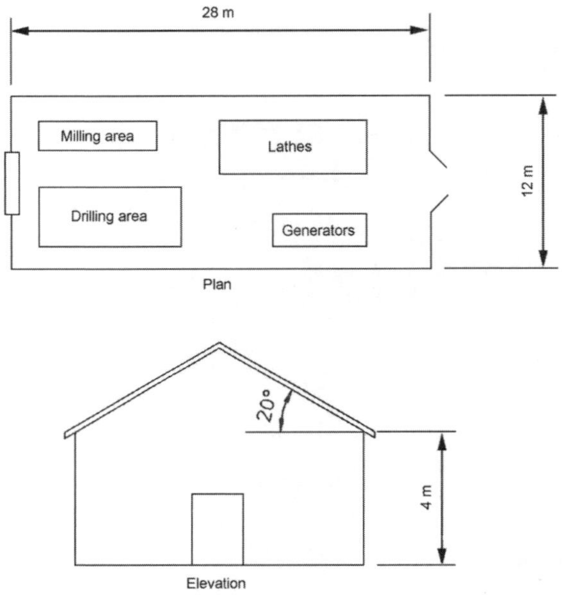

Figure 21.15 Roof extract unit extraction example

Thus the volume is determined by:

Length (L) x width (W) x height (H) = volume

Dimensions are - 28 m x 12 m x 5 m = 1680 m^3

c) The next item to recognise is the air change rate. Over the years the data in Table 21.1 has become a suggested reference for air change rates.

Situation	Air changes per hour	Situation	Air changes per hour
Assembly halls	4 - 6	Furnace rooms	30 - 60
Bakeries	15 -30	Garages	6 - 10
Banks	2 - 4	Hospital wards	6 - 8
Bathrooms	6 - 8	Hospital treatment rooms	6 - 8

Situation	Air changes per hour	Situation	Air changes per hour
Bars	6 - 8	Kitchens for restaurants	13 - 30
Boiler houses	15 - 30	Laboratories	4 - 6
Cafes	8 - 12	Laundries	10 - 15
Canteens	8 - 12	Libraries	2 - 4
♣ Churches	1 - 10	Offices	1 - 6
♦ Cinemas	6 - 10	Paint shops	30 - 60
Classrooms	2 - 4	Residences	1 - 2
Cleaners	15 - 30	♦ Restaurants	10 - 15
Dance halls	8 - 12	Storage areas	1 - 2
Domestic kitchens	10 - 15	Swimming baths	15 - 30
Dyers	15 - 30	♦ Theatres	6 - 10
Engine rooms	15 - 30	Workshops	6 - 10
Foundries	30 - 60		

Note: ♦ General requirements are 28 m^3/h (8 l/s) per person minimum in public places, this figure increases if smoking is allowed.

♣ Dependent on height of building and number of persons.

Table 21.1 Recommended air changes per hour

From Table 21.1 it can be seen that for a general workshop the rate varies between 6 and 10 air changes per hour. The most convenient rate to calculate on is 8, midway through the range. (Later on, we will see that when selecting the fans, the actual air change rate given will vary, usually upward).

d) The formula for obtaining the volume flow of air is:

Ventilation rate (m^3/sec) =

$$\frac{\text{volume of building } (m^3) \times \text{air exchanges per hour}}{3600}$$

Our example:

Volume of building = 1680 m^3

Air changes per hour = 8

3600 is a constant

$$\therefore \quad \frac{1680 \times 8}{3600} = 3.73 \text{ m}^3/\text{s}$$

e) We know that we need a fan or fans to be able to produce/at least 3.73 m^3/s to achieve 8 air changes per hour.

f) The next step is to choose the correct type of fan

For industrial purposes there are three basic types:-

- Roof mounted
- Wall mounted
- Duct mounted

g) Both wall and roof mounted options offer the best degree of flexibility in terms of positioning and controllability. One should avoid choosing a single fan to provide the ventilation, due to the inflexibility and the problem it would cause in leaving areas of the building without any air movement.

h) When roof extract units are considered, then the rough guide is to use one unit per 10 metres of length of building. (The same is true when considering wall mounted units). However, if very high air change rates are required then the 10 metre rule may need to be modified.

i) Before considering where to position the extract fans, careful consideration is needed to ensure that the air inlets to the building are also taken into account. Normally these are not powered.

Again the best rule of thumb is the area of such inlets should equal twice the area of the extract fan.

Example:

A 500 mm diameter fan has an area of 0.2 m^2,
so the air inlets have an area of at least 0.4 m^2.

An important point to remember is to site inlets sensitively so that they do not cause draughts to the personnel in the building. Usually the inlets need to be situated at low level to allow for cross ventilation (see Figure 21.16). The higher the inlets are mounted, then higher levels of air stagnation (no air movement) may occur, at low levels.

j) The mounting of the roof extract units should follow the examples in Figures 21.17 and 21.18. A choice can be made

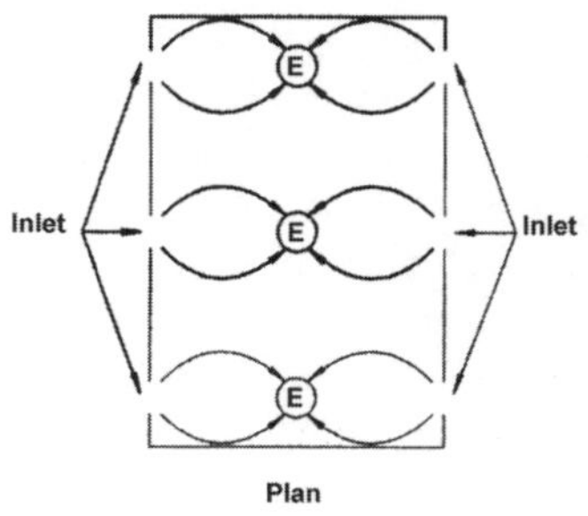

Figure 21.16 Siting of low level inlets

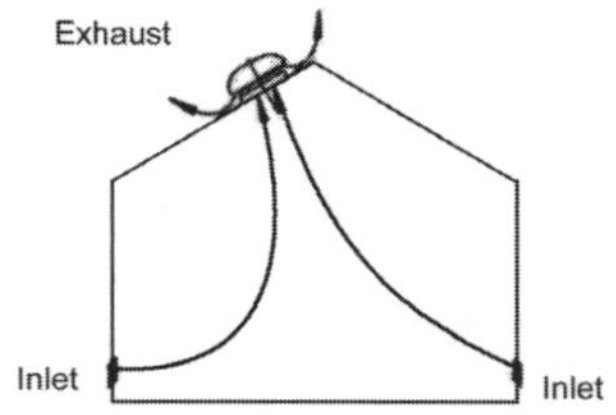

Figure 21.17 Mounting of roof extract option

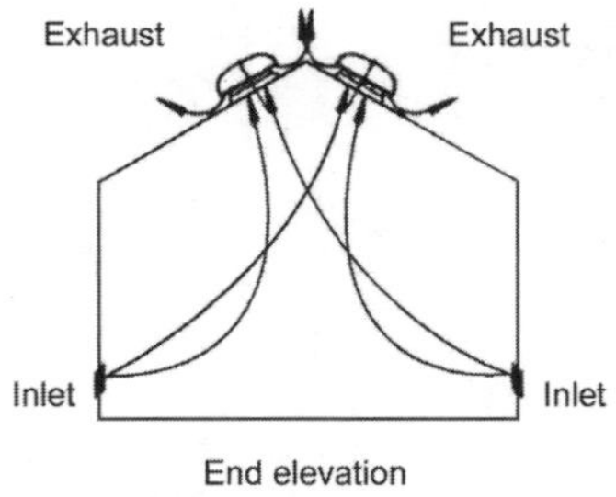

Figure 21.18 Mounting of roof extract option

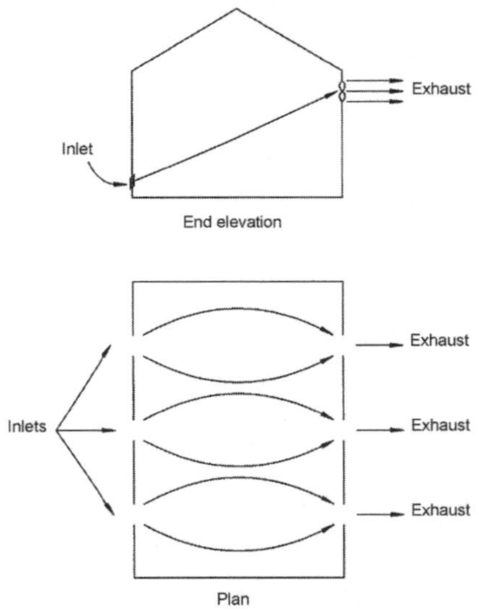

Figure 21.19 Location of wall mounted fans

to situate the units on one side of the roof as near to the apex as possible (Figure 21.17) or one can alternate as in (Figure 21.18), remembering that the units need to be as close as possible to the apex of the roof.

k) The wall mounted fans should be located on one wall; as high as possible, with the inlets set in the opposing wall give in a cross-ventilation system (see Figure 21.19).

21.2.7 Sizing the fans

21.2.7.1 Wall mounted

The unit needs to be complete and fully assembled with a wall plate, and motor compliance guard.

Using the earlier rule of one fan per 10 m, then at about 30 m length, 3 fans should suffice. From a typical standard range catalogue, we need to be looking for a volume per fan of about 1.25 m^3/s. The best choice would be a 450 mm diameter fan running at 1370 rpm, which for one particular manufacturer gives 1.71 m^3/s.

Do not worry too much about the extra flowrate; you will simply achieve 10 air changes per hour. By using closely matched speed controllers the flowrate can easily be reduced.

21.2.7.2 Roof mounted

Using roof extract units, the number necessary would be the same as the wall mounted requirement. e.g. three units being sufficient for the project.

Once again 450 mm units running at 1370 rpm for this manufacturer each give 1.38 m^3/s of volume flow. Again this is more than the 1.25 m^3/s per fan necessary to give 8 changes of air per hour. Then, by using the same speed controllers as the wall mounted system, a reduction in flowrate can easily be achieved.

So, by following these few simple steps, affordable ventilation is possible.

21.2.8 Construction

21.2.8.1 Cowl and base

The cowl and base should be manufactured from stainless steel, aluminium or cold pressed glass reinforced polyester resin to ensure long life and resistance to the severest weather conditions. If the cowls are made from such a resin, then they should contain an inbuilt ultra violet stabiliser to ensure that they do not fade during intensive sunlight. The method of construction described will ensure a material which is strong, light weight and which offers excellent properties against atmospheric corrosion. All fasteners and fittings should be of stainless steel.

21.2.8.2 Motors

Electric motors should have the following features:

- Totally enclosed, single or three-phase AC as appropriate
- Rated for continuous running in ambient temperatures up to 50°C
- Squirrel cage induction type for direct on-line starting
- All should have Class F insulation
- Pre-lubricated with high quality grease - re-lubricate after 30,000 hrs or 5 years
- Should have excellent speed control characteristics, capable of regulation to 20-30% of full speed
- Ratings to comply with BS 5000 Part 99 and IEC 60034-1
- Protection to IP55 - IEC 60034-5
- Overheat protection as standard on all single phase motors. To be available as a cost option on others

Flameproof and/or two speed motors may be necessary for some applications according to zone and also the variation in summer/winter duties.

21.2.8.3 Mountings

Fan support arms should be of mild steel resiliently attached to the base.

Mounting positions

- The roof extract units should be designed to operate efficiently when mounted horizontally or on a pitched roof up to an angle of 30°.
- Purlin boxes, soaker sheets, and direct mounting sheets for most popular roof profiles should be available for all sizes of units.
- Curb mounts - it is normal for the building contractor to fabricate this item.

21.2.8.4 Anti-backdraught shutters

Aluminium anti-backdraught shutters should be fitted as standard. The use of these shutters does not reduce the fan unit performance and will improve weathering under extreme conditions.

Shutter features:

- On side discharge units the shutters are opened by increased airflow and closed on flow reduction.
- On vertical discharge units the shutters are opened by increased airflow and closed by stainless steel springs.
- Synthetic buffers should be fitted to the units to ensure quiet operation.

21.2.8.5 Ancillaries

The following ancillaries may be necessary according to the application:

Inlet guards — Closes off the aperture within the ceiling.

Bird guards — Prevents entry of foreign objects via the discharge aperture.

Security bars — Manufactured with double row of steel bards for added protection.

Motorised dampers — Heat loss a critical factor - mount into ceiling aperture. Manufactured in aluminium, opposed multi-leaf blades.

Acoustic curbs — To meet critical noise criteria, splitter type specially developed to fit the units.

Speed controllers — Solid state electronic or auto-transformer.

21.2.9 Input units

It is not always desirable to have low-level inlets set in walls, especially where a building is full of equipment. Extract ventilation without a power supply will mean that the pressure within the building will be below atmospheric. To reduce leakage at doors and other openings, input units mounted on the roof may therefore be a solution to the problem. These units are similar to the extract fans, with of course, a reversal in the airflow direction.

Where the building is not full of heat generating equipment and where there are occupants whose environment needs to be heated then air heaters are often added. There may also be the need for this supply air to be cleaned to maintain an acceptable indoor air quality. In such cases, various types of filler are added.

21.2.10 High temperature smoke venting

The major cause of deaths at a fire is from the hot toxic smoke, rather than from the fire itself. Control and essential removal of this smoke from a building is therefore a vital component in any fire protection scheme. As our knowledge of the behaviour of fire increases, the traditional methods of exhausting the fire smoke, via natural venting have proved inadequate and systems using more positive and readily controlled fan-powered units are now frequently favoured.

Unlike normal ventilation systems, extraction rates for fire smoke venting have little to do with the size of the room. The amount of **smoke** produced depends largely on the size of the fire. As the smoke plume rises, surrounding cool air is entrained into the plume and becomes so well mixed with the hot smoky products of combustion as to form an inseparable component of the smoke. See Figure 21.20.

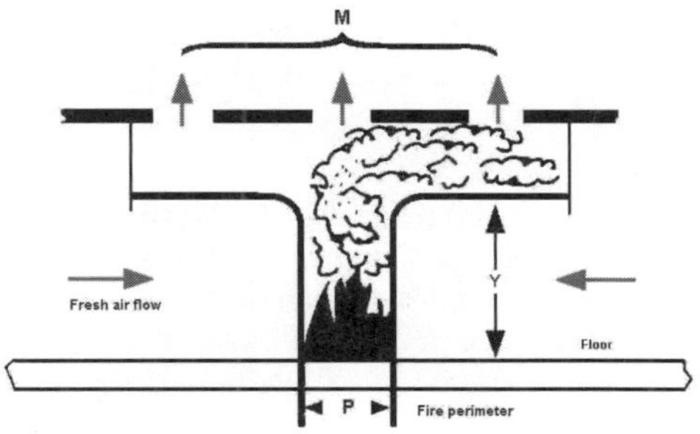

Figure 21.20 Smoke production

The quantity of smoke produced by a fire will depend on three factors:

- The perimeter of the fire.
- The temperature of the flames in the plume.
- The effective height of the column of hot gases above the fire.

At flame temperatures of around 800°C and ambient air temperatures around 17°C (density 1.22 kg/m^3), the production of smoke from a fire can be obtained from the simple expression:

$$M = 0.19\ P\ Y^{1.5} \qquad \text{Equ 21.3}$$

where:

M = mass of smoke produced (kg/s)

P = perimeter of the fire (m)

Y = height of the smoke layer (m)

The temperature of the smoke can be calculated using the formula:

$$\theta = \frac{Q_s}{M} \qquad \text{Equ 21.4}$$

where:

θ = temperature of the smoke above ambient (°C)

Q_S = heat carried by the smoke (kW)

M = mass of smoke production (kg/s)

Early smoke venting systems were designed around single storey low ceiling factory buildings. Here, the height through which the smoke rises is small. Hence, the smoke remains hot enough so that, when combined with a relatively deep smoke layer, it provides the buoyancy required to force it through the natural smoke vents provided.

In larger and more complicated buildings and especially for atriums and shopping malls, there is often insufficient buoyancy. During the critical early stages of a fire, when people are escaping, the smoke may be too cool to form a stable layer.

The relative merits of natural ventilation and fan powered ventilation are summarised in Table 21.2.

Method	Advantages	Disadvantage
Natural	1) Lightweight (if aluminium) 2) Self-regulating 3) Easy to retrofit or reuse 4) Operate at high temperatures 5) Units in non-fire zones can provide replacement air	1) Easily affected by wind 2) Require large areas of inlet 3) Large openings on roof 4) "Cool" smoke a problem 5) Material distortion 6) Not acceptable to all approving authorities
Powered	1) Guaranteed exhaust rate 2) Few smaller openings in roof 3) Can handle "Cool" smoke 4) Small area of inlet required 5) Can be used with ducting 6) Can be sited away from risk area 7) Will provide normal ventilation for building (2 speed)	1) Weight can cause a problem 2) Electrical supply and wiring 3) Retrofit not always possible 4) Expensive if high temperature (above 400°C)

Table 21.2 Merits of natural and fan-powered ventilation

21.2.10.1 Extractor fan requirements

The requirements for an extractor fan in a fire smoke venting system can be listed as follows:

a) To extract the hot smoky gases for a sufficient period of time to enable occupants to escape from the building.

b) To keep the building free of smoke for long enough to assist the firemen in locating the seat of the fire (will usually do this whilst performing requirement a) above.

c) If possible, to assist in clearing the residue smoke from the building, after the fire has been extinguished.

d) To provide the normal ventilation requirements of the building.

e) To extract the cold smoke during the early critical stages of a fire.

Because of the dilution which takes place, the smoke temperature is much lower than might be anticipated. It is often however extremely toxic. Nevertheless it is necessary to design the mechanical and electrical parts to withstand temperatures which, as far as the UK London fire authorities are concerned, reach 300°C, to allow sufficient time for building evacuation.

And also for the fire to be fought, this temperature has to be withstood during operation for one hour. Requirements in other countries differ. In all cases it is essential that the electrical supply is from an independently protected source, connected to the fan motor by fire resistant cable such as Pyrotenax.

It might be thought that "plastic" roof cowls would be unsuitable for such fans. This however is not the case and especially with vertical discharge propeller units the hot gases are kept away from the fibreglass reinforced plastic and the units continue to function for the requisite time. A typical system is shown in Figure 21.21 for which the design parameters were a sprinkled building with a design fire size of 3 m x 3 m with a heat output of 5 MW.

21.2.11 Conclusions

The use of extraction fans has continued to increase over the last few years despite their apparent simplicity. It is recognised that they have many advantages where a free floor area is desired and where the use of the building may change. Contrary to natural ventilation, they are independent of the weather and the

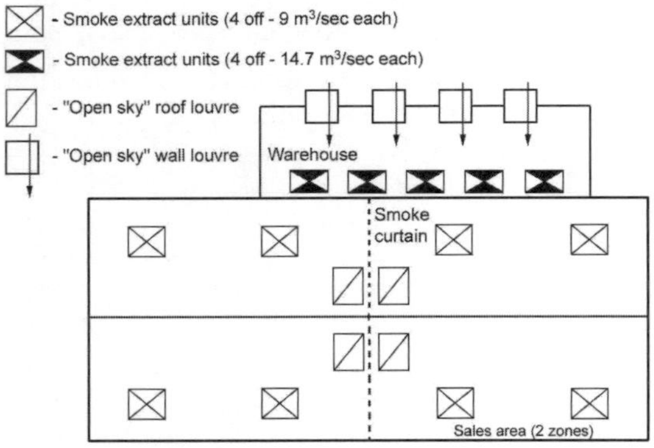

Design parameters

Sprinklered building
Design fire size - 3 m x 3 m x 5 MW

	Sales area	Warehouse
Height of clear layer	5 m	8 m
Smoke extract rates per zone	36 m^3/sec	58.6 m^3/sec
Maximim smoke temp	203 °C	109 °C
Fan specification	Category H.T. 300/0.5 (300 °C for 1/2 hour)	Category H.T. 105/5 (150 °C for 5 hours)
Units selected	4	4

Figure 21.21 Typical smoke extract system

ambient air conditions. The rate of extract can be tailored to suit the building usage and the number of openings in the roof or walls can be reduced. Modern units, incorporating mixed flow or centrifugal impellers, can overcome the resistance of ducting systems and filters to ensure good indoor air quality.

Where there is a desire to reduce the power used to a minimum, then hybrid systems should be considered. These would use natural or passive ventilation when ambient conditions favour it, but powered ventilation would be available when the outside air was warm and there was an absence of wind. It would also be of use for smoke ventilation, if the units were appropriately rated.

21.3 Residential ventilation

21.3.1 The UK situation

For many years we have been accustomed to providing some form of ventilation in two particular areas of a home. These particular applications are characterised by high levels of humidity and are as follows:

- Bathroom extract
- Cooker extract

Under previous conditions, these two areas were perhaps unique in British houses and therefore required the provision of extract ventilation with or without recirculation.

In recent years, under UK Government encouragement, the air tightness of domestic residences has improved considerably, to the extent that trickle ventilation openings and airbricks now have to be incorporated into new houses to improve the natural ventilation performance.

21.3.2 The situation elsewhere

In the USA and Commonwealth countries such as Australia and New Zealand, the situation is somewhat different and mechanical ventilation systems have been developed to overcome problems associated with mould growth due to the high ambient humidities at certain parts of the daily cycle.

Ventilation was also seen as a means of controlling volatile organic compounds, radon emissions from brickwork in granite and similar areas, body odours etc. Whole house ventilation has therefore become almost the "norm" in all new accommodation in these countries and this has included the roof spaces, attics, etc.

21.3.3 Introduction of the new part F Building Regulations

Part F (Ventilation) of the Building Regulations for England & Wales is currently under review by the UK Government. This is as a result of changes to Part L which includes minimising uncontrolled air leakage through the building envelope, which, although contributing to the ventilation, can result in draughts as well as wasted energy. However, sealing up the building can mean that there is not adequate ventilation, and the new Part F due to take effect from January 1st 2006 will deal with designing adequate ventilation systems to maintain good indoor air quality, without relying on air leakage into the building. The UK Government's approach may be summed up as "build tight and ventilate right".

With the introduction of the new part F and similar legislation in other European countries there will undoubtedly be a similar trend to whole house ventilation in the UK. Essentially the strategy has been outlined in BRE Digest 398, which is a best value approach already common in flats in the UK.

Essentially the concept is to provide simultaneous low rate extraction from at least kitchens, bathrooms, shower rooms, utility rooms and WCs and this is ducted to a central extract fan. Extract grilles are positioned in these various rooms as appropriate but should be as close as practicable to the actual source of water vapour. The extracted air should then be discharged outside the building via a single duct terminating in an outlet duct or cowl. The more sophisticated systems therefore incorporate a heat exchanger to minimise the heat losses to atmosphere.

21.3.4 Air tightness of dwellings

The dwelling should be as airtight as practicable to ensure economic operation. Currently, the achievable lower limit is a mean, background air infiltration rate of 0.2 air changes per hour (ach) for a conventional brick-built cavity-walled house, but this can be bettered in a timber-frame construction.

Generally speaking, a house with a mean natural air infiltration of 0.2 ach will have a higher rate if "pumped up" by an external fan in accordance with the accepted method of test. At an applied pressure difference of 50 Pa this should not however exceed 0.4 ach.

21.3.5 Air flowrate and air distribution

The total extract air flowrate during normal operation of the system, should be equivalent to between 0.5 and 0.7 ach based on the whole dwelling volume, less an allowance for background natural infiltration.

A facility to boost the air extract rate from the kitchen during periods of cooking, and from the bathroom during washing is highly desirable. It is suggested that an increase in extract airflow rate of 50% in a single room, or 25% for the whole system, would be a reasonable minimum.

21.3.6 System controls

A master on/off switch should be mounted on or near the fan unit to isolate the system electrically during cleaning and maintenance. A variable fan-speed control facility to boost the air extract rate from the kitchen during periods of cooking, and from the bathroom during washing is highly desirable.

21.3.7 Noise

Noise generated by continuous, operation should be controlled. Suggested reasonable levels are shown in Table 21.3. The CIBSE Guide B5, is also worth studying. These levels are for normal operation; they do not apply to systems under boost operation. They should not be exceeded anywhere that is normally accessible to occupants. The noise should be steady and contain no distinguishable tonal or impulsive characteristics. Noise caused by air flowing in the ductwork, (regenerated noise) can be reduced by minimising velocities and pressure loves across duct components and by having duct terminals, branches and bends well spaced.

Room classification	Suggested design background noise level dbA
Bedrooms	below 30
Living rooms	below 35
Dining rooms	below 35
Kitchens, bathrooms, utility rooms, circulation areas	below 45

Table 21.3 Background noise levels from mechanical services

21.3.8 Fan siting

The fan may be sited anywhere in the dwelling provided there is adequate access for cleaning and maintenance, and that any noise produced by the system will not disturb the occupants or their neighbours.

21.3.9 Dwelling characteristics

With mechanical ventilation, outdoor air must be admitted into the dwelling to replace extracted air. Dwellings which are not airtight will have a combination of controlled ventilation, provided by the fan system, and uncontrolled infiltration through air leakage paths all over the structure. With a leakage rate of 4 ach at 50 Pa, it is probably sufficient to rely on structural air leakage. If the bottom edges of internal doors clear the floor surface by 5 to 8 mm, there is likely to be a sufficient opening for air movement.

21.3.10 Ductwork

Ventilation ducts should be sized to give average air velocities below 4 m/s during normal operation to minimise noise. Higher air velocities are usually acceptable for boost operation.

Non-circular ducting of equivalent cross-sectional area may also be used. Size the duct system and terminal fittings, taking into account the airflow rate and pressure performance available from the fan unit. All joints should be sealed properly to avoid air leaks into or out of the ducts. Vertical exhaust ducts should be fitted with condensate traps.

Horizontal exhaust ducts should slope away from fans to prevent condensate running back and so reduce the risk of contacting live parts. Where ducts pass through an outside wall or attic floor, they should be carefully sealed to the building envelope to maintain the air tightness of the dwelling.

21.3.11 Duct terminal fittings

The terminal fittings should be durable, easy to clean and capable of passing the required airflow rate, at the available pressure drop, without generating excessive noise. Extract grilles should incorporate a dust filter and, if fitted in a cooker hood, a grease filter.

Extract grilles should be positioned so that they clear air from as much of the room as practical, ideally as far from the internal door and as high as possible. Above a shower cubicle or bath is a good place in the bathroom.

The extract-air outlet should always be outside the building (for example through a roof, wall or soffit) and away from any noise-sensitive areas such as bedroom windows. Outlet fittings should have a louvre, cowl or similar device to prevent rain, birds, large insects and rodents from entering the duct system.

21.3.12 Fire precautions

Ducting within the kitchen and connected to a cooker hood should be made of a non-flammable material such as steel. A fire damper is essential in all installations to close off the cooker hood's air extraction opening; it should be as close as possible to the hood.

Fire dampers are desirable in other kitchen extract terminals. Quick-acting fire dampers, such as a spring-loaded or gravity operated flap with thermal release, are preferred. If metal extract ducting is used, a fire damper and/or thermal fan cut-out switch should be fitted in the air extract system upstream of the fan unit. If plastic ducting is used for extract, fire dampers must, in addition, be fitted wherever the duct passes through any floor and through those ceilings and internal walls, which are required to be fire resisting. Fire regulations may impose requirements additional to those given here.

21.3.13 Cleaning and maintenance

Cleaning intervals depend largely upon the location and effectiveness of the system's air filters. Air filters and extract grilles will probably need to be cleaned at least two or three times a year. Fan impellers can be inspected, and cleaned if necessary annually. Cooker hood grease filters may need cleaning monthly to prevent contamination of the ductwork. The main source of information on cleaning and maintenance is the manufacturer's instructions.

21.3.13 Window opening and summer operation

Fan systems are usually operated continuously because they provide most of the ventilation in the dwelling. All habitable and service rooms on an external wall (except sanitary accommodation separate from a bathroom) should have one or more rapid ventilation openings, such as an openable window.

21.3.14 The fan and motor unit

Readers, who diligently read Chapter 13, will have noted that the efficiency of small AC electric motors can be disastrously low. There has therefore been a definite move to DC motorised fans for these systems. Modern electronic design has enabled a 240 volt single phase 50 Hz AC to be easily transformed to 12 or 24 volt DC.

Direct current motors run cooler, prolonging the life of insulating materials, lubricants and bearings. Electronically commutated DC motors can have the control module mounted "on board" to reduce the power and control wiring to a minimum, (Figure 21.22).

Figure 21.22 Typical DC fan
Courtesy of Vent-Axia Ltd

21.3.15 Fan mounting boxes

Fans can be mounted in boxes, which can if necessary be acoustically lined. The sides of the box can be provided with a number of spigots to which the ducting from the various rooms can be attached. Filters can also be included and a further spigot on the outlet side enables the discharge ducting to be fitted. Figure 21.23 illustrates a typical ventilation system installation.

Figure 21.23 Whole house ventilation system
Courtesy of Vent-Axia Ltd

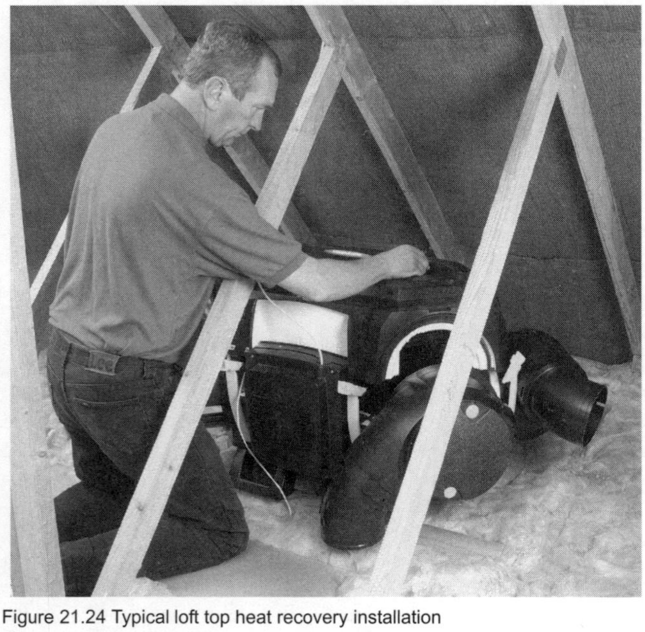

Figure 21.24 Typical loft top heat recovery installation
Courtesy of Vent-Axia Ltd

21.3.16 Heat recovery

Air-to-air heat exchangers have now been developed in the small sizes required for residential ventilation. These enable the heat which would otherwise be lost to atmosphere to be recovered and re-used for house heating. Typical installations can be disassembled and assembled on walls, ceilings and loft areas of roof voids with limited access. (Figure 21.24)

21.3.17 Conclusions

Whilst it might be thought that the provision of such systems will increase the household electricity bill, this is far from the case. By careful control, overall usage can fall as controlled ventilation, lower leakage and heat recovery all contribute to improving household efficiency.

21.4 Tunnel ventilation

21.4.1 Introduction

Ventilation of tunnels is necessary to remove pollution or heat and to control smoke in the event of fire. Pollution in road tunnels is caused by the emission of carbon monoxide, diesel smoke and the oxides of nitrogen from vehicles. In rail tunnels, it can come from locomotive diesel smoke while, in metros, heat removal requirements are far greater than those for providing fresh air for people.

Heat removal does not normally affect the ventilation requirements of a road tunnel as the calculated rate is far less than the ventilation rate required for removal of pollution. Heat removal can, however, affect the ventilation rate needed for rail tunnels — when a train is running in a long tunnel, air may be pushed through the tunnel at nearly the same speed as the train. This causes a local temperature increase around the train and overheating of the engine.

Control of the effects of fire is extremely important in tunnels. A crucial aspect is determining the design fire size as this will influence the ventilation requirements. In a road tunnel, the source of fire is the vehicles, i.e. the fuel and cargo and the flammability of the vehicle itself. In rail tunnels, the situation is similar, but there is also the possibility of electrical fires. For urban metros, electrical fires are important, as are general fires that can occur in buildings.

21.4.2 Ventilation and smoke control in metros

Ventilation of metros can be divided into two areas: tunnels and stations. At stations, heat generated by people, electrical equipment and trains needs to be removed. This is done by normal HVAC methods, by supply and extract ducts, by through-ventilation or by air conditioning. Tunnels are ventilated by longitudinal methods to remove heat from the trains and to avoid excessive air movement caused by trains.

For smoke control in stations, the system would be similar to that in a normal building except that access to external atmosphere for smoke extraction or escape is more restricted. The station has to be divided into compartments from which smoke can be extracted using high temperature fans and ducting.

Smoke should be extracted as close to the source of fire as possible and so it is important to know the design fire size and the amount and temperature of the smoke that will be emitted. The fire theories developed originally by the UK Fire Research Station (FRS),can be used for this purpose. The smoke temperature calculated should be the maximum. If the smoke is diluted with fresh air the temperature will drop. A difficult problem to overcome is the natural buoyancy of the smoke, which causes it to rise even against opposing airflow.

For tunnel ventilation, total heat input from the trains is used to calculate the ventilation requirements. Heat is usually removed by extracting air from the tunnels, with replacement air coming partly from the station and also from supply fans close to the stations or from natural vent shafts. Natural vents or fans are also often used to reduce the airflow caused by the piston effect of the train.

For smoke control in a tunnel, the principle is to provide sufficient velocity to blow the smoke away from escape routes. This is normally not difficult as the fires are usually small, about 15MW. Research has established the rate at which smoke is produced and travels away from a fire. If air velocity exceeds this figure, the smoke will move away from the fire. For a 15MW fire, 3m/s in the tunnel cross-section is sufficient to ensure this. The fans are usually fully reversible to enable the closest escape route to be used.

Although large areas are conventionally ventilated by using large fans to supply and extract air, jet fans can also be used (see Figure 21.25). These induce airflow by providing a high velocity jet of air, which diffuses and entrains the air through which it passes. The energy of the jet converts to static pressure, which causes a general air movement. Jet fans can be used in tunnels or large spaces such as covered stations. See Figure 21.26.

21.4.3 Ventilation of mainline rail tunnels

Rail tunnels are generally simpler to ventilate than metro tunnels. There is usually a single tunnel open at each end, with a central extract/supply shaft. As these tunnels are not usually connected to others, there is no need for complex network calculations involving movement of trains or combinations of fans.

In most tunnels everyday ventilation is provided by the passage of trains. However, in some tunnels, ventilation will be required to remove diesel smoke or to ensure sufficient air movement past the trains to cool them. In all but the shortest tunnels, there should be ventilation for controlling smoke in case of fire.

Figure 21.25 A 2500 mm reversible main ventilating fan for Western Harbour Crossing Tunnel, Hong Kong

Figure 21.26 Three 1250 mm jet fans used for ventilating the Nanterre-Orgeval Tunnel, France

For rail tunnels, the only practical ventilation system is longitudinal air movement along the tunnels. This can be provided either by large, reversible fans extracting/supplying at the centre of a tunnel, or by jet fans. If air is extracted or supplied at the centre of a tunnel, however, it enters or leaves by the easiest path, usually through the empty tunnel, not the section containing the train. So the relative system resistances must be taken into account when calculating the flow past the train.

With a jet fan system, air enters through one portal and exits through the other so all the air has to pass the train. The fans can be 100 % reversible to direct airflow away from the escape route. However, with a longitudinal system, if there is a fire in the centre of the train, the smoke has to be directed past parts of the train where there may be passengers.

Smoke control is again based on providing sufficient velocity to ensure that smoke only goes to one side of the fire, and on choosing the direction of flow to provide safety for the greatest number of people. Fans used for smoke control in rail tunnels would normally be rated for 150°C or 250°C for one hour.

21.4.4 Road tunnel ventilation

Road tunnel ventilation rates are based either on pollution control for comfort or the control of smoke in case of a fire. For comfort, the rate depends on the levels of carbon monoxide, diesel smoke or the oxides of nitrogen.

The pollutant emission levels of vehicles can be obtained from the handbook issued by the World Road Association (previously known as PIARC), which is formed from representatives of many nations. The information is in the form of base data with factors that are used to represent the differing conditions in each country. This base is then modified for vehicle speed, the gradient of the road and the altitude of the tunnel.

It is also necessary to know the number and weight of vehicles in the tunnel. The number of vehicles is calculated from the length of the tunnel, the number of lanes and the number of vehicles/kilometre/lane. This information, from an earlier edition of the PIARC handbook, is shown in Figure 21.27.

Revised figures presented at the Montreal Conference of PIARC in 1995 and later conferences give lower levels based on the improved emission rates of modern vehicles. For global

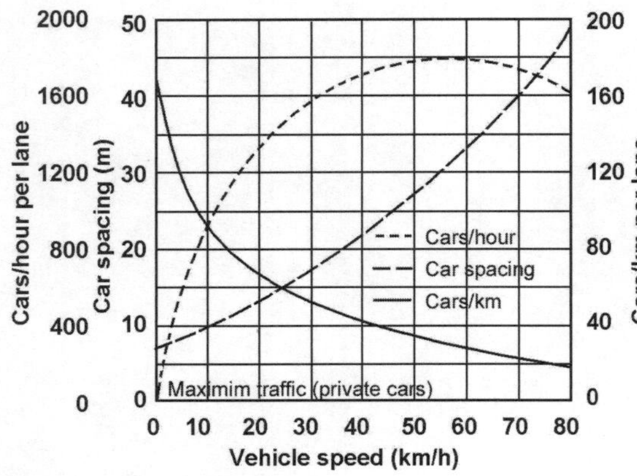

Figure 21.27 Maximum vehicle statistics

installations, and taking into account the emissions from older vehicles, the rates shown are still recommended for tunnels in any country other than the developed world.

The ventilation rate required is calculated from the number of vehicles in the tunnel multiplied by the weight of the vehicles, times the specific emission (CO, NO_X, and smoke), divided by the required increase in the level of emission (CO, NO_X and smoke). Each emission will give a different ventilation rate, so the highest must be used.

21.4.4.1 Dealing with the poisonous gases

Carbon monoxide is a short-term poison that is easily absorbed into the body, but is also quickly given up. The normal levels of dilution designed for are 100 to 250 ppm. Diesel smoke obscures vision and therefore the levels have to be set to ensure sufficient visibility to drive safely. The allowable levels of obscurity are between 0.005 m^{-1} for high speed conditions to 0.009 m^{-1} for lower speeds. The actual level set will depend on the country, speed, traffic conditions and lighting. Oxides of nitrogen are long-term poisons and are not normally important unless background levels are already high.

When establishing the ventilation rate for these gases, allowance must be made for the pollution level of the "fresh air". Any air supplied to the tunnel may already be partially polluted. Examples are: a city centre or twin tunnels, when the discharge air from one tunnel may partially recirculate into the inlet of the tunnel in the opposite direction. If the air is already polluted it could lead to a large increase in the required ventilation rate.

21.4.4.2 Control of smoke and hot gases

In recent years the effects of fires in tunnels has received considerable attention, following some well-publicised catastrophes. The ventilation necessary to control smoke and hot gases in the event of fire has been the subject of much research.

Two methods may be used. The smoke is either extracted at high level and removed through a ducted system, or air is blown towards the fire to force the smoke and hot gases in the desired direction. Tests have been carried out by PIARC to establish the volume needed in either of the cases described above.

21.4.5 Ventilation systems

For road tunnels, a number of types of ventilation system can be used, either singly or in combination.

- Fully transverse
- Semi-transverse
- Mixed
- Longitudinal

21.4.5.1 Fully transverse system

A fully transverse system has ducted supply and extract air. The supply is normally at low level from a duct underneath the roadway, so that the fresh air mixes with the exhaust fumes from the vehicles. The polluted air, which is buoyant, is extracted at high level, normally through a ducted system above the roadway, see Figure 21.28. It is technically the most exact system as it is not affected by wind pressure on the portals or traffic movement. It is normally used on long, congested or two-way tunnels. The costs of the civil construction and the mechanical equipment is high and the air distribution needs to be balanced carefully.

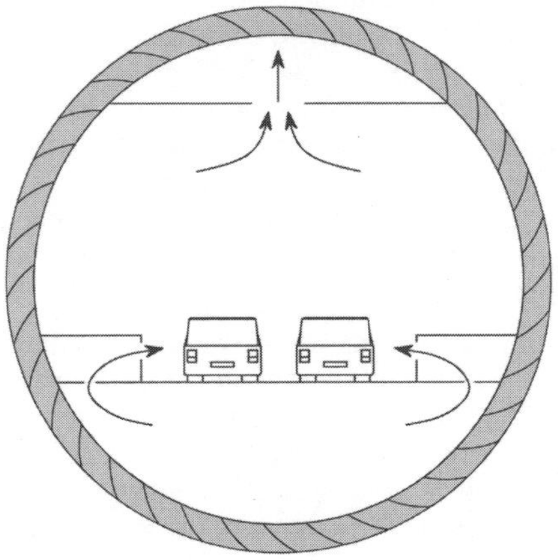

Figure 21.28 Fully transverse ventilation system

The fans used would usually be large axial flow fans operating in parallel, normally of fixed pitch design, with the total volume being controlled by switching fans on and off. Additional volume steps and reduced running costs can be achieved by using two-speed fans. Variable pitch and variable speed fans can also be used. The fans can be installed with their shafts vertical or horizontal. Splitter silencers are usually needed on both the system and atmospheric side of the fans. Two-speed fans can offer lower noise levels at night, when ventilation needs are normally lower.

When a fire occurs in a tunnel without longitudinal movement of air through the tunnel, it draws in fresh air from either side and produces a plume of smoke and hot gases. The plume stratifies above the fresh air and spread sideways along the tunnel, typically for 300m on each side of the fire. Beyond this distance, it cools sufficiently to lose its buoyancy and begins to mix with the fresh air being drawn in by the fire.

For effective smoke control, smoke must be extracted before mixing occurs. The system may extract air from the correct point, but it may not extract sufficient to capture all the smoke produced by the fire. The volume of air removed from close to the fire must therefore be increased. Often, additional extract points with dampers are fitted to the system. Those near to the fire can then be opened and the others closed to ensure air close to the fire is extracted.

If the tunnel is for one-way traffic, it is preferable to blow smoke and hot gases away from stationary vehicles. To achieve this, the supply and extract fans in different parts of the tunnel can be reversed to create the required longitudinal velocity.

17.5.5.2 Semi-transverse system

This system is similar to both the supply or extract system of a fully transverse system. If it is an extract system, (see Figure 21.29) the air is extracted at high level and the fresh air enters at the portals and is extracted through grilles at high level. The air becomes progressively more polluted towards the centre of the tunnel, so the ventilation rate must be increased to compensate.

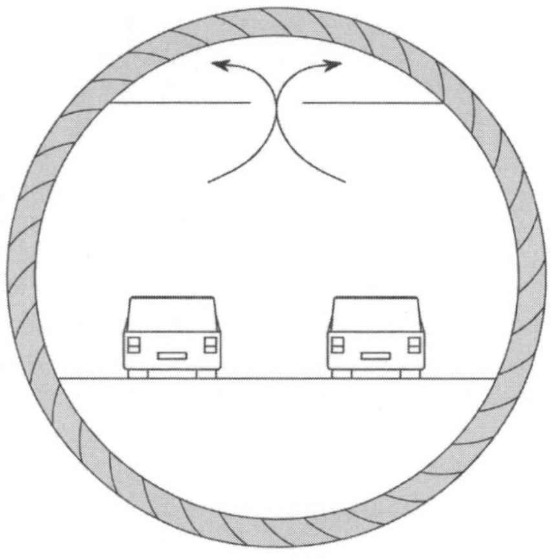

Figure 21.29 Semi-transverse (extract) ventilation system

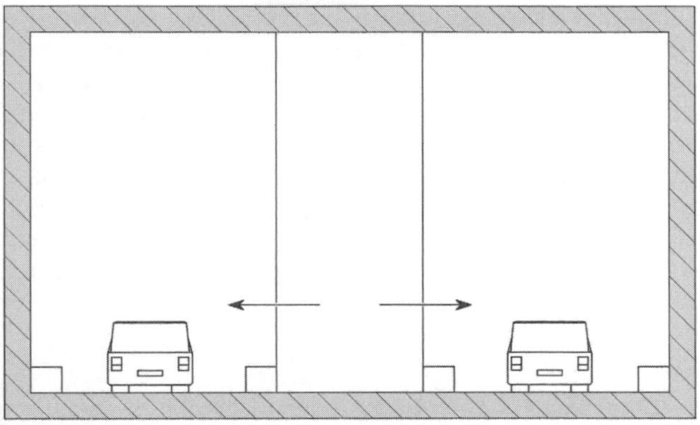

Figure 21.30 Semi-transverse (supply) ventilation system

In a supply system (see Figure 21.30), fresh air is supplied at low level and polluted air exits from the portals. All the air would be at the maximum allowable pollution level so there is no need for the ventilation rate to be increased. The air ducts can be above or below the road surface, or at the side of the carriageways.

A semi-transverse system is technically less exact than a fully transverse system as it relies on longitudinal air movement along the tunnel, which will be affected by wind pressure on the portals and by the movement of traffic. The ventilation rate, therefore, has to be increased to take into account the reduction of airflow in a particular direction due to these effects.

A semi-transverse system would be used for long, congested or two-way tunnels, although it may be limited by a maximum velocity at the portals. To reduce the portal velocity, a semi-transverse system may be used for the end sections of a tunnel with a fully transverse system for the centre section. Like the fully transverse system, a semi-transverse system can be expensive to construct and install.

For smoke control, a semi-transverse system has to operate in the same way as a fully transverse system, with extraction at high level. If it is a supply system, the fans must be reversible and able to exhaust from high level, with dampers to close off the low-level grilles. Fan installations are as for fully transverse systems but, as there is a longitudinal movement of air along the tunnel, the resulting pressure loss must be added to the loss in the ducted system.

21.4.5.3 Mixed system

A supply duct may feed part of the length of a tunnel section and an extract duct draw from the remaining length, with longitudinal flow between sections. In long tunnels through mountains, a longitudinal system can be used provided there are shafts to the surface to supply and extract air at regular intervals.

21.4.5.4 Longitudinal system

In a longitudinal ventilation system, air movement is along the length of the tunnel so there is no need fo r distribution ducts in the tunnel. The air can enter at one portal and leave at another, or be supplied or extracted from locations in the tunnel. Air movement can be induced either using large fans at these locations, by jet fans, or a combination of both types.

All the polluted air from a longitudinally ventilated tunnel exits from the portals. Large fans can be installed at these points to extract and discharge the air at high level. If the design velocity in the tunnel is very low, the fresh air may not be turbulent enough to dilute exhaust fumes from one car before they enter the air intake of the next. However, if the ventilation rate is designed to cope with fire, this is unlikely to occur.

If the tunnel is in an exposed position, the wind effects on the portals may be higher than the pressure that can be overcome using jet fans. In this case, the fans can be 100 % reversible to ensure that the airflow through the tunnel can follow the same direction as the wind pressure. The ventilation rate is determined either from the rate needed to dilute pollutants, or to blow smoke away from a fire.

Jet fans are usually installed in multiples spaced along the tunnel. Fan running time can be low because the air flow induced by traffic movement often provides sufficient ventilation. The fans are, therefore, sequenced to ensure uniform operating hours. It is important to note that the airflow through the tunnel is due to the pressure rise caused by the jet fan, which creates thrust by ejecting a jet of high velocity air. As this air decelerates, it transfers its energy to the general flowrate, causing a pressure increase equal to the fan thrust divided by the cross-sectional area of the tunnel. This pressure pushes the airflow through the tunnel and overcomes the drop in tunnel pressure.

As the deceleration of the air occurs gradually, if the longitudinal distance between fans is insufficient the deceleration will be incomplete and the increased velocity will affect the performance of the next set of fans. It is common to have ten tunnel diameters between sets of fans to eliminate this problem.

Other effects to be taken into account are the velocity in the tunnel and the proximity of the fan to the walls and ceiling. Fan thrust is measured under still conditions and is due to the change of momentum of air passing through the fan. If the air is already moving at the inlet to the fan, the change of momentum is reduced. Figure 21.31 shows the magnitude of this effect.

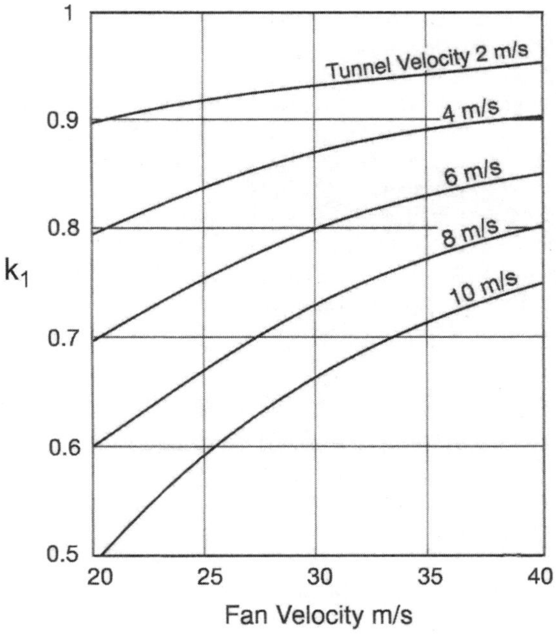

Figure 21.31 Effect of tunnel air velocity on fan thrust

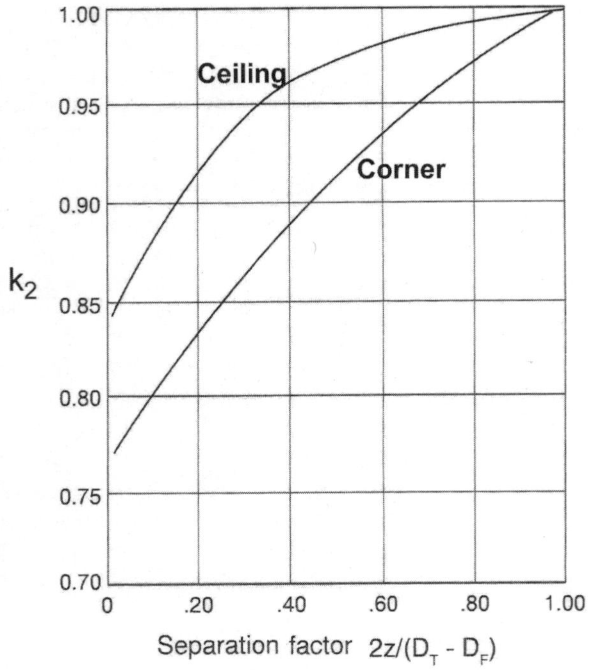

Figure 21.32 Tunnel fan installation factors

If a jet fan is installed close to a wall, there will be an additional loss due to the friction between the jet and the wall. PIARC suggested that the approximate effect of this is as shown in Figure 21.32. If the fan is close to a wall and a ceiling, the effect must be applied for both.

Reversible fans are slightly less efficient and slightly noisier than uni-directional fans but they are more flexible. They allow one-way tunnels to be used for two-way traffic when required and the normal direction of airflow to be reversed in fire conditions.

For tunnel ventilation schemes, jet fans offer lower capital and operating costs. They are often used with other systems, an example being to provide airflow at the required velocity in the event of fire or to provide the first levels of ventilation at reduced operating costs. Jet fans provide the performance required by a longitudinal ventilation system, which is the system generally accepted within Europe as the designer's first choice.

However, tunnel length can be a limiting factor as there is a practical maximum for the air velocity in the tunnel. For safety, this should be below 10 m/s; velocities above 7 m/s are rare. This system is also not satisfactory for tunnels of over 300 m with urban two-way traffic, unless there are emergency exhaust fans for smoke venting and protecting passenger escape routes. Potential fire loads of 10 MW and greater precluded the use of earlier designs of jet fans where the average air temperature could exceed 300 °C. But, in the last few years, these fans have been tested and certified for operation at 400 °C for two hours. This will enable people to be evacuated safely before fan failure.

21.4.6 Axial flow fans for vehicular tunnels

The most common type of axial flow fan has an impeller designed to operate mainly in the forward direction but is capable of being reversed. In order to improve performance in the reverse direction it does not have guide vanes. Such a fan would give approximately 65% of the forward flow in reverse and 50% of the pressure development. If the performance in reverse is less important, fans can be fitted with guide vanes to increase pressure development.

When air passes through a fan, the impeller turns the air and gives it a rotating component. The guide vanes turn this component into the axial direction with a regain of the kinetic energy from the rotating velocity. The increase in pressure depends on the type of fan but it could be 5-50%. The performance in reverse would typically be 55% of the volume in the forward direction, with 35% of the pressure.

If a fan is needed to give the same performance in both directions, this can be achieved in two ways. If first cost is the most important criterion, a reversible impeller can be manufactured by turning alternate blades through 180°. In this case, the blades running in the forward direction do most of the work but there is additional drag from the blades operating in reverse. This type of fan produces about 85% of the volume of a standard fan without guide vanes and 70% of the pressure. If performance and efficiency are important, a fan with 100% reversible blade sections is selected. This fan gives about 90% of the volume of a standard fan and 85% of the pressure.

A 100% reversible section consists of two top surfaces of an aerofoil section back-to-back in the reverse direction. The top surfaces of an aerofoil section creates nearly all the lift of the section by reason of the accelerated flow, which gives a negative pressure. When the fan operates in one direction, the top surface in the forward direction produces its normal lift although there will be additional drag dues to the extra thickness at the trailing edge from the reversed top surface. When the fan operates in the reverse direction, the other surface operates in its normal direction and an identical performance is produced. This type of section increases the fan efficiency by at least 5% compared to a fan with reversed blades.

The other type of fan commonly used for tunnels is a contra-rotating axial flow fan. This consists of two similar fans operating in series with the impellers rotating in opposite directions. The first impeller rotates the air in one direction and the second in the opposite direction so that it leaves the fan nearly axially. Thus, the contra-rotating fan develops up to three times the total pressure of a single fan at a very high efficiency.

21.4.6.1 Flowrate control

The technical considerations for flowrate control are detailed in Chapter 6. Flowrate control for tunnels is not required to make the ventilation system operate correctly. It is, however, used to reduce the operating costs of the tunnel since only four to six steps of ventilation are necessary.

The methods of flowrate control are:

- Variable speed
- Variable pitch
- Fans in parallel

Variable speed

The variable speed fan is a simple concept and usually employs induction motors driven by inverters. This method gives a good reduction of power with volume because, as a first approximation, the power reduces with the cube of volume. However, there is an initial penalty in the additional losses in the motor because the supply from the inverter is not truly sinusoidal and there is the loss in the inverter itself. The latter is an extra heat input into the plant room, which must be removed by ventilation.

The use of an inverter reduces starting current and so gives the best ratio of starting current to starting time. Care must be taken over the design of the system and input and output filters incorporated as necessary to eliminate problems with electro-magnetic compatibility and avoid affecting the mains supply. Also, larger inverters in some cases cost more than the complete fan.

Operating at a reduced speed reduces bearing wear and the temperature rise of a motor, although the non-sinusoidal wave form does give rise to additional magnetic forces in the winding.

The reliability of the winding, therefore, would not necessarily be improved by this decrease in temperature. The noise level of the impeller will be reduced by about 17dB at half speed, but additional magnetic noise may be apparent, particularly at the lowest speeds.

Variable pitch

A variable pitch fan is fitted with an electric, pneumatic or hydraulic actuator, which enables the impeller pitch angle to be changed while the fan is running. The fan volume flow can thus be reduced by reducing the pitch angle. Again, the first approximation to power reduction is that it reduces by the cube of the change of flowrate; however, the fan is normally selected for high efficiency at the maximum duty point and so, as the flowrate is reduced, so is the fan efficiency. The sound level reduces as the volume is initially reduced but can increase again at very low angles when the fan efficiency is very low.

This type of fan requires more maintenance and, because of the complexity of the design, is less reliable than a fixed pitch fan. The cost of a variable pitch fan would be between that of a fixed pitch fan and a fixed pitch fan plus inverter. Operating at reduced pitch angle and power will reduce the temperature rise of the motor and increase the life of the motor windings. Starting at reduced pitch angle will reduce the run-up time, particularly at reduced voltage.

Fans in parallel

Flowrate control can be most simply achieved by using more than one fan in parallel, i.e. the volume flow is divided between the number of fans used, all operating at the same pressure. The flowrate is controlled by switching off fans. As each fan is switched off, the flowrate through the total system reduces and so does the pressure. The fans therefore move down their fan characteristic curves actually giving less pressure but more flow from each.

The power does not reduce directly with the volume cubed as the losses in the parallel part of the installation (the fan, damper and connecting duct work) do not reduce because the volume through this section does not reduce. But if two-speed motors are used with full and three-quarter speeds, the power on the lower speed will be only 42% of that on high speed for a volume of 75%. Subsequent volume and power reductions can be made from this low level.

The other advantage of using a number of fans in parallel is that the fans are smaller, more standard and can be factory assembled and shipped ready to install. A higher percentage of stand-by in case of a fan failure is also achieved.

21.4.7 Calculation of jet tunnel fan requirements

21.4.7.1 Fresh air requirements

Fresh air required will depend on the following factors:

- The specified maximum permitted carbon monoxide and diesel smoke level. (Nitrous oxides have not been considered to be significant in the past, but in recent times levels are now being specified.)
- The number of vehicles/hour and the number of these which are diesel-fuelled
- Speed of traffic
- Gradients
- Altitude

In practice, it is accepted that the maximum air requirements occur when traffic is heavily congested at a speed of 10-15km/h. The figures in Tables 21.4 and 21.5 are based on vehicle emission data published in 1987 by PIARC following the XVIIIth World Road Congress.

Type of tunnel	CO at peak traffic (ppm) congested traffic or standstill
Urban tunnels (used to capacity)	
Daily congestion	100 - 150
Seldom congested	250
Inter-urban tunnels (highway or mountain)	250

Table 21.4 Recommended CO levels (PIARC 1987)

	Permissible visibility limited
Type of tunnel	K_{lim} m^{-1}
Urban tunnel with dense rapid traffic	0.005
Congested traffic	0.009
When K = 0.12 m^{-1} the tunnel must be closed	

Table 21.5 Recommended visibility limits (PIARC 1987)

The recommended CO levels are shown (see also Figures 21.33 and 21.34). The figures in Table 21.5 are also based on diesel vehicle emission data published in 1987 by PIARC. The

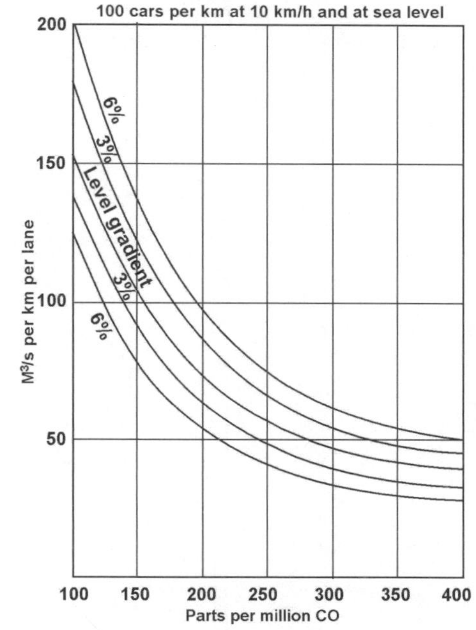

Figure 21.33 Fresh air requirements for CO dilution at sea level

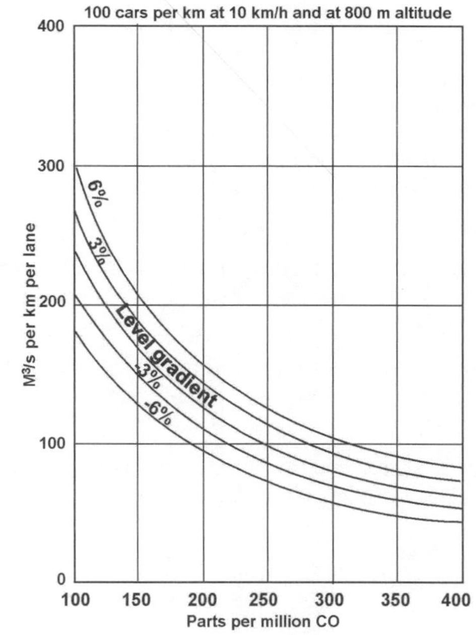

Figure 21.34 Fresh air requirements for CO dilution at 800 m altitude

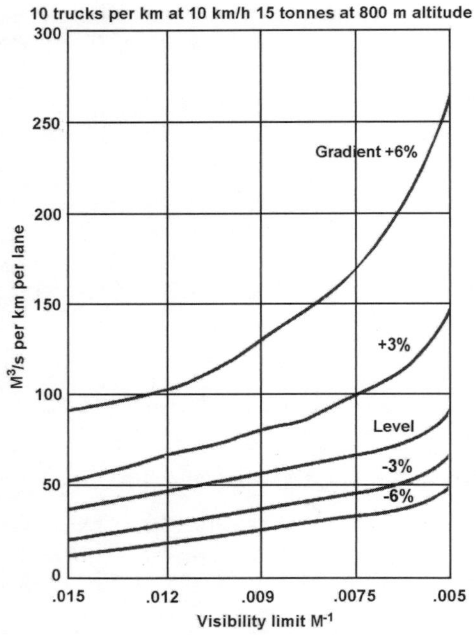

Figure 21.35 Fresh air requirements for smoke dilution at 800 m altitude

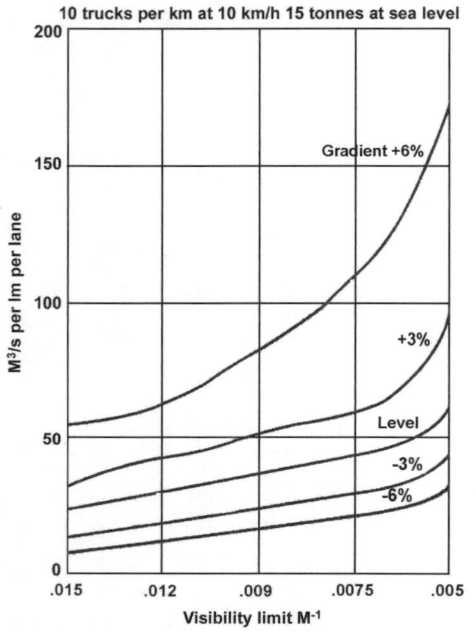

Figure 21.36 Fresh air requirements for smoke dilution at sea level

charts assume congested conditions with 10% of the vehicles having diesel engines, and having an average weight of 15 tonnes. (Figures 21.35 and 21.36 give the recommended fresh air requirements for smoke dilution.)

21.4.7.2 Tunnel thrust requirements

When the airflow requirements have been established it is necessary to calculate how much thrust is required to overcome the resistance of:

- **Inlet and outlet loss of the tunnel** The combined loss is generally assumed to be about 1.5 times the tunnel air dynamic pressure.
- **Tunnel surface friction** The loss associated with suspended fittings and road direction signs.

$$\text{Pressure loss} = \frac{1}{2}\rho\, V_T^{\,2}\, f \frac{L}{D_H} \qquad \text{Equ 21.5}$$

where:

V_T = tunnel velocity (m/s)

L = length of tunnel (m)

D_H = hydraulic diameter (m) (circular equivalent to tunnel cross-section)

f = friction factor

ρ = air density (kg/m³)

The value of "f" is generally taken as 0.025, but can vary from 0.02 to 0.04 depending on surface roughness and physical obstructions like lights, signs, etc.

There will be drag resistance where the traffic speed is lower than the average tunnel velocity in a single-direction tunnel. In a two-way tunnel, the traffic speed in the opposite direction to the tunnel air velocity must also be considered. Wind or temperature and barometer pressure difference between tunnel entry and exit must be considered.

The actual total thrust required from the fans is:

$$\text{Total thrust} = P \times A_T \ (\text{Newtons})$$

where:

P = summation of pressures (Pa)

A_T = tunnel area (m²)

The rating of a jet fan is commonly identified in terms of thrust applied to the air. The basic thrust rating is equal to the change of momentum between the fan inlet and outlet which is the product of the mass flow and some average velocity.

$$\text{Theoretical thrust} = \text{air density} \times \text{air volume} \times \text{air velocity}$$

$$= \rho \frac{Q^2}{A} \qquad \text{Equ 21.6}$$

where:

ρ = air density (kg/m³)

Q = airflow volume (m³/s)

A = fan outlet area (m²)

In the simplified equation 21.6 the nominal outlet velocity has been used, i.e. the fan flowrate divided by the fan outlet area. With a jet fan this is far from correct as the velocity varies considerably at the outlet plane (Figure 21.37).

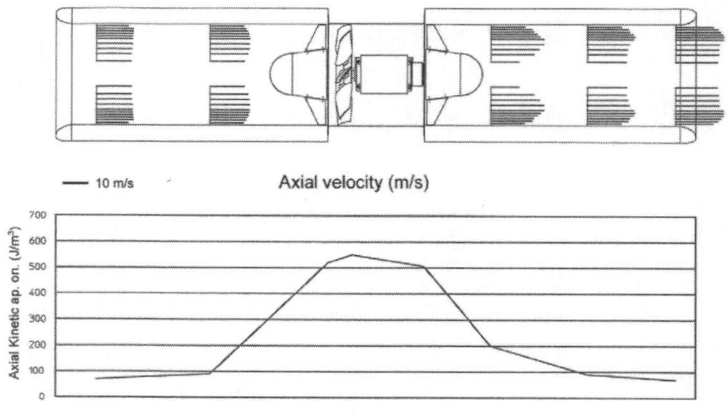

Figure 21.37 Axial velocity profile along a jet fan

Strictly speaking, the local thrust at varying radii should be integrated. Immediately on the fan outlet this is difficult and the thrust is therefore measured on a rig in accordance with ISO 13350.

Measured thrusts have been found to vary between about 85% and 105% of this theoretical value, depending on the blade design and resultant velocity distribution, distance from impeller to silencer outlet, effects of swirl, etc.

21.4.7.3 Entry and exit pressure losses

The entry and exit loss, $P_{EN\,EX.}$ is normally assumed to equate to about 1.5 times the tunnel air dynamic pressure. With careful design by the use of a "streamlined bellmouth" opening to the tunnel and gradual diffusion at the exit, these losses can be reduced.

$$p_{dT} = \frac{1}{2}\rho\, v_T^{\,2} \qquad \text{Equ 21.7}$$

where:

P_{dT} = tunnel air dynamic pressure (Pa)

V_T = average tunnel air velocity (m/s)

ρ = air density (kg/m³)

note:

$$v_T = \frac{q_T}{A_T}$$

where:

q_T = tunnel air volume flowrate (m³/s)

A_T = tunnel cross sectional area (m²)

17.4.7.4 Traffic drag or resistance

There will be a drag resistance in a single direction tunnel where the vehicle speeds are lower than the average tunnel air velocity. In a two-way tunnel, the traffic speed in the opposite direction to the tunnel air velocity must also be considered.

In recent times it has been recognised that jet fans can be used in the event of fire for the control of smoke. Depending on the predicted fire size, the resultant drag from a large number of stationary vehicles could be higher than that for moving traffic, when there would be fewer vehicles in the tunnel.

Traffic drag loss in a bi-directional tunnel may be approximated as:

$$p_{DRAG} = C_d \cdot \frac{A_v}{A_T} \cdot \frac{1}{2}\rho \left[\begin{array}{l} (N_{C1} + N_{T1})(v_{v1} + v_T)^2 \\ -(N_{C2} + N_{T2})|v_{v2} - v_T|(v_{V2} - v_T) \end{array} \right] \qquad \text{Equ 21.8}$$

where:

P_{DRAG} = pressure loss due to traffic drag (Pa)

C_d = vehicle coefficient of drag (1.0)

A_v = frontal area of vehicles (cars 2 m², trucks 6 m²) (m²)

N_{C1} = number of cars in tunnel moving against airflow

N_{T1} = number of trucks in tunnel moving against airflow.

N_{C2} = number of cars in tunnel moving with airflow

N_{T2} = number of trucks in tunnel moving with airflow

v_{v1} = vehicle speed of traffic moving against airflow (m/s)

v_{v2} = vehicle speed of traffic moving with airflow (m/s)

In a unidirectional tunnel the second term within the square brackets becomes zero.

21.4.7.5 Ambient conditions

At the tunnel entry and exit, ambient conditions may differ especially where the tunnel is long and/or the altitude changes as in mountainous environments. There may then be differences in wind velocity/direction, air temperature and barometric pressure, leading to stack effects either adding to or detracting from the tunnel resistance. The difference in barometric pressure at the two tunnel ends, if measured in Pascals is simply added to the system loss. This effect is known as p_{stack}.

21.4.7.6 Tunnel surface friction

Tunnel surface friction, together with the loss associated with suspended fittings such as lighting, road direction signs etc needs to be established. This pressure loss can be calculated from:

$$p_L = \frac{1}{2}\rho\, v_T^{\,2}\, f \frac{L}{D_h} \qquad \text{Equ 21.9}$$

where:

v_T = average tunnel air velocity (m/s)

L = length of tunnel (m)

D_h = hydraulic diameter (m)

f = friction factor

ρ = air density (kg/m³)

note:

$$D_h = \frac{4A_T}{P_T}$$

where:

P_T = tunnel periphery at a section (m)

A_T = cross-sectional area of the tunnel (m²)

The value of "f" can vary from about 0.02 minimum to 0.04 maximum. It is dependent on the surface roughness of the tunnel surfaces and the size and number of the tunnel fittings. In the absence of information to the contrary, a reasonable figure to use is f = 0.025

Total tunnel thrust - T_T

The total thrust required from the jet fan is numerically equal to the losses in the tunnel

$$T_T = p_T\, A_T \ \text{(N)}$$

where p_T is the summation of the pressure losses in items 1.0 to 4.0 i.e.,

$$p_T = p_{ENEX} + p_{DRAG} \pm p_{STACK} + p_L \qquad \text{Equ 21.10}$$

Jet fan thrust

To assist in the design of longitudinal ventilation systems by jet fans, it is convenient to rate them in terms of the thrust applied to the air, which they develop.

The basic thrust is equal to the change of air momentum between the fan inlet and outlet. This is the product of the mass air flowrate and the "average" air velocity at the fan inlet/outlet. The theoretical fan thrust is given by:

$$T_m = \text{air density} \times \text{air volume flowrate} \times \text{air velocity}$$

$$= \rho\, q_{vF}\, v_F$$

$$= \frac{\rho\, q_{vF}^{\,2}}{A_F} \ \text{(N)} \qquad \text{Equ 21.11}$$

where:

q_{VF} = fan air volume flowrate (m³/s)

V_F = average velocity at fan outlet (m/s)

A_F = cross-sectional area of fan (m²)

It should be noted, however, that this formula is only correct for a uniform velocity. The velocity profile at the fan outlet is far from even. The degree of distortion is very much dependent on the fan design particularly the hub to tip ratio on the impeller, the basis on which the blades have been designed (free, forced or arbitrary vortex) effectiveness of fairings, motor obstruction etc.

The measured thrust is obtained from tests carried out in accordance with ISO 13350. It varies from between 0.85 and 1.05 times the value of the "theoretical" thrust. Other designs have been tested with values as low as 0.65 times the "theoretical" thrust.

The total thrust developed by a number of fans in a tunnel is the sum of the individual thrusts. Fans may be located in groups operating in parallel, or in series spaced lengthwise along the tunnel, or any combination of the two.

General working rules are that fans in series should be spaced at ten or more tunnel diameters apart. Alternatively the spacing (m) can be taken as equal to the fan dynamic pressure (Pa) ÷ 10. Fans in parallel should have a minimum distance between centres of 2.0 times fan diameter.

The above rules are of necessity approximate only. More accurate calculations require knowledge of the Craya-Curlet Number for detailed tunnel designs.

The number of fans required:

$$N_F = \frac{T_T}{T_i} \text{ to the next whole number} \qquad \text{Equ 21.12}$$

where:

T_T = total tunnel thrust (N)

T_i = installed fan thrust (N)

When installed in the tunnel the actual thrust transmitted to the tunnel air will be less than that measured under the laboratory conditions specified in ISO 13350. Thus:

$$\text{Installed fan thrust } T_i = T_m\, k_1\, k_2\, k_3 \qquad \text{Equ 21.13}$$

k_1 is a correction coefficient based on the fact that the tunnel air velocity "offloads" the fan as compared with still air conditions. This may be obtained from Figure 21.31.

k_2 is a correction coefficient based on the knowledge that as the fans are eccentrically placed in the tunnel adjacent to one or two surfaces, some of the air will attach itself to the wall and or roof, rather than be directed into the main flow. This effect will be more severe the closer the fans are to these surfaces.

Figure 21.32 is for the case of no inclination i.e., the jet fan is parallel to the tunnel axis. In Figure 21.32:

z = distance of jet axis to tunnel wall or ceiling

D_F = jet fan diameter

D_T = tunnel diameter (can use hydraulic diameter for rectangular tunnel too)

Note: The corner factor applies to a fan installed equal distances from wall and ceiling.

The horizontal axis is more complex than most graphs of this coefficient reflecting the effect of both tunnel and jet fan diameters. It is developed from work carried out at London South Bank University.

k_3 as found from Figure 21.38 is a correction coefficient based on the knowledge that a small jet fan inclination of the fan can improve the installation performance.

Here, a family of curves have been presented for 4 separation factors (note the separation factor is the horizontal axis on Figure 21.32).

As expected, for the smallest separation, the jet clings to the wall, even at high inclination. The best result from jet fan inclination is achieved at a separation factor of 0.16, where an inclination of about 7 degrees gives an increase in thrust of 10%.

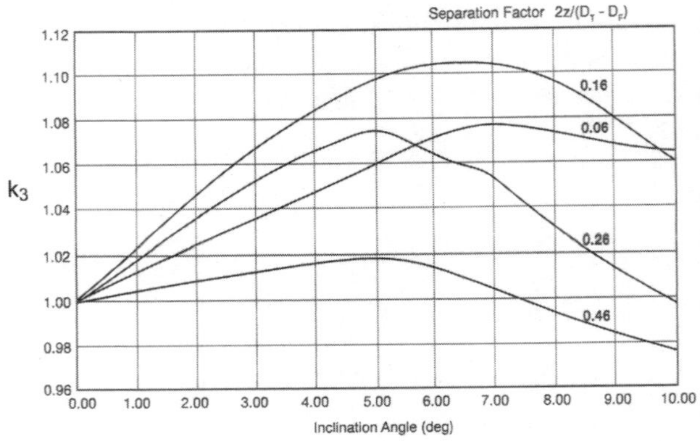

Figure 21.38 Effect of jet inclination on fan thrust

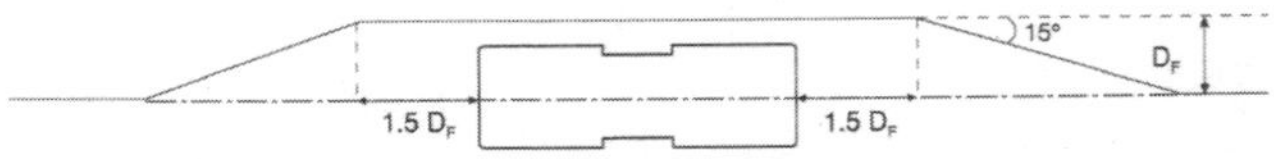

Figure 21.39 Typical tunnel niche

Clearly the optimum inclination angle differs with separation factor, increasing with decreasing separation.

Other factors which can affect the installed thrust capability are:

a) How close the first fan is mounted to the tunnel entry portal.

b) How close the last fan is mounted to the tunnel exit portal.

c) In immersed tube or cut and cover tunnels, it is common to have limited headroom and to locate fans in areas that are locally heightened i.e., niches.

For the typical construction shown in Figure 21.39. The combined coefficient for k_2. k_3 for the jet fan installed with the centre of the discharge on the ceiling line is as follows:

Inclination angle (°)	$k_2 \cdot k_3$
0	0.82
5	0.88
10	0.93
15	0.90

21.4.7.7 Testing for performance

Air flow

The volume flowrates are measured using complete units, i.e. the fan with silencers and bellmouths fitted, but with the inlet bellmouths replaced by inlet measuring cones in accordance with ISO 5801. Single measurements are taken and correspond to the volume flow close to zero static pressure. The velocities are derived from the area calculated from the inlet/outlet diameter.

Thrust

To measure the thrust, the jetfoil units are mounted on a test rig. This consists of a platform, supported by low friction linear bearings, mounted on a frame. The bearings constrain the movement to that in the direction of the axis of rotation of the fan. The platform is restrained by a load cell which measures the force exerted by the fan.

The rig is installed centrally in a large building to ensure that the circulating velocities are low and that there are minimal effects from the proximity of walls, ceiling and floors. The rig is levelled and the force exerted by the fan measured when the thrust and power reading have stabilised. This arrangement is one of the methods approved in ISO 13350.

21.4.7.8 "Real" thrust requirements

The total thrust developed by a number of fans in a tunnel is usually calculated as the sum of the individual thrusts. Fans may be located in parallel groups but should be spaced a number of tunnel diameters apart length-ways to ensure that one fan does not affect the other. Alternatively, an empirical rule is that the spacing can be taken as:

spacing (m) = fan dynamic pressure (mm)

Mounted in a tunnel, the fan will be off-loaded by the tunnel velocity, which will reduce the thrust available as seen in Figure 21.31. The higher the outlet velocity, the less is the fall-off in thrust. However, a high outlet velocity fan has a lower rating of Newtons/kilowatt. The distance of the fan from the tunnel wall or ceiling will also have an effect as seen in Figure 21.32. While work carried out in the late 1990s indicates that the ratio of fan diameter to tunnel diameter will also be of importance (Figure 21.40).

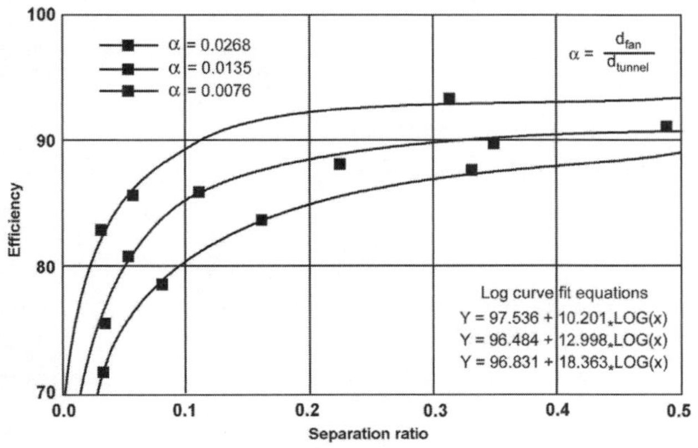

Figure 21.40 Variation in efficiency with separation ratio

$$\text{Installed fan thrust} = N_1\left[1-\frac{V_T}{V_F}\right]K_2 = N_1K_1K_2 \qquad \text{Equ 21.14}$$

where

N_1	=	thrust measured under laboratory conditions (Newtons)
V_T	=	average tunnel velocity (m/s)
V_F	=	average fan outlet velocity (m/s)
K_1	=	velocity coefficient (see Figure 21.31)
K_2	=	proximity of fan to tunnel wall coefficient (see Figure 21.32)
K_3	=	inclination of fan to tunnel coefficient (see Figure 21.38)

Other factors which can influence the result are:

- The distance apart of adjacent fans
- The distance between successive fans down the length of the tunnel
- How close the first fan is mounted to the tunnel entry portal
- How close to the exit portal the last fan is mounted

Fans would typically be located as shown in Figure 21.41. Where the tunnel must cater for traffic in either direction, fully reversible fans may be necessary, with blades designed to give equal flow for either rotation. Uni-directional fans may be acceptable and these can give about 60% of the forward thrust in reverse.

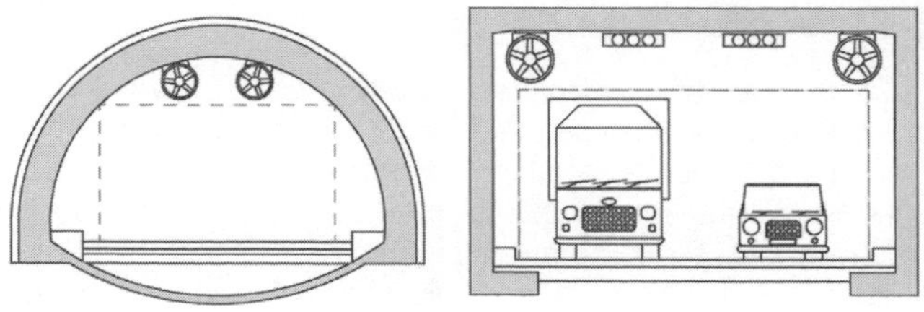

Figure 21.41 Typical fan locations

It is apparent that there is a degree of empiricism in the calculation of such systems. They work, but there are a number of "coefficients" used which disguise the lack of knowledge. In the major fan companies, these have been the result of extensive research programmes. They are unique to a particular design of fan and its consequent velocity profile.

Manufacturers look on them as closely guarded secrets and they are applicable to their product range. The alternative, and equally valid, statement is that they are unknown for some companies.

21.4.7.9 Guidelines for jet tunnel fan selection

When selecting jet fans the following guidelines should be followed:

- Choose the largest fans that can be fitted. Larger fans give a higher ratio of thrust to capital and installation costs than smaller fans.
- For the lowest operating cost, choose a low speed and/or pitch angle. The ratio of power to thrust is directly related to the fan outlet velocity, so for any given thrust requirements, the higher the velocity the higher the power consumption. Reducing the thrust for a given size of fan increases the number of fans and hence the capital cost, unless using a lower velocity also means that the length of silencers can be reduced of the silencers eliminated completely.
- The installation cost can be reduced by providing local niches or installing fans in car-only lanes.

21.4.8 Ventilation during construction

Ventilation is needed for construction as well as operation of tunnels. The requirements are for:

- sufficient fresh air for the people working in the tunnel
- removal of pollution from diesel powered vehicles
- removal of heat
- removal of dust caused by blasting.

The requirement for fresh air given in BS 6164 for working in tunnels which are freely ventilated, (i.e., not working in compressed air) is 9 m^3/min/m^2 of tunnel face. To this is added 1.9 m^3/min per kilowatt of diesel powered vehicles.

Fresh air for heat removal depends on heat input and allowable temperature. If the ventilation rate required is too high, cold water cooling systems can be used. The ventilation rate for blasting depends on the rate at which the fumes need to be cleared. 2 m^3/min/kg of gelignite will clear the face in 20 minutes. The ventilation system used consists of flexible ducting at high level in the tunnel and axial fans fitted as close to the inlet portal as possible to reduce cabling costs. The ducting is used to supply fresh air close to the face, particularly important when blasting.

The air becomes progressively polluted as it moves away from the face, so any air added to the tunnel reduces the pollution, which is highest at the exit from the portal. The pressure loss of the ducting varies with its design from 0.01 to 0.03 x velocity pressure x length/diameter. It is important to use the manufacturer's pressure loss data.

Multi-stage axial fans are ideal for this application. As the length of the ducting is increased, additional fans can be added to increase the pressure capability. By this means it is necessary only to install the number of fans needed to provide the required pressure, thus reducing operating and capital costs to a minimum. It is normally possible to select fans of a diameter similar to the ducting, which gives a very simple installation. The axial fans are light-weight and simple to support at high level in the tunnel. Fans can be added as required when the actual system pressure loss is known.

21.5 Drying

21.5.1 Introduction

A drying system is often necessary in industry:

- to reduce the moisture within a material to improve it
- to make industrial processes more efficient
- to recover the moisture where this has value.

Moisture may be present in a solid material in various forms as:

- surface moisture
- absorbed moisture
- water of crystallisation
- liquid in which a solid is in suspension or solution.

21.5.2 Moisture content

The amount of moisture present may be mathematically presented by the moisture content (m.c.), calculated in either of two ways:

- on a dry basis

$$\text{m.c.} = \frac{\text{weight of moisture}}{\text{weight of dry stock}}$$

- on a wet basis

$$\text{m.c.} = \frac{\text{weight of moisture}}{\text{weight of moisture } + \text{dry stock}}$$

In the past, moisture content was determined by weighing before and after drying in an oven:

a) at 100°C for about 5 hours

b) at 155°C for about ¼ hour (Carter-Simon oven)

depending on the ability of the material to withstand the particular temperature. With advances in electronics, however, such methods have been largely superseded by meters which measure changes in the electrical conductivity of a material with moisture content and can be calibrated accordingly.

21.5.3 Equilibrium moisture content

If a material is completely dried, it will regain moisture on contact with ambient air to an amount which will be dependent on the material, the air temperature and the air relative humidity. It will settle at some value, for this given set of conditions, known as the equilibrium moisture content (emc). It should also dry naturally in air to this value. Typical values may be obtained from Figure 21.42.

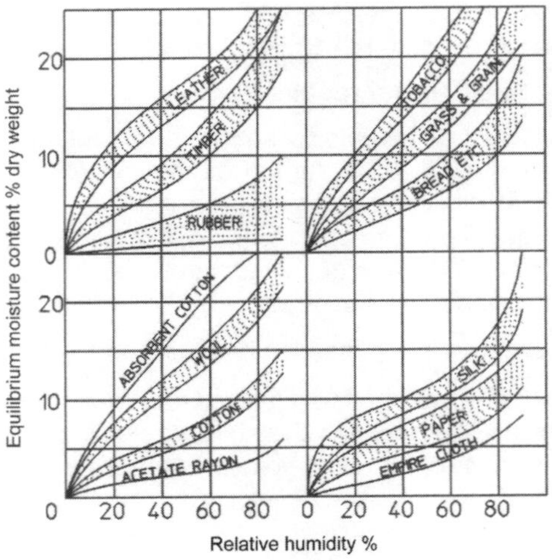

Figure 21.42 Typical equilibrium moisture contents

21.5.4 Methods of removing moisture

There are a number of different ways of removing moisture from a material:

a) Compression (squeezing)

b) Centrifuging (spinning)

(these two methods are only possible down to a moderate m.c.)

c) Air movement (heated or ambient temperature)

d) Application of heat (air movement will be necessary to remove the moisture)

e) Vacuum drying

f) Freeze drying (often in a vacuum)

g) Electro osmosis

For the purposes of *Fans & Ventilation* we are interested in method c), which employs fans, and where appropriate, method d) which uses heat, assisted by fans.

21.5.5 The drying of solids in air

There are 3 parts to this process:

1. Drying from a surface saturated with moisture – drying is then at a constant rate.
2. Unsaturated surface drying – this is a fairly short period when dry patches appear and the drying rate falls uniformly.
3. A second falling rate period in which the rate of drying is controlled by the rate of internal diffusion.

21.5.6 Critical moisture content

It should be observed that the critical moisture content is that at which unsaturated surface drying commences i.e., when the first dry spots appear. This varies from material to material as shown in Table 21.6.

Material	Critical moisture content
Pottery before firing	14 to 16%
Rubber	10%
Tea	174%
Leather	90 to 125%
Paper	33 to 70%

Table 21.6 Critical moisture contents of various materials

The whole process is shown diagrammatically in Figure 21.43.

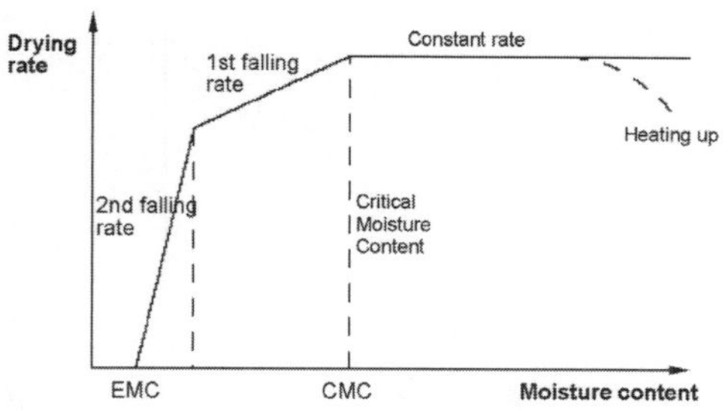

Figure 21.43 Drying process

21.5.7 Rate of drying

The rate of drying in each of the 3 parts of the drying process outlined in Section 21.5.5, may be calculated as follows (referring also to Figure 21.43):

1. Constant rate

Heat required to evaporate dM of moisture = dM·L

where:

L = latent heat

Heat supplied by the air = $kA\cdot\Delta\theta\cdot dt$

where:

k = overall heat transfer coefficient

A = surface area

$\Delta\theta$ = temperature difference between the air and the surface

$dM\cdot L = KA\cdot\Delta\theta\cdot dt$

or the rate of drying:

$$\frac{dM}{dt} = \frac{kA\cdot\Delta\theta}{L} \qquad \text{Equ 21.17}$$

For drying of solids in air with heat supplied by convection only, the surface of the material will be very nearly at the bulb temperature and $\Delta\theta$ becomes the wet bulb depression, and then k = convective heat transfer coefficient.

When drying materials in trays, with the air flow parallel to the surface:

$k - 0.0128\ G^{0.8}$

where:

G = mass velocity = ρv

When the airflow is perpendicular to the surface:

$k = 0/37\ G^{0.37}$

2. First falling rate

This is mostly controlled by the same process, since the surface drying is almost certainly well in excess of the diffusion rate. Calculations for this period are generally unnecessary in view of the relatively short time involved.

3. Second falling rate

The rate is controlled mainly by the rate of diffusion of moisture from the inside of the material to the outside surface. Air velocity is of relatively little importance as it does not accelerate the rate of diffusion.

The application of heat is probably therefore of greater importance, the air movement serving mainly to carry away the moisture reaching the surface. It has been suggested that the "soakage" equation can be used to indicate the processes involved:

$$\left(\frac{dM}{dt}\right)\text{fall} = k'\frac{d^2M}{dS^2} \qquad \text{Equ 21.18}$$

where:

M = mass of moisture

S = semi (half) thickness

k′ = constant depending on the rate of diffusion

For thin slabs and long drying times:

$$\left(\frac{dM}{dt}\right)\text{fall} = \frac{-k''}{S^2}(m - m_e)$$

where:

m = moisture content

m_e = equilibrium moisture content

$$\int_{m_i}^{m}\frac{-dM}{m - m_e} = \int_0^t \frac{k''}{S_2}\,dt$$

$$-\left[\log_e(m - m_e) - \log_e(m_i - m_e)\right] = \frac{k''t}{S^2}$$

$$\log_e\frac{m_i - m_e}{m - m_e} = \frac{k''t}{S^2}$$

or

$$t = \frac{S^2}{k''}\log_e\frac{m_i - m_e}{m - m_e} \qquad \text{Equ 21.19}$$

where:

m_i = initial moisture content

m = final moisture content

Since k″ is a constant, drying time is proportional to (thickness)^2.

Although values of k″ are not readily available, this expression is useful for finding, from known conditions, the drying times for other conditions, but the same material.

Full-scale plant may therefore be designed from data obtained with pilot plants. It should be noted that k″ may vary if the stock temperature is changed. This expression holds for homogeneous solids such as rubber, soap, gelatine, glue etc., but it is not so accurate for granular materials such as sand, paint pigments etc., probably due to the different way in which the moisture is released.

It has therefore been suggested that:

$$-\left(\frac{dM}{dt}\right)_{fall} = \frac{k'''}{S}\left(\frac{dM}{dt}\right)_{const}\frac{m - m_e}{m_i - m_e}$$

or

$$t = (m_i - m_e)\log_e\frac{m_i - m_e}{m - m_e}\frac{S}{k'''} \qquad \text{Equ 21.20}$$

i.e. drying time $\propto$ to thickness

Note: In calculations thickness maybe used in most cases to replace semi-thickness S.

21.5.7.1 Example

A material has an initial moisture content of 20% and is dried to 14% in 20 hours, the equilibrium moisture content of the material being 10% and the sample having a thickness of 6 mm.

How long would you expect it to take to dry a 12 mm thick material to 12% moisture content assuming:

a) a homogenous material?

b) a granular material?

Solution a)

$$\left(\frac{a_2}{a_1}\right)^2 \frac{\log_e\left(\frac{m_i - m_e}{m - m_e}\right)_2}{\log_e\left(\frac{m_i - m_e}{m - m_e}\right)_1} = \frac{t_2}{t_1}$$

$$\left(\frac{12}{6}\right)^2 \frac{\log_e\left(\frac{20-10}{12-10}\right)_2}{\log_e\left(\frac{20-10}{14-10}\right)_1} = \frac{t_2}{20}$$

$$4\,\frac{\log_e 5}{\log_e 2\cdot 5} = \frac{t_2}{20}$$

$$\text{or } t_2 = 80\,\frac{\log_e 5}{\log_e 2\cdot 5}$$

$$= 80\,\frac{0.699}{0.3979}$$

$$= 140 \text{ hours}$$

Solution b)

$$\left(\frac{S_2}{S_1}\right) \frac{\log_e\left(\frac{m_i - m_e}{m - m_e}\right)_2}{\log_e\left(\frac{m_i - m_e}{m - m_e}\right)_1} = \frac{t_2}{t_1}$$

$$\left(\frac{12}{6}\right) \frac{\log_e\left(\frac{20-10}{12-10}\right)}{\log_e\left(\frac{20-10}{14-10}\right)} = \frac{t_2}{20}$$

$$2\,\frac{\log_e 5}{\log_e 2\cdot 5} = \frac{t_2}{20}$$

$$\text{or } t_2 = 40\,\frac{0.699}{0.3979}$$

$$= 70 \text{ hours}$$

21.5.8 Elementary psychrometry

1. Air used directly and discharged

The psychrometry for such a system is illustrated in Figure 21.44.

In this case conditions A and B may be determined but point C cannot. The maximum amount of moisture which the air can absorb is $W_d - W_b$. In practice only $W_c - W_b$ is absorbed and

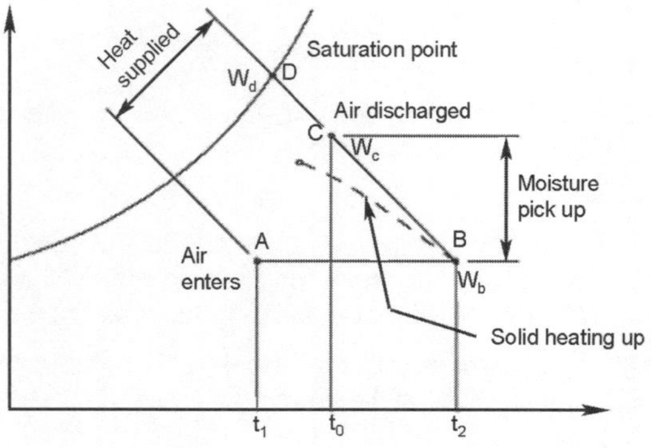

Figure 21.44 Psychrometry for direct system

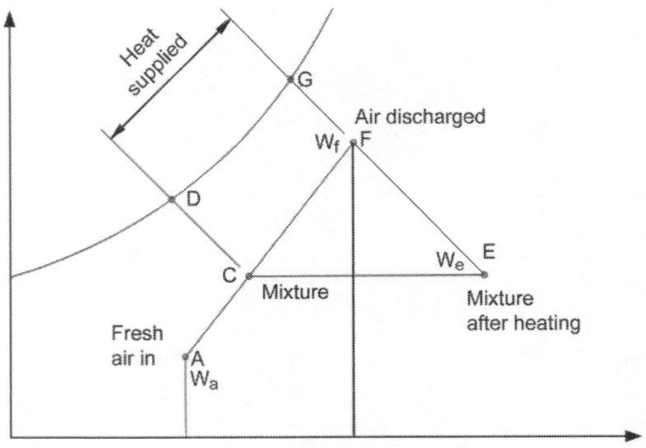

Figure 21.45 Psychrometry for recirculated system

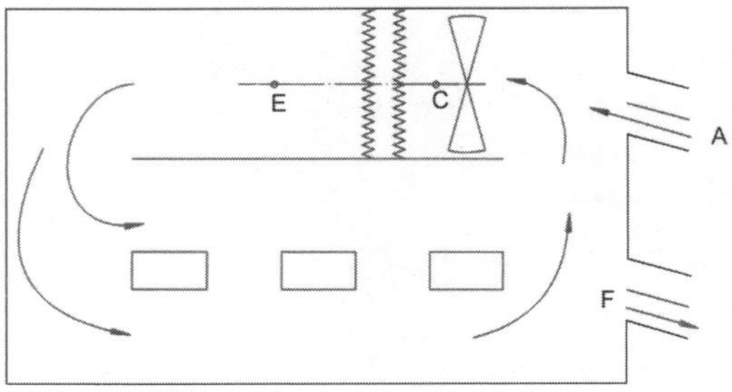

Figure 21.46 Fresh/recirculated air system

$\frac{W_c - W_b}{W_d - Wb}$ may be between 0.1 and 0.7. Since it is desirable only to supply latent heat of evaporation, discharge at condition C is generally wasteful.

2. Use of recirculated air

The psychrometry for such a system is illustrated in Figure 21.45 and a typical system is shown in Figure 21.46.

Here fresh air at A is mixed with recirculated air at condition F giving condition C. Condition F may be set by means of controls to a higher moisture content than without recirculation.

Thus, for a given volume of fresh air a greater amount of evaporation takes place. Moisture pick up = W_f - W_a.

Heat requirements

1. Heat given to the air = $G\,C_{air}\,(t_{out} - t_{in})$
2. Latent heat of evaporation = $G\,(W_{out} - W_{in})\,L$
3. Heat given to solid = $M\,C_{solid}\,(t_{solid\ out} - t_{solid\ in})$
4. Heat given to moisture in solid = $RM\,(t_{solid\ out} - t_{solid\ in})$
   $t_{solid\ out}$ = wet bulb temperature of air leaving dryer
5. Heat loss from dryer

21.5.9 Practical drying systems

There are two main classification of dryers:

- **Direct dryer** — material is heated by direct contact with the heating medium e.g. hot air.
- **Indirect dryer** — material is indirectly heated by medium via conduction or radiation.

We are primarily concerned with direct dryers

- Batch dryers — Suitable for clay ware, powders and foodstuffs, see Figure 21.47.
- Continuous dryers — continuous dryers for sheet, Figures 21.48 and 21.49.

For drying of leather sheet 27 to 50 °C is typical, see Figure 21.50.

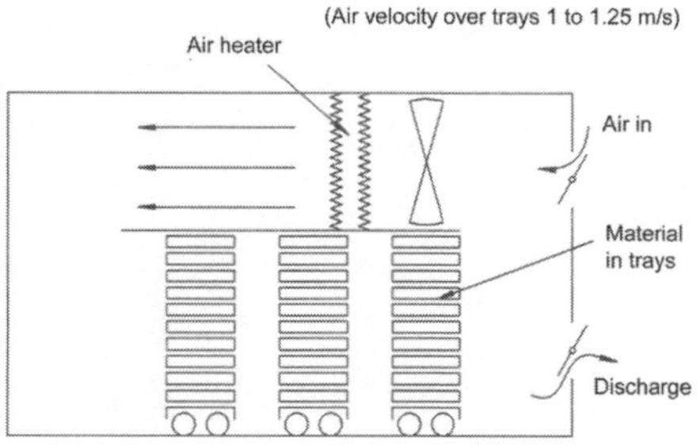

Figure 21.47 Batch dryer

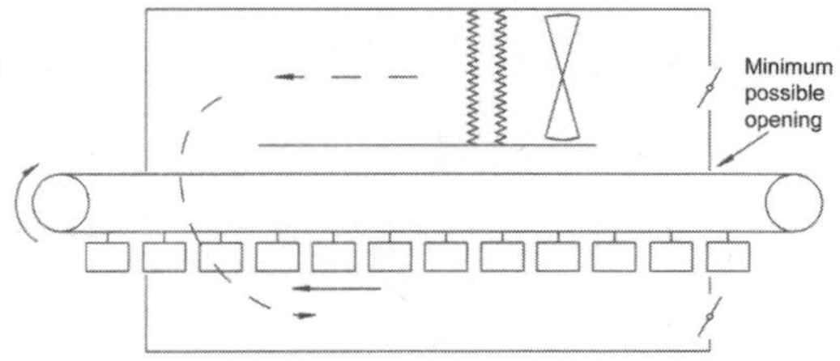

Figure 21.48 Continuous dryer

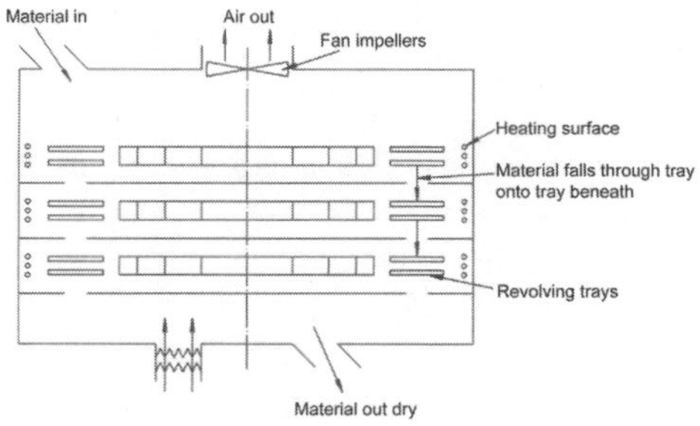

Figure 21.49 Turbo continuous dryer for sheet materials

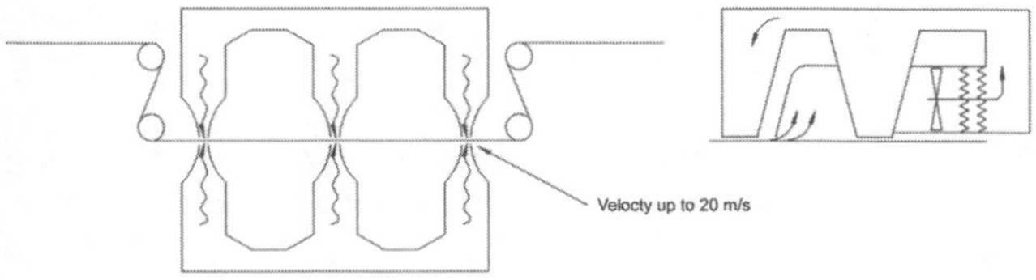

Figure 21.50 Stentar drying method for woven material

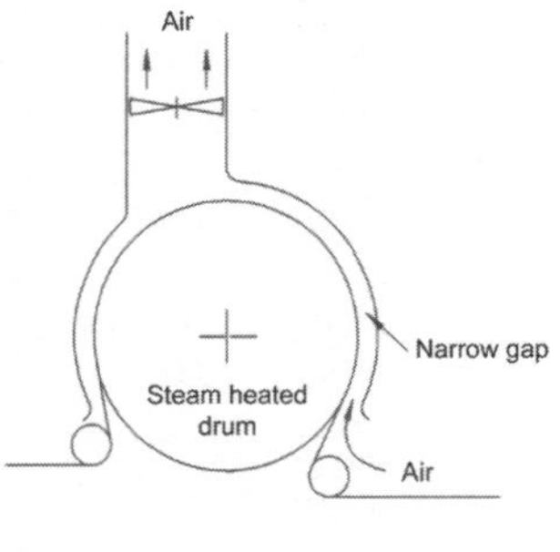

Figure 21.51 Callendar dryer

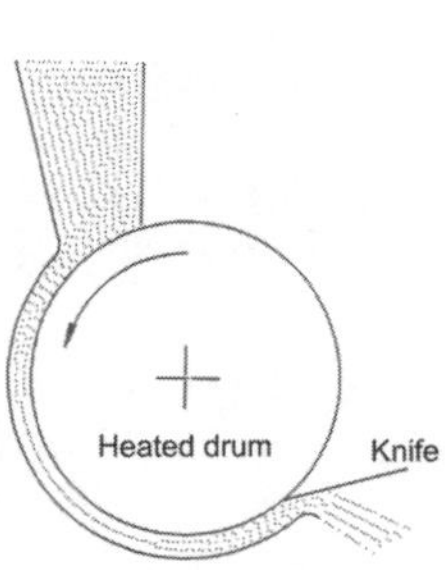

Figure 21.52 Paste or slurry dryer

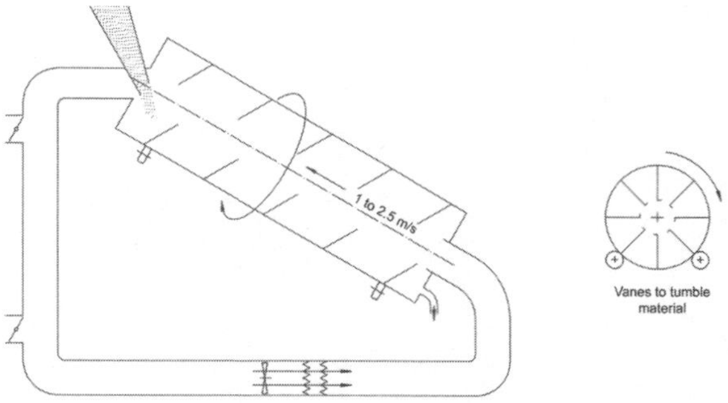

Figure 21.53 Rotary drum dryer

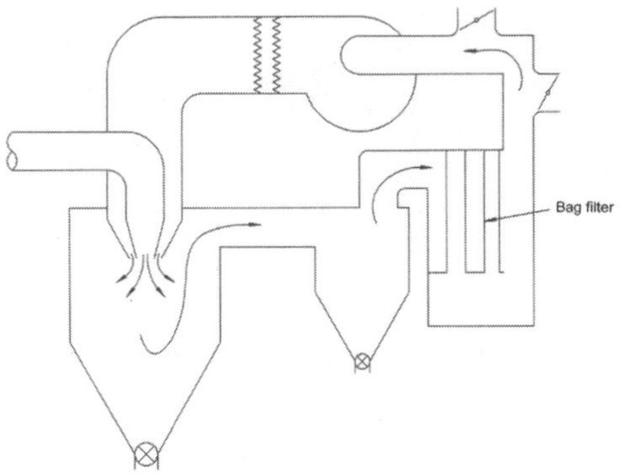

Figure 21.54 Spray dryer

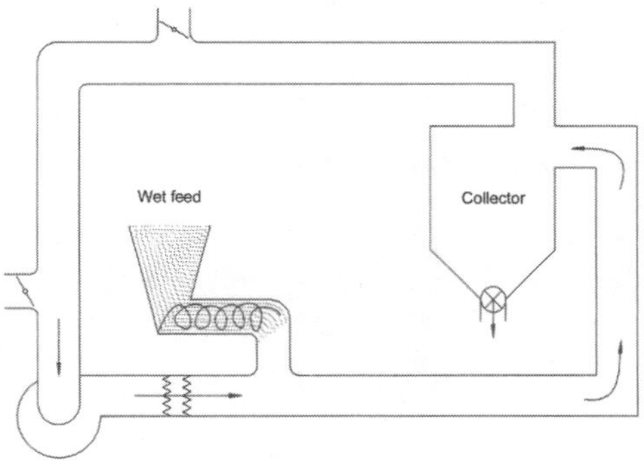

Figure 21.55 Pneumatic or flash dryer

In the "Shirley" accelerated dryer, which uses a rotating drum, air velocities of 15-20 m/s, ft/min are employed.

Other types of dryer are illustrated in Figures 21.51 to 21.55.

In Figure 21.53, instant drying is achieved because of very intimate contact between solution and heated air.

Rotary type atomisers probably give better control of particle size, dried egg, mild, detergents etc.

21.6 Mechanical draught

21.6.1 Introduction

Gas, oil and coal fired boilers are used extensively to provide heating, hot water and steam for process applications, and in the generation of electrical power. In all cases where a fuel is being burnt, fans are used to provide combustion air, to transport exhaust gases and in some cases used to transport and deliver the fuel (coal) into the furnace.

Fans are used to deliver the air used for combustion into the furnace, and are known as forced draught (FD) fans. In smaller boiler units FD may be the only fan in the system. In larger units additional fans are also installed which transport the exhaust gases out of the boiler unit. These are known as induced

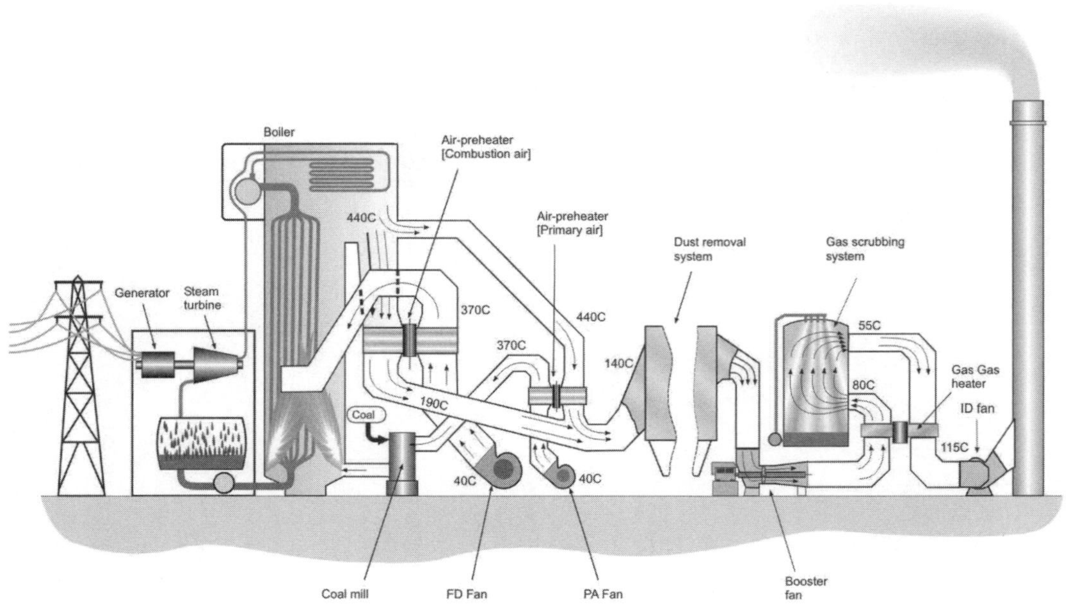

Figure 21.56 The position of various fans in a typical water-tube boiler plant
Courtesy of R Mulholland

draught (ID) fans and normally handle hot gas at around 140°C. On larger coal fired plant additional fans, called primary air (PA) fans are used to transport the powdered coal, typically the constituency of talcum powder, into the furnace. In older plants, the alternative was mill exhausters, down stream of the pulverising mill. These were modified paddle bladed centrifugal fans with heavy duty casings and impellers.

For industrial boilers the absorbed power of an individual fan would be in the range of 100 to 10,000 kW depending on the size of the plant. The fans, which are used to transport the air and exhaust gas through the boiler furnace and ducts, are sized not only to cater for the flow requirements, but also to overcome the pressure drop through the system. Figure 21.56 shows the position of the various fans in a typical water-tube boiler plant.

The three main types of fans are as follows:

Forced draught fan (FD)

This fan draws atmospheric air and delivers it, within a ducting system, to the combustion furnace. In larger plants the air is supplied into a heat exchanger where the air is preheated, normally to around 300°C, before entering the furnace. Both centrifugal and axial fans can be used.

To cater for changes in the boiler loads, in the case of centrifugal units the fan output is varied by the use of variable geometry inlet vanes or variable speed. On axial fans variable pitch, in motion, blade adjustment is used to regulate the airflow.

Induced draught fan (ID)

This fan is situated at the opposite end of the boiler ducting system from the FD fans and handles combustion gases, normally at around 140°C. The fans are specially designed to cope with the higher gas temperatures and in some cases must be able to cope with erosive dust.

Again the fan output must be adjustable to cater for the variable output of the boiler unit. Because of the lower density ID fans are larger than the FD fans but in many cases look similar. Both centrifugal and axial fans can be used, however when dust particles are present the axial fan blades require a more elaborate erosion protection system.

Primary air fan (PA)

This type of fan is only found on a coal-fired plant and its main function is to transport powdered coal into the furnace. It draws atmospheric air and supplies it to a heat exchanger where temperature is normally raised to around 300°C. The hot air then passes through the coal grinding plant where it picks up the coal dust and transports it on to the furnace.

Although the fans are significantly smaller than the FD and ID fans, the requirement to provide a much higher air pressure means that they still absorb significant power. As a result of the higher pressure requirement, the duty is more suited to a centrifugal fan.

21.6.2 Combustion

The most economical use of fuel has received the attention of manufacturers, designers and government agencies for many years. Older readers will perhaps remember those essential books of their youth - *The Efficient Use of Fuel* and *The Efficient Use of Steam*. Since that time, of course, the cost of fuel has increased enormously, whilst the need to reduce carbon dioxide emissions is now seen as an aid to self-preservation.

Defined in simple terms, combustion is the chemical composition of oxygen with combustible material such as carbon, hydrogen and, if unavoidable, sulphur. Oxygen is of course a constituent of the air around us. Under normal ambient conditions, air contains about 21% oxygen by weight. The remaining 79% however is almost entirely composed of nitrogen which to all intents and purposes is inert.

Before combustion actually takes place, a solid fuel must be heated to ignition temperature. The volatile gases in combination with the oxygen in the air supply then burn, and by increasing the temperature of the remaining material, ignite the fixed carbon. This is converted into carbon monoxide or carbon dioxide, according to the amount of oxygen present. Any non-combustible material remains as ash. It should be noted that pulverised coal generally burns firstly, by the formation of carbon monoxide (and other volatile distillates) and then further to carbon dioxide.

For liquid fuels the combustion process is simpler. They are soon converted into gaseous compounds, which burn very much as gases proper.

Gaseous fuels burn immediately and do not have the severe problems of an ash residue. They may however produce significant quantities of moisture in the form of water vapour.

When fuels are burned, the whole of the heat produced cannot be used. Apart from furnace radiation losses, some of the heat is taken up by the products of combustion.

Perhaps most importantly there will be a considerable loss due to the amount of excess air used in an endeavour to obtain complete combustion. For this reason alone the use of mechanical draught is now almost universal.

The theory of combustion is comparatively simple. It is much more difficult to apply in practice, however. Properly mixing air in the correct proportions with fuels and combustible gases to obtain complete combustion is not easy. Often the quantity of air delivered to the furnace is far in excess of that theoretically required. Although excess air means a loss of boiler efficiency, it is often necessary to ensure complete combustion, the amount depending on the quality, quantity and size of the fuel burnt. When fan draught is used, the air supply can be closely regulated and controlled.

21.6.3 Operating advantages

Whilst the numbers of boiler plant continuing to use natural draught alone is now very small, it is as well to remember the advantages that are obtained with the use of mechanical draught fans:

- increased boiler output and reduced heat losses via the chimney
- lower grade and less costly fuel may be used
- exact adjustment of draught to boiler load requirements
- improved combustion obtainable which with proper firing will reduce smoke emissions
- permits the addition of heat recuperating equipment such as economisers and air pre-heaters to reduce exit gas temperatures and therefore heat losses.

21.6.4 Determining the correct fan duty

The carbon dioxide percentage in the flue gases at exit from the boiler is a measure of the excess air admitted for combustion. It is dependent on the average maximum theoretical CO_2 % of the particular fuel being burnt and the method of firing. Table 21.7 gives the general range of boiler operating conditions.

It may be assumed that the lower CO_2 % corresponds to "good" combustion whilst the higher figure is 'very good'. Medium percentages would be 8 to 10% for coal, whilst 5 to 8% would be considered "poor". It will be noted that the figures for oil are closer to their theoretical maximum, reflecting increased ease of obtaining good combustion with this fuel.

Type of boiler	Fuel and firing	Boiler efficiency %	Operating CO_2 range %	Average maximum theoretical CO_2 %
Water-tube	Pulverised fuel	85 to 88	12.5 to 15	18.5
	Coal - stoker	77 to 84	11 to 13.5	
	Oil	82 to 86	11.5 to 12.5	15.4
	Wood	75 to 82	11 to 14.5	20.2
Shell	Coal - hand	60 to 68	9 to 11	18.5
	Coal - stoker	68 to 75	10 to 13	
	Oil	70 to 77	11 to 12	15.4

Table 21.7 General range of boiler operating conditions

For more detailed information concerning particular boiler types and other fuels you should consult the boiler manufacturers.

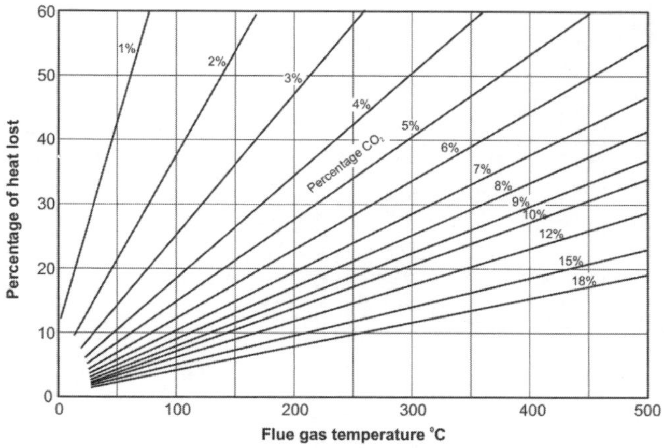

Figure 21.57 Percentage of CO_2 at various gas temperatures

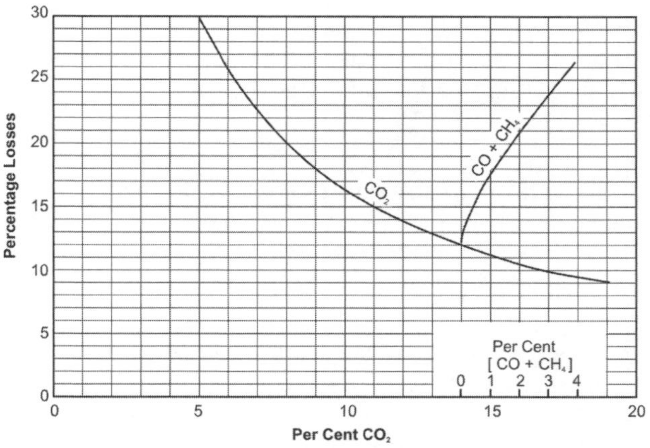

Figure 21.58 The effect of varying percentages of CO_2 with flue gases

Note: With modern boiler types such as the condensing boiler, efficiencies greater than those specified above are possible.

Neglecting the losses due to radiation and the unburnt fuel in the ash, the effect of an increase in the percentage of CO_2 at various gas temperatures can be seen from Figure 21.57.

Figure 21.58 shows the effect of varying percentages of CO_2 with flue gases at a temperature of about 200°C, a usual figure where economisers are in use. This figure also indicates the rapid increase in heat loss when the combustion is incomplete, as indicated by the presence of carbon monoxide and hydrocarbons. The air supply should not therefore be reduced to such an extent that any appreciable amount of CO is present, so even under the best conditions, the percentage of CO_2 cannot easily exceed about 14% with coal.

Figure 21.59 shows approximately the relation between draught and the rate of combustion for various types of fuels **burnt on ordinary grates**. Higher values of draught are re-

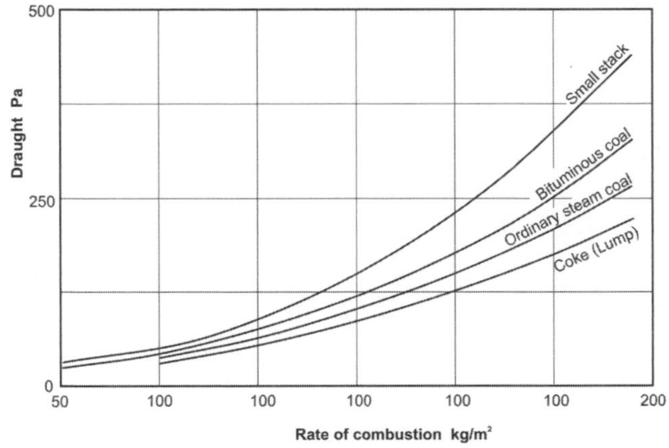

Figure 21.59 The relation between draught and the rate of combustion for various types of fuels

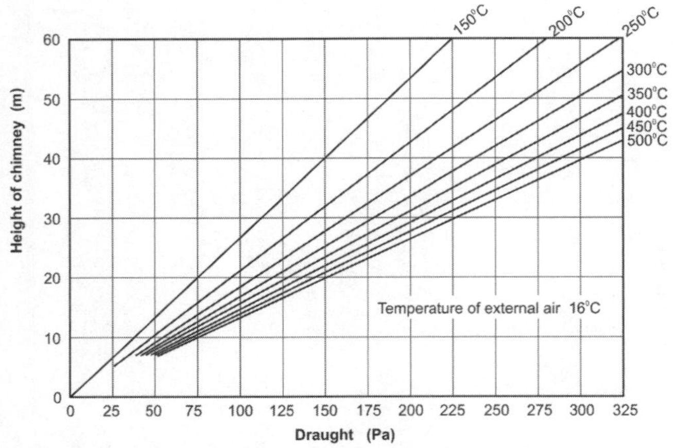

Figure 21.60 Height of chimney required to give natural draught

quired for chain grates, fluidized beds etc. Figure 21.60 shows the height of chimney required theoretically to give natural draught up to 325 Pa for various flue gas temperatures at normal altitudes.

From these two diagrams it will be seen that the highest rates of combustion are impossible without chimneys of considerable height or with high gas temperatures. It will be seen from Figure 21.57 that this would cause an excessive loss of useful heat.

Allowances must be made for infiltration into the boiler and flue system and the increase in volume of flue gases due to moisture in the fuel. Air and gas volume flowrates so obtained do not include margins for the infiltration and fuel moisture mentioned in Table 21.8.

Thus when calculating the flue gas flowrate to be handled by an induced draught fan, it is customary to increase the flow by 15% to allow for overload. This is the design duty.

Allowance for Infiltration	% addition	Allowance for moisture	% addition
Poor brick flues	15	Bituminous coal	3
Good brick flues	5	Fuel oil	4.5
Average steel flues	3.5	Dry wood	10
Rotary air heater	7.5		
Grit Collector	1		

Table 21.8 Margins for infiltration and fuel moisture

21.6.5 Combustion air and flue gases

21.6.5.1 Volumetric flowrates

The nomogram in Figure 21.61 provides an easy means of obtaining a close approximation of the weight and volume flowrate of air required for combustion of the fuel and the volumetric flowrate of flue gases.

If the design efficiency of the boiler and the average operating percentage can be obtained with reasonable accuracy, the flow of gases to be handled by the induced draught fan may be determined sufficiently accurately for sizing purposes.

In estimating these quantities it may be possible to obtain from site the data required. If, however, the information regarding operating conditions is unreliable or scanty, it is possible to use the appropriate data from Table 21.7.

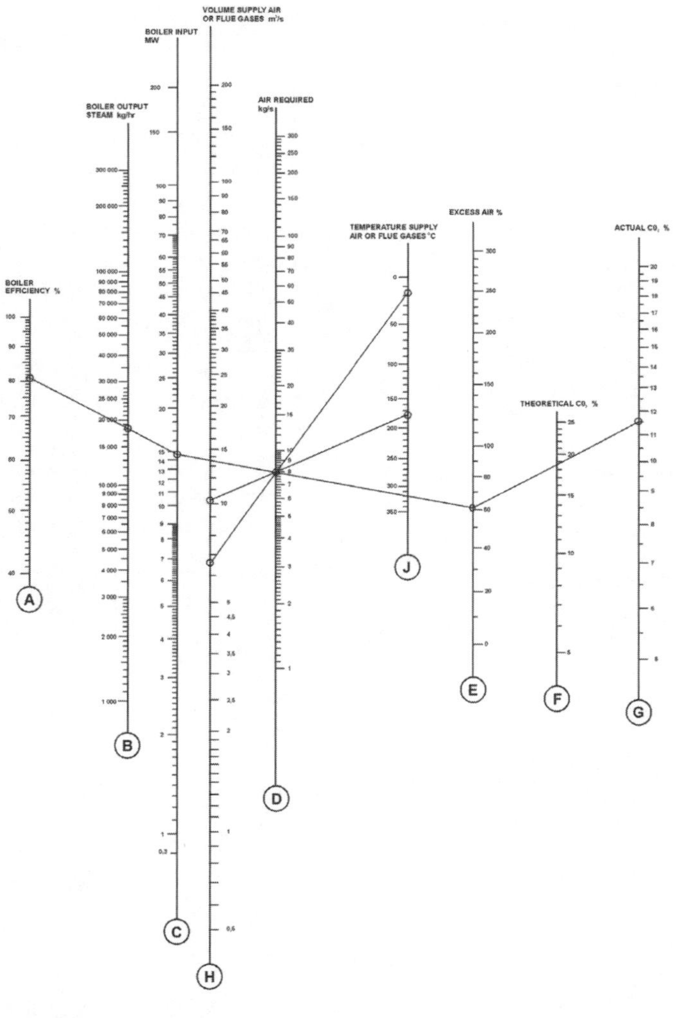

Figure 21.61 Nomogram for combustion air and gas volumetric flowrate

21.6.5.2 Use of the nomogram

The method of using the nomogram is detailed below, but for illustration an example of a particular installation has been chosen, this being shown on the nomogram by appropriate lines. It uses the data, as in Table 21.9.

Boiler efficiency	80%
Boiler evaporation at maximum continuous rating	18116 kg/hr
Operating CO_2	11.5%
Maximum theoretical CO_2 (for average coal)	18.5%
Temperature of air entering boiler	16°C
Temperature of flue gases	177°C

Table 21.9 Boiler installation details

Using the nomogram in Figure 21.61:

1) Join point on A (80%) to point on B (18,116 kg/hr) extending line to cut C (Boiler Input - 14654 kW).
2) Join point on G (11.5%) to point on F (18.5%) extending to cut E (Excess Air - 60%, see note below).
3) Join intersection on C to intersection on E cutting D. (Weight of air required for combustion - 485 kg/s).

Then to obtain volume of air required for combustion:

4) Join intersection on D to point on J (16 °C), extending line to cut H giving volume of (6.61 m^3/s).

Next, to obtain volume of flue gases (dry products):

4a) Join same intersection on D to point J (177 °C), extending line to cut H giving volume of flue gases (10.2 m^3/s).

Note: In the example the excess air amounts to 60% and it will be noted that if the percentage of excess air is known it is unnecessary to plot the operating and theoretical CO_2 points.

Given the boiler to be coal-fired and furnished with brick flues in moderately good condition, the allowances to be made are:

Infiltration into flues, say 10%,

Moisture in coal 3.0%,

plus allowances for overload etc. 15%,

thus:

the total volume of the gas becomes:

$$10.2 \times 1.1 \times 1.03 \times 1.15 = 13.3 \ \ m^3/s$$

21.7 Dust and fume extraction

21.7.1 Introduction

The intention of any local extract system is to reduce what could be a danger due to the pollution of the atmosphere by some mechanical or chemical process. It is imperative that the contaminant be captured as close as possible to its source by moving a mass of air across possible escape routes.

An alternative is to reduce the concentration of the contaminant below the danger level by introducing a large quantity of clean air and removing a corresponding amount of dirty air from the affected area. Local exhaust will, however, give more positive control than this so-called dilution method.

21.7.2 Types of extract system

There are many different types of extract system, among which the following are examples:

- Wood refuse plants - to keep saws, planes and other machines working
- Fume exhaust plants - to extract fumes, which would be harmful to health, if allowed to escape
- Kitchen extract system - to remove unpleasant odours and steam
- Recovery plant - where dust produced in some mechanical process such as machining, milling or polishing, has a value
- Atmospheric pollution prevention.

21.7.3 Components of an extract system

All systems will use ductwork, fans and hoods to capture and transport the contaminants. It may also be necessary to incorporate some sort of air cleaner or dust collector. The cost of these items can be considerable and it is therefore essential to ensure efficient capture of the contaminants so that the amount of extract air may be minimized. This is especially important where dangerous fumes have to be extracted. Some fumes may act as cumulative poisons, whilst some mixtures of fine dust particles and air become explosive and must be avoided.

21.7.4 Categories of particles to be extracted

There are two main categories of particle to be considered:

- Fine dusts, fumes, vapours and smokes which can generally be dealt with by air movement.
- Heavier particles which need special precautions and should be caught in their trajectory.

Exhaust hoods have to be located in the path of the particles.

In both cases, it may be advantageous to use both blowing as well as exhaust opening i.e. a so-called "push-pull" system. This can help to reduce factory heating loads by decreasing the number of air changes. The blower air can be cold. Air can be recirculated into occupied spaces but only if the dust or fume is non-toxic, the collection efficiency is high and there are no small particles which could elude collection.

21.7.5 General design considerations

Circular cross-section ducting is preferred, as rectangular ducts produce lower velocities in their corners, leading to dust deposition and build-up. The inside of the duct should be kept as smooth as possible and free from projections. A means of identifying blockages and removing them by inspection doors and clean-out hatches should be incorporated.

The solid particles are not assumed to affect the flow in any way as the usual mixtures are from 0.1 m^3/s of air per kg of dust for the heaviest exhaust duties down to about 10 m^3/s of air per kg of dust for grinders etc. On a volume basis this is about 5000 : 1 down to 500000 : 1. It is normally possible to ignore the effects of the compressibility of the air, although system pressures are usually much higher than those experienced in air conditioning etc.

21.7.6 Motion of fine particles, fumes and vapours

In fine particle control, an open inlet may be modelled on the assumption that it approximates to a point source, see Figure 21.62. (See also Chapter 3, Figure 3.36.)

Design may necessitate that the exhaust opening may need to be some distance from the source of dust emission, see Figure 21.63. For example, this could help the operator of a grinder. The force to capture a particle of known physical characteristics

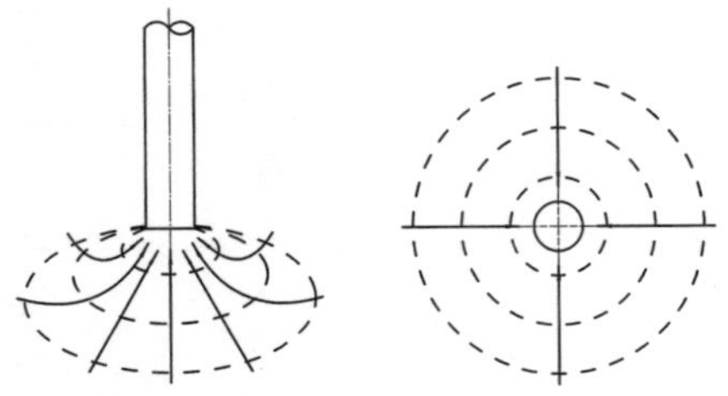

Figure 21.62 Point source approximation

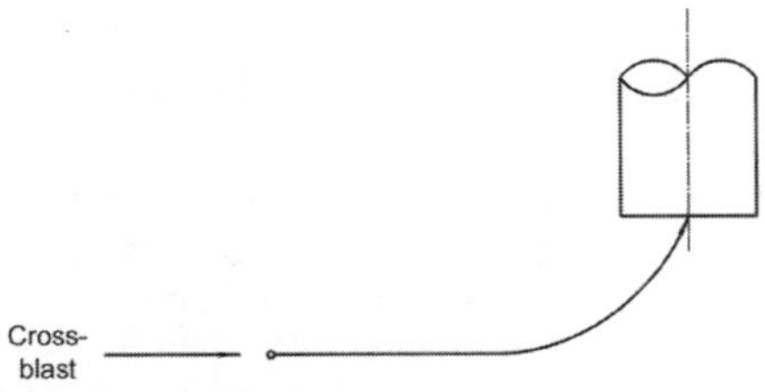

Figure 21.63 Extract at distance from emission

can be calculated, and from this the extract velocity can also be calculated.

It is often possible to use a cross-blast to advantage by positioning the extracting opening somewhat downstream of the point of dust emission, as in Figure 21.64.

Instead of a point, the source of emission may be an area. An example of this is a pickling tank. In such cases the exhaust hood should overlap the area as shown in Figure 21.65.

The minimum angle of the hood should be 35°. Any condensation will then run down the surface of the hood. Any lesser angle and the condensate will "rain" back on the work and operator beneath. The efficiency of exhaust opening can be increased by flanging, see Figure 21.66.

If the flange is brought out to the 5% velocity contour, then the volumetric flowrate is reduced to 70% of the value without this addition. Thus the duct size may be reduced and hence with the reduction in flowrate, there is also a reduction in fan size, motor power and running costs.

Where a tank is against a wall than the extract flowrate may be reduced to 75% of that normally required for the same efficiency of collection with the same cross-draught. This latter, however, may also be reduced giving even greater savings. The ultimate of this approach is the spray booth, illustrated in Figure 21.67.

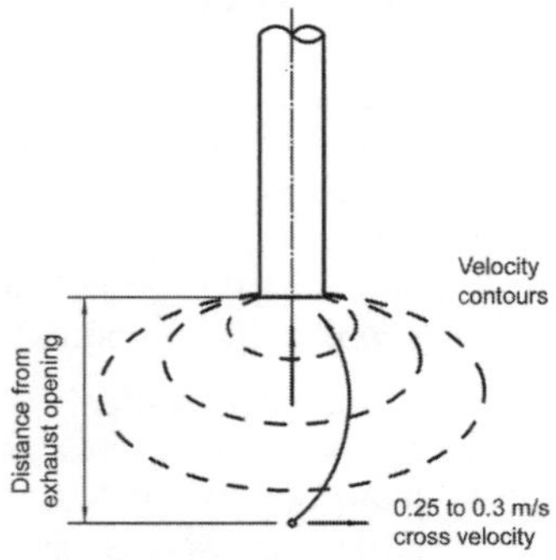

Figure 21.64 Position of extract when cross-blast present

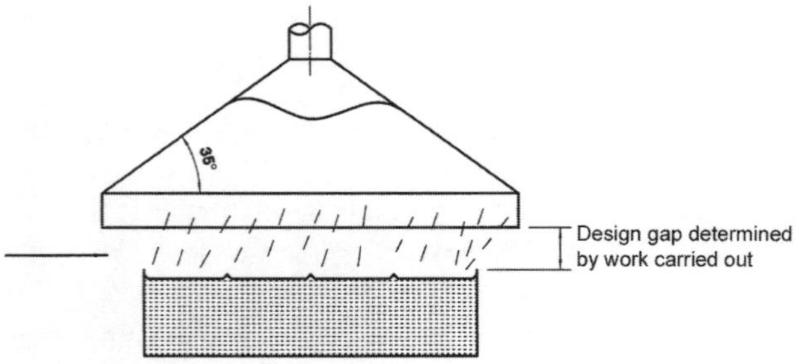

Figure 21.65 Exhaust hood overlapping tank emission

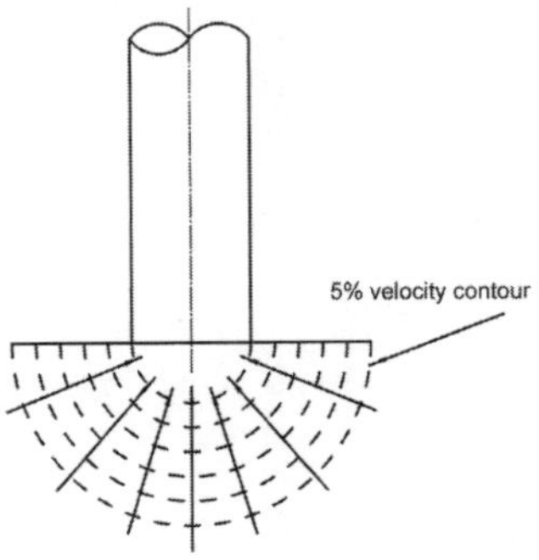

Figure 21.66 Exhaust opening with flange

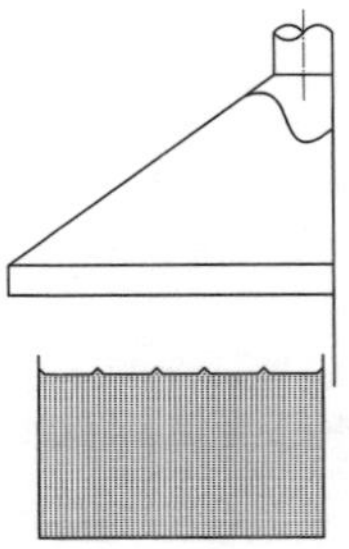

Figure 21.67 Exhaust hood against a wall

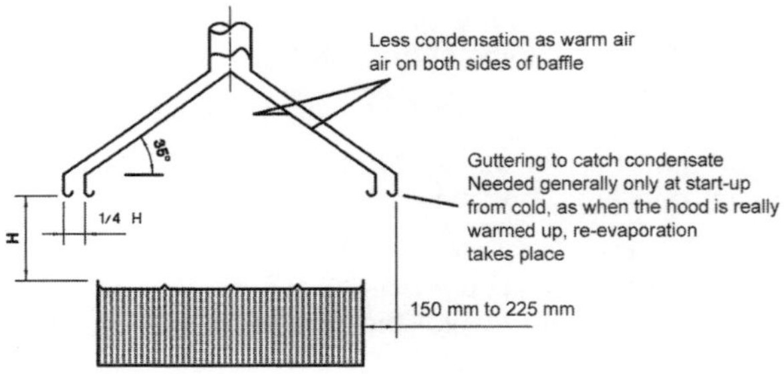

Figure 21.68 Baffle type hood

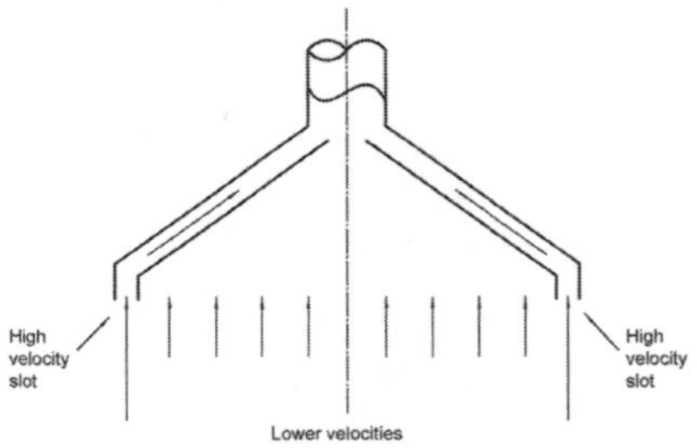

Figure 21.69 Double canopy hood

Where small volumes of fumes are produced, a large hood with corresponding large volume flowrates might be considered uneconomic. The design might then be modified to the baffle type shown in Figure 21.68.

Where larger volumes of fumes are produced, a double canopy hood can effect savings in extract flowrate. See Figure 21.69.

Large hoods may necessitate more than one opening to reduce height, see Figure 21.70.

Distances from hood periphery to extract opening must all be equal to maintain an even airflow, see Figure 21.71.

In long hoods a solution is often a long box or plenum chamber above the hood. This has a longitudinal slot in its underside. As the pressure loss across the slot is large, it will tend to be the same for all the hood. The loss in the box is by comparison

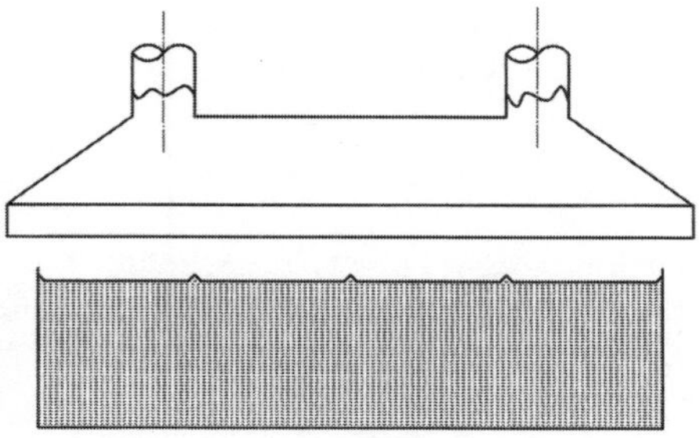

Figure 21.70 Large hood with double extract

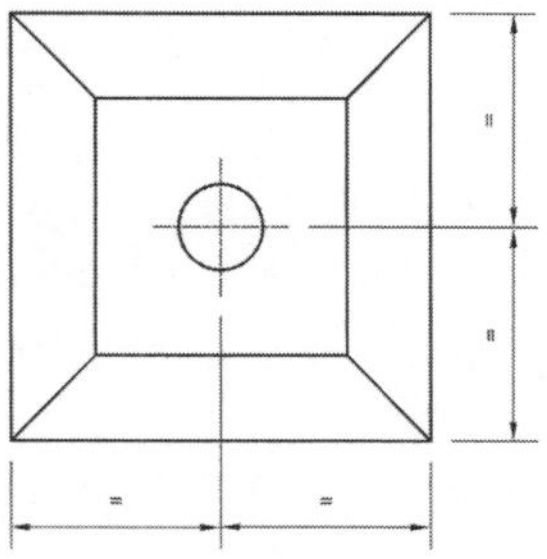

Figure 21.71 Equalisation of distance from hood periphery to extract opening

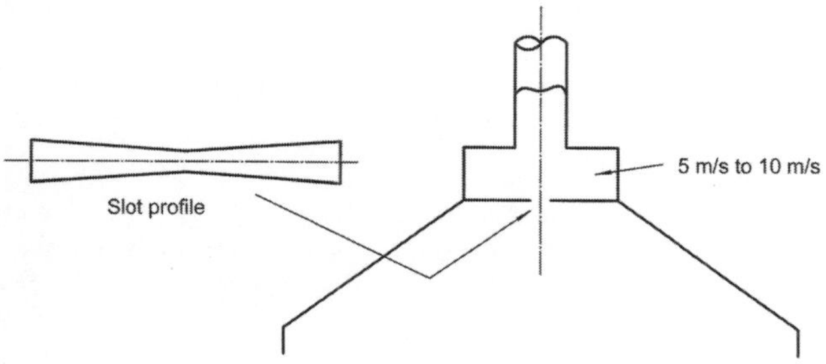

Figure 21.72. Slotted plenum extract

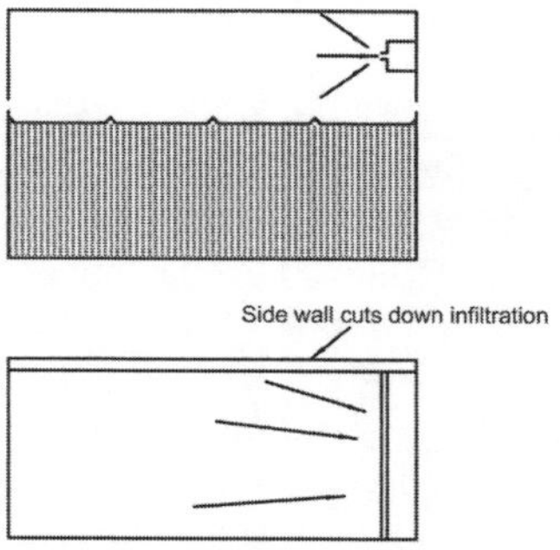

Figure 21.73 Side extract

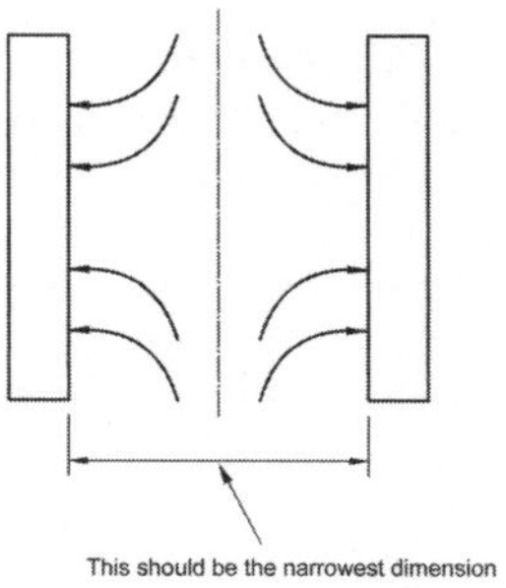

Figure 21.74 Double extract at sides

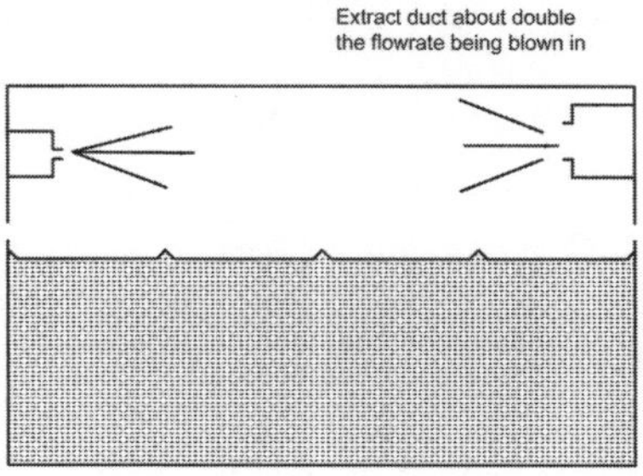

Figure 21.75 "Push-pull" system

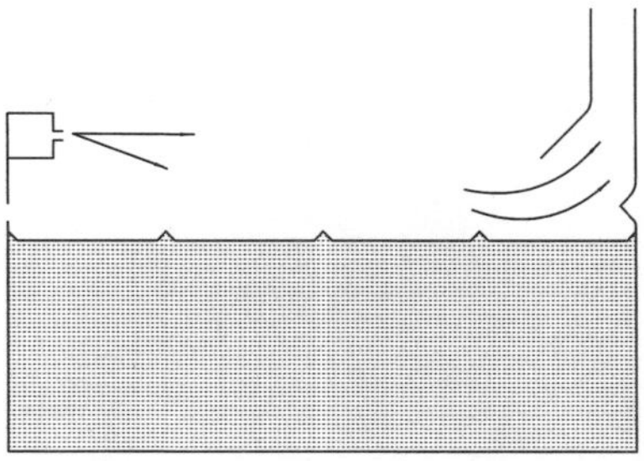

Figure 21.76 Modified "push-pull" system

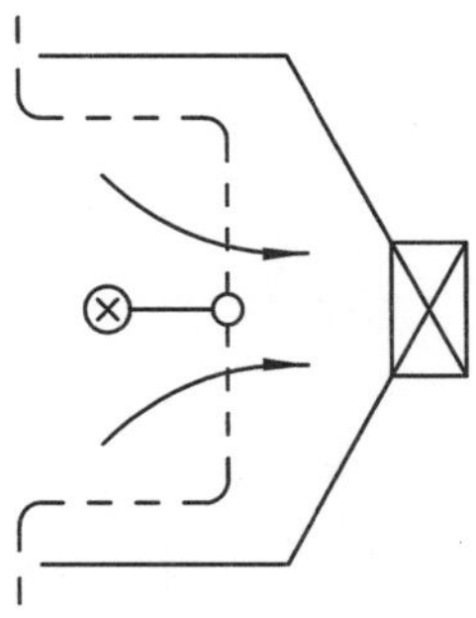

Figure 21.77 Spray booth extract

small and negligible from any part of the slot. To maintain exactly equal losses, the slot might theoretically be shaped as in Figure 21.72.

All the refinements in hood design detailed above tend to increase both the first cost and weight. In an age when energy costs and greenhouse effects were not so important as today they were therefore largely ignored.

Weight **was** an important consideration as hoods have usually to be suspended from roof trusses. Cleaning is also more difficult with a baffled or double hood. Nevertheless these problems can be overcome and it behoves us all to consider again these solutions when energy costs are so important.

Interior lighting to a large hood may be necessary and here bulkhead water-tight light fittings should be used. Smooth exteriors to these should be chosen, so that they may be easily wiped clean of grease, dust etc. If sheet metal ducts are used then earthing points for the lights should also be fitted.

In some industries, such as the automotive, a monorail system is needed above the tank and it is difficult to fit an extract hood. Side extract may then be the only possible solution as illustrated in Figure 21.73.

This may be arranged at either side if the tank is wide, Figure 21.74.

Alternatively a "push-pull" system may be employed, see Figure 21.75.

Where one end of the tank can be blanked off, then a modified form can be used. (Figure 21.76.)

This method is effective up to about 1 m wide with velocities of 10 m/s. Slots should be about 35 mm to 50 mm wide. Again the slots should be tapered along their length to maintain an equal pressure loss, the widest point being furthest from the extraction duct. Spray booths incorporate many of these features, albeit in a different configuration, (Figure 21.77).

21.7.7 Dust features

When heavier particles such as e.g. wood chips are being extracted, then duct wall thicknesses should be increased and bends offset slightly at their beginning, where possible, to decrease damage, see Figure 21.78. Damage can lead to leakage into extract ducts with consequent loss of suction and possible increase in fan power.

"Lobster backed" bends of less than 250mm diameter should be made up in 15° segments. Larger diameters should be in less than 15° segments to maintain the approximation to a curve. The inside of ducts must be smooth and care should be taken at gaskets in flanged ductwork to ensure that there are no lips.

With condensable fumes, low points in duct systems should be fitted with drain points. A drain should also be incorporated in the fan casing, see Figure 21.79. All bends should have an easy radius with a bend ratio R/D of 2 or more. Ducts should generally be not less than 100 mm diameter to obviate clogging, although this size should be used with discretion. Air tightness is important, as due to the higher pressure losses, leakage may be quite large, "robbing" the furthest sections of the system.

If dampers must be used then they should be of the slide type, coming from the top of the duct so that no dust build-up occurs, as shown in Figure 21.80.

Sweep-up points should be self-closing and airtight. It is not normal to allow for the air quantity passing through them as they are only open for short periods. Electrical continuity should be arranged across flanged sections. Sparking would otherwise occur across gaskets due to the build up of static electricity.

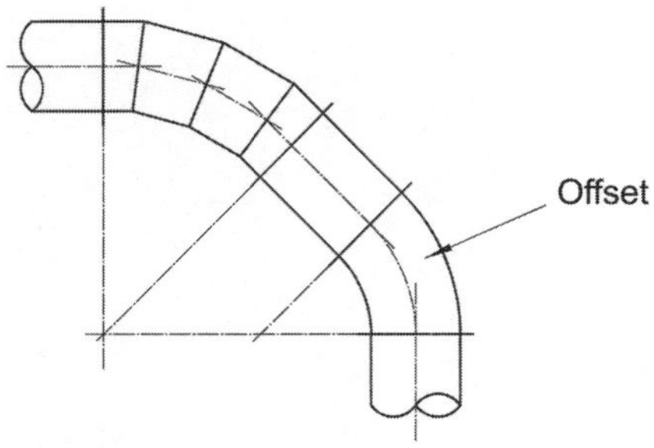

Figure 21.78 Offset bends

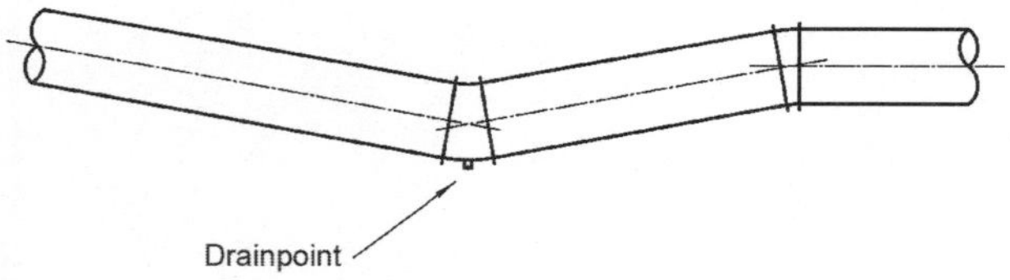

Figure 21.79 Drain point positioning

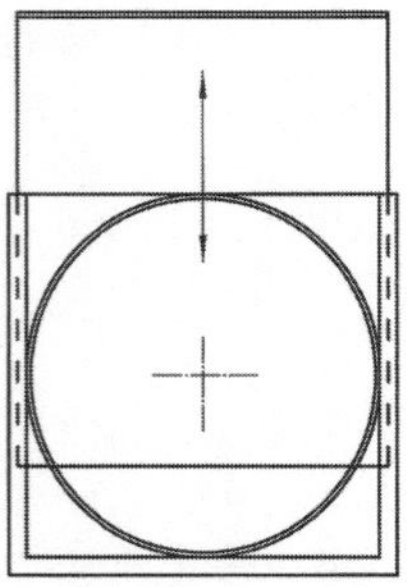

Figure 21.80 Positioning of slide dampers

21.7.8 Balancing of duct systems

In any dust or fume extract plant it is essential to "balance" the system to ensure that the design extract flowrates are achieved and that all points have adequate suction without overloading the fan-driving motor.

The alternative strategies are as follows:

1. Size the fan and all branches on a given extract velocity without allowing for any balancing. In fact the design flowrate will be achieved and the motor may be overloaded. No proper control over the extract from individual machines will be achieved.
2. Balance branches by one of these methods:

a) Using dampers or blast gates or internal cones of a suitable design

b) Re-arrange machines or re-route ductwork to equalise the pressure loss to all extract points

c) Reduce the size of branch ducts so that they have an appropriately high and equal pressure loss at the design flowrate.

3. Size as in method 1, but then calculate or measure the actual air quantities flowing and then determine the new and correctly fan pressure.

Of the above, 2 c) is preferred as flowrates are controlled without the need for dissipating energy across dampers or cones. The velocities in short legs may however exceed the notional design figure.

21.8 Explosive atmospheres

21.8.1 Introduction

On the 1st July 2003 the ATEX Directive of the European Union became a part of UK law. Whilst its provisions can, and do, affect all equipment used in hazardous areas, this Section concentrates on its requirements for fans. They are, after all, the prime mover in ventilation and air conditioning systems.

Unfortunately there is a lot of ignorance as to its intentions and requirements. Even some fan manufacturers appear to be unaware of its effect on them. Some have been heard to say that it only applies to the electric driving motor - NOT TRUE. Others appear to believe that it has no effect on materials of construction, running clearances between stationary and rotating parts, bearing selection and so on. They couldn't be more wrong!

Just as unfortunate has been the reaction of many designers and ventilation system users. You know the sort of response - "Oh heck, another piece of legislation to worry about. Put something in the specification like must comply with the ATEX Directive".

YOU CAN'T DO IT!

The Directive requires that the customer **and** the manufacturer **each** carefully consider their response to the particular problem. The main duties of the end user (who will no doubt appraise his plant designer) are:

- to prevent the formation of explosive atmospheres
- to make an assessment of explosion risks
- to categorise the work place area and divide it into appropriate zones
- to select appropriate products according to the zone

- to prepare an explosion protection document
- to identify hazardous areas and "sign-post" them with warnings

Having completed all this, he is only then in a position to approach a fan supplier with an appropriate enquiry giving all the relevant information.

And here another set of problems raises its ugly head. The European Commission "mandated" CEN (Comité Europėan de Normalisation) to produce Standards covering the requirements for all types of machinery. Fans were assigned to a committee (what else?) designated CEN/TC305/WG2/SC1, which comprised representatives from a number of European Standards bodies including BSI, DIN, AFNOR, UNI and SIS. The author was privileged (?) to lead the UK delegation.

21.8.2 The need for a Standard

If a piece of equipment, such as a fan, has been manufactured to a relevant mandated Standard, then it is deemed to comply with the Directive. Engineers are much happier working with Standards than the Europeanised legalese in which the Directive is couched! The committee set to work with a will and produced a draft within the allotted timescale. There were arguments, some bitter, between DIN (Germany) and the rest of the nations represented, principally over clearances and material pairings between rotating and stationary parts. However, eventually an amicable compromise was reached, and the draft sent upstream to its parent CEN/TC305/WG2.

Whilst apparently endorsing its technical content, the draft was rejected as it was not in the format approved for all ATEX mandated standards. Evidently it was felt that it more closely resembled a standard produced for the Machinery Directive. It was then rewritten in the correct format, which gave a further opportunity for delegates to re-raise objections that they might still have! The fan manufacturers would have preferred the "machinery" format as it enabled easy comparisons to be made between "standard" and "explosion proof" fans.

21.8.3 Zone classification and fan categories

Users of flameproof electrical equipment will have been familiar for a number of years with the zone classification and gas grouping used to delineate the required features. These are mirrored in the draft fan Standard pr EN14986 by three categories suffixed with a G or D to identify a gas or dust mixed with the air.

It must be emphasised that the choice of category is ultimately the user's, but in the absence of specific information, Table 21.10 would be used by fan manufacturers. The user may need to make an assessment based on his knowledge of the fan site. A fan sited in the middle of the Sahara desert, well away from other habitation, might not be the same risk as if it were sited close to a school, for example.

In Zone	Applicable category	If designed for
0	1G	gas/air mixture or vapour/air mixture or mist/air mixture
1	1G or 2G	gas/air mixture or vapour/air mixture or mist/air mixture
2	1G or 2G or 3G	gas/air mixture or vapour/air mixture or mist/air mixture
20	1D	dust/air mixture
21	1D or 2D	dust/air mixture
22	1D or 2D or 3D	dust/air mixture

Table 21.10 Zone classification and applicable fan category

It has been established that there shall never be more than one category difference (step lower) for the outside of a fan than for inside the fan casing and that for a ducted fan located in an unventilated room, the same category shall be applied for the outside and the inside fo the fan casing

These provisos have been established to take account of situations for example where the fan is handling a more dangerous gas but is located in a safer area.

21.8.4 prEN 14986 – contents of this draft Standard

Fans for operation in all such atmospheres have to be of a rigid design. This requirement is considered as fulfilled for casings, support structures, guards, protective devices and other external parts, if the deformation resulting from a single impact test at the most vulnerable point is so small that the moving parts do not come into contact with the casing (see EN13463-1). All impellers, shafts, bearing, pulleys, cooling discs, etc, have to be positioned by positive locking devices.

Fan casings must be of substantially gas-tight construction (defined as category E from ISO 13349:1999 Table 4). A gas tight seal at the shaft entry will be necessary where there is a difference between the gas inside and around the fan. The material pairings between stationary and rotating parts must be taken from a comprehensive list, (see Table 21.11) it being noted that not all plastics are necessarily flameproof.

It must be understood that the pairings in the Table are "hedged" with a number of requirements and footnotes. It is essential to refer to prEN 14986 for full details. Impellers have to be rigid with all calculated stresses less than 2/3 of the yield stress.

Item	Material (1)	Material (2)	Category	
			3	2 and 1
1	Leaded brass CuZn39Pb	Carbon or stainless steel or cast iron	yes	yes
2	Copper	Carbon or stainless steel or cast iron	yes	yes
3	Tin	Carbon or stainless steel or cast iron	yes	yes
4	Aluminium alloy	Aluminium alloy	yes	yes
5	Aluminium alloy	Naval brass CuZn39Sn	yes	yes
6	Aluminium alloy	Leaded brass CuZnPb3/CuZn39Pb	yes	yes
7	Nickel based alloy	Nickel based alloy	yes	yes
8	Stainless steel	Stainless steel	yes	yes
9	Any other steel alloy or cast iron	Any other steel alloy or cast iron	yes	yes
10	Any steel alloy	CuZn37	yes	no
11	Plastic	Plastic	yes	yes
12	Plastic	Naval brass CuZn39Sn	yes	yes
13	Plastic	Aluminium alloy	yes	yes
14	Plastic	Nickel based alloy or nickel based steel alloy	yes	yes
15	Plastic	Leaded brass CuZnPb3	yes	yes
16	Plastic	Any steel alloy or cast iron	yes	yes
17	Plastic	Stainless steel	yes	yes
18	Rubber	Any steel alloy or cast iron	yes	yes
19	Rubber coated metal	Rubber coated metal	yes	yes

Table 21.11 Permissible material pairings for gas explosion groups 11A and 11B

Category 3 fans must be designed for easy inspection and cleaning. There must be a clearance of 1% of the possible contact diameters between rotating and stationary parts with a minimum of 3 mm and a maximum of 20 mm. This requirement does not of course apply to shaft seals where the rubbing speed is very low. Maximum temperature of surfaces shall be less than 75% of the gas ignition temperature. The L_{10} bearing life shall be greater than 20,000 hours. Direct drive is preferred, although belt drives may be used with appropriate precautions. These fans may be self-certified by the manufacturer.

Category 2 fans are generally similar to Category 3 but L_{10} bearing lives have to be greater than 40,000 hours. Belt drives are not allowed. Casings must be continuously welded with gaskets at all openings and splits. Design documentation, risk assessment data etc. must be deposited (in a sealed envelope) with a Notified Body. This will be opened and used in evidence in the even of any accident etc.

Category 1G fans, typically used for atmospheres containing hydrogen or acetylene, are generally similar to Category 2 but also require flame arresters on the inlet and outlet or internal arresters. These have to be tested and witnessed by a third party (Notified Body). Gas tightness also has to be witness tested.

Where the fan is also handling dust, i.e. Category 1D there will also have to be apparatus for frequent automatic cleaning of arresters. An alternative is to fit quick acting leakproof dampers on the inlet and outlet (closing in microseconds).

Electric motors in the airstream are **not** acceptable in any Category 1 fan, but pneumatic or hydraulic motors **may** be possible with appropriate safeguards.

It need hardly be added that whilst a Category 3 fan may have only a relatively small price premium compared with a "standard" fan, that for Category 2 and Category 1 fans will be considerable. But then the penalties for non-compliance can also be considerable!

21.8.5 Clearances between rotating and stationary parts

Of the above requirements, perhaps that specifying the clearance between the running and stationary parts has created the most anguish. Some manufacturers have chosen to ignore it, whilst others perhaps do not appreciate its significance.

The effects of such increased clearances are especially severe on the performance of axial flow fans. These depend, for the development of high pressures, on a minimum tip gap between the blade periphery and the circular casing. The magnitude of this degradation in performance is very much dependent on the blade design. So called forced-vortex blades (which have a large blade chord at the tip when compared with the chord at the hub) are severely affected. Free vortex blades (which have

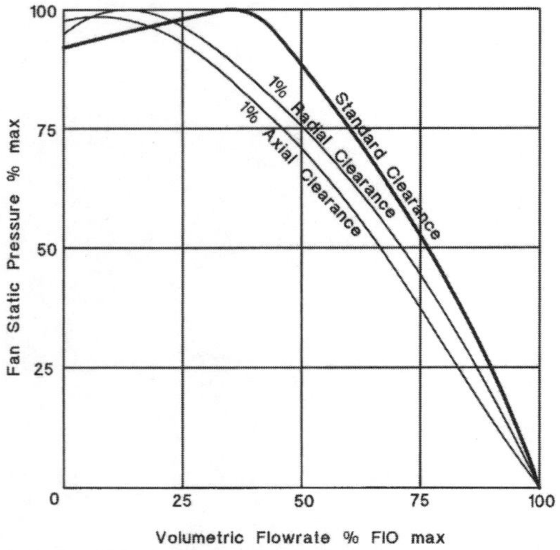

Figure 21.81 Influence of tip clearance on the pressure against flowrate curve of a typical axial flow fan

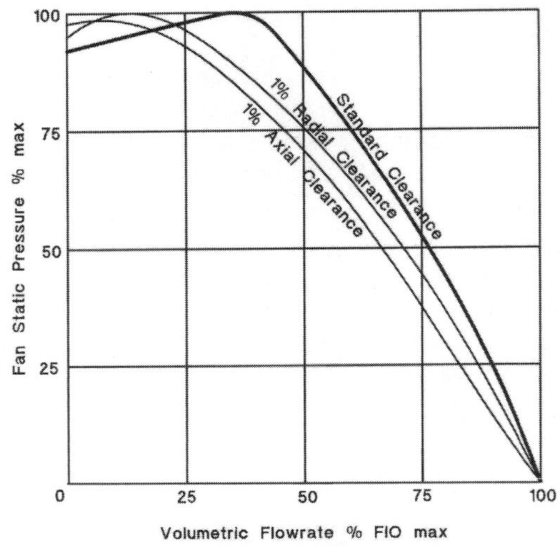

Figure 21.82 Influence of clearance at inlet on the pressure against flowrate curve of a typical backward bladed centrifugal fan

minimum chord at the tip and a very large chord at the hub with considerable twist) are less affected. For a given duty, they are however usually larger in any case.

However, the majority of axial flow fans have arbitrary vortex blades. Figure 21.81 shows the typical performance of such fans with different tip gaps. It will be instantly appreciated how important this factor is. A normal tip gap for a "standard" fan would be 0.2 to 0.3%. For a fan complying with ATEX it is 1%.

It might be thought that centrifugal fans would be unaffected. This is certainly not the case, especially with high efficiency fans where there is an overlap between the inlet cone and the impeller shroud. The dimension of the gap between the lip of the venturi inlet cone and the impeller shroud is again critical to minimize recirculation and thus maintain efficiency. In this case Figure 21.82 shows the typical degradation according to the orientation of the clearance.

21.8.6 Actions required by manufacturers and users

So can we ignore the ATEX Directive until the finally approved standard is published? The answer is very definitely **NO**. Ever since 1st July 2003 we have had to comply. Most of the Notified Bodies have taken the stance that the last published version of the draft represents the agreed state of the art as known to the reputable manufacturers. It should therefore be followed. In any event we have the non-specific EN 13463 Standards to which all products must comply in the absence of a product specific standard.

Purchasers of explosion proof fans are recommended to obtain the position statement on ATEX of the UK Fan Manufacturers Association (FMA). They should also consider carefully the purchase of such fans and make sure that they are not laying themselves open to future trouble by ignoring both their duties and the manufacturer's obligations. Such fans should not be chosen simply on first cost considerations - its too dangerous on so many counts. Perhaps they should ask what features the fan manufacturer has included, what tip gaps his performance is based on and what material combinations he has used. If he is evasive, hesitant or says these questions are not important, be very wary.

The draft standard prEN14986 has at the time of writing been in the public domain for over 6 months, during which time comments were invited. It perhaps demonstrates the widespread interest generated to note that over 50 such letters were received by the committee. They reflected the fact that there are quite a number of manufacturers with an interest in this market. There was little commonality in design for explosion protection before work on this Standard started, a general reluctance to change existing designs, a scepticism about some of the scientific justification for particular requirements, and some who do not appreciate the need for the Standard to follow the logic of the Directive, and the basic requirements standard for non-electrical equipment, EN 13463 part 1.

Considerable comment the scope was received. Some felt there was confusion that the Standard tried to address only the mechanical parts of fans, and yet almost all fans were sold or used with a motor. There was also felt to be confusion that electric motors often had integral fans that might not meet the requirements of this Standard.

There was pressure not to exclude Group IIC gases from its scope, and this was done, but it was realised later that there was not the technical basis for designing a safe fan for this expanded scope, and the requirements for the small number of gases in Group IIC might not always be the same. So there are now references later to requirements specific to hydrogen service.

There were also discussions about the working temperature range. The wide range in the enquiry draft arises from alignment with mining service, where a 60°C ambient is possible. However, many types of electric motor are only available off the shelf with an ambient range up to 40°C. This conflict will probably need to be resolved by marking many fans with the actual intended range of service temperatures and may require X marking of a large number of fans.

21.8.7 Probable changes to prEN 14986

It is apparent that in the past many manufacturers had no data about the leakage rates either from joints in the casing, or from around the fan shaft. The draft Standard now will provide essentially 3 options:

- If the casing is not designed or tested to be leak tight, the fan should have the same category inside as out.
- If casing construction requirements are met, with effectively sealed joints, there can be a one category difference between inside and out.
- If the fan is additionally tested, and the user is provided with measured information about the leakage rate around the shaft seal, then there can be a 2 category difference between inside and out. This only applies of course to fans with a ducted inlet and outlet.

The section on gas temperatures has been reworded, mainly with the intention that is clearer.

Material pairings, predictably again, generated much discussion, with some strong views about what was or was not acceptable. To some degree this reflects a belief that in the past appropriate material pairings were the principle means by which a standard fan was modified for an ATEX type application. As now drafted, that is nor longer the case; many other features also being relevant.

To some degree the difference of opinion arises from fragmentary information; there are no complete data sets about the risk, for all material pairings, for all gases, and probably never will be. Even if all the experimental work was done, it would have to be recognised that there is no sharp line between safe and unsafe.

The consequence was that the committee chairman rightly declared that the table in the public enquiry draft was already a compromise; and no progress would be made if a debate were reopened on all issues. So changes are limited. The use of aluminium inlet cones was however definitely "banned".

Footnotes to the table regarding paint were split, to prevent aluminium being used where rust could be present, and also to prevent iron oxide in the paint, where aluminium was used for construction. For stainless steel, Germany presented data showing the chrome content could be safely reduced from 18.5% to 18.1 % and this was accepted.

A note about different plastics not being automatically permissible was deleted, but it was not normative, and should not create new problems.

The reference to a test in ISO 1210 was changed to making it an option rather than a requirement.

A proposal to provide different minimum thicknesses for linings where these are used for category 2 and category 3 fans was accepted.

For category 1 fans, the German Notified Bodies have certified some small designs for vapour recovery plants, particularly in petrol handling facilities. Safety is based on a very strong, rigid casing; flame arrestors on the inlet and outlet, and full testing.

They proposed to remove a requirement for vibration monitoring. The UK argued strongly against this, as vibration monitoring seems a good way of identifying incipient faults, before a fan fails in a way that creates an ignition risk. The counter argument is that with flame arrestors and an explosion resistant casing, full testing shows the fan remains safe, even if the impeller does disintegrate. The compromise text agreed now says the manufacturer must make a special ignition hazard assessment concerning this aspect, for category 1 fans.

Vibration monitoring is retained as a requirement for category 2D fans, because of the risk of dust deposits on moving parts.

Markings

This generated strong views, about how the category number inside and out should be identified. Some wanted a scheme that says (for instance) cat 2 (I) and cat 3(o).

However, the EU Commission guide, currently under revision, has an elaborate extension to the current marking scheme, and would in this case mean the category would be 2/3. Should CEN automatically fall in line with the Commission? Probably CENELEC would not be willing to do this. This is a power struggle, and the outcome cannot be predicted in the market and the standards committees. The author's own view is that you cannot squeeze all the necessary information for safe installation into any sensible marking scheme, and when it comes to explosion safety, people really must read the instructions.

Annexes

Annex A will make clear the distinction between routine tests and type tests.

21.8.8 Conclusions

It would seem that all discussion is now at an end and finality has been reached. We can look forward to the publication of EN14986 in the near future. It must be emphasised that there will be some important changes from prEN14986 the most recent document in the public domain. Whilst the Sections above give the flavour of the discussions, the copyright naturally resides with CEN. Interested parties are therefore strongly recommended to purchase EN14986 as soon as it is published.

21.9 Pneumatic conveying

21.9.1 Introduction

A conveying system of any kind is used to transport a material from one particular place to another. For the chosen method of pneumatic conveying, air is used as the transport medium. It is applicable to most granular materials, be they in that form naturally or have to be pulverised or crushed. In this system, energy is needed to accelerate the material from rest, to lift it as required and to overcome the losses due to friction.

In pneumatic conveying installations, the pressure losses due to friction include not only those due to the sliding of the material along horizontal ducts and around the walls of bends, but also those due to the air itself. For the purposes of this Section, it is, of course, assumed that the driving force is supplied by a fan.

The ratio of material flowrate to air flowrate will not normally exceed 6 : 1 and even this will require a very high pressure fan. So called "dense phase" systems, where the material proceeds essentially as a "plug" along the duct to which is applied a pressure difference usually incorporate either Rootes blowers or compressors of the rotary or reciprocating type. The theory developed in this Section is not applicable to these.

21.9.2 The basis of a design

There is probably no part of fan engineering where "know how" is more important than pneumatic conveying. It has been claimed to be a "black art" rather than a science. Indeed "trial and error" methods (especially the latter) may be necessary before a satisfactory solution in terms of economy and operating convenience is reached.

21.9.3 Conveying velocities

These are usually based on values obtained through sometimes bitter experience. Those given in Table 21.12 are if anything conservative, which, as in life, is not a bad thing. However it is necessary to develop values based on elementary fluid mechanics.

Material	Conveying velocity v_c m/s
Dry sawdust	15
Pulverised coal	20
Oats	22
Crushed limestone	25
Barley	27
Granulated salt	27
Maize	28
Wheat	27
Cement	35
Sand	35

Table 21.12 Typical conveying velocities

21.9.3.1 Vertical velocity

For any particular granule, a floating velocity can be derived from the formula:

$$\nu_f = \sqrt{\frac{2g\rho_p V_p}{C_D \rho_a A_p}} \qquad \text{Equ 21.21}$$

Where

ν_f = floating velocity (m/s)

g = acceleration due to gravity (m/s^2)

ρ_p = particle density (kg/m^3)

ρ_a = air density (kg/m^3)

A_p = frontal area of particle (m^2)

C_D = coefficient of drag

V_p = particle volume (m^3)

For spherical particle this can be reduced to:

$$\nu_f = \sqrt{\frac{4_g \rho_p d_p}{3C_D \rho_a}} \qquad \text{Equ 21.22}$$

where:

d_p = particle diameter (m)

The coefficient of drag C_D for a particle with sharp edges is virtually unaffected by Reynolds number and for most shapes is approximately 1.0. for rounded bodies such as spheres, ellipses, long cylinders etc, the value will vary from 0.5 to 1.0 in the order given.

21.9.3.2 Horizontal velocity

The particles are seen to bounce along the bottom of the duct and spin with high velocity. The "Magnus Effect" (when any rotating object placed in a moving airstream produces circulation and aerodynamic lift) occurs when:

$$\rho_F \, g \, V_p = C_L A_p \cdot \tfrac{1}{2} \rho_F \nu_H{}^2$$

or:

$$\nu_H = \sqrt{\frac{V_p}{A_p} \cdot \frac{\rho_p}{\rho_F} \cdot \frac{2_g}{C_L}} \qquad \text{Equ 21.23}$$

where:

ν_H = horizontal velocity (m/s)

However, in practice it is found that the velocity calculated for vertical conveying is adequate for horizontal conveying. Much of the work on pneumatic conveying by fan systems has been carried out on cereal grains of various types where the bulk density and specific gravity are known, (see Table 21.13).

Description	Bulk density kg/m^3	Specific gravity
Black rape	592	0.865
Oats	416 - 513	1.25
Barley	592 - 689	1.3
Maize	592	1.34
Wheat	609	1.3

Table 21.13 Bulk density and specific gravity of grains

Bulk density should not be confused with density. It includes all voids between adjacent grains when in repose and is therefore a lower figure.

The dimensions of certain grains are given in Table 21.14.

Description	Dimensions mm x mm
Argentine wheat	6 x 3
British wheat	6.5 x 3.5
Oats	8 to 13 x 1 to 4
Barley	8 to 14 x 1 to 5

Table 21.14 Typical dimensions of grains

It seems to have become the practice in low pressure fan systems to make an initial design based on a value of R between 1 and 2, where $R = \frac{\text{weight flow of material}}{\text{weight flow of air}}$. Larger values of R may require less power but higher fan pressures (see Section 21.9.4.2).

An alternative semi-empirical approach to the calculation of conveying velocity is obtained from the equation

$$v_c = k_v\sqrt{S_p d}$$

where:

k_v	=	given in Table 21.15
S_p	=	specific gravity of the particle
d	=	diameter of the particle (mm)
v_c	=	conveying velocity (m/s)

Description	k_v	C_D
Sphere	23	0.47
Rounded	19	0.70
Average	17	0.93
Elongated	16	0.05
Flattened	14	0.04

Table 21.15 Values of k_v for typical particles

21.9.4 Pressure losses

The total pressure loss in a pneumatic conveying system may be obtained by summating the pressure losses due to various factors and detailed below:

p_{fr}	=	friction of air alone for horizontal & vertical straight ducts (Pa)
p_{acc}	=	acceleration of the particles (Pa)
p_L	=	the lift in vertical ducts (Pa)
p_r	=	resistance of particles to the air (Pa)
p_c	=	particle collision and wall friction (Pa)
p_b	=	loss of forward momentum at bends (Pa)
p_{pfd}	=	loss in particle feed device (Pa)
p_{ps}	=	loss in particle separator (Pa)

Thus:

the fan pressure p_F is obtained from

$$p_F = p_A + p_L + p_r + p_c + p_b + p_{pfd} + p_{ps} \quad \text{Pa} \qquad \text{Equ 21.14}$$

21.9.4.1 Pressure loss due to air alone

$$p_A = \frac{0.02l}{d} \times \frac{1}{2}\rho v_c^{\ 2}$$

where:

l	=	length of duct (m)
d	=	diameter of duct (m)
ρ	=	air density (kg/m³)
v_c	=	conveying velocity ,relative velocity + material velocity (m/s)

Bends usually have a large radius i.e. at least 6 times the duct diameter. Their equivalent loss in straight duct may be reasonably taken.

21.9.4.2 Pressure loss due to the particles

a) To accelerate the particles:

The force required is equal to the rate of change of momentum i.e.

$$p_{acc}A = G\int dv$$

where:

A	=	duct cross-section area (m²)
G	=	weight flow of material (kg/s)
v	=	material velocity (m/s)

Now:

$$R = \frac{\text{weight flow of material}}{\text{weight flow of air}}$$

Thus:

$$G = RW_f = R_{\gamma f}Q = R_{\gamma f}(v_r + v_m)A = R_{\gamma f}v_c A$$

or

$$p_{acc} = \frac{R_{\gamma f}v_c A v_m}{gA}$$

$$= R_{pf}v_m(v_r + v_m)$$

$$= RK \cdot \frac{1}{2}\rho v_c^{\ 2} \qquad \text{Equ 21.15}$$

where:

$$K = \frac{2v_m}{v_m + v_r}$$

b) To lift the particles in vertical ducts:

Power supplied by air = Power required for lift

or

$$p_L Q = GH = R_{\gamma f}QH$$

or

$$p_L = R_{\gamma f}H$$

where:

H	=	total loft of material (m)
gf	=	weight/unit volume of conveying air (kg/m³)

c) Resistance of particles to the air:

$$p_r A = C_D A_n \frac{1}{2}\rho v_r^{\ 2}$$

where:

A_n	=	total projected area of particles
vr	=	velocity of particles relative to air

i.e.

$$C_D = \frac{\text{drag force}}{A\frac{1}{2}\rho v_r^{\ 2}}$$

$$G = \gamma_p \Sigma V$$

$$= \gamma_p A_n f v_m$$

where:

Σv	=	total volume/unit time (m³/s)
f	=	$\frac{V}{A}$ for a single particle
A_n	=	$\frac{R_{\gamma f}Q}{f_{\gamma p}v_m}$

This calculation is virtually impossible to make in practice as f is usually unknown. For horizontal conveying, Gasterstadt found that the pressure loss due to particle resistance

$$p_r = 1 + kR \times \text{loss due to air alone} \qquad \text{Equ 21.16}$$

The factor k is obtained from Table 21.16.

Vm + vr = vc m/s	k
15.8	0.43
17.8	0.36
26.6	0.30

Table 21.16 Values of k against conveying velocity

d) Particle collision and wall friction:

In collision some particle momentum is lost, but the exact amount may only be found by experiment. H.E. Rose and H. E. Barnach found that

$$p_c = k \times \frac{1}{2}\rho v_c^{\,2}$$

where:

$$k \quad = \quad 0.00022\frac{L}{D}\sqrt{\frac{S_p}{S_f}} \times R$$

and

S_p = specific gravity of particle

S_f = specific gravity of air

Putting some values to the above

$$\sqrt{\frac{S_p}{S_f}} = \sqrt{\frac{1.3 \times 1000}{1.2}} = 32.9$$

thus:

$$k = 0.0072\ \frac{L}{D} \cdot R$$

i.e. approximately 15 to 30% of the loss for air alone

e) Loss of forward momentum at bends:

$p_b = c \times$ loss due to acceleration

Since bends are gradual, the particles perhaps lose about half of their forward momentum i.e. C = 0.5 and

$$p_b = 0.5RK \cdot \frac{1}{2}\rho v_c^{\,2} \qquad \text{Equ 21.27}$$

21.9.5 Types of conveying system

a) **Direct**

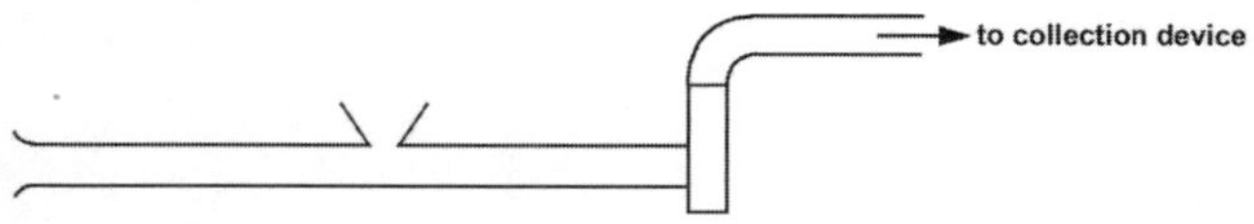

Figure 21.83 Direct system

b) **Suction**

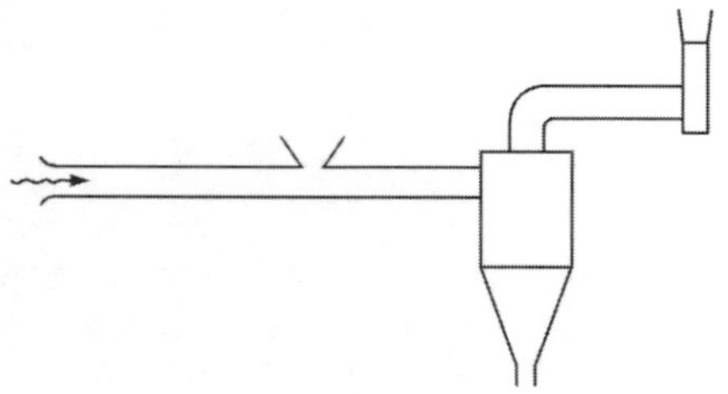

Figure 21.84 Suction system

c) **Screw feed**

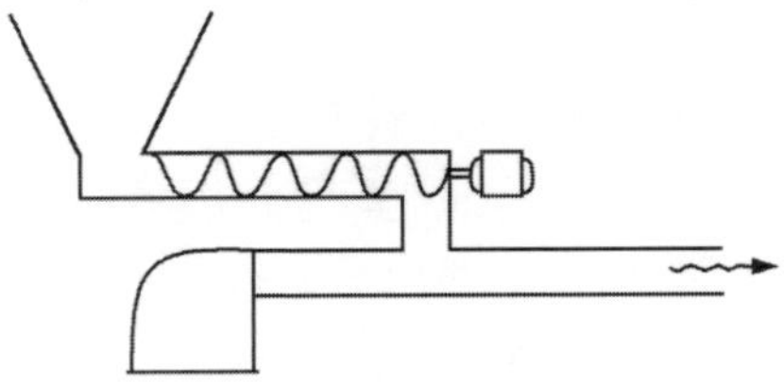

Figure 21.85 Screw feed system

d) **Injector**

In the injector system, in order to prevent blowback at the point of entry of materials, the static pressure at this point must be equal to or below atmospheric pressure, (see Figure 21.87).

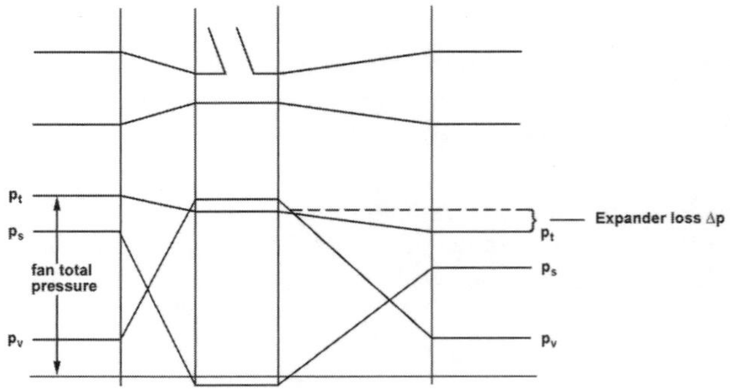

Figure 21.87 Injector system design showing pressures at various points

Throat total pressure = system total pressure + loss in expander

or

$$p_{vt} + p_{st} = kp_{vt} + p_L$$

i.e.

$$(1-k)p_{vt} = p_L - p_{st}$$

but

$$ps_t = 0 \text{ or } -\Delta_p(\text{if below atmospheric pressure})$$

$$p_{vt} = \frac{p_L}{1-k} \quad \left(\text{or } \frac{p_L + \Delta_p}{1-k}\right)$$

$$\text{the throat area} = \frac{Q}{v_t}$$

More information on Venturi injectors was given by Segler in his book *Pneumatic Grain Conveying*, this including values of k.

21.10 Bibliography

No ifs or butts, Dr Andrew Geens and Dr Max Graham, Building Services Journal, March 2005.

Proposed amendment of Approved Document F of the Building Regulations for England & Wales: Ventilation, ODPM (Office of the Deputy Prime Minister), Building Regulations Division, UK. www.odpm.gov.uk/approved documents.

BRE Digest 398, *Continuous mechanical ventilation in dwellings: design, installation and operation.* Sept 1, 1994, ISBN: 851256414.

CIBSE Guide B5, *Noise and vibration control for HVAC (2002)*.

FRS, the fire division of BRE, Garston, Watford WD25 9XX, UK telephone 01923 664100 fax 01923 664910 e-mail: frs@bre.co.uk.

World Road Association (PIARC), previously known as Permanent International Association of Road Congress, La Grande

Arche, Paroi nord, niveau 8, F-92055 La Defense Cedex France Tel: 01 47 96 81 21, Fax: 01 49 00 02 02, Email: piarc@wanadoo.fr, www.piarc.org.

ISO 13350:1999, *Industrial fans — Performance testing of jet fans.*

Centre for Tunnel Aerodynamics Research, London South Bank University, 103 Borough Road, London SE1 0AA UK.

ISO 5801:1997, *Industrial fans — Performance testing using standardized.*

BS 6164:2001, *Code of practice for safety in tunnelling in the construction industry.*

Circular Air Jets, T Djeridane, M Amielh, F Anselmet, and L Fulachier, (1993). Experimental investigation of the near-field region of variable density turbulent jets. Proc. 5th Int. Symposium. on refined flow modelling and turbulence measurement, Paris.

The Efficient Use of Steam, Oliver Lyle (HMSO, London 1947); *The Efficient Use of Fuel* HMSO, London.

Equipment and Protective systems intended for use in Potentially Explosive Atmospheres (ATEX) Directive 94/9/EC. (The Directive has been mandatory from 1st July 2003; it is named after the French "ATmosphere EXplosible".)

The Machinery Directive 98/37/EC provides the regulatory basis for the harmonisation of the essential health and safety requirements for machinery at European Union level.

prEN 14986, *Design of fans working in potentially explosive atmospheres.*

CEN/TC 305, Standard under development.

ISO 13349:1999, *Industrial fans - Vocabulary and definitions of categories.*

Position statement on fans intended for use in Potentially Explosive Atmospheres; Conformity with the ATEX Directive 94/9/EC. UK Fan Manufacturers Association (FMA) the specialist fan group within the HEVAC Association – part of FETA. Tel: +44 0118 940 3416, www.feta.co.uk.

BS EN 13463-1:2001, *Non-electrical equipment for potentially explosive atmospheres. Basic method and requirements.*

Pneumatic Grain Conveying with special reference to agricultural application, G Segler, published by author 1951, Braunschweig (Brunnswick).

22 Units, conversions, standards and preferred numbers

The modern SI system has been adopted in legislation by practically every country in the world. There are, however, extensive changeover problems and many different kinds of data are still only available in units which do not conform to the SI system. Chapter 22 attempts to bridge the difficulties experienced when dealing with quantities expressed in older units.

Almost without exception it is recommended that the quantities be converted directly to SI units; calculation can then be carried out using the coherent SI system. However, it is important that all results of calculations should be reviewed by someone fully conversant with the system of units and the practical values encountered.

The "Système International d'Unités" (SI) is therefore the universally accepted system of units. To undertake any calculations requires a knowledge of units and their correct usage. It is fortunate that the basic electrical units were defined in terms of the metric units from which the SI system evolved and for fans this has some unique advantages compared to other unit systems.

This Chapter outlines the basis of the SI system and the recognised ways of expressing quantities. Guidance is giving on how to check equations for the consistency of units, also a table of conversion factors is included for units likely to be met in drive systems calculations.

To ensure standardisation across national borders, most motors are manufactured to standard dimensions and sometimes standard output powers. This standardisation is based on a preferred number series which is explained in detail. The use of preferred numbers has, of recent years, been extended to the sizing of fans.

Special Note: Some American publications have great difficulties with SI units. Treat all SI values quoted with caution.

Conversion factors for hardness and toughness testing, which are not defined by SI are also given.

Contents:

22.1 SI, The International System of Units

SI, Système Internationale d'Unités, the international measurement unit system, is not a completely new system. It is based on an earlier metric system and is coming more and more into world-wide use. The SI system is now systematically constructed to cover in practice all scientific, technical and daily requirements and is subject to international agreement. This means that it is now possible to apply the SI system uniformly throughout the world. A measurement system which is suitable for all technical and scientific purposes has to fulfil many requirements. Some of the basic requirements which SI satisfies are consistency, consequential applicability, coherence, the convenient expression of multiples and sub-multiples over a wide range of numerical values and accuracy.

Consistency means that each unit shall represent one, and only one, quantity.

Consequential applicability means that each quantity shall be measured in one, and only one, unit.

Coherence means that all units for every quantity shall be compatible so as to eliminate the need for arbitrary conversion factors in calculations involving related quantities.

Convenient expression of multiples and sub-multiples means the convenient multiplication of units to enable the use of practical numerical values within a particular application.

Accuracy means that the base units shall be precisely derived and defined. Six of the seven base units are thus determined from distinct precisely defined physical phenomena, the seventh, the kilogram, is determined by one standard body which is held in Paris.

In 1971 the Council of Ministers of the EEC ratified a Directive, 71/354/EEC, on units which committed all member states to amend legislation to authorise SI units within 18 months of that date and to implement all provisions of the Directive within a further five years. An amending Directive, 76/770/EEC, legislates the obligations. The Units of Measurement Directives place non-SI units into four chapters A to D.

Chapter A prescribes those units which are for permanent use and member states are obliged to authorise them in their laws by 21 April, 1978.

Chapter B contains a list of all units which member states have undertaken to cease to authorise in their laws with effect from 31 December, 1977.

Chapter C contains a list of units which member states have undertaken to cease to authorise in their laws with effect from 31 December, 1979.

Chapter D covers remaining units and some other units and will be reviewed before 31 December, 1979.

The formal content of the SI is determined and authorised by the General Conference of Weights and Measures (CGPM) and, for more detailed descriptions of the System reference should be made to BS 3763 and SI - The International System of Units published by HMSO. However, the basic advice to industry on the use of SI is now contained in ISO 1000, BS 5555.

The SI system includes three classes of units: 1) base units 2) supplementary units 3) derived units.

22.1.1 Brief history of unit systems

Although often called the metric system, the SI system essentially has more basic units and overcomes problems encountered during the development of a consistent system of units to serve all science and engineering functions. The metric system was introduced at the beginning of the nineteenth century based on the unit of length being the metre. The unit of mass followed and together with the unit of time formed the basic metric system. In 1873 the British Association for the Advancement of Science agreed on the use of the centimetre, gramme and second as the basic units for scientific work (the CGS system) but engineering within the United Kingdom had been well established using the British units of feet, pounds and seconds (the FPS system). As electrical experiments took place it was the metric system that became the basis for units peculiar to the electrical sciences and many basic electrical units were added to the CGS system.

An international authority on the metric system was established in 1875 with the Bureaux International des Poids et Mesures at Sèvres defining the units of length, mass and time as the metre, kilogramme and second respectively (the MKS system) as these units were more convenient than the CGS system. The basic units of length, mass and time are insufficient to cover electrical units and consequently units employing the permeability and permittivity of free space became necessary. Confusion also arose because of the links with the CGS system with, in particular, the use of p (the ratio of the circumference of a circle to its diameter) appearing in equations which were usually not associated with circles.

To overcome the complications the International Electrotechnical Commission (IEC) rationalized the units in 1950 and adopted as a fourth basic unit the unit of electrical current, the ampere (the MKSA system). This had been suggested about a half century earlier by an Italian professor called Giorgi and this system of units was consequently named the Giorgi System. Although this system of units covered electrical engineering it did not cover all branches of science and consequently the Conference Générale des Poids et Measures (CGPM) in 1954 agreed a rationalized and coherent system of units which became in 1960 the SI system.

After some subsequent additions the system now has seven basic units as follows:

length	metre (m)
mass	kilogramme (kg)
time	second (s)
electric current	Ampere (A)
light	candela (cd)
temperature	Kelvin (K)
substance	mole (mol)

In addition the following supplementary units are in use:

plane angle	radian (rad)
solid angle	steradian (sr)

The supplementary units are both ratios and therefore have no basic units.

The SI system is based around the seven basic units and the two supplementary units together with derived units for the more commonly used quantities and a series of prefixes used for the formation of multiples and submultiples.

Actual temperatures are normally ex-pressed in Celsius units (°C) and temperature differences in Kelvin units (K).

The Council of Ministers of the European Economic Community (EEC) ratified Directive 71/354/EEC in 1971 calling for all member states to amend legislation to authorize SI units and this was followed in 1976 by Directive 76/770/EEC to legalize the obligations. British Standard BS 5555 gives further details of SI units and their use.

22.1.2 Method of expressing symbols and numbers

The following rules apply to symbols for units

The symbol should be lower case unless the unit is derived from a proper name in which case the symbol should be upper case (or the first letter upper case if more than one letter), for example, metre - m, Ampere - A, Hertz - Hz

The symbol should not contain a final full stop, for example, m not m.

The symbol should remain unaltered in the plural, for example, m not ms

If multiple symbols are required they should be separated by a space if confusion can occur, for example, kg m/s^2 not kgm/s^2. If a pair of units are each represented by a single letter they are not separated if the absence of a space is not likely to cause confusion, for example Nm

The symbol should be reduced to its simplest expression, for example, W not J/s, Js^{-1}, kg m^2/s^3 or kg m^2 s^{-3}

Where more than one symbol is required and division is involved use a solidus or superscript, do not use more than one solidus, for example, use m/s^2 or m s^{2,} do not use m/s/s.

Large numbers should be written with the digits grouped in threes and with a '.' or ',' used to denote the decimal place, for example, 12 345 678, 0.000 012, 0,012 3. Use 10 000 not 10,000 to denote ten thousand, for example. One exception to this rule is that numbers with only four digits and without a decimal point do not normally have the space after the first digit, for example 1234 not 1 234.

Alternatively very large or very small numbers can be represented in exponential form, that is a number multiplied by a factor in the form of 10 to a positive or negative power. For example 100 is equivalent to 10^2 and 0.001 is equivalent to 10^{-3}. Therefore 1 234 000 can be expressed in the form 1.234 x 10^6 and 0.0123 can be expressed as 1.23 x 10^{-2}.

There is a further group of specialised units which are primarily for use within astronomy and physics.

22.1.3 Multiples of SI units

The prefixes in Table 22.1 are used to form names and symbols of multiples and subdivisions of the SI units. The symbol of a prefix is considered to be combined with the unit symbol for the base unit, supplementary unit or derived unit to which it is directly attached, forming with it a symbol for a new unit which can be provided with a positive or negative exponent and which can be combined with other unit symbols to form symbols for compound units.

Factor by which the unit is multiplied	Prefix		Example
	name	symbol	
10^{24}	yotta	Y	
10^{21}	zetta	Z	
10^{18}	exa	E	
10^{15}	peta	P	
10^{12}	tera	T	1 terrajoule = 1TJ
10^{9}	giga	G	1 gigawatt = 1GM
10^{6}	mega	M	1 megavolt= 1 MV
10^{3}	kilo	k	1 kilometre= 1 km
10^{2}	hecto	h	1 hectogram = 1 hg
10^{1}	deca	da	1 decalumen = 1 dalm
10^{-1}	deci	d	1 decimetre = 1 dm
10^{-2}	centi	c	1 centimetre = 1 cm
10^{-3}	milli	m	1 milligram = 1 mg
10^{-6}	micro		1 microgram = 1 m
10^{-9}	nano	n	1 nanohenry = 1 nH
10^{-12}	pico	p	1 picofarad = 1 pF
10^{-15}	femto	f	1 femtometre = 1 fm
10^{-18}	atto	a	1 attosecond = 1 as

Table 22.1 Multiples of SI units

Whenever possible units should be multiples or submultiples of 3 - hecto, deca, deci and centi therefore should not normally be used.

The prefix symbol should appear immediately before the basic symbol, for example mA for milliampere. It should be noted that mm^3 means (0.001 m)^3 not 0.001 m^3 and that mm^{-1} means (10^{-3} m)$^{-1}$ not 10^{-3} m^{-1}. The use of dk should be avoided as this may cause confusion, for example, dkg is decagramme not deci killogramme.

It is normal practice to use millimetre (mm) as the unit of length on engineering drawings.

22.1.4 Derived units

These are expressed in terms of base units and/or supplementary units by multiplication and division according to the laws of physics relating the various quantities, see Table 22.2.

Quantity	Name of derived SI unit	Symbol	Expressed in terms of base or supplementary units
frequency	Hertz	Hz	1 Hz = 1/s
force	Newton	N	1 N = 1 kg m/s^2
pressure, stress	Pascal	Pa	1 Pa = 1 N/m^2
energy, work, heat	Joule	J	1 J = 1 Nm
power	Watt	W	1 W = 1 J/s
electric charge, quantity of electricity	Coulomb	C	1 C = 1 A s
electric potential	Volt	V	1 V = 1 J/C = 1 W/A
electric capacitance	Farad	F	1 F = 1 C/V
electric resistance	Ohm		1 = V/A
electric conductance	Siemens	S	1 S = 1/
magnetic flux	Weber	Wb	1 Wb = 1 V s
magnetic flux density	Tesla	T	1 T = 1 Wb/m^2
inductance	Henry	H	1 H = 1 Wb/A
luminous flux	lumen	lm	1 lm = 1 cd sr
illuminance	lux	lx	1lx = 1 lm/m^2
radioactivity	Becquerel	Bq	1 Bq = 1/s
absorbed dose	Gray	Gy	1 Gy = 1 J/kg

Table 22.2 Some derived SI units having special names

Non SI units. There are certain units not included in SI which cannot, for a variety of reasons, be eliminated, despite the fact that these can, in principle, be expressed in SI units. Some of the non-SI units which may be used together with the SI units and their multiples and are recognised by the CIPM, Comité International des Poids et Mesures, are shown in Table 22.3.

Conductance is sometimes used as the reciprocal of resistance for which the unit is the Siemen (S), also referred to as mho.

Quantity	Name of unit	Unit symbol	Definition
time	minute hour day	min h d	1 min = 60 sec 1 h = 60 min 1 d = 24 h
plane angle	degree minute second	°	1° = (p/180) rad 1 = (1/60)° 1 = (1/60)
volume	litre	l	1 l = 1 dm^3
mass	tonne	t	1 t = 10^3 kg
pressure of fluid	bar	bar	1 bar = 10^{5}Pa

Table 22.3 Non-SI units

22.1.5 Checking units in equations

It is sometimes useful to check equations by checking whether the units are consistent on each side of the equals sign. In particular force and mass are often confused, for example, a mass of 1 kg produces a force due to gravity of 1 kg multiplied by the acceleration due to gravity (9.807 m/s^2) to give 9.807 N but this is often referred to as 1 kilogramme force (1 kgf) - if the units of kilogramme force are taken as the same as kilogramme then errors will result. Checking the consistency of the units should immediately indicate whether an error is present.

It is normal practice to replace the units with the letters M for mass, L for length and T for time when checking units and this avoids any consideration of multiples and submultiples that are not relevant when checking for consistency. For example if a constant torque is applied to a rotating mass, the time to change the speed by a given amount is given by:

$$t \quad \frac{Jn}{M} \qquad \text{Equ 22.1}$$

where

t = time for speed change

J = inertia of rotating mass

n = speed change

M = torque applied

The units of J may be kg m^2 which can be expressed as ML^2. The units of n may be revolutions per second or radians per second for which revolutions and radians have no basic units giving the units of n as 1/s which translates to 1/T. The units of torque may be Nm where N can be replaced by the basic units kg m/s^2 giving (ML/T^2) L = ML^2/T^2 representing torque. A mistake could be made at this stage if torque is given as kgf m and then the units are translated as ML. Substituting the correct M, L and T expressions into equation 22.1 gives:

$$t \quad \frac{ML^2 \quad 1/T}{ML^2/T^2} \quad T \qquad \text{Equ 22.2}$$

which shows the units are consistent. If the error above had been made the unit of time would be given as L/T and this would have been a clear indication that something was amiss.

A more complicated example, involving both electrical current (A) and temperature (K) units, is the rate of temperature rise of a current carrying conductor assuming all the heat generated is stored within the conductor, which is given by:

$$\frac{\quad}{t} \quad \frac{D^2}{dc} \qquad \text{Equ 22.3}$$

where

= temperature rise

t = time the current flows

D = current density

= conductor resistivity

d = density of conductor

c = specific heat of conductor

The units of D may be A/m^2 which can be expressed as A/L^2. The units of are usually expressed in terms of a resistance across the faces of a cube , for example, m^2/m or m. is a derived unit equivalent to a voltage divided by a current. In turn voltage is power divided by current, power is work divided by time, work is force multiplied by distance and force is mass multiplied by acceleration. Repeated substitutions then leads to the basic units of resistivity as equivalent to ML^3/T^{3}A^2. The unit of density is equivalent to M/L^3. Specific heat is usually ex-pressed in terms of J/kg K and therefore after substituting for J the unit of c is equivalent to L/T^{2}K. Substituting in equation 22.3 gives the units as:

$$\frac{\quad}{t} \quad \frac{\left(A/L^2\right)^2 \quad ML^3/T^3A^2}{M/L^3 \quad L^2/T^2K} \quad \frac{K}{T} \qquad \text{Equ 22.4}$$

which gives consistent units for the rate of temperature rise.

22.2 Conversion factors for SI units

A number of electrical terms evolved from the CGS system may still be encountered as well as units in the FPS system and other special units. In general it is best to convert units to the SI system before inserting quantities into equations as this will avoid problems involving conversion factors in the equations themselves. The engineering units which may be encountered in other systems of units are listed in Tables 22.4 to 22.6 with the conversion factor to give SI units.

Quantity	Unit	Conversion factor	SI unit
Length	cable	219.456	m
	chain (Gunter's)	20.116 8	
	chain (Ramden's)	30.48	
	fathom	1.828 8	
	feet	0.304 8	
	furlong	201.168	
	inch	0.025 4	
	micron	1 x 10^{-6}	
	mile (nautical Brit)	1.853 184 x 10^3	
	mile (nautical Int)	1.852 x 10^3	
	mile (statute)	1.609 34 x 10^3	
	mil	2.54 x 10^{-5}	
	rod	5.029 2	
	yard	0.914 4	
Area	acre	4.046 856 x 10^3	m^2
	are	100	
	centare	1	
	hectare	1 x 10^4	
Volume	barrel (Brit)	0.163 65	m^3
	barrel (US petrol)	0.158 98	
	barrel (US dry)	0.115 63	
	barrel (US liquid)	0.119 24	
	gallon (Brit) (Imp)	4.545 9 x 10^{-3}	
	gallon (US dry)	4.404 8 x 10^{-3}	
	gallon (US liquid)	3.785 3 x 10^{-3}	
	litre	1 x 10^{-3}	
	ounce (Brit fluid)	2.841 225 x 10^{-5}	
	ounce (US fluid)	2.957 373 x 10^{-5}	
	pint (Brit) (Imp)	5.682 4 x 10^{-4}	
	pint (US dry)	5.505 95 x 10^{-4}	
	pint (US liquid)	4.731 63 x 10^{-4}	
	quart (Brit) (Imp)	1.136 49 x 10^{-3}	
	quart (US dry)	1.101 19 x 10^{-3}	
	quart (US liquid)	9.463 2 x 10^{-4}	

Table 22.4 Conversion factors for length, area and volume

To obtain the quantity in SI units, the original units should be multiplied by the conversion factor, for example, to convert a length of 10 inch to metre units multiply by 0.025 4 giving 0.254 m. To convert from one non-SI unit to another non-SI unit, first convert to the SI unit by multiplying by the conversion factor and then convert to the other non-SI unit by dividing by the conversion factor for the other unit. If multiple units are involved, convert each separately by multiplying by the appropriate conversion factor for any units with positive powers and dividing by the appropriate conversion factors for any units with negative powers. For example, to convert lb/in^3 (or lb in^{-3}) to SI units multiple by 0.453 6 to convert lb to kg and divide by 0.025 4^3 to convert in^{-3} to m^{-3} giving 27 680 kg m^{-3} (or 27680kg/m^3).

To convert Fahrenheit to Celsius or centigrade subtract 32 and then multiply by 5/9

To convert Fahrenheit to Rankine add 459.67

To convert Celsius and centigrade to Kelvin add 273.15

To convert Rankine to Kelvin multiply by 5/9

For temperature conversion to Kelvin (K) the following should be noted:

Absolute zero	=	0 Kelvin
		0 Rankine
		-273.15 Celsius
		-273.15 centigrade
		-459.67 Fahrenheit
Ice point	=	273.15 Kelvin
		491.67 Rankine
		0 Celsius
		0 centigrade
		32 Fahrenheit

Quantity	Unit	Conversion factor	SI unit
Energy and Work	British Thermal Unit (Btu)	1.054 35 x 10^3	J
	Btu (IST)	1.055 04 x 10^3	
	Btu (mean)	1.055 87 x 10^3	
	Btu (39 F)	1.059 66 x 10^3	
	Btu (60 F)	1.054 68 x 10^3	
	calorie, gm (cal, gm)	4.184	
	cal, gm (mean)	4.19	
	cal, gm (15 C)	4.186	
	cal, gm (20 C)	4.182	
	calorie, kg (cal, kg)	4.184 x 10^3	
	cal, kg (mean)	4.19 x 10^3	
	centigrade heat unit (15°C)	1.898 3 x 10^3	
	erg	1 x 10^{-7}	
	foot-pound	1.355 82	
Force	dyne	1 x 10^{-5}	N
	kilogramme force	9.807	
	poundal	0.138 255	
Mass	hundredweight (long)	50.802 3	kg
	hundredweight (short)	45.359 2	
	ounce (troy)	0.031 103	
	ounce (avdp)	0.028 35	
	pounds (troy)	0.373 241	
	pounds (avdp)	0.453 592	
	slugs	14.593 9	
	ton (long)	1.106 046 x 10^3	
	ton (short)	9.071 847 x 10^2	
	ton (metric) (tonne)	1 x 10^3	
Power	cheval-vapeur	735.499	W
	horsepower	746	
	horsepower (boiler)	9.809 5 x 10^3	
	horsepower (metric)	735.499	

Quantity	Unit	Conversion factor	SI unit
Pressure	atmosphere (atm)	1.103 25 x 10^5	Pa
	bar	1 x 10^5	
	inch of mercury (32°F) (in Hg)	3.386 4 x 10^3	
	inch of mercury (60°F) (in Hg)	3.376 1 x 10^3	
	inch of water (4°C) (in H_2O)	0.248 7	
	mm of mercury (0°C)	133.313	
	torr	133.313	
Velocity	knot	0.514 4	m/s
Viscosity (dynamic)	centipoise (cp)	1 x 10^{-3}	Ns/m^2
	poise (p)	0.1	
Viscosity (kinematic)	centistoke (cSt)	1 x 10^{-6}	m^2/s
	Stoke (St)	1 x 10^{-4}	

Table 22.5 Conversion factors for quantities related to mechanics

Quantity	Unit	Conversion factor	SI unit
Electric charge	Faraday (chem)	9.649 x 10^4	C
	Faraday (phys)	9.652 x 10^4	
Magnetic flux	line	1 x 10^{-8}	Wb
	Maxwell	1 x 10^{-8}	
Magnetic flux density	Gauss	1 x 10^{-4}	T
Magnetizing force	Oersted	79.577 47	A-turns/m
Magnetomotive force	Gilbert	0.795 77	A-turns

Table 22.6 Conversion factors for quantities

Detailed conversion information is now given for a number of different quantities.

22.2.1 Plane angle

Quantity designation:

SI unit: radian, rad.

Normal multiple units: mrad,   rad.

Example: 2 rad = 2 × 57.2958 = 114.5916°

rad	... g gon, grade	... ° degree	... ′ minute	... ″ second	angular mil
1	63.6620	57.2958	3.43775 × 103	0.206265 × 106	1.00268 × 103
15.7080 $\cdot 10^{-3}$	1	0.9	54	3.24 × 103	15.75
17.4533 $\cdot 10^{-3}$	1.11111	1	60	3.6 × 103	17.5
0.290888 $\cdot 10^{-3}$	18.5185 $\cdot 10^{-3}$	16.6667 $\cdot 10^{-3}$	1	60	0.291667
4.84814 $\cdot 10^{-6}$	0.308642 $\cdot 10^{-3}$	0.277778 $\cdot 10^{-3}$	16.6667 $\cdot 10^{-3}$	1	4.86111 $\cdot 10^{-3}$
0.997331 $\cdot 10^{-3}$	63.4921 $\cdot 10^{-3}$	57.1429 $\cdot 10^{-3}$	3.42857	205.714	1

grade (g) (or gon), 1 g = 1 gon =   /200 rad.

1° =   /180 rad.

For some purposes, the angular mil is taken to be 10^{-3} rad. The figures shown here are based on the concept that an angular mil is equal to 360/6400 degrees.

NOTE: 1 grade (... g) = 1 /100 of a right angle

22.2.2 Length

Quantity designation l.

SI unit: metre (m).

Normal multiple units: km, cm, mm, .

Example: 3 in = 3 × 25.4 × 10^{-3} = 76.2 × 10^{-3} m

metre m	inch in	foot ft	yard yd	mile
1	39.370	3.2808	1.0936	0.62137 · 10^{-3}
25.4 · 10^{-3}	1	83.333 · 10^{-3}	27.778 · 10^{-3}	15.783 · 10^{-6}
0.3048	12	1	0.33333	0.18939 · 10^{-3}
0.9144	36	3	1	0.56818 · 10^{-3}
1.6093 · 10^{3}	63.36 ·10^{3}	5.28 · 10^{3}	1.76 · 10^{3}	1
1.852 · 10^{3}	72.913 · 10^{3}	6.0761 · 10^{3}	2.0254 · 10^{3}	1.1508

1 nautical mile = 6080 ft = 1853.184 m

1 Å, 1 Ångström = 10^{-10} m

1 astronomic unit = 0.1496 · 10^{12} m

1 light year = 9.4605 · 10^{15} m

1 parsec = 30.857 · 10^{12} m

22.2.3 Area

Quantity designation: A.

SI unit: square metre (m^2).

Normal multiple units: km^2, dm^2, cm^2, mm^2.

Example: 4 ft^2 = 4 ·92.903 ·10^{-3}= 0.371612 m^2

m^2	in^2	ft^2	yd^2	acre	square mile $mile^2$
1	1.5500 · 10^{3}	10.764	1.1960	0.24710 · 10^{-3}	0.38610 · 10^{-6}
0.64516 · 10^{-3}	1	6.9444 · 10^{-3}	0.77161 · 10^{-3}	0.15942 · 10^{-6}	0.24910 · 10^{-9}
92.903 · 10^{-3}	144	1	0.11111	22.957 · 10^{-6}	35-870 · 10^{-9}
0.83613	1.296 · 10^{3}	9	1	0.20661 · 10^{-3}	0.32283 · 10^{-6}
4.0469 · 10^{-3}	6.2726 · 10^{6}	43.56 · 10^{3}	4.84 · 10^{3}	1	1.5625 · 10^{-3}
2.5900 · 10^{6}	4.0145 · 10^{9}	27.878 · 10^{6}	3.0976 · 10^{6}	640	1

1 acre = 100 m^2

1 hectare = 100 acres = 10000 m^2

22.2.4 Volume

Quantity designation: V.

SI unit: cubic metre (m^3).

Normal multiple units: dm^3, cm^3, mm^3.

Non SI unit: litre (l): 1 l = 0.001 m^3 = 1 dm^3.

Normal multiple units: cl, ml.

Example:

5 US gallon = 5 · 3.7854 · 10^{-3} = 18.927 · 10^{-3} m^3 = 18.927 l

m^3	in^3	ft^3	yd^3	UK gallon	US gallon
1	61.024 · 10^{3}	35.315	1.3080	219.97	264.17
16.387 · 10^{-6}	1	0.57870 · 10^{-3}	21.434 · 10^{-6}	3.6046 · 10^{-3}	4.3290 · 10^{-3}
28.317 · 10^{-3}	1.728 · 10^{3}	1	37.037 · 10^{-3}	6.2288	7.4805
0.76456	46.656 · 10^{3}	27	1	168.18	201.97
4.5461 · 10^{-3}	277.42	0.16054	5.9461 · 10^{-3}	1	1.2010
3.7854 · 10^{-3}	231	0.13368	4.9511 · 10^{-3}	0.83268	1

gross (register) tonnage used in shipping
1 ton = 100 ft^3 = 2.83168 m^3

1 UK fluid ounce, fl oz = 28.4131 cm^3

1 US fluid ounce, fl oz = 29.5735 cm^3

22.2.5 Time

Quantity designation: t.

SI unit: second (s).

Normal multiple units: ms, s, ns.

Non SI units: day (d), hour (h), minute (min).

Example: 100000 s = 100000/3600 = 27.778 h

s	min	h	d (day)	week
1	16.6667 · 10^{-3}	0.277778 · 10^{-3}	11.5741 ·10^{-6}	1.65344 · 10^{-6}
60	1	16.6667 · 10^{-3}	0.694444 · 10^{-3}	99.2063 · 10^{-6}
3.6 · 10^{3}	60	1	41.6667 · 10^{-3}	5.95238 · 10^{-3}
86.4 · 10^{3}	1.44 · 10^{3}	24	1	0.142857
604.8 · 10^{3}	10.08 · 10^{3}	168	7	1

1 tropical year = 31556925.974 s = 365.24219878 d

1 sidereal year = 31558150 s

1 calendar year = 365 d = 8760 h

22.2.6 Linear velocity

Quantity designation: v.

SI unit: metre per second (m/s).

Normal multiple units: km/h.

m/s	km/h	ft/s	mile/h
1	3.6	3.2808	2.2369
0.27778	1	0.91134	0.62137
0.3048	1.0973	1	0.68182
0.44704	1.6093	1.4667	1
0.51444	1.852	1.6878	1.1508

1 knot = 1 nautical mile per hour = 1.853 km/h

22.2.7 Linear acceleration

Quantity designation: a.

SI unit: metre per second squared (m/s^2).

m/s^2	cm/s^2	ft/s^2	in/s^2	g
1	100	3.2808	39.37	0.10197
10 × 10-3	1	32.808 · 10-3	393.7 · 10^{-3}	1.0197 · 10^{-3}
0.3048	30.48	1	12	31.081 · 10^{-3}

m/s^2	cm/s^2	ft/s^2	in/s^2	g
25.4×10^{-3}	2.54	$83.33 \cdot 10^{-3}$	1	$2.59 \cdot 10^{-3}$
9.80665	980.665	32.174	386.09	1

22.2.8 Angular velocity

Quantity designation: .

SI unit: radian per second (rad/s).

The SI and Imperial units are identical.

Angular velocity is normally calculated from N revolutions/s by 2 N

22.2.9 Angular acceleration

Quantity designation: a.

SI unit: radian per second squared (rad/s^2)

The SI and Imperial units are identical.

22.2.10 Mass

Quantity designation: m.

SI unit: kilogram (kg).

Normal multiple units: g, g, mg.

Non SI unit: tonne (t) = 1000 kg.

kg	lb (pound)	oz (ounce)	hundred-weight cwt (UK)	ton (UK)
1	2.2046	35.274	$19.684 \cdot 10^{-3}$	$0.98421 \cdot 10^{-3}$
0.45359	1	16	$8.9286 \cdot 10^{-3}$	$0.44643 \cdot 10^{-3}$
14.594	32.174	514.79	$0.28727 \cdot 10^{-3}$	$14.363 \cdot 10^{-3}$
28.350×10^{-3}	62.5×10^{-3}	1	$0.55804 \cdot 10^{-3}$	$27.902 \cdot 10^{-6}$
50.802	112	$1.792 \cdot 10^{3}$	1	$50 \cdot 10^{-3}$
$1.0161 \cdot 10^{3}$	$2.24 \cdot 10^{3}$	$35.84 \cdot 10^{-3}$	20	1

oz = ounce, also called ounce avoirdupois

1 ounce troy = $31.1035 \cdot 10^{-3}$ kg

cwt = hundredweight

USA cwt = 100 lb   USA ton = 2000 lb

22.2.11 Density

Also called specific weight.

Quantity designation: .

SI unit: kilogram per cubic metre (kg/m^3)

Non SI units: kg/dm^3, g/cm^3.

kg/m^3	g/cm^3	lb/in^3	lb/ft^3
1	10^{-3}	$36.127 \cdot 10^{-6}$	$62.428 \cdot 10^{-3}$
10^{3}	1	$36.127 \cdot 10^{-3}$	62.428
$27.680 \cdot 10^{3}$	27.680	1	$1.728 \cdot 10^{3}$
16.019	$16.019 \cdot 10^{-3}$	$0.57870 \cdot 10^{-3}$	1

The term specific gravity or relative density is also used and is the ratio of the mass of a given volume of substance to the mass of an equal volume of water at temperature of 4 °C and a pressure of 101.325 kPa absolute. The density of water at 4 °C and 101.325 kPa absolute is 1000.02 kg/m^3.

22.2.12 Force

Quantity designation: F.

SI unit: newton (N).

Normal multiple units: MN, kN.

N	dyn	Kilogram-force, kgf kilopond,	pound-force lbf
1	$0.1 \cdot 10^{6}$	0.10197	0.22481
$10 \cdot 10^{-6}$	1	$1.0197 \cdot 10^{-6}$	$2.2841 \cdot 10^{-6}$
9.8066	$0.98066 \cdot 10^{6}$	1	2.2046
4.4482	$0.44482 \cdot 10^{6}$	0.45359	1

22.2.13 Torque

Quantity designation: T.

SI unit: Newton metre (Nm).

Normal multiple units: MNm, kNm.

Nm	kgf m	lbf × in	lbf × ft
1	0.10197	8.8508	0.73756
9.8066	1	86.796	7.2330
0.11299	$11.521 \cdot 10^{-3}$	1	$83.333 \cdot 10^{-3}$
1.3558	0.13826	12	1

Torque, power and speed are related by the formula: P = 2 NT

22.2.14 Pressure, stress

Quantity designation: p, , .

SI unit: Pascal (Pa), 1 Pa = 1 N/m^2.

Normal multiple units: GPa, MPa, kPa and for stress: MN/m^2, N/m^2, N/mm^2.

Pa	bar	kgf/cm^2 technical atmos	kgf/mm^2	Torr (≈ mm Hg)	standard atm	lbf/in^2 (psi)
1	$10 \cdot 10^{-6}$	$10.197 \cdot 10^{-6}$	$0.10197 \cdot 10^{6}$	$7.5006 \cdot 10^{-3}$	$9.8692 \cdot 10^{-6}$	$0.14504 \cdot 10^{-3}$
$100 \cdot 10^{3}$	1	1.0197	$10.197 \cdot 10^{-3}$	750.06	0.98692	14.504
$98.066 \cdot 10^{3}$	0.98066	1	$10 \cdot 10^{-3}$	735.56	0.96784	14.223
$9.8066 \cdot 10^{6}$	98.066	100	1	$73.556 \cdot 10^{3}$	96.784	$1.4223 \cdot 10^{3}$
133.32	$1.3332 \cdot 10^{-3}$	$1.3595 \cdot 10^{-3}$	$13.595 \cdot 10^{-6}$	1	$1.3158 \cdot 10^{-3}$	$19.337 \cdot 10^{-3}$
$101.32 \cdot 10^{3}$	1.0132	1.0332	$10.332 \cdot 10^{-3}$	760	1	14.696
$6.8948 \cdot 10^{3}$	$68.948 \cdot 10^{-3}$	$70.307 \cdot 10^{-3}$	$0.70307 \cdot 10^{-3}$	51.715	$68.046 \cdot 10^{-3}$	1

The preferred pressure unit for fan applications is the "bar".

1 mm H_2O 9.81 Pa

1 in H_2O 249.09 Pa

1 in Hg 3386.4 Pa

1 ata = 1 technical atmosphere (absolute)

1 atu = 1 technical atmosphere (gauge)

22.2.15 Dynamic viscosity

Quantity designation: .

SI unit: Pascal second Pa s.

Normal multiple units mPa s (= cP).

N s/m²	N s/mm²	P (poise)	cP m Pa s
1	10^{-6}	10	10^3
10^6	1	10×10^6	10^9
0.1	0.1×10^{-6}	1	100
10^{-3}	10^{-9}	10×10^{-3}	1

22.2.16 Kinematic viscosity

Quantity designation: .

SI unit: square metres per second (m²/s).

Normal multiple units: mm²/s.

For conversion to other units of viscosity see nomogram below.

m²/s	mm²/s cSt	St (Stoke)
1	10^6	$10 \cdot 10^3$
10^{-6}	1	$10 \cdot 10^{-3}$
$0.1 \cdot 10^{-3}$	100	1

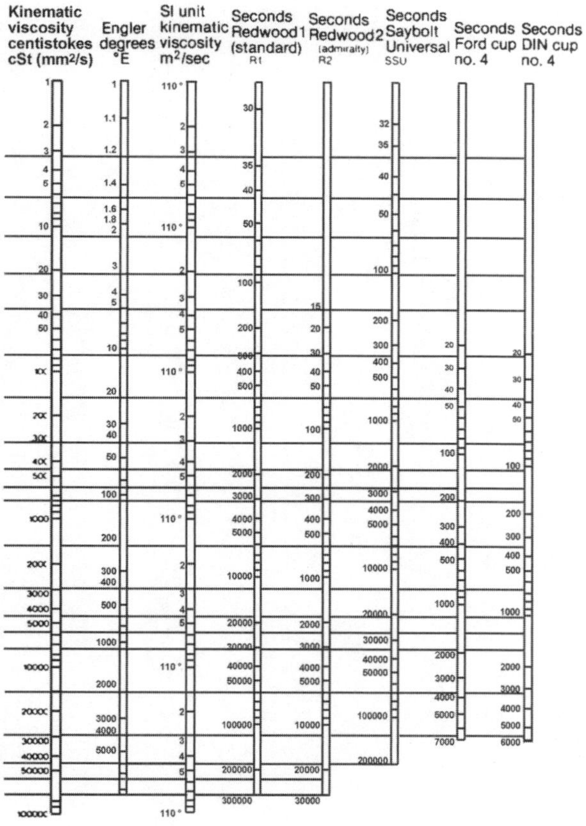

	cSt x	°E x	R1 x	SSU x
Kinematic viscosity cSt	1	7.58	0.247	0.216
Engler degrees °E	0.132	1	0.0326	0.0285
Seconds, Redwood 1, R1	4.05	30.7	1	0.887
Seconds, Saybolt Universal SSU	4.62	35.11	1.14	1

The above factors apply for values above 60cSt

22.2.17 Energy

Quantity designations: E, W, Q, L, U depending upon the type of energy.

SI unit: Joule (J).

Normal multiple units: TJ, GJ, MJ, kJ, mJ.

Since 1 J = 1 Nm = 1 Ws then Nm and Ws can also be used for all types of energy. The unit Joule should, however, be used for expressing all types of energy.

1 erg = 0.1×10^{-6} J.

J Joule	kWh kilowatt hour	kgf m kilogram-force metre	kcal kilo-calorie	ch h metric horse-power hour	ft. lbf (foot pound-force)	Btu (British thermal unit)
1	$0.27778 \cdot 10^{-6}$	0.10197	$0.23885 \cdot 10^{-3}$	$0.37767 \cdot 10^{-6}$	0.73756	$0.94782 \cdot 10^{-3}$
3.6 · 106	1	$0.36710 \cdot 10^6$	859.85	1.3596	$2.6552 \cdot 10^6$	$3.4121 \cdot 10^3$
9.8066	$2.7241 \cdot 10^{-6}$	1	$2.3423 \cdot 10^{-3}$	$3.7037 \cdot 10^{-6}$	7.2330	$9.2949 \cdot 10^{-3}$
$4.1868 \cdot 10^3$	$1.163 \cdot 10^{-3}$	426.94	1	$1.5812 \cdot 10^{-3}$	$3.0880 \cdot 10^3$	3.9683
$2.6478 \cdot 10^6$	0.73550	$0.27 \cdot 10^6$	632.42	1	$1.9529 \cdot 10^6$	$2.5096 \cdot 10^3$
1.3558	$0.37662 \cdot 10^{-6}$	0.13826	$0.32383 \cdot 10^{-3}$	$0.51206 \cdot 10^{-6}$	1	$1.2851 \cdot 10^{-3}$
$1.0551 \cdot 10^3$	$0.29307 \cdot 10^{-3}$	107.59	0.25200	$0.39847 \cdot 10^{-3}$	778.17	1

22.2.18 Power

Quantity designation: P.

SI unit: Watt (W).

Normal multiple units: GW, MW, kW, mW, W.

W	kgf m/s	kcal/s	kcal/h	ch metric horse-power	hp horse-power	ft lbf/s	Btu/h
1	0.10197	$0.23885 \cdot 10^{-3}$	0.85985	$1.3596 \cdot 10^{-3}$	$1.3410 \cdot 10^{-3}$	0.73756	3.4121
9.8066	1	$2.3423 \cdot 10^{-3}$	8.4322	$13.333 \cdot 10^{-3}$	$13.151 \cdot 10^{-3}$	7.2330	33.462
$4.1868 \cdot 10^3$	426.94	1	$3.6 \cdot 10^3$	5.6925	5.6146	$3.0880 \cdot 10^3$	$14.286 \cdot 10^3$
1.163	0.11859	$0.27778 \cdot 10^{-3}$	1	$1.5812 \cdot 10^{-3}$	$1.5596 \cdot 10^{-3}$	0.85779	3.9683
735.50	75	0.17567	632.42	1	0.98632	542.48	$2.5096 \cdot 10^3$
745.70	76.040	0.17811	641.19	1.0139	1	550	$2.5444 \cdot 10^3$
1.3558	0.13826	$0.32383 \cdot 10^{-3}$	1.1658	$1.8434 \cdot 10^{-3}$	$1.8182 \cdot 10^{-3}$	1	4.6262
0.29307	$29.885 \cdot 10^{-3}$	$69.999 \cdot 10^{-6}$	0.25200	$0.39847 \cdot 10^{-3}$	$0.39302 \cdot 10^{-3}$	0.21616	1

22.2.19 Flow

Quantity designation: q_V.

SI unit: cubic metre per second (m³/s).

Non-SI units: l/s, ml/s, m³/h.

I gallon/ min Igpm	US gallon/ min USgpm	barrel/ day bpd	m³/s	m³/h	l/s	l/min
1	1.2009	41.175	$75.768 \cdot 10^{-6}$	$272.76 \cdot 10^{-3}$	$75.768 \cdot 10^{-3}$	4.5461
0.83268	1	34.286	$63.09 \cdot 10^{-6}$	$227.12 \cdot 10^{-3}$	$63.09 \cdot 10^{-3}$	3.7854
$24.286 \cdot 10^{-3}$	$29.167 \cdot 10^{-3}$	1	$1.84 \cdot 10^{-6}$	$6.6244 \cdot 10^{-3}$	$1.84 \cdot 10^{-3}$	$110.41 \cdot 10^{-3}$
$13.198 \cdot 10^{3}$	$15.85 \cdot 10^{3}$	$543.44 \cdot 10^{3}$	1	3600	1000	60000
3.6662	4.4029	150.95	$277.78 \cdot 10^{-6}$	1	$277.78 \cdot 10^{-3}$	16.667
13.198	15.85	543.44	$1.0 \cdot 10^{-3}$	3.6	1	60
$219.97 \cdot 10^{-3}$	$264.17 \cdot 10^{-3}$	9.0573	$16.667 \cdot 10^{-6}$	$60 \cdot 10^{-3}$	$16.667 \cdot 10^{-3}$	1

1 barrel = 42 US gallon

22.2.20 Temperature

Absolute temperature: Quantity designation: T.

SI unit: Kelvin (K).

Temperature: Quantity designation: t.

unit degree Celsius (Centigrade) (°C).

	Kelvin*** scale	Celsius scale	Rankine scale	Fahrenheit** scale	Physical relation-ship
Relative temperature value	0 K	- 273.15 °C	0 °R	- 459.67 °F	Absolute zero
	273.15 K	0 °C	491.67 °R	32 °F	Melting point of ice*
	273.16 K	0.01 °C	491.688 °R	32.018 °F	Triple point of water*
	373.15 K	100 °C	671.67 °R	212 °F	Boiling point of water*
Relative temperature intervals (diffs)	1 K 0.55556 K	1 °C 0.55556 °C	1.8 °R 1 °R	1.8 °F 1 °F	

* For defined conditions

** Value in °C = $\frac{1}{1.8}$ (value in °F -32) = (value in K -273.15)

*** Value in K = $\frac{5}{9}$ × (value in °R)

22.3 Other conversion factors

22.3.1 Hardness

Hardness is not defined within the SI system. The following table can be used for conversion between the popular systems used.

Tensile strength	Vickers hardness (F ≥ 98 N)	Brinell hardness	Rockwell hardness	
N/mm²	HV	BHN	HR_B	HR_C
200	63	60		
210	65	62		
220	69	66		
225	70	67		
230	72	68		
240	75	71		
250	79	75		
255	80	76		
260	82	78		
270	85	81	41	
280	88	84	45	
285	90	86	48	
290	91	87	49	
300	94	89	51	
305	95	90	52	
310	97	92	54	
320	100	95	56	
30	103	98	58	
335	105	100	59	
340	107	102	60	
350	110	105	62	
360	113	107	63.5	
370	115	109	64.5	
380	119	113	66	
385	120	114	67	
390	122	116	67.5	
400	125	119	69	
410	128	122	70	
415	130	124	71	
420	132	125	72	
430	135	128	73	
440	138	131	74	
450	140	133	75	
460	143	136	76.5	
465	145	138	77	
470	147	140	77.5	
480	150	143	78.5	
490	153	145	79.5	
495	155	147	80	
500	157	149	81	
510	160	152	81.5	
520	163	155	82.5	
530	165	157	83	
540	168	160	84.5	
545	170	162	85	
550	172	163	85.5	
560	175	166	86	
570	178	169	86.5	
575	180	171	87	
580	181	172		
590	184	175	88	
595	185	176		
600	187	178	89	
610	190	181	89.5	
620	193	184	90	
625	195	185	91	
630	197	187	91.5	
640	200	190	92	
650	203	193	92.5	
660	205	195		
670	208	198	93	
675	210	199	93.5	
680	212	201		
690	215	204	94	
700	219	208		
705	220	209	95	
710	222	211	95.5	
720	225	214	96	
730	228	216		
740	230	219	96.5	
750	233	221	97	
755	235	223	97.5	
760	237	225		
770	240	228	98	
780	243	231		21
785	245	233		
790	247	235		
800	250	238	99	22
810	253	240	99.5	
820	255	242		23

Tensile strength	Vickers hardness (F ≥ 98 N)	Brinell hardness	Rockwell hardness	
N/mm²	HV	BHN	HR_B	HR_C
830	258	245		
835	260	247		24
840	262	249		
850	265	252		
860	268	255		25
865	270	257		
870	272	258		26
880	275	261		
890	278	264		
900	280	266		27
910	283	269		
915	285	271		
920	287	273		28
930	290	276		
940	293	278		29
950	295	280		
960	299	284		
965	300	285		
970	302	287		30
980	305	290		
990	308	293		
995	310	295		31
1000	311	296		
1010	314	299		
1020	317	301		32
1030	320	304		
1040	323	307		
1050	327	311		
1060	330	314		33
1070	333	316		
1080	336	319		34
1090	339	322		
1095	340	323		
1100	342	325		
1110	345	328		35
1120	349	332		
1125	350	333		
1130	352	334		
1140	355	337		
1150	358	340		36
1155	360	342		
1160	361	343		
1170	364	346		
1180	367	349		37
1190	370	352		
1200	373	354		
1210	376	357		38
1220	380	361		
1230	382	363		
1240	385	366		39
1250	388	369		
1255	390	371		
1260	392	372		
1270	394	374		40
1280	397	377		
1290	400	380		
1300	403	383		41
1310	407	387		
1320	410	390		
1330	413	393		42
1340	417	396		
1350	420	399		
1360	423	402		
1370	426	405		43
1380	429	408		
1385	430	409		
1390	431	410		
1400	434	413		
1410	437	415		44
1420	440	418		
1430	443	421		
1440	446	424		
1450	449	427		
1455	450	428		45
1460	452	429		

Tensile strength	Vickers hardness (F ≥ 98 N)	Brinell hardness	Rockwell hardness	
N/mm²	HV	BHN	HR_B	HR_C
1470	455	432		
1480	458	435		
1485	460	437		
1490	461	438		46
1500	464	441		
1510	467	444		
1520	470	447		
1530	473	449		
1540	476	452		47
1550	479	455		
1555	480	456		
1560	481	457		
1570	484	460		
1580	486	462		48
1590	489	465		
1595	490	466		
1600	491	467		
1610	494	470		
1620	497	472		
1630	500	475		49
1640	503	478		
1650	506	481		
1660	509	483		
1665	510	485		
1670	511	486		
1680	514	488		50

Values based on DIN 50150.

22.3.2 Material toughness

Material toughness is not defined by SI. Most materials are assessed by conducting impact testing. Two differing test methods can be used with various sizes and styles of test specimen. The following table can be used as a **guide** to the relative toughness of the various tests:

Charpy V notch kgm/cm²	Charpy V notch ft lb	Charpy V notch Joule	Izod ft lb
0.4	2.3	3.1	2.5
0.9	5.2	7	6.4
1.5			10.8
2.2			16
3.1	18	24.4	21.5
4.1	23.8	32.2	27.8
5.2	30	40.6	34.1
6.5	37.7	51	40.4
8.0	46.4	62.9	46.7
9.4	54.5	73.9	53
10.9	63	85.4	59.3
12.6			65.6
14.1	82	111	71.9
15.8			78.2
17.7	102	138	84.5
19.4			90.8
21.1	122	165	97.1
23.0	134	182	103.4

22.4 Preferred numbers

22.4.1 General

Standardisation is necessary if products are to be sold across national frontiers without problems of installation and operation. This applies in particular to dimensions to ensure that products can be purchased to a standard with the knowledge that it will fit in place of the same basic product manufactured by a different company or indeed manufactured in a different country. The general principle can also be applied to output ratings and in turn standard ratings can be related to a set of standard principle dimensions. The electric motor industry was one of the first to adopt standard dimensions and ratings for a wide range of products and this is universally accepted world-wide. The same basic principle has more recently been extended to other products, for example gearboxes of particular interest to fans. There is also considerable standardisation throughout industry on parts that are used in the make up of fans and a good example is the Standards applicable to fixings such as bolts and screws. With respect to the principle dimensions of fans, a preferred number series forms the basis.

The problem of deciding what range of sizes to produce for any particular design ought to be considered at an early stage of its life cycle. Unfortunately, in the past, this was frequently not the case, and manufacturers often produced unique sizes to suit a favourite customer, without consideration of its effect on the competitive advantages it might have for other applications.

The demand of consumers for an infinite range of choice has to be balanced against any increase in the cost of production, stocking and distribution, which might result from the introduction of too many types and sizes.

Long experience within the industry has shown, however, that sizes following a geometric progression can satisfy demand, as they are in harmony with the fan laws.

22.4.2 Preferred number series

Preferred numbers used for fans are therefore based on a geometric series where each number has a common ration between it and the adjacent number. This suits both large and small numbers in a particular range. Table 22.7 shows preferred numbers in the R20 and R40 series. These series give roughly a 12% and 6% increase respectively to the next highest number in the series.

R20	R40	R20	R40	R20	R40
1	1	10	10	100	100
	1.06		10.6		106
1.12	1.12	11.2	11.2	112	112
	1.18		11.8		118
1.25	1.25	12.5	12.5	125	125
	1.32		13.2		132
1.4	1.4	14.	14	140	140
	1.5		15		150
1.6	1.6	16	16	160	160
	1.7		17		170
1.8	1.8	18	18	180	180
	1.9		19		190
2	2	20	20	200	200
	2.12		21.2		212
2.24	2.24	22.4	22.4	224	224
	2.36		23.6		236
2.5	2.5	25	25	250	250
	2.65		26.5		265
2.8	2.8	28	28	280	280
	3		30		300
3.15	3.15	31.5	31.5	315	315
	3.35		33.5		335
3.55	3.55	35.5	35.5	355	355
	3.75		37.5		375
4	4	40	40	400	400
	4.25		42.5		425
4.5	4.5	45	45	450	450
	4.75		47.5		475
5	5	50	50	500	500
	5.3		53		530
5.6	5.6	56	56	560	560
	6		60		600
6.3	6.3	63	63	630	630
	6.7		67		670
7.1	7.1	71	71	710	710
	7.5		75		750
8	8	80	80	800	800
	8.5		85		850
9	9	90	90	900	900
	9.5		95		950
				1000	1000

Table 22.7 Preferred numbers in R20 and R40 series

As previously stated, the fan laws indicate that a geometric progression in sizes is to be desired whilst BS 2045 1965 and PD 6481 1977 favour Preferred Numbers. These are arbitrarily rounded off values derived from a geometric series having one of the common ratios shown in Table 22.8.

Series	R5	R10	R20	R40	R80
Ratio	$\sqrt[5]{10}$	$\sqrt[10]{10}$	$\sqrt[20]{10}$	$\sqrt[40]{10}$	$\sqrt[80]{10}$
=	1.58	1.26	1.12	1.06	1.03

Table 22.8 Renard series ratios

Further information is given in ISO497. Subsequently, Eurovent produced their document 1/2 for the sizes of fan circular flanges and this first formally recognised the R20 series for a range of standard light duty fans. This was then adopted internationally as ISO 6580.

More recently a Standard, ISO 13350 covering fan sizes, circular flanges and rectangular outlet/inlet flanges, has been produced, again based on an R20 series and covering fan sizes from 100 mm to 2000 mm. Customers are urged to specify this Standard, so that the situation current in the motor world, will eventually be replicated in the fan world. A level playing field may then be achieved and competitive products will be fairly compared.

These sizes are now widely used in Germany for both axial and centrifugal fans. They are also dominant in the UK axial fan market.

100	112	125	140	160	180	200	224	250	280	315	355
400	450	500	560	630	710	800	900	1000	1120	1250	1400
1600	1800	2000									

Table 22.9 R20 series for fan sizes and flanges as standardized in ISO 6580 and ISO 13350

By using the rounded numbers, the divergence from the theoretical value nowhere exceeds 1.22% as can be seen by reference to Table 22.10. The similarity with the standard frame sizes of electric motors, although the latter have used a frame 132 from the R40 series in preference to the adjacent sizes from the R20 series, is striking.

Theoretical values		Renard series (rounded)	Percentage difference between R20 series and calculated values
Basic logarithm	Calculated value	R20	
2.00	100.00	100	0

Theoretical values		Renard series (rounded)	Percentage difference between R20 series and calculated values
Basic logarithm	Calculated value	R20	
2.05	112.20	112	-0.18
2.10	125.89	125	-0.71
2.15	141.25	140	-0.88
2.20	158.49	160	+0.95
2.25	177.83	180	+1.22
2.30	199.53	200	+0.24
2.35	223.87	224	+0.06
2.40	251.19	250	-0.47
2.45	281.84	280	-0.65
2.50	316.23	315	-.039
2.55	354.81	355	+0.05
2.60	398.11	400	+0.47
2.65	446.68	450	+0.74
2.70	501.19	500	-0.24
2.75	562.34	560	-0.42
2.80	630.96	630	-0.15
2.85	707.95	710	+0.29
2.90	794.33	800	+0.71
2.95	891.25	900	+0.98
3.00	1000.0	1000	0
3.05	1122.0	1120	-0.18
3.10	1258.9	1250	-0.71
3.15	1412.5	1400	-0.88
3.20	1584.9	1600	+0.95
3.25	1778.3	1800	+1.22
3.30	1995.3	2000	+0.24

Table 22.10 Calculation of R20 series

22.4 Normal quantities and units used in fan technology

The fan and air movement industries within the English speaking world have for many years used a number of specialised units. These have been based on the imperial system, adapted as necessary. Often an arbitrary choice was made, for example pressure measured in inches water gauge.

Table 22.11 gives a number of conversion factors designed to assist those who are unfamiliar with the magnitude of the SI units. They will also be useful in converting values from earlier textbooks, catalogues, and other data.

Quantity	Imperial unit	SI unit	Conversion factors
Volume flowrate	cfm (ft^3/min)	cubic metres per second (m^3/s)	4.7195×10^{-4}
	cfm (ft^3/min)	litres per second (l/s)	4.7195×10^{-1}
	cu sec (ft^3/s)	cubic metres per second (m^3/s)	2.8316×10^{-2}
Pressure	inches w.g.	Pascal (Pa or N/m^2)	2.4909×10^{2}
	inches w.g.	kiloPascal (kPa)	2.4909×10^{-1}
	inches w.g.	millibar (mbar)	2.4909
	inches H.g.	kiloPascal (kPa)	3.3864
Power	hp (bhp or ahp)	Watt (W or J/s)	7.4570×10^{2}
	hp	kiloWatt (kW)	7.4570×10^{-1}
Torque (5)	lbf-in	Newton metre (Nm)	1.1298×10^{-1}
	lbf-ft	Newton metre (Nm)	1.3558
Density	lb/ft^3	kilogram per cubic metre (kg/m^3)	1.6018×10
Tip speed Outlet velocity or Duct velocity	fpm (ft/min)	metres per second (m/s)	5.0800×10^{-3}
	fps (ft/sec)	metres per second (m/s)	3.0480×10^{-1}
	mph (miles/hr)	metres per second (m/s)	4.4704×10^{-1}
Rotational speed (2)	rpm (rev/min)	revolutions per second (rev/s)	1.6667×10^{-2}
Dimensions	inches	millimetres (mm)	2.5400×10
	feet	metre (m)	3.0480×10^{-1}
	thou (mil) = 0.001 in	micrometre (m)	2.5400×10
Moment of inertia (6)	lb-ft^2	kilogram metre squared (kg m^2)	4.2140×10^{-2}
	slug-ft^2	kilogram metre squared (kg m^2)	1.3558
Stress (5)	lbf-ft^2	Pascal (Pa or N/m^2)	6.8948×10^{3}
	ton f-in^2	megaPascal (MPa)	1.5444×10
Energy (work or heat equivalent)	Therm	megaJoule (MJ)	1.0551×10^{-2}
	hp hr (horsepower hour)	megaJoule (MJ)	2.6845
	Btu (British thermal unit)	kiloJoule (kJ)	1.0551
	ft-lbf	Joule (J)	1.3558
	kW hr	megaJoule (MJ)	3.6000
Temperature (3)	°F	Kelvin	(°F+ 459.67) 1.8

Table 22.11 Metric and Imperial conversion factors

Notes to Table 22.11:

1. The choice of the appropriate multiple or sub-multiple of an SI unit is governed by convenience. The multiple chosen for a particular application should be the one which will lead to numerical values within a practical range (ie kiloPascal for pressure, kilowatts for power, and megaPascal for stress).
2. The second is the SI base unit of time, although outside SI the minute has been recognised by CIPM as necessary to retain for use because of its practical importance. The use of rev/min for rotational speed is still therefore continued.
3. The Kelvin is the SI base unit of thermodynamic temperature and is preferred for most scientific and technological purposes. The degree Celsius (°C) is acceptable for practical applications.
4. Multiply Imperial unit by this factor to obtain SI Standard, except the Kelvin temperature.
5. Great care must be taken in the conversion of these units. In the Imperial system the pound force or weight lbf (mass x acceleration due to gravity) was often loosely referred to as 'lb'.
6. For reasons as in (5) above inertia was often given as $\frac{W}{g}k^2$ ie slug/ft^2.

Multiples:

Name	Symbol	Factor
micro		10^{-6}
milli	m	10^{-3}
kilo	k	10^{3}
mega	M	10^{6}

Examples:

a 5712 c.f.m. = $5712 \times 4.7195 \times 10^{-4}$

= 2.6958 m^3/s

b 20.6 c.f.m.	=	$20.6 \times 4.7198 \times 10^{-1}$
	=	9.7228 l/s
c 1.35 in.w.g.	=	$1.35 \times 2.4909 \times 10^{2}$
	=	336.27 Pa
d 40.6 in.w.g.	=	$40.6 \times 2.4909 \times 10^{-1}$
	=	10.11 kPa

and so on.

The USA continues to use such units based on the Imperial system. In view of their dominant position in the air conditioning market, such units will have widespread currency for many years to come. It is essential, therefore, for engineers to be conversant with the conversion factors used for translating between the two systems.

22.6 Bibliography

BS 5555:1981, ISO 1000-1981, *Specification for SI units and recommendations for the use of their multiples and of certain other units.*

ISO 6580:2005, *General-purpose industrial fans — Circular flanges – Dimensions.*

ISO 497:1973, *Guide to the choice of series of preferred numbers and of series containing more rounded values of preferred numbers.*

23 Useful fan terms translated

A

abrasion	abrasion	Abrieb
abrasion resisting fan	ventilateur pour gaz chargé de poussières abrasives	Ventilator zur Förderung von schleifendem Staub
access or inspection door	porte de visite	Inspektionsdeckel
acidity	acidité	Azicität
addition	juxtaposition	Anlagerung
aeration	ventilation, aération	Belüftung
air extracting fans	ventilateurs pour extraction d'air	Ventilatoren, luftabsaugend
ammeter	ampèremètre	Amperemeter
amount	quantite	Menge
analysis	analyse	Analyse
anti-vibration mountings	dispositifs antivibratiles	Schwingungsdämpfer
approximate	approché	Angenähert
arc	arc, courbe, coude	Bogen
asbestos	asbeste, amiante	Asbest
asbestos fibre	fibre d'amiante	Asbestfaser
asbestos-cement	amiante-ciment	Asbestzement
atmosphere	atmosphére	Atmosiphäre
atomic or nuclear energy	énergie atomique ou nucléaire	Atomenergie, Kernenergie
axial fans	ventilateurs axiaux	Axialventilatoren
axial-flow fan	ventilateur hélicoïde	Axialventilator
axial-flow-indirect drive	hélicoïde - entraínment indirect	Axial - Indirekter Antrieb
axial-flow-long casing-guide vane-direct drive	hélicoïde - enveloppe longue - distributeur - entraínment direct	Axial - Langes Gehäuse - Leitschaufein - direkter Antrieb
axial-flow-multistage - indirect drive	hélicoïde - multi-étages - entraínment indirect	Axial - Mehrstufig - indirekter Antrieb
axial-flow-propeller fan	hélicoïde - "propeller fan"	Axial - Für Einbau in einer Öffnung
axial-flow-shielded motor (bifurcated) - direct drive	hélicoïde - moteur protégé - entraínment direct	Axial - Motor mit Fremdkühlung - direkter Antrieb
axial-flow-short casing - direct drive	hélicoïde - enveloppe courte - entraínment direct	Axial - Kurzes Gehäuse - direkter Antrieb
axis	axe	Achse
axis of rotation	axe de rotation	Drehachse
axle	essieu	Welle

B

backplate	un disque arrière de roue	Deckscheibe
backpressure	contre-presslon	Gegendruck
backward curved or inclined	aubes courbées ou inclinées vers l'arrière	Rückwärts gekrümmt oder geneigt
ball bearing	roulement à billes	Kugellager
base	fondation	Fundament
base angles	equerres de base	Fußwinkel
base plate	plaque de fondation, socle	Grundplatte
baseframe	chassis support	Grandrahmen
bearing	palier	Lager
bearing bracket	plaque support du palier, flasques de palier	Lagerkonsole, Lagerschilde
bearing cover	couvercle à palier	Lager-deckel
bearing pedestal	support-palier	Lagerbock
bearing stool	tabouret-palier	Lagerträger
bearing supports	support-paliers	Haltestreben für Lager
bed	fondation	Fundament
belt drive	commande à courroie	Riemenantrieb
bend	arc, courbe	Bogen
bid	offre, devis	Angebot
bin	réservoir, bac, caisse	Behälter
blade inlet edge	bord d'attaque de l'aube	Eintrittskante der Laufradschaufel
blade leading edge	bord d'attaque de l'aube	Eintrittskante der Laufradschaufel
blade root	pied de pale	Fuß der Laufradschaufel
blade tip	bord de fuite de l'aube, bord périphérique	Austrittskante der Laufradschaufel, Spitze der Laufradschaufel
blade trailing edge	bord de fuite de l'aube	Austrittskante der Laufradschaufel
blades	aubes (ou pales)	Laufradschaufein
bolt	vis	Schraube
breakdown	arrêt de service, trouble	Betrebsstörung
cross-flow fan	ventilateur tangentiel	Questromventilator
building	construction	Konstruktion, Aufbau
building site	terrain à bâtir, chantier	Baustelle
bushing	douille, boite, mançhon	Buchse, Mute, Hulse
butterfly valve	robinet modérateur	Drosselventil

C

calibration	étalonnage, tarage	Eichung
capacity	capacité de transport	Förderleistung
carbon	charbon, carbone	Kohle
carbonic acid	acide carbonique	Kohlensäure
casing backplate	flasque arrière	Gehäuserückwand
casing coverplate	flanc démontable	Gehäusedeckel
casing drain	purge de volute	Ablaufstutzen
casing inlet sideplate	flasque avant	Gehäusevorderwand
casing stiffeners	renforts d'enveloppe	Gehäuseverstärkungen
casing wear ring	bagues d'usure fixes	Spaltringe
castillated not	écrou à créneaux	Kronenmutter
cavitation	cavitation	Kavitation
centre fairing	carénage	Motorverkleidung
centrifugal fan	ventilateur centrifuge	radialventilator
centrifugal-backward curved-indirect drive	centrifuge-aubes à courbées vers l'arrière - entraínment indirect	Radial-Rückwärts gekrümmt-indirekter Antrieb
centrifugal-double inlet	centrifuge-double ouïe d'aspiration	Radial-Zweiseitig saugend
centrifugal-forward curved-direct drive	centrifuge-aubes à courbées vers l'avant - entraínment direct	Radial-Vorwärts gekrümmt-direkter Antrieb
centrifugal-multistage	centrifuge-multi-étage	Radial-Mehrstufig
centrifugal-paddle blade-indirect drive	centrifuge-aubes radiales-roue sans disque-entraínment indirect	Radial-Laufrad ohne Deckscheibe und Boden-indirekter Antrieb
centrifugal-two stage with duct connection (duplex)	centrifuge-deux étages séparéspar un conduit intermédiaire (duplex)	Radial-Zweistufig mit Verbindung durch Leitung
centrifugal-vane control-coupled drive	centrifuge-commande par aubage - entraínment par accouplement	Radial-Drallregler-Antrieb über Kupplung
chain	châine	Kette
chain drive	transmission par châine	Kettengetriebe
chassis	châssis	Fahrgestell
chemistry	chimie	Chemie
chlorine	chlore	Chlor
chlorine dioxide	bioxyde de chlore	Chlordioxid
circuit diagram	schema de connexions	Schaltbild
circulation	circulation	Umlauf
circulating fan	ventilateur brasseur d'air	Umwälzventilator
cleaning	lavage, rinçage, curage	Spolung
clockwise rotation	tourne "vers la droite"	Rechtsdrehend
clutch	embrayage	Kupplung
coal	charbon, carbone	Kohle
coefficient of elongation	coéfficient d'allongement	Dehnungskoeffizient

coefficient of expansion	coéfficient de dilatation	Ausdehnungskoeffizient
combination baseplate	socle commun	Sockel für Motor und Lager
companion flange	contre-bride	Gegenflansch
compensation nut	écrou compensateur	Ausgleichsmutter
compressed air	air comprimé	Druckluft
compression spring	ressort de compression	Druckfeder
connecting rod	bielle	Schubstange
connecting rod bolts	boulon de bielle	Schubstangenschrauben
connecting rod cap	tête de bielle	Schubstangendeckel
connecting rod half brg bottom	coussinet de bielle inférieur	Unteres Schubstanglager
connecting rod half brg top	coussinet de bielle supérieur	Oberes Schubstanglager
connecting rod nuts	écrous de bielle	Schubstangenmuttern
connection	assemblage, raccord	Anschluss
construction	construction	Konstruktion, Aufbau
container	réservoir, bac, caisse	Behälter
contamination	impureté, contamination	Verunreinigung
continuous	continu, à action continue	Kontinuierlich
control systems, automatic, heating, ventilation and air conditioning (HVAC)	systèmes de contrôle automatique pour chauffage, ventilation et climatisation d'air	Steuerungssysteme, automatisch, für Heizungsanlagen, Lüftungsanlagen und Klimaanlagen
control valve	soupape de reglage	Regelventil
controller	régulateur	Regler
conveying fan	ventilateur pour transport pneumatique	Transportventilator
coolant	produit refroidisseur, réfrigerant	Kuhlmittel
cooling disc (or cooling impeller) guard	protection du disque (ou de la turbine) de refroidissement	Schutzgitter für Kühlscheibe (oder -laufrad)
cooling disc (or impeller)	disque (ou turbine) de refroidisssement	Kühlscheibe (oder -laufrad)
corrosion	corrosion	Korrosion
corrosion resistance	résistance à la corrosion	Korrosionsbeständigkeit
corrosion resisting fan	ventilateur pour gaz corrosif	Korrosionssicherer Ventilator
counter-clockwise rotation	tourne "vers la gauche"	Linksdrehend
counterpressure	contre-presslon	Gegendruck
coupling	embrayage, accouplement	Kupplung
coupling guard	protection de l'accouplement	Kupplungsschutz
cover	couvercle	Deckel
crankcase	bâti	Gestellblock
crankshaft	arbre vilebrequin	Kurbelwelle
cross-flow	tangentiel	Querstrom
cross-flow fan	ventilateur tangentiel	Questromventilator
crosshead	crosse	Kreuzkopf
crosshead pin	tourillon de crosse	Kreuzkopfzapfen
crosshead pin bearing	coussinet de crosse	Kreuzkopflager
cross-section	section transversale	Querschnitt
curve	courbe	Kurve
customer service	service pour les clients	Kundenservice
cut-off	bec	Gehäusezunge
cut-out	interrupteur	Schalter
cylinder block	corps du cylindre	Zylinderblock

D

damper control	commande par registre	Regulierung durch Drosselklappe
DC motor	moteur à courant continu	Gleichstrommotor
dead weight	poids mort	Eigengewicht
deaeration	ventilation	Entlüftung
decarbonisation	décarbonisation	Entcarbonisierung
delivery	capacité de transport	Förderleistung
density	densité, poids spécifique	Dichte
depression	depression, vide partiel	Unterdruck
diagram	diagramme	Diagramm
diaphragm plate	platine	Wandplatte
dimension	dimension	Abmessung
direct drive	entraînment direct	Direkter Antrieb
direction of rotation	sens de rotation	Drehsinn des laufrades
disc valve	soupape à siége plan	Tellerventil
discharge	écoulernent	Ausfluß Ausströmung
discharge rate	débit	Förderstrom
dosing	dosage	Dosierung
downstream centre fairing	carénage aval	Abströmhaube
downstream guide vane	aube directirce aval	Nachleitschaufel
downstream guide vanes (a set)	redresseur	Nachleitapparat
drain	écoulernent, effluent	Ablauf
drive belt(s)	courroie(s) d'entraïnment	Antriebsriernen
drive guard	carter de protection d'entraïnment	Riemenschutz
driveshaft	essieu moteur	Antriebsachse
driving clutch	embrayage	Antriebskupplung
driving gear	pignon de commande	Antriebsrad
driving side	côté de la commande	Antriebsseite
ductility	extensibilité, ductilité	Dehnbarkelt
dust fan	ventilateur pour gaz poussiéreux	Staubventilator

E

efficiency	rendement	Wirkungsgrad
elasticity	extensibilité, ductilité	Dehnbarkeit
elbow	arc, courbe	Bogen
elimination	élimination, separation	Ausscheiden
elongation	allongement	Expansion
end cover	couvercle de fermeture	Abschlußdeckel
energy	énergie	Energie
engaging piece	entraineur	Mitnehmer
engine	machine, engin	Maschine
erection	montage	Montierung
example of application	exemple d'application	Anwendungsbeispiel
exhauster	désaérateur	Entlüfter
expansion	expansion	Expansion
extended cut-off	bec prolongé	Zungenblech

F

factory	fabrique, usine, ateliers	Fabrik
fairing supports	supports de carénage	Haltestreben für Verkleidungen
fan casing	enveloppe du ventilateur	Ventilatorengehäuse
fan coil units	ventilateurs à serpentins	Klimakonvektoren
fan inlet	ouïe d'aspiration du ventilateur	Eintrittsöffnung des Ventilators
fan outlet	ouïe de refoulement du ventilateur	Austrittsöffnung des Ventilators
fan pulley	poulie du ventilateur	Ventilatorriemenscheibe
fans for corrosive atmospheres	ventilateurs pour atmosphères corrosives	Ventilatoren für aggressive Umgebungen
fans, portable, industrial, electric	ventilateurs portatifs industriels	Ventilatoren, elektrisch, tragbar, industriell
fault	arrêt de service, trouble	Betriebsstörung
faulty installation	défault de montage	Einbaufehler
feather	ressort	Feder
feed	arrivée	Zufluß
filter	filtre	Fitter
filter area	surface du filtre	Filterfläche
flameproof fan	ventilateur ininflammable, antidéflagrant	Explosionsgeschützter Ventilator, Ventilator, feuerfest
flange	bride	Flansch

flow	courant	Strömung
flow measuring and control equipment	matériel de mesure et de contrôle de débit	Durchflussmessgeräte und Durchflussregler
foot or feet	pied(s)	Fuß oder Füße
force	énergie	Energie
forced-feed lubrication	graissage sous pression	Druckschmierung
foundation	fondation	Fundament
free inlet fan	ventilateur refoulant	Frei ansaugender Ventilator
free outlet fan	ventilateur aspirant	Frei ausblasender Ventilator
friction	friction	Reibung
fulcrum	axe de rotation	Drehachse
fully ducted fan	ventilateur aspirant-refoulant	Ventilator für beidseitigen Leitungsanschluß

G

gasket	bague d'etanchéité	Dichtungsring
gastight fan	ventilateur étanche	Gasdichter Ventilator
gate valve	soupape à coulisse	Schieber
gauge	mesurer	Messen
gauging	étalonnage, tarage	Eichung
gear transmission	transmission	Getriebe
gear wheel	roue dentée	Zahmad
gearbox	carter d'engrenages	Getrielbegehäuse
general purpose fan	ventilateur courant	Ventilator für allgemeine Zwecke
generator	chaudiére à vapeur	Dampfkessel
gland	presse-étoupe	Stopfbuchse
grade	grosseur de grain	Komgröße
graduation	graduation, gamme	Skala
graph	diagramme	Diagramm
grey cast iron	fonte grise, fonte moulée	Grauguß
guide vane	aube directrice	Leitschaufel
guide vanes (a set)	aubage directeur	Leitapparat

H

high efficiency fans	ventilateurs à haut rendement	Hochleistungsventilatoren
high pressure fans	ventilateurs à haute pression	Ventilatoren, Hochdruck
high temperature fans	ventilateurs pour hautes températures	Ventilatoren für hohe Temperaturen
hot gas fan	ventilateur pour gaz chauds	Heißgasventilator
hub	moyeu	Nabe
hub boss	bossage central du moyeu	Nabenkörper
hub disc	disc de moyeu	Nabenscheibe
hub rim	jante de moyeu	Nabenkranz
hub spider	croisillon de moyeu	Nabenstern

I

impeller	roue	Laufrad
impeller backplate	disque arrière de roue	Laufradboden
impeller centreplate	disque central de roue	Gemeinsamer Laufradboden
impeller endplate	disque latèral de roue	Endscheibes des Laufrades
impeller inlet clearance	jeu à l'entrée de la roue	Laufradspalt am Eintritt
impeller intermediate shroud	disque intermèdiaire de roue	Zwischenscheibe oder Zwischenring des Laufrades
impeller shroud	disque avant de roue	Deckscheibe oder Deckring des Laufrades
impeller tip clearance	jeu périphérique de la roue	Laufradspalt
impeller tip diameter	diamètre de roue	Laufraddurchmesser
impeller-side guard	protection côte roue	Laufradseitiges Schutzgitter
impermeable	imperméable	Undurchlässig
impurity	impureté, contamination	Verunreinigung
induction motor	moteur asynchrone	Asynchronmotor
inflow	arrivée	Zufluß
initial velocity	vitesse initiale	Anfangsgeschwindigkeit
inlet box	caisson d'aspiration	Saugkasten
inlet flange	bride d'aspiration	Eintrittsflansch
inlet guard	protection à l'aspiration	Schutzgitter am Eintritt
inlet spigot	manchette d'aspiration	Eintrittsstutzen
inlet vane control	inclineur	Drallregler am Eintritt
input side	côté de la commande	Antriebsseite
installation	installation, assemblage	Einbau
installation dimension	dimension de montage	Einbaumaß
insulation	isolation	Isolierung
interconnecting duct	conduit intermèdiaire	Verbindungsleitung
internal diameter	diametre interieur	Innendurchmesser
iron	fer	Eisen
issue	éccoulernent	Ablauf, Auströmung

J

jet fan	ventilateur de jet	Strahlventilator
journal bearing	palier de l'arbre, palier lisse	Kurbellager, Gleitlager

L

laminar	laminaire	Laminar
leakage	coulage, fuite	Leckage
line voltage	tension du résau	Netzspannung
loading	charge	Belastung
low pressure fans	ventilateurs à basse pression	Ventilatoren, Niederdruck
lubricant	lubrifiant	Schmiermittel

M

machine	machine	Maschine
magnitude	taille, dimension	Größe
main shaft	arbre principal	Hauptwelle
maintenance	entretien, soin	Instandhaltung, Wartung
manual operation	service manuel	Handbetrieb
manufacturing tolerance	tolérance de fabrication	Fertigungstoleranz
marine fans	ventilateurs pour usage marin	Schiffsventilatoren
material (of construction)	materiel	Werkstoff
material defect	defectuosité	Materialfehler
measure	mesurer	Messen
measuring	dosage	Dosierung
measuring error	erreur de mesure	Meßfehler
mechanical seal	joint mécanique	Gleitringdichtung
medium pressure fans	ventilateurs à moyenne pression	Ventilatoren, Mitteldruck
metallic coating	recouvrement métallique	Metallüberzug
method	operation, méthode	Verfahren
mill	fabrique, usine, ateliers	Fabrik
mine ventilation fans	ventilateurs et souffleurs pour mines	Ventilatoren und Gebläse für Bergwerke
mixed-flow	hélico-centrifuge	Halbaxial
mixed-flow fan	ventilateur hélico-centrifuge	Halbaxialventilator
modular system	système modulaire	Modular Baukastensystem
monitoring	contrôle, surveillance	Oberwachung
motor (engine)	moteur	Motor
motor arms	bras support de moteur	Haltearme für Motor
motor bracket	plaque support du palier	Motorkonsole
motor or other prime mover	moteur ou autre dispositif d'entraïnment	Elektromotor oder andere Antriebsmaschine
motor pulley	poulie du moteur	Motorriemenscheibe
motor stool	tabouret-moteur	Motorbock
motor supports	supports du moteur	Haltestreben für Motor
motor-side guard	protection côte moteur	Motorseitiges Schutzgitter
mounting	installation, montage	Einbau, Montage
mounting lugs	pattes de fixation	Befestigungslaschen
mounting ring (wall flange)	bride de fixation	Wandraing (Wandbefestigungsflansch)
multi-stage	roue multicanal	Mehrstufig

multi-stage fan	ventilateur multi-étage	Mehrstufig Ventilator
N		
non-clogging fan	ventilateur conçu pour éviter l'engorgement	Verschmitzungssicherer Ventilator
nozzle	porte-vent, buse, tuyau	Duse
nuclear power industry fans	ventilateurs et soufflantes pour l'industrie nucléaire	Ventilatoren und Gebläse für die Nuklearindustrie
number of revolutions	nombre de tours	Drehzahl
nut	écrou	Mutter
O		
offer	offre, devis	Angebot
oil	huile	Öl
oil level gauge	jauge de niveau d'huile	Ölstandanzeiger
operating instructions	instructions de service	Bedienungsvorschrift
operating temperature	température de fonctionnement	Betriebstemperatur
orifice	orifice	Blende
oscillation	oscillation, vibration	Schwindung
outlet expander	diffuseur au refoulement	Austrittsdiffusor
outlet flange	bride de refoulement	Austrittsflansch
outlet position	position de l'ouïe	Stellung der Austrittsöffnung
outlet reducer	convergent au refoulement	Austrittskonfusor
outlet spigot	manchette de refoulement	Austrittsstutzen
outlet transformer	pièce de transformation au refoulement	Übergangstück am Austritt
output	débit nominal	Nennleistung
overflow	trop-plein	Oberlauf
overload	surcharge	Oberlastung
overload protection	protection contre les surcharges	Oberlastungsschutz
oxidation	oxydation	Oxydation
P		
packing	joint, garniture, étoupage	Dichtung
packing fluid	liquide obturant	Sperrflüssigkeit
panel	panneau de distribution	Schalttafel
parallel operation	marche en paralléle	Parallellbetrieb
partition fan	ventilateur de paroi	Wand - oder Dachventilator
peak load	charge maximale	Spitzenbelastung
performance curve	courbe de performance	Leistungsdiagramm
pilot plant	usine de recherche	Versuchsanlage
pinion	pignon	Ritzel
piping	tuyauterie	Rohrleitung
piston	piston	Kolben
piston displacement	cylindre	Hubraum
piston rod	tige de piston	Kolbebstange
plain bearing	palier lisse	Gleitlager
plan view	projection horizontale	Grundriß
plant	fabrique, usine, ateliers	Fabrik
plunger	piston, plongeur	Plunger
power	énergie	Energie
power absorption	puissance absorbée	Leistungsaufnahme
power consumption	puissance nécessaire	Kraftbedarf
power demand	puissance nécessaire	Kraftbedarf
power input	puissance absorbée	Leistungsautnahme
powered roof ventilator	ventilateur de toiture	Dachventilator
process	operation, méthode	Verfahren
Q		
quantity	quantité, debit	Menge
R		
radial fans	ventilateurs radiaux	Radialventilatoren
radial/forward curved	radiale aubes courbées vers l'avant	Vorwärts gekrümmt
rate	vitesse, vélocité	Geschwindigkeit
rated capacity	débit nominal	Mennleistung
refrigeration industry fans	ventilateurs pour industrie du froid	Ventilatoren für die Kälteindustrie
risk of fracture	danger de rupture	Bruchgefahr
roller bearing	roulement à rouleaux	Rollenlager
S		
saddle clip	collier de prise	Anbohrschelle
safety factor	coéfficient de securité	Sicherheitsfaktor
scale	graduation, gamme	Skala
scroll plate	volute	Gehäusemantel
seal	joint, garniture, étoupage	Dichtung
sealing liquid	liquide obturant	Sperrflûssigkeit
sealing properties	properiété d'étanchéité	Dichtungseigenschaft
sealing ring	bague d'étanchéité	Dichtungsring
sealing sleeve	joint d'étanchéité	Dichtungsmanschette
service condition	condition de fonctionnement	Betriebsbedingung
service life	durée de service	Betriebsdauer
servicing	entretien	Instandhaltung
shaft	arbre	Welle
shaft extension	bout d'arbre	Wellenende
shaft guard	protection de l'arbre	Wellenschutz
shaft seal	dispositif d'étanchéite sur l'arbre	Wellendichtung
shaped inlet	pavillon d'aspiration	Einströmdüse
shroud	un disque avant	Laufradboden
size	taille, dimension	Größe
sleeve	manchon	Muffe, Hulse
solder	souder	Löten
solidification	solidification	Erstarrung
sound reduction equipment	matériel d'insonorisation	Vorrichtungen zur Schalldämpfung
soundproofing systems for ducts	systèmes d'insonorisation pour conduits	Schallschutzsysteme für Rohrleitungen
source	source	Quelle
space required	encombrement	Raumbedarf
spare	pièce de rechange	Ersatzteil
spare part list	liste des pièces de rechange	Ersatzteilliste
sparkproof fan	ventilateur antidéflagrant	Funkengeschützter Ventilator
special designs	exécutions spéciales	Sonderausführungen
special purpose fan	ventilateur spécial	Ventilator für spezielle Zwecke
specific gravity	densité, poids spécifique	Dichte
speed	vitesse, velocité	Geschwindigkeit
speed range	gamme de tours	Drehzahlbereich
split pins	goupilles	Splinte
spring	ressort, source	Feder
stainless	inoxydable	Rostfrei, Nichtrostend
standard design	exécution standard	Standardausführung
starting	mise en mouvement	Inbetriebnahme
start-up	mise en mouvement	Inbetriebnahme
strain	charge	Belastung
strainer	filtre	Filter
strength	solidité	Festigkeit
stress	charge	Belastung
supervision	contrôl surveillance	Oberwachung
supply	assemblage, raccord	Anschluß
surface	surface	Olberfläche, Fläche
surface finish	état de surface	Oberflägchen-beschaffenheit
switch	interrupteur	Schalter
switchboard	panneau de distribution	Schalttafel
T		
tachometer	compte-tours, tachomètre	Drehzahlmesser
tank	réservoir, bac, caisse	Behälter
tapping sleeve	collier de prise	Anbohrschelle

temperature	temperature	Temperatur
tender	offre, devis	Angebot
tensile strength	résistance à la tradion	Zugestigkeit
tension	tension	Spannung
testing of materials	essai des matériaux	Materialprufung
three-phase AC motor	moteur à courant triphasé	Drehstrommotor
throughput	debit	Durchflußmenge
time	temps	Zeit
transmission housing	carter de transmission	Antriebsgehäuse
transmission shaft	arbre d'entrâinement	Antriebswelle
tunnel ventilating fans	ventilateurs pour tunnels	Ventilatoren für Tunnelbelüftungen

U

upstream centre fairing	carénage amont	Anströmhaube
upstream guide vane	aube directrice amont	Vorleitschauel
upstream guide vanes (a set)	distributeur	Vorleitapparat

V

vane control	commande par aubage	Drallregulierung
variable pitch control	commande par variation de pas	Laufradschaufel-verstellung
variable speed control	commande par variation de vitesse	Drehzahlreguliering
variable-pitch (VP) & variable speed fans	ventilateurs à pas variable et à vitesse variable	Ventilatoren regelbar
vee-belt	courroie trapezoidale	Keilriemen
vee-belt pulley	pouilie à gorge	Keilriemenscheibe
velocity	vitesse, vélocité	Geschwindigkeit
vent	désaérateur	Entlûfter
ventilation	ventilation, aération	Entlüftung, Belüftung
vibration	oscillation, vibration	Schwindung
viscosity	viscosité	Viskositet, Zäghigkeit
volume	quantité, volume	Menge, Volumen

W

wear	ponçage, abrasion	Abrieb
weight	poids	Gewicht
weld	souder	Schweißen
wet gas fan	ventilateur pour gaz humides	Naßgasventilator
working condition	condition de fonctionnement	Betriebsbedingung
works/plant	fabrique, usine, ateliers	Fabrik

24 Guide to manufacturers and suppliers

The classification guide summarises the various fan types, covering their differing styles, sizes and basic principles of operation. All definitions are in accordance with ISO13349 (BS 848 Part 8).

The guide has been categorized in a particular way to impose strict boundary limits on fan types and the operating conditions available, with the specific aim of simplifying the choice of supplier from the users' point of view.

The guide covers all fan types, followed by ancillary products and services. Trade names are comprehensively listed too. It is preceded by the names and addresses and contact details of all companies appearing in the classification guide. These are listed alphabetically, by country.

It is strongly recommended that direct contact with the relevant companies is made to ensure that their details are clarified wherever necessary.

Contents:

24.1 Introduction

The classification guide summarises the various types of fans according to their differing styles and sizes and basic principles of operation.

Within the fan industry there are certain established practices by which fans are sometimes named according to their design or construction. This may also be according to the field of use or particular application.

Despite the fact that the means of designation and description are not always strictly logical it is often obvious to both user and manufacturer what is intended by a particular main group.

One reason for the classification guide being categorised in this way has been to impose strict boundary limits – with the express aim of simplifying the choice of supplier from the user's point of view.

Classification

Classification has been based primarily on the fan function i.e.

- Ducted fans
- Partition fans
- Open fans

Followed by — Fluid path and impeller blading

Then —

The Arrangement, e.g.for ducted fans — direct or vee belt drive for centrifugal or radial flow fans.

The Arrangement for axial, propeller, mixed flow or ring shaped fans etc. — tube axial, vane axial, contra rotating, reversible or bifurcated, and direct or vee belt drive.

Operating conditions

Reference is also made to the range of operating conditions offered by a manufacturer or supplier. This relates to the entire range of fans supplied and will not therefore apply to every fan type under which the company is listed.

It is strongly recommended that direct contact with the relevant companies is made to ensure that their details are clarified wherever necessary.

The operating conditions are defined as follows:

A — General purpose

B — Domestic

Special purpose:

C — Hot gas

D — Smoke venting

E — Wet gas

F — Gas tight

G — Dust handling

H — Conveyance/Transport

J — Abrasion resistant

K — Non clogging

L — Corrosive use

M — Aerospace

N — Marine

P — Military vehicles

Names and Addresses — Section 24.2

This Section has been based on a questionnaire sent to a wide range of selected manufacturers and suppliers of fans. Where possible the information supplied has been used. Full company and contact details are given. Companies are listed alphabetically, by country of origin.

Fan types — Section 24.3

This important Section has also been based on the same questionnaire. Discussions have been held with many of the companies to ensure wherever possible, that their activities were correctly interpreted. There will however, inevitably be some overlapping due to limitations of category descriptions and space, and the information is for guidance only.

The user is advised to check with each manufacturer and supplier for specific information. Companies are listed alphabetically within this Section, followed by the letter reference for the operating conditions available, then with their country of origin.

Ancillary products and services — Section 24.4

Also based on the same questionnaire, this Section lists companies alphabetically under the relevant product or service and with their country of origin.

Trade names — Section 24.5

This Section has been compiled similarly. It lists companies alphabetically under the relevant trade name and with their country of origin.

24.2 Names and addresses

AUSTRIA (A)

Helios Ventilatoren GmbH
Siemensstrasse 15A
A-6023 Innsbruck
Austria
Tel: 0512 26 59 88
Fax: 0512 26 59 88 85
E-Mail: info@heliosventilatoren.at
Web: www.heliosventilatoren.at

BELGIUM (B)

Almeco NV-SA
Rue de la Royenne
B-7700 Mouscron
Belgium
Tel: 056 854080
Fax: 056 854081
E-Mail: info@almeco.be
Web: www.almeco.be

Alstom Belgium SA Energy-MTM
Leuvensesteenweg 474
BE - 2812 Muiizen
Belgium
Tel: 015 450011
Fax: 015 423337

De Raedt SA
Chaussée de Namur 66
B-1400 Nivelles
Belgium
Tel: 067 89 23 23
Fax: 067 8923 29
E-Mail: nd@deraedt.be
Web: www.deraedt.be

Gates Europe nv
Dr. Carlierlaan 30
B-9320 Erembodegem
Belgium
Tel: 053 76 27 11
Fax: 053 76 27 13
E-Mail: ptindustrial@gates.com
Web: www.gates.com/europe/pti

Toussaint Nyssenne SA
Clos du Chemin Creux 6C
Holleweggaarde
B-1030 Brussels
Belgium
Tel: 087 30 69 11
Fax: 087 31 44 76
E-Mail: henri.schippers@touny.com
Web: www.touny.com

BRAZIL (BR)

Howden Ventiladores Ltda
Rua Dr Sodré 122 - 2° andar
CEP 04535-110-Vila Olimpia-SP
San Paulo
Brazil
Tel: 011 3089 0044
Fax: 011 3089 0067
E-Mail: howden@howden.com.br
Web: www.howden.com

CANADA (CAN)

Leader Fan Industries Ltd
130 Claireville Drive
Toronto
Ontario
M9W 5Y3
Canada
Tel: 416 675 4700
Fax: 416 675 4707
E-Mail: jgs@leaderfan.com
Web: www.leaderfan.com

TLT Co-Vent Fans Inc
1381 Hocquart Street
St. Bruno
Montreal
J3V 6B5
Canada
Tel: 450 441 3233
Fax: 450 441 2189
E-Mail: jcohen@coventfans.ca
Web: www.coventfans.ca

Universal Fan & Blower Ltd
30 Barker's Lane
Bloomfield
Ontario
K0K 1G0
Canada
Tel: 0613 393 3267
Fax: 0613 393 1937
E-Mail: info@universalfan.com
Web: www.universalfan.com

DENMARK (DK)

B. Bille A/S
Rugmarken 34
DK-3520 Farum
Denmark
Tel: 044 95 68 11
Fax: 044 95 66 49
E-Mail: mail@bbille.com
Web: www.bbille.com

Bruel and Kjaer
DK-2850 Nærum
Denmark
Tel: 04580 0500
Fax: 04580 1405
E-Mail: info@bksv.com
Web: www.bksv.co.uk

Danfoss Drives A/S
Ulsnæs 1
DK-6300 Gråsten
Denmark
Tel: 07488 2222
Fax: 07465 2580
Web: www.danfoss.com

Exhausto A/S
Odensevej 76
DK-5550 Langeskov
Denmark
Tel: 0656 61234
Fax: 0656 61200
E-Mail: exhausto@exhausto.dk
Web: www.exhausto.com

Multi-Wing International a·s
Staktoften 16
DK-2950 Vedbaek
Denmark
Tel: 045 89 01 33
Fax: 045 89 31 33
E-Mail: info@multi-wing.com
Web: www.multi-wing.com

FINLAND (FIN)

ABB Drives
HVAC
PO Box 184
FIN-00381 Helsinki
Finland
Tel: 01 022 11
Fax: 01 022 22330
E-Mail: mikko.lonnberg@fi.abb.com
Web: www.abb.com/fi

Koja Oy
PO Box 351
FIN-33101 Tampere
Finland
Tel: 03 282 5111
Fax: 03 282 5401
E-Mail: vesa.sorkio@koja.fi
Web: www.koja.fi

Oy Swegon AB
Munkinmäentie 1
FIN-02400 Kirkkonummi
Finland
Tel: 09 221 981
Fax: 09 221 98200
E-Mail: info@swegon.fi
Web: www.swegon.fi

Vallox Oy
Myllykyläntie 9-11
FIN-32200 Loimaa
Finland
Tel: 02 763 6300
Fax: 02 763 1539
E-Mail: markku.maenpaa@vallox.com
Web: www.vallox.com

FRANCE (F)

Airap
5/7 Avenue Ferdinand Buisson
F-75016 Paris
France
Tel: 01 46 20 37 20
Fax: 01 46 20 34 13
E-Mail: ch.rappoport@airap.fr
Web: www.airap.fr

Aldes Aeraulique
20, Boulevard Joliot Curie
F-69694 Vénissieux Cedex
France
Tel: 04 78 77 15 15
Fax: 04 78 76 15 97
E-Mail: aldes@aldes.com
Web: www.aldes.com

Gates France SARL
BP 37 Zone Industrielle

F-95380 Louvres
France
Tel: 01 34 47 41 41
Fax: 01 34 72 60 54
E-Mail: ptindustrial@gates.com
Web: www.gates.com/europe/pti

Groupe Leader
68 Boulevard Jules Durand
F-76056 Le Havre Cedex
France
Tel: 02 35 53 05 75
Fax: 02 35 53 16 32
E-Mail: sb@groupe-leader.fr
Web: www.groupe-leader.fr

Helios Ventilateurs Sarl
ZI La Fosse à la Barbière
2 rue Louis Saillant
F-93605 Aulnay sous Bois Cedex
France
Tel: 01 48 65 75 61
Fax: 01 48 67 28 53
E-Mail: info@helios-fr.com
Web: www.helios-fr.com

Howden Sirocco SA
19 rue de la Ladrie
BP 125
F-59653 Villeneuve d'Ascq Cedex
France
Tel: 03 28 33 32 30
Fax: 03 28 33 32 31
E-Mail: sales@howdensirocco.com
Web: www.howden.com

Nicotra France SA
8 Chemins des Mûriers - ZI mi-plaine
F-69745 Genas-Cedex
France
Tel: 04 72 79 01 20
Fax: 04 72 79 01 21
E-Mail: nicotra.france@wanadoo.fr
Web: www.nicotra.it

Sardou SA
18 rue du Sauvoy
F-77165 Saint-Soupplets
France
Tel: 01 60 01 03 67
Fax: 0160 01 03 34
E-Mail: info@sardou.net
Web: www.sardou.net

Timken Europe
France
Tel: 03 89 2144 44
Fax: 03 89 2145 99
E-Mail: andre.van.der.merwe@timken.com
Web: www.timken.com
See also: The Timken Company, USA

Vortice France
72 rue Baratte
F-94106 Cholet Saint Maur Cedex
France
Tel: 01 55 12 50 00
Fax: 01 55 12 50 01
E-Mail: contact@vortice-france.com
Web: www.vortice.com

GERMANY (D)

Gates GmbH
Haus Gravener Strasse 191-193
D-40764 Langenfeld
Germany
Tel: 02173 5108 0
Fax: 02173 795 150
E-Mail: ptindustrial@gates.com
Web: www.gates.com/europe/pti

Gebhardt Ventilatoren
Gebhardstrasse 19-25
D-74638 Waldenburg
Germany
Tel: 07942 1010
Fax: 07942 101170
E-Mail: info@gebhardt.de
Web: www.gebhardt.de

Gewa-Werth GmbH
Rheinische Strasse 1
D-58332 Schwelm
Germany
Tel: 023 36 2820
Fax: 023 36 10595
E-Mail: info@gewa-werth.de
Web: www.gewa-werth.de

Helios Ventilatoren
Lupfenstrasse 8
D-78056 VS - Schwenningen
Germany
Tel: 07720 6060
Fax: 07720 606166
E-Mail: info@heliosventilatoren.de
Web: www.heliosventilatoren.de

Hofmann Maschinen- und Anlagenbau GmbH
Altrheinstrasse 11
D-67550 Worms
Germany
Tel: 06242 9040
Fax: 06242 904286
E-Mail: info@hofmannmaschinen.com
Web: www.hofmannmaschinen.com

Karl Klein Ventilatorenbau GmbH
Waldstrasse 24
D-73773 Aichwald
Germany
Tel: 0711 369 060
Fax: 0711 369 0650
E-Mail: info@karl-klein.de
Web: www.karl-klein.de

Mayr GmbH + Co KG
Eichenstrasse 1
D-87665 Mauerstetten
Germany
Tel: 08341 8040
Fax: 08341 804421
E-Mail: info@mayr.de
Web: www.mayr.de

Mietzsch GmbH
Grossenhainer Strasse 137
D-01129 Dresden
Germany
Tel: 0351 84330
Fax: 0351 8433160
E-Mail: mietzsch.lufttechnik@t-online.de
Web: www.mietzsch.de

Nicotra GmbH
Weissenfelder Strasse 2
D-85551 Kirchheim-München
Germany
Tel: 089 9006920
Fax: 089 90069210
E-Mail: info@nicotra.de
Web: www.nicotra.de

NTN Wälzlager (Europa) GmbH
Max-Plank-Strasse 23
D-40669 Erkrath
Germany
Tel: 0211 25080
Fax: 0211 2508400
E-Mail: sales-de@ntn-europe.com
Web: www.ntn-europe.com

Piller Industrieventilatoren GmbH
Nienhagener Strasse 4-6
D-37186 Moringen
Germany
Tel: 05554 201 0
Fax: 05554 201 271
E-Mail: info@piller.de
Web: www.piller.de

PM°DM Precision Motors Deutsche Minebea GmbH
Auf Herdenen 10
D-78052 Villingen-Schwenningen
Germany
Tel: 07721 9970
Fax: 07721 9970249
E-Mail: andras.lelkes@nmb-minebea.com
Web: www.pmdm.de

Prüftechnik AG
Oskar-Messter-Strasse 19-21
D-85737 Ismaning
Germany
Tel: 089 99 61 60
Fax: 089 99 61 62 00
E-Mail: info@pruftechnik.com
Web: www.pruftechnik.com

Rosenberg Ventilatoren GmbH
Maybachstrasse 1/9
D-74653 Kunzlesau
Germany
Tel: 07940 1420
Fax: 07940 142125
E-Mail: info@rosenberg-gmbh.com
Web: www.rosenberg-gmbh.com

Schenck RoTec GmbH
Landwehrstrasse 55
D-64293 Darmstadt
Germany
Tel: 06151 322311
Fax: 06151 322315
E-Mail: rotec@schenck.net
Web: www.schenck-rotec.de

TLT- Turbo GmbH
Gleiwitestrasse 7
D-66482 Zweibruecken
Germany
Tel: 06332 808 0
Fax: 06332 808 267
E-Mail: tlt@tlt.de
Web: www.tlt.de

Turbowerke Meissen Howden GmbH
Naundorfer Strasse 4

D-01640 Coswig
Germany
Tel: 03523 940
Fax: 03523 94265
E-Mail: turbowerke@howdenindustrial.de
Web: www.howden.com

VEM motors GmbH
Carl Friedrich Gauss Strasse 1
D-38855 Wernigerode
Germany
Tel: 039 43680
Fax: 039 4368 2120
E-Mail: motors@vem-group.com
Web: www.vem-group.com

Witt & Sohn AG
Wuppermanstrasse 6-10
D-25421 Pinneberg
Germany
Tel: 04101 70070
Fax: 04101 700762
E-Mail: witt@wittfan.de
Web: www.wittfan.de

GREECE (GR)

Fyrogenis SA
20th Km National Road 1
GR-145 69 Athens
Greece
Tel: 0210 813 6301
Fax: 0210 813 5301
E-Mail: exp.dpt@fyrogenis.gr
Web: www.fyrogenis.gr

ITALY (I)

Aertecnica Croci Srl
Via Ticinese 8
I-28050 Pombia (NO)
Italy
Tel: 0321 956498
Fax: 0321 957259
E-Mail: aertecnica@interbusiness.it
Web: www.aercroci.com

Boldrocchi Srl
Via Trento e Trieste, 93
I-20046 Biassono
Italy
Tel: 039 22021
Fax: 039 2754200
E-Mail: boldrocchi@boldrocchi.it
Web: www.boldrocchi.it

CBI Service Srl
Viale dell'Industria 22
I-20040 Cambiago
Italy
Tel: 02 95308400
Fax: 02 95308391
E-Mail: info@industriecbi.it
Web: www.industriecbi.it

Cofimco SpA
Via Gramsci 62
I-28050 Pombia (Novara)
Italy
Tel: 0321 968311
Fax: 0321 958992
E-Mail: info@cofimco.com
Web: www.cofimco.com

Comefri SpA
Via Buja 3
I-33010 Magnano in Riviera (Udine)
Italy
Tel: 0432 798 811
Fax: 0432 798 890
E-Mail: info@comefri.com
Web: www.comefri.com

Dynair Srl
Via Napoleone Tirale 1
I-25017 Lonato (BS)
Italy
Tel: 030 9913575
Fax: 030 9913766
E-Mail: commexportd@dynair.it
Web: www.dynair.it

O Erre SpA
Via delle Industrie 25
I-20050 Mezzago
Italy
Tel: 039 627460
Fax: 039 6022440
E-Mail: oerre@oerre.it
Web: www.oerre.it

Euroventilatori SpA
Via Risorgimento 90
I-36070 S. Pietro Mussolino (Vicenza)
Italy
Tel: 0444 472472
Fax: 0444 472418
E-Mail: info@euroventilatori-int.it
Web: www.euroventilatori-int.com

F C R SpA
Via Enrico Fermi 3
I-20092 Cinsicello Balsamo
Italy
Tel: 0261 7981
Fax: 0261 798300
E-Mail: fcr@fcr.it
Web: www.fcr.it

Ferrari Ventilatori Industriali SpA
Via Marchetti 28
I-36071 Arzignano (VI)
Italy
Tel: 0444 471100
Fax: 0444 471105
E-Mail: info@ferrariventilatori.it
Web: www.ferrariventilatori.it

Gates Srl
Via Senigallia 18
(Int. 2 - Blocco A - Edificio 1)
I-20161 Milano (MI)
Italy
Tel: 02 662 16 21
Fax: 02 645 86 36
E-Mail: ptindustrial@gates.com
Web: www.gates.com/europe/pti

Marzorati Ventilazione Srl
Via Varese 38
I-21047 Saronno (VA)
Italy
Tel: 02 967 01633
Fax: 02 96701419
E-Mail: info@marzorativentilazione.com
Web: www.marzorativentilazione.com

Nicotra SpA
Via Modena 18
I-24040 Zingonia
Italy
Tel: 035873111
Fax: 035884319
E-Mail: info@nicotra.it
Web: www.nicotra.it

Technik SpA
Via dei Lavoratori 78
I-20092 Cinisello Balsamo (MI)
Italy
Tel: 02 660761
Fax: 02 66076329
E-Mail: info@tecnik.it
Web: www.tecnik.it

Termotecnica Pericoli Srl
PO Box 262
I-17031 Albenga
Italy
Tel: 0182 589006
Fax: 0182 589005
E-Mail: termotecnica@pericoli.com
Web: www.pericoli.com

Vortice Elettrosociali SpA
Strada Cerca
2 Frazione di Zoate
I-20067 Tribiano (Milan)
Italy
Tel: 02 9069935 6
Fax: 02 9069931 4
E-Mail: export@vortice-italy.com
Web: www.vortice.com

THE NETHERLANDS (NL)

ACT-RX Technology Europe
The Netherlands
E-Mail: info@arx-europe.com
Web: www.arx-europe.com

Almeco Nederland BV
Gemsstraat 12
1338 KG Almere
The Netherlands
Tel: 036 5292212
Fax: 036 5292925
E-Mail: info@almeco.nl
Web: www.almeco.be

MBL (Europe) BV
Energieweg 1-3
2382 NA Zoeterwoude
The Netherlands
Tel: 71-5899264
Fax: 71-5895062
E-Mail: info@mitsuboshi.co.jp
Web: www.mitsuboshi.com

Naaykens' Luchttechnische Apparatenbouw BV
PO Box 2233
5001 CE Tilburg
The Netherlands
Tel: 013 5425002
Fax: 013 5359885
E-Mail: info@naaykens.com
Web: www.naaykens.com

POLAND (PL)

Swegon Sp. z o.o.
ul. Owocowa 23

PL-62-080 Tarnowo Podgórne
Poland
Tel: 0816 8700
Fax: 0814 6354
E-Mail: poznan@swegon.pl
Web: www.swegon.pl

SERBIA & MONTE NEGRO (SEM)

Minel Kotlogradnja AD
Uralska 3
11060 Belgrade
Serbia & Monte Negro
Tel: 011 2783 222
Fax: 011 2781 597
E-Mail: office@minel-kotlogradnja.co.yu
Web: www.minel-kotlogradnja.co.yu

SLOVENIA (SLO)

Rotomatika Fans d.o.o.
Spodnja Kanomlja 23
SL-5281 Spodnja Idrija
Slovenia
Tel: 05 37 56 000
Fax: 05 37 56 517
E-Mail: sales@rotomatika-fans.si
Web: www.rotomatika.si

SPAIN (E)

Casals Cardona Industries SA
F. Casablancas, 24
E-08243 Manresa
Spain
Tel: 0938 748 480
Fax: 0938 757 668
E-Mail: tecnium@tecnium.es
Web: www.tecnium.es

Conductaire SA
Vereda de los Barros
P.I. Ventorro del Cano
E-28925 Alcorcon (Madrid)
Spain
Tel: 091 6324980
Fax: 091 632 1950
E-Mail: conductaire@nexo.es
Web: www.conductaire.com

Gates SA
Polígono Industrial
Les Malloles
E-08660 Balsareny
Spain
Tel: 093 877 7000
Fax: 093877 70 39
E-Mail: ptindustrial@gates.com
Web: www.gates.com/europe/pti

GER SA
Ctra de Valencia Km 6,300
Naves 12, 13 y 14
E-50410 Cuarta De Huerva (Zaragoza)
Spain
Tel: 0976 503558
Fax: 097 6504486
E-Mail: compras@gersa.com
Web: www.gersa.com

Koolclima SL
Políg. ind. "Uranga"
C/ Montecarlo 14
E-28942 Fuenlabrada (Madrid)
Spain
Tel: 091 600 2900
Fax: 091 607 0309
E-Mail: general@koolclima.com
Web: www.koolair.es

Nicotra Espana SA
Ctra Alcala - Villar del Olmo M-204 km 2.830
E-28810 Villalbilla (Madrid)
Spain
Tel: 091 8846110
Fax: 091 8859450
E-Mail: info@nicotra.es
Web: www.nicotra.es

Novovent SA
Josep Finestres, 9
E-08030 Barcelona
Spain
Tel: 093 278 8277
Fax: 093 278 8267
E-Mail: export@novovent.com
Web: www.novovent.com

Sodeca SA
Ctra. de Berga Km. 0.7
E-08580 Sant Quirze de Besora (Barcelona)
Spain
Tel: 093 8529111
Fax: 093 8529042
E-Mail: comercial@sodeca.com
Web: www.sodeca.com

Tecnivel SA
Calle Leo 5
E-28007 Madrid
Spain
Tel: 091 557 1130
Fax: 091 557 0917
E-Mail: tecnivel@tecnigrupo.com
Web: www.tecnigrupo.com

Termoven SA
c/ Isabel Colbrand
10-12. 5º Local 163-164
E-28050 Madrid
Spain
Tel: 091 358 9926
Fax: 091 358 8509
E-Mail: comercial@termoven.es
Web: www.termoven.es

Tradair SA
c/ Puerto de Pajares 32
E-28919 Leganés-Madrid
Spain
Tel: 091 428 2180
Fax: 091 341 1297
E-Mail: info@tradair.es
Web: www.tradair.es

Vemair
Calle Tuerca 25
Parque Industrial Santa Ana
E-28529 Madrid
Spain
Tel: 091301 1116
Fax: 091666 4611
E-Mail: correo@vemair.com
Web: www.vemair.com

Ventiladores Chaysol SA
Avenida Alcotanes 45
E-28320 Pinto (Madrid)
Spain
Tel: 091 692 8470
Fax: 091 692 8471
E-Mail: consultas@chaysol.com
Web: www.chaysol.com

SWEDEN (SE)

GIA SwedVent
PO Box 59
SE-772 22 Grangesberg
Sweden
Tel: 0240 797 00
Fax: 0240 797 25
E-Mail: info@gia.se
Web: www.gia.se

C A Östberg AB
Industrigatan 2
SE-775 35 Avesta
Sweden
Tel: 0226 86000
Fax: 0226 86003
Web: www.ostbeg.com

Removex AB
Mofallavägen 6
SE-696 75 Åmmeberg
Sweden
Tel: 07515 34070
Fax: 0583 34070
E-Mail: info@removex.se
Web: www.removex.se

SPM Instrument AB
Box 504
SE-645 25 Strängnäs
Sweden
Tel: 0152 22500
Fax: 0152 15075
E-Mail: info@spminstrument.se
Web: www.spminstrument.com

Swegon AB
Frejgatan 14
SE-535 30 Kvänum
Sweden
Tel: 0512 32 200
Fax: 0512 32 300
E-Mail: info@swegon.se
Web: www.swegon.se

Systemair AB
Industrivägen 3
SE-739 30 Skinnskatteberg
Sweden
Tel: 0222 440 00
Fax: 0222 44099
E-Mail: mailbox@systemair.se
Web: www.systemair.se

SWITZERLAND (CH)

Colasit AG
Faulenbachweg 63
CH-3700 Spiez
Switzerland
Tel: 033 655 6161
Fax: 033 654 8161
E-Mail: info@colasit.ch
Web: www.colasit.ch

Helios Ventilatoren AG
Steinackerstrsse 36
CH-8902 Urdorf-Zürich
Switzerland
Tel: 01 735 36 36
Fax: 01-735 36 37
E-Mail: info@helios.ch
Web: www.helios.ch

TURKEY (TU)

FITA Teknik
Ahmet Vefikpasa Cad 36
TR-34280 Çapa - Istanbul
Turkey
Tel: 0212 5864613
Fax: 0212 5881500
E-Mail: info@fitateknik.com
Web: www.fitateknik.com

Imas AS
Atatürk Organize Sanayi Bölgesi 10006 Sokak
No:29
TR-35620 Büyük Çigli - Izmir
Turkey
Tel: 0232 376 8700
Fax: 0232 376 8576
E-Mail: imas_makina@superonline.com
Web: www.imasklima.com.tr

Selnikel
Karaca Sokak 19
TR-06610 Gaziosmanpasa - Ankara
Turkey
Tel: 0312 442 7950
Fax: 0312 441 1314
E-Mail: info@selnikel.com.tr
Web: www.selnikel.com.tr

Sönmez Metal
Acarlar Is Merk. F Blok Kat : 6
TR-34805 Kavacik-Beykoz - Istanbul
Turkey
Tel: 0425 50 00
Fax: 0425 50 10
E-Mail: info@sonmezmetal.com
Web: www.sonmezmetal.com

Termas AS
Koresehitleri
Cad Mithat Unlu Sok No 12
Zincirlikuyu
Istanbul
Turkey
Tel: 0212 2666046

Tetisan Ltd
Tunc Caddesi
Has Sanayi Sitesi A Blok
TR-34850 Hadimkoy-Istanbul
Turkey
Tel: 0212 6232015
Fax: 0212 623 2017
E-Mail: info@tetisan.com
Web: www.tetisan.com

UNITED KINGDOM (UK)

Advanced Air (UK) Ltd
Burrell Way
Thetford
Norfolk
IP 24 3WB
United Kingdom
Tel: 01842 765657
Fax: 01842 753493
E-Mail: sales@advancedair.co.uk
Web: www.advancedair.co.uk

Advanced Design Technology Ltd
Monticello House
45 Russell Square
London
WC1B 4JP
United Kingdom
Tel: 020 7907 4715
Fax: 020 7907 4711
E-Mail: info@adtechnology.co.uk
Web: www.adtechnology.co.uk

AEG-Lafert Electric Motors Ltd
Electra House
Electra Way
Crewe
Cheshire
CW1 6QL
United Kingdom
Tel: 01270 270022
Fax: 01270 270023
E-Mail: info.aegmotors@lafert.com
Web: www.lafert.com

AHR International Ltd
70 Park Crescent
Elstree
Hertfordshire
WD6 3PU
United Kingdom
Tel: 020 8207 0930
Fax: 020 8207 0689
E-Mail: sales@ahrinternational.com
Web: www.ahrinternational.com

Air Control Industries Ltd
Silver Street
Chard
Somerset
TA20 2AE
United Kingdom
Tel: 01460 67171
Fax: 01460 61700
E-Mail: sales@air-con.co.uk
Web: www.air-con.co.uk

Airflow Developments Ltd
Lancaster Road
Cressex Business Park
High Wycombe
HP12 3QP
United Kingdom
Tel: 01494 525252
Fax: 01494 461073
E-Mail: info@airflow.co.uk
Web: www.airflow.com

Airflow Products Ltd
Underhill Lane
Sheffield
S6 1NL
United Kingdom
Tel: 0114 2327788
Fax: 0114 2327799
E-Mail: sales@airflow-group.com
Web: www.airflow-group.com

Airscrew Ltd
111 Windmill Road
Sunbury on Thames
Middlesex
TW16 7EF
United Kingdom
Tel: 01932 765822
Fax: 01932 761
E-Mail: mail@airscrew.co.uk
Web: www.airscrew.co.uk

Alfa Fans Ltd
Unit 7, Green Lane
Cannock
Staffordshire
WS11 OJJ
United Kingdom
Tel: 01543 572553
Fax: 01543 462393
E-Mail: sales@alfafans.co.uk
Web: www.alfafans.co.uk

Alldays Peacock & Co Ltd
Winterstoke Road
Weston-super-Mare
BS23 3YS
United Kingdom
Tel: 01934 636263
Fax: 01934 623727
E-Mail: apco@alldays-peacock-co.com
Web: www.apco1650.demon.co.uk

Allianz Cornhill International
Haslemere Road
Liphook
Hampshire
GU30 7UN
United Kingdom
Tel: 01428 722407
Fax: 01428 724824
E-Mail: marketing@allianzcornhill engineering.co.uk
Web: www.allianzcornhillengineering.co.uk

APMG Ltd
Mount Skip Lane
Little Hulton
Manchester
M38 9AL
United Kingdom
Tel: 0161 799 2200
Fax: 0161 799 2270
E-Mail: enquiries@apmg.co.uk
Web: www.apmg.co.uk

Applied Energy Products Ltd
Morley Way
Peterborough
PE2 9JJ
United Kingdom
Tel: 01733 456789
Fax: 01733 310606
Web: www.applied-energy.com

Axair Fans UK Ltd
Lowfield Drive
Centre 500
Wolstanton
Newcastle-under-Lyme,
ST5 0UU
United Kingdom
Tel: 01782 349430
Fax: 01782 349439
E-Mail: sales@axair-fans.co.uk
Web: www.axair-fans.co.uk

Beatson Fans & Motors Ltd
17-35 Mowbray Street
Sheffield
S3 8EN
United Kingdom
Tel: 0114 276 8088
Fax: 0114 275 8172
E-Mail: enquiries@beatson.co.uk
Web: www.beatson.co.uk

Biddle Air Systems Ltd
St. Mary's Road
Nuneaton
Warwickshire
CV11 5AU
United Kingdom
Tel: 024 7638 4233
Fax: 024 7637 3621
E-Mail: info@biddle-air.co.uk
Web: www.biddle-air.co.uk

B.O.B. Stevenson Ltd
Coleman Street
Derby
DE24 8NN
United Kingdom
Tel: 01332 574112
Fax: 01322 757286
E-Mail: sales@bobstevenson.co.uk
Web: www.bobstevenson.co.uk

Bri-Mac Engineering Ltd
Stambermill Works
Bagley Street
Lye
Stourbridge
DY9 7AR
United Kingdom
Tel: 01384 423030
Fax: 01384 422774
E-Mail: sales@bri-mac.co.uk
Web: www.bri-mac.co.uk

Brook Crompton
St Thomas' Road
Huddersfield
HD1 3LJ
United Kingdom
Tel: 01484 557200
Fax: 01484 557201
E-Mail: csc@brook-crompton.com
Web: www.brook-crompton.com

Brown Group Ltd
Lordswood Industrial Estate
Chatham
Kent
ME5 8UD
United Kingdom
Tel: 01634 687141
Fax: 1634 686347
E-Mail: mail@browngroupltd.com
Web: www.browngroupltd.com

Bruel and Kjaer UK Ltd
Bedford House
Rutherford Close
Stevenage
SG1 2ND
United Kingdom
Tel: 01438 739000
Fax: 01438 739099
E-Mail: ukinfo@bksv.com
Web: www.bksv.co.uk

BSI Product Services
Maylands Avenue
Hemel Hempstead
Hertfordshire
HP2 4SQ
United Kingdom
Tel: 01442 278607
Fax: 01422 278630
E-Mail: product.services@bsi-global.com
Web: www.bsi-global.com

Bureau Veritas
224-226 Tower Bridge Road
London
SE1 2TX
United Kingdom
Tel: 020 7550 8900
Fax: 020 7403 1590
Web: www.bureauveritas.com

CE-Air International
Newton Moor Industrial Estate
Hyde
Cheshire
SK14 4LG
United Kingdom
Tel: 0161 368 1476
Fax: 0161 367 8145
E-Mail: sales@ceair.co.uk
Web: www.ceair.co.uk

CEMB Hofmann (UK) Ltd
Unit 1 Longwood Road
Trafford Park
Manchester
M17 1PZ
United Kingdom
Tel: 0161 8723123
Fax: 161 877 9967
E-Mail: balancing@cembhofmann.co.uk
Web: www.cembhofmann.co.uk

Central Fans - Colasit Ltd
Unit 19-20
New Meadow Road
Redditch
Worcestershire
B98 8YW
United Kingdom
Tel: 01527 517200
Fax: 01527 517195
E-Mail: mail@central-fans.co.uk
Web: www.central-fans.co.uk

Colt International Ltd
New Lane
Havant
Hampshire
PO9 2LY
United Kingdom
Tel: 023 9245 1111
Fax: 023 9245 4220
E-Mail: info@coltgroup.com
Web: www.coltgroup.com

Comair Rotron Europe Ltd
Unit 9, The IO Centre
Nash Road
Park Farm North
Redditch
Worcestershire
B98 7AS
United Kingdom
Tel: 01527 520525
Fax: 01527 520565
E-Mail: sales@comairrotroneurope.com
Web: www.comairrotroneurope.com

Cooper Roller Bearings Co Ltd
Wisbech Road
King's Lynn
Norfolk
PE30 5JX
United Kingdom
Tel: 01553 767667
Fax: 01553 660494
E-Mail: coopersalesuk@kaydon.com
Web: www.cooperbearings.com

Criptic-Arvis Ltd
Croft Grange Works
Bridge Park Road
Thurmaston
Leicester
LE4 8BL
United Kingdom
Tel: 0116 2609700
Fax: 0116 2640147
E-Mail: sales@arvis.co.uk
Web: www.arvis.co.uk

Danfoss Ltd
Capswood
Oxford Road
Denham
Buckinghamshire
UB8 4LH
United Kingdom
Tel: 0870 608 0008
Fax: 0870 608 0009
Web: www.danfoss.com

Delrac Ltd
128 Malden Road
New Malden
Surrey
KT3 6DD
United Kingdom
Tel: 0208 3369000
Fax: 0208 942 0110
E-Mail: sales@delrac-london.co.uk

Direct Bearings & Power Transmissions Ltd
19 Patricia Way
Pysons Road Industrial Estate
Broadstairs
Kent
CT10 2LF
United Kingdom
Tel: 01843 600200
Fax: 01843 600210
E-Mail: sales@directbearings.co.uk
Web: www.directbearings.co.uk

DNV
Cromarty House
67-72 Regent Quay
Aberdeen
AB11 5AR
United Kingdom
Tel: 01224 335000
Fax: 01224 593311
E-Mail: aberdeen.techqual@dnv.com
Web: www.dnv.com

Domus Ventilation Ltd
Bearwalden Business Park
Royston Road

Wendens Ambo
Saffron Walden
Essex
CB11 3TL
United Kingdom
Tel: 01799 540602
Fax: 01799 541143
E-Mail: info@domusventilation.com
Web: www.domusventilation.com

Dynamic Air Products Ltd
1 Hurricane Close
Old Sarum Business Park
Salisbury
SP4 6LG
United Kingdom
Tel: 01722 416070
Fax: 01722 416069
E-Mail: info@dynamic-air-products.co.uk
Web: www.dynamic-air-products.co.uk

ebm-papst UK Ltd
Chelmsford Business Park
Chelmsford
CM2 5EZ
United Kingdom
Tel: 01245 468555
Fax: 01245 466336
E-Mail: sales@uk.ebmpapst.com
Web: www.ebmpapst.co.uk

Elta Fans Ltd
17 Barnes Wallis Road
Segensworth Industrial Estate
Fareham
Hampshire
PO15 5ST
United Kingdom
Tel: 01489 566500
Fax: 01489 566555
E-Mail: mailbox@eltafans.co.uk
Web: www.eltafans.com

Encon Air Systems Ltd
31 Quarry Park Close
Charter Gate
Moulton Park Industrial Estate
Northampton
NN3 6QB
United Kingdom
Tel: 01604 494187
Fax: 01604 645848
E-Mail: sales@encon-air.co.uk
Web: www.encon-air.co.uk

G. English Electronics Ltd
Unit 8, io Centre
The Royal Arsenal
Woolwich
London
SE18 6SR
United Kingdom
Tel: 020 8855 0991
Fax: 020 8854 5563
E-Mail: info@gelec.co.uk
Web: www.gelec.co.uk

European Thermodynamics Ltd
3 Kingsley Business Park
New Road
Kibworth Beauchamp
Leicestershire
LE8 0LE
United Kingdom
Tel: 0116 279 6899
Fax: 0116 276 3490
E-Mail: info@etdyn.com
Web: www.etdyn.com

Exhausto Ltd
Unit 3 Lancaster Court
Coronation Road
Cressex Business Park
High Wycombe
Buckinghamshire
HP12 3TD
United Kingdom
Tel: 01494 465166
Fax: 01494 465163
E-Mail: info@exhausto.co.uk
Web: www.exhausto.co.uk

Fan Engineering (Midlands) Ltd
19 Sandy Way
Tamworth
B77 4EX
United Kingdom
Tel: 01827 57000
Fax: 01827 64641
E-Mail: enquiries@fanengineering.co.uk
Web: www.fanengineering.co.uk

Fans & Blowers Ltd
Walrow Industrial Estate
Highbridge
Somerset
TA9 4AG
United Kingdom
Tel: 01278 784004
Fax: 01278 786910
E-Mail: fab-sales@btconnect.com
Web: www.fansandblowers.com

Fenner Drives
Hudson Road
Leeds
LS9 7DF
United Kingdom
Tel: 0113 249 3486
Fax: 0113 248 9656
E-Mail: drives@fenner.com
Web: www.fenner.com

Fläkt Woods Ltd
Tufnell Way
Colchester Essex
CO4 5AR
United Kingdom
Tel: 01206 544122
Fax: 01206 574434
E-Mail: info.uk@flaktwoods.com
Web: www.flaktwoods.com

Flamgard Engineering Ltd
Units 2 & 3 Pontnewynydd Industrial Estate
Pontnewynydd
Pontypool
Torfaen
NP4 6YW
United Kingdom
Tel: 01495 757347
Fax: 01495 755443
E-Mail: sales@flamgard.co.uk
Web: www.flamgard.co.uk

Flender Power Transmission Ltd
Thornbury Works
Leeds Road
Thornbury
Bradford
BD3 7EB
United Kingdom
Tel: 01274 657700
Fax: 01274 669836
E-Mail: info@flender-power.co.uk
Web: www.flender-power.co.uk

Fluent Europe Ltd
Sheffield Business Park
Europa Link
Sheffield
S9 1XU
United Kingdom
Tel: 0114 2818888
Fax: 0114 2818818
E-Mail: info@fluent.co.uk
Web: www.fluent.co.uk

Gamak Motors Ltd
Claycliffe Business Park
Barugh Green Road
Barugh Green
Barnsley
S75 1JU
United Kingdom
Tel: 01226 382727
Fax: 01266 38370
E-Mail: sales@gamakmotors.co.uk
Web: www.gamakmotors.co.uk

Gates Power Transmission Ltd
Tinwald Downs Road
Heath Hall
Dumfries
DG1 1TS
United Kingdom
Tel: 01387 242000
Fax: 01387 242010
E-Mail: ptindustrial@gates.com
Web: www.gates.com/europe/pti

Greenmount Fans (North) Ltd
Unit 8
Saville Street
Lancashire
BL2 1BY
United Kingdom
Tel: 01204 364362
Fax: 01204 364368
E-Mail: sales@greenmountfans.co.uk
Web: www.greenmountfans.co.uk

Greenwood Air Management Ltd
Brookside Industrial Estate
Rustington
West Sussex
BN16 3LH
United Kingdom
Tel: 01903 771021
Fax: 01903 782398
E-Mail: info@greenwood.co.uk
Web: www.greenwood.co.uk

Gulfoke Ltd
New Coach House
21 Grange Way
Colchester
CO2 8HF
United Kingdom
Tel: 01206 506555
Fax: 01206 871224
E-Mail: gulfoke@parrot.co.uk
Web: www.parrot.co.uk

Halifax Fan Ltd
Mistral Works
Unit 1, Brookfoot Business Park
Elland Road
Brighouse
HD6 2SD
United Kingdom
Tel: 01484 475123
Fax: 01484 475122
E-Mail: sales@halifax-fan.co.uk
Web: www.halifax-fan.co.uk

Helios Ventilation Systems Ltd
5 Crown Gate
Wyncollis Road
Colchester
CO4 9HZ
United Kingdom
Tel: 01206 228500
Fax: 01206 228500
E-Mail: sales@heliosfans.co.uk
Web: www.heliosfans.co.uk

Howden Buffalo
United Kingdom
See Howden Industrial UK

Howden Industrial UK
Old Govan Road
Renfrew
PA4 8XJ
United Kingdom
Tel: 0141 885 7500
Fax: 0141 885 7555
E-Mail: industrial.uk@howden.com
Web: www.howden.com

HRP Ltd
Rougham Industrial Estate
Rougham
Suffolk
IP30 9ND
United Kingdom
Tel: 01359 271131
Fax: 01359 271132
E-Mail: rougham@hrpltd.co.uk
Web: www.hrponline.co.uk

HSB Inspection Quality
Cairo House
Greenacres Road
Waterhead
Oldham
OL6 8DB
United Kingdom
Tel: 0845 345 5670
Fax: 0845 345 5680
E-Mail: alan.hulme@hsbeil.com
Web: www.hsbiq.com

Imofa UK Ltd
New Coach House
21 Grange Way
Colchester
CO2 8HF
United Kingdom
Tel: 01206 505909
Fax: 01206 794095
E-Mail: sales@imofa.co.uk
Web: www.imofa.co.uk

INA Bearing Company Ltd/FAG
Forge Lane
Minworth
Sutton Coldfield
West Midlands
B76 1AP
United Kingdom
Tel: 0121 351 3833
Fax: 0121 351 7686
E-Mail: ina.bearing@uk.ina.com
Web: www.uk.ina.com

IRD UK Ltd
Unit B4, Brymail One Estate
River Lane
Saltney
Chester
CH4 8RG
United Kingdom
Tel: 01244 682222
Fax: 01244 677977
E-Mail: sales@irdbalancing.com
Web: www.irdbalancing.com

Kiloheat Ltd
Enterprise Way
Edenbridge
Kent
TN8 6HF
United Kingdom
Tel: 01732 866000
Fax: 01732 866370
E-Mail: info@kiloheat.co.uk
Web: www.kiloheat.co.uk

Lenze Ltd
Caxton Road
Bedford
MK41 0HT
United Kingdom
Tel: 01234 321321
Fax: 01234 261815
E-Mail: sales@lenze.co.uk
Web: www.lenze.co.uk

Lloyd's Register
71 Fenchurch Street
London
EC3M 4BS
United Kingdom
Tel: 020 7423 2892
Fax: 020 7423 1525
E-Mail: etbs@lr.org
Web: www.lr.org

The London Fan Company Ltd
75-81 Stirling Road
London
W3 8DJ
United Kingdom
Tel: 020 8992 6923
Fax: 020 8992 6928
E-Mail: sales@londonfan.co.uk
Web: www.londonfan.co.uk

Lord Corporation (Europe) Ltd
Unit 30 Stretford Motorway Estate
Barton Dock Road
Stretford
Manchester
M32 0ZH
United Kingdom
Tel: 0161 865 8048
Fax: 0161 865 0096
E-Mail: ed_auslander@lord.com
Web: www.lordcorporation.co.uk

MAN Acoustics Ltd
Walrow Industrial Estate
Highbridge
Somerset
TA9 4AG
United Kingdom
Tel: 01278 789335
Fax: 01278 785613
E-Mail: sales@man-acoustics.com
Web: www.man-acoustics.com

Marstair Ltd
Armytage Road
Brighouse
West Yorkshire
HD6 1QF
United Kingdom
Tel: 01484 405600
Fax: 01484 405620
E-Mail: info@marstair.com
Web: www.marstair.com

Matthews & Yates Ltd
Peartree Road
Stanway
Colchester
CO8 0LD
United Kingdom
Tel: 01206 543311
Fax: 01206 760497
E-Mail: sales@matthews-yates.co.uk
Web: www.matthews-yates.co.uk

Metrico International Ltd
Unit 2A, Brymau 3 Industrial Estate
River Lane
Saltney
Chester
CH4 8RQ
United Kingdom
Tel: 01244 677878
Fax: 01244 677080
E-Mail: sales@metrico.co.uk
Web: www.metrico.co.uk

Michell Bearings
Scotswood Road
Newcastle upon Tyne
NE15 6LL
United Kingdom
Tel: 0191 273 0291
Fax: 0191 272 2787
E-Mail: humble-g@michellbearings.co.uk
Web: www.michellbearings.co.uk

National Physical Laboratory
Queens Road
Teddington
Middlesex
TW11 0LW
United Kingdom
Tel: 020 8943 6880
Fax: 020 8943 6458
E-Mail: enquiry@npl.co.uk
Web: www.npl.co.uk

Nicotra UK Ltd
Unit D
Parkgate Business Park
Rail Mill Way
Rotherham
S62 6JQ
United Kingdom
Tel: 01709 780760
Fax: 01709 780762

E-Mail: brian@nicotra.co.uk
Web: www.nicotra.co.uk

NMB Minebea (UK) Ltd
1 Sterling Centre
Eastern Road
Bracknell
RG12 2PW
United Kingdom
Tel: 01522 500933
Fax: 01522 696485
E-Mail: graham.anderson@nmb.co.uk
Web: www.nmb-europe.com

Northern Fan Supplies
Unit E1 Longford Trading Estate
Thomas Street
Stretford
Manchester
M32 0JT
United Kingdom
Tel: 0161 864 1777
Fax: 0161 864 2777
E-Mail: nfs@nfan.co.uk
Web: www.nfan.co.uk

Northey Technologies Ltd
Nortech House
Allens Lane
Poole
Dorset
BH16 5DG
United Kingdom
Tel: 01202 668600
Fax: 01202 668500
E-Mail: info@northey.net
Web: www.northey.net

NTN Bearings (UK) Ltd
Wellington Crescent
Fradley Park
Fradley
Kichfield
WS13 8RZ
United Kingdom
Tel: 01543 445000
Fax: 01543 445035
E-Mail: sales-uk@ntn-europe.com
Web: www.ntn-europe.com

The Nuaire Group
Western Industrial Estate
Caerphilly
CF83 1NA
United Kingdom
Tel: 029 2088 5911
Fax: 029 2088 7033
E-Mail: info@nuaire.co.uk
Web: www.nuaire.co.uk

Ondrives Ltd
Foxwood Industrial Park
Chesterfield
Derbyshire
S41 9RN
United Kingdom
Tel: 01246 455500
Fax: 01246 455522
E-Mail: sales@ondrives.com
Web: www.ondrives.com

Oriental Motor (UK) Ltd
Unit 5 Faraday Office Park
Rankine Road
Basingstoke
Hampshire
RG24 8AG
United Kingdom
Tel: 01256 347090
Fax: 01256 347099
E-Mail: info@oriental-motor.co.uk
Web: www.oriental-motor.co.uk

PCA Engineers Ltd
Homer House
Sibthorp Street
Lincoln
LN5 7SB
United Kingdom
Tel: 01522 530106
Fax: 01522 511703
E-Mail: info@pcaeng.co.uk
Web: www.pcaeng.co.uk

Prüftechnik Ltd
Burton Road
Streethay
Lichfield
Staffordshire
WS13 8LN
United Kingdom
Tel: 01543 417722
Fax: 01543 417723
E-Mail: info@pruftechnik.co.uk
Web: www.pruftechnik.co.uk

Remco Products Ltd
Eastmead Industrial Estate
Lavant
Chicheste
West Sussex
PO18 0DB
United Kingdom
Tel: 01243 528414
Fax: 01243 532127
E-Mail: sales@remco.co.uk
Web: www.remco.co.uk

Rencol Tolerance Rings Ltd
Unit 16 Concorde Road
Patchway
Bristol
BS34 5TB
United Kingdom
Tel: 0117 938 1700
Fax: 0117 915 7982
E-Mail: admin@rencol.co.uk
Web: www.rencol.co.uk

R H F Fans Ltd
Unit 2 Ferrous Way
Gilchrist Road
Northbank Industrial Estate
Irlam
Manchester
M44 5FS
United Kingdom
Tel: 0161 776 6400
Fax: 0161 775 6566
E-Mail: sales@rhf-fans.co.uk
Web: www.rhf-fans.co.uk

Rockwell Automation
Pitfield
Kiln Farm
Milton Keynes
MK11 3DR
United Kingdom
Tel: 0870 2425004
Fax: 01980 261917
Web: www.ra.rockwell.com

Roof Units
Fleming Way
Crawley
West Sussex
RH10 9YX
United Kingdom
Tel: 01293 441570
Fax: 01293 534898
E-Mail: ru@roofunitsltd.co.uk

Royal & SunAlliance, Engineering Business
Royal & SunAlliance, Engineering Business
1st Floor
17 York Street
Manchester
M2 3RS
United Kingdom
Tel: 0161 235 3090
Fax: 0161 235 3702
E-Mail: engineering.consultancy@uk.royalsun.com

Schenck Balancing & Diagnostic Systems
Lombard Way
Banbury
Oxfordshire
OX16 4TX
United Kingdom
Tel: 01295 251122
Fax: 01295 252111
E-Mail: rotec@schenck.co.uk
Web: www.schenck.co.uk

Secomak Ltd
502 Honeypot Lane
Stanmore
Middlesex
HA7 1JR
United Kingdom
Tel: 020 8952 5566
Fax: 020 8952 6983
E-Mail: info@secomak.com
Web: www.secomak.com

SGS United Kingdom Ltd
SGS House
Johns Lane
Tividale
Warley
West Midlands
B69 3HX
United Kingdom
Tel: 021 520 6454
Fax: 021 522 3532
E-Mail: mark.homer@sgs.com
Web: www.sgs.com

Silver Box Fans Ltd
Unit 13, Shaftsbury Industrial Estate
Letchworth
SG6 1HE
United Kingdom
Tel: 01462 481051
Fax: 01462 481126
E-Mail: clive@silverbox.co.uk
Web: www.silverbox.co.uk

SKF (UK) Ltd
Sundon Park Road
Luton

LU3 3BL
United Kingdom
Tel: 01582 490049
Fax: 01582 848091
E-Mail: marketing.uk@skf.com
Web: www.skf.com

A O Smith Electrical Products Ltd
PO Box 8
Marshall Way
Gainsborough
Lincolnshire
DN21 1XU
United Kingdom
Tel: 01427 614141
Fax: 1427 617513
E-Mail: sales@aosmithepc.co.uk
Web: www.aosmithelectricalproducts.co.uk

Soler + Palau Ltd
19-23 Betts Avenue
Martlesham Heath
Ipswich
IP5 3RH
United Kingdom
Tel: 01473 626277
Fax: 01473 610468
E-Mail: eugene@solerandpalau.co.uk
Web: ww.solerandpalau.co.uk

Sound Research Laboratories Ltd
Holbrook House
Little Waldingfield
Sudbury
Suffolk
CO10 0TH
United Kingdom
Tel: 01787 247595
Fax: 01787 248420
E-Mail: marketing@soundresearch.co.uk
Web: www.soundresearch.co.uk

SPM Instrument UK Ltd
Suite 12, Hardmans Business Centre
New Hall Hey Road
Rawtenstall
Lancashire
BB4 6HH
United Kingdom
Tel: 01706 835 331
Fax: 01706 260 640
E-Mail: info@spminstrument.co.uk
Web: www.pminstrument.co.uk

Standard & Pochin Ltd
Units 6 & 7
Westminster Road
Wareham
Dorset
BH20 4SP
United Kingdom
Tel: 01929 554311
Fax: 01929 556726
E-Mail: sales@standardandpochin.co.uk
Web: www.standardandpochin.co.uk

Stockbridge Airco Ltd
Blossom Street Works
Blossom Street
Ancoats
Manchester
M4 6AE
United Kingdom
Tel: 0161 236 9314
Fax: 0161 228 0009
E-Mail: fans@stockbridge-airco.com
Web: www.stockbridge-airco.com

Swegon Ltd
Essex House
Astra Centre
Edinburgh Way
Harlow
Essex
CM20 2BN
United Kingdom
Tel: 01279 416087
Fax: 01279 416076
E-Mail: sales@swegon.co.uk
Web: www.swegon.co.uk

Systemair Ltd
Pharaoh House
Arnolde Close
Medway City Estate
Rochester
Kent
ME2 4SP
United Kingdom
Tel: 01634 735000
Fax: 01634 735001
E-Mail: mailbox@systemair.co.uk
Web: www.systemair.co.uk

Teco Electric Europe Ltd
Teco Building
Marshall Stevens Way
Trafford Park
Manchester
M17 1PP
United Kingdom
Tel: 0161 877 8025
Fax: 0161 877 8030
E-Mail: enquiries@teco.co.uk
Web: www.teco.co.uk

Torin Ltd
Greenbridge
Swindon
Witshire
SN3 3JB
United Kingdom
Tel: 01793 524291
Fax: 01793 486570
E-Mail: sales@torin-sifan.com
Web: www.torin-sifan.com

Tyco Electronics-Crompton Small Motors
Wheatley Hall Road
Doncaster
DN2 4NB
United Kingdom
Tel: 01302 812712
Fax: 01302 634738
E-Mail: andy.j.hall@tycoelectronics.com
Web: www.cromptonsmallmotors.com

Ubbink (UK) Ltd
Borough Road
Brackley
Northamptonshire
NN13 7TB
United Kingdom
Tel: 01280 700211
Fax: 01280 705332
E-Mail: info@ubbink.co.uk
Web: www.ubbink.co.uk

Vectaire Ltd
Lincoln Road
Cressex Business Park
High Wycombe
HP12 3RH
United Kingdom
Tel: 01494 522333
Fax: 01494 522337
E-Mail: sales@vectaire.co.uk
Web: www.vectaire.co.uk

Vent-Axia Ltd
Fleming Way
Crawley
West Sussex
RH10 9YX
United Kingdom
Tel: 01293 526062
Fax: 01293 552375
E-Mail: info@vent-axia.com
Web: www.vent-axia.com

Vortice Ltd
Beeches House
Esatren Avenue
Burton on Trent
Staffordshire
DE13 0BB
United Kingdom
Tel: 01283 492949
Fax: 01283 544121
E-Mail: getit@vortice.ltd.uk
Web: www.vortice.ltd.uk

WEG Electric Motors (UK) Ltd
28-29 Walkers Road
Manorside Industrial Estate
North Moons Moat
Redditch
Worcestershire
B98 9HE
United Kingdom
Tel: 01527 596748
Fax: 01527 591133
E-Mail: sales@wegelectricmotors.co.uk
Web: www.weg.com.br

Witt UK
Fan Systems Group
Hollyoak Works, Rochdale Road
Greetland
Halifax
HX4 8HB
United Kingdom
Tel: 01422 378131
Fax: 01422 378672
E-Mail: sales@fansystems.co.uk
Web: www.fansystems.co.uk

Woodcock & Wilson Ltd
Airstream Works
Blackmoorfoot Road
Crosland Hill
Huddersfield
HD4 7AA
United Kingdom
Tel: 01484 462777
Fax: 01484 462888
E-Mail: sales@fanmanufacturers.com
Web: www.anmanufacturers.com

Wyko Industrial Services
Amber Way
Halesowen

West Midlands
B62 8WG
United Kingdom
Tel: 0121 508 6341
Fax: 0121 508 6333
E-Mail: marketing@wyko.co.uk
Web: www.wyko.co.uk

USA (US)

ACME Engineering & Manufacturing Corp
PO Box 978
Muskogee
Oklahoma 74402
United States
Tel: 918 682 7791
Fax: 918 682 0134
E-Mail: lbuddrus@acmefan.com
Web: www.acmefan.com

Fairbrother & Associates Inc
11001 Falls Road
Lutherville
Maryland 21093
United States
Tel: 410 828 8484
Fax: 410 828 8492
E-Mail: fairbrotherassoc@worldnet.att.net

Greenheck
PO Box 410
Schofield
Wisconsin 54476-0410
United States
Tel: 715 359 6171
Fax: 715 355 6484
E-Mail: terry.radtke@greenheck.com
Web: www.greenheck.com

Hartzell Fan Inc
910 South Downing Street
Piqua
Ohio 45356
United States
Tel: 937 773 7411
Fax: 937 773 8994
E-Mail: info@hartzellfan.com
Web: www.hartzellfan.com

Howden Buffalo Inc
2029 W DeKalb Street
Camden
South Carolina 29020
United States
Tel: 803 713 2200
Fax: 803 713 2222
E-Mail: sales@howdenbuffalo.com
Web: www.howden.com

Loren Cook Company
PO Box 4047
Springfield
Missouri 65808
United States
Tel: 417 869 6474
Fax: 417 862 3820
E-Mail: sales@lorencook.com
Web: www.lorencook.com

Robinson Industries Inc
PO Box 100
Zelienople
Pennsylvania 16063
United States
Tel: 724 452 6121
Fax: 724 452 0388
E-Mail: jay.henderson@robinsonfans.com
Web: www.robinsonfans.com

The Timken Company
1835 Dueber Avenue SW
PO Box 6932
Canton
Ohio 44706-0932
United States
Tel: 330 438 3000
Fax: 330 471 4388
Web: www.timken.com

TLT-Babcock Inc
260 Springside Drive
Akron
Ohio 44333
United States
Tel: 330 867 8540
Fax: 330 869 4819
E-Mail: jgray@tltbabcock.com
Web: www.tltbabcock.com

24.3 Fan types

			Ducted fans														
			Centrifugal or radial flow														
					Backward						Radial						
	Country Code	Range of operating conditions offered — see Code definitions in Section 24.1	Forward curved		Inclined		Curved		Aerofoil		Tipped		Shrouded		Open paddle	Backplated paddle	
			Direct drive	Vee belt drive	Direct drive	Vee belt drive	Direct drive	Vee belt drive	Direct drive	Vee belt drive	Direct drive	Vee belt drive	Direct drive	Vee belt drive	Vee belt drive	Direct drive	Vee belt drive
ACME Engineering & Manufacturing Corp.	US	A	●		●	●	●	●	●		●		●	●	●	●	●
ACT-RX Technology Europe	NL	A, B	●										●				
Aertecnica Croci Srl	I	G			●	●	●	●									
Air Control Industries Ltd	UK	A, C, H	●		●		●						●				
Airap	F	A, C, D, E, F	●	●													
Airflow Developments Ltd	UK	A, B, C, F	●														
Airflow Products Ltd	UK	A, C, G, J, K															
Airscrew Ltd	UK	A, H, M, N, P	●						●								
Alfa Fans Ltd	UK	A, G	●														
Alldays Peacock & Co Ltd	UK	A, G					●				●				●	●	
Almeco Nederland BV	NL	A	●	●	●	●	●	●	●	●	●	●	●	●	●	●	●
Almeco NV-SA	B	A	●	●	●	●	●	●	●	●	●	●	●	●	●	●	●
APMG Ltd	UK	L	●	●	●	●							●	●			
Applied Energy Products Ltd	UK	A	●	●						●							
Axair Fans UK Ltd	UK	A	●				●									●	
Beatson Fans & Motors Ltd	UK	A														●	
B Bille A/S	DK	A, C, D, E, F, G,H, J, K			●	●	●	●							●	●	●
B.O.B. Stevenson Ltd	UK	A			●	●							●	●			
Boldrocchi Srl	I	A, C, G, J					●		●		●		●				
Casals Cardona Industries SA	E	A, D, F			●	●	●	●							●	●	●
CE-Air International	UK	A															
Central Fans - Colasit Ltd	UK	E, F, G, H, J, K	●	●	●	●	●	●									
Cofimco SpA	I	A															
Colasit AG	SE	E, F, G, H, J, K	●	●	●	●	●	●									
Colt International Ltd	UK	A, C															
Comair Rotron Europe Ltd	UK	A					●									●	
Comefri SpA	I	A	●	●			●	●		●			●	●			
De Raedt SA	B	A, B, C, D, E, F, G, J, K			●	●	●	●	●	●							
Domus Ventilation Ltd	UK	B															
Dynair Srl	I	A, B, C	●				●										
ebm-papst UK Ltd	UK	A, B	●				●										

Ducted fans										Partition fans								Open fans					
										Powered roof ventilators				Wall mounting				Recirculating		Main cooler		Jet tunnel	Unit
Axial flow					Propeller		Mixed or compound flows		Ring shaped	Axial flow	Propeller	Centrifugal	Mixed	Axial flow	Propeller	Centrifugal	Mixed	Axial flow	Axial flow	Axial flow	Axial flow	Axial flow	Cross flow
Aerofoil section					Sheet blading					Aerofoil section	Sheet blading			Aerofoil section	Sheet blading			Aerofoil section	Sheet blading	Aerofoil section	Sheet blading	Aerofoil section	Forward curved
Tube axial	Vane axial	Contra rotating	Reversible	Bifurcated	Tube axial	Bifurcated	Direct dirve	Vee belt drive	Direct drive														
												●				●							
●																							
●																							
●		●	●	●						●	●												
														●		●	●						
●																							
●	●						●	●								●	●						
				●		●						●											
●	●	●	●	●						●		●				●			●		●		●
●	●	●	●	●						●		●				●			●		●		●
				●																			
●					●					●				●						●			
●		●	●	●						●				●									
					●																		
●				●				●															
	●																						
●	●		●	●						●				●				●		●			
										●				●				●	●	●			
	●									●				●									
●																							
●		●										●											
							●							●		●	●						
												●		●		●							
●							●																

24.3 Fan types

	Country Code	Range of operating conditions offered — see Code definitions in Section 24.1	Ducted fans														
			Centrifugal or radial flow														
			Forward curved		Backward						Radial						
					Inclined		Curved		Aerofoil		Tipped		Shrouded		Open paddle	Backplated paddle	
			Direct drive	Vee belt drive	Direct drive	Vee belt drive	Direct drive	Vee belt drive	Direct drive	Vee belt drive	Direct drive	Vee belt drive	Direct drive	Vee belt drive	Vee belt drive	Direct drive	Vee belt drive
Elta Fans Ltd	UK	A, C															
G. English Electronics Ltd	UK	A															
O Erre SpA	I	A, B, L															
European Thermo-dynamics Ltd	UK	A, B	●				●										
Euroventilatori SpA	I	A, D, G, J, K	●	●	●	●	●	●	●	●	●	●	●	●	●	●	●
Exhausto A/S	DK	A, B, C, D	●				●										
Exhausto Ltd	UK	A, B, C, D	●				●										
Fairbrother & Associates Inc	US	A, G, H, J	●		●								●		●		
Fans & Blowers Ltd	UK	A, C, E, F, G, H, J, K	●	●	●	●	●	●									
Ferrari Ventilatori Industriali SpA	I	A, C, D, E, F	●	●	●	●	●	●	●	●	●	●	●	●			
Fläkt Woods Ltd	UK	A, C, G	●	●	●	●	●	●	●	●						●	●
Flamgard Engineering Ltd	UK	A	●	●			●	●									
Gebhardt Ventilatoren	D	A, C, D, F	●	●	●	●	●	●	●	●							
Gewa-Werth GmbH	D	A, B, C, D	●				●										
Greenheck	US	A, B, D			●	●											
Greenmount Fans (North) Ltd	UK	A	●	●	●	●	●	●								●	●
Greenwood Air Management Ltd	UK	A, B	●				●										
Gulfoke Ltd	UK	A	●	●			●	●									
Halifax Fan Ltd	UK	A, C, D, E, F, G, J, K	●	●	●	●	●	●	●	●	●	●	●	●	●	●	●
Hartzell Fan Inc	US	A, D, L					●	●									
Helios Ventilateurs Sarl	F	A	●				●										
Helios Ventilation Systems Ltd	UK	A	●				●										
Helios Ventilatoren	D	A	●				●										
Helios Ventilatoren AG	SE	A	●				●										
Helios Ventilatoren GmbH	A	A	●				●										
Howden Buffalo Inc	US	A, C, D, F, G, J, K			●	●	●	●	●	●	●	●					
Howden Industrial UK	UK	A, C, D, E, F, G, J, K	●		●	●	●	●	●	●	●	●			●		
Howden Sirocco SA	F	A, C, D, E, F, G, J, K	●		●	●	●	●	●	●	●	●			●		
Howden Ventiladores Ltda	BR	A, C, D, E, F, G, J, K	●		●	●	●	●	●	●	●	●			●		
Imofa UK Ltd	UK	A	●	●			●	●									

Ducted fans										Partition fans								Open fans					
										Powered roof ventilators				Wall mounting				Recirculating		Main cooler		Jet tunnel	Unit
Axial flow					Propeller		Mixed or compound flows		Ring shaped	Axial flow	Propeller	Centrifugal	Mixed	Axial flow	Propeller	Centrifugal	Mixed	Axial flow	Axial flow	Axial flow	Axial flow	Axial flow	Cross flow
Aerofoil section					Sheet blading					Aerofoil section	Sheet blading			Aerofoil section	Sheet blading			Aerofoil section	Sheet blading	Aerofoil section	Sheet blading	Aerofoil section	Forward curved
Tube axial	Vane axial	Contra rotating	Reversible	Bifurcated	Tube axial	Bifurcated	Direct dirve	Vee belt drive	Direct drive														
●				●			●			●			●	●								●	
●	●	●	●	●																			
	●			●										●	●	●	●		●				
●																							
●	●	●	●	●	●	●	●	●	●														
	●											●				●							
●																							
●	●	●	●	●	●			●		●	●	●	●	●	●							●	
●		●	●	●			●																
●												●		●									
	●											●				●							
	●																						
	●		●																				
●										●				●	●								
●		●	●	●			●																
						●																	
	●				●					●		●				●	●						
●							●			●		●		●								●	
●							●			●		●		●								●	
●							●			●		●		●								●	
●							●			●		●		●								●	
●							●			●		●		●								●	
	●		●																			●	
	●		●	●			●	●															
	●		●	●			●	●															
	●		●	●			●	●															
●	●	●	●				●																

24.3 Fan types

	Country Code	Range of operating conditions offered — see Code definitions in Section 24.1	Ducted fans														
			Centrifugal or radial flow														
			Forward curved		Backward						Radial						
					Inclined		Curved		Aerofoil		Tipped		Shrouded		Open paddle	Backplated paddle	
			Direct drive	Vee belt drive	Direct drive	Vee belt drive	Direct drive	Vee belt drive	Direct drive	Vee belt drive	Direct drive	Vee belt drive	Direct drive	Vee belt drive	Vee belt drive	Direct drive	Vee belt drive
Kiloheat Ltd	UK	A, C, D, G, J, K			●	●	●	●	●	●				●	●		
Karl Klein Ventilatorenbau GmbH	D	A									●	●	●	●	●	●	●
Koolclima SL	E	A	●	●			●	●									
Leader Fan Industries Ltd	CAN	A														●	
The London Fan Company Ltd	UK	A	●	●													
Loren Cook Company	US	A, B															
Marzorati Ventilazione Srl	I	A, B	●	●													
Matthews & Yates Ltd	UK	A, C, D, F, J				●				●							
MBL (Europe) BV	NL	A		●		●		●		●		●		●	●		●
Metrico International Ltd	UK	A, C, D, G, K		●	●	●									●		
Mietzsch GmbH	D	A, C, D, F	●	●	●	●	●	●	●	●	●	●	●	●			
Minel Kotlogradnja AD	SEM	A, D	●	●	●	●	●	●	●	●	●	●	●	●	●	●	●
Multi-Wing International a·s	DK	A															
Naaykens' Luchttechnische Apparatenbouw BV	NL	A	●	●	●	●	●	●	●	●	●	●	●	●	●	●	●
Nicotra Espana SA	E	A	●	●				●									
Nicotra France SA	F	A	●	●				●									
Nicotra GmbH	D	A	●	●				●									
Nicotra SpA	I	A	●	●				●									
Nicotra UK Ltd	UK	A	●	●				●									
NMB Minebea (UK) Ltd	UK	A	●				●		●								
Northern Fan Supplies	UK	A, D			●	●	●	●							●	●	●
Northey Technologies Ltd	UK	A	●	●													
Novovent SA	E	A, B, C, D, E, F, H	●				●										
The Nuaire Group	UK	A, B, D															
C A Östberg AB	SE	A	●				●										
Piller Industrieventilatoren GmbH	D	A, C, E, F, G, H, J, K			●	●	●	●					●	●		●	●
PM°DM Precision Motors Deutsche Minebea GmbH	D	A	●				●		●								
Remco Products Ltd	UK	A															
Removex AB	SE	A, D															
R H F Fans Ltd	UK	A	●	●			●	●								●	●

Ducted fans										Partition fans								Open fans					
										Powered roof ventilators				Wall mounting				Recirculating		Main cooler		Jet tunnel	Unit
Axial flow					Propeller		Mixed or compound flows		Ring shaped	Axial flow	Propeller	Centrifugal	Mixed	Axial flow	Propeller	Centrifugal	Mixed	Axial flow	Axial flow	Axial flow	Axial flow	Axial flow	Cross flow
Aerofoil section					Sheet blading					Aerofoil section	Sheet blading			Aerofoil section	Sheet blading			Aerofoil section	Sheet blading	Aerofoil section	Sheet blading	Aerofoil section	Forward curved
Tube axial	Vane axial	Contra rotating	Reversible	Bifurcated	Tube axial	Bifurcated	Direct dirve	Vee belt drive	Direct drive														
●																							
●			●							●	●			●	●				●				
●		●	●	●		●				●	●	●		●	●			●		●			
	●				●					●		●				●					●		
●				●						●		●	●									●	
								●															
●				●						●				●									
●												●		●									
●	●	●	●	●																			
●	●	●	●	●						●				●									
										●		●		●									
										●		●		●									
										●		●		●									
										●		●		●									
										●		●		●									
●							●																
						●	●	●				●	●		●	●							
			●																				
●	●									●		●		●	●								
	●		●	●							●				●	●	●						
												●				●							
●							●																
															●	●			●		●		
●	●	●	●	●	●			●		●	●	●	●	●	●							●	
●				●																			

24.3 Fan types

	Country Code	Range of operating conditions offered — see Code definitions in Section 24.1	Ducted fans														
			Centrifugal or radial flow														
			Forward curved		Backward						Radial						
					Inclined		Curved		Aerofoil		Tipped		Shrouded		Open paddle	Backplated paddle	
			Direct drive	Vee belt drive	Direct drive	Vee belt drive	Direct drive	Vee belt drive	Direct drive	Vee belt drive	Direct drive	Vee belt drive	Direct drive	Vee belt drive	Vee belt drive	Direct drive	Vee belt drive
Robinson Industries Inc	US	A			●	●		●	●		●	●			●	●	●
Roof Units	UK	A	●	●	●	●	●	●	●	●	●	●	●	●	●	●	●
Rosenberg Ventilatoren GmbH	D	A, B	●				●						●				
Rotomatika Fans d.o.o.	SL	A	●		●		●		●				●				
Sardou SA	F	A, B	●	●	●	●	●	●	●	●	●	●	●	●	●	●	
Secomak Ltd	UK	A, C, D, F, H	●														
Selnikel	TU	A, B, C, J			●	●											
Silver Box Fans Ltd	UK	A	●				●										
Sodeca SA	E	A, B, D															
Soler + Palau Ltd	UK	A	●	●	●												
Standard & Pochin Ltd	UK	A,F,G,J,K	●	●	●	●	●	●	●	●	●	●	●	●	●	●	●
Stockbridge Airco Ltd	UK	A, C, F, G, J			●	●		●		●							
Tecnifan SA	E	A, C	●	●			●	●	●	●							
Termotecnica Pericoli Srl	I	A, C, G	●				●										
TLT Co-Vent Fans Inc	CAN	A, C, D, F															
TLT- Turbo GmbH	D	A, C, D, F															
TLT-Babcock Inc	US	A, C, D, F															
Torin Ltd	UK	A	●	●	●	●	●	●	●	●	●	●	●	●			
Turbowerke Meissen Howden GmbH	D	A, C, D, E, F, G, J, K	●		●	●	●	●	●	●	●	●			●		
Ubbink (UK) Ltd	UK	B															
Universal Fan & Blower Ltd	CAN	A, E, F, L		●		●		●						●		●	●
Vectaire Ltd	UK	A, C	●		●		●	●									
Vent-Axia Ltd	UK	A	●	●	●	●	●	●	●	●	●	●	●	●	●	●	●
Vortice Elettrosociali SpA	I	A, B	●				●										
Vortice France	F	A, B	●				●										
Vortice Ltd	UK	A, B	●				●										
Witt & Sohn AG	D	A, C, D, F, G, K	●	●			●	●									
Witt UK	UK	A, C, D, E, F, G, H, J, K	●	●	●	●	●	●	●	●	●	●	●	●	●	●	●
Woodcock & Wilson Ltd	UK	A, C, D, E, F, G, H, J, K			●	●	●	●	●	●	●	●			●	●	

Ducted fans										Partition fans								Open fans					
										Powered roof ventilators				Wall mounting				Recirculating		Main cooler		Jet tunnel	Unit
Axial flow					Propeller		Mixed or compound flows		Ring shaped	Axial flow	Propeller	Centrifugal	Mixed	Axial flow	Propeller	Centrifugal	Mixed	Axial flow	Axial flow	Axial flow	Axial flow	Axial flow	Cross flow
Aerofoil section					Sheet blading					Aerofoil section	Sheet blading			Aerofoil section	Sheet blading			Aerofoil section	Sheet blading	Aerofoil section	Sheet blading	Aerofoil section	Forward curved
Tube axial	Vane axial	Contra rotating	Reversible	Bifurcated	Tube axial	Bifurcated	Direct dirve	Vee belt drive	Direct drive														
	●		●																				
●	●	●	●	●	●	●	●	●	●	●	●	●	●	●	●	●	●	●	●	●	●	●	●
●					●							●		●							●		
	●		●				●																
																●	●	●	●		●	●	●
●	●	●	●	●										●	●	●	●						
●											●	●		●						●			
●	●			●	●	●								●	●								
	●	●																					
	●											●				●							
	●		●							●				●								●	
	●		●							●				●								●	
	●		●							●				●								●	
●	●	●	●		●																		
	●		●	●			●	●															
●														●		●		●					
●	●																						
●				●			●			●		●		●		●							
●	●	●	●	●	●	●	●	●	●	●	●	●	●	●	●	●	●	●	●	●	●	●	●
			●																				
			●																				
			●																				
	●		●	●																		●	
●	●		●																			●	
		●	●	●																			

24.4 Ancillary products and services

ACOUSTIC BOXES

Fläkt Woods Ltd	UK
MAN Acoustics Ltd	UK
Roof Units	UK
Vent-Axia Ltd	UK

AERODYNAMIC DESIGN

ACT-RX Technology Europe	NL
European Thermodynamics Ltd	UK
Fläkt Woods Ltd	UK

AIR DISTRIBUTION PRODUCTS

Advanced Air (UK) Ltd	UK
CBI Service Srl	I
Conductaire SA	E
Delrac Ltd	UK
Encon Air Systems Ltd	UK
Fläkt Woods Ltd	UK
Howden Buffalo Inc	US
Howden Industrial UK	UK
Howden Sirocco SA	F
Howden Ventiladores Ltda	BR
Roof Units	UK
Technik SpA	I
Tetisan Ltd	TU
Turbowerke Meissen Howden GmbH	D
Vent-Axia Ltd	UK

AIR HANDLING UNITS

Alstom Belgium SA Energy-MTM	B
Delrac Ltd	UK
Dynamic Air Products Ltd	UK
FITA Teknik	TU
Fläkt Woods Ltd	UK
Fyrogenis SA	GR
GER SA	E
HRP Ltd	UK
Imas AS	TU
Koja Oy	FIN
Koolclima SL	E
Roof Units	UK
Rosenberg Ventilatoren GmbH	D
Sönmez Metal	TU
Oy Swegon Ab	FIN
Swegon AB	SE
Swegon Ltd	UK
Swegon Sp. z o.o.	PL
Systemair AB	SE
Systemair Ltd	UK
Technik SpA	I
Tecnivel SA	E
Termas AS	TU
Termotecnica Pericoli Srl	I
Termoven SA	E
Tetisan Ltd	TU
Toussaint Nyssenne SA	B
Tradair SA	E
Vallox Oy	FIN
Vent-Axia Ltd	UK
Ventiladores Chaysol SA	E

ANTI-VIBRATION MOUNTS

Airscrew Ltd	UK
APMG Ltd	UK
Applied Energy Products Ltd	UK
Axair Fans UK Ltd	UK
Brown Group Ltd	UK
CE-Air International	UK
De Raedt SA	B
Elta Fans Ltd	UK
European Thermodynamics Ltd	UK
Exhausto Ltd	UK
Fairbrother & Associates Inc	US
Fläkt Woods Ltd	UK
Gebhardt Ventilatoren	D
Greenmount Fans (North) Ltd	UK
Halifax Fan Ltd	UK
Howden Buffalo Inc	US
Howden Industrial UK	UK
Howden Sirocco SA	F
Howden Ventiladores Ltda	BR
Imofa UK Ltd	UK
Ondrives Ltd	UK
Remco Products Ltd	UK
Removex AB	SE
Rencol Tolerance Rings Ltd	UK
Sardou SA	F
Secomak Ltd	UK
Standard & Pochin Ltd	UK
Tecnifan SA	E
TLT Co-Vent Fans Inc	CAN
TLT- Turbo GmbH	D
TLT-Babcock Inc	US
Turbowerke Meissen Howden GmbH	D
Vectaire Ltd	UK
Witt & Sohn AG	D
Witt UK	UK
Wyko Industrial Services	UK

BALANCING & DIAGNOSTIC EQUIPMENT

CEMB Hofmann (UK) Ltd	UK
Hofmann Maschinen- und Anlagenbau GmbH	D
Howden Buffalo Inc	US
Howden Industrial UK	UK
Howden Sirocco SA	F
Howden Ventiladores Ltda	BR
IRD UK Ltd	UK
Lord Corporation (Europe) Ltd	UK
Schenck Balancing & Diagnostic Systems	UK
Schenck RoTec GmbH	D
Turbowerke Meissen Howden GmbH	D

BALL/ROLLER BEARINGS

AHR International Ltd	UK
Airscrew Ltd	UK
APMG Ltd	UK
Applied Energy Products Ltd	UK
Boldrocchi Srl	I
Bri-Mac Engineering Ltd	UK
Brown Group Ltd	UK
Comair Rotron Europe Ltd	UK
Cooper Roller Bearings Co Ltd	UK
Criptic-Arvis Ltd	UK
De Raedt SA	B
Direct Bearings & Power Transmissions Ltd	UK
Domus Ventilation Ltd	UK
Gebhardt Ventilatoren	D
Greenwood Air Management Ltd	UK
Halifax Fan Ltd	UK
Howden Buffalo Inc	US
Howden Industrial UK	UK
Howden Sirocco SA	F
Howden Ventiladores Ltda	BR
INA Bearing Company Ltd	UK
Kiloheat Ltd	UK
Michell Bearings	UK
Nicotra Espana SA	E
Nicotra France SA	F
Nicotra GmbH	D
Nicotra SpA	I
Nicotra UK Ltd	UK
NMB Minebea (UK) Ltd	UK
NTN Bearings (UK) Ltd	UK
NTN Wälzlager (Europa) GmbH	D
Ondrives Ltd	UK
Piller Industrieventilatoren GmbH	D
PM°DM Precision Motors Deutsche Minebea GmbH	D
Rosenberg Ventilatoren GmbH	D
Sardou SA	F
Silver Box Fans Ltd	UK
SKF (UK) Ltd	UK
Standard & Pochin Ltd	UK
The Timken Company	US
Timken Europe	F
TLT Co-Vent Fans Inc	CAN
TLT- Turbo GmbH	D

TLT-Babcock Inc	US
Turbowerke Meissen Howden GmbH	D
Vortice Elettrosociali SpA	I
Vortice France	F
Vortice Ltd	UK
Witt & Sohn AG	D
Witt UK	UK
Wyko Industrial Services	UK

BEARING HOUSINGS

Bri-Mac Engineering Ltd	UK
Howden Buffalo Inc	US
Howden Industrial UK	UK
Howden Sirocco SA	F
Howden Ventiladores Ltda	BR
Turbowerke Meissen Howden GmbH	D

CERTIFICATION SERVICES

Allianz Cornhill International	UK
BSI Product Services	UK
Bureau Veritas	UK
DNV	UK
Lloyd's Register	UK
Royal & SunAlliance, Engineering Business	UK
SGS United Kingdom Ltd	UK

CFD (COMPUTATIONAL FLUID DYNAMICS)

ACT-RX Technology Europe	NL
Advanced Design Technology Ltd	UK
Fläkt Woods Ltd	UK
Fluent Europe Ltd	UK
PCA Engineers Ltd	UK

COMBINATION BASEFRAMES

Direct Bearings & Power Transmissions Ltd	UK
Fans & Blowers Ltd	UK
Gebhardt Ventilatoren	D
Halifax Fan Ltd	UK
Howden Buffalo Inc	US
Howden Industrial UK	UK
Howden Sirocco SA	F
Howden Ventiladores Ltda	BR
Imofa UK Ltd	UK
Kiloheat Ltd	UK
Matthews & Yates Ltd	UK
Novovent SA	E
Standard & Pochin Ltd	UK
Turbowerke Meissen Howden GmbH	D
Universal Fan & Blower Ltd	CAN
Witt & Sohn AG	D
Witt UK	UK

CONDITION MONITORING

Howden Buffalo Inc	US
Howden Industrial UK	UK
Howden Sirocco SA	F
Howden Ventiladores Ltda	BR
Rockwell Automation Ltd	UK
SPM Instrument AB	SE
SPM Instrument UK Ltd	UK
Turbowerke Meissen Howden GmbH	D

CONSULTANCY SERVICES

ACT-RX Technology Europe	NL
Advanced Design Technology Ltd	UK
European Thermodynamics Ltd	UK
PCA Engineers Ltd	UK

CONTROLLERS - SOFT START

ABB Drives	FIN
CE-Air International	UK
Danfoss Drives A/S	DK
Danfoss Ltd	UK
De Raedt SA	B
Gamak Motors Ltd	UK
Halifax Fan Ltd	UK
Standard & Pochin Ltd	UK
Weg Electric Motors (UK) Ltd	UK
Witt & Sohn AG	D

CONTROLLERS - VARIABLE FREQUENCY

ABB Drives	FIN
Axair Fans UK Ltd	UK
Brown Group Ltd	UK
Danfoss Drives A/S	DK
Danfoss Ltd	UK
De Raedt SA	B
Exhausto Ltd	UK
Fläkt Woods Ltd	UK
Flender Power Transmission Ltd	UK
Gamak Motors Ltd	UK
Gebhardt Ventilatoren	D
Halifax Fan Ltd	UK
Howden Buffalo Inc	US
Howden Industrial UK	UK
Howden Sirocco SA	F
Howden Ventiladores Ltda	BR
Imofa UK Ltd	UK
Lenze Ltd	UK
Metrico International Ltd	UK
Northey Technologies Ltd	UK
Novovent SA	E
Rosenberg Ventilatoren GmbH	D
Teco Electric Europe Ltd	UK
The Timken Company	US
Timken Europe	F
Turbowerke Meissen Howden GmbH	D
Weg Electric Motors (UK) Ltd	UK
Witt & Sohn AG	D
Witt UK	UK
Wyko Industrial Services	UK

CONTROLLERS - VARIABLE VOLTAGE

ABB Drives	FIN
Axair Fans UK Ltd	UK
Brown Group Ltd	UK
ebm-papst UK Ltd	UK
Elta Fans Ltd	UK
Exhausto A/S	DK
Exhausto Ltd	UK
Gamak Motors Ltd	UK
Gebhardt Ventilatoren	D
Halifax Fan Ltd	UK
Helios Ventilateurs Sarl	F
Helios Ventilation Systems Ltd	UK
Helios Ventilatoren	D
Helios Ventilatoren AG	SE
Helios Ventilatoren GmbH	A
Matthews & Yates Ltd	UK
C A Östberg AB	SE
Remco Products Ltd	UK
Rosenberg Ventilatoren GmbH	D
Soler + Palau Ltd	UK
Weg Electric Motors (UK) Ltd	UK
Wyko Industrial Services	UK

COOLING DISCS

Airap	F
Cooper Roller Bearings Co Ltd	UK
Gebhardt Ventilatoren	D
Halifax Fan Ltd	UK
Howden Buffalo Inc	US
Howden Industrial UK	UK
Howden Sirocco SA	F
Howden Ventiladores Ltda	BR
Kiloheat Ltd	UK
Piller Industrieventilatoren GmbH	D
Selnikel	TU
Standard & Pochin Ltd	UK
TLT Co-Vent Fans Inc	CAN
TLT- Turbo GmbH	D
TLT-Babcock Inc	US
Turbowerke Meissen Howden GmbH	D
Witt & Sohn AG	D
Witt UK	UK

DAMPERS

Advanced Air (UK) Ltd	UK
FITA Teknik	TU
Fläkt Woods Ltd	UK
Flamgard Engineering Ltd	UK
Greenheck	US
Greenmount Fans (North) Ltd	UK
Howden Buffalo Inc	US
Howden Industrial UK	UK
Howden Sirocco SA	F

Howden Ventiladores Ltda BR
Koolclima SL E
MAN Acoustics Ltd UK
Removex AB SE
Schenck Balancing & Diagnostic Systems UK
Turbowerke Meissen Howden GmbH D

DESIGN SOFTWARE

Advanced Design Technology Ltd UK
Fläkt Woods Ltd UK
Howden Buffalo Inc US
Howden Industrial UK UK
Howden Sirocco SA F
Howden Ventiladores Ltda BR
PCA Engineers Ltd UK
Turbowerke Meissen Howden GmbH D

DRAIN POINTS

Axair Fans UK Ltd UK
Fans & Blowers Ltd UK
Gebhardt Ventilatoren D
Imofa UK Ltd UK
Standard & Pochin Ltd UK
Universal Fan & Blower Ltd CAN
Witt & Sohn AG D

ELECTRIC MOTORS - AIRSTREAM RATED

ABB Drives FIN
AEG-Lafert Electric Motors Ltd UK
Airscrew Ltd UK
Brook Crompton UK
Gamak Motors Ltd UK
Howden Buffalo Inc US
Howden Industrial UK UK
Howden Sirocco SA F
Howden Ventiladores Ltda BR
Matthews & Yates Ltd UK
Oriental Motor (UK) Ltd UK
Remco Products Ltd UK
Sardou SA F
A O Smith Electrical Products Ltd UK
Standard & Pochin Ltd UK
Teco Electric Europe Ltd UK
Turbowerke Meissen Howden GmbH D
Weg Electric Motors (UK) Ltd UK
Witt & Sohn AG D
Witt UK UK

ELECTRIC MOTORS - CAPACITOR START

ABB Drives FIN
AEG-Lafert Electric Motors Ltd UK
Brook Crompton UK
ebm-papst UK Ltd UK
Fläkt Woods Ltd UK
Gamak Motors Ltd UK
Gebhardt Ventilatoren D
Halifax Fan Ltd UK
Oriental Motor (UK) Ltd UK
Remco Products Ltd UK
Sardou SA F
Soler + Palau Ltd UK
Standard & Pochin Ltd UK
Tyco Electronics-Crompton Small Motors UK
Weg Electric Motors (UK) Ltd UK
Witt & Sohn AG D

ELECTRIC MOTORS - DC

Airscrew Ltd UK
Brook Crompton UK
ebm-papst UK Ltd UK
Gamak Motors Ltd UK
Gebhardt Ventilatoren D
Halifax Fan Ltd UK
Mayr GmbH + Co KG D
NMB Minebea (UK) Ltd UK
Oriental Motor (UK) Ltd UK
PM°DM Precision Motors Deutsche Minebea GmbH D
Rosenberg Ventilatoren GmbH D
Sardou SA F
Soler + Palau Ltd UK
Teco Electric Europe Ltd UK
Torin Ltd UK
Tyco Electronics-Crompton Small Motors UK
Witt & Sohn AG D

ELECTRIC MOTORS - FOOT MOUNTING TEFV

AEG-Lafert Electric Motors Ltd UK
Axair Fans UK Ltd UK
Beatson Fans & Motors Ltd UK
Brook Crompton UK
CE-Air International UK
De Raedt SA B
Flender Power Transmission Ltd UK
Gamak Motors Ltd UK
Halifax Fan Ltd UK
Howden Buffalo Inc US
Howden Industrial UK UK
Howden Sirocco SA F
Howden Ventiladores Ltda BR
Oriental Motor (UK) Ltd UK
Standard & Pochin Ltd UK
Teco Electric Europe Ltd UK
Turbowerke Meissen Howden GmbH D
Weg Electric Motors (UK) Ltd UK
Witt UK UK

ELECTRIC MOTORS - "INSIDE OUT"

ABB Drives FIN
Axair Fans UK Ltd UK
Gebhardt Ventilatoren D
Helios Ventilateurs Sarl F
Helios Ventilation Systems Ltd UK
Helios Ventilatoren D
Helios Ventilatoren AG SE
Helios Ventilatoren GmbH A
Kiloheat Ltd UK
NMB Minebea (UK) Ltd UK
PM°DM Precision Motors Deutsche Minebea GmbH D
A O Smith Electrical Products Ltd UK

ELECTRIC MOTORS - SINGLE PHASE

AEG-Lafert Electric Motors Ltd UK
Airap F
Axair Fans UK Ltd UK
Brook Crompton UK
CE-Air International UK
Domus Ventilation Ltd UK
Elta Fans Ltd UK
Fairbrother & Associates Inc US
Fans & Blowers Ltd UK
Fläkt Woods Ltd UK
Gamak Motors Ltd UK
Gebhardt Ventilatoren D
Halifax Fan Ltd UK
Helios Ventilateurs Sarl F
Helios Ventilation Systems Ltd UK
Helios Ventilatoren D
Helios Ventilatoren AG SE
Helios Ventilatoren GmbH A
Metrico International Ltd UK
Oriental Motor (UK) Ltd UK
C A Östberg AB SE
Sardou SA F
A O Smith Electrical Products Ltd UK
Standard & Pochin Ltd UK
Teco Electric Europe Ltd UK
Tyco Electronics-Crompton Small Motors UK
VEM motors GmbH D
Vortice Elettrosociali SpA I
Vortice France F
Vortice Ltd UK
Weg Electric Motors (UK) Ltd UK
Witt & Sohn AG D

ELECTRIC MOTORS - SQUIRREL CAGE

ABB Drives FIN
AEG-Lafert Electric Motors Ltd UK
Axair Fans UK Ltd UK
Brook Crompton UK
Gamak Motors Ltd UK
Halifax Fan Ltd UK

Howden Buffalo Inc	US
Howden Industrial UK	UK
Howden Sirocco SA	F
Howden Ventiladores Ltda	BR
Oriental Motor (UK) Ltd	UK
Remco Products Ltd	UK
Sardou SA	F
Standard & Pochin Ltd	UK
Teco Electric Europe Ltd	UK
Turbowerke Meissen Howden GmbH	D
VEM motors GmbH	D
Weg Electric Motors (UK) Ltd	UK
Witt & Sohn AG	D
Witt UK	UK

ELECTRIC MOTORS - THREE PHASE

ABB Drives	FIN
Airap	F
Axair Fans UK Ltd	UK
Brown Group Ltd	UK
Fairbrother & Associates Inc	US
Fans & Blowers Ltd	UK
Flender Power Transmission Ltd	UK
Gebhardt Ventilatoren	D
Halifax Fan Ltd	UK
Helios Ventilateurs Sarl	F
Helios Ventilation Systems Ltd	UK
Helios Ventilatoren	D
Helios Ventilatoren AG	SE
Helios Ventilatoren GmbH	A
Howden Buffalo Inc	US
Howden Industrial UK	UK
Howden Sirocco SA	F
Howden Ventiladores Ltda	BR
Kiloheat Ltd	UK
Matthews & Yates Ltd	UK
Mayr GmbH + Co KG	D
C A Östberg AB	SE
Remco Products Ltd	UK
Rosenberg Ventilatoren GmbH	D
Sardou SA	F
Secomak Ltd	UK
A O Smith Electrical Products Ltd	UK
Standard & Pochin Ltd	UK
Turbowerke Meissen Howden GmbH	D
Tyco Electronics-Crompton Small Motors	UK
VEM motors GmbH	D
Vortice Elettrosociali SpA	I
Vortice France	F
Vortice Ltd	UK
Witt & Sohn AG	D
Witt UK	UK
Wyko Industrial Services	UK

EXTENDED LUBRICATORS

CE-Air International	UK
Gebhardt Ventilatoren	D
Halifax Fan Ltd	UK
Howden Buffalo Inc	US
Howden Industrial UK	UK
Howden Sirocco SA	F
Howden Ventiladores Ltda	BR
Kiloheat Ltd	UK
Piller Industrieventilatoren GmbH	D
Standard & Pochin Ltd	UK
Turbowerke Meissen Howden GmbH	D
Witt & Sohn AG	D
Witt UK	UK

FAN COIL UNITS

Advanced Air (UK) Ltd	UK
Biddle Air Systems Ltd	UK
Delrac Ltd	UK
Fläkt Woods Ltd	UK
GER SA	E
HRP Ltd	UK
Imas AS	TU
Koja Oy	FIN
Koolclima SL	E
Marstair Ltd	UK
Marzorati Ventilazione Srl	I
Sönmez Metal	TU
Oy Swegon Ab	FIN
Swegon AB	SE
Swegon Ltd	UK
Swegon Sp. z o.o.	PL
Technik SpA	I
Tecnivel SA	E
Termas AS	TU
Termoven SA	E
Tetisan Ltd	TU
Vemair	E

FLEXIBLE CONNECTORS

Airap	F
APMG Ltd	UK
Applied Energy Products Ltd	UK
Axair Fans UK Ltd	UK
Elta Fans Ltd	UK
Exhausto A/S	DK
Fläkt Woods Ltd	UK
Gebhardt Ventilatoren	D
Halifax Fan Ltd	UK
Helios Ventilateurs Sarl	F
Helios Ventilation Systems Ltd	UK
Helios Ventilatoren	D
Helios Ventilatoren AG	SE
Helios Ventilatoren GmbH	A
Howden Buffalo Inc	US
Howden Industrial UK	UK
Howden Sirocco SA	F
Howden Ventiladores Ltda	BR
Matthews & Yates Ltd	UK
Nicotra Espana SA	E
Nicotra France SA	F
Nicotra GmbH	D
Nicotra SpA	I
Nicotra UK Ltd	UK
Novovent SA	E
Ondrives Ltd	UK
Rencol Tolerance Rings Ltd	UK
Rosenberg Ventilatoren GmbH	D
Sardou SA	F
Secomak Ltd	UK
Selnikel	TU
Silver Box Fans Ltd	UK
Soler + Palau Ltd	UK
Standard & Pochin Ltd	UK
Turbowerke Meissen Howden GmbH	D
Vectaire Ltd	UK
Witt & Sohn AG	D
Witt UK	UK

FLEXIBLE COUPLINGS

Gates Europe nv	B
Gates France SARL	F
Gates GmbH	D
Gates Power Transmission Ltd	UK
Gates SA	E
Gates Srl	I
Howden Buffalo Inc	US
Howden Industrial UK	UK
Howden Sirocco SA	F
Howden Ventiladores Ltda	BR
Turbowerke Meissen Howden GmbH	D

INLET BOXES

De Raedt SA	B
Fläkt Woods Ltd	UK
Halifax Fan Ltd	UK
Howden Buffalo Inc	US
Howden Industrial UK	UK
Howden Sirocco SA	F
Howden Ventiladores Ltda	BR
Imofa UK Ltd	UK
Piller Industrieventilatoren GmbH	D
Rosenberg Ventilatoren GmbH	D
Selnikel	TU
TLT Co-Vent Fans Inc	CAN
TLT- Turbo GmbH	D
TLT-Babcock Inc	US
Turbowerke Meissen Howden GmbH	D
Universal Fan & Blower Ltd	CAN
Witt & Sohn AG	D

Witt UK	UK

INLET/OUTLET GUARDS

Airscrew Ltd	UK
Axair Fans UK Ltd	UK
Elta Fans Ltd	UK
European Thermodynamics Ltd	UK
Fläkt Woods Ltd	UK
Gebhardt Ventilatoren	D
Howden Buffalo Inc	US
Howden Industrial UK	UK
Howden Sirocco SA	F
Howden Ventiladores Ltda	BR
Imofa UK Ltd	UK
Nicotra Espana SA	E
Nicotra France SA	F
Nicotra GmbH	D
Nicotra SpA	I
Nicotra UK Ltd	UK
Novovent SA	E
Selnikel	TU
Soler + Palau Ltd	UK
Standard & Pochin Ltd	UK
TLT Co-Vent Fans Inc	CAN
TLT- Turbo GmbH	D
TLT-Babcock Inc	US
Turbowerke Meissen Howden GmbH	D
Vectaire Ltd	UK
Vortice Elettrosociali SpA	I
Vortice France	F
Vortice Ltd	UK
Witt & Sohn AG	D
Witt UK	UK

INSPECTION DOORS

Airap	F
APMG Ltd	UK
Elta Fans Ltd	UK
Fans & Blowers Ltd	UK
Gebhardt Ventilatoren	D
Halifax Fan Ltd	UK
Howden Buffalo Inc	US
Howden Industrial UK	UK
Howden Sirocco SA	F
Howden Ventiladores Ltda	BR
Imofa UK Ltd	UK
Nicotra Espana SA	E
Nicotra France SA	F
Nicotra GmbH	D
Nicotra SpA	I
Nicotra UK Ltd	UK
Piller Industrieventilatoren GmbH	D
Selnikel	TU
Standard & Pochin Ltd	UK
Turbowerke Meissen Howden GmbH	D
Universal Fan & Blower Ltd	CAN
Witt & Sohn AG	D
Witt UK	UK

INSPECTION SERVICES

Allianz Cornhill International	UK
BSI Product Services	UK
Bureau Veritas	UK
DNV	UK
Howden Buffalo Inc	US
Howden Industrial UK	UK
Howden Sirocco SA	F
Howden Ventiladores Ltda	BR
HSB Inspection Quality	UK
Lloyd's Register	UK
Royal & SunAlliance, Engineering Business	UK
SGS United Kingdom Ltd	UK
Turbowerke Meissen Howden GmbH	D

MATCHING FLANGES

APMG Ltd	UK
Axair Fans UK Ltd	UK
Fläkt Woods Ltd	UK
Halifax Fan Ltd	UK
Helios Ventilateurs Sarl	F
Helios Ventilation Systems Ltd	UK
Helios Ventilatoren	D
Helios Ventilatoren AG	SE
Helios Ventilatoren GmbH	A
Howden Buffalo Inc	US
Howden Industrial UK	UK
Howden Sirocco SA	F
Howden Ventiladores Ltda	BR
Imofa UK Ltd	UK
Matthews & Yates Ltd	UK
Nicotra Espana SA	E
Nicotra France SA	F
Nicotra GmbH	D
Nicotra SpA	I
Nicotra UK Ltd	UK
Novovent SA	E
Secomak Ltd	UK
Selnikel	TU
Soler + Palau Ltd	UK
Standard & Pochin Ltd	UK
TLT Co-Vent Fans Inc	CAN
TLT- Turbo GmbH	D
TLT-Babcock Inc	US
Turbowerke Meissen Howden GmbH	D
Vectaire Ltd	UK
Witt UK	UK

MECHANICAL DESIGN, ANALYSIS AND TROUBLE SHOOTING

Howden Buffalo Inc	US
Howden Industrial UK	UK
Howden Sirocco SA	F
Howden Ventiladores Ltda	BR
Turbowerke Meissen Howden GmbH	D

MOUNTING FEET (FOR AXIAL FLOW FANS ETC)

Airap	F
CE-Air International	UK
Elta Fans Ltd	UK
Fläkt Woods Ltd	UK
Gebhardt Ventilatoren	D
Halifax Fan Ltd	UK
Helios Ventilateurs Sarl	F
Helios Ventilation Systems Ltd	UK
Helios Ventilatoren	D
Helios Ventilatoren AG	SE
Helios Ventilatoren GmbH	A
Howden Buffalo Inc	US
Howden Industrial UK	UK
Howden Sirocco SA	F
Howden Ventiladores Ltda	BR
Matthews & Yates Ltd	UK
Metrico International Ltd	UK
Novovent SA	E
Silver Box Fans Ltd	UK
Soler + Palau Ltd	UK
Standard & Pochin Ltd	UK
Turbowerke Meissen Howden GmbH	D
Universal Fan & Blower Ltd	CAN
Vectaire Ltd	UK
Witt & Sohn AG	D
Witt UK	UK

NOISE AND VIBRATION MEASUREMENT AND ANALYSIS

Bruel and Kjaer	DK
Bruel and Kjaer UK Ltd	UK
Howden Buffalo Inc	US
Howden Industrial UK	UK
Howden Sirocco SA	F
Howden Ventiladores Ltda	BR
Sound Research Laboratories Ltd	UK
Turbowerke Meissen Howden GmbH	D

PILLOW BLOCKS

Bri-Mac Engineering Ltd	UK
Howden Buffalo Inc	US
Howden Industrial UK	UK
Howden Sirocco SA	F
Howden Ventiladores Ltda	BR
Turbowerke Meissen Howden GmbH	D

PORTABLE BALANCING EQUIPMENT

Howden Buffalo Inc	US
Howden Industrial UK	UK
Howden Sirocco SA	F
Howden Ventiladores Ltda	BR

IRD UK Ltd	UK
Turbowerke Meissen Howden GmbH	D

ROTOR BALANCING MACHINES

Howden Buffalo Inc	US
Howden Industrial UK	UK
Howden Sirocco SA	F
Howden Ventiladores Ltda	BR
IRD UK Ltd	UK
Turbowerke Meissen Howden GmbH	D

SHAFT ALIGNMENT

Howden Buffalo Inc	US
Howden Industrial UK	UK
Howden Sirocco SA	F
Howden Ventiladores Ltda	BR
Prüftechnik AG	D
Prütechnik Ltd	UK
Turbowerke Meissen Howden GmbH	D

SHAFT SEALS

Airap	F
Airscrew Ltd	UK
Bri-Mac Engineering Ltd	UK
Brown Group Ltd	UK
Criptic-Arvis Ltd	UK
Fairbrother & Associates Inc	US
Gebhardt Ventilatoren	D
Greenmount Fans (North) Ltd	UK
Halifax Fan Ltd	UK
Howden Buffalo Inc	US
Howden Industrial UK	UK
Howden Sirocco SA	F
Howden Ventiladores Ltda	BR
Kiloheat Ltd	UK
Piller Industrieventilatoren GmbH	D
Standard & Pochin Ltd	UK
Turbowerke Meissen Howden GmbH	D
Universal Fan & Blower Ltd	CAN
Witt & Sohn AG	D
Witt UK	UK
Wyko Industrial Services	UK

SILENCERS

Fläkt Woods Ltd	UK
Howden Buffalo Inc	US
Howden Buffalo Inc	US
Howden Industrial UK	UK
Howden Industrial UK	UK
Howden Sirocco SA	F
Howden Sirocco SA	F
Howden Ventiladores Ltda	BR
Howden Ventiladores Ltda	BR
MAN Acoustics Ltd	UK
Northey Technologies Ltd	UK
Removex AB	SE
Turbowerke Meissen Howden GmbH	D

SLEEVE BEARINGS

AHR International Ltd	UK
APMG Ltd	UK
Boldrocchi Srl	I
Bri-Mac Engineering Ltd	UK
Comair Rotron Europe Ltd	UK
Criptic-Arvis Ltd	UK
De Raedt SA	B
Direct Bearings & Power Transmissions Ltd	UK
Domus Ventilation Ltd	UK
Greenwood Air Management Ltd	UK
Halifax Fan Ltd	UK
Howden Buffalo Inc	US
Howden Industrial UK	UK
Howden Sirocco SA	F
Howden Ventiladores Ltda	BR
INA Bearing Company Ltd	UK
Nicotra Espana SA	E
Nicotra France SA	F
Nicotra GmbH	D
Nicotra SpA	I
Nicotra UK Ltd	UK
Piller Industrieventilatoren GmbH	D
Sardou SA	F
Standard & Pochin Ltd	UK
TLT Co-Vent Fans Inc	CAN
TLT- Turbo GmbH	D
TLT-Babcock Inc	US
Turbowerke Meissen Howden GmbH	D
Witt UK	UK

SPLIT CASINGS

APMG Ltd	UK
B.O.B. Stevenson Ltd	UK
Boldrocchi Srl	I
Fans & Blowers Ltd	UK
Gebhardt Ventilatoren	D
Halifax Fan Ltd	UK
Howden Buffalo Inc	US
Howden Industrial UK	UK
Howden Sirocco SA	F
Howden Ventiladores Ltda	BR
Nicotra Espana SA	E
Nicotra France SA	F
Nicotra GmbH	D
Nicotra SpA	I
Nicotra UK Ltd	UK
Piller Industrieventilatoren GmbH	D
Standard & Pochin Ltd	UK
TLT Co-Vent Fans Inc	CAN
TLT- Turbo GmbH	D
TLT-Babcock Inc	US
Turbowerke Meissen Howden GmbH	D
Witt & Sohn AG	D
Witt UK	UK

SYNCHRONOUS BELTS

Gates Europe nv	B
Gates France SARL	F
Gates GmbH	D
Gates Power Transmission Ltd	UK
Gates SA	E
Gates Srl	I

TESTING

Fläkt Woods Ltd	UK
Howden Buffalo Inc	US
Howden Industrial UK	UK
Howden Sirocco SA	F
Howden Ventiladores Ltda	BR
National Physical Laboratory	UK
Royal & SunAlliance, Engineering Business	UK
Turbowerke Meissen Howden GmbH	D

TILTING THRUST PADS

Howden Buffalo Inc	US
Howden Industrial UK	UK
Howden Sirocco SA	F
Howden Ventiladores Ltda	BR
Michell Bearings	UK
Turbowerke Meissen Howden GmbH	D

VEE BELT DRIVES AND GUARDS

APMG Ltd	UK
Brown Group Ltd	UK
CE-Air International	UK
Fans & Blowers Ltd	UK
Fenner Drives	UK
Gates Europe nv	B
Gates France SARL	F
Gates GmbH	D
Gates Power Transmission Ltd	UK
Gates SA	E
Gates Srl	I
Gebhardt Ventilatoren	D
Halifax Fan Ltd	UK
Howden Buffalo Inc	US
Howden Industrial UK	UK
Howden Sirocco SA	F
Howden Ventiladores Ltda	BR
Imofa UK Ltd	UK
Kiloheat Ltd	UK
MBL (Europe) BV	NL
Novovent SA	E
Sardou SA	F
Selnikel	TU
Standard & Pochin Ltd	UK
Turbowerke Meissen Howden GmbH	D
Universal Fan & Blower Ltd	CAN
Witt & Sohn AG	D

Witt UK	UK
Wyko Industrial Services	UK

VENTILATION SYSTEMS

Aldes Aeraulique	F
Encon Air Systems Ltd	UK
Fläkt Woods Ltd	UK
GER SA	E
GIA SwedVent	SE
Greenheck	US
Groupe Leader	F
Hartzell Fan Inc	US
HRP Ltd	UK
Imas AS	TU
Koja Oy	FIN
Loren Cook Company	US
Marzorati Ventilazione Srl	I
Removex AB	SE
Rotomatika Fans d.o.o.	SL
Sönmez Metal	TU
Oy Swegon Ab	FIN
Swegon AB	SE
Swegon Ltd	UK
Swegon Sp. z o.o.	PL
Termas AS	TU
Termoven SA	E
Tradair SA	E
Ventiladores Chaysol SA	E

24.5 Trade Names

ARX
ACT-RX fans are branded as ARX
ACT-RX Technology Europe NL

Aerex
Axial fans
De Raedt SA B

Aeroto
Industrial noise protection
Boldrocchi Srl I

Airap
Industrial axial and centrifugal fans
Airap F

Alcosa
Extensive range of centrifugal industrial fans
B.O.B. Stevenson Ltd UK

Arpex
Torsionally rigid couplings
Flender Power Transmission Ltd UK

Arvis
Bearing housings and plummer blocks for rolling element bearings
Criptic-Arvis Ltd UK

Attenuators
Attenuators for noise reduction of industrial centrifugal fans
B Bille A/S DK

BFM Type B56
Fractional power motors
Beatson Fans & Motors Ltd UK

Belair
Intelligent domestic ventilation system
Gebhart Ventilatoren D

Breeza
Impellers
The London Fan Company Ltd UK

Breezax
Impellers
The London Fan Company Ltd UK

Brooks roof units
Roof extract units mixed flow and centrifugal
Matthews & Yates UK

Buffalo
(Became part of Howden)
Howden Industrial UK UK

CBM fans
Range of forward curved OEM centrifugal fans
Soler & Palau Ltd UK

CK
Centrifugal inline duct-fan with circular connection
C A Östberg AB SE

CMM System
Reliable and cost-effective condition monitoring modules
SPM Instrument AB SE

Carter
(Became part of Howden)
Howden Industrial UK UK

Centrifugal fans
Centrifugal fans for industrial purposes
B Bille A/S DK

Centripal EV
Centrifugal fans
Fläkt Woods Ltd UK

Ceradyna
A simple and highly reliable patented bearing system
ACT-RX Technology Europe NL

Checo
Fans and louvres
Ce-Air International UK

Chimney Draft
Chimney fans - RS/RSV
Exhausto A/S DK

Chittom
Axial fans
Ce-Air International UK

Chittom Equipment
Axial flow fans manual/autovariable
Ce-Air International UK

Christy Hunt
Hunt Saroe self-aligning ring oiling plain bearing units bronze/white metal
Criptic-Arvis Ltd UK

Cincinnati Fan
Air moving and material handling fans. Direct and belt driven
Fairbrother & Associates Inc US

Colasit R
Plastic fan for aggressive and corrosive gases
Colasit AG CH

Comfortventilation
Centrifugal or radial fans - BESB/BESF, Roof fans - DTH/DTV, Wall fans - VVR
Exhausto A/S DK

Compact
Range of commercial plate and duct fans for general ventilation
Elta Fans Ltd UK

Compact
Range of HVAC and OEM plate and cased axial fans
Soler & Palau Ltd UK

Cooper
Split roller bearings and seals
Cooper Roller Bearings Co Ltd UK

Croft
Rigid split ring oiler plummer block
Criptic-Arvis Ltd UK

Crompton Small Motors
Electric motors
Tyco Electronics - Crompton Small Motors UK

Cyclone
Range of FCC VCB and V aerofoil centrifugal fans
Matthews & Yates UK

Dantop
Roof fans
Exhausto Ltd UK

De Cardenas Ventilatione
Heavy duty fans
Boldrocchi Srl I

De Raedt
Axial and centrifugal fans
De Raedt SA B

Discfan
Round slim axial bathroom fan
Domus Ventilation Ltd UK

Durafan
Corrosion resistant plastic fans
APMG Ltd UK

EBV System
Exhausto building ventilation
Exhausto Ltd UK

Ecofit
External rotor motor fans
Axair Fans UK Ltd UK

Ecofit
Commercial fans
Greenwood Air Management Ltd UK

Ecofit
Fans, motors
Rosenberg Ventilatoren GmbH D

Ecologia
Air pollution control
Boldrocchi Srl I

Elf
Economy axial domestic fan
Domus Ventilation Ltd UK

Engart
(Became part of Howden)
Howden Industrial UK UK

Enviromental
Central ventilation system
Greenwood Air Management Ltd UK

Etri
Fans, motors
Rosenberg Ventilatoren GmbH D

Exhausto
Chimney fans
Exhausto Ltd UK

Fan Systems
Centrifugal fans
Witt UK UK

Fantraxx
Commercial and industrial ventilation products
Leader Fan Industries Ltd CAN

Flomax
Tube axial range of fans
NMB Minebea UK Ltd UK

Fumex
Emergency smoke extraction motor
Brook Crompton UK

G W Axial
Axial flow fans, aerofoil section impellers
Matthews & Yates UK

Genovent
Centrifugal roof extract fan
Gebhart Ventilatoren D

Gip
Silent high efficeiency fans
Sardou SA F

HM horizontal balancing machine
Universal balancing machine for a wide range of rotors
Schenck Rotec GmbH D

HT Horizontal b.m.
Hard bearing h.m. in a table-top version, best suited for small workpieces such as miniature fans or complete assemblies
Schenck Rotec GmbH D

Hunt Saroe
Split self-aligning ring oiling enclosed self lubricating bronze fan bearings
Criptic-Arvis Ltd UK

IRD Balancing
Supply and on-going servicing of fan balancing equipment
IRD UK Ltd UK

IRD Balancing
Balancing consulting services and support
IRD UK Ltd UK

IRE
Insulated centrifugal inline ductfan with circular and rectangular connection
C A Östberg AB SE

Imofa
Centrifugal and axial flow fans and ventilation equipment
Gulfoke Ltd UK

Imofa
Centrifugal and axial flow fans and ventilation equipment
Imofa UK Ltd UK

Imofa UK
Centrifugal fan and axial flow fans, air handling units, ventilation equipment, control panels
Imofa UK Ltd UK

JM Aerofoil
Axial flow fans
Fläkt Woods Ltd UK

Jetblack
Blow-off gun, blower driven
Air Control Industries Ltd UK

Kiloheat fans
Specialists in air movement
Kiloheat Ltd UK

Leaderfan
Positive pressure blowers
Leader Fan Industries Ltd CAN

Leonova TM
Portable, multifunctional instrument for bearing and lubrication condition monitoring, vibration analysis
SPM Instrument AB SE

Liberator
Powered smoke and heat extract fan
Colt International Ltd UK

Lined
In-lined mixed flow extractor fans for duct mounting, 14 models with and without timers
Vortice Ltd UK

MG4
Stand-alone measuring unit for continuous condition monitoring
SPM Instrument AB SE

MS 30
Fans for trains and buses, 10dB less than classic fans
Sardou SA F

Mitsuboshi
Vee and fan belts
MBL (Europe) BV B

Mitsuboshi
Timing belts
MBL (Europe) BV B

Modulair
Bespoke light industrial ventilation solutions
Applied Energy Products Ltd UK

Multi-wing
Axial impellers
Multi-Wing International A/S DK

Multiflow
Range of mixed flow general purpose fans
Elta Fans Ltd UK

Neupex
Flexible shaft couplings
Flender Power Transmission Ltd UK

Novovent
Ventilation and air conditioning systems
Novovent SA E

Nutlink
Detachable vee belt
BTL UK

Nutlink
Detachable vee belt
Fenner Drives UK

PSB
Car parking ventilation and smoke extraction axial fans
Witt UK UK

Piller
The centre of motion
Piller Industrieventilatoren GmbH D

Poly ChainR GT2
Polyurethane synchronous belt for extremely powerful industrial drives
Gates Power Transmission Ltd UK

PowergripR GT3
Powerful rubber synchronous belt
Gates Power Transmission Ltd UK

Punto
Splashproof axial extract fans used to extract air either directly to the outside or into ducting up to 4 m long
Vortice Ltd UK

Quad-Power II
Gates top of the range raw edge narrow section vee-belt
Gates Power Transmission Ltd UK

RK
Centrifugal inline duct fan with circular and rectangular connection
C A Östberg AB SE

Radax
Mixed flow ducted fan range
Helios Ventilation Systems Ltd UK

Recyclair R
Waste air scrubber
Colasit AG CH

Remco
FHP motorsand fans for replacement use
Remco Products Ltd UK

Rencol Tolerance Ring
Spring steel fasteners
Rencol Tolerance Rings UK

Roof Units
Ventilation
Roof Units UK

Rosenberg
External rotor motor fans
Axair Fans UK Ltd UK

Rosenberg
External rotor motor fans
Axair Fans UK Ltd UK

Rosenberg
Air handling units, fans, motors
Rosenberg Ventilatoren GmbH D

Rotavent
High performance fan with aerofoil blades
Gebhart Ventilatoren D

Rotorex
Range of special OEM sickle blade axial fans
Soler & Palau Ltd UK

Rupex
Pin and buffer flexible couplings
Flender Power Transmission Ltd UK

S E A T
Plastic fans
Axair Fans UK Ltd UK

Sanitary fans
Sanitary fans for pharmaceutical purposes
B Bille A/S DK

Sardou
Low noise fans for trains, cars, industry, air and space
Sardou SA F

Select
Domestic fans
Greenwood Air Management Ltd UK

Selnikel
Industrial centrifugal fans
Selnikel TK

Silver Box Fans
Manufacture of blowers for inflatables
Silver Box Fans Ltd UK

Smokevent
Range of axial flow fan units for smoke use, tested to EN 12101-3 Standard
Elta Fans Ltd UK

Smokex®
Emergency smoke extract fan
Vectaire Ltd UK

Static balancing M/c series ESA-ESF
Measurement of the static unbalance of discshaped rotors, e.g. fans
Schenck Rotec GmbH D

Stevenson
Heavy duty industrial fan range
B.O.B. Stevenson Ltd UK

Stevenson
Heavy duty industrial fan range
B.O.B. Stevenson Ltd UK

Sunon
AC and DC industrial fans range
G. English Electronics Ltd UK

Superlink
Detachable wedge belt
BTL UK

Superlink
Detachable wedge belt
Fenner Drives UK

Technifan
Industrial fans
Technifan SA E

Tendo-AC
Three-phase synchronous servo motors
Mayr GmbH & Co KG D

Tendo-PM
Permanent magnetic motors and servo motors
Mayr GmbH & Co KG D

Thermoair
Ventilation equipment, air handling units, unit air heaters
Gulfoke Ltd UK

Thermoair
Ventilation equipment, air handling units, unit air heaters
Imofa UK Ltd UK

Tornado
Powered heat extract fan
Colt International Ltd UK

Turbodesign [1]
3D inviscid inverse design code for turbomachinery
Advanced Design Technology Ltd UK

Turbodesign [2]
3D transonic viscous inverse design code for transonic axial fans, compressors and turbines
Advanced Design Technology Ltd UK

Turngrove
Axial & bifurcated fans for building services & heating & ventilating applications
B.O.B. Stevenson Ltd UK

VLT R
Frequency converters
Danfoss Drives A/S DK

Vario
Axial supply and exhaust fans with manual or automatic shutters
Vortice Ltd UK

Vent-Axia
Ventilation
Vent-Axia Ltd UK

Vent-Axia
Heating
Vent-Axia Ltd UK

Vent-Axia
Air conditioning
Vent-Axia Ltd UK

Ventlight
Combined shower fan and light kit
Domus Ventilation Ltd UK

W Range
3-phase motor for safe or hazardous areas
Brook Crompton UK

W21 Line
TEFC + TEAO electric motors
WEG Electric Motors (UK) Ltd UK

Whirlwind
Vertical discharge powered heat extract fan
Colt International Ltd UK

Witt UK
Motor bases and anti-vibration mounts
Witt UK UK

Xodus
Range of fans suitable for kitchens, bathrooms & utility rooms with an IP25 rating
Applied Energy Products Ltd UK

Xpelair
Full range of fans and ventilation solutions
Applied Energy Products Ltd UK

Xpelair
A market leader in domestic, commercial & light industrial ventilation
Applied Energy Products Ltd UK

Xplus
Comprehensive range of light industrial fans designed for high performance and ease of installation
Applied Energy Products Ltd UK

25 Reference Index

The reference index contains a large number of key words used within the industry. It lists the page numbers on which the key words are used. The list of contents at the start of each chapter also provides a useful guide.

N

O

P

Q

R

T

U

V

Acknowledgments

The publishers wish to acknowledge the help and assistance of the following organisations in supplying data, photographs and illustrations, and where appropriate, permission to reproduce material from their own publications.

ABB Drives
AMCA (Air Movement & Control Association International, Inc.)
ANSI (American National Standards Institute)
BSI (British Standards Institution)
CEN (Commité Européen de Normalisation)
CIBSE (The Chartered Institution of Building Services Engineers)
Eurovent (European Committee of Air Handling and Refrigeration Equipment)
FETA (Federation of Environmental Trade Associations)
Fläkt Woods Ltd
FMA (Fan Manufacturers Association)
HEVAC (Heating Ventilating & Air Conditioning Manufacturers Association)
INA Bearing Company Ltd
ISO (International Standards Organisation)
Michell Bearings
PM°DM Precision Motors Deutsche Minebea GmbH
Prüftechnik Ltd
Rockwell Automation Ltd
Schenck RoTec GmbH
SKF (UK) Ltd
SPM Instrument AB
The News Ltd, Sydney, Australia
Vent-Axia Ltd

Index to advertisers